H. Ehrig H.-J. Kreowski
G. Rozenberg (Eds.)

Graph Grammars and Their Application to Computer Science

4th International Workshop
Bremen, Germany, March 5-9, 1990
Proceedings

Springer-Verlag
Berlin Heidelberg New York
London Paris Tokyo
Hong Kong Barcelona
Budapest

Series Editors

Gerhard Goos
GMD Forschungsstelle
Universität Karlsruhe
Vincenz-Priessnitz-Straße 1
W-7500 Karlsruhe, FRG

Juris Hartmanis
Department of Computer Science
Cornell University
Upson Hall
Ithaca, NY 14853, USA

Volume Editors

Hartmut Ehrig
Technische Universität Berlin, Fachbereich Informatik
Franklinstraße 28/29, W-1000 Berlin 10, FRG

Hans-Jörg Kreowski
Universität Bremen, Fachbereich Mathematik/Informatik
Bibliothekstraße, W-2800 Bremen 33, FRG

Grzegorz Rozenberg
Universiteit Leiden, Faculteit de Wiskunde en Informatica
Niels Bohrweg 1, 2300 RA Leiden, The Netherlands

CR Subject Classification (1991): F.4.2-3, I.1.1, I.2.4, I.5.1, J.3

ISBN 3-540-54478-X Springer-Verlag Berlin Heidelberg New York
ISBN 0-387-54478-X Springer-Verlag New York Berlin Heidelberg

Printed in Germany

Typesetting: Camera ready by author
Printing and binding: Druckhaus Beltz, Hemsbach/Bergstr.
2145/3140-543210 - Printed on acid-free paper

Preface

This volume consists of papers selected from the contributions to the *Fourth International Workshop on Graph Grammars and Their Application to Computer Science* which took place in Bremen, March 5 - 9, 1990. The workshop had 86 participants from 18 countries in 4 continents. The program contained 4 tutorial talks, 36 technical presentations, 4 system demonstrations, a panel discussion and an open-problems session. The organization of the workshop was supported by the ESPRIT Basic Research Working Group *Computing by Graph Transformation.*

The research area of graph grammars is theoretically attractive and well motivated by various applications. More than 20 years ago, the concept of a graph grammar was introduced by A. Rosenfeld in the U.S.A. as a formulation of some problems in pattern recognition and image processing as well as by H.J. Schneider in Germany as a method for data type specification. Since then, researchers from all over the world have contributed steadily to the field. This volume as well as the proceedings of the previous three workshops in Bad Honnef 1978 (published as Lecture Notes in Computer Science 73), in Osnabrück 1982 (Lecture Notes in Computer Science 153) and in Warrenton, Virginia, 1986 (Lecture Notes in Computer Science 291) provide a rich record of the development of the field.

This volume is again intended as a source of information for researchers active in the area as well as for scientists who would like to know more about graph grammars. We think that through this volume the reader can get a good idea of the state of the art of graph grammars, and she/he can recognize the newest trends.

The volume is organized in five sections. The first section contains three short tutorials on hyperedge replacement, node label controlled graph grammars and the algebraic approach based on double and single pushouts as well as a note on the algebraic and the logic description of graph languages. Most of the technical contributions are closely related to at least one of these four graph grammar approaches. The second section is a collection of statements concerning the future trends in the area of graph grammars and potential applications. The third (short) section consists of four system descriptions. The fourth section provides the technical contributions. The topics of the papers cover foundations, algorithmic and implementational aspects, and various issues from application areas like concurrent computing, functional and logic programming, computer graphics, artificial intelligence and biology. In the last section, the description of the ESPRIT Basic Research Working Group *Computing by Graph Transformation* is given.

We are grateful to all who helped us in reviewing the submitted papers. The referees were: M. Bauderon, M. de Boer, L. Bonsiepen, F.J. Brandenburg, H. Bunke, M. Chytil, B. Courcelle, M. Dauchet, F. Drewes, J. Engelfriet, G. Engels, P. Fitzhorn, F.D. Fracchia, H. Göttler, A. Habel, F. Hinz, D. Janssens, K.P. Jantke, J.R. Kennaway, C. Kim, H.-P. Kriegel, C. Lautemann, M. Löwe, H. Lück, J. Lück, B. Mayoh, M. Nagl, F. Nake, F. Parisi-Presicce, A. Paz, D. Plump, P. Prusinkiewicz, J.-C. Raoult, F. Rossi, H. Schneider, A. Schürr, R. Siromoney, R. Sleep, W. Vogler, E. Wanke, E. Welzl, J. Winkowski. In particular, we would like to thank B. Courcelle, M. Nagl and A. Rosenfeld as the members of the advisory board of the workshop for their support in organizing the workshop and editing the proceedings. We are grateful to Annegret Habel for helping with the local organization. Finally, we gladly acknowledge the financial support by the Commission of the European Communities, and the University of Bremen.

July 1991

Hartmut Ehrig (Berlin)
Hans-Jörg Kreowski (Bremen)
Grzegorz Rozenberg (Leiden)

Contents

Part 1

Part 2

Part 3

Part 4

Part 5

A Note on Hyperedge Replacement

Frank Drewes, Hans–Jörg Kreowski
Fachbereich Mathematik und Informatik
Universität Bremen
Postfach 33 04 40
W–2800 BREMEN

ABSTRACT: In this note, we recall the basic features of hyperedge replacement as one of the most elementary and frequently used concepts of graph rewriting. Moreover, we discuss the Contextfreeness Lemma for derivations in hyperedge–replacement grammars.

Keywords: hyperedge replacement, contextfreeness, bounded treewidth.

CONTENTS

0 Introduction

Hyperedge replacement is one of the most elementary and frequently used concepts of graph transformation with the characteristics of context–free rewriting. Introduced in the early seventies by Feder [Fed 71] and Pavlidis [Pav 72] (under various pseudonyms), it has been intensively and systematically studied in the eighties (starting with the special case of edge replacement) by Bauderon and Courcelle [BC 87, Cou 87b, Cou 87a], Habel, Kreowski and Vogler [Kre 79, HK 87a, HK 87b, HKV 89a, HKV 89b], Lengauer and Wanke [LW 88] and others.

A hyperedge is an atomic item with a label and a fixed number of ordered tentacles. It can be attached to any kind of structure coming with a set of nodes in such a way that each tentacle is attached to a node. So the hyperedge controls a sequence of nodes, called the attachment, and can play the role of a placeholder, which may be replaced by some other structure eventually. In such a replacement step the hyperedge disappears, and the replacing structure is added and suitably connected to the original structure. For this purpose, the replacing structure is assumed to have a sequence of external nodes which are fused with the attachment of the replaced hyperedge. The replacement of a hyperedge by a structure can be iterated if the original structure or the replacing one are decorated by other hyperedges.

The aim of this note is to recall the basic features of hyperedge replacement where we employ ordinary undirected and unlabelled graphs as underlying structures of interest. Moreover, we discuss the Contextfreeness Lemma for derivations in hyperedge-replacement grammars as one of the main results in the theory of hyperedge replacement. We demonstrate the use of this result by proving the correctness of a grammar which generates all graphs of treewidth k (for some $k \geq 2$).

Readers who want to learn more about hyperedge replacement are refered to the literature and, in particular, Habel's thesis [Hab 89], which presents the area of hyperedge replacement in a neat, systematic and nearly complete way (see also the introductory note by Habel and Kreowski [HK 87c]).

1 Hyperedge–decorated graphs

In this section, we recall the basic notions hyperedge replacement and hyperedge–replacement grammars are built upon. We choose ordinary unlabelled undirected graphs without loops and multiple edges as basic structures of interest. To generate sets of graphs, these graphs become equipped with *hyperedges* each of which has got a label and is attached to several nodes of the graph. Further, every such *(hyperedge-)decorated* graph has got a sequence of so-called *external* nodes. Every hyperedge of a decorated graph serves as a placeholder for a graph or — recursively — for another decorated graph, which can replace the hyperedge by fusing its external nodes with those the hyperedge is attached to.

1.1 Definition (graph)
A(n unlabeled, undirected) *graph* is a pair $G = (V, E)$ where V is a set of *nodes* (or *vertices*) and E is a set of 2–elements subsets of V, called *edges*. The set of all graphs is denoted by $\mathcal{G}_0$. □

1.2 Definition (decorated graph)
Let N be a set of *nonterminal labels*. A (hyperedge–)*decorated graph* (over N) is a system $H = (V, E, Y, lab, att, ext)$ where $(V, E) \in \mathcal{G}_0$, Y is a set of *hyperedges*, $lab : Y \to N$ is a mapping, called the *labelling*, $att : Y \to V^*$ is a mapping, called the *attachment*, and $ext \in V^*$, called the (sequence of) *external nodes*. □

Remarks

1. The components V, E, Y, lab, att, and ext of H are also denoted by V_H,E_H,Y_H,lab_H, att_H, and ext_H, respectively.
2. The length of ext_H is called the *type* of H and denoted by *type*(H). If *type*$(H) = m$, H is called an m*–graph*.
3. The length of $att_H(y)$ for $y \in Y_H$ is called the *type* of y, denoted by *type*(y). If *type*$(y) = n$, y is called an n*–edge*.
4. Note that ext as well as $att(y)$, for $y \in Y_H$, may contain nodes twice or more.
5. The class of all decorated graphs over N is denoted by $\mathcal{G}(N)$. The class of all m–graphs over N is denoted by $\mathcal{G}_m(N)$.
6. A graph can be seen as a 0–graph without hyperedges. In this sense, $\mathcal{G}_0 \subseteq \mathcal{G}_0(N) \subseteq \mathcal{G}(N)$. Moreover, the subclass of $\mathcal{G}_m(N)$, for $m \geq 0$, consisting of all m–graphs without hyperedges is denoted by $\mathcal{G}_m$ while $\mathcal{G}$ denotes the union of all $\mathcal{G}_m$, for $m \geq 0$. In the description of m–graphs without hyperedges we will drop the components Y, lab and att.
7. Another particular case is a decorated graph with a single hyperedge where external nodes and attachment coincide, i.e., $H = (V, \emptyset, \{y\}, lab, att, ext)$ with $ext = att(y)$. Such an H is called a *handle*. If $V = \{1, \ldots, n\}$, $lab(y) = A$, and $att(y) = ext = 1 \ldots n$, H is denoted by $(A, n)^\bullet$. □

Two decorated graphs are said to be isomorphic if there is an isomorphism between the underlying graphs preserving external nodes, and there is a bijective mapping between their hyperedges which is consistent with the graph isomorphism. Formally, this looks as follows.

1.3 Definition (isomorphic decorated graphs)
Let $H, H' \in \mathcal{G}(N)$. Then H and H' are *isomorphic*, denoted by $H \cong H'$, if there are bijective mappings $f : V_H \to V_{H'}$ and $g : Y_H \to Y_{H'}$ such that $E_{H'} = \{\{f(x), f(y)\} \mid \{x, y\} \in E_H\}$, $lab_H(y) = lab_{H'}(g(y))$ and $f^*(att_H(y)) = att_{H'}(g(y))$ for all $y \in Y_H$, and $f^*(ext_H) = ext_{H'}$, where f^* is the natural extension of f to sequences. □

1.4 Example (hyperedge–decorated graph)
The picture below shows a decorated graph consisting of a graph on six vertices having six edges, together with four hyperedges. Its external nodes are drawn as filled circles (whereas all others are unfilled).

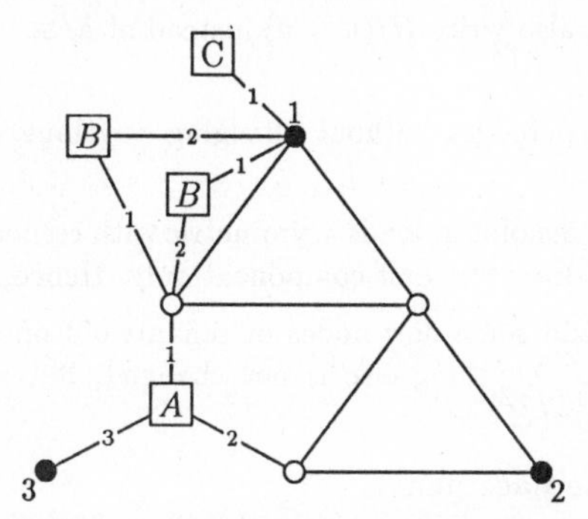

The hyperedges are labelled A, B, and C, and the order on their tentacles is indicated by their numbers. Similarly, we provide numbers for the external nodes indicating the ordering on ext. The type of this decorated graph as well as that of the A–labelled hyperedge is 3, while the hyperedges labelled with B are of type 2 and the one labelled with C is of type 1. □

2 Hyperedge replacement

In this section, we will make precise what it means to replace a hyperedge e of a decorated graph by another decorated graph G. As we mentioned above, this procedure can intuitively be seen as substituting G for e in such a way that the external nodes of G become fused with the nodes e was attached to. In fact, we will give a definition which is a bit more general, because it allows to replace several hyperedges in one step. In order to be able to give a concise definition for this, three simple operations hyperedge replacement is based on are needed: Hyperedge removal, disjoint union of decorated graphs, and nodes fusion (as a special case of nodes–set exchange).

2.1 Definition (operations on decorated graphs)
Let $H, H' \in \mathcal{G}(N)$.

1. Let $B \subseteq Y_H$. Then the *removal* of B from H yields the decorated graph

$$H - B = (V_H, E_H, Y_H - B, lab, att, ext_H)$$

with $lab(y) = lab_H(y)$ and $att(y) = att_H(y)$ for all $y \in Y_H - B$.

2. The *disjoint union* of H and H' is the decorated graph

$$H + H' = (V_H \mathring{\cup} V_{H'}, E_H \mathring{\cup} E_{H'}, Y_H \mathring{\cup} Y_{H'}, lab, att, ext_H)$$

with $lab(y) = lab_H(y)$ and $att(y) = att_H(y)$ for all $y \in Y_H$ and $lab(y) = lab_{H'}(y)$ and $att(y) = att_{H'}(y)$ for all $y \in Y_{H'}$. Here $\mathring{\cup}$ denotes the disjoint union of sets.

For $H_1, \ldots, H_n \in \mathcal{G}(N)$ $\sum_{i=1}^{i} H_i$ means $H_1 + H_2 + \ldots + H_n$.

3. Let $f : V_H \to V$ be a mapping for some set V. Then the *nodes–set exchange* in H through f yields the decorated graph
$$H/f = (V, E, Y_H, lab_H, att, ext)$$
with $E = \{\{f(x), f(y)\} \mid \{x,y\} \in E_H, f(x) \neq f(y)\}$, $att(y) = f^*(att_H(y))$ for all $y \in Y_H$, and $ext = f^*(ext_H)$, where f^* is the natural extension of f to sequences.
4. Let δ be an equivalence relation on V_H, let V be the corresponding quotient set (of V_H through δ) and $nat : V_H \to V$ the natural mapping. Then H/nat is called the *nodes fusion* according to δ, and is also denoted by H/δ. If, $u = x_1 \cdots x_n$ and $v = y_1 \cdots y_n$ ($x_i, y_i \in V_H$ for $i = 1, \ldots, n$, $n \geq 0$, where $n = 0$ means $u = \lambda = v$) and $\equiv$ is the equivalence relation on V_H generated by $x_i \equiv y_i$ for $i = 1, \ldots, n$, we also write $H/(u = v)$ instead of $h/\equiv$. □

Remarks

1. Removal removes some hyperedges without changing anything else. In particular, we have $type(H - B) = type(H)$.
2. It should be noted that the disjoint union is asymmetric with respect to the choice of the external nodes, which are borrowed from the first component only. Hence, $type(H + H') = type(H)$.
3. Nodes–set exchange may add some new nodes or rename old ones (including identifications of originally different nodes). Anything else is not changed, but adapted to the exchange. In particular, $type(H/f) = type(H)$. □

2.2 Definition (hyperedge replacement)
Let $H \in \mathcal{G}(N)$ and $B = \{y_1, \ldots, y_n\} \subseteq Y_H$. Let $repl : B \to \mathcal{G}(N)$ be a mapping with $type(y_i) = type(repl(y_i))$ for $i = 1, \ldots, n$. Then the *replacement* of B in H through *repl* yields the decorated graph REPLACE$(H, repl) = ((H - B) + \sum_{i=1}^{n} repl(y_i)) / (att = ext)$ with $att = att(y_1) \cdots att(y_n)$ and $ext = ext_{repl(y_1)} \cdots ext_{repl(y_n)}$. □

Remarks

1. Hyperedge replacement is a simple construction where some hyperedges are removed, the associated decorated graphs are added disjointly and their external nodes are fused with the corresponding nodes formerly attached to the replaced hyperedges.
2. Note that the component graphs replacing hyperedges are fully embedded into the resulting graph where their external nodes loose this status and may additionally be fused with other nodes.
3. If a mapping $repl : B \to \mathcal{G}(N)$ meets the assumption of the definition, it will be referred to as *well–typed.*
4. In the special case where $B = \{y\}$ and $repl(y) = R$, we may denote $REPLACE(H, repl)$ by $H \otimes_y R$. □

2.3 Example (hyperedge replacement)
Consider again the decorated graph G given in Example 1.4.

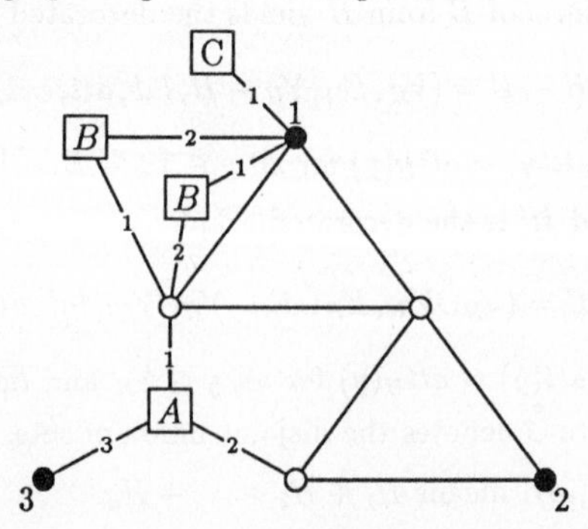

If we replace the hyperedge labelled with A by G itself, we get

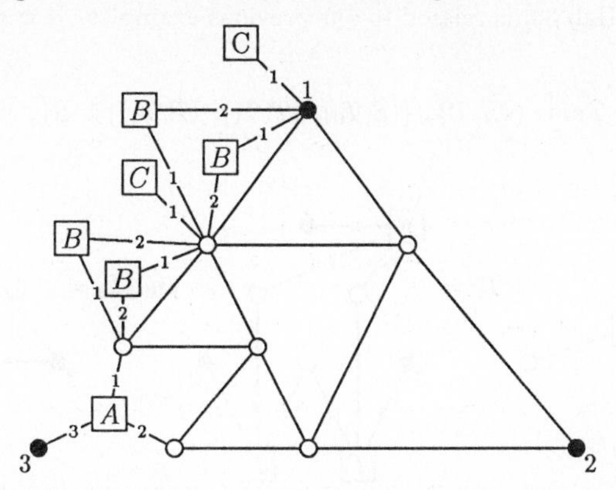

Observe that G cannot replace the other two hyperedges, because this would not be well–typed. □

3 Hyperedge–replacement grammars and languages

We will now define hyperedge–replacement grammars by means of hyperedge replacement. As usual in the context of formal languages, *derivations* will be defined through composing single derivation steps, so–called *direct derivations.* Such a direct derivation is given by a hyperedge replacement if the label of each replaced hyperedge and the replacing decorated graph form a production of the grammar. The graphs generated by such a grammar are obtained by starting from the handle $(S,0)^\bullet$, where S is the start symbol of the grammar, and iterating derivation steps until all hyperedges are replaced.

3.1 Definition (hyperedge–replacement grammar and language)

1. A *production* (over N) is a pair $p = (A, R)$ with $A \in N$ and $R \in \mathcal{G}(N)$. A is called the *left–hand side* of p and denoted by $lhs(p)$. R is called the *right–hand side* and denoted by $rhs(p)$.
2. Let $H \in \mathcal{G}(N)$ and $B \subseteq Y_H$. Let P be a set of productions. A mapping $b : B \to P$ is called a *base* in H if $lab_H(y) = lhs(b(y))$ and $type(y) = type(rhs(b(y)))$ for all $y \in B$.
3. Let $H, H' \in \mathcal{G}(N)$ and $b : B \to P$ be a base in H. Then H *directly derives* H' *through* b if $H' \cong \text{REPLACE}(H, repl)$ with $repl : B \to \mathcal{G}(N)$, $repl(y) = rhs(b(y))$ for all $y \in B$. A direct derivation is denoted by $H \underset{b}{\Longrightarrow} H'$ or $H \underset{P}{\Longrightarrow} H'$. If $B = \{y\}$ and $b(y) = p$ in particular, we may also write $H \underset{y,p}{\Longrightarrow} H'$ or $H \underset{p}{\Longrightarrow} H'$.
4. A sequence of direct derivations of the form $H_0 \underset{P}{\Longrightarrow} H_1 \underset{P}{\Longrightarrow} \ldots \underset{P}{\Longrightarrow} H_k$ is called a *derivation* (of length k) from H_0 to H_k and is denoted by $H_0 \overset{k}{\underset{P}{\Longrightarrow}} H_k$. If, moreover, $H \cong H'$, we call this a derivation (of length 0) from H to H' and denote it by $H \overset{0}{\underset{P}{\Longrightarrow}} H'$. If, in $H \overset{k}{\underset{P}{\Longrightarrow}} H'$ for some $k \geq 0$, the length k does not matter, we write $H \overset{*}{\underset{P}{\Longrightarrow}} H'$. We will omit P if it is clear from the context.
5. A *hyperedge–replacement grammar* is a system $\text{HRG} = (N, P, Z)$ where N is a set of *nonterminals*, P is a finite set of *productions* (over N), and $Z \in \mathcal{G}_0(N)$.
6. Given such a grammar $\text{HRG} = (N, P, Z)$, the *generated graph language* consists of all graphs derivable from Z: $L(\text{HRG}) = \{G \in \mathcal{G}_0 \mid Z \overset{*}{\underset{P}{\Longrightarrow}} G\}$. If $Z = (S,0)^\bullet$ for some $S \in N$, we may denote HRG by (N, P, S).
7. The class of all hyperedge–replacement grammars will be denoted by $\mathcal{HRG}$. □

3.2 Example (hyperedge–replacement grammar)
An example taken from [Hab 89] is related to our previous examples. It is given by the hyperedge-replacement grammar

$$Tri = (\{S,T\},\ \{(S,T_0),\ (T,T_1),\ (T,T_2)\}),\ S),$$

where

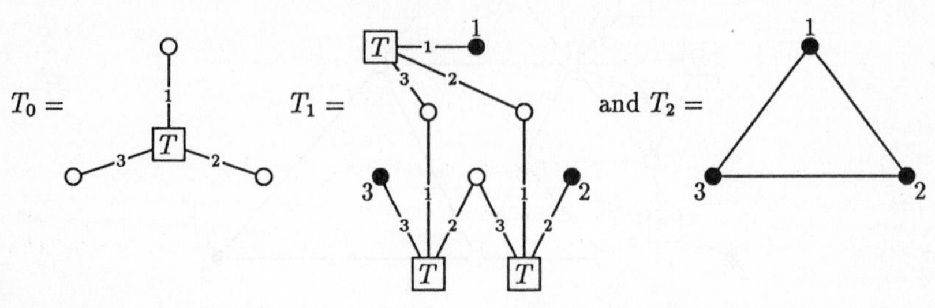

It generates a set of graphs which could be called "irregular Sierpinski triangles". For example, the Sierpinski triangles

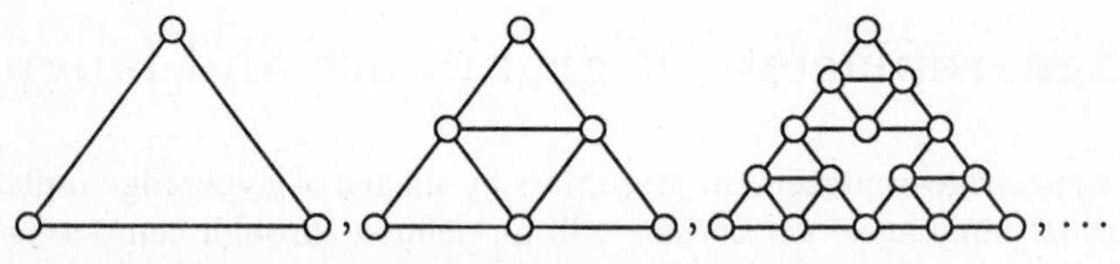

are generated, but also, for instance, the graph

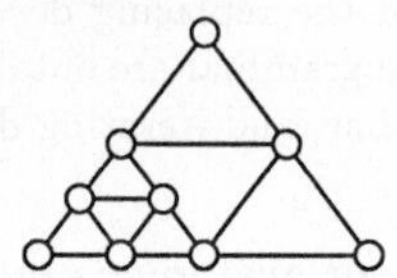

is generated by the derivation

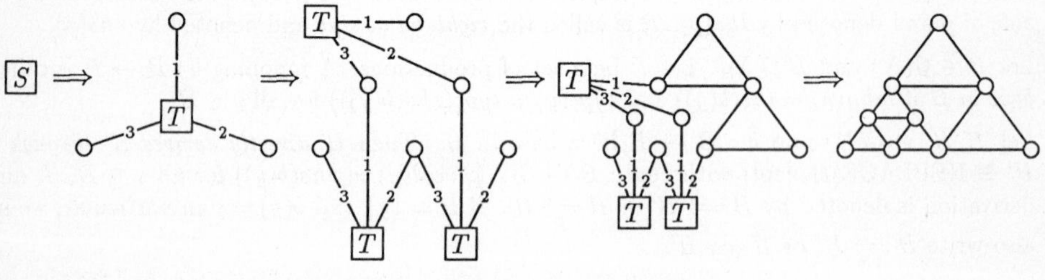

□

4 The Contextfreeness Lemma

In this section, we will concentrate on one of the most significant properties of hyperedge–replacement grammars: their contextfreeness. Replacements of different hyperedges cannot interfere with each other, except for some node fusions. Consequently, each derivation starting in a decorated graph H_0 can be restricted to any decorated graph K_0 embeddable into H_0. In particular, we may consider the fibres of a derivation, being the restrictions to the handles induced by the hyperedges of H_0. The most interesting fact about these fibres is that the original derivation can be reconstructed from all of them together. This result on hyperedge replacement is called *"Contextfreeness Lemma"*. A recursive version of it yields decompositions of graphs derived from handles into smaller graphs derived from handles. More precisely, each graph derived from a handle coincides with the right-hand side of a production where all hyperedges are replaced by derived graphs.

4.1 Lemma (Contextfreeness Lemma)
Let $HRG = (N, P, Z) \in \mathcal{HRG}$, $A \in N$, $G \in \mathcal{G}_k$, for some $k \geq 0$, and $n \geq 0$.

Then there is a derivation $s = \left((A,k)^\bullet \overset{n+1}{\underset{P}{\Longrightarrow}} G\right)$ if and only if there is some $(A, R) \in P$ and, for each $y \in Y_R$, there exists a derivation of the form $s(y) = \left((lab_R(y), type(y))^\bullet \overset{n_y}{\underset{P}{\Longrightarrow}} G(y)\right)$, where $n_y \leq n$, such that $G \cong \mathrm{REPLACE}(R, repl)$ with $repl(y) = G(y)$, for all $y \in Y_R$. □

Remarks

1. Because s starts in a handle, it decomposes always into $(A, R) \in P$ and $R \overset{n}{\underset{P}{\Longrightarrow}} G$. The latter derivation can be restricted to the handles of R, leading to the derivations $s(y)$. Vice versa, such fibres can jointly be embedded into R, yielding the graph $\mathrm{REPLACE}(R, repl)$ with $repl(y) = G(y)$, for $y \in Y_R$.
2. In the situation of the Contextfreeness Lemma, the graph $G(y)$ for $y \in Y_R$ will be called the *y-component* of G. □

4.2 Example (Use of the Contextfreeness Lemma)
Let k be an arbitrary, but fixed natural number. We call an m-edge y *well-formed* if its attachment contains m distinct vertices, i.e., for all $i, j \in \{1, \ldots, m\}$, $i \neq j$ implies $att(y)_i \neq att(y)_j$. An m-graph is said to be well-formed if it has m distinct external nodes and all its hyperedges are well-formed.

In order to demonstrate the use of the Contextfreeness Lemma with a non-trivial example, let us take the class of graphs of *tree-width* at most k. This notion was introduced by Robertson and Seymour [RS 86]. Many hard graph problems have been shown to become easy (i.e., at least polynomial time decidable) if they are restricted to graphs of bounded tree-width (see Arnborg [Arn 85] and Lautemann [Lau 90], and also Bodlaender [Bod 86, Bod 88a, Bod 88b], Arnborg, Lagergreen and Seese [ALS 88], Courcelle [Cou 88], and Lautemann [Lau 88]). Let A be some fixed label, and let $\mathcal{P}$ be the set of all productions (A, R), where R is well-formed and $|V_R| = k + 1$. A graph G is of tree-width $\leq k$ if $|V_G| \leq k$ or there is a derivation $(A, 0)^\bullet \overset{*}{\underset{\mathcal{P}}{\Longrightarrow}} G$.[1] Let

$$Tw_k = \{G \in \mathcal{G}_0 | G \text{ is of tree-width } \leq k \text{ and } |V_G| \geq k\}.$$

(We could as well drop the restriction $|V_G| \geq k$, but this way the example becomes nicer, and the class of graphs with $|V_G| < k$ is finite, anyway.) Observe that the definition for Tw_k yields no hyperedge-replacement grammar, since $\mathcal{P}$ is infinite because — in contrast to edges — hyperedges may be parallel. (Here, two hyperedges e, e' of a decorated graph G are said to be parallel if they are attached to the same set of nodes, ie, $\{att(e)_1, \ldots, att(e)_{type(e)}\} = \{att(e')_1, \ldots, att(e')_{type(e')}\}$.)

Let, for a label A and natural numbers n_1, n_2, $Complete(A, n_1, n_2)$ denote some well-formed, decorated graph $G \in \mathcal{G}_k(\{A\})$, such that $|V_G| = n_1$, $E_G = \emptyset$, and Y_G is a maximum set of non-parallel, well-formed, n_2-edges. (The order on $att(y)$ we choose for any $y \in Y_G$ will not matter in the following.)

We define $HRG = (\{\boxed{k}, \boxed{?}, S\}, P_{HRG} = \{Start, NewNode, Comp, EdgeNo, EdgeYes\}, S)$ with

$$\begin{aligned}
Start &= (S, U((\boxed{k}, k)^\bullet)) \\
&\quad \text{where } U(G),\ G \in \mathcal{G}(N), \text{ forgets the external nodes of } G \\
NewNode &= (\boxed{k}, Complete(\boxed{k}, k+1, k)) \\
Comp &= (\boxed{k}, Complete(\boxed{?}, k, 2)) \\
EdgeNo &= (\boxed{?}, 1\bullet \qquad \bullet 2) \\
EdgeYes &= (\boxed{?}, 1\bullet\!\!-\!\!\!-\!\!\!-\!\!\bullet 2).
\end{aligned}$$

[1] Actually, this is not the original definition (cf. [RS 86]), which does not use hyperedge replacement, but the class of graphs it defines is the same.

By definition, the first production to be applied is always *Start*, yielding the decorated graph $(\boxed{k}, k)^\bullet$ without external nodes. So we may (and will) forget about this production and start with $(\boxed{k}, k)^\bullet$ in all considerations that follow. Production *NewNode* allows us to replace repeatedly $\boxed{k}$-labelled hyperedges by decorated graphs on $k+1$ vertices having no edges, but a maximum set of non-parallel, well-formed k-edges all labelled $\boxed{k}$. Production *Comp* then replaces a $\boxed{k}$-labelled hyperedge by a complete graph on k vertices where the edges are represented by 2-edges labelled $\boxed{?}$, and *EdgeNo* and *EdgeYes* finally allow one either to delete such a $\boxed{?}$-labelled 2-edge or to replace it by a "normal" edge.

We will use the Contextfreeness Lemma in order to show the following.

Main Claim. $L(HRG) = Tw_k$. □

$L(HRG) \subseteq Tw_k$ is easy to show. For the other direction, if we have a graph $G \in Tw_k$ then either $|V_G| = k$ or a derivation must exist which yields G and does only use productions of $\mathcal{P}$. So, essentially, the problems we have to cope with are the following ones.

(1) The productions in $\mathcal{P}$ contain parallel hyperedges. Since the productions of HRG do not (and in particular *NewNode*, which is the important production of HRG, does not), we will have got to prove that from one $\boxed{k}$-labelled hyperedge we can generate the same graphs as from any number of parallel $\boxed{k}$-labelled hyperedges.

(2) The right-hand side R of a production in $\mathcal{P}$ may contain hyperedges of type $|V_R|$, but the right-hand side of *NewNode* does not contain a $(k+1)$-edge, so it must be shown that these hyperedges are not really needed.

(3) Finally, we will have got to show that it does not matter if some right-hand side contains hyperedges attached to fewer than k nodes, although $rhs(NewNode)$ contains only k-edges.

For (1) and (2) we will prove seperate claims, and the proof of the main claim will then essentially deal with (3).

Before we treat (1) and (2), observe that an easy application of the Contextfreeness Lemma proves that every graph on k nodes can be derived from $(\boxed{k}, k)^\bullet$. More exactly, for $G \in \mathcal{G}_k$ and $|V_G| = k$ we have $(\boxed{k}, k)^\bullet \underset{P_{HRG}}{\overset{*}{\Longrightarrow}} G$. This is because $(\boxed{k}, k)^\bullet$ directly derives $Complete(\boxed{?}, k, 2)$ through production *Comp*, $REPLACE(Complete(\boxed{?}, k, 2), repl) = G$ with

$$repl(y) = \begin{cases} 1\bullet\!\!-\!\!\!-\!\!\!-\!\!\bullet 2 & \text{if } \{att(y)_1, att(y)_2\} \in E_G \\ 1\bullet \quad\quad \bullet 2 & \text{otherwise,} \end{cases}$$

and furthermore $(\boxed{?}, 2)^\bullet \underset{\{EdgeYes\}}{\Longrightarrow} 1\bullet\!\!-\!\!\!-\!\!\!-\!\!\bullet 2$ as well as $(\boxed{?}, 2)^\bullet \underset{\{EdgeNo\}}{\Longrightarrow} 1\bullet \quad\quad \bullet 2$, so an application of the Contextfreeness Lemma yields $(\boxed{k}, k)^\bullet \underset{P_{HRG}}{\overset{*}{\Longrightarrow}} G$.

Claim 1. For all $G_1, G_2 \in \mathcal{G}_k$

$$(\boxed{k}, k)^\bullet \underset{P_{HRG}}{\overset{*}{\Longrightarrow}} G_1 \text{ and } (\boxed{k}, k)^\bullet \underset{P_{HRG}}{\overset{*}{\Longrightarrow}} G_2 \text{ implies } (\boxed{k}, k)^\bullet \underset{P_{HRG}}{\overset{*}{\Longrightarrow}} (G_1 + G_2)/(ext_{G_1} = ext_{G_2}).$$

□

Proof. Let $G = (G_1 + G_2)/(ext_{G_1} = ext_{G_2})$. We proceed by induction on the number of occurrences of production *NewNode* in the derivations for G_1 and G_2. If neither derivation contains an application of *NewNode* then $|V_G| = k$, so the claim follows from what we saw above. Otherwise let, w.l.o.g., the derivation for G_1 start with *NewNode*. (Of course, since the other productions do not contain any $\boxed{k}$-labelled hyperedge in their right-hand sides, a derivation containing *NewNode* must start with it.) By contextfreeness, we have $G_1 = REPLACE(rhs(NewNode), repl_1)$ with $repl_1(y) = G_1(y)$ and

$s_1(y) = (\boxed{k}, k)^\bullet \underset{P_{HRG}}{\overset{n_y}{\Longrightarrow}} G_1(y)$ for all $y \in Y_{rhs(NewNode)}$. Now, if y_0 is the hyperedge of $rhs(NewNode)$ attached to the nodes of $ext_{rhs(NewNode)}$, we have $G = REPLACE(rhs(NewNode), repl)$ with

$$repl(y) = \begin{cases} repl_1(y) & \text{if } y \neq y_0 \\ (G_1(y_0) + G_2)/(ext_{G_1(y_0)} = ext_{G_2}) & \text{otherwise.} \end{cases}$$

But the fibre $s_1(y_0)$, i.e., the derivation yielding $G_1(y_0)$, contains fewer applications of *NewNode* than that for G_1, so by induction hypothesis

$$(\boxed{k}, k)^\bullet \underset{P_{HRG}}{\overset{*}{\Longrightarrow}} (G_1(y_0) + G_2)/(ext_{G_1(y_0)} = ext_{G_2}),$$

hence by contextfreeness $(\boxed{k}, k)^\bullet \underset{P_{HRG}}{\overset{*}{\Longrightarrow}} G$. □

Claim 2. If $(A, l)^\bullet \underset{\mathcal{P}}{\overset{*}{\Longrightarrow}} G$ then there is a derivation $(A, l)^\bullet \underset{\mathcal{P}}{\overset{*}{\Longrightarrow}} G$, such that for each production $(A, R) \in \mathcal{P}$ used in this derivation and all $y \in Y_R$ we have $type(y) \leq k$. □

Proof. Let $(A, l)^\bullet \underset{\mathcal{P}}{\overset{n}{\Longrightarrow}} G$, for some $l \leq k + 1$. We show by induction on n that the claim holds true. Because of the Contextfreeness Lemma we have $G = REPLACE(R, repl)$ for some production $(A, R) \in \mathcal{P}$, and with y–components $G(y)$. By induction hypothesis we may assume that each of the fibres $(lab(y), type(y))^\bullet \underset{\mathcal{P}}{\overset{n_y}{\Longrightarrow}} G(y)$ has got the claimed property, since $n_y < n$. If $y_0 \in Y_R$ and $type(y_0) = k + 1$ consider the fibre $s(y_0)$. Applying the Contextfreeness Lemma once again we get $G(y_0) = REPLACE(R', repl'))$ for some production $(A, R') \in \mathcal{P}$, and with y–components $G(y_0)(y)$. Since every node of R becomes fused with one of R' in $R_1 = REPLACE(R, y_0 \mapsto R')$, we have $|V_{R_1}| = |V_{R'}| = k + 1$. But then $(A, R_1) \in \mathcal{P}$ and, since $G = REPLACE(R_1, repl_1)$ with $repl_1(y) = \begin{cases} repl(y) & \text{if } y \in Y_R - \{y_0\} \\ repl'(y) & \text{otherwise,} \end{cases}$ there is a derivation as claimed, using the production (A, R_1) instead of the two productions (A, R) and (A, R'). □

Proof of the main claim.

"$\subseteq$". We omit this part, because it is very easy. By contextfreeness, we can collaps the final applications of *Comp*, *EdgeNo* and *EdgeYes* into the last application of *NewNode*, if $|V_G| > k$. (Observe that *NewNode* has the property we required.)

"$\supseteq$". Let $G \in Tw_k$. By Claim 2 we may work with $\mathcal{P}_{\leq k} = \{(A, R) \in \mathcal{P} | type(y) \leq k \text{ for all } y \in Y_R\}$ instead of $\mathcal{P}$. Since we have Claim 1 in order to be able to treat parallel hyperedges occurring in Y_R, the major difference between the productions of $\mathcal{P}_{\leq k}$ and *NewNode* is that those of $\mathcal{P}_{\leq k}$ may contain hyperedges whose type is less than k. On the other hand we might make every hyperedge become a (well-formed) k–edge by adding arbitrary nodes to its attachment, since all right–hand sides contain enough nodes. This means we are almost ready if we can show that, for $l \leq k$, $(A, l)^\bullet \underset{\mathcal{P}_{\leq k}}{\overset{n}{\Longrightarrow}} G$ implies

$$(A, k)^\bullet \underset{\mathcal{P}_k}{\overset{*}{\Longrightarrow}} \overline{G} = (V_G \mathring{\cup} \{l + 1, \ldots, k\}, E_G, Y_G, lab_G, att_G, ext_G \circ l + 1 \ldots k),$$

where $\mathcal{P}_k = \{(A, R) \in \mathcal{P} | type(y) = k \text{ for all } y \in Y_R\}$. We proceed by induction on n. Let $G = REPLACE(R, repl)$ for a production $(A, R) \in \mathcal{P}_{\leq k}$ and with y–components $G(y)$. By induction hypothesis for each $y \in Y_H$ we have $(lab(y), k)^\bullet \underset{\mathcal{P}_k}{\overset{*}{\Longrightarrow}} \overline{G(y)}$, so $G = REPLACE(R', repl')$, where R' is any decorated graph which can be obtained from R by replacing every hyperedge y by a k–hyperedge y' such that the attachment of y is a prefix of that of y', and where $repl'(y') = \overline{G(y)}$ for all $y' \in Y_{R'}$. By definition of $REPLACE$ we have $\overline{G} = REPLACE(\overline{R'}, repl')$. So we are ready if we can construct a set $P \subset \mathcal{P}_k$ with $(A, k)^\bullet \underset{P}{\overset{*}{\Longrightarrow}} \overline{R'}$. Let, w.l.o.g., $V_R = \{1, \ldots, k+1\}$ and $ext_R = 1 \ldots l$. We set $P = \{P_{l+1}, \ldots, P_{k+1}\}$ with

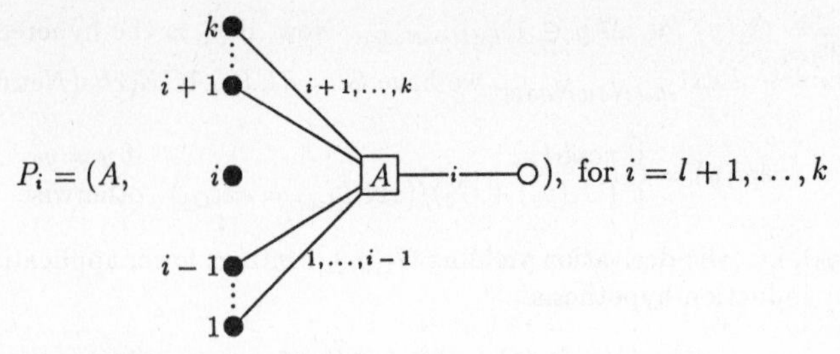

and $P_{k+1} = (A, (V_{\overline{R'}}, E_{\overline{R'}}, Y_{\overline{R'}}, lab_{\overline{R'}}, att_{\overline{R'}}, 1 \ldots k))$.

It is easy to show that $(A, k)^{\bullet} \overset{*}{\underset{P}{\Longrightarrow}} \overline{R'}$.

Now, our productions look almost like *NewNode*, and we could complete the proof using Claim 1 in order to treat parallel hyperedges. Additionally, we would have to use the facts that

- the ordering on $att(y)$, for any $y \in Y_{rhs(NewNode)}$, does not matter, ie, if G is derivable then also $(V_G, E_G, Y_G, lab_G, att_G, \pi(ext_G))$, for π a permutation, and
- we may "remove" each $\boxed{k}$-labelled hyperedge by a derivation $(\boxed{k}, k)^{\bullet} \overset{*}{\underset{P_{HRG}}{\Longrightarrow}} G$ with $G \in \mathcal{G}_k$ such that $|V_G| = k$ and $|E_G| = 0$.

We end the discussion of this example here, because all interesting things have already been done. The example should have been sufficient to illustrate the use and importance of the Contextfreeness Lemma. Just imagine an ad hoc proof for the main claim which does not use the lemma — it is supposed to become rather lengthy and (even more) difficult to read. □

5 Concluding remarks

We hope that readers of the note have got a first insight into hyperedge replacement and have become curious to learn more. Indeed, the theory of hyperedge replacement is comparatively rich with structural results corresponding to the properties of contextfree Chomsky–languages and decidability results concerning graph–theoretic properties of the members of the generated languages. As mentioned in the introduction, the details can be found in the literature (see, e.g., [BC 87, Cou 87a, Cou 87b, Hab 89, HK 87b, HKV 89a, HKV 89b, LW 88]).

References

[ALS 88] S. Arnborg, J. Lagergren, D. Seese: *Which problems are easy for tree-decomposable graphs? Lecture Notes in Computer Science* 317, 38–51, 1988.

[Arn 85] S. Arnborg: *Efficient algorithms for combinatorial problems on graphs with bounded decomposability — a survey. BIT* 25, 2–23, 1985.

[BC 87] M. Bauderon, B. Courcelle: *Graph expressions and graph rewriting. Mathematical Systems Theory* 20, 83–127, 1987.

[Bod 86] H. L. Bodlaender: *Classes of graphs with bounded tree–width.* Report RUU–CS–86–22, Rijkskuniversiteit Utrecht, vakgroep informatica, 1986.

[Bod 88a] H. L. Bodlaender: *Dynamic programming on graphs with bonded tree–width. Lecture Notes in Computer Science* 317, 105–118, 1988.

[Bod 88b] H. L. Bodlaender: *NC–algorithms for graphs with small tree-width.* Report RUU-CS-88-4, Rijskuniversiteit Utrecht, vakgroep informatica, 1988.

[Cou 87a] B. Courcelle: *On context-free sets of graphs and their monadic second-order theory.* *Lecture Notes in Computer Science* 291, 133–146, 1987.

[Cou 87b] B. Courcelle: *A representation of graphs by algebraic expressions and its use for graph rewriting systems.* *Lecture Notes in Computer Science* 291, 112–132, 1987.

[Cou 88] B. Courcelle: *The monadic second-order logic of graphs, III: tree-width, forbidden minors and complexity issues.* Preprint, Université Bordeaux 1, 1988.

[Fed 71] J. Feder: *Plex languages.* *Information Science* 3, 225–241, 1971.

[Hab 89] A. Habel: *Hyperedge Replacement: Grammars and Languages.* Dissertation, Univ. Bremen, 1989.

[HK 87a] A. Habel, H.-J. Kreowski: *Characteristics of graph languages generated by edge replacement.* *Theoretical Computer Science* 51, 81–115, 1987.

[HK 87b] A. Habel, H.–J. Kreowski: *May we introduce to you: hyperedge replacement.* *Lecture Notes in Computer Science* 291, 15–26, 1987.

[HK 87c] A. Habel, H.–J. Kreowski: *Some structural aspects of hypergraph languages generated by hyperedge replacement.* *Lecture Notes in Computer Science* 247, 207–219, 1987.

[HKV 89a] A. Habel, H.–J. Kreowski, W. Vogler: *Decidable boundedness problems for hyperedge replacement graph grammars.* *Lecture Notes in Computer Science* 351, 275–289, 1989. Proc. TAPSOFT 89.

[HKV 89b] A. Habel, H.–J. Kreowski, W. Vogler: *Metatheorems for decision problems on hyperedge replacement graph languages.* *Acta Informatica* 26, 657–677, 1989.

[Kre 79] H.-J. Kreowski: *A pumping lemma for context-free graph languages.* *Lecture Notes in Computer Science* 73, 270–283, 1979.

[Lau 88] C. Lautemann: *Decomposition trees: structured graph representation and efficient algorithms.* *Lecture Notes in Computer Science* 299, 28–39, 1988. Proc. CAAP 88.

[Lau 90] C. Lautemann: *The complexity of graph languages generated by hyperedge replacement.* *Acta Informatica* 27, 399–421, 1990.

[LW 88] T. Lengauer, E. Wanke: *Efficient analysis of graph properties on context-free graph languages.* *Lecture Notes in Computer Science* 317, 379–393, 1988.

[Pav 72] T. Pavlidis: *Linear and context-free graph grammars.* *Journal of the ACM* 19(1), 11–23, 1972.

[RS 86] N. Robertson, P.D. Seymour: *Graph minors II. Algorithmic aspects of tree width.* *Journal of Algorithms* 7, 309–322, 1986.

Graph grammars based on node rewriting: an introduction to NLC graph grammars

Joost Engelfriet   Grzegorz Rozenberg

Department of Computer Science, Leiden University
P.O.Box 9512, 2300 RA Leiden, The Netherlands

ABSTRACT. An elementary introduction to the notion of an NLC graph grammar is given, and several of its extensions and variations are discussed in a systematic way. Simple concepts are considered rather than technical details.

Keywords: graph grammar, node rewriting, node label controlled

CONTENTS

1. Node rewriting and the NLC methodology

One way of rewriting a graph h into a graph h′ is to *replace* a subgraph m of h by a graph d and to *embed* d into the remainder of h, i.e., into the graph that remains after removing m from h; h′ is the resulting graph. We say that h is the *host graph*, m is the *mother graph*, and d is the *daughter graph*. Note that the removal of m from h includes the removal of all edges of h that are incident with nodes of m. The embedding process connects d to the remainder of h by establishing edges between certain nodes of d and certain nodes of the remainder of h.

Many approaches to graph grammars are based on this way of rewriting. In the restricted case of *node rewriting* the mother graph consists of one node only, a "local unit" of the host graph. Thus, if also the embedding is done in a "local" fashion (by connecting the daughter graph to host nodes that are "close" to the mother node), then a node rewriting step is a local graph transformation. The iteration of such node rewriting steps leads to a *global* transformation of a graph into a graph that is based on *local* transformations. This is the underlying idea of graph grammars based on node rewriting. In such a grammar the replacement of nodes is specified by a finite number of *productions* and the embedding mechanism is specified by a finite number of *connection instructions*. Often these finitely many productions and connection instructions are combined in a finite number of rewriting rules, where each rewriting rule consists of one production and a finite number of connection instructions. The rewriting rules are the main constituent of the graph grammar, which is a finite specification of a graph rewriting system.

A typical example of a node-rewriting mechanism is the *Node Label Controlled* mechanism, or NLC mechanism. In the NLC framework one rewrites undirected node-labeled graphs. The replacing productions are node-replacing productions and the embedding connection instructions connect the daughter graph to the neighbourhood of the mother node - hence the rewriting process is completely local. In the NLC approach "everything" is based on node labels. A *production* is of the form $A \longrightarrow d$, where A is a (nonterminal) node label, and d is an undirected graph with (terminal or nonterminal) node labels. Such a production can be applied to <u>any</u> node m in the host graph that has label A (i.e., there are no application conditions); its application results in the replacement of mother node m by daughter graph d. A *connection instruction* is of the form (δ,μ), where both δ and μ are (terminal or nonterminal) node labels. The meaning of such an instruction is that the embedding process should establish an edge between <u>each</u> node labeled δ in the daughter graph d and <u>each</u> node labeled μ in the neighbourhood of the mother node m. Since a connection instruction is an ordered pair, a finite number of connection instructions is a relation, called a *connection relation*.

Example 1. Consider the production $A \longrightarrow d$ in Fig.1, and consider the connection instructions (a,c), (b,b), and (A,b), where we assume that A is a nonterminal node label and a,b,c are terminal node labels. Consider also the host graph h in Fig.2, and let m be the unique node of h with label A. The application of $A \longrightarrow d$ to mother node m is shown in two steps in Figs.3 and 4. Fig.3 shows the result of replacing m by d; m is removed, together with its three incident edges, and d is added (disjointly) to the remainder of h. Fig.4 shows the resulting graph h′, i.e., the result of embedding d in the remainder of h; the five edges that are added by the embedding process are drawn fatter than the other edges. Thus, h is transformed into h′ by the application of $A \longrightarrow d$ to m. □

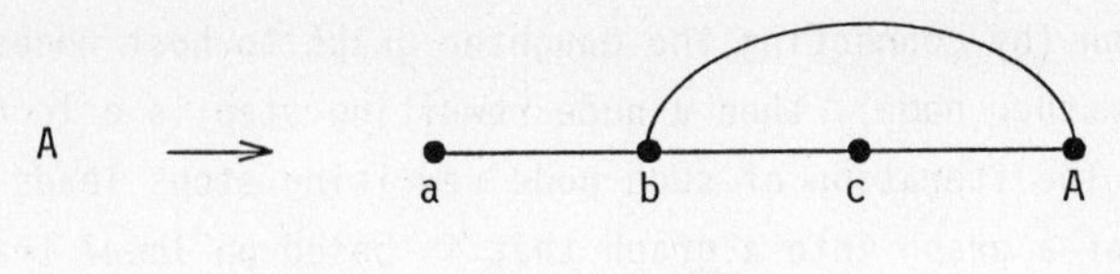

Figure 1

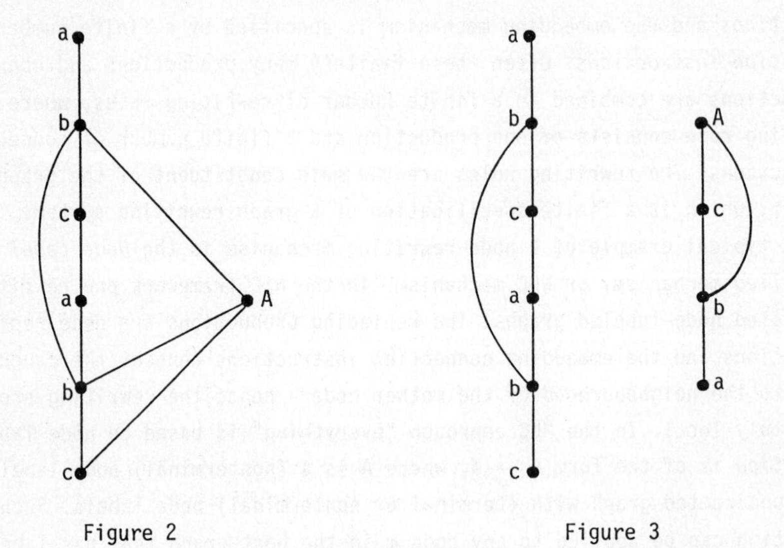

Figure 2

Figure 3

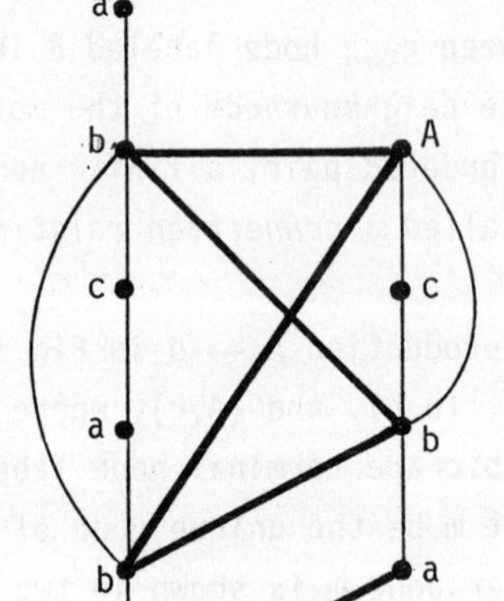

Figure 4

An *NLC graph grammar* is a system $G = (\Sigma,\Delta,z,P,C)$ where $\Sigma-\Delta$ and Δ (with $\Delta \subseteq \Sigma$) are the alphabets of nonterminal and terminal node labels, respectively, z is the initial graph (over Σ), P is a finite set of productions, and C is a connection relation, i.e., a binary relation over Σ. As usual, the graph language generated by G is $L(G) = \{g \in GR(\Delta) \mid z \Longrightarrow^* g\}$, where $GR(\Delta)$ is the set of undirected graphs with node labels in Δ, $\Longrightarrow$ represents one rewriting step, and $\Longrightarrow^*$ represents a derivation, i.e., a sequence of rewriting steps.

To simplify the description of certain graphs, in examples, let us view a string $\alpha_1\alpha_2\cdots\alpha_n$, with $\alpha_i \in \Delta$ for some alphabet Δ and $n \geq 1$, as an undirected node-labeled graph with nodes $x_1,x_2,\ldots,x_n$ and edges $\{x_i,x_{i+1}\}$ for every $1 \leq i \leq n-1$, such that x_i has label α_i for every $1 \leq i \leq n$. In particular, for $\alpha \in \Delta$, α also denotes the graph with one α-labeled node and no edges.

Example 2. Let us consider an NLC graph grammar $G = (\Sigma,\Delta,z,P,C)$ such that L(G) is the set of all strings in $(abc)^+$ with additional edges between all nodes with label b. G is defined as follows: $\Sigma = \{A,a,b,c\}$, $\Delta = \{a,b,c\}$, $z = A$, P consists of the production in Fig.1 and the production $A \longrightarrow abc$, and $C = \{(a,c), (b,b), (A,b)\}$. The intermediate graphs derived by this grammar are the strings in $(abc)^*A$ with additional edges between all nodes with label b or label A. One rewriting step of G was considered in Example 1. □

NLC graph grammars are one attempt to define a class of "context-free" graph grammars, i.e., graph grammars that are similar to the usual context-free grammars for strings. Such an attempt is useful if one wishes to carry over the nice properties of context-free grammars to the case of graphs. In particular, "context-free" graph grammars are meant to be a framework for the description of recursively defined properties of graphs. NLC graph grammars are context-free in the sense that they are completely local and have no application conditions. However, in general, they do not have the context-free property that the result of a derivation is independent of the order in which the productions are applied, a property that guarantees the existence of derivation trees (which embody the recursive nature of context-free grammars). A graph grammar that <u>has</u> this property is said to be *order-independent* or *confluent* (not to be confused with the related but different notion of confluence in term rewriting systems).

Example 3. Consider an NLC grammar G with initial graph S and productions $S \longrightarrow AB$, $A \longrightarrow a$ and $B \longrightarrow b$ (where S,A,B are nonterminal node labels and a,b are terminal node labels). Viewed as a context-free string grammar, $L(G) = \{ab\}$ and G has exactly one derivation tree. Suppose now that G has connection relation $C = \{(a,B),(b,a)\}$. Then $L(G) = \{ab,g\}$ where g is the graph with two nodes, labeled a and b, and no edges. Note that, intuitively, G still has one derivation tree, but the two ways of traversing this tree produce different graphs: ab is produced when $A \longrightarrow a$ is applied before $B \longrightarrow b$, whereas g is produced when they are applied in the reverse order. □

One way to solve the problem of non-confluence is simply to restrict attention to confluent NLC (or C-NLC) graph grammars: confluence is a decidable property. Another way is to put natural structural restrictions on the NLC grammar that guarantee confluence. An attractive example of such a restriction is the "boundary" restriction: in a *boundary* NLC (or B-NLC) graph grammar no two nodes with a nonterminal label are connected by an edge (in the right-hand sides of productions and in the initial graph). In fact, an important feature of NLC rewriting is the following: if, in a derivation, one obtains an intermediate graph with two nodes x and y that are <u>not</u> connected by an edge, then, whatever will happen later to these nodes in the derivation, no node descending from x will ever be connected to a node descending from y: connections can be broken, but cannot be re-established. In a boundary NLC grammar this feature ensures that in every intermediate graph nodes with a nonterminal label are unconnected, and this in turn ensures that the grammar is confluent. It should be clear that the grammar of Example 2 is a B-NLC grammmar; in fact, it is even linear: there is at most one node with a nonterminal label in every derived graph.

2. Extensions and variations

The NLC approach described in the previous section can be extended and modified in various ways, which is the reason that, today, NLC is a generic name for a methodology of graph rewriting based on node replacement and "local" embedding. Here we discuss some of these extensions and variations, which are also called "NLC-like" graph grammars.

2.1. It is often convenient to be able to refer in the connection instructions *directly* to nodes in the daughter graph rather than through their labels. Hence each connection instruction will now be of the form (x,μ), where x is a node in the daughter graph (i.e., in one of the right-hand sides of the productions of the grammar), and, as before, μ is a node label. The meaning of this instruction is that node x should be connected to each node labeled μ in the neighbourhood of the mother node. This gives us a convenient way of distinguishing between individual nodes in the daughter graph.

Example 4. Consider the daughter graph d in Fig.5, and suppose that some mother node m is replaced by d. Although x and y have the same label a, if (x,b) is a connection instruction, but (y,b) is not, then x will be connected to all b-labeled neighbours of m, but y will not! □

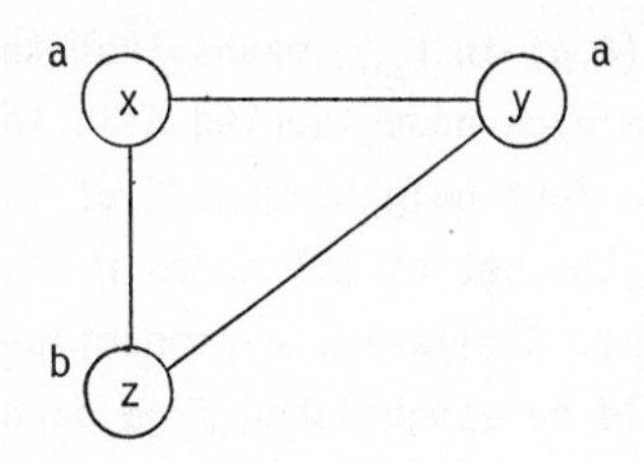

Figure 5

NLC-like grammars with this type of connection instructions are called *NCE graph grammars*, i.e., graph grammars *with Neighbourhood Controlled Embedding*. This acronym stresses the locality of the embedding process, which NCE grammars "inherit" from NLC grammars. Of course, NCE grammars are still "node label controlled" as far as replacement is concerned; the embedding process is "node label controlled" with respect to the neighbourhood of the mother node only, because the nodes of the daughter graph are accessed directly. NCE graph grammars generate the same class of graph languages as NLC graph grammars, but this is not always true when combined with some of the variations to be discussed.

In the formal definition of an NCE graph grammar one could stick to one connection relation and require the right-hand sides of productions to be mutually disjoint. However, it is more natural to define an NCE grammar as a system (Σ,Δ,z,R), such that Σ, Δ, and z are as before, R is a finite set of "rewriting rules", and each rewriting rule is a pair $(A \longrightarrow d, C)$ where $A \longrightarrow d$ is a production and C is a connection relation "for d", i.e., $C \subseteq V_d \times \Sigma$ (where V_d is the set of nodes of d).

One might think of extending the connection instructions to be of the form (x,y), where x is a node of the daughter graph and y is a neighbour of the mother node. However, this does not work in general, because the number of neighbours of the mother node may be unbounded, which makes it impossible to access individual neighbours. A way to distinguish between neighbours in a better way than just through their labels will be discussed later.

2.2. It is natural to extend the domain of graph rewriting to graphs more general than undirected node-labeled graphs. In particular we will discuss directed graphs and graphs with edge labels.

To extend the NLC approach to directed node-labeled graphs is quite easy. Rather than just one connection relation C we use two connection relations, C_{in} and C_{out}, to deal with the incoming edges and the outgoing edges of the mother node, respectively. These connection relations are used in an obvious way. Thus, a connection instruction (δ,μ) in C_{out} means that the embedding process should establish an edge from each node labeled δ in the daughter graph d to each node labeled μ in the "out-neighbourhood" of the mother node m (where the out-neighbourhood of m is the set of all nodes n for which there is an edge from m to n in the host graph). Similarly, a connection instruction (δ,μ) in C_{in} means that an edge should be established from each node labeled μ in the "in-neighbourhood" of m to each node labeled δ in the daughter graph. Note that the grammar is direction preserving, in the sense that edges leading to m (from m, respectively) are "replaced" by edges leading to d (from d, respectively). It is not difficult to think of a variation in which directions may be changed: one could take four connection relations $C_{i,j}$, with $i,j \in \{in,out\}$.

As observed before, neighbours of the mother node can be distinguished by the embedding process of an NLC grammar when they have distinct labels only. The extension to directed graphs already gives some more discerning power: the grammar can also distinguish between "out-neighbours" and "in-neighbours" of the mother node. Extending the NLC approach to (undirected) graphs that, in addition to node labels, also have edge labels gives even more discerning power in the neighbourhood of the mother node.

Example 5. For the host graph in Fig.6, when replacing the mother node m, the embedding process can distinguish between neighbours x and y (both labeled a), because x is a p-neighbour of m (i.e., is connected to m by an edge with label p) and y is a q-neighbour of m. Note that it is still impossible to distinguish between neighbours x and u. □

Thus, intuitively, the neighbours of the mother node m are divided into several distinct types, depending on the label of the edge that connects the neighbour to m. A natural, and useful, idea is to allow the embedding process to change the type of these neighbours: when the embedding process connects a daughter node x to a neighbour n of m, the type of n with respect to x may differ from its type with respect to m. This leads to connection instructions of the form (δ,q,μ,p), where p and q are edge labels, and δ and μ are node labels as before. The meaning of this connection instruction is that the

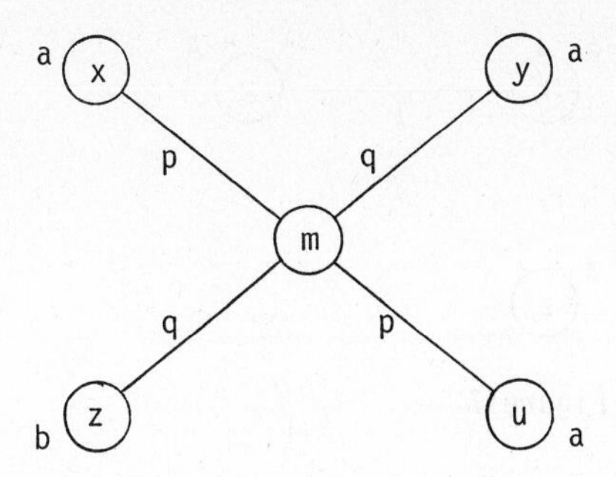

Figure 6

embedding process should establish an edge with label q between each δ-labeled node in the daughter graph and each μ-labeled p-neighbour of the mother node.

Even if one is not interested in generating graphs with edge labels, the edge labels can be used as a natural additional control of the rewriting and embedding process during the derivation of graphs. In many cases this facilitates the construction and the understanding of the grammar.

We use a small letter d to indicate that an NLC-like graph grammar generates directed graphs, and a small letter e to indicate that the generated graphs have edge labels in addition to node labels. Thus, one has, e.g., dNLC grammars, eNCE grammars, and edNLC grammars.

Example 6. To show the use of dynamically changing edge labels, we consider an eNCE grammar G that generates all strings in a^+ with additional edges $\{x_i, x_{i+2}\}$ for every i; all edges have label q. Note that the connection instructions of an eNCE grammar have the form (x,q,μ,p), where x is a daughter node, p,q are edge labels, and μ is a node label. $G = (\Sigma,\Delta,\Omega,z,R)$ is defined as follows: $\Sigma = \{A,a\}$, $\Delta = \{a\}$, the edge label alphabet Ω is $\{q,1,2\}$, $z = A$, and R consists of the two rewriting rules $(A \longrightarrow d_1,C_1)$ and $(A \longrightarrow d_2,C_2)$, where $A \longrightarrow d_1$ and $A \longrightarrow d_2$ are shown in Fig.7, $C_1 = \{(x,q,a,1), (x,q,a,2), (y,2,a,1)\}$ and $C_2 = \{(z,q,a,1), (z,q,a,2)\}$. G generates intermediate graphs of the form shown in Fig.8. The edge labels 1 and 2 indicate the first and second node, respectively, "to the left of" the mother node. Application of the rewriting rule $(A \longrightarrow d_1,C_1)$ leads to a change of type of the first neighbour of S: after application it becomes the second neighbour of S; this change of type is caused by the connection instruction (y,2,a,1). Note that G is a C-eNCE grammar, because it is confluent; in fact, it is even a B-eNCE grammar, and even a linear eNCE grammar. □

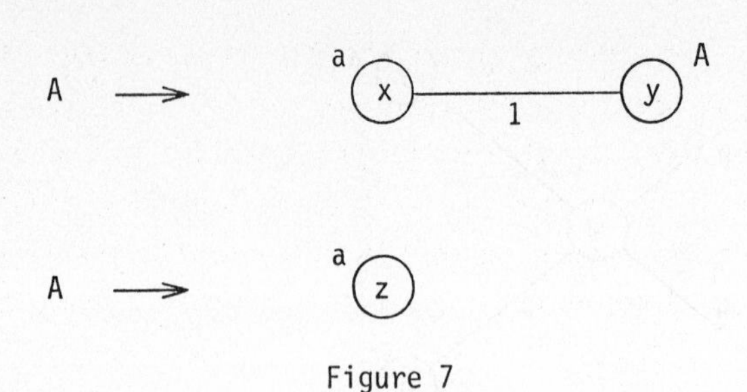

Figure 7

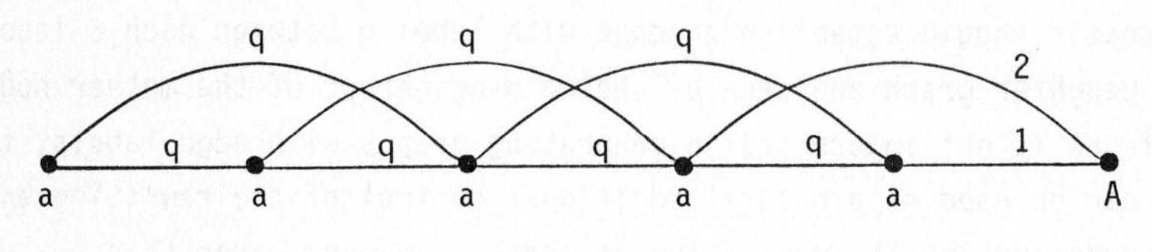

Figure 8

2.3. Combining all features discussed above naturally leads to the class of edNCE graph grammars. Each rewriting rule of such a grammar is of the form $(A \rightarrow d, C_{in}, C_{out})$, and each connection instruction is of the form (x,q,μ,p), where x is a node of d, p,q are edge labels, and μ is a node label (as in Example 6). If this instruction is an element of, say, C_{out}, then it is interpreted as follows: the embedding process should establish an edge with label q from node x of d to each μ-labeled p-neighbour in the "out-neighbourhood" of m. It can be shown, for edNCE grammars, that the information concerning the labels of neighbours can be coded into the labels of the connecting edges. Thus, one might in fact assume the connection instructions to be of the simpler form (x,q,p) and disregard the label μ of the neighbour of the mother node.

As argued in Section 1, of particular interest are the C-edNCE and the B-edNCE grammars: the confluent edNCE grammars and the subclass of boundary edNCE grammars.

In going from the class of NLC grammars to the class of edNCE grammars we have clearly made the embedding mechanism much more involved. Thus, since in analyzing derivations of edNCE grammars one has to keep track additionally of edge labels, directions, and individual nodes, it would seem that it is more involved to analyze edNCE grammars than NLC grammars. However, there is a definite trade-off here: it turns out that the additional features of the edNCE grammar can often be used in a straightforward, easily understandable fashion to show that they have certain desirable properties that are not possessed by NLC grammars. As a simple example, every B-edNCE grammar has an equivalent B-edNCE grammar in Chomsky Normal Form (appropriately defined); the analogous result does not hold for B-NLC grammars. As another example, it turns out that the class of C-edNCE graph languages can be characterized in several different, natural ways, as opposed to the class of C-NLC languages.

3. Bibliographical comments

NLC graph grammars were introduced in [JR1], and investigated in, e.g., [JR2, JR3, EMR, JRW, EJKR, KR, EL1, HM]. Boundary NLC grammars were introduced in [RW1], and studied, e.g., in [RW2, RW3, Wel1/2, Vog]. The notion of confluence for graph grammars in general was stressed in [Cou], where also confluent NLC grammars were investigated. In [Jef] the slightly more general notion of order independence was discussed. Another subclass of the class of C-NLC grammars was considered in [JR6, Cou]: the neighbourhood-uniform NLC grammars.

NCE graph grammars were introduced in [JR5] and used, e.g., in [ELR1]. The extension to directed graphs, i.e., the dNLC graph grammar, was introduced in [JR4] and investigated, e.g., in [AR, ARE, AER]. The extension to graphs with edge labels, i.e., the eNLC and edNLC graph grammar, was first considered in [JRV]. The eNCE and edNCE graph grammars were introduced in [Bra1, Kau1/2] and [ELW, EL2], independently, and studied, e.g., in [EL3, ELR2, ER]. In particular, the class of C-edNCE grammars is investigated in [Kau1, Bra2, Sch, Eng1/2, CER]. Some other extensions of NLC grammars are considered in [GJRT, MR1/2].

References

[AER] IJ.J.Aalbersberg, J.Engelfriet, G.Rozenberg; The complexity of regular DNLC graph languages, JCSS 40 (1990), 376-404

[AR] IJ.J.Aalbersberg, G.Rozenberg; Traces, dependency graphs and DNLC grammars, Discrete Appl. Math. 11 (1985), 299-306

[ARE] IJ.J.Aalbersberg, G.Rozenberg, A.Ehrenfeucht; On the membership problem for regular DNLC grammars, Discrete Appl. Math. 13 (1986), 79-85

[Bra1] F.J.Brandenburg; On partially ordered graph grammars, in [ENRR], 99-111
[Bra2] F.J.Brandenburg; On polynomial time graph grammars, Proc. STACS 88, Lecture Notes in Computer Science 294, Springer-verlag, Berlin, 227-236
[Cou] B.Courcelle; An axiomatic definition of context-free rewriting and its application to NLC graph grammars, Theor. Comput. Sci. 55 (1987), 141-181
[CER] B.Courcelle, J.Engelfriet, G.Rozenberg; Context-free handle-rewriting hypergraph grammars, this Volume
[EMR] A.Ehrenfeucht, M.G.Main, G.Rozenberg; Restrictions on NLC graph grammars, Theor. Comput. Sci. 31 (1984), 211-223
[EJKR] H.Ehrig, D.Janssens, H.-J.Kreowski, G.Rozenberg; Concurrency of node-label controlled graph transformations, Report 82-38, University of Antwerp, U.I.A., 1982
[ENRR] H.Ehrig, M.Nagl, G.Rozenberg, A.Rosenfeld (eds.); "Graph-Grammars and their Application to Computer Science", Lecture Notes in Computer Science 291, Springer-Verlag, Berlin, 1987
[Eng1] J.Engelfriet; Context-free NCE graph grammars, Proc. FCT '89, Lecture Notes in Computer Science 380, Springer-Verlag, Berlin, 1989, 148-161
[Eng2] J.Engelfriet; A characterization of context-free NCE graph languages by monadic second-order logic on trees, this Volume
[EL1] J.Engelfriet, G.Leih; Nonterminal bounded NLC graph grammars, Theor. Comput. Sci. 59 (1988), 309-315
[EL2] J.Engelfriet, G.Leih; Linear graph grammars: power and complexity, Inform. and Comput. 81 (1989), 88-121
[EL3] J.Engelfriet, G.Leih; Complexity of boundary graph languages, RAIRO Theoretical Informatics and Applications 24 (1990), 267-274
[ELR1] J.Engelfriet, G.Leih, G.Rozenberg; Apex graph grammars and attribute grammars, Acta Informatica 25 (1988), 537-571
[ELR2] J.Engelfriet, G.Leih, G.Rozenberg; Nonterminal separation in graph grammars, Report 88-29, Leiden University; to appear in Theor. Comput. Sci.
[ELW] J.Engelfriet, G.Leih, E.Welzl; Boundary graph grammars with dynamic edge relabeling, JCSS 40 (1990), 307-345
[ER] J.Engelfriet, G.Rozenberg; A comparison of boundary graph grammars and context-free hypergraph grammars, Inform. and Comput. 84 (1990), 163-206
[GJRT] H.J.Genrich, D.Janssens, G.Rozenberg, P.S.Thiagarajan; Generalized handle grammars and their relation to Petri nets, EIK 4 (1984), 179-206
[HM] J.Hoffmann, M.G.Main; Results on NLC grammars with one-letter terminal alphabets, Theor. Comput. Sci. 73 (1990), 279-294
[JR1] D.Janssens, G.Rozenberg; On the structure of node-label-controlled graph languages, Information Sciences 20 (1980), 191-216
[JR2] D.Janssens, G.Rozenberg; Restrictions, extensions, and variations of NLC grammars, Information Sciences 20 (1980), 217-244
[JR3] D.Janssens, G.Rozenberg; Decision problems for node label controlled graph grammars, JCSS 22 (1981), 144-177
[JR4] D.Janssens, G.Rozenberg; A characterization of context-free string languages by directed node-label controlled graph grammars, Acta Informatica 16 (1981), 63-85
[JR5] D.Janssens, G.Rozenberg; Graph grammars with neighbourhood-controlled embedding, Theor. Comput. Sci. 21 (1982), 55-74
[JR6] D.Janssens, G.Rozenberg; Neighborhood-uniform NLC grammars, Computer Vision, Graphics, and Image Processing 35 (1986), 131-151
[JRV] D.Janssens, G.Rozenberg, R.Verraedt; On sequential and parallel node-rewriting graph grammars, Computer Graphics and Image Processing 18 (1982), 279-304
[JRW] D.Janssens, G.Rozenberg, E.Welzl; The bounded degree problem for NLC grammars is decidable, JCSS 33 (1986), 415-422
[Jef] J.Jeffs; Embedding rule independent theory of graph grammars, in [ENRR], 299-308
[Kau1] M.Kaul; "Syntaxanalyse von Graphen bei Präzedenz-Graph-Grammatiken", Dissertation, Universität Osnabrück, 1985
[Kau2] M.Kaul; Practical applications of precedence graph grammars, in

[ENRR], 326-342
[KR] H.-J.Kreowski, G.Rozenberg; Note on node-rewriting graph grammars, Inf. Proc. Letters 18 (1984), 21-24
[MR1] M.G.Main, G.Rozenberg; Handle NLC grammars and r.e. languages, JCSS 35 (1987), 192-205
[MR2] M.G.Main, G.Rozenberg; Edge-label controlled graph grammars, JCSS 40 (1990), 188-228
[RW1] G.Rozenberg, E.Welzl; Boundary NLC graph grammars - basic definitions, normal forms, and complexity, Inform. and Control 69 (1986), 136-167
[RW2] G.Rozenberg, E.Welzl; Graph theoretic closure properties of the family of boundary NLC graph languages, Acta Informatica 23 (1986), 289-309
[RW3] G.Rozenberg, E.Welzl; Combinatorial properties of boundary NLC graph languages, Discrete Appl. Math. 16 (1987), 59-73
[Sch] R.Schuster; Graphgrammatiken und Grapheinbettungen: Algorithmen und Komplexität, Report MIP-8711, Passau, 1987
[Vog] W.Vogler; On hyperedge replacement and BNLC graph grammars, in Graph-Theoretic Concepts in Computer Science WG '89, Lecture Notes in Computer Science 411, Springer-Verlag, Berlin, 78-93
[Wel1] E.Welzl; Boundary NLC and partition controlled graph grammars, in [ENRR], 593-609
[Wel2] E.Welzl; On the set of all subgraphs of the graphs in a boundary NLC graph language, in "The Book of L" (G.Rozenberg, A.Salomaa, eds.), Springer-Verlag, Berlin, 1986, 445-459

Tutorial Introduction to the Algebraic Approach of Graph Grammars Based on Double and Single Pushouts

HARTMUT EHRIG, MARTIN KORFF, MICHAEL LÖWE

Technical University of Berlin
Department of Computer Science
Franklinstraße 28/29, D-1000 Berlin 10

Abstract. The gluing construction on which the algebraic notion of a derivation is based operationally provides a simple and intuitive understanding of graph rewriting. Inheriting the powerful toolbox of category theory, its abstract version as a (single resp. double) pushout leads to highly compact and elegant proofs especially for the basic constructions of sequential and parallel independent derivations as well as for concurrent and amalgamated productions respectively.

Contents

1 General Format of Productions and Direct Derivations

Generally, the rewriting of a graph G, via some rule p is done by first deleting some part DEL (from G) and then adding a new part ADD, finally resulting into a derived graph H. Most often a rule p shall only be applicable if G additionally contains some context K_L, which is essentially to be kept, i.e. rewriting means to replace some part $L = DEL \cup K_L$ by another one $R = K_R \cup ADD$:

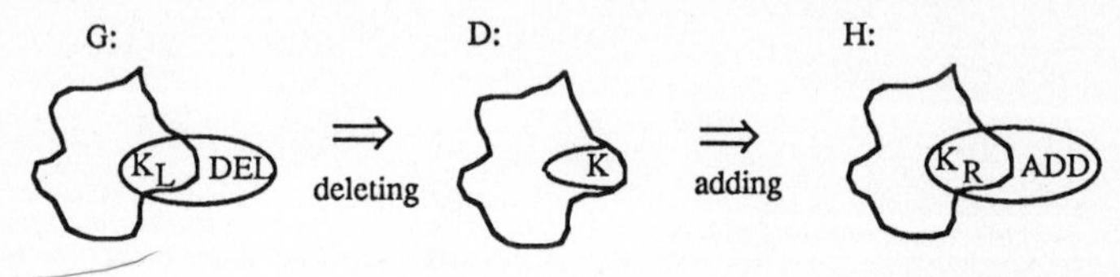

1.1 Production

Consequently a *production rule* p is a pair of graphs (L, R), called left and right hand side resp., together with designated graphs $K_L \subseteq L$ and $K_R \subseteq R$, the gluing points, which are in one-to-one correspondence to each other.

A rule p can be applied to a graph G not only if L is a subgraph of G, as shown in the picture above, but also by allowing that items K_L may be mapped onto the same item in G. The corresponding match, the *occurrence map*, is then said to be non-injective.

1.2 Gluing Condition

The dangling points $DANGLING$ are those points in L (considered as subgraph of G) which are source or target of arcs in $G - L$. The identification points $IDENTIFICATION$ are those points in L which are identified in G. The gluing points $GLUING$ of L are given by K_L. The condition

$$DANGLING \cup IDENTIFICATION \subseteq GLUING$$

is called gluing condition and assures that the context D in Step 2 below is indeed a graph.

1.3 Direct Derivations

A *direct derivation* from a mother graph G to a daughter graph H is obtained in three steps:

Step 1: Match the left hand side L of the production with a subgraph of the mother graph G and check the gluing condition (see 1.2).

Step 2: Consider the left hand side L except of its gluing points K_L, i.e. $L - K_L$, and remove it from the mother graph G. The result is a context graph D which still contains the gluing points K_L.

Step 3: Add the right hand side R to the context graph D by gluing together the gluing points K_L of D with corresponding gluing points K_R of R. The result is the daughter graph H.

The algebraic approach of graph rewriting is of extreme simplicity and clarity since the deletion of items as well as the embedding of the part to be added must explicitly be specified: each item of L to be deleted must be matched on an item of G exclusively by its own (Identification condition). Moreover vertices in G to be deleted are not allowed to have edges which are not specified in L as candidates for deletion themselves (Dangling condition).

2 An Introductory Example

Illustrating the essential features of the approach the following example has been designed as small as possible.

2.1 Production

Consider the production (L, R) where the gluing points K_L of L and K_R of R are designated by the numbers 1, 2 and 3 attached to items (i.e. nodes and arcs) in L and R respectively. The bijective correspondence between K_L and K_R is given by corresponding numbers:

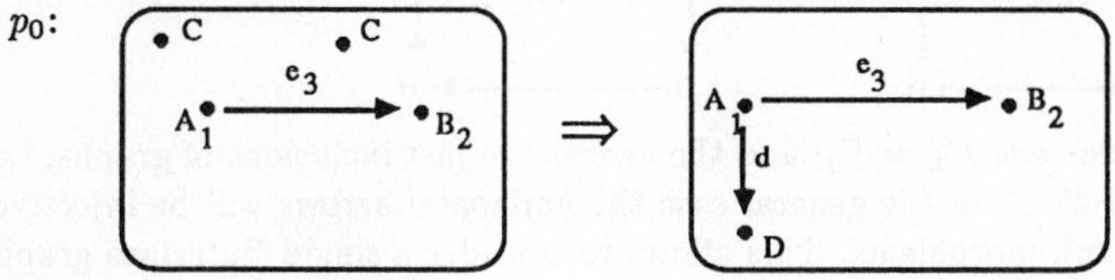

2.2 Direct Derivation

Consider the following mother graph

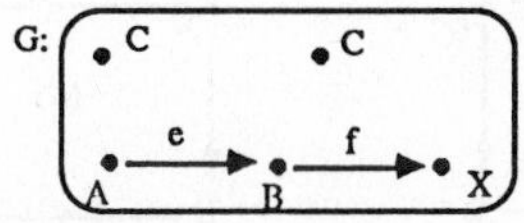

We are going to construct a direct derivation from G to a daughter graph H using the production above:

Step 1: All items in G up to the arc labelled f and its target node labelled X are matched by the left hand side L. The gluing condition is satisfied because the boundary (dangling) point 2 of L is included in the gluing points $K_L = \{1, 2, 3\}$ of L.

Step 2: Removing $L - K_L$ from G leads to the following context graph D which still contains the designated gluing points K_L of L.

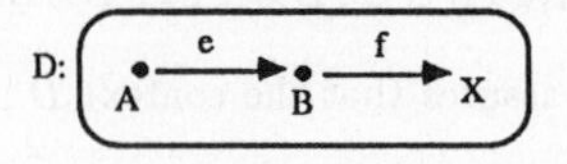

Remark:

▹ Note that if we construct the context D as $G - (L - K_L)$ on nodes and arcs separately, we cannot expect that D is in general a graph unless we make sure that the dangling points are still included in D. But whenever the gluing condition 1.2 is satisfied these points are gluing points and hence they are still included in the *graph* $G - (L - K_L)$.

Step 3: The result of step 3 is the following daughter graph H obtained as gluing of D and R at corresponding gluing points 1, 2 and 3:

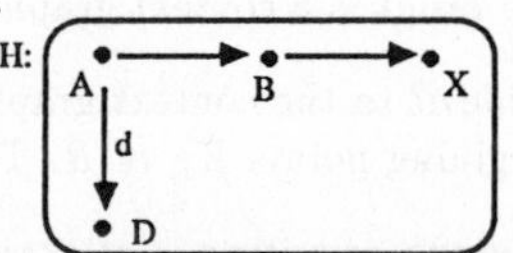

Note, that p_0 is not applicable to G' or G'' due to a violated gluing condition (There are non-gluing items (the 'C's) which are *identification* resp. *dangling* points):

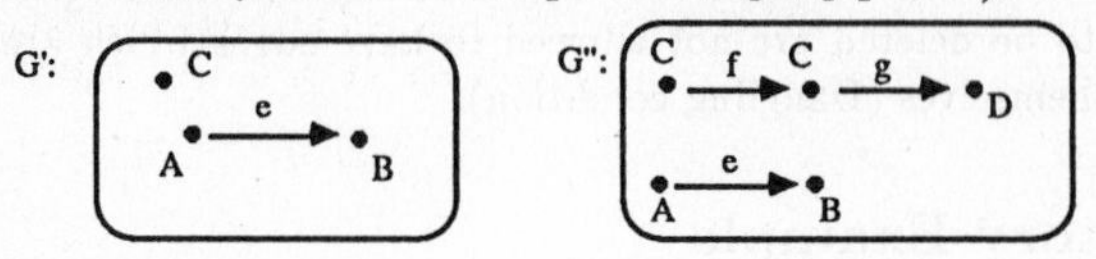

3 Towards an Algebraic Version

The graphs (L, R) of the production, the designated gluing points $K_L \subseteq L$ and $K_R \subseteq R$, the mother graph G, context graph D, and daughter graph H can be combined in the following diagrams:

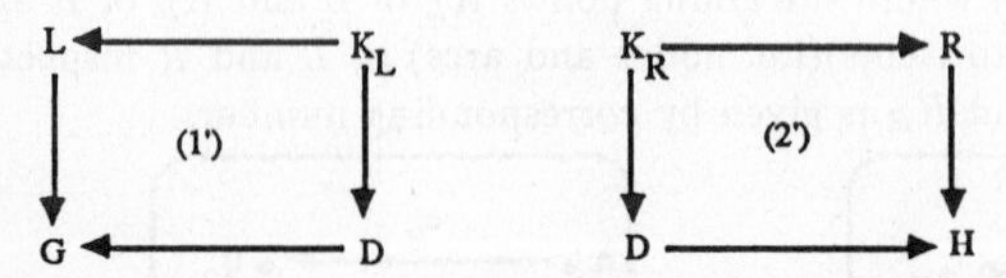

where in the simplest case we have $K_L = K_R$ and the arrows are just inclusions of graphs, i.e. $L \rightarrow G$ means that L is a subgraph of G. In the general case the horizontal arrows will be injective and the vertical arrows arbitrary graph morphisms. This allows to consider a single "interface graph" K and two injective graph morphisms $K \longleftarrow L$ and $K \longrightarrow R$ such that the diagrams (1') and (2') above can be combined to the following diagrams (1) and (2):

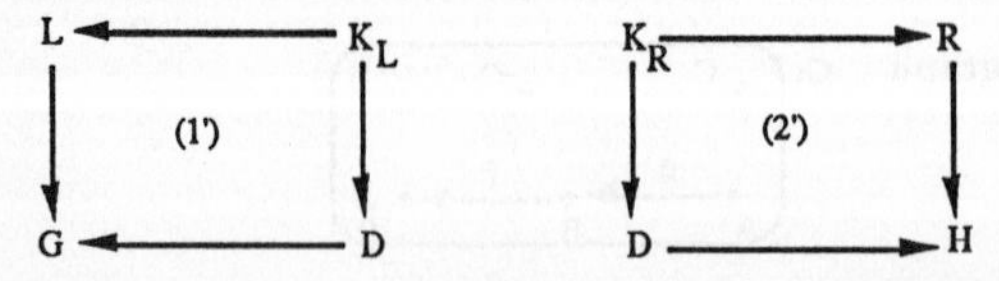

By construction in step 3, H is the gluing of R and D along K, which means that for each k in K the images of k in R and D w.r.t. $K \longrightarrow R$ resp. $K \longrightarrow D$ are glued together in H. As a result we obtain also graph morphisms $D \longrightarrow H$ and $R \longrightarrow H$ and the diagram (2) is a "gluing diagram" in the sense defined in 6. The construction of D in steps 1 and 2 implies that also (1) becomes a gluing diagram, i.e. G can be considered as the gluing of L and D along K. This leads to a simple symmetric definition of productions and direct derivations:

4 Algebraic Version of Productions and Direct Derivations

In the following we give a precise notion of productions and direct derivations using the notions "graph", "graph morphism", and "gluing diagram" which are formally defined in section 5, 6 and 7 below.

4.1 Productions

In addition to the pair (L, R) we consider a third graph K of gluing points together with injective graph morphisms $l : K \longrightarrow L$ and $r : K \longrightarrow R$. Hence a *production* p is given by the following graphs and graph morphisms:

$$p = (L \xleftarrow{l} K \xrightarrow{r} R)$$

Remarks:

- ▷ We obtain subgraphs $K_L = l(K) \subseteq L$ and $K_R = r(K) \subseteq R$ and a bijective correspondence between K_L, K_R and K via l and r and hence also between K_L and K_R as required in 1.1.
- ▷ Considering only the generative power of productions it would be sufficient to take a discrete graph K (i.e. without arcs) but in connection with independence of productions and derivations (see 8.8) it is important to allow also arcs as gluing points in K and hence in K_L and K_R.

4.2 Direct Derivations

A production p (as in 3.1) *can be applied* to a mother graph G if there is a graph morphism $g : L \rightarrow G$, the *occurrence map*, such that there is the following gluing diagram (1):

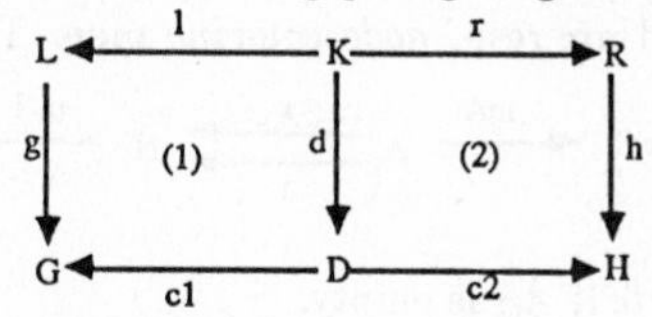

In this case, we always obtain the gluing diagram (2) which uniquely defines the graph H (up to isomorphism); we then call this a *direct derivation* from G to the daughter graph H w.r.t. p and g

$$(p, g) : G \Longrightarrow H \quad \text{briefly} \quad p : G \Longrightarrow H$$

Remarks:

- ▷ This definition of a direct derivation $p : G \Longrightarrow H$ based on gluing diagrams (1) and (2) is symmetric. This means that the production

$$p^{-1} = (R \xleftarrow{r} K \xrightarrow{l} L)$$

defines a direct derivation $p^{-1} : H \Longrightarrow G$ using the same gluing diagrams (1) and (2) in opposite order:

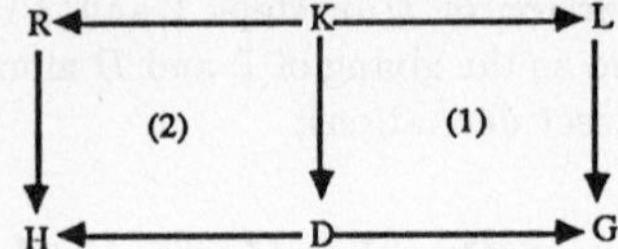

▷ The definition of a direct derivation in 1.2 does not seem to be symmetric at first glance. But it turns out to be equivalent to our algebraic version above. We show how step 1 to 3 are reflected in our definition:

Step 1: The match of L with a subgraph of G is given by the graph morphism $g : L \longrightarrow G$. The gluing condition is satisfied if and only if diagram (1) becomes a gluing diagram (see remark 4 in 6.).

Step 2: Given $l : K \longrightarrow L$ and $g : L \longrightarrow G$ the context graph D is constructed as $D = G - g(L - l(K))$ on nodes and arcs separately. The gluing condition makes sure that source and target of each arc in D are also nodes in D. The graph morphisms $d : K \longrightarrow D$ and $c1 : D \longrightarrow G$ are defined by $d(x) = gl(x)$ and $c1(y) = y$ for all items x in K and y in D.

Step 3: H is constructed as the gluing of R and D along K leading to the gluing diagram (2).

▷ The algebraic version of direct derivations avoids to require the gluing condition, because it is a consequence of (1) being a gluing diagram.

5 Definition of Graphs and Graph Morphisms

1. Let $C = (C_A, C_N)$ be a pair of sets, called pair of *color alphabets* for arcs and nodes respectively, which will be fixed in the following.

2. A *(colored) graph* $G = (A, N, s, t, mA, mN)$ consists of sets A, N, called set of arcs and nodes respectively, and mappings $s : A \to N, t : A \to N$, called *source* resp. *target map*, $mA : A \to C_A, mN : N \to C_N$, called *arc* resp. *node coloring map*. These data can be summarized in the diagram

$$G: C_A \xleftarrow{mA} A \underset{t}{\overset{s}{\rightrightarrows}} N \xrightarrow{mN} C_N$$

3. A graph G is called *discrete* if A_G is empty.

4. Graph G' is called *subgraph* of G if $A' \subseteq A$, $N' \subseteq N$ and all the mappings s', t', mA' and mN' are restrictions of the corresponding ones from G.

5. Given two graphs G and G' a *graph morphism* $f : G \to G'$, f or $G \to G'$ for short, is a pair of maps $f = (f_A : A \to A', f_N : N \to N')$ such that $f_N \circ s = s' \circ f_A, f_N \circ t = t' \circ f_A, mA' \circ f_A = mA$ and $mN' \circ f_N = mN$, i.e. the following diagram commutes for source and target mappings separately:

$$\begin{array}{ccccccc} & & A & \underset{t}{\overset{s}{\rightrightarrows}} & N & & \\ & \swarrow^{mA} & \downarrow f_A & & \downarrow f_N & \searrow^{mN} & \\ C_A & & & & & & C_N \\ & \nwarrow_{mA'} & A' & \underset{t'}{\overset{s'}{\rightrightarrows}} & N' & \nearrow_{mN'} & \end{array}$$

A graph morphism $f = (f_A, f_N)$ is called *injective* resp. *surjective* if both f_A and f_N are injective resp. surjective mappings. If $f : G \to G'$ is injective and surjective it is called an *isomorphism*, and there is also an inverse isomorphism $f' : G' \to G$.

6. The *composition* $f' \circ f : G \to G''$ of two graph morphisms $f = (f_A, f_N) : G \to G'$ and $f' = (f'_A, f'_N) : G' \to G''$ is defined by $f' \circ f = (f'_A \circ f_A, f'_N \circ f_N)$.

7. Graphs and graph morphisms as above are defining a category in the sense of category theory, called the *category of (colored) graphs.*

6 Explicit Version of Gluing and Gluing Diagram

Given graph morphisms $b : K \to B$ and $d : K \to D$ with injective b the gluing of B and D along K (more precisely along b and d) is given by the following graph G below. Moreover we obtain graph morphisms $c : D \to G$ and $g : B \to G$ leading to the following diagram (1), called *gluing diagram:*

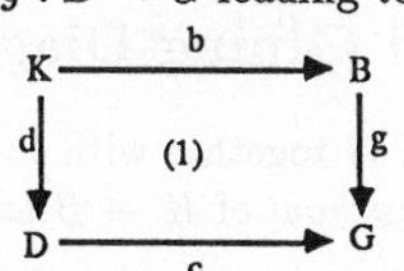

1. $G = D + (B - bK)$ for the sets of arcs and nodes respectively where s_G, t_G for all $a \in G_A$ and mG are defined by:
   $s_G(a) =$ **if** $a \in D_A$ **then** $s_D(a)$
   else if $s_B(a) \in (B - bK)_N$ **then** $s_B(a)$
   else dk **where** $s_B(a) = bk$ **for** $k \in K_N$
   $t_G(a)$ is defined similar replacing "s" by "t" simultaneously
   $m_G(x) =$ **if** $x \in D$ **then** $m_D(x)$ **else** $m_B(x)$ **for all** $x \in G$.

2. $c : D \to G$ is the inclusion from D into G.

3. $g : B \to G$ is defined for all $x \in B$ by
   $g(x) =$ **if** $x \in (B - bK)$ **then** x **else** dk **where** $x = bk$ **for** $k \in K$.

Note that s_G, t_G and g are well-defined because b is injective.

Remarks:

▷ The gluing of G, B and D along K is also defined for non-injective b. In this case G is defined on nodes and arcs by the following quotient

$$G = (B + D)_{/\equiv}$$

where $\equiv$ is the equivalence generated by the relation $\approx$ on $B + D$ defined by $b(k) \approx d(k)$ for each k in K.

▷ The gluing G above can be constructed by the following iterative procedure:

procedure gluing **input parameters** (B, D, K, b, d) **output parameter** (G)
begin $G := B + D$ <disjoint union>
for all nodes and arcs k **in** K **do**
$G := identify(G, bk, dk)$
end

where $identify(G, bk, dk)$ is a procedure which identifies the items bk and dk in the graph G.

▷ An algebraic version of the gluing construction and gluing diagram — which is much more suitable for proofs than the explicit construction above — will be given in section 7.

- ▹ If we have given injective $b : K \to B$ and arbitrary $g : B \to G$ we can construct a context graph D leading to a gluing diagram (1) as above if and only if the following gluing condition is satisfied:

$$BOUNDARY \subseteq GLUING$$

where $BOUNDARY$ and $GLUING$ are subgraphs of B defined by:

$$\begin{array}{ll} GLUING = bk & (= \mathrm{K1\ in\ 1.1}) \\ DANGLING = \{x \in B_N \mid \exists a \in (G - gB) : g(x) = s_G(a) \text{ or } g(x) = t_G(a)\} & \\ IDENTIFICATION = \{x \in B \mid \exists y \in B : x \neq y \text{ and } g(x) = g(y)\} & \\ BOUNDARY = DANGLING \cup IDENTIFICATION & \end{array}$$

7 Algebraic Version of Gluing and Gluing Diagram

Given graph morphism $K \to B$ and $K \to D$ a graph G together with two graph morphisms $B \to G$ and $D \to G$ is called *gluing* of B and D along K or *pushout* of $K \to B$ and $K \to D$ if we have

1. *(Commutativity)*: $K \to B \to G = K \to D \to G$, and
2. *(Universal Property)*: For all graphs G' and graph morphisms $B \to G'$ and $D \to G'$ satisfying $K \to B \to G' = K \to D \to G'$ there is a unique graph morphism $G \to G'$ such that

$$B \to G \to G' = B \to G' \text{ and } D \to G \to G' = D \to G'$$

The situation can be illustrated by the following diagram

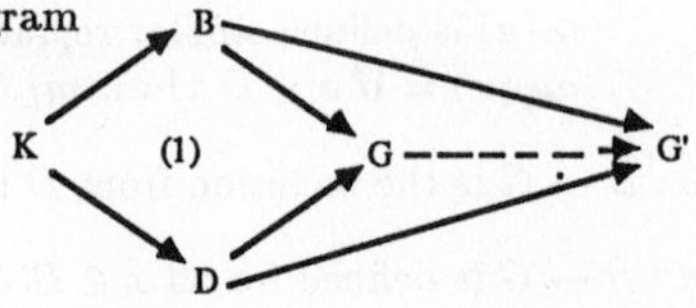

In general diagram (1) is called *gluing diagram* or *pushout* if it satisfies the properties 1. and 2. above.

Interpretation. The commutativity means that the items of B and D coming from the "interface" graph K are identified in G. On the other hand we want to make sure that no other items of B and D are glued together and that G does not contain other items which are not coming from B or D. These both requirements are expressed by the universal property of G where G is compared with any other G' satisfying a similar commutativity as G.

Remarks:

- ▹ The gluing graph G is uniquely determined up to isomorphism by the commutativity and universal property above. This is a general property of pushouts in any category.
- ▹ The explicit versions of gluing and gluing diagrams in 6. above are satisfying the properties 1. and 2. above and hence they are special cases of the algebraic version. On the other hand the uniqueness of G up to isomorphism implies that for each algebraic gluing graph G (resp. gluing diagram (1)) there is an isomorphic explicit gluing graph G' (resp. gluing diagram (1')). In this sense the explicit and the algebraic version are equivalent up to isomorphism. It is advisable to use the explicit version for explicit constructions and the algebraic version for proofs of theorems in the algebraic approach.
- ▹ A fundamental lemma about pushouts in any category - and hence also for gluing diagrams of graphs - states that pushouts can be composed horizontally as well as vertically: if (1), (2), (3)

and (4) are pushouts

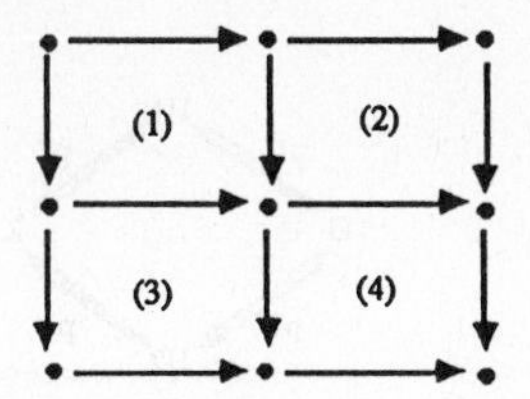

then also (1) + (2), (3) + (4), (1) + (3), (2) + (4) and hence also (1) + (2) + (3) + (4) are pushouts. To prove this using the explicit version of gluing diagrams would be very tedious.

8 Essentials of the Algebraic Approach

The essentials of the algebraic approach in contrast to other approaches are the following:

8.1 Rule applications via graph morphisms. Application of productions to graphs are given by graph morphisms $g : B1 \rightarrow G$,the occurrence maps, which may even be noninjective. Non-injectivity is essential for considering parallel productions where the left hand sides are allowed to overlap in the mother graph G.

8.2 The embedding mechanism. Given an occurrence map $g : L \rightarrow G$, an embedding is possible, i.e. additional arcs between vertices v in $g(L)$ and other vertices of G, provided v is (the image of) a gluing item k (along $K \rightarrow L \rightarrow G$). Since gluing items are preserved during a derivation this avoids to define an additional embedding mechanism.

8.4 Double Gluing. A direct derivation is given by a pair of gluing diagrams, also called *double gluing*, which is conceptually simple and symmetric.

The procedure to define graph derivations, graph grammars and graph languages once we have productions and direct derivations is almost the same than in other approaches:

8.5 Derivations. Graph derivations are sequences of $n \geq 0$ direct derivations.

8.6 Graph Grammar. A graph grammar $GG = (C,T,PROD,START)$ consists of a pair of color alphabets $C = (C_A, C_N)$, a pair of terminal color alphabets $T = (T_A, T_N)$ included in C, a finite set of productions $PROD$ and a startgraph $START$.

8.7 Graph Language. The graph language $L(GG)$ is defined to be the set of all terminal colored graphs G s.t. there exists a graph derivation from $START$ to G.

8.8 Pushouts. Gluing diagrams are pushouts in the category of graphs which allow to apply several well-known techniques and results from category theory to graph derivations and graph grammars.

This applies also to the following essential constructions of new productions and derivations from old ones concerning parallelism, concurrency and amalgamation and corresponding results.

8.9 Church-Rosser Property. Given direct derivations $p1 : G \Longrightarrow H1$ and $p2 : G \Longrightarrow H2$ which are *parallel independent* (i.e. the overlap of the left hand sides of $p1$ and $p2$ in G must be included in the intersection of the corresponding interface graphs K_1 and K_2 in G) we are able to construct direct derivations $p2 : H1 \Longrightarrow X$ and $p1 : H2 \Longrightarrow X$ with same graph X. Moreover, the sequences $G \Longrightarrow H1 \Longrightarrow X$ via $(p1,p2)$ and $G \Longrightarrow H2 \Longrightarrow X$ via $(p2,p1)$ are *sequential independent* (i.e. the overlap of the right hand side of the first production with the left hand side of the second production consists of gluing points, only).

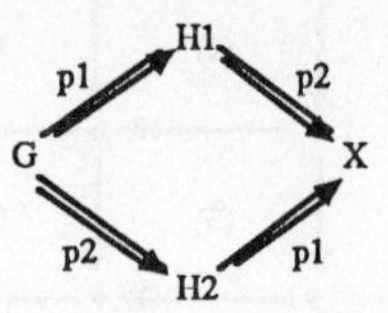

Vice versa given a sequential independent sequence $G \Longrightarrow H1 \Longrightarrow X$ via $(p1,p2)$ or $G \Longrightarrow H2 \Longrightarrow X$ via $(p2,p1)$ we are able to construct the other sequence with same G and X such that $G \Longrightarrow H1$ and $G \Longrightarrow H2$ become parallel independent.

8.10 Parallelism. Given $p1 = (L1 \longleftarrow K1 \longrightarrow R1)$ and $p2 = (L2 \longleftarrow K2 \longrightarrow R2)$ we are able to define a parallel production

$$p1 + p2 = (L1 + L2 \longleftarrow K1 + K2 \longrightarrow R1 + R2).$$

If $G \Longrightarrow H1$ via $p1$ and $G \Longrightarrow H2$ via $p2$ are parallel independent derivations there is also a parallel derivation $G \Longrightarrow X$ via $p1 + p2$ with same graph X as in 8. above. Vice versa given any parallel derivation $G \Longrightarrow X$ via $p1 + p2$ there are sequential independent derivation sequences $G \Longrightarrow H1 \Longrightarrow X$ via $(p1,p2)$ and $G \Longrightarrow H2 \Longrightarrow X$ via $(p2,p2)$ as in 8. above.

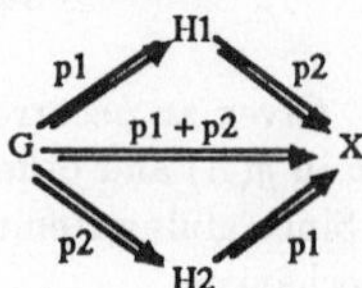

Remark:

- ▹ Application of the parallel production $p1 + p2$ above is possible if and only if the derivation $G \Longrightarrow H1$ via $p1$ and $G \Longrightarrow H2$ via $p2$ are parallel independent if and only if the sequences $G \Longrightarrow H1 \Longrightarrow X$ via $(p1,p2)$ is sequential independent.

  Problem 1: What can be done if $G \Longrightarrow H2 \Longrightarrow X$ via $(p1,p2)$ is not sequential independent, i.e. production $p2$ needs some items which are generated by $p1$ or $p2$ removes items which are used by $p1$?

  Problem 2: What can be done if $G \Longrightarrow H1$ via $p1$ and $G \Longrightarrow H2$ via $p2$ are not parallel independent, i.e. $p1$ and $p2$ may delete the same items of G?

  These problems are solved below:

8.11 Concurrency. In the case of problem 1 above we are able to construct a new production $p3 = p1 \oplus_R p2$, called sequential composition or *concurrent production of* $p1$ *and* $p2$ *via* R, where R is the overlapping part of $p1$ and $p2$ in $H1$. The *concurrency theorem* states that there is a 1-1-correspondence between applications of the concurrent production $G \Longrightarrow X$ via $p3$ and sequences $G \Longrightarrow H1 \Longrightarrow X$ via $(p1,p2)$ where $p1$ and $p2$ are at least overlapping in R.

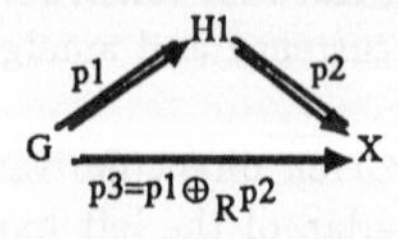

8.12 Amalgamation. In the case of problem 2 above let us assume that the overlap of $p1$ and $p2$ in G consists of common gluing or deleting items, i.e. items which are preserved or deleted by both productions in the same way (overlap of items which are preserved by one production and deleted by the other one is not allowed). In this case the productions $p1$ and $p2$ are called *amalgamable* and we are able to construct a *common subproduction* r of $p1$ and $p2$ and also an *amalgamated production* $p4 = p1 \oplus_r p2$ which is the union of $p1$ and $p2$ with shared subpart r (in fact a gluing resp. pushout in a category of graph productions).

Moreover, we are able to construct productions $p1'$ and $p2'$ corresponding to $p1$ and $p2$ without the effect of r. The *amalgamation theorem* states in the case of given $G \Longrightarrow H1$ via $p1$ and $G \Longrightarrow H2$ via $p2$ which are amalgamable we are able to construct derivations $H1 \Longrightarrow X$ via $p2'$, $H2 \Longrightarrow X$ via $p1'$ and $G \Longrightarrow X$ via $p4 = p1 \oplus_r p2$ with same graph X.

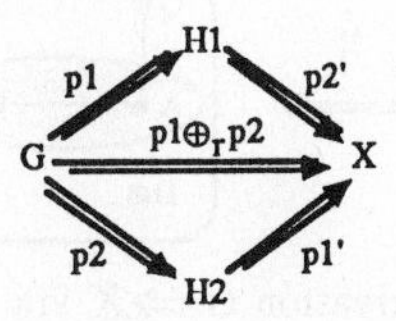

Vice versa given $G \Longrightarrow X$ via $p4$ we are able to construct the derivation sequences $G \Longrightarrow H1 \Longrightarrow X$ via $(p1, p2')$ and $G \Longrightarrow H2 \Longrightarrow X$ via $(p2, p1')$.

Note that in contrast to 8.9 above in this diagram the single derivations are in general neither sequential nor parallel independent.

9 Examples

Let us consider the following productions where we adopt the notation of 2.1 for corresponding gluing points.

9.1 Church-Rosser-Property

Productions $p1$ and $p2$ applied to G below are parallel independent. Application of 8.8 leads to

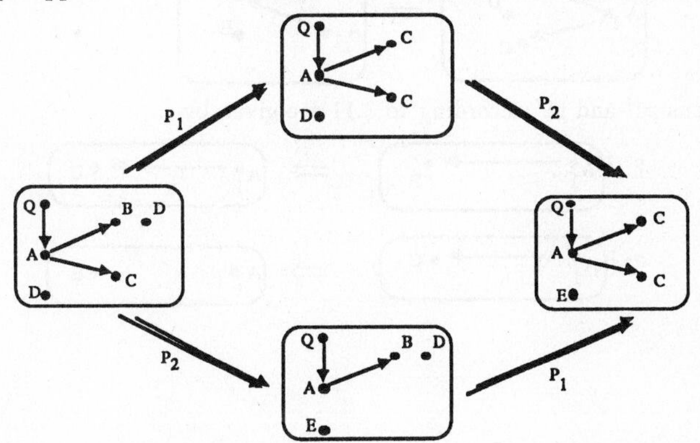

9.2 Parallelism

The parallel production $p1 + p2$ (see 8.9) given by the disjoint union of the left and right hand sides respectively leads in one direct derivation step from the left graph G to the right graph X in 9.1. Note that the graph morphism $g1 + g2 : (L1 + L2)$. G is non-injective due to the overlapping of the left hand sides of $p1$ and $p2$ in G.

9.3 Concurrency

The following derivation sequence $G \Longrightarrow H1 \Longrightarrow X$ via $(p1, p2)$ is not sequential independent. The overlapping of $p1, p2$ in $H1$ corresponding to the relation R is shown dottedly outlined.

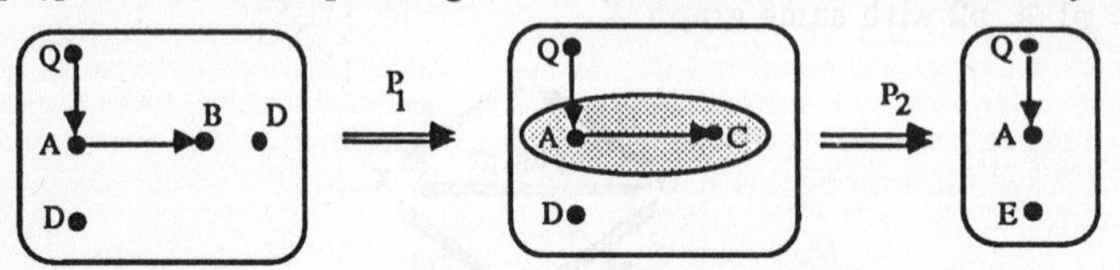

But due to 8.10 there is a direct derivation $G \Longrightarrow X$ via the following concurrent production $p1 \oplus_R p2$:

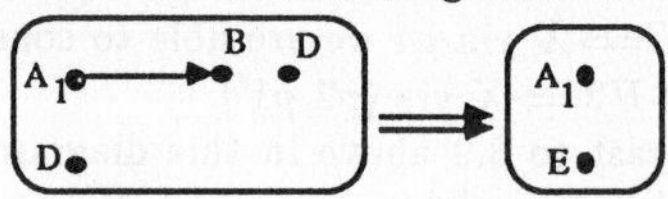

9.4 Amalgamation

The productions $p1$ and $p2$ applied to

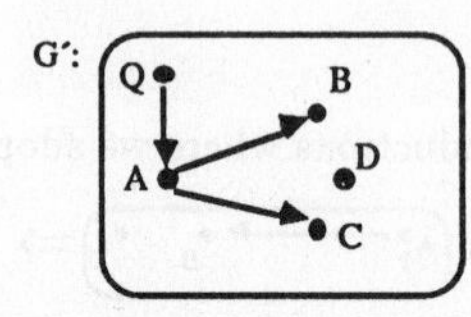

are not parallel independent but amalgamable. Both productions particularly overlap in an item to be deleted: the node labelled D.

The *common subproduction* of $p1$ and $p2$ is: 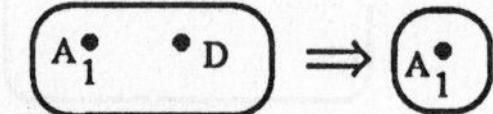

The *amalgamated production* is given by $p1 \oplus_r p2$:

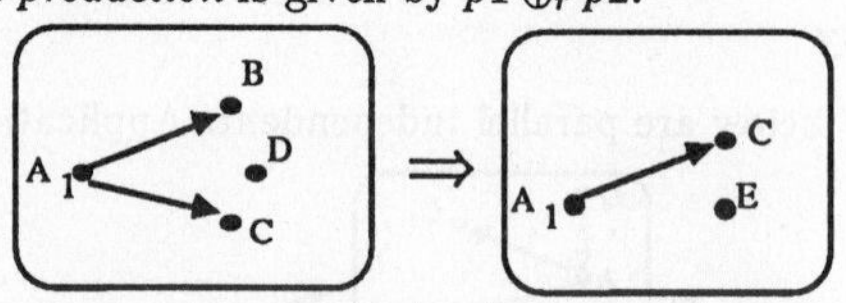

The productions $p1'$ and $p2'$ according to 8.11 are given by

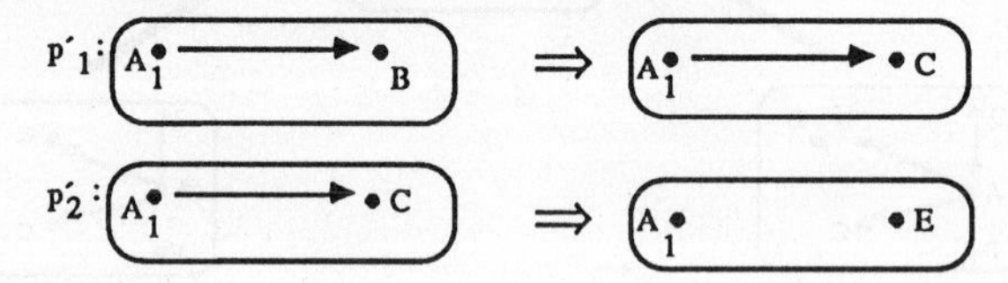

And the corresponding derivations according to 8.11 are given by

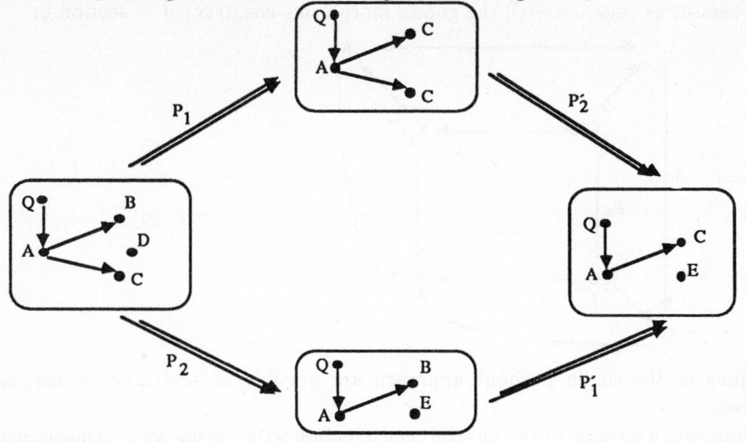

10 Single Pushout Approach

In some application contexts, productions should be applicable although the gluing condition is not satisfied. In this case, all dangling edges are going to be deleted by production application. Almost all operational approaches to graph transformation realize this idea.

Operational behaviour of this kind can be obtained in the algebraic approach if we interpret a pair of injective morphisms $(l : K \to L, r : K \to R)$ as a single partial morphism $r' : L \to R$ and perform direct derivations as a single pushout construction in the category of graphs and partial graph morphisms: A *partial graph morphism* $r : L \to R$ is a morphism from some subgraph L_r of L to R. A *single pushout production* is an injective partial graph morphism. The application of a production $r : L \to R$ to a mother graph G is given by a total morphism $g : L \to G$ (the *redex*). The *direct derivation* from G to the daughter graph H is given by the pushout of r and g:

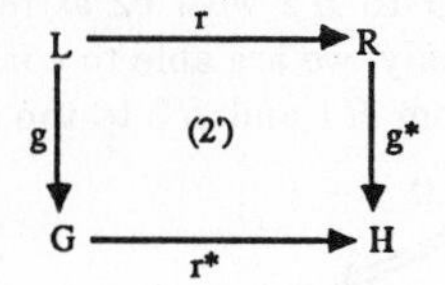

Given r and g, $H, r*$, and $g*$ can be explicitly constructed in four steps:

Step 1: (Construction of the gluing graph) The gluing objects K of r at g are all objects in L_r which satisfy:

(1) For all $x \in K$ and $y \in L$: $r(x) = r(y)$ or $g(x) = g(y)$ $\overset{\text{implies}}{\Longrightarrow}$ $y \in K$.

(2) For all arcs $a \in K$: $s_L(a), t_L(a) \in K$

Step 2: (Construction of G_{r*} and R_{g*}, Deletion step). Erase from G all objects which are images of non-gluing items w.r.t. g and all arcs, whose source or target nodes are erased. Symmetrically, treat R. The remainders are G_{r*} resp. R_{g*}.

Step 3: (Gluing) Now construct H as the gluing (as given in 6.) of G_{r*} and R_{g*} w.r.t. the total morphisms $r|_K : K \to R_{g*}$ and $g|_K : K \to G_{r*}$.

Step 4: (Construction of r* and g*) The domain of $r*$ is G_{r*} and the domain of $g*$ is R_{g*}. The morphisms themselves coincide with the gluing morphisms constructed in section 6:

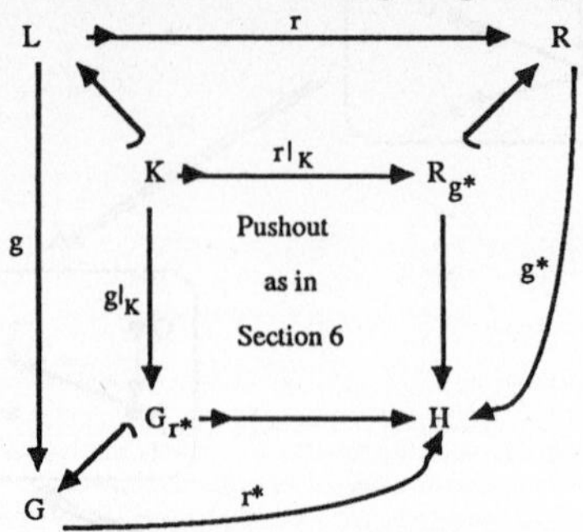

Direct derivations in the single pushout approach are possible at arbitrary redices, no gluing condition is required.

Analogously to section 6 remark 1., the pushout construction with partial morphisms is also defined if r is not injective. Even if g is partial the pushout exists.

Single pushout derivations generalize double pushout derivations (as given in sections 4–7) in the following sense: The translation of a production $p = (l : K \to L, r : K \to R)$ to a partial morphism is given by $T(p) : L \to R$ such that $L_{T(p)} = l(K)$ and $T(p)|_{L_{T(p)}} = r \circ l^{-1}$. If G can be derived to H with p at a redex g in the double pushout framework, $T(p)$ applied to G at g yields H as well. Vice versa, a production $r : L \to R$ in the single pushout setting can be translated to $t(r) = (l : L_r \to L, r|_{L_r} : L_r \to R)$ where l is the inclusion of L_r into L. Now suppose G can be derived to H with r at g, and g satisfies the gluing condition for $t(r)$, then $t(r)$ applied to G at g provides H as well.

Due to the abstract notion for direct derivations in the single pushout approach (which is even simpler than the corresponding notion in the double pushout framework), it is possible to reobtain almost all results w.r.t. Church-Rosser-properties, parallelism, concurrency, and amalgamation. (The proofs are even simpler.) Consider for example the Church-Rosser-Property: Given two direct derivations from G to $H1$ with $r1$ at redex $g1$ and from G to $H2$ with $r2$ at redex $g2$ which are parallel independent, i.e. $g1$ and $g2$ overlap in gluing points only, we are able to construct redices $f1$ for $r1$ in $H2$ and $f2$ for $r2$ in $H1$ which lead to derivations from $H1$ and $H2$ to the same daughter graph H:

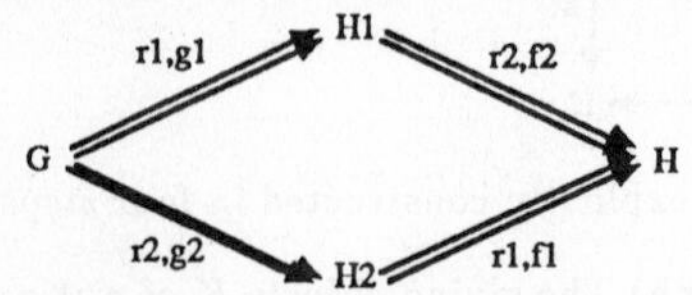

Note that the notion of gluing points in this context is given by the first step of the pushout construction with partial morphisms. Thus, the gluing points are not statically given by a production but they depend on the actual redex at which the production is applied.

Due to the more general contexts in which single pushout rules are applicable, some results are weaker than in the double pushout framework. For example: Having the Church Rosser property for $r1$ and $r2$, the parallel rule $r1 + r2$ provides the same derived graph H. However, parallel rules at arbitrary redices cannot be sequentialized in all cases, since redices without gluing condition allow identification on non-gluing items (compare 8.9).

10.1 Examples

Since the single generalizes the double pushout approach, all the examples in section 9.1 can be considered as examples for the single pushout approach, too.

The increased expressive power results from the fact that the the gluing condition needs not to be satisfied. In order to demonstrate this we will refer to the introductory example from section 2. Applying the single pushout rule p_0 to the graph G' requires that the *identification condition* is violated. The application of p_0 to the graph G'' implies that the *dangling condition* is violated. We obtain the derived graphs H' resp. H'':

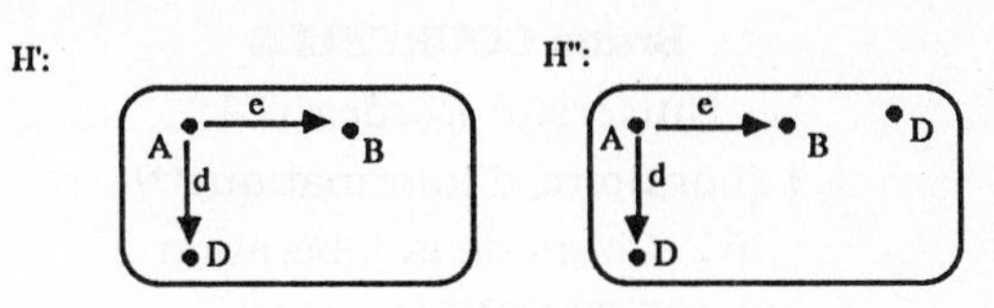

References

In the following we give some key references to the algebraic approach in general and to the material presented in this tutorial.

The algebraic approach was introduced in [EPS 73], a detailed introduction is given in [Ehr 79], while concepts of parallelism and amalgamation are studied in full detail in [Kr 77] and [BFH 87] respectively.

In [EKMRW 81] it is shown how the algebraic approach can be extended from graphs to more general structures, including partial graph, partially colored graphs and relational structures.

The idea of single pushout tranformations has been introduced by [Rao 84] and has been refined by [Ken 87]. [Löw 90] provides a comprehensive introduction to the single pushout approach investigating parallelism, concurrency, and amalgamation in this framework.

[BFH 87] P. Boehm, H.-R. Fonio, A. Habel: : *"Amalgamation of Graph Transformations: A Synchronization Mechanism,"* Journal of Computer and System Sciences 34 (1987), p. 377 – 408

[EPS 73] H. Ehrig, M. Pfender, H.J. Schneider: : "Graph Grammars: An Algebraic Approach, " Proc. IEEE Conf. SWAT'73, Iowa City 1973, p. 167 – 180

[Ehr 79] H. Ehrig: : *"Introduction to the Algebraic Theory of Graph Grammars,"* LNCS 73 (1979), p. 1 – 69

[EKMRW 81] H. Ehrig, H.-J. Kreowski, A. Maggiolo-Schettini, B. Rosen, J. Winkowski:: *"Transformation of Structures: An Algebraic Approach,"* Math. Syst. Theory 14 (1981), p. 305 - 334

[Ken 87] R. Kennaway: : *"On "On Graph Rewriting,"* Theoretical Computer Science 52 (1987), p. 37 – 58

[Kr 77] H.-J. Kreowski: : *"Manipulation von Graphmanipulationen, "* Ph.D. Thesis, TU Berlin 1977

[Löw 90] M. Löwe:: *"Algebraic Approach to Graph Transformation Based on Single Pushout Derivations,"* Technical Report TU Berlin No 90-5, 1990 (short version to appear in Proc. of WG'90)

[Rao 84] J.L. Raoult:: *"On Graph Rewriting"* Theoretical Computer Science 32 (1984), p. 1 – 24

THE LOGICAL EXPRESSION OF GRAPH PROPERTIES
(Abstract)

Bruno COURCELLE
Université Bordeaux-I
Laboratoire d'Informatique(+)
351, Cours de la Libération
33405 TALENCE - Cedex -
France

A graph can be considered as a logical structure in a fairly natural way: the domain of the structure is the set of vertices, and a binary relation R represents edges, in such a way that R(x,y) holds iff there is an edge linking x to y. It follows that logical formulas can express graph properties, and that they can also define sets of graphs: a formula defines the set of graphs where it holds if one considers the graph as a logical structure.

Investigating the expressive power of logical languages, and the properties of sets of graphs in relation with the syntax of logical formulas defining them has proved fruitful as shown by the following list of results.

1. Fagin has proved that a graph property is NP iff it is expressible by an existential second-order formula. Many other logical characterizations of complexity classes have been found. This research is motivated by the design of efficient query languages for relational data bases. We refer the reader to the surveys of Immerman [10], and Kannellakis [11] on this topic.

2. If a graph property can be expressed in monadic second-order logic, (Courcelle[4,5]), or in extended monadic second-order logic (Arnborg et al. [1]), then its validity can decided in polynomial or even linear time in some cases, on graphs of tree-width at most k, or on graphs generated by a fixed hyperedge-replacement graph-grammar. Such grammars, are context-free, in a precise sense; defined in Courcelle [3].

Notes :(+) Laboratoire associé au CNRS. Email : courcell@geocub.greco-prog.fr. This work has been supported by the ESPRIT-Basic Research Action 3299 ("Computing by Graph Transformation").

3. Monadic second-order definable sets of graphs behave w.r.t. context-free sets of graphs like regular languages w.r.t. context-free languages. In particular, the graphs of a context-free set that satisfy a monadic second-order formula form a context-free set. The monadic theory of a context-free set of graphs is decidable. See Courcelle [3,4,5].

4. There are two variants of monadic second-order logic, according to whether quantified variables can denote sets of edges or not. (See Courcelle [9], for the relevant definitions.) Some monadic second-order properties are not expressible without quantifications on sets of edges. This question is the subject of Courcelle [8].

5. Infinite equational graphs extend infinite regular trees in a natural way. In other words, they are generated by hyperedge replacement grammars, such that each nonterminal is the left handside of a single production, by means of infinite derivations sequences. Their monadic second-order theory is decidable, and they can be characterized, up to isomorphism, by monadic-second order formulas. See Courcelle[6,7].

References

[1] ARNBORG S., LAGERGREN J., SEESE D., Problems easy for tree-decomposable graphs,ICALP 1988, L N C S 317,pp.38-51.

[2] BAUDERON M., Infinite hypergraphs, Theor. Comput. Sci., to appear.(Extended abstract in this volume).

[3] COURCELLE B., An axiomatic definition of context-free rewriting and its application to NLC graph grammars, Theoretical Computer Science 55 (1987) 141-181.

[4] COURCELLE B., Graph rewriting : An algebraic and logic approach, in "Handbook of Theoretical Computer.Science Volume B", J. Van Leeuwen ed., Elsevier,1990, pp.193-242

[5] COURCELLE B., The monadic second-order logic of graphs I, recognizable sets of finite graphs. Information and Computation 85 (1990) 12-75.

[6] COURCELLE B., The monadic second-order logic of graphs II, Infinite graphs of bounded width, Mathematical Systems Theory, 21(1989)187-221.

[7] COURCELLE B., The monadic second-order logic of graphs IV, Definability properties of equational graphs, Annals Pure Applied Logic 49(1990)193-255.

[8] COURCELLE B., The monadic second order logic of graphs VI : On several representations of graphs by relational structures, Report 89-99, (see also Logic in Computer Science 1990, Philadelphia).

[9] COURCELLE B., Graphs as relational structures: an algebraic and logical approach, this volume.

[10] IMMERMAN N., Languages that capture complexity classes,SIAM J.Comput. 16(1987)760-777.

[11] KANNELLAKIS P., Elements of relational database theory, same volume as [4], pp.1073-1156.

Panel Discussion:
The Use of Graph Grammars in Applications

At the fourth Graph Grammars Workshop, Manfred Nagl organized and chaired a panel discussion on the Use of Graph Grammars in Applications. Most of the statements are documented by panelists and other contributors in this section in alphabetical order of the authors.

Graph grammars as a modelling tool in developmental biology

Martin J.M. de Boer
University of Regina

Map L-systems have been successfully applied in the past decade to the study of cell division patterns and plant morphogenesis. They are used to describe the spatial-temporal organization of cell division in a formal and compact way and provide a framework for computer simulation.

Map L-systems for theoretical cell division patterns provide clues for division patterns in real organisms, which are very often difficult to observe in great detail. From the map L-systems the combinatory of the cell division patterns can be investigated, and wall and cell growth functions can be determined (De Boer and Lindenmayer, 1987). The scope of the patterns that can be generated by map L-systems of a particular size and the architecture of the generated structures (archetypes) have been investigated exhaustively (Lück and Lück, 1986).

The relationship between cell division pattern and shape formation has been studied for particular developmental systems by computer simulation based on map L-systems (De Boer, this volume) and realistic visualization and animation of the development of cellular structures has been performed (Fracchia and Prusinkiewicz, this volume).

In order to study the three-dimensional organization of cell division patterns extensions of map L-systems to cellwork L-systems for the generation of three-dimensional cell patterns have been made (Lindenmayer, 1984) and have been worked out in more detail recently (Fracchia and Prusinkiewicz, this volume; Lück and Lück, this volume).

Considerable work in the area of graph grammars must be done in order to make map L-systems more widely accessible and applicable. First of all, extensions of the map DOL-systems to *probabilistic* and *interactive* map L-systems are necessary in order to simulate cell division patterns more realistically and in order to simulate the control of cell division pattern by cell-cell interaction, respectively. Probabilistic or interactive control on the wall (edge) level (Nakamura et al. 1986) seems unworkable and too difficult to interpret biologically. Assuming that all patterns are modifications of regular groundplans that can be generated by map DOL-systems, the most suitable formalisms would be map DOL-systems that have control on the wall (edge) level and on top of it incorporate probabilistic and/or interactive control on the cell (region) level.

Standardization of the construction of map L-systems for the characterization of observed patterns seems necessary in order to improve the accessibility and suitability of the method to developmental biologists. It is desirable to construct a smallest map DOL-system that emphasizes the underlying recursive nature of the pattern and for suitable probabilistic parameters matches the observed patterns.

Finally, it must be strived for to make the formalisms simpler and interactive computer programs must be written for them. It has been suggested (by Hans-Jörg Kreowski and Brian Mayoh, personal communication) that hyperedge replacement grammars and map L-systems share common features. Considering this will hopefully lead to a cross fertilization and perhaps even to a convergence of the two types of graph grammars.

References

M.J.M. De Boer and A. Lindenmayer. 1987. Map OL-systems with edge label control: Comparison of marker and cyclic systems. Pages 378-392 in H. Ehrig, M. Nagl, G. Rozenberg and A. Rosenfeld, eds. Graph Grammars and their Application to Computer Science. Lecture Notes in Computer Science 291. Springer-Verlag, Berlin.

M.J.M. De Boer. Modelling and simulation of the development of cellular layers with map Lindenmayer systems. This volume.

F.D. Fracchia and P. Prusinkiewicz. Physically-based graphical interpretation of map L-systems and cellwork L-systems. This volume.

A. Lindenmayer. 1984. Models for plant tissue development with cell division orientation regulated by preprophase bands of microtubules. Differentiation 26: 1-10.

H.B. Lück and J. Lück. 1986. Unconventional leaves. (An application of map OL-systems to biology). Pages 275-289 in G. Rozenberg and A. Salomaa, eds. The Book of L. Springer-Verlag, Berlin.

J. Lück and H.B. Lück. Double-wall cellwork systems for plant meristems. This volume.

A. Nakamura, A. Lindenmayer and K. Aizawa. 1986. Some systems for map generation. Pages 323-332 in G. Rozenberg and A. Salomaa, eds. The Book of L. Springer-Verlag, Berlin.

Graph grammars - a useful tool for pattern recognition?

H. Bunke
University of Berne

Pattern recognition has been an active area of research for more than thirty years and a number of useful concepts have found their way into commercial applications in the meantime. Important application areas of pattern recognition include character and printed document recognition, robot vision, industrial inspection, speech understanding, interpretation of medical signals like ECG, and many others. The most important data which are processed and interpreted by pattern recognition systems include greylevel-, color- and range-images, human speech, and other time-dependent one-dimensional signals.

The task of any pattern recognition system is to automatically infer the meaning of unknown patterns, which are presented to the system as input. In order to characterize a simple pattern it is offen sufficient to classify it. For example, the meaning of the printed character "5" is sufficiently detailed described by giving its class name "five". In case of more complex patterns it is usually necessary to describe an unknown input by a symbolic representation consisting of elementary parts together with attributes and relations.

Over the years a large number of different pattern recognition methods have been proposed. They can be categorized into statistical, artificial neural network based, artificial intelligence based, artificial intelligence based, and structural approaches [1]. Statistical and neural network based methods are similar to each other. They are primarily applied for the purpose of pattern classification. An unknown pattern is represented by a point in an n-dimensional feature space and the recognition procedure is based on a partition of this feature space into mutually disjoint regions representing a pattern class, each. Statistical and neural network based approaches are insensitive to noise and computationally inexpensive. On the other hand, there is a number of unsolved problems like feature selection and classifier training. In artificial intelligence based approaches, pattern classes are represented by explicitly storing knowledge about them. Applying this knowledge to the features extracted from an unknown pattern, recognition is accomplished by drawing domain specific inferences or logical conclusions. Artificial intelligence approaches are superior to statistical or neural network based methods if highly complex patterns are involved and if the explicit knowledge of a human expert about the patterns under study is required for recognition rather than pattern properties that are implicitly inherent to a sample set. On the other hand, artificial intelligence methods usually have a high computational time complexity and they fail if no explicit knowledge about pattern classes is available. Finally, structural approaches are somehow in the middle between statistical and artificial intelligence based methods. They emphasize pattern structure, i.e. relationships that exist between the elementary components of a pattern and the hierarchical composition of complex patterns from their parts. The basic idea in structural pattern recognition is to compare an unknown pattern with repre-sentatives of the different classes. These representatives are either directly given by a finite number of prototypes or indirectly by means of a grammar.

The block diagram of a structural pattern recognition system which is based on graph grammars is shown in Fig. 1. An unknown pattern, for example, a grey-level image is preprocessed and transformed into a structural representation in the first processing phase. Then this structural representation is fed into a parser and analyzed according to a graph grammar that represents the different pattern classes. In case of successful analysis we get as recognition result not only the class of the unknown input but also its derivation according to the grammar rules, which is equivalent to a structural description.

Graph grammars as a tool for structural pattern recognition have a number of advantages. First, graph grammars is a two-dimensional representation formalism in its very nature as opposed to one-dimensional approaches, like string grammars. This is beneficial if we deal with problems where two-dimensionality is important, for example, in the analysis of pictorial patterns. Secondly, graphs are a universal representation formalism, particularly when augmented by attributes. So we can expect that a large class of pattern recognition problems can be conveniently modeled by means of graphs and graph grammars. Finally, graph

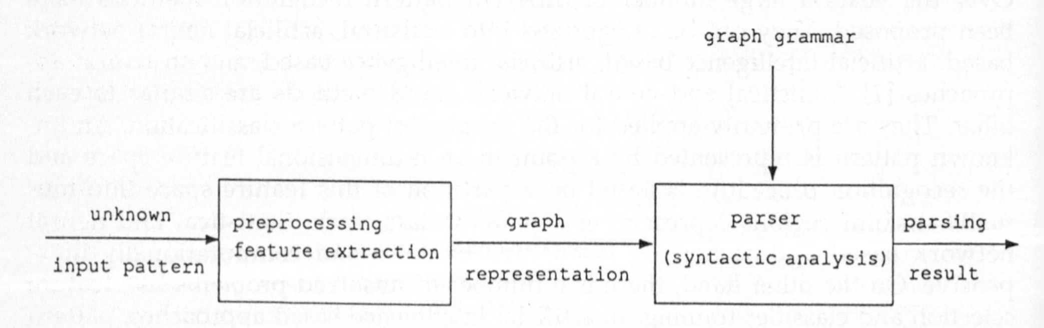

Fig. 1: Block diagram of a structural pattern recognition system based on graph grammars

grammars allow the compact representation of a large, possibly infinite number of pattern representatives using a much smaller, always finite number of rules.

On the other hand, the application of graph grammars to problems in pattern recognition suffers from a number of drawbacks. First, there is the inference problem. Unlike in statistical or neural network based pattern recognition where a number of learning algorithms are available which automatically infer a classifier from a sample set there aren't such inference algorithms for graph grammars – at least not for real world applications. Secondly, there is the parsing problem. As it can be seen in Fig. 1, the core of any structural pattern recognition system relying on graph grammars is a parser. Unfortunately, parsing of graph grammars is a field that has not been very intensively studied. Only few parsing algorithms for graph grammars are known at the time being and those parsers that are able to handle graph languages useful for practical applications have a very high computational complexity. Finally, and most seriously, there is the noise and distortions problem. In any real world application the data at the input of a pattern recognition system are noisy and distorted. This means that the graph representation, which is obtained after preprocessing and feature extraction and given to the parser, is often not an element of the language generated by the graph grammar. Nevertheless, the parser has to try to reconstruct its undistorted version, i.e. that element of the graph language that is most similar to the distorted input. Otherwise, recognition would fail. For the case of string grammars, error correcting parsers fulfilling the task of recognizing distorted inputs not belonging to the language generated by the grammar have been introduced many years ago. For graph grammars no such parser has ever been developed. Unfortunately, it is to be expected thatsuch parsers, if they were developed, suffer from an almost prohibitive high time or space complexity.

In the balance it turns out that the usefulness of graph grammars as a tool for pattern recognition is rather limited. Nevertheless, there are some applications where graph grammars have been successfully applied. An example is circuit diagram recognition [2]. The problem of parsing has been avoided in [2] by using the graph grammar in a generative way, i.e. by applying the productions in a forward directed way [3]. Due to a number of topological and geometric constraints the productions of the graph grammar described in [2] work in a deterministic way. Other examples of successful applications of graph grammars in pattern recognition are described in [4][5].

In conclusion, graph grammars are just one particular method among many others in pattern recognition. There are severe disadvantages making the successful application of graph grammar in pattern recognition difficult. Nevertheless, there seem to be some classes of problems where graph grammars can be very successfully used. It can be expected that the importance of graph grammars in pattern recognition will not significantly increase in this decade. But it is certainly possible that more niches will be found where graph grammars will turn out superior to other approaches. The high computational complexity of parsing and the lack of know-how in error correcting parsing are the most serious obstacles for the further propagation of graph grammars in pattern recognition. So any new results in these areas could eventually lead to an increased interest in graph grammars as a tool for pattern recognition.

References

[1] H. Bunke: Hybrid pattern recognitions methods: In: H. Bunke, A. Sanfeliu (eds.:) Syntactic and structural pattern recognition – theory and applications, World Scientific Publ. Co., Singapore, 1990, 307-347.

[2] H. Bunke: Graph grammars as a generative tool in image understanding In: H. Ehrig, M. Nagl, G. Rozenberg, A. Rosenfeld (eds.:) Graph-grammars and their application to computer science, Springer Lecture Notes 153, 1983, 8-19.

[3] Bunke, T. Glauser, T.-H. Tran: An efficient implementation of graph grammars based on the RETE matching algorithm, in this volume.

[4] Pfaltz: Web grammars and picture description. Comp. Graphics and Image Processing 1, 1972, 193-219.

[5] M. Brayer: Web grammars and their application to pattern recognition. Technical Report TR-EE 75-1, Purdue University, West Lafayette/IN, 1975.

Graph grammars as tools in applications work ?

J. Cuny
University of Massachusetts

I don't think I said very much in the discussion. What little I did say was in response to the question "Why aren't graph grammars more widely accepted as tools in applications work?"

I have run into a number of skeptics in the parallel programming community. For the most part, they are willing to concede that graph grammars can provide a framework for the specification of interprocess communication structures, but they do not believe that they will be of use in providing more sophisticated kinds of support. I have been asked "Now that you've used graph grammars to guide the development of your editor, can't you throw away all its grammatical underpinnings?" The answer to this question is "Yes" unless the grammatical structure of graph families can be exploited by efficient algorithms to support more of the programming process. These algorithms must be useful for graphs with hundreds or even thousands of nodes, graphs that are prohibitively expensive to generate for analysis. Specifically, in the context of large graphs:

Can we provide information that might be useful in mapping: Is the graph connected? Does it have bounded degree? Does it have fixed degree?

Can we provide information that is useful in debugging: Is a given pair of process nodes mutually reachable? How many distinct node types occur? Is it possible to identify "boundary" nodes?

Can we provide information that is useful in visualization: Does the grammatical structure suggest a coherent layout? Is it possible to generate partial views of the graph? Does the graph have repetitive structures that can be deleted for the purpose of display?

The specification of communication structures is relatively easy (at least in retrospect). Until we can solve some of the harder questions in programming support, we will not get a lot of converts.

Modelling of biological structures

F. David Fracchia
University of Regina

The focus of research conducted by the Computer Graphics Group at the University of Regina is the modeling of biological structures. Our objectives are the study of morphogenesis – the development of form in organisms – and the simulation of such development. Visualization of biological structures and simulation of developmental processes are among the most exciting applications of computer graphics. In this respect, we have found graph grammars to be a very powerful tool.

An inherent feature of graph grammars and related concepts, such as bracketed L-systems and map L-systems, is the definition of complex objects by a generative process using a relatively small number of simple rules. In the graphics literature this effect is referred to as database amplification. Although a generative mechanism can be applied to many areas, such as the modeling of fractal mountains, definition of space-filling curves and creation of floor plan layouts, it is particularly useful in the modeling of living organisms since not only the final structure, but also the process of development are captured by the grammar. Because segments of living organisms develop concurrently (many cell divisions may occur at the same time), the parallel application of grammar rules is more appropriate than sequential rewriting. The intermediate results from each derivation step are useful in illustrating consecutive stages of organism development. The generation of developmental stages beyond those recorded by observations lends to the predictive value of the model. Since L-systems describe the process of growth, they appear to be well suited for the animation of plant development. The problem is that subsequent frames must be relatively similar to each other to provide a good impression of continuous growth. By introducing continuous parameters, we have overcome the limitations of the originally discrete formalism.

Models based on L-systems are also useful for the validation of scientific hypertheses. For example, the modeling of fern gametophytes supports the hypothesis that cell division patterns play the decisive role in the formation of their overall shape.

Our research has resulted in the generation of many flowering plant structures and cellular tissues, and also has revealed problems open for further study. One problem is concerned with the construction of rules which capture the development of a specific organism. In other words, given observations of an organism at various stages of growth, the problem is to infer the productions, including the necessary geometric information (very important), that fully describe its development. Another problem deals with the computational complexity of generative algorithms. For example, what would be the differences in time and space complexity between the generation of a context-sensitive branching structure represented by a string, versus a tree representation? Finally, we need more detailed biological data to construct models of development in three dimensions.

Usefulness of graph grammars in applications

Herbert Göttler
Universität Mainz

We are discussing here the usefullness of graph grammars in applications. What I have heard till now from the members of the panel, were different ideas what kind of application areas this could be. I bet, an unprejudiced listener would consider some of the areas not being really "applications" but rather "esoteric theoretical stuff". I do not object, everyone should have his/her playground. For my present playground, the development of syntax-directed graphical editors and CAD-systems, even in an industrial environment, I believe graph grammars to be useful concepts. Otherwise I would not be here. That's why the situation has a touch of surrealism to me when we waste our time to listen to mutual affirmations like: "Graph grammars are great for xyz".

Besides this workshop, I've been attending the last two ones. If I'm not completely wrong, I think I've seen most of the people here on these workshops, too, and hardly new ones, except for the students of the "hard core members", perhaps. I wonder – but not just today – why graph grammars still do not get the attention they deserve although they are around now for twenty years? What could the reasons be that there is such a bad resonance? These are questions which we should ask us within this panel, too. I've got two main reasons why graph grammars have this life in the shadow.

Firstly, the usefulness of a primarily theoretical concept like graph grammars are best proved by successful "real" application, not by toy problems. Are there real projects around whose solutions were derived best by graph grammars? Not many, I believe. We have to look for better realistic problems!

A second reason why we have difficulties to get disciples is the disgusting appearance of graph grammars to the newcomer. I have already been told this by some people. There are many different notations around and I do not see work being done to change the situation. After twenty years, a field of research like ours should be able to supply a good introductory textbook for interested persons, discussing the pros and cons of the different approaches to graph rewriting, and should provide an acceptable notation! If we had a unified system for the description of most of the important aspects of graph rewriting, I think, we would get fresh blood, new ideas and more interesting problems.

We should do something soon to prevent graph grammars from perishing!

Applying graph grammars to software engineering

Simon M. Kaplan
University of Illinois at Urbana-Champaign*

This position paper concerns application of graph grammars to software engineering problems of various kinds. This idea is of course not new, as graphs with replicated patterns are found throughout the software engineering field in various guises, and the proceedings of all the graph grammar workshops contain many papers explaining how graph grammar theory can be applied to various software engineering and related applications ranging from internal structures in programming environments to representation mechanisms in artificial intelligence applications.

The purpose of this paper is to outline some potentially interesting new directions for application and suggest some ways of making the field of graph

---

* *This research supported in part by the National Science Foundation under grant CCR-8809479, by the Center for Supercomputing Research and Development at the University of Illinois and by AT&T through the Illinois Software Engineering Project.*
Author's Address: Department of Computer Science, University of Illinois, 1304 W. Springfield, Urbana, IL 61801 USA. Email: kaplan@cs.uiuc.edu.

grammars more accessible to the general computer science community thereby (we hope) increasing the penetration of graph grammars.

The areas where graph grammar applications have been investigated most thoroughly thus far are those where the graphs (and grammars) are hidden inside the system, such as graph rewriting systems or programming environments for conventional languages which use graph grammars as a way of controlling internal structures. I think there is a need for graph grammars to "break out" of this mold and be used in environments where the graphs and grammars are directly accessible to and manipulated by the user. There are several application domains in which this is potentially useful.

The first is the area of *visual languages*. People often prepare for a programming task (especially concurrent programming tasks) by drawing "prototypical graph pictures" which exemplify the state of the running system. Our experience with the Garp and Δ systems is that these can often be seen as graph grammars, or can easily be turned into productions. The user has then written his program at a much higher level of abstraction because the pictures are closer to what he has in mind. Designing a visual language based around graph rewriting and building an associated environment seems an excellent potential application of graph grammar technology.

The second is the area of *hypertext systems*. Hypertext is a way of presenting information to a user in a nonlinear way, for example allowing the user to browse an encyclopedia by association rather than from A through to Z. They are thus potentially very useful as a medium for presenting all kinds of complicated interlinked structures to the user, for example program code and associated documentation. In practice, though, it turns out that current hypertext systems make it rather difficult to determine how to structure the hypertext links among nodes, and provide no good abstraction mechanisms for supporting such structures. A hypertext system is just a graph with considerable replicated (possibly at different levels of abstraction) structure. Thus there is potentially a wide range of applications of graph grammar based ideas to making the control of hypertext structures more tractable (in the sense of easier for users to use, rather than the more theoretical meaning of the word).

Both of these applications have a distinct advantage from the viewpoint of one wishing (as I do) to export the ideas in the graph grammar community to other areas in computer science: *they are new*. What this means is that we do not have to compete with existing formalisms or approaches, but can use a graph grammar based approach from the start.

If we want to expand the application of graph grammars, we have to do something to make the field more accessible. When I give talks on graph grammar applications to software engineering (mostly to applications in the area of concurrent specifications and languages) I find the audience has little difficulty grasping the basic ideas and seeing how the grammars "work". However, when presented with the proceedings of a graph grammar workshop, many people are "turned off". I believe there are several reasons for this. There are too many forms – algebraic, expression, NLC, hyperedge, many other specialized subspecies – and it is difficult for the non-initiate to see how they are related. There is no "universal sugar" in which to write graph productions (an analog to the λ calculus). And there is no single place (or small set of places) where one can go to find an introduction to the theory and a range of applications which allow the reader to become familiar enough with the ideas to try applying them himself.

In conclusion, to increase the visibility of graph grammars we really need to do two things: (a) continue to find new and interesting applications ourselves, and show that the graph grammar approach pays dividends in these areas, and (b) continue to evolve the field so that universal notations and simple ways of explaining the ideas emerge and can be taught to others.

Applied graph transformation

Hans-Jörg Kreowski
Universität Bremen

The area of graph grammars and graph transformation is strongly related to various directions of potential applications. This has been, is and will be the major stimulus for developing the field.

In the past

In the late 60s and the early 70s graph grammars were introduced and investigated by Feder, Milgram, Montanari, Pavlidis, Pfaltz, Rosenfeld, Schneider and others to tackle problems in

- pattern recognition and picture processing,
- specification of data structure, and
- semantics of programming languages.

I was not involved in the business in those days. Thus my knowledge and opinions are mainly based on the early papers. They present very interesting considerations, that are still up to date. The problems we are facing today have not changed much since then. I can only guess why the breakthrough of graph grammars did not take place in that period already. On one hand, the available computer systems were not appropriate. They did not run fast enough, their storage was too small, and they did not support interaction sufficiently. On the other hand, many applications require recognition procedures. Unfortunately, the parsing and membership problems for graph languages were underestimated. In the first two decades of the history of graph grammars, only a few and quite restricted solutions came up. They failed to meet the expectations and necessities.

In the present

The situation today is more promising. During the last decade, the foundations .of graph grammars have been tremendously improved. The algorithmic aspects have been intensively studied with some success. Significant progress has be made concerning the relationship of graph transformation and certain applicational issues like

- evaluation of algebraic functional and logical expressions,
- software specification environments,
- pattern generation and design of rule-based systems,
- parallelism and concurrency.

Moreover, a number of experimental graph grammar tools are implemented. In particular, some recognition algorithms of acceptable usefulness are available meanwhile.

In the future

Will the future bring the breakthrough? In my opinion, there is a fair chance. Graphs and graph-like structures are quite often suitable representations of complex objects as they occur in all kinds of data processing problems. Graphs combine the advantage of intuitive understanding with mathematical feasibility. Rule-basedness is considered to be a powerful and manageable method for specifying the dynamic behaviour of systems. Putting graphs and rules together yields graph grammars. The graph grammar approach to system specification and the design of algorithms on complex objects embodies three actual key conceptions of computing: rule-basedness, visualization and proof methods. I expect that graph transformation will be shown to be a suitable base for

- the development of integrated specification environments,
- fractal-pattern processing,
- the design of expert systems.

The interest in the applicability of graph transformation will grow with the quality of interactive tools with graphical interface - an exciting perspective.

Graph grammars for knowledge representation

Brian Mayoh
Aarhus University

Graph grammars have an important role to play in the areas of computer science where the careful representation of knowledge is crucial. Databases, simulation, planning, and expert systems are areas, in which the representation of knowledge requires careful definition and use of graphic notation: semantic nets, frames, type inheritance hierarchies and other kinds of conceptual structures. Practitioners in such areas do in fact use graphs of various kinds, but theoreticians seem to prefer "one dimensional" logical representations. Many expressive logics have been devised, but logical inference does not always correspond to "reasonable use of represented knowledge".

Maybe logical programming should be replaced by graphical programming. Graph productions seem to capture better the heuristic nature of reasoning, just as graphs seem to capture the imprecision and uncertainty of represented knowledge better than logical formulas. Programming languages should be graphical in two senses: programs should be graphs, not strings of symbols; data structures should be graphs, not glorified lists or arrays. In traditional imperative languages one calls procedures with particular actual parameters; in functional programming languages one applies functions to particular arguments; in graphical programming languages one should apply productions to particular graphs. More work should be done on graphical languages, exploiting recent advances in parallel and distributed computing, visual languages, expert system shell designs, and graph rewriting.

But there is also theoretical work to be done. We need a better understanding of various kinds of graphs. Do we understand dynamic graphs, where the edges represent functions, processes, or events? Do we understand reflexive graphs, where vertices or edges can themselves be graphs? What are the graph equivalents of modal (temporal or epistemic), non-monotonic, higher order and metalogics? What are the graph equivalents of the many logical treatments of uncertainty: default logic, preferences, fuzzy logic, circumcription etc.? Note that statistical treatments of uncertainty have recently become graphical; there is a growing literature on influence diagrams.

All in all it seems that graph grammars have a bright future in "knowledge based" areas of computer science.

Graph grammars which are suitable for applications

Manfred Nagl
RWTH Aachen

Since more than ten years [Na 79] we are working with *graph grammars* for *certain applications*. All of them can be subsumed under the title "intelligent and integrated interactive tools". In the area of software development environments our experiences stem from the IPSEN, but also from the PROGRESS project (cf. e.g. [NS 90], [Sc 90]). Minor applications have been text systems and library recherche systems. Therefore, I would like to argue from this range of applications although there are serious indications that the experience we got so far and the problems we detected are also valid in other areas like concurrency [Ja 86], parallelism [Br 88], and others.

In all these applications graph grammars are used for internal, high-level programming. This means, that the *effects* of concrete (e.g. tools of a software development environment) or abstract tools (formal specification of semantics) on internal data structures are *operationally specified* by a language for manipulating attributed node and edge labelled directed graphs, one essential of which is rewriting by rules. From these specifications the software architecture and the implementation of tools can be derived.

For any kind of operational description, may we call it specification or very high-level programming, we need a framework within which we can conveniently program or specify. This convenience makes it necessary that different *programming/specification paradigms* are supported.

One essential is that the specification language is as *declarative* as possible, therefore reducing the operational part in order to avoid overspecification. In PROGRESS we introduce a schema part which declares the node types, edge types, and attribute equations of a graph class. The node types are instances of an object-oriented class hierarchy. The attribute equations can be seen as a data flow oriented element of the language. The rules can also be regarded as a descriptive way to define operations. *Imperative* language features are used to build up complex transactions from other ones, namely elementary transactions like application of rewriting rules or subgraph tests, and of other transactions.

This procedure to *offer different* specification/programming *paradigms* has to be *extended*. Similarities, expressed by an object-oriented class hierarchy, need not and should not be restricted to the internal structures of nodes, but must be extended to graph classes. The imperative style of programming has to be improved by offering also language elements to express independent or parallel execution. For applications in the area of concurrency elements for synchronization or event handling have to be introduced. For complex and/or distributed applications suitable mechanisms for the integration of different graph and/or specification parts have to be offered (what we call specification in the large).

Up to now we have used graph grammars only as an abstract tool for various applications, i.e. a conceptual tool to be used in a paper and pencil mode. Of course, in order to have a breakthrough for grammars there must also be *tools* for *building up, maintaining,* and *validating* graph grammar specifications. Building up and maintaining, is done by using an editor/analyser tool, validating by an analyser, an instrumentor and an execution tool. A remarkable step in the direction of such an integrated graph grammar specification environment is found in [NS 90].

Besides these tools to handle graph grammar specifications there must also be some help for the complex *translation* from a graph grammar specification *to an efficient implementation* in one of the conventional programming languages. As, at the moment, there is no efficient automatic and complete translation for a graph grammar specification available, this translation has to be carried out at least in parts by hand. However, the translation process can be partially automated. In any case, some basic tools are necessary as, for example, a general purpose graph (object) storage.

Graph grammars are not widely accepted at this moment. There are two main competitors, namely data modelling languages from the data base area and attributed tree grammars from compiler construction. In order to have graph grammars widely populated the advantages of *graph grammars* have to be made clear, i.e. that they are *better for modelling*. Graph grammars offer mechanisms to specify complex operations which cannot be found in the data modelling discussion. Compared to attributed tree grammar approaches, graph grammars can describe the whole structural knowledge of a certain application area by the same graph-theoretic framework. Therefore, they do not distinguish between the context free structure, expressed by trees, and the non context free structure, expressed by attributes, as it is done by attributed grammars.

Compared to the two main competitors the *implementation* and *translation tools* of graph grammars have to be *improved*, as aketched above. This means, that a basic software layer suitable for graph grammar implementations has to be developed in order to be used at different sites. Furthermore, integrated graph grammar specification environments and, finally, transformation tools for getting an implementation from a graph grammar specification (compilers, generator tools) have to be developed.

References

[Na 79] M. Nagl: Graph Grammars: Theory, Applications, Implementation (in German), Wiesbaden: Vieweg (1979).

[NS 90] M. Nagl, A. Schürr: A Specification Environment for Graph Grammars, this volume.

[Sc 90] A. Schürr: PROGRESS – A VHL–Language Based on Graph Grammars, this volume.

[Br 88] Th. Brandes: Formal Methods for Specifying Automatic Parallelization (in German), Doctoral Thesis, Heidelberg: Hüthig Verlag (1988).

[Ja 86] M. Jackel: Formal Specification of Ada's Concurrent Constructs by Graph Grammars, Doctoral Thesis, University of Osnabrück (1986).

Usefulness and visibility of graph grammars

Azaria Paz
Technion - Israel Institute of Technology

While the theory of graph grammars has developed nicely and substantially in the past several years, it is still considered and studied as an extension and generalization of the theory of string-grammars.

It is my opinion that we should look for properties of graph grammars which are characteristic of graph grammars and are not a generalization of known properties of string grammars. Thus it might be interesting to define graph grammar whose production rules are based on the combining of two already generated graphs into a new graph according to some preasigned specifications (e.g. from G1=(V1,E1) and G2=(V2,E2) generate G3=(V1xV2,E3), where e=[(v1,v2),(w1,w2)] is in E3 if either (v1,v2) is in E1 or (w1,w2) is in E2). Such approaches may prove usefull when dealing with the generation of very large and very regular graphs (as presented in the talk of professor Cuny).

With regard to the usefullness and visibility of graph grammars.

As with any other theory one can find many areas to which graph grammars are not relevant, but if here are some areas to which it is relevant then it serves a purpose. In general scientists do not always solve problems which ought to be solved but rather solve problems they can solve. Thus most problems in real life are nonlinear but lineartity in much more investigated. A theory is certainly usefull if it is an eye opener, if it provides new prospectives and if it helps towards a better understanding of various fenomena.

Applications of graph grammars and directions for research

M.R. Sleep [1]
University of East Anglia

During the 1990 Graph Grammar workshop held at Bremen, a number of participants were asked to state briefly their views on potential applications and future directions of research into Graph Grammars, and later invited to provide a short written version of the statement. This is my written contribution. It contains at least the spirit of the statement I made during the workshop, and some remarks about present and future research.

The statement

1. I prefer to dicuss the applications of *Graph Rewriting* rather that what seems to me the narrower term *Graph Grammars*.

2. Applications *are* important, not only because they help justify the expenditure of resources on particular areas of research, but also because they provide real challenges to research workers. Even when there is no political or economic pressure for considering applications, judicious choice of some real problem can often provide invaluable guidelines for otherwise somewhat arbitrary design decisions concerning both theoretical and practical constructions.

3. However, it is also important to ensure that the applications are *realistic*. If the main motivation for doing a particular piece of research is intellectual curiosity, then it is probably wiser in the long run to recognise this rather than invent artificial applications.

4. One very fruitful area for the application of Graph Rewriting to information technology is contributing to the sound definition and efficient implementation of *standards*. Whether proprietory or otherwise, standards increasingly dominate information technolgoy, whether for languages (ranging from very high level through to very low level) to machines (ranging from proprietary standard "system" architectures such as IBM PC down to microprocessor chips and below).

5. Good standards are not only well specified, they are also *practicable*. This secondary requirement means that involvement in standards activities drives not only the research into using graph rewriting for specifications, but also to provide an operational semantics (computing by graph transformation).

[1] Partially supported by ESPRIT Basic Research Action 3074 (Semagraph) and ESPRIT project 2025 (European Declarative System).

Present directions for research in graph rewriting

The remarks below reflect the author's interest and concerns, and should be regarded more as demonstrating the potential richness of the field rather than as a comprehensive survey.

The most coherent "school" of graph rewriting is undoubtedly the "Graph Grammar" (Gra Gra) school founded by Rosenfeld and others. [Ehrig 89] gives an excellent overview. This school has traditionally stressed "high formalism" through - for example - the use of category theory and the invention and development of the double pushout construction.

In addition there is a considerable body of knowledge about *term* or *tree* rewriting (see for example [Klop 90]). This knowledge has not yet adequately been related to graph rewriting. There are surprises here for those versed in term (or tree) rewriting: for example conventional "tree rewriting" does *not* turn out to be a special case of graph rewriting - at least not with any of the present models of graph rewriting. One has to work quite hard to avoid the sharing implicit in graph rewriting models.

In parallel with the advances made by the Gra Gra school there have been more pragmatic developments using some operational definition of "graph rewriting". These activities generally lack the coherence and sound formal underpining of the Gra Gra school, primarily because they view graph manipulation simply as an implementation technique. However, they have led to significant practical advances. Thus [Peyton Jones 87a] describes the use of a number of ad-hoc graph rewriting techniques for the implementation of functional languages, and [Peyton Jones 87b, Watson 87] describe the design of a parallel graph rewriting machine.

Finally, there is growing interest in the design and implementation of programming languages based on some form of parallel rewriting. LEAN [Bar 89] and Dactl [Glauert 88] are two early examples of parallel graph rewriting languages. Less obviously, I would consider some of the "committed choice" logic languages such are PARLOG and Concurrent Prolog to be languages of parallel graph rewriting, albeit of a restricted sort. Similarly the family of "actor" languages could be seen as languages of parallel graph rewriting. Techniques for translation of functional and logic languages to pure graph rewriting models are described in [Hammond 88].

The various approaches outlined above have different strengths and weaknesses, which happily tend to be complementary. One of the tasks of ESPRIT Basic Research Actions no. 3074 Semagraph [Sleep 89] and no. 3299 Gra Gra [Ehrig 89] is to relate these various approaches. Some success can already be reported: the work of [Staples 80] and Barendregt et al. [Bar 87] establishes precise notions simulating term rewriting using graphs (introducing the notion of *term graph rewriting),* and the work of Hoffmann and Plump [Hoffmann 89] relates term rewriting to an appropriate category of graphs and the classical double pushout construction. Recent work of Kennaway [Kennaway 90] relates a single pushout construction [Raoult 84, Kennaway 87] to the classical double pushout model of graph rewriting.

Conclusions and future directions

I believe that the results mentioned above are early prizes from what will turn out to be a very rich seam of intellectual endeavour justified both for its foundational nature and its potential applications. My main conclusion is that we have only begun to tap the European strengths in graph rewriting, both theoretical and practical. The problem will be to find the good research frameworks for integrating and harmonising work in this area, and here I believe that future projects should contain a strong element of applications drive.

References

[Bar 87] Barendregt, H.P., van Eekelen, M.C.J.D., Glauert, J.R.W., Kennaway, J.R., Plasmeijer, M.J., and Sleep, M.R., 1987, "Term graph rewriting", Proc. PARLE conference, Lecture Notes in Computer Science, 259, 141-158, Springer.

[Bar 89] Barendregt, H.P., van Eekelen, M.C.J.D., Glauert, J.R.W., Kennaway, J.R., Plasmeijer, M.J. and Sleep, M.R., 1989, "LEAN an Intermediate Language based on Graph Rewriting". Journal of Parallel Computing 9, North Holland.

[Ehrig 89] Ehrig, H., Löwe, M., eds., 1989, "Computing by Graph Transformation" Report no. 89/14, Technical University Berlin.

[Glauert 88] Glauert, J.R.W., Kennaway, J.R. and Sleep, M.R., 1988, "Final Specification of Dactl", University of East Anglia report SYS-C88-11.

[Hammond 88] Hammond, K. and Papadopoulos, G.A., 1988, "Parallel implementations of declarative languages based on graph rewriting", Proc. Alvey Technical Conference.

[Hoffmann 89] Hoffmann, B. and Plump, D., 1989, "Jungle Evaluation for Efficient Term Rewriting", Lecture Notes in Computer Science, 343, Springer.

[Kennaway 87] Kennaway, J.R., 1987, "On 'On graph rewritings'", Theor. Comp. Sci., 52, 37-58.

[Kennaway 90] Kennaway, J.R., 1987, "Graph rewriting in some categories of partial morphisms", University of East Anglia Report 1990.

[Klop 90] Klop, J.W., 1990, "Term rewriting systems", Handbook of Logic in Computer Science eds., Abramsky S., Gabbay D. and Maibaum T., Oxfort University Press.

[Peyton Jones 87a] Peyton Jones, S.L., 1987, "The Implementation of Functional Languages", Prentice-Hall, London.

[Peyton Jones 87b] Peyton Jones, S.L., Clack, C.D. and Salkild, J., 1987, "GRIP: a parallel graph reduction machine", ICL Technical Journal, 5, 595-599.

[Raoult 84] Raoult, J.C., 1984, "On graph rewritings", Theor. Comp. Sci., 32, 1-24.

[Sleep 89] Sleep, M.R., 1989, "The semantics and pragmatics of generalized graph rewriting", (SemaGraph) ESPRIT Basic Research Action 3074 Invited paper, EATCS Bulletin No. 39 Oct. 1989.

[Staples 80] Staples, J., 1980, "Computation on graph-like expressions", Theor. Comp. Sci., 10, 171-185.

[Watson 87] Watson, I., Sargeant, J., Watson, P. and Woods, V., 1987, "Flagship computational models and machine architecture", ICL Technical Journal, 5, 555-594.

$Graph^{Ed}$: An Interactive Tool For Developing Graph Grammars

Michael Himsolt
Lehrstuhl für Informatik, Universität Passau
Innstraße 33, Postfach 2540, D-8390 Passau, FRG
himsolt@unipas.fmi.uni-passau.de

Abstract

$Graph^{Ed}$ is a powerful interactive editor for drawing and manipulating graphs and graph grammars. It helps designing graphs, networks, data structures, entity relationship diagrams, petri nets, electrical circuits, VLSI circuits, flowcharts or even arbitrary diagrams. Its unique feature is the ability to handle graph grammars. They are useful both for theoretical investigations and as parametrisized graph macros.

1. Basics

The primitives of graphs are nodes, edges and labels. Each of them is handled in a flexible manner by $Graph^{Ed}$.

Nodes may have arbitrary color, size and shape. $Graph^{Ed}$ provides both built-in styles and user-defined icons for node shape. There are five optional ways to attatch edges to nodes.

Edges between nodes are drawn either as straight lines or arbitrary polylines, using one of six linestyles. Like nodes, edges may be colored. The arrow of an edge in a directed graph may also be changed in size and shape. $Graph^{Ed}$ can handle self-loops and more than one edge between two nodes.

Labels are arbitrary texts using various, user-definable fonts. They are automatically attatched to the nodes or edges. In the case of nodes, there are five possible placements, namely centred or in one of the corners.

2. The User Interface

Graph^Ed's object oriented user interface provides all functions that are necessary for flexible manipulations of graphs. It supports insertion, deletion and change of nodes and edges including their labels. The user may bundle nodes together (like subgraphs) and move, manipulate or delete them in common. Cut & paste operations are also available.

There may be up to 16 windows, and each of them may contain an arbitrary number of graphs. Further utilities include a grid raster for tidier drawing of graphs and printing to postscript printers.

Attributes of nodes, edges, graphs and graph grammars are set up using special pop-up windows. There are other windows to change Graph^Ed's configuration, for example to add or delete fonts. A file selector is also included.

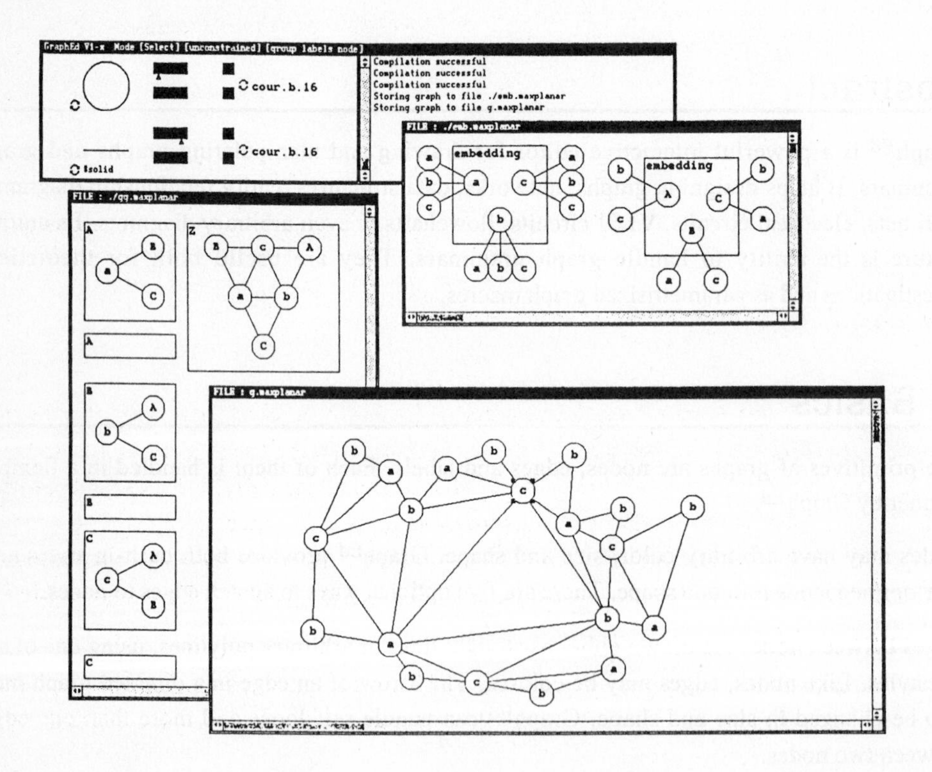

All these functions are fully mouse- and menudriven and harmonized with the SunTools environment. All operations are kept as simple and intuitive as possible, and working with Graph^{Ed} can be learned in just a few hours. Most actions may be performed either with the menu or with keystrokes.

3. Graph Grammars

Graph^{Ed}'s most advanced feature are graph grammars. Graph^{Ed} can currently handle three types of graph grammars : 1-NCE, NLC, and BNLC (see [JanRoz 82], [ENRR 86]). NLC and BNLC are mostly useful for theoretical investigations, whereas 1-NCE is more suitable for practical issues, due to a more flexible embedding mechanism.

NCE-1 graph grammar productions may be given using an intuitive graphical representation : the left side is just a (very big) node with the right side drawn inside. The embedding rules are drawn around the left side :

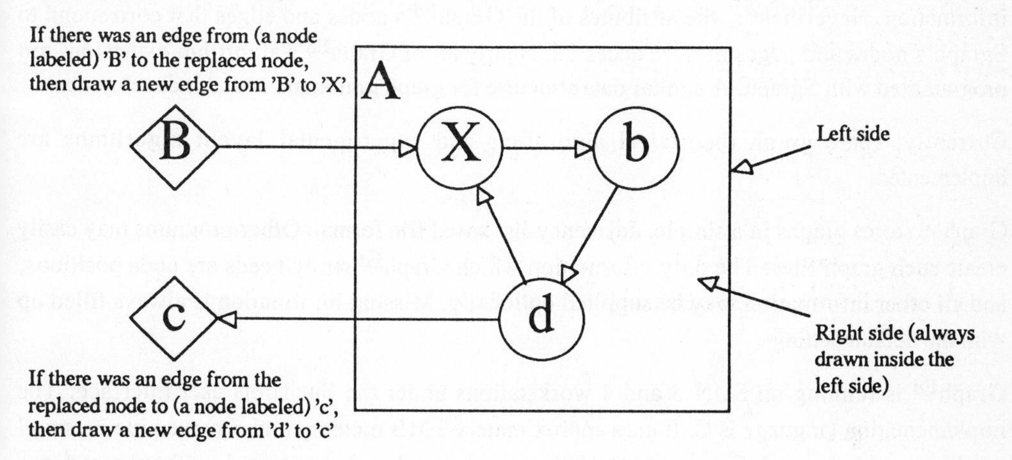

NLC and BNLC graph grammar productions are slightly different from that. Basically, there is a single embedding relation common to all productions. Nodes of the right side can only be distinguished by their labels, whereas 1-NCE is able to address individual nodes. Thus, the productions contain no embedding information. Instead, there ia a special 'embedding' form to specify the global embedding relation. Its syntax is similar to the one of the 1-NCE embeddings described above. For convenience, NLC-style global embedding relations are also available for 1-NCE graph grammars.

All operations available for graphs may also be used for graph grammars. Additionaly, the user may define special node and edge styles for the left side and the embedding rules and apply them with a 'pretty print' operation.

Using graph grammars is easy : set a 'current' production, select a node and choose 'apply' from Graph^{Ed}'s menu.

Graph^{Ed}'s graph grammar facilities are built as an interpreter : a production is in fact a special graph. Graph^{Ed} partitions this graph into left side, right side and embedding information. To apply a production on a node, it is replaced by a copy of the right side and reconnected according to the embedding rules. This implementation makes it particularly easy to extend Graph^{Ed} to other graph grammar models. There is nothing more to do than to write a 'parser' for the production and a procedure for the 'apply' operation.

4. User-defined Algorithms

Graph^{Ed} provides two options to include user-defined algorithms written in C : they may either use Graph^{Ed}'s own data structures or a simplified structure called Sgraph. Sgraph is easier to use and more flexible, since it is not neccessary to maintain drawing or user interface information. Nevertheless, the attributes of the Graph^{Ed}'s nodes and edges that correspond to Sgraph's nodes and edges may be accessed. Nearly all of Graph^{Ed}'s algorithm extensions are programmed with Sgraph. A similar data structure for graph grammars is under development.

Currently, some graph theoretical algorithms and experimental layout algorithms are implemented.

Graph^{Ed} stores graphs in a simple, adjacency-list based file format. Other programs may easily create such graph files. The only information which Graph^{Ed} really needs are node positions, and all other information may be supplied optionally. Missing information is always filled up with the default settings.

Graph^{Ed} is running on SUN 3 and 4 workstations under the SunTools user interface. The implementation language is C. It uses approximately 2 MB memory and can even run on small machines at high speed. Graphs up to 1000 nodes have already been used, without any decay of speed.

5. Future Directions

Graph^{Ed} will be extended in various directions. Besides further enhancements of the user interface, handling of derivation trees will be added to the graph grammars. We will include more layout algorithms, especially some which use the graph grammar derivation trees as additional structural information. Porting to the X-Windows environment is also under consideration.

References

[ENRR 86] H. Ehrig, M.Nagl, G. Rozenberg, A. Rosenfeld (Eds) : Graph-Grammars and Their Application to Computer Science, Lecture Notes in Computer Science 291, Springer Verlag Berlin, Heidelberg, New York 1987

[Graph^{Ed} 88] Michael Himsolt, Entwicklung eines Grapheneditors, Diplomarbeit, Universität Passau, 1988

[Graph^{Ed} 89] Michael Himsolt, Graph^{Ed} : An Interactive Graph Editor, in : STACS 89, Lecture Notes in Computer Science 349, Springer-Verlag Berlin, Heidelberg, New York 1989

[Graph^{Ed} 90] Michael Himsolt, Graph^{Ed} User Manual, to appear

[JanRoz 82] D. Janssens, G. Rozenberg, Graph grammars with neighbourhood-controlled embedding, Theoretical Computer Science 21, 1982

Presentation of the
IPSEN–Environment:
An Integrated and Incremental Project Support ENviroment

Andy Schürr

Lehrstuhl für Informatik III
Aachen University of Technology
Ahornstraße 55, D–5100 Aachen,
West Germany

The **IPSEN** system (Integrated and Incremental Project Support ENvironment) was intended from its very beginning to cover and to integrate most of the activities occurring in the software life–cycle. Within the presented IPSEN prototype, this **integration of tools** comprises project management, the design and analysis of software architectures, coding, analysis, execution, test and evaluation of module implementations (Modula–2 specific), and last but not least the production of software documentations.

Besides integration in the sense of **smooth interaction** of tools with a **uniform user interface**, we consider at least the following additional aspects of integration:

- **Complex consistency relations** are introduced not only within, but also between different software documents, like a system's software architecture and its accompanying documentation. These interdocument relations allow for the creation of **integrated sets of documents**, describing a software project.
- For the design and implementation of the IPSEN system we used uniform modeling and implementation mechanisms based on the model of **attributed graphs** and on the formalism of **PRO**grammed Graph **RE**writing **S**y**S**tems (cf. "**PROGRESS**: A VHL–Language Based on Graph Grammars", this vol.).

Incrementally working tools take into account that software development has to be seen as an incremental evolutionary process. Tools with fine–grained incremental operations provide fast and direct response to user interactions, showing inconsistencies within and between documents as soon as possible.

The presented prototype consists of about 135.000 lines of Modula–2 code and is running on SUN 3 workstations.

Presentation of the

PROGRESS-Editor:

A text-oriented hybrid editor for PROgrammed Graph REwriting SyStems

Andy Schürr
Lehrstuhl für Informatik III
Aachen University of Technology
Ahornstraße 55, D-5100 Aachen,
West Germany

The language **PROGRESS** (c.f. "PROGRESS: A VHL-Language based on Graph Grammars", this vol.) is, as far as we know, the first **strongly typed** language which is based on the concepts of **PRO**grammed **G**raph **RE**writing **S**y**S**tems. Currently, an **integrated programming support environment** for this language, including a syntax-directed editor and an interpreter, is under construction (cf. "A Specification Environment for Graph Grammars", this vol.).

The design and implementation of this environment is based on our previous experiences with the realization of the **I**ntegrated **P**roject **S**upport **EN**vironment **IPSEN**. Thus, **attributed graphs** are used for the internal representation of documents (in this case: PROGRESS programs), and the language PROGRESS has been used to "specify" parts of the PROGRESS enviroment's implementation (**bootstrapping**).

Up to now, the implementation of the first version of a **syntax-directed, text-oriented hybrid editor** has been finished. This editor consists of the following main components: (1) A menue-driven, **syntax-directed editor,** (2) an integrated emacs-like **text editor**, (3) an incrementally working **multiple-entry parser**, (4) an incrementally working **unparser** (pretty-printer), (5) a **layout editor**, (5) and a demand-driven, incrementally working **type-checker**.

The presented hybrid editor's implementation consists of about 140.000 lines of Modula-2 code and is running on SUN 3 workstations.

PLEXUS: Tools for Analyzing Graph Grammars

Egon Wanke

Universität-Gesamthochschule-Paderborn
Fachbereich 17 - Mathematatik/Informatik
Warburger Straße 100
D-4790 Paderborn

The library system:

PLEXUS is a library system that has been developed for the implementation of graph algorithms for certain graph grammars, see also [Wan88]. It supports the implementation of algorithms for graph grammars in that the embedding is done by a one to one identification of special vertices. It provides a very comfortable data type for graphs of such graph models. Nevertheless, the major part of PLEXUS concerns usually defined graphs represented by adjacency lists. Graphs are allowed to be undirected, directed, or mixed, i.e., having directed and undirected edges. Hyperedges, multiple edges, ordered edges, and self-loops are also available. Terminal and nonterminal vertices could be equipped with geometrical positions and users specified informations.

PLEXUS is based on a very general dynamic object-oriented approach. Graph objects and sets of graph objects are not realized by integers or static arrays, respectively, but by dynamic data structures. Thus, it is possible to create and destroy vertices and edges during the processing of a graph. The underlying data types are called *object*, *set*, and *sequence*. Each graph, each vertex, and each edge is realized by an *object*. Each graph has a *set* of vertices and a *set* of edges. The incident vertices of an edge are stored in a *set* associated with the edge. Each nonterminal vertex is associated with a *sequence* of terminal vertices. The functions and data structures provided by the PLEXUS library system should support a fast and clean implementation of graph algorithms that are easy to debug. All available functions run in asymptotically optimal time.

Tools for analyzing graph grammars:

PLEXUS includes a tool that illustrates the processing of context-free graph grammars. In such a processing the tool determines whether or not the language of a context-free graph grammar contains a graph that has a property Π. The processing works as follows, see also [LW88]: Let R be the set of productions of a context-free graph grammar Γ, and let R' be an empty set.

In the first step, we consider all productions $(X, G) \in R$ that have a terminal right hand side. That is, the graph G on the right hand side has no nonterminal objects. For each such production (X, G), we generate an equivalent production (X, G^b) in R'. The productions (X, G) and (X, G^b) are equivalent with respect to Π. That is, if

$$H \Rightarrow_{n,(X,G)} H_1$$

is a derivation step in that the production (X, G) is applied to the nonterminal object n of any graph H and if

$$H \Rightarrow_{n,(X,G^b)} H_2$$

is a derivation step in that the production (X, G^b) is applied to the nonterminal object n of H, then H_1 has property Π if and only if H_2 has property Π. Here, the graphs G^b are as small as possible. If R' has two equivalent productions with respect to Π, we remove one of them from R'.

In each further step, we consider all productions $(X, G) \in R$ that have a right hand side G with some nonterminal objects. We try to apply all productions from R' to all nonterminal objects of G. For each resulting terminal graph G', a *new* production (X, G^b) in R' will be generated. The productions (X, G') and (X, G^b) are equivalent with respect to Π. The graphs G^b are as small as possible. If R' has two equivalent productions with respect to Π, we remove one of them from R'.

This process is continued until we have found a production (X, G) in R' such that G has property Π and X is the axiom of Γ—in this case the language of Γ has a graph with property Π—or no new productions will be generated in R'—in this case the language of Γ has no graph with property Π. The processing always terminates if the number of equivalent productions is finite.

Since PLEXUS supports geometrical positions of terminal and nonterminal vertices, the tool can illustrate each step of the processing on a graphics screen. Currently, the tool is implemented to deal with connectivity properties.

Implementation:
The system is written in the C programming language under the UNIX[1] environment on a SUN-3/50 Workstation[2]. All graphics routines are based on the Sun-View[3] system. The entire efficiency of PLEXUS is only available in a C like environment.

Acknowledgment:
This project has been supported by grants Le-291/1-1,2,3,4 from Deutsche Forschungsgemeinschaft. We would like to thank Elof Frank and Markus Kellermeyer for their high quality programming.

References

[LW88] T. Lengauer and E. Wanke. Efficient analysis of graph properties on context-free graph languages. In T. Lepistö and A. Salomaa, editors, *Proceedings of ICALP '88*, volume 317 of *Lecture Notes in Computer Science*, pages 379–393. Springer-Verlag, Berlin/New York, 1988.

[Wan88] E. Wanke. Plexus: A system for implementing hierarchical graph algorithms. In R. Cori and M. Wirsing, editors, *Proceedings of STACS '88*, volume 294 of *Lecture Notes in Computer Science*, pages 401–402. Springer-Verlag, Berlin/New York, 1988. System demonstration.

---

[1]UNIX is a trademark.

[2]SUN Workstation is a trademark.

[3]Sun-View is a trademark.

AN ALGEBRAIC THEORY OF GRAPH REDUCTION

Stefan Arnborg *
The Royal Institute of Technology
NADA, KTH, S-100 44 Stockholm, Sweden

Bruno Courcelle†
Bordeaux-1 University
Laboratoire d'Informatique (associé au CNRS),
351 cours de la Libération, 33405, Talence, France.

Andrzej Proskurowski
University of Oregon
CIS department, Eugene, Oregon 97403, USA

Detlef Seese
Akademie der Wissenschaften
Karl-Weierstraß Institut für Mathematik,
Mohrenstr. 39, PF 1304, 1086 Berlin, Germany

ABSTRACT: We show how membership in classes of graphs definable in monadic second order logic and of bounded treewidth can be decided by finite sets of terminating reduction rules. The method is constructive in the sense that we describe an algorithm which will produce, from a formula in monadic second order logic and an integer k such that the class defined by the formula is of treewidth $\leq k$, a set of rewrite rules that reduces any member of the class to one of finitely many graphs, in a number of steps bounded by the size of the graph. This reduction system corresponds to an algorithm that runs in time linear in the size of the graph.
Keywords: Graph, algebra, reduction.

CONTENTS

0 Introduction

There are several ways to define sets of finite graphs by finite devices. The main ones are graph-grammars[23] (and in particular context-free ones, see, e.g., Courcelle[18,19,20]), logical formulas (and in particular monadic second-order ones, see Courcelle[16,17,18,19,21] and Arnborg et al. [5,6]), by forbidden configurations (and in particular forbidden minors, see, *e.g.*, Robertson and Seymour [33], Arnborg, Proskurowski and Corneil [9]), and finally by reduction.

*Supported by NFR and STU.
†Supported by the "Programme de Recherches Coordonnées: Mathematiques et Informatique" and the ESPRIT-BRA project 3299 "Computing by Graph Transformations".

We say that a set of graphs L is defined by *reduction* if a graph rewriting relation $\rightarrow_R$ is given, together with a finite set K of acceptable graphs, so that, given any graph G, every sequence of $\rightarrow_R$ - rewritings terminates (with a graph called a *normal form* of G) and the graph so obtained is in K if and only if G is in L (thus every normal form of G in L is in K and no normal form of G not in L is in K). Classical examples concern trees, series-parallel graphs, flowcharts (Hecht and Ullmann [25]). As an example, consider a graph. Remove a pendant edge with its degree one vertex. Repeat this removal until a graph with no pendant edge is obtained. This graph is an isolated vertex if and only if the original graph was a tree.

In this paper we present an algebraic theory of graph reduction. Our main theorem says that every set of graphs of bounded treewidth that is defined by a monadic second-order formula is definable by reduction. In addition, the pair (R, K) can be effectively obtained. Although the construction method is intractable in general, it can be applied to specific cases of interest, and we provide examples concerning outerplanar graphs and partial 3-trees. The proofs omitted here appear in the full version [4].

For the classes of trees, forests (partial 1-trees), two-terminal series-parallel graphs, partial 2-trees, and partial 3-trees, membership can be decided with help of terminating sets of reduction rules [7,36]. Moreover, using these reduction rules gives an embedding of a partial k-tree in a (full) k-tree for $k = 1,2,3$, and this embedding is produced in time linear in the size of the given graph [7,27,29]. The straightforward generalization of this method does not work beyond $k = 3$ [26]. Once an embedding in a k-tree is given for a graph, many combinatorial problems can be solved in time linear in the size of the graph [5,6,8,11,37] (the constant of proportionality depends however on the value of k). Since no linear time embedding algorithm is known for arbitrary k [1], this problem is in a sense the bottleneck for fast solution of a large number of combinatorial optimization problems on partial k-trees. The method presented here will make it possible to solve some of these problems in linear time without access to a k-tree embedding, tree-decomposition, elimination order or parse tree.

We will develop the theory of rewriting and reduction systems in a universal algebra setting. We consider subsets of the carrier of an algebra that are defined as a union of equivalence classes of a congruence with finitely many classes. Such sets can be characterized with a rewriting system allowing a value $f[l]$ to be replaced by $f[r]$, whenever $f[\,]$ is a context and (l, r) is a pair in a finite list of rewrite rules. In general, this rewriting relation will be neither computable nor Noetherian and thus of little value algorithmically. But if there is an integer valued *size* function defined for the carrier of the algebra obeying certain monotonicity properties and such that there are only finitely many objects of each size, then the rewriting system can be modified to a reduction system which is guaranteed to reduce the size of the object on each rewrite step. A reduction system will always rewrite an object from the desired set to one of finitely many representatives of its congruence class, in a number of rewrite steps bounded by the size of the object. We will apply the universal algebraic framework to algebras of graphs, as developed in [10,16,17,18,19,20], in such a way that the reduction relation is easily computable on the graph itself and not only on the expression evaluating to the graph. The algebra of graphs that we shall use here is related to the construction of k-terminal recursive graph families [37,38]. We have a sequence of domains $(\mathbf{G}_i)_{i=0}^{\infty}$ where the domain $\mathbf{G}_i$ is the family of i-sourced graphs, *i.e.*, of graphs with a sequence of i distinguished vertices or *sources*. We need only two operations to combine graphs of $\mathbf{G}_i$ into a new graph of $\mathbf{G}_i$, namely the generalized *parallel* and *series* composition. We also need a set of 'lifting' operators to add sources to a given graph and basic nullary operators to introduce vertices and edges. The operator sets for i-sourced graphs give rise to an infinite sequence of finite signatures, $(F_k)_0^{\infty}$, with $F_k \subset F_{k+1}$, where the subalgebra generated by the signature F_k is the class of graphs of treewidth at most k.

[1] An $O(n^2)$ approximate embedding algorithm was developed by Robertson and Seymour[31,32]. Various improvements are possible, see, *e.g.* Courcelle[17] and Bodlaender[14]. A probabilistic and approximate algorithm with $O(n \log n)$ performance was developed by Matousek and Thomas[29]

Based on our algebra of i-sourced graphs it is easy to define a rewriting relation on graphs, by matching the left-hand side of a pair in a given graph and replacing the matched subgraph by the right-hand side of the pair. This leads to a rewriting relation that can be implemented locally on a graph and thus more efficiently than the rewriting relation that depends on all global parses of a graph.

The monadic second order predicate logic was used by Courcelle [16,17,18,20,19] and Arnborg, Lagergren and Seese [5,6] as a powerful tool to formalize graph properties (hereafter called *MS-definable* properties) easily decided on families of graphs of bounded treewidth.

We claim that membership in every MS-definable set of graphs of bounded treewidth can be decided in linear time. We give applications to some families of outerplanar and planar graphs.

1 Basic graph-theoretic and algebraic definitions

1.1 Graphs

We will consider graphs which are unoriented loop-free multigraphs given by finite disjoint sets V and E of vertices and edges, respectively, and the incidence relation $I \subset V \times E$ that is constrained to make every edge incident to exactly two vertices. If there is no single vertex that disconnects a graph, the graph is called *non-separable.* A set of vertices, $S \subset V$, is a *separator* in a graph G if the removal of S and the edges incident to its vertices disconnects G. We use other standard concepts of graph theory presented by Bondy and Murty [22]. We assume also familiarity with the theory of partial k-trees, *i.e.*, partial graphs of k-trees. A graph G is of treewidth at most k if a family $\{X_n\}_{n\in N}$ of vertex subsets of G can be arranged as nodes in a tree T so that those nodes containing an arbitrary given vertex induce a subtree of T (*i.e.*, a connected subgraph of T), every pair of adjacent vertices share membership of some X_n, and $|X_n| \leq k+1$ for all $n \in N$. Such an arrangement is called a *tree-decomposition of width at most* k. The class of partial k-trees is exactly the class of graphs of treewidth at most k[37].

1.2 Algebras

Let $\mathcal{S}$ be a finite set of sorts. A set F is a finite $\mathcal{S}$-sorted signature if F is finite and every f in F has a profile $s_1 \times \cdots \times s_\beta \rightarrow s$, where β is nonnegative and finite (it may be zero, which corresponds to constant or nullary operators), and $s_1, \ldots, s_\beta, s$ are all in $\mathcal{S}$.

A tuple $M = ((M_s)_{s\in\mathcal{S}}, (f_M)_{f\in F})$, where $M_s \cap M_{s'} = \emptyset$ if $s \neq s'$ and f_M is a total mapping $M_{s_1} \times \cdots \times M_{s_\beta} \rightarrow M_s$ whenever f is of profile $s_1 \times \cdots \times s_\beta \rightarrow s$, is an F-algebra. The sets M_s, $s \in \mathcal{S}$ are its *domains.* We let also $M = \cup_{s\in\mathcal{S}} M_s$, as customary (note the ambiguous use of M).

We denote by $M(F)$ the initial F-algebra (term algebra over F), and by $h_M : M(F) \rightarrow M$ the unique homomorphism associated with M. Let $t \in M(F, \{x_1, \ldots, x_n\})$ be a linear term with variables $x_1, \ldots, x_n$ (*i.e.*, each variable x_i occurs exactly once and has a fixed sort $\sigma(x_i)$). Let $d_i \in M_{\sigma(x_i)}$ for $m < i \leq n$. Then we have a mapping f of profile $\sigma(x_1) \times \cdots \times \sigma(x_m) \rightarrow \sigma(t)$ defined by $f(a_1, \ldots, a_m) = t[a_1, \ldots, a_m, d_{m+1}, \ldots, d_n]$. These mappings will be called *derived operations* or, if $m = 1$, *contexts.* If in addition $n = 1$, *i.e.*, if t is an expression over F with exactly one occurrence of a variable, then we say that the context is *generated by* F. Note that, if $h_M(M(F))$ is a proper subset of M, then some contexts may not be generated by F.

An equivalence relation $\approx$ is *stable* under the operations of F if, for every f_M of M and pairs $(v_i, u_i)_{i=1}^{\beta}$, if $v_i \approx u_i$ for $0 < i \leq \beta$, then $f_M(v_1, \ldots, v_\beta) \approx f_M(u_1, \ldots, u_\beta)$. $\approx$ is *finite* if in addition it has a finite number of equivalence classes. The class of an element d is denoted $[d]$ when the intended equivalence is clear from the context.

A *congruence* on M is an equivalence relation $\approx$ on M such that

(i) any two elements equivalent under $\approx$ are of the same sort and

(ii) $\approx$ is stable under the operations of M (and, as a consequence, under the derived operations of M).

Let $L \subset M$. We denote by $\sim_L$ the congruence on M defined by: $m \sim_L m'$ if and only if, for every context $f[\ ]$, $f[m] \in L$ if and only if $f[m'] \in L$. We say that L is *generated* by F if $L \subset h_M(M(F))$, *i.e.*, if every member of L can be written as an expression over F. We say that L is *M-recognizable* if L is a union of classes of a finite congruence on M. The following lemma is proved in Courcelle[16]:

Lemma 1.1 *Let M be an F-algebra and $L \subset M_s$ for some $s \in \mathcal{S}$. The following conditions are equivalent:*

(1) L is M-recognizable.

(2) $\sim_L$ is finite.

(3) $L = h^{-1}(C)$ where $h : x \mapsto [x]_{\sim_L}$ is an F-homomorphism and $C = \{[x]_{\sim_L} \mid x \in L\}$.

2 Rewriting systems on an algebra

A rewriting system R on an algebra M of signature F is a finite list of pairs of elements of M, $R = \{(l_i, r_i)_{i \in I}\}$, where each r_i is of the same sort as the corresponding l_i. We write $m \to_{(R,F)} m'$ if and only if $m = f[l_i]$, $m' = f[r_i]$ and $m \neq m'$, for some context f and some $i \in I$. Let $\to^*_{(R,F)}$ be the reflexive transitive closure of the relation $\to_{(R,F)}$. For a subset M_0 of M and a rewriting system R, we let $L_w((R,F), M_0)$ be the set $\{m \mid$ there is an m' such that $m \to^*_{(R,F)} m'$ and $m' \in M_0\}$. When the set of operations F is clear from the context, we write $\to_R$ and $L_w(R, M_0)$. We say that $L_w((R,F), M_0)$ is the set *weakly defined* by (R,F) and M_0.

Proposition 2.2 *Let $L \subset M$ be M-recognizable and generated by F. Then $L = L_w(R, M_0)$ for some rewriting system R and some finite subset M_0 of L.*

The rewriting system R of proposition 2.2 can be obtained as follows: for every congruence class $[x]_{\sim_L}$ introduce a representative $\overline{x}$. Let $D = \{\overline{x} \mid x \in M\}$. For every operator $f \in F$ and sort-compatible operands from D, if d is the result of applying f to the operands and $d \neq \overline{d}$, then add $(d, \overline{d})$ to R. An explicit algorithm will follow from some assumptions on computability and termination of arbitrary sequences of rewritings.

The relation $\to_R$ is said to be *Noetherian* if there is no infinite sequence $m_0, m_1, m_2, \ldots$ such that $m_i \to_R m_{i+1}$ for all $i \geq 0$. An element m' such that $m' \to_{(R,F)} m''$ for no m'' is called *(R,F)-irreducible* or if (R,F) is clear from the context, *irreducible*. If $m \to^*_{(R,F)} m'$ and m' is irreducible, then m' is called a *(R,F)-normal form* of m or just a *normal form*.

Let R be a rewriting system and M_0 be a finite subset of M. We say that $((R,F), M_0)$ *defines* L and we write this $L = L((R,F), M_0)$ if the following conditions hold:

(i) $L = L_w((R,F), M_0)$,

(ii) $\to_{(R,F)}$ is Noetherian,

(iii) M_0 is a set of (R,F)-irreducibles;

(iv) For every m, and for every (R,F)-normal form m' of m, either $m' \in M_0$ and $m \in L$, or $m' \notin M_0$ and $m \notin L$

We now consider how to find a Noetherian $\rightarrow_R$ in an important special case. A *size function* on an algebra M is a mapping $m \mapsto |m|$ associating a non-negative integer $|m| \in \mathrm{N}$ with every $m \in M$. We assume that there are only finitely many elements in M of each given size and that if $|m| < |m'|$ then $|f[m]| < |f[m']|$ for every context $f[\,]$.

We call a rewriting system R a *reduction system* if $|r| < |l|$ for every pair (l, r) in R. Obviously, a reduction system is Noetherian. It produces a normal form of m in at most $|m|$ steps.

Proposition 2.3 *If M has a size function, then every recognizable set L in M is defined by $((R', F), M_0)$ for some reduction system R' and some finite set M_0.*

Proof. Consider a finite congruence $\approx$ such that L is the union of some of its classes. For each congruence class $[m]$, select one element $\overline{m}$ of minimum size as representative of the class and let D_m be the finite set of elements in the class that are of minimum size. Let D be the union of the sets D_m. For every (sort-compatible) combination of operation f_M and operands $d_1, \ldots, d_\beta$ from D, consider $d = f_M(d_1, \ldots, d_\beta)$ such that $c[d] \in L$ for some context c. If $d \notin D$, then put in R' the pair $(d, \overline{d})$. Since D contains all smallest elements of $[d]$, $|\overline{d}| < |d|$ and the system R' will be a reduction system. Also, all normal forms of elements generated by F are in the finite set D. Let $M_0 = D \cap L$. Since the rewriting relation $\rightarrow_{(R',F)}$ respects the congruence, $L_w((R', F), M_0) = [M_0]$, and this set is equal to L since M_0 is the set of normal forms in L. Thus, L is defined by $((R', F), M_0)$. ∎

3 Graph reductions

A comprehensive account of algebras defined to construct graphs is given in [10]. We will consider an algebraic definition of unlabeled, undirected multigraphs. It is easy to extend or modify this algebra to vertex and edge labeled graphs, simple graphs, directed graphs, hypergraphs, and combinations of these.

3.1 Generating partial k-trees

Let $\mathcal{S}_k$ be the set $\{g_0, \ldots, g_k\}$, where g_i is the sort of *i-sourced graphs.* A *concrete i-sourced graph* is an unoriented multigraph and a sequence of i distinct vertices, $G(V, E, I, s)$, where V and E are finite disjoint sets (the vertices and edges, respectively), $I \subset V \times E$ is the incidence relation required to make every edge incident to exactly two vertices, and $s: \{1, \ldots, i\} \rightarrow V$ is the injective map indicating the jth source for $1 \leq j \leq i$. An i-sourced graph is the isomorphism class of a concrete i-sourced graph. The graphs of sort g_i form the domain G_i of i-sourced graphs.

Consider a sequence of sets of operations, $(D_i)_{i=0}^{\infty}$. For $i \geq 0$, the set D_i consists of the following operators:

P_i : $\mathrm{G}_i \times \mathrm{G}_i \rightarrow \mathrm{G}_i$; the parallel composition of two i-sourced graphs. It is obtained by fusing corresponding sources of the two i-sourced graphs. P_0 is the special case of disjoint union of two graphs.

l_i^j : $\mathrm{G}_{i-1} \rightarrow \mathrm{G}_i$ for $1 \leq j \leq i$, the lifting of an $(i-1)$-sourced graph to an i-sourced graph by insertion of a new isolated source vertex at position j among the sources.

r_i : $\mathrm{G}_i \rightarrow \mathrm{G}_{i-1}$, $i \geq 1$ removes the last element from the sequence of sources of an i-sourced graph (but keeps the corresponding vertex).

S_i : $\mathbf{G}_i^i = \mathbf{G}_i \times \cdots \times \mathbf{G}_i \to \mathbf{G}_i$; the series composition of i i-sourced graphs making them into a new i-sourced graph. $S_i(G_1, \ldots, G_i)$ can be defined in terms of the operators l_{i+1}^j, r_{i+1} and P_{i+1} as $r_{i+1}(P_{i+1}(l_{i+1}^1(G_1), P_{i+1}(l_{i+1}^2(G_2), \ldots, l_{i+1}^i(G_i)\ldots)))$, for $i \geq 2$. Intuitively, each of the operands will have $i-1$ of its i sources identified with $i-1$ sources of the result and one with an $(i+1)$th vertex that is subsequently not regarded as a source. For $i=1$, $S_1(G_1)$ is defined to have a new vertex which is a source and is connected by an edge with the source of the argument (which is not a source of the result). Thus, $S_1(G_1) = r_2(P_2(e_2, l_2^1(G_1)))$, where e_2 is defined below.

$e_i \in \mathbf{G}_i$ for $i = 1, 2$ is a nullary operator which evaluates to an edge with its two end-vertices. One ($i = 1$) or both ($i = 2$) of the end-vertices are sources. (Note that $e_1 = r_2(e_2)$.)

$\mathbf{i} \in \mathbf{G}_i$ for $i = 0$, is the empty graph $\mathbf{0}$. For $i > 0$, $\mathbf{i}$ is a derived operator, evaluating to the i-sourced edgeless graph $l_i^1(\ldots l_0^1(\mathbf{0})\ldots)$.

r_i^* : $\mathbf{G}_i \to \mathbf{G}_0$ is introduced as a derived operator, $r_i^*(G) = r_i(r_{i-1}(\ldots r_1(G)\ldots))$, *i.e.*, r_i^* is the operator that removes all sources (as sources, not as vertices) from an i-sourced graph.

Thus, for instance, D_0 consists of the operators $\mathbf{0}$ and P_0, D_1 consists of P_1, S_1, l_1^1, r_1, $\mathbf{1}$ and e_1. D_2 consists of the ordinary parallel and series operations P_2 and S_2, as well as l_2^1, l_2^2, r_2, and e_2.

In the case we are interested in labeled graphs or hypergraphs, more operators would be needed to define labels and hyperedges of different sizes, see [10]. We let the signature F_k be the union of operator sets D_i for $i = 0, \ldots, k$. We let F_∞ be the union of the signatures F_k. We shall denote by M_k the F_k-algebra with set of sorts $\{g_0, \ldots, g_k\}$ and domains $\{\mathbf{G}_0, \ldots, \mathbf{G}_k\}$. Sometimes, we will use the term *parse* of a graph G over a signature F to denote an expression over F that evaluates to G.

The following proposition shows that F_k generates a proper subset of the domains of M_k.

Proposition 3.4 *F_k generates the class of graphs of tree-width at most k.*

The Proposition will follow from these two lemmas, proven in [4]:

Lemma 3.5 *For any term $t \in M(F_k)$ evaluating to a graph G, and every partial graph H of G, there is a term $t' \in M(F_k)$ which evaluates to H.*

Lemma 3.6 *For a K_k-subgraph induced by vertex set K in any k-tree T, there is a term $t \in M(F_k)$ which evaluates to the graph T with the vertices in K as sources, in any given order.*

Proof. (of Proposition 3.4) By the previous two Lemmas, every partial k-tree is generated by F_k. For an expression $t \in M(F_k)$ we can produce a tree-decomposition of width at most k for $h_M(t)$ as follows: The tree of the decomposition is the undirected tree T corresponding to the parse tree t. The vertex set X_n for node n of T consists of the sources of the value of the subexpression corresponding to n, except if the operation of n is e_1 or S_i, $1 \leq i \leq k$. In the latter case, the set X_n is the union of the sources of the operands of n (these are merged by the operator S_i, so there are $i+1$ vertices for such a node). In the former case, both end points of the edge will be in the corresponding vertex set. It is easy to show by induction over the structure of t that this gives a tree-decomposition of width at most k for $h_M(t)$. ∎

Corollary 3.7 *An i-sourced graph G_i has a tree-decomposition of width at most k such that all sources of G_i are in one vertex set of the tree-decomposition, if and only if G_i is generated by F_k.*

Courcelle [16,18,19,21] defines a *recognizable* set of graphs L to be a set definable as the union of classes of a congruence with finitely many congruence classes of every sort, *w.r.t.* the infinite signature F_∞. This is essentially the finiteness notion mentioned by Lengauer and Wanke[28].

Basic definitions concerning expressing graph properties in Monadic Second Order Logic are given in Courcelle [16,19,20,21] and Arnborg et al. [5]. As a simple example, consider two-colorability of graphs. It is defined by the second formula below which uses the definition of adjacency expressed by the first formula:

$$\begin{aligned} Adj(x,y) &\equiv \exists e\, I(x,e) \wedge I(y,e) \\ Twocol &\equiv \exists X \exists Y\, \forall x V(x) \rightarrow ((x \in X \vee x \in Y) \wedge \neg(x \in X \wedge x \in Y) \\ &\quad \wedge \neg \exists x_1 \exists x_2\, ((x_1 \in X \leftrightarrow x_2 \in X) \wedge Adj(x_1, x_2)) \end{aligned}$$

Theorem 3.8 (Courcelle[16]) *Every monadic second order definable set of graphs is recognizable.*

Courcelle [16] proved the above theorem by exhibiting a finite congruence over M_k fine enough to discriminate all sets definable with a formula of height at most h (the height of a logical formula is the largest level of nesting of quantifiers). Under this congruence, two graphs G and G' are congruent if, for every monadic second order formula ϕ with height at most h, $G \models \phi$ if and only if $G' \models \phi$. We can explain the finiteness of the resulting congruence by observing (and proving, *e.g.*, by induction over h) that there are only finitely many such formulas which cannot be proven equal using renamings of quantified variables and the laws of propositional calculus.

3.2 Graph rewriting systems.

Graph rewriting systems can be either defined as concrete substitution mechanisms (by which a subgraph of a graph is replaced by another graph) or by a rewriting on the algebra of graphs. These two aspects are investigated and shown equivalent in [10]. The graph rewriting systems introduced below are simpler than the most general ones considered in [10], but are sufficient for our purposes.

A *graph rewriting system* is a finite set S of pairs of graphs such that two graphs in any pair are of the same sort. We associate with S a binary relation $\Rightarrow_S$ on (0-sourced) graphs defined as follows:

$G \Rightarrow_S G'$ if and only if, for some pair (H, H') in S with H and H' of some sort g_i, there is an i-sourced graph K such that $G = r_i^*(P_i(H,K))$ and $G' = r_i^*(P_i(H',K))$.

This means that $\overline{H}$, isomorphic to H, is a subgraph of G such that the edges and vertices of G not in $\overline{H}$ and the vertices corresponding to sources of H span a subgraph $\overline{K}$ of G, isomorphic to K, (that we shall call the *context* of $\overline{H}$ in G), with the property that the vertices common to $\overline{H}$ and $\overline{K}$ correspond to the sources of H and K in the considered isomorphisms. Intuitively, $G \Rightarrow_S G'$ means that if G can be expressed as a parallel composition of H, a left side in S, and a context K, then H can be replaced by H' (the corresponding right side in S) in that context to form G'.

Proposition 3.9 *Let S be a graph rewriting system and let k be the smallest number such that every graph occurring among the pairs in S is generated by F_k. For every $k' \geq k$, for all graphs G and G', $G \Rightarrow_S G'$ if and only if $G \rightarrow_{(S,F_{k'})} G'$.*

We can take the total number of edges and vertices as the size of a graph. Now the notion of a graph *reduction* system follows from that of a rewriting system in Section 2.

Theorem 3.10 *Let L be a recognizable set of graphs of treewidth at most k. Then $L = L(R, M_0)$ for some graph reductiong system R and some finite set $M_0 \subset L$.*

Proof. It follows from Proposition 3.4 that L is generated by F_k. We consider the results in Section 3 for the F_k-algebra M_k. This algebra has a size function. So it follows from Proposition 2.3 that $L = L(R, M_0)$ for some finite reduction system R and some finite subset M_0 of L. ∎

Remark. The sets of graphs L as in Theorem 3.10 are definable by hyperedge replacement graph grammars (CFHR). This follows from the closure property of CFHR languages *w.r.t.* intersection with recognizable languages and from the fact that the set of graphs with tree-width at most k is CFHR. Some CFHR sets of graphs are definable by reduction without being recognizable. An example can be constructed from the non-recognizable CF (string) language $\{a^n b^n | n > 1\}$ which is defined by the reduction system $\{a^2 b^2 \rightarrow ab\}$ with the accepting word ab.

It will often be convenient to consider a graph algebra generating only i-sourced graphs, for some i, and consider such a graph 'equivalent' to the corresponding graph with sources removed by the operator r_i^*. We can find representatives (or all minimum size members) of all congruence classes of type g_0 if we have them for type g_i: If S_i is a set containing a minimum size representative (or every minimum size member) from each class of type g_i, then the set S contains a minimum size representative (all minimum size members) from each class of g_0, where S is the union of the set $r_i^*(S_i)$ of graphs from S_i with sources removed, and the set of graphs with fewer than i vertices. The latter set is not finite, since there is no limit on the multiplicity of edges. But in all our examples, a minimum size member of a class has a finite bound (actually, 1 or 2) on the multiplicity of an edge.

We are now ready to state an algorithmic result. The proof is in [4].

Theorem 3.11 *Membership in every monadic second order definable set of graphs of bounded treewidth can be decided in time linear in the number of vertices on a RAM with the uniform cost measure.*

Corollary 3.12 *For every k, membership in the class of partial k-trees is decidable in linear time.*

Remark. The corresponding algorithms are known only for $k \leq 3$.

3.3 Effectiveness of the main result

If we are given a recognizable set L of graphs, we can only construct the reduction system for L if we have $\sim_L$ or one of its finite refinements available. It turns out that this is not necessarily the case. As an example, each minor-closed class which excludes a planar graph is recognizable and of bounded treewidth, but an algorithm that constructs representatives of the congruence classes or decides congruence with respect to a given such property is not known.

We can construct the reduction system R and the set M_0 of Theorem 3.10 if L is defined by a known MS-formula φ and if we know an upper bound on the treewidth of graphs in L (but we have no method to decide the existence of, or to compute, such a bound from φ). By the proof of Theorem 3.8 given in [16] one can construct from φ a family of finite sets of formulae $\{\Phi_i\}_{i=0}^k$, with the following property: Let $G \approx G'$ if and only if G and G' are of the same type i, $0 \leq i \leq k$ and if for every $\psi \in \Phi_i$ we have $G \models \psi$ if and only if $G' \models \psi$. This equivalence relation is a decidable congruence *w.r.t.* F_k, and it is a refinement of $\sim_L$. We can thus construct the (finite) set X of minimum size graphs generated by F_k in each of the (finitely many) congruence classes, since every minimum size graph in a class must be produced by an operator with minimum size arguments. It is now easy to produce R and M_0 using the procedure indicated in the proof of Theorem 3.10 and 2.3.

Suppose, on the other hand, that we do not know a bound on the treewidth implied by φ, but we are given L as $\{G \mid G \models \varphi, \text{treewidth of } G \text{ is at most } k\}$. We know by results of Robertson and Seymour [33,31,32] and Courcelle [17,20] that the class of graphs of treewidth at most k is MS-definable (by means of a set of forbidden minors) but the corresponding MS-formula, Θ_k, is not known for $k \geq 4$. So L is MS-definable (with the formula $\varphi \wedge \Theta_k$) but we cannot find R and M_0 of Theorem 3.10. It appears difficult to find Θ_k – once it is available one could effectively find the minimal forbidden minors for partial k-trees. It is not enough to know an algorithm deciding the membership in L to be able to construct $\sim_L$. One must also know at least an upper bound on the number of equivalence classes of $\sim_L$, see also Lengauer and Wanke[28]. Fellows and Langston[24] show how to find the minimal forbidden minors from an algorithm deciding $\sim_L$ or one of its finite refinements, for a minor-closed family L and when a tree-width bound is known for these minors.

In the examples that follow, we do not derive the classes automatically. Instead, we start with the nullary operators and generate new values using combinations of old values as arguments to the operators of the algebra. Each time a new value has been obtained, we must decide if it is congruent by $\sim_L$ to one previously obtained, and if so generate a reduction rule. Here it is fatal to conclude that two graphs are congruent when they are not, but the opposite mistake only results in more classes than strictly necessary. The congruences must ultimately be proved, but often it turns out that there is a small set of congruent pairs that generates all congruences. Only if one infinitely often fails to identify two congruent values as such does this procedure fail to terminate.

4 Applications

4.1 Outerplanar graphs

Let us follow up with an example, to our knowledge not considered in the literature, possibly because of its relative simplicity. Here, the class of interest is that of **biconnected** (*i.e.*, non-separable) **outerplanar graphs**, P. A graph is outerplanar if and only if it has an embedding in the plane such that all the vertices lie on the border of the unbounded region of the plane ('the outer mesh'). These graphs are isomorphic to convex n-gons with non-intersecting chords. All extant algorithms for recognition of these structures (see, for instance [34,12]) use the fact that these graphs are Hamiltonian. Below, we present a reduction system based on an algebra generating a superclass of biconnected outerplanar graphs.

Consider the class of 2-sourced graphs $\mathbf{G}_2$, the nullary operation e_2 evaluating to a single 2-sourced edge, and two operations $P_2, S_2 : \mathbf{G}_2 \times \mathbf{G}_2 \rightarrow \mathbf{G}_2$, denoting respectively the parallel and the series operation with the usual interpretation. We also have the operation r_2^*, source removal, which can only be the outermost operator since no operator takes a 0-sourced graph as argument. For this reason we can restrict our attention to the congruence classes of type g_2 and derive those in g_0 by projection. Let us take the declared constant graph e_2 as the representative of the equivalence class 1. Applied to arguments from class 1, the two operations result in, respectively, the two-edge, two-vertex graph, which belongs also to class 1, and the path of length 2, the representative of class 2. $P_2(1,2)$ and $P_2(2,2)$ define the classes 3 and 4, respectively. We take $S_2(1,4)$ as representative of the class 5 that contains every graph that can not be a subgraph of a graph in P. The complete tables for these two operations follow. The P_2 operator is commutative, so we give only half the table. The S_2 operation is not commutative in the graph algebra (the sources should be regarded as an ordered set), but it is so in the quotient algebra as can be seen from the operation table. Script entries correspond to new class representatives (*e.g.*, $P_2(1,2)$ is the representative of class 3) non-script entries generate reduction rules, by the construction of Proposition 2.2.

P_2	1	2	3	4	5
1	1	*3*	3	4	5
2		*4*	4	5	5
3			4	5	5
4				5	5
5					5

S_2	1	2	3	4	5
1	*2*	2	2	*5*	5
2	2	2	2	5	5
3	2	2	2	5	5
4	*5*	5	5	5	5
5	5	5	5	5	5

The major task in every such example is to show the equivalence of the result of some operation with some earlier encountered graph. It is clear that an edge can be added or removed parallel to an existing one without violating outerplanarity or biconnectedness. Similarly, one can add an edge between vertices with a common degree 2 neighbor, if the local context shows that the added edge cannot be essential for the outer mesh of the graph. Since any two-path has to be included in the outer mesh, the path can be extended by another degree 2 vertex. Below, we give the five representatives of the equivalence classes of $\sim_P$ of type g_2 in figure 1. In this and the following figures, we indicate source 1 by the leftmost unfilled vertex and source 2 by the rightmost such vertex. The pairs of the reduction system, referring to the operations giving rise to new reduction rules (according to proposition 2.2) is shown in figure 2. Note that we do not need rules with both sides in class 5, and that some redundant rules are not included. As an example, the rule $P_2(3,3) \to 4$ is redundant because we have

$$
\begin{aligned}
P_2(3,3) &= P_2(P_2(1,2),P_2(1,2))) && \text{(definitions)}\\
&= P_2(P_2(1,1),P_2(2,2)) && \text{(since } P_2 \text{ is associative and commutative)}\\
&\to P_2(1,P_2(2,2)) && \text{(by the first rule of Figure 2)}\\
&\to P_2(2,2)=4 && \text{(by the second rule of Figure 2)}
\end{aligned}
$$

M_0 is the set of 0-sourced graphs obtainable as a normal form of a minimum size member of one of the accepted classes, with sources removed, and the graphs K_0 and K_1. In this case, 1, 3 and 4 are the accepted classes, and they all have unique minimum size elements. But the representative of class 4 with sources removed is the cycle of length 4, C_4, which is rewritten by the third rule to C_3 which is the representative of class 3. Thus, in this case M_0 will be $\{C_3, K_2, K_1, K_0\}$; the last two being the only graphs in the class not generated by the operators considered.

4.2 Partial 3-trees

We will now refer to the reduction system presented in [7] and describe it from the new perspective of an algebra generating exactly the class of partial 3-trees.

The representatives for the equivalence classes are the 8 subgraphs of the triangle between three sources ($\mathbf{3}$, $c_1 = l_3^1(e_2)$, $c_2 = l_3^2(e_2)$, $c_3 = l_3^3(e_2)$, $p_1 = P_3(c_2,c_3)$, $p_2 = P_3(c_1,c_3)$, $p_3 = P_3(c_1,c_2)$, $\Delta = P_3(c_1,p_1)$) and $s = S_3(c_1,c_2,c_1)$, the vertex of degree three adjacent to three sources.

The parallel composition of s with itself leads to the buddy reduction rule (cf. [7]). The series composition of three instances of s (recall that $S_3 : G_3 \times G_3 \times G_3 \to G_3$) leads to the cube rule. The parallel composition of s with the one-edge graphs leads to the triangle rule and the isolated vertex, pendant edge, parallel and series rules are obtained from nontrivial series compositions of the smaller graphs.

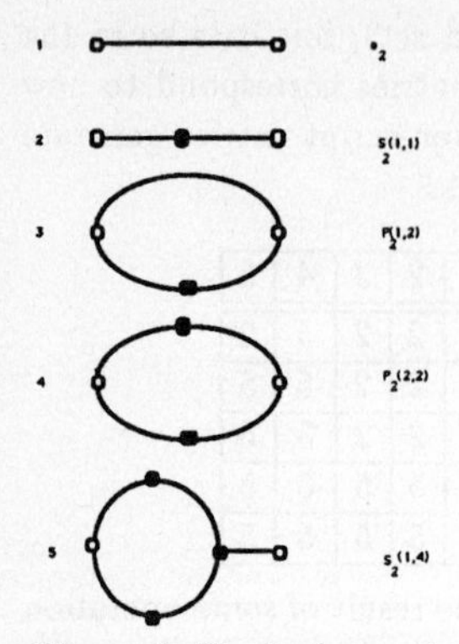

Figure 1: Equivalence class representatives for biconnected outerplanar graphs.

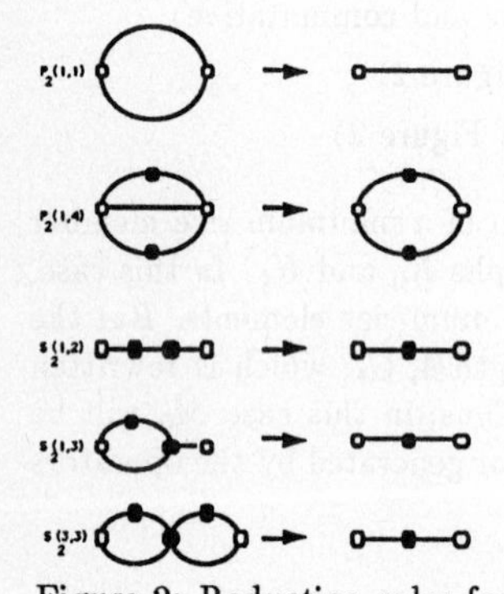

Figure 2: Reduction rules for biconnected outerplanar graphs.

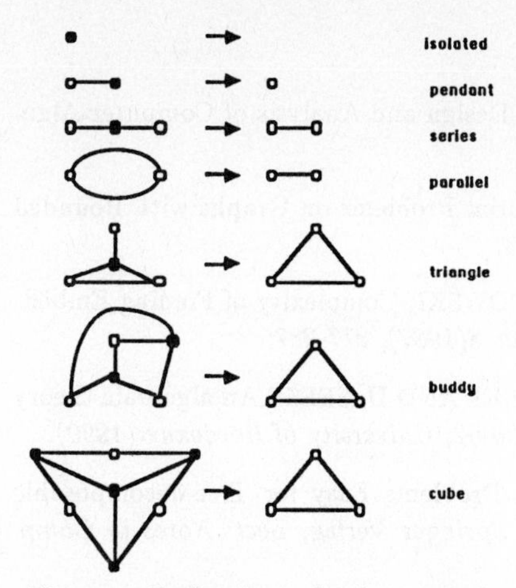

Figure 3: Reduction rules for partial 3-trees.

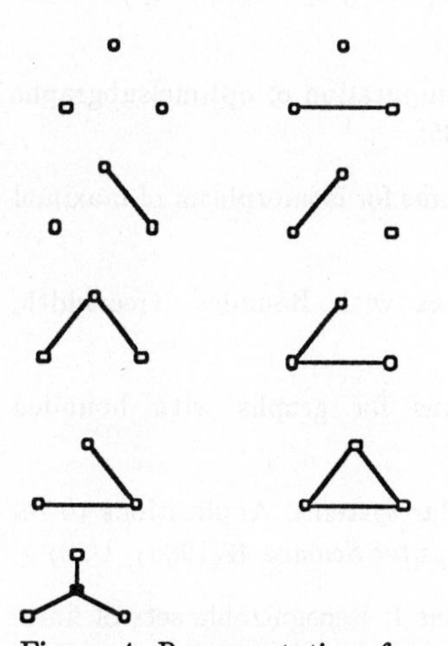

Figure 4: Representatives for congruence classes for partial 3-trees.

References

[1] A.V. AHO, J.E. HOPCROFT AND J.D. ULLMAN, Design and Analysis of Computer Algorithms *Addison-Wesley 1972.*

[2] S. ARNBORG, Efficient Algorithms for Combinatorial Problems on Graphs with Bounded Decomposability — A Survey, *BIT 25 (1985), 2-33;*

[3] S. ARNBORG, D.G. CORNEIL AND A. PROSKUROWSKI, Complexity of Finding Embeddings in a k-tree, *SIAM J. Alg. and Discr. Methods 8(1987), 277-287*;

[4] S. ARNBORG, B. COURCELLE, A. PROSKUROWSKI, AND D. SEESE,An algebraic theory of graph reduction, *Technical Report LaBRI TR 90-02, University of Bordeaux (1990).*

[5] S. ARNBORG, J. LAGERGREN AND D. SEESE, Problems Easy for Tree-decomposable graphs (extended abstract). *Proc. 15 th ICALP, Springer Verlag, Lect. Notes in Comp. Sc.317 (1988) 38-51*;

[6] S. ARNBORG, J. LAGERGREN AND D. SEESE, Problems Easy for Tree-decomposable graphs *to appear, J. of Algorithms.*

[7] S. ARNBORG AND A. PROSKUROWSKI, Characterization and Recognition of Partial 3-trees, *SIAM J. Alg. and Discr. Methods 7(1986), 305-314*;

[8] S. ARNBORG AND A. PROSKUROWSKI, Linear Time Algorithms for NP-hard Problems on Graphs Embedded in k-trees, *Discr. Appl. Math. 23(1989) 11-24;*

[9] S. ARNBORG, A. PROSKUROWSKI AND D.G. CORNEIL, Forbidden minors characterization of partial 3-trees , *Discrete Math., to appear*

[10] M. BAUDERON AND B. COURCELLE, Graph expressions and graph rewritings, *Mathematical Systems Theory 20*(1987), 83-127;

[11] J.A. BERN, E. LAWLER AND A. WONG, Linear time computation of optimalsubgraphs of decomposable graphs, *J. of Algorithms 8* (1987), 216-235;

[12] T. BEYER, W. JONES AND S. MITCHELL, Linear algorithms for isomorphism of maximal outerplanar graphs, *JACM 26(4), Oct.1979, 603-610*;

[13] H.L. BODLAENDER, Dynamic Programming on Graphs with Bounded Tree-width, *MIT/LCS/TR-394, MIT 1987.*

[14] H.L. BODLAENDER, Improved self-reduction algorithms for graphs with bounded treewidth. *RUU-CS-88-29, University of Utrecht 1988.*

[15] B. COURCELLE, Equivalence and transformation of regular systems. Applications to recursive program schemes and grammars, *Theoretical Computer Science 42*(1986), 1-22;

[16] B. COURCELLE, The monadic second order logic of graphs I: Recognizable sets of finite graphs, *Information and Computation* **85** (1990) 12-75;

[17] B. COURCELLE, The monadic second order logic of graphs III: Tree-width, forbidden minors, and complexity issues, Report I-8852, Bordeaux-1 University (1988);

[18] B. COURCELLE, Graph rewriting: an algebraic and logical approach, in "Handbook of Theoretical Computer Science",Volume B, J.B. van Leeuwen, Ed. *Elsevier,* 194-242.

[19] B. COURCELLE, Some applications of logic, universal algebra and of category theory to the theory of graph transformations, *Bulletin of the EATCS 36, October 1988, 161-218.*

[20] B. COURCELLE, The monadic second-order logic of graphs: Definable sets of finite graphs, *LNCS 344 (1989) 30-53.*

[21] B. COURCELLE, Graphs as relational structures; an algebraic and logical approach, *this volume.*

[22] J.A. BONDY AND U.S.R. MURTY, *Graph Theory with Applications*, North Holland (1976);

[23] H. EHRIG, M. NAGL, G. ROZENBERG AND A. ROSENFELD, (Eds.), *Proceedings of the 3rd international workshop on Graph Grammars and their Application to Computer Science, Springer Verlag, Lect. Notes in Comp. Sc.291*

[24] M. FELLOWS and M. LANGSTON, An analogue of the Myhill-Nerode theorem and its use in computing finite basis characterizations, *FOCS 1989 520-525*

[25] M. HECHT AND J. ULLMANN, Flow graph reducibility, *SIAM J. Comp. 1(1972)188-202*;

[26] J. LAGERGREN, manuscript (1987);

[27] Y. KAJITANI, A. ISHIZUKA AND S. UENO, Characterization of partial 3-trees in terms of 3 structures, *Graphs and Combinatorics* 2(1986) 233-246.

[28] T. LENGAUER AND E. WANKE, Efficient analysis of graph properties on context-free graph languages, *ICALP 88, LNCS 317, (1988), 379-393*;

[29] J. MATOUŠEK AND R. THOMAS, Algorithms finding tree-decompositions of graphs, manuscript (1988);

[30] N. ROBERTSON AND P.D. SEYMOUR, Some new results on the well-quasi ordering of graphs, *Annals of Discrete Mathematics 23*(1987), 343-354;

[31] N. ROBERTSON AND P.D. SEYMOUR, Graph Minors X Preprint.

[32] N. ROBERTSON AND P.D. SEYMOUR, Graph Minors XIII, The Disjoint Path Problem Preprint.

[33] N. ROBERTSON AND P.D. SEYMOUR, Graph Minors XIV, Wagners conjecture, Preprint.

[34] M.M. SYSLO, Linear time Algorithm for Coding Outerplanar Graphs, Institute of Computer Science, Wroclaw University, Raport Nr N-20 (1977);

[35] J.W.THATCHER, J.B.WRIGHT, Generalized Finite Automata Theory with an Application to a Decision Problem in Second-Order Logic, *Mathematical Systems Theory 2(1968), 57-81.*

[36] A. WALD AND C.J. COLBOURN, Steiner Trees, Partial 2-trees, and Minimum IFI Networks, *Networks 13 (1983), 159-167*;

[37] T.V. WIMER, Linear algorithms on k-terminal graphs, *PhD. thesis, Clemson University, August 1987*;

[38] T.V.WIMER, S.T.HEDETNIEMI, AND R.LASKAR, A methodology for constructing linear graph algorithms, DCS, Clemson University, September 1985.

Programming with Very Large Graphs[1]

Duane A. Bailey
Department of Computer Science
Williams College, Williamstown MA 01267

Janice E. Cuny
Charles D. Fisher
Department of Computer and Information Science
University of Massachusetts, Amherst MA 01003

Abstract: The ParaGraph graph editor is a tool for specifying the graphical structure of parallel algorithms. Based on an extended formalism of Aggregate Rewriting Graph Grammars, it is an improvement on existing techniques for describing the families of regular, scalable communication graphs. We expect that ParaGraph will prove useful as a testbed for new techniques for describing, visualizing and analyzing the structure of *very large graphs*. This work describes ongoing formal (and practical) efforts to make ParaGraph an a effective tool for specifying massive parallelism.

Keywords: graph editors, massively parallel programming, graph visualization, very large graphs

CONTENTS

1. Introduction

Massively parallel programs are most naturally represented as graphs. Graphs often accompany the description of algorithms in the literature and they are often used as the basis for parallel program visualization and animation[12, 24]. As a result, several programming environments for message-passing architectures now support the view that *a parallel program is an annotated graph*[3, 5, 23]. Nodes in this *programgraph* represent processes, edges represent potential communication channels, and annotations textually define process code and its relation to the communication structure. This view reduces the disparity between the designer's conceptualization of his algorithm and its implementation. For massively parallel algorithms, however, it requires the development of support tools for the manipulation of very large graphs.

Massive parallelism most easily achieved by the composition of similar processes communicating across regularly structured networks (see Figure 1). Programmers exploit this regularity by

[1]The Parallel Programming Environments Project at the University of Massachusetts is supported by the Office of Naval Research under contract N000014-84-K-0647 and by the National Science Foundation under grants DCR-8500332 and CCR-8712410.

designing, implementing, and debugging their programs in-the-small and then scaling upward to obtain the performance of massive parallelism. Even production parallel programs require rescaling to reflect problem size constraints or hardware availability. Thus support for the use of very large programgraphs must include support for scaling. Current environments provide this only to a limited extent through multiple instantiation of nodes[6, 21], replication operators[25], and textual descriptions[4, 20]. We report here on the ParaGraph editor which provides comprehensive support for modeling parallel programs as scalable graphs.

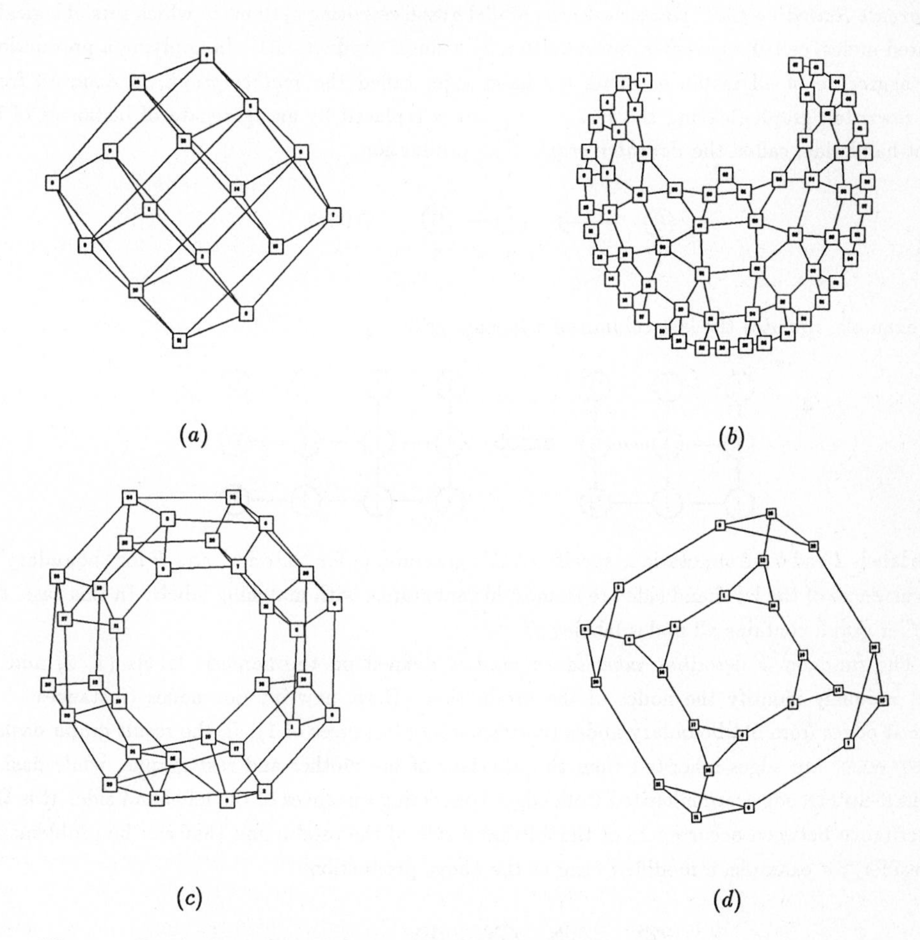

Figure 1: Graphs frequently used in parallel computation: the binary cube (a), x-tree (b), torus (c), cube connected cycles (d). Each was constructed by ParaGraph from smaller members of the same family.

ParaGraph[2, 3] uses graph rewriting to provide scalability. The editing operations of ParaGraph are based on *Aggregate Rewriting (AR)* Graph Grammars[1], a formalism specifically designed to facilitate the specification of the *graph families* typical of parallel computation. ParaGraph has been successfully used within our environment to support parallel program specification and visualization, but we have identified two areas that need improvement. The first is its interface which is unsuitable for naive users; the second is its support for visualization of very large graphs. These areas are the subject of this discussion.

In Section 2, we review AR graph grammars. In Section 3, we describe the ParaGraph editor and its current role in the development of parallel programs. In Sections 4, we consider improvements to our interface and, in Section 5, we propose some methods for visualizing large graphs. In Section 6, we summarize the contributions of this work.

2. Aggregate Rewriting Graph Grammars

Aggregate Rewriting (AR) Grammars are parallel graph rewriting systems in which sets of logically related nodes, called *aggregates*, are rewritten by a single production[1]. In applying a production, the aggregate of all instances of its left-hand side, called the *mother graph*, is removed from the rewritten graph (leaving the *rest graph*) and is replaced by an aggregate of instances of its right-hand side, called the *daughter graph*. The production,

$$\underset{1}{\text{(B)}} \longrightarrow \underset{2}{\text{(I)}} - \underset{3}{\text{(B)}} \qquad \phi(2)=1$$

for example, rewrites the last column of a labeled grid:

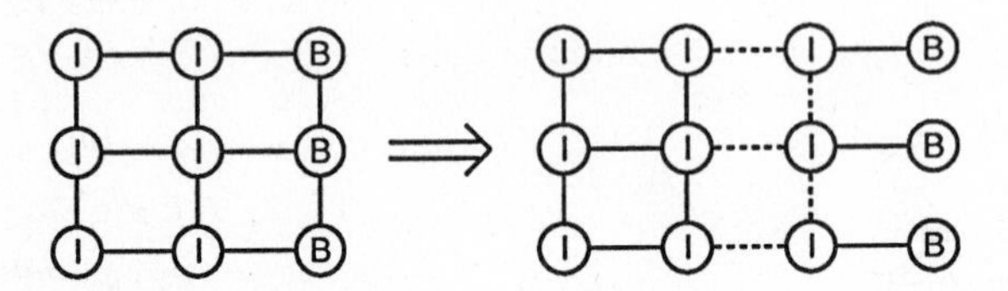

The labels I and b are annotations specific to this grammar (I for "interior" and B for "boundary"). Occurrences of the left-hand side are isomorphic subgraphs with matching labels. In this case, the mother graph contains all nodes labeled B.

The function ϕ describes *inheritance*, and is defined on the numeric labels (1, 2, and 3) that uniquely identify the nodes of the production. Here, new interior nodes (instances of 2) inherit edges from old boundary nodes (corresponding instances of 1). In the result graph dashed EAST-WEST are edges inherited from the interface of the mother and rest graphs, while dashed NORTH-SOUTH edges are inherited from edges connecting instances of the left-hand side. It is this inheritance between occurrences of the left-hand side of the production that can be problematic. Consider, for example, a modified form of the above production:

$$\underset{1}{\text{(B)}} \longrightarrow \underset{2}{\text{(I)}} - \underset{3}{\text{(B)}} \qquad \phi(2)=1, \quad \phi(3)=1$$

What should be the image of an edge between rewritten boundary nodes? If all possible pairs of the replacing node are connected, the original edges are propagated as crossbars (dashed):

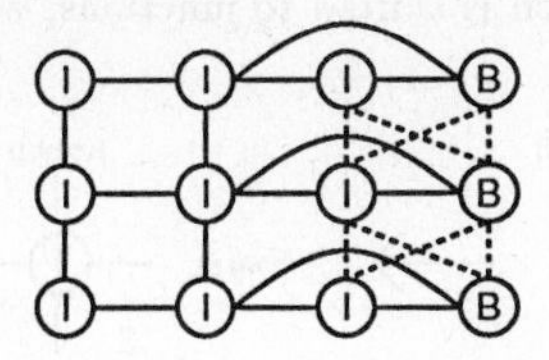

Crossbars, however, rarely occur in parallel systems. If the edge is not propagated, however, the communication graph is potentially disconnected, leaving processes isolated. AR graph grammars strike a balance by introducing the notion of *partitioning*. Nodes of a production's right-hand side are partitioned, inducing a partitioning of nodes in the daughter graph and a partitioning of the inheritance function. Pairs of inheriting daughter graph nodes are then connected only if they reside in the same partition. Thus, if we partition nodes 2 and 3 (with a similar partitioning of ϕ) we have

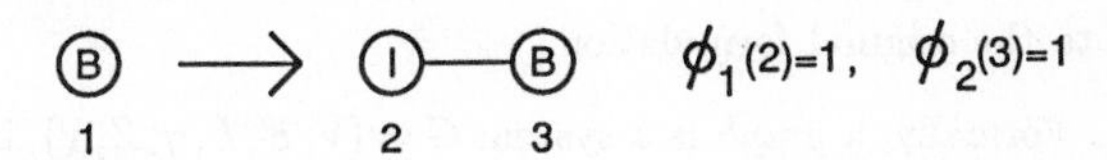

which produces the following transformation in the grid

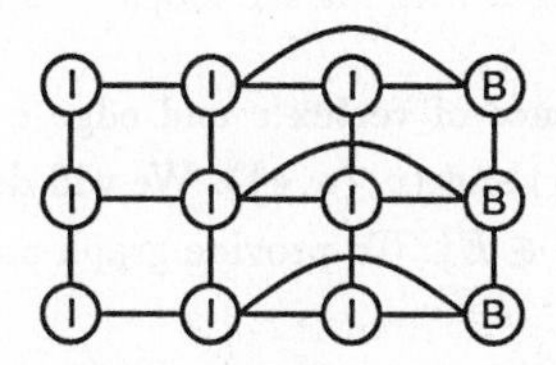

Partitioning allows the *copying of aggregates* (here, a column), which is crucial to the description of the communication structures typical of highly parallel computation.

Our example production, however, still does not result in the desired addition of a single column of nodes to form a larger grid. Such a transformation would require that instances of the new boundary node inherit NORTH-SOUTH but not EAST-WEST connections. Inheritance of *some* of a parent's connections was not possible in our original formulation of AR grammars, but experience has led us to modify the formalism to support *partial inheritance* of edges.

To specify partial inheritance a production must distinguish among edges incident on mother graph nodes. A number of possible mechanisms for this exist (e.g. use of adjacent node labels[14, 15, 16]) but our application motivates a different approach. In our graphs edges incident upon a node represent a process's communication channels in the corresponding program. In process code, programmers must use names to correctly distinguish these channels, either by naming them globally (as in Occam[13] and CSP[11]), or locally (as in Poker[22] or Simon[9]). The latter approach — in which processes refer to logical port names without reference to the neighboring node — promotes modularity and reusability by allowing the programmer to write position-independent. We introduce the notion of a *junction* to provide the programmer with a natural mechanism for specifying logical port names and to provide us with a formal mechanism for distinguishing edges. Each edge has two junctions, which are labeled and rewritten in a manner similar to nodes.

When the junctions of our production are explicitly labeled (**north**, **south**, and **west**) and the domain of the inheritance function is shifted to junctions, we have

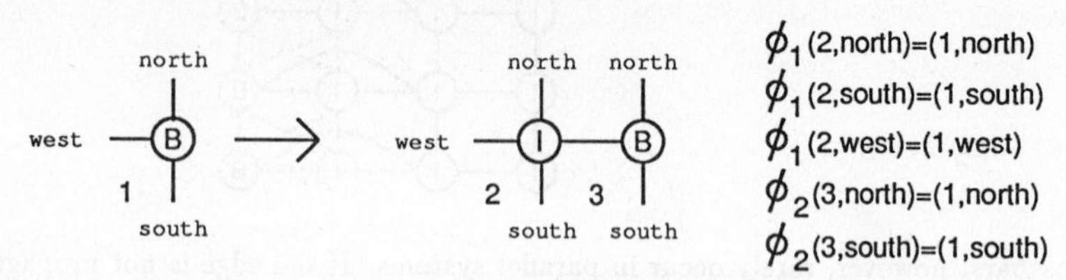

This production allows instances of node 2 to inherit all edges while instances of node 3 inherit just those with junctions labeled **north** or **south**. Junctions participate only partially in occurrence matching: where the corresponding edges are missing from the mother graph (as, for example, at the top and bottom boundary nodes) they are not required for matching. This mechanism aids in uniform rewriting of graphs that are nearly symmetric.

Note that partial inheritance results in a refinement of the original AR grammars: if all junctions have the same label and partitions of ϕ do not distinguish junctions, the extended AR grammars reduce to the original formulation.

The Formalism. Formally, a *graph* is a system $G = (V, E, L, \gamma, Z, \chi)$, in which

V is a finite set of vertices, and

E is a set of undirected edges without self-loops — a set of two element sets on V.

The junction denoting the incidence of vertex v and edge e is denoted (v, e). Each edge $\{u, v\}$, then, has two junctions, $(u, \{u, v\})$ and $(v, \{u, v\})$. We will denote the set of junctions on a graph $J = \{(u, \{u, v\}) \mid u, v \in V,\ \{u, v\} \in E\}$. To provide graph annotation, we note that

L is a set of vertex labels,

$\gamma : V \to L$ is a total function mapping vertices to their labels,

Z is a set of junction labels, and

$\chi : (V \times E) \to Z$ is a total function on J mapping junctions to their labels.

Graphs G and H are *isomorphic* if there exists a bijection $\iota_V : V_G \to V_H$ which induces the natural bijection between E_G and E_H. An *occurrence of G in H* is a subgraph G' of H which is isomorphic to G. We assume that this isomorphism preserves the labels of vertices and junctions.

An *aggregate rewriting production* $\mathcal{P} = (M, D, \phi = \sum_i \phi_i)$ rewrites the mother graph (the aggregate of occurrences of M), producing the daughter graph (the disjoint union of corresponding instances of D) under the direction of an *inheritance function* $\phi : V_D \times Z_D \to V_M \times Z_M$. The inheritance function is a partial map that indicates the labeled junctions in M that will provide connecting edges for some of the junctions of D. We restrict ϕ by requiring that a node in D inherit from at most one node in M, and that each junction of M be inherited by some junction of D. We express the partitioning of inheritance by a fixed partitioning of ϕ into one or more ϕ_i. The trivial partitioning $\phi_1 = \phi$ is used when no logical partitioning of ϕ is desired.

We now describe the mechanics of the parallel rewrite rule. First the mother graph is removed from the host graph, leaving the rest graph. The *interface* is the set of edges that are not found in either the mother or rest graph. Next, the daughter graph is constructed by forming the disjoint union of copies of D, one for each instance of M in the mother graph; we assume that the instances of M and D within the mother and daughter graphs are similarly indexed. Finally, the daughter graph is sewn to the rest graph by reconstructing the interface edges using the following rules:

- *Edges incident to the rest graph and the mother graph.* Suppose edge $e = \{u, v\}$ is incident to the rest graph at node$u$ and an occurrence of M in the mother graph at node v. An edge is introduced between u and all nodes v' with a junction (labeled z') inheriting from junction $(v, \{u, v\})$. The junctions of this new edge are labeled $\chi((u, \{u, v\}))$ at $(u, \{u, v'\})$ and z' at $(v', \{u, v'\})$.

- *Edges incident to two instances of the left graph.* Suppose the edge $e = \{u, v\}$ is incident to the n-th and m-th occurrences of M in the mother graph at junctions labeled z_n and z_m respectively. Then an edge is introduced between all inheriting pairs of junctions (labeled z'_n and z'_m, respectively) on nodes $u'_n \in V_{D_n}$ and $v'_m \in V_{D_m}$ — provided the junctions reside in the same partition. Each new edge has junction labels z'_n at $(u'_n, \{u'_n, v'_m\})$ and z'_m at $(v'_m, \{u'_n, v'_m\})$.

We now consider the analogy between AR graph grammar derivations and the growth of parallel programs using ParaGraph.

3. Designing Very Large Programs

The ParaGraph editor is a central specification tool in a parallel programming environment that we have built for simulated, local-memory, MIMD architectures. The simulator, like many parallel processors, executes a collection of independent processes. As the degree of parallelism increases, it becomes less feasible to tailor and distribute processes by hand. ParaGraph, therefore, must automate distribution of code in a manner that exploits the similarity of processes and the symmetries of the underlying communication topology.

We describe ParaGraph by tracing the development of a simple application: a vector-summation algorithm. The algorithm reads a vector of values at the leaves of a binary tree and reduces it to its sum as it is passed upward toward the root. Leaves read an external value, while nonleaves feed the sum of two incoming values upward, until the total exits the root.

The Small Program. The first step toward creating the program is to specify its smallest instance. This involves (1) drawing its communication graph, (2) annotating the processes, and (3) developing one or more code bodies based on the process annotations.

Consider the snapshot of a specification of a tree summation program, shown in the monochrome screendump of Figure 2. The *view* of the graph labeled `tree-0` (upper left) shows the smallest version of the program, consisting of a single leaf process. Annotation of graph components is performed through typed labels, or *attributes*; in practice, processes usually require multiple labels. Here, the labels are shown in the *scorecard* beside the `tree-0` view: `code_filename` gives the name of the code to be run by this process (`tree.c`) and `process_name` gives the logical name of the process (`leaf`). They are *explicit* attributes because they are specified and interpreted by the user. Other attributes (e.g. `degree`) are *implicit* because they are declared and maintained

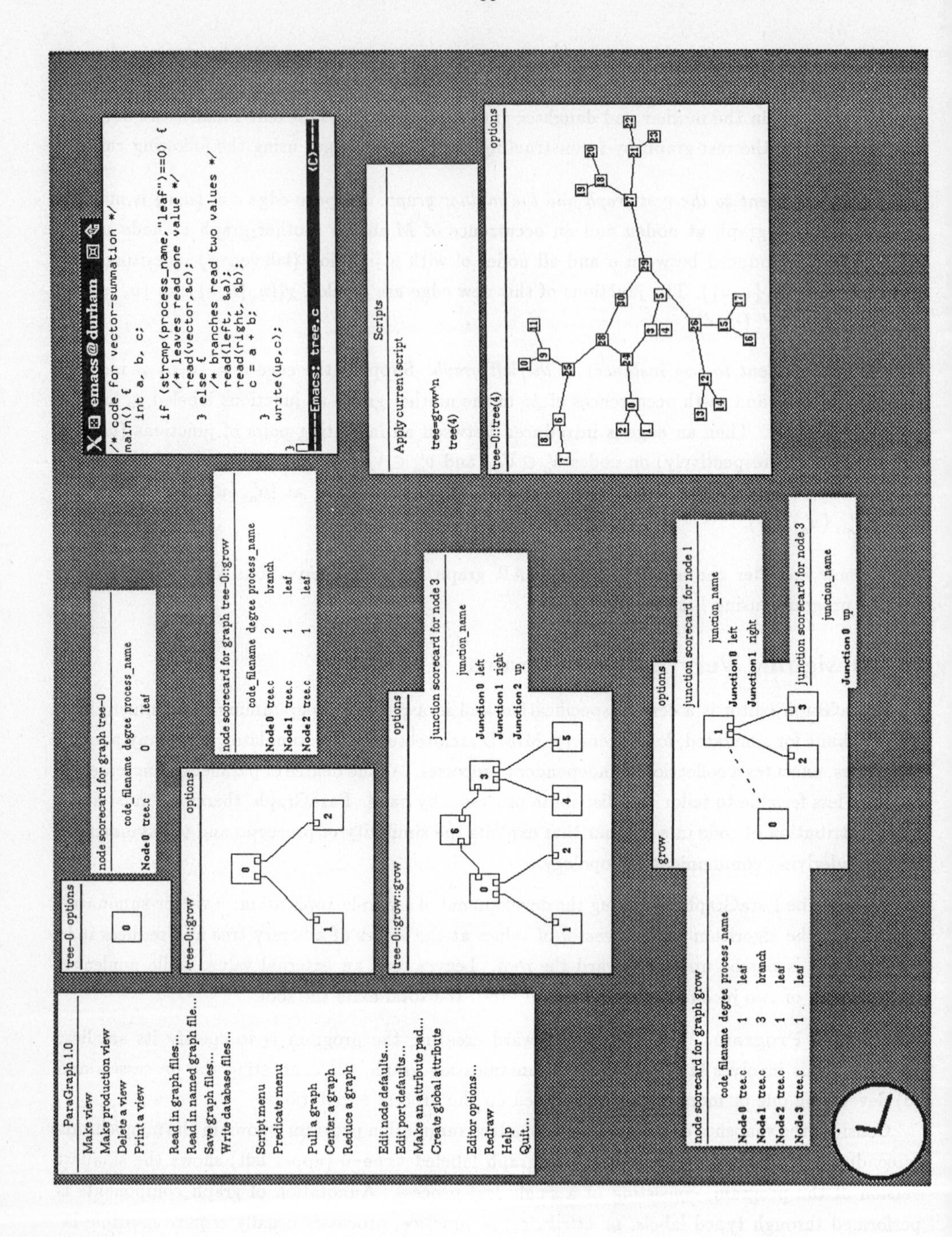

Figure 2: Screen dump of a ParaGraph session.

automatically by the editor. All attribute values are available as preprocessor macros within each process' code.

To generate code suitable for execution, the programmer dumps the graph into a *program database* that is passed to a *code distributor*, which uses preprocessor macros and graph structure to generate code tailored to each process. In the tree summation program, for example, `tree.c` (upper right) is used as the basis for both leaf and nonleaf process code. The attribute `process_name` is used to distinguish between types of processes at runtime: leaf nodes get their initial value from a port `vector` while branch nodes read and sum their children's values. All nodes pass their values upward through port up. Ports not referenced in the graph refer to process-specific files on secondary storage. Once code is generated for each process, the entire program is compiled, linked, downloaded and executed.

The Large Program. In its current state our program has only a single node and makes no use of parallelism! The construction of a large program involves (1) specifying the production which transforms an instance of the programgraph into the next larger member of the family, and (2) successively rewriting the graph with the production until the desired graph is generated.

In this example, the production labeled `grow` is all that is needed. The left side is `node 0` which is rewritten to a structure consisting of nodes `1`, `2`, and `3`. Inheritance is shown by dashed lines (`node 1` inherits from `node 0`) with line color indicating partitioning (not needed here). All rewriting in the productions is based on user-defined attributes. To provide flexibility in rewriting, annotations on the production are expressions that may contain free variables. Occurrences of the left-hand side are isomorphic subgraphs whose labels match with consistent bindings of the free variables. These bindings are then used to establish the context of label rewriting in the corresponding instances of the right-hand side.

This production simultaneously rewrites all nodes with code name `tree.c` and process name `leaf`. If it is applied once to `tree-0`, the three node graph in view `tree-0::grow` results; if it is applied twice, the seven node graph in view `tree-0::grow::grow` results. Note that the labels on these last two windows indicate the "heritage" of the tree structures they contain.

Once a graph has grown to more than one node, it needs junction labels to support interprocess communication. These labels (`left`, `right` and `up`) are introduced during rewriting. They are shown here as junction titles in scorecards associated with the nodes of the production: `node 1` has two junctions named `left` and `right`; nodes `2` and `3` each have a junction named labeled `up`. During rewriting, the daughter graph inherits junctions from the mother graph. Thus, the junction scorecard for `node 3` of `tree-0::grow::grow` has accumulated `left`, `right`, *and* `up` junctions.

Often the order of production application must be controlled in order generate only members of a logical family of topologies. In ParaGraph this is accomplished, in a manner similar to that of programmed graph grammars[7], with a *script*. A script is a parameterized expression defined over the set of production names. Concatenation is implicit, and quantified iteration is indicated by $\wedge$ followed by a parameter that determines the number of iterations used in a specific derivation. Thus, `grow∧n` applies production `grow` n times; application to `tree-0` with `n=` 4 results in the view `tree-0::tree(4)`.

ParaGraph allows the programmer to work in an environment that directly supports the necessary communication abstractions of parallel programming by enabling him to concisely specify scalable, annotated graphs. The information from these graphs — including generated information on layout — is available to other programming tools. Currently, for example, we use ParaGraph specifications as the basis for the displays produced by an animating debugger[12].

Our experience with ParaGraph's use, however, has led us to conclude that its interface is difficult for most programmers. We discuss our design for an improved interface in the next section.

4. Drafting Very Large Graphs

While the concept of growing a large programgraph from a smaller instance is not difficulty, the use of ParaGraph requires an understanding of graph grammar mechanisms foreign to most programmers. We are therefore designing a simplified interface[8] in which transformations are based on more familiar graphical operations. Our experience with ParaGraph indicates that two operations form a useful basis for the new interface: *copy* which duplicates a subgraph and *replace* which replaces single nodes. (Both operations, in fact, correspond to node rewriting graph grammar productions.)

We will attempt to convey our approach and demonstrate its feasibility with a simple example of replacement: the growth of an x-tree, a topology of Figure 1. Given an x-tree, the next instance of the family is the result of replacing all leaf nodes in a similar manner, thus the user need only specify one prototypical three node replacement (Figure 3a). The dashed square represents the node to be replaced, the replacing subgraph is drawn within the borders of the square (as in GraphEd[10]), and neighbors will be drawn drawn surrounding the borders. Junctions on the replaced node are represented as shaded strips along its perimeter. The interface assumes that all nodes have the same set of potential junctions and as edges are drawn corresponding junctions are induced on all nodes.

To indicate inheritance of neighbor connections an edge is drawn from the junctions of the inner graph (the daughter junction) to the junctions of the original node (the mother junction) (Figure 3b). Since we are only replacing leaf nodes the bottom and corner junctions are not used.

The addition of these edges results in the display of dashed squares representing the neighbors adjacent to the rewritten node. The neighbor above is an interior node that remains untouched by this replacement. The left and right neighbors, however, represent sibling leaves that will be replaced concurrently. For these nodes we must specify the junction in the neighbor's replacement graph which supports the opposite end of the inherited channel. As we draw through a junction on the original node, its neighbor is automatically expanded to allow us to identify that junction (Figure 3c).

All changes are automatically *tessellated* throughout the picture : completing the connection indicated by the arrow causes the similar connections to be inferred *from* the neighbor on the right. This reduces the redundant specifications and enforces correct construction of the underlying AR production. Drawing another edge finalizes the transformation (Figure 3d). Note that the completed edges in this drawing represent *actual edges* rather than inheritance.

While many aspects of the production can be deduced from user actions, partitioning is more difficult. If the underlying AR production for this drawing is to correctly generate an x-tree, the editor must infer partitioning to prevent crossbars between replacements for adjacent leaf nodes. To do this the interface identifies the most conservative partitioning consistent with the edges

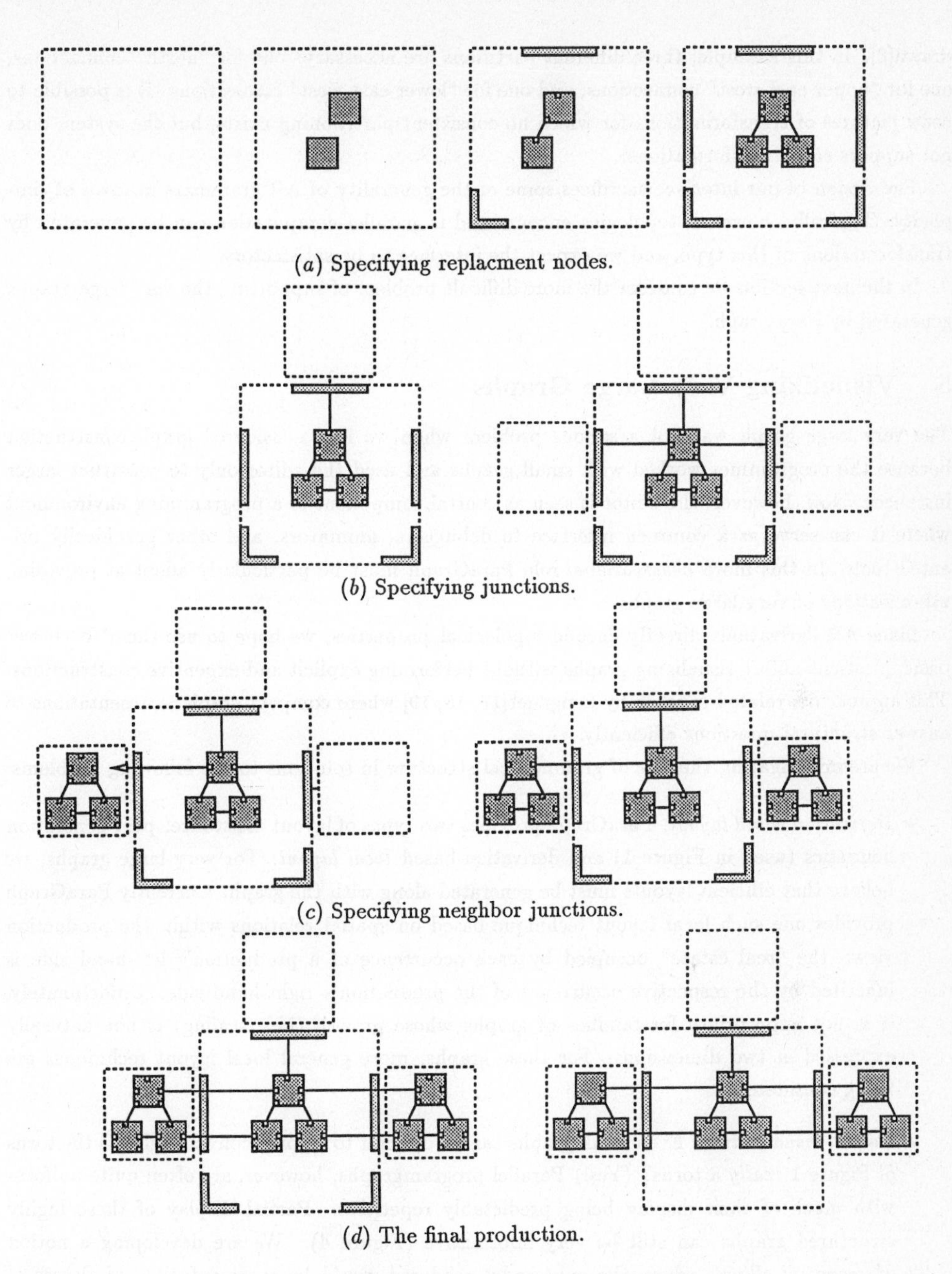

(*a*) Specifying replacment nodes.

(*b*) Specifying junctions.

(*c*) Specifying neighbor junctions.

(*d*) The final production.

Figure 3: Production specification for x-tree growth.

drawn[8]. In this example, three different partitions are necessary: one for "north" connections, one for "upper east-west" connections, and one for "lower east-west" connections. It is possible to draw pictures of transformations for which no consistent partitioning exists, but the system does not support such transformations.

The design of our interface sacrifices some of the generality of AR grammars in favor of simplicity. Typically, however, topologies encountered in parallel computation can be generated by transformations of this type, and we expect the interface to be satisfactory.

In the next section we consider the more difficult problem of supporting the very large graphs generated by ParaGraph.

5. Visualizing Very Large Graphs

The very large graph was not a serious problem when we first considered graph construction because the programmer worked with small graphs and used the editor only to construct larger instances. Now, however, the editor is seen as central component of a programming environment where it can serve as a common interface to debuggers, animators, and other graphically oriented tools. In this more *observational* role ParaGraph must be particularly adept at providing *visualizations* of very large graphs.

Since AR derivations directly encode topological properties, we hope to use them to answer basic questions about visualizing graphs without performing explicit and expensive constructions. This approach is related to work by Lengauer[17, 18, 19] where compact graph representations to answer structural questions efficiently.

We are investigating the role of grammatical structure in solutions to the following problems:

- ⋆ *Derivation based layout.* ParaGraph provides two types of layout assistance: post-generation heuristics (used in Figure 1) and derivation-based *local layout*. For very large graphs, we believe that efficient layouts must be generated along with the graph. Currently ParaGraph provides one such local layout technique based on spatial relations within the production view: the "real estate" occupied by each occurrence of a production's left-hand side is inherited by the respective occurrence of the production's right-hand side. Unfortunately, it is not appropriate for families of graphs whose growth (i.e. scaling) is not naturally expressed in two dimensions. For these graphs, more general local layout techniques are being considered.

- ⋆ *Partial visualization.* Even small graphs can be difficult to display convincingly: Is the torus of Figure 1 *really* a torus? (Yes!) Parallel programgraphs, however, are often quite uniform with much of their display being predictably repetitive. Partial display of these highly structured graphs can still be very informative (Figure 4). We are developing a notion of *graphical ellipsis* where the only nodes rendered would be representatives of classes of nodes (`branch`, or `leaf`), or nodes introduced early in the derivation (the first node labeled `branch`), or those nodes that represent the region active in growth (any node labeled `leaf`).

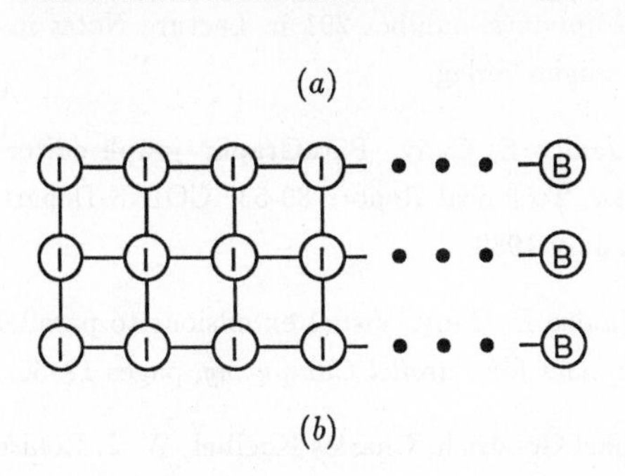

Figure 4: An example of (a) a derivation and (b) a structure represented using ellipses.

⋆ *Lazy evaluation.* In the large programgraph, the programmer will often focus attention on a small (topological or logical) neighborhood of nodes. If the rendering of the graph can be restricted to that neighborhood, the cost and complexity of the display can be substantially reduced. While many editors provide spatial "scrolling" of graphs, we are considering a formal method for lazy evaluation which generalizes this concept by using grammatical structure to provide demand-based-derivation of small portions of a graph.

We expect that grammatical structure will provide information useful in the visualization of large graphs and we believe that ParaGraph provides a natural framework for testing the effectiveness of such techniques.

6. Conclusions

The ParaGraph editor is a tool for specifying scalable graph descriptions based on the mechanisms of aggregate rewriting graph grammars. We have been able to incorporate ParaGraph into our parallel programming environment where it supports the specification of massively parallel programs as annotated graphs. This approach reduces the disparity between a programmer's concept of his algorithm and his code and it provides visual information that can be exploited by other programming tools. We are now in the process of designing more compact (albeit, less expressive) representations of productions that will provide a domain-specific view of graph rewriting that is more natural to parallel programmers.

Initially, this project was motivated by the difficulty of accurately constructing scalable graph families. However, ParaGraph's success in graph construction has exposed the more fundamental difficulty of manipulating very large graphs for visualization. Fortunately, we believe that ParaGraph provides a useful platform for the evaluation of derivation-based solutions to these challenging problems.

References

1. Duane A. Bailey and Janice E. Cuny. Graph grammar based specification of interconnection structures for massively parallel computation. In *Proceedings of the Third International Workshop on Graph Grammars*, number 291 in Lecture Notes in Computer Science, pages 73–85, Berlin, 1987. Springer-Verlag.

2. Duane A. Bailey and Janice E. Cuny. ParaGraph: graph editor support for parallel programming environments. Technical Report 89-53, COINS Department, University of Massachusetts at Amherst, July 1989.

3. Duane A. Bailey and Janice E. Cuny. Visual extensions to parallel programming languages. In *Languages and Compilers for Parallel Computing*, pages 17–36. The MIT Press, 1990.

4. Francine Berman, Michael Goodrich, Charles Koelbel, W. J. Robison III, and Karen Showell. Prep-P: A mapping preprocessor for CHiP architectures. In *1985 International Conference on Parallel Processing*, pages 731–733, August 1985.

5. J. C. Browne. *Software Engineering of Parallel Programs in a Computationally Oriented Display Environment*, pages 75–94. The MIT Press, 1990.

6. J.C. Browne, M. Azam, and S. Sobek. CODE: A unified approach to parallel programming. *IEEE Software*, pages 10–19, July 1989.

7. Horst Bunke. *Graph-Grammars and Their Application to Computer Science and Biology*, pages 155–166. Number 73 in Lecture Notes in Computer Science. Springer-Verlag, Berlin, 1979.

8. Charles D. Fisher. Approaches to specifying aggregate rewriting graph grammar productions. Technical Report 90-49, Univ. of Massachusetts, Amherst, MA, June 1990.

9. Richard M. Fujimoto. Simon: Simulator of multicomputer networks. Technical Report UCB/CSD 83/140, UC Berkeley, Berkeley, August 1983.

10. Michael Himsolt. GraphEd user manual. Technical report, Universitat Passau, Passau, 1990.

11. C.A.R. Hoare. Communicating sequential processes. *Communications of the ACM*, 21(8):666–677, 1978.

12. Alfred A. Hough and Janice E. Cuny. Initial experiences with a pattern-oriented parallel debugger. In *ACM SIGPLAN and SIGOPS Workshop on Parallel and Distributed Debugging*, May 1988.

13. INMOS. *Occam Programming Manual.* INMOS, Ltd., Bristol, England, 1983.

14. D. Janssens and G. Rozenberg. Restrictions, extensions, and variations of NLC grammars. *Information Sciences*, 20:217–244, 1980.

15. D. Janssens and G. Rozenberg. Graph grammars with neighbourhood–controlled embedding. *Theoretical Computer Science*, 21:55–74, 1982.

16. D. Janssens and G. Rozenberg. Graph grammars with node-label controlled rewriting and embedding. In Volker Claus, Hartmut Ehrig, and Grzegorz Rozenberg, editors, *Graph-Grammars and Their Application to Computer Science and Biology*, number 153 in Lecture Notes in Computer Science, pages 186–205, Berlin, 1982. Springer-Verlag.

17. Thomas Lengauer. The complexity of compacting hierarchically specified layouts. In *Conference on Foundations of Computer Science*, pages 358–368, 1982.

18. Thomas Lengauer. Efficient solution of connectivity problems on hierarchically defined graphs. Technical Report Bericht Nr. 24, Universitat-Gesamthochschule Paderborn, Paderborn, West Germany, 1985.

19. Thomas Lengauer. Hierachical planarity testing algorithms. Technical Report Bericht Nr. 25, Universitat-Gesamthochschule Paderborn, Paderborn, West Germany, 1985.

20. Hungwen Li, Ching-Chy Wang, and Mark Lavin. Structured process: A new language attribute for better interaction of parallel architecture and algorithm. In *1985 International Conference on Parallel Processing*, pages 247–254, August 1985.

21. James Purtilo, Daniel A. Reed, and Dirk C. Grunwald. Environments for prototyping parallel algorithms. In *1987 International Conference on Parallel Processing*, pages 431–438, 1987.

22. Lawrence Snyder. Introduction to the configurable, highly parallel computer. *Computer*, 15(1):47–56, 1982.

23. Lawrence Snyder. Parallel programming and the Poker programming environment. *Computer*, 17(7):27–37, 1984.

24. David Socha, Mary L. Bailey, and David Notkin. Voyeur: Graphical views of parallel programs. In *SIGPLAN Workshop on Parallel and Distributed Debugging*, pages 206–215, May 1988.

25. M. Wohlert. Spectral user's manual. Technical report.

DESCRIBING GÖTTLER'S OPERATIONAL GRAPH GRAMMARS WITH PUSHOUTS

Klaus Barthelmann
Joh.-Gutenberg-Universität, FB Mathematik (Informatik),
Staudingerweg 9, D-6500 Mainz, Germany

ABSTRACT: We shall show how the kind of graph grammars invented by Göttler [8, 9, 10] can be defined in categorical terms. Derivations can then be carried out in the framework of [6]. This translation enables us to review the definitions which were given with implementations in mind. Furthermore it may suggest a way to add expressive power to the algebraic approach. And, hopefully, some theorems carry over between the algebraic approach and special cases (notably NLC [12, 16] or NCE graph grammars [13]) of the operational graph grammars considered in this paper.

Keywords: CR-classification: S.4.2 grammars and rewriting systems; additional keywords: graph grammars

Contents

0 Introduction

The well-known *algebraic approach* of graph grammars is described, for example, in [6, 4, 5]. Its elegance and generality come from the fact that it only defines *whether* a given graph can be derived from another graph by applying a production. Some additional effort is required to determine *how* the result of the application can be obtained. Therefore the algebraic approach is a good starting point in defining more powerful derivation mechanisms.

Operational graph grammars, in contrast, emphasize the effective execution of derivation steps. Manipulations on graphs are supposed to be carried out by computers in an efficient way. In most cases the embedding of generated parts into the rest of a graph can be more flexible than in the algebraic case, which fixes the neighbourhood of removed vertices.

Among the most well-known representatives of operational graph grammars are NLC (*node label controlled*) [12, 16] or NCE (*neighbourhood controlled embedding*) [13], and grammars with set-theoretic operators [14, 15]. With respect to the expressive power, Göttler's approach [8, 9, 10] lies somewhere in between (although this may not be obvious at first sight since different formalisms are involved). That seems to be a good compromise between expense and usefulness for many applications. The method offers another advantage: Productions themselves are depicted as graphs. In this way the illustration of morphisms (by numbers) and the multiple drawing of the gluing graph, as in the algebraic approach, are avoided. A graphical representation seems preferable to a linear notation also.

In the rest of this paper, by "operational graph grammars" we refer to Göttler's approach.

The paper is organized as follows: First the reader is reminded of the basic notions in the algebraic approach. Furthermore the category to work with is introduced. Then, an introduction to Göttler's approach follows which might be unfamiliar to the reader. The main section is divided into three parts. First, we show that the effect of an operational production in a given situation can as well be achieved by an algebraic one. As situations change, we need different algebraic productions. So there is a family of algebraic productions related to one operational production. After some technical terms are introduced, this relation can be expressed formally: We require the existence of some morphisms forming pushouts and satisfying additional restrictions.

The reader is assumed to be familiar with some basic notions from category theory (see, for example, [17]).

1 The Algebraic Approach

Definition 1.1 (Production) *A* production *consists of three graphs and two morphisms between them:* $p = ({}'B \xleftarrow{{}'b} K \xrightarrow{b'} B')$. *We call* $'B$ *the* lefthand side, B' *the* righthand side *and* K *the* gluing graph *of the production.*

Definition 1.2 (Derivation) *Graph* H *can be* derived *from graph* G *by applying production* p $(G \stackrel{p}{\Longrightarrow} H)$ *iff there exist a* context graph D *and additional morphisms such that the following diagram consists of two pushouts:*

$$\begin{array}{ccccc} {}'B & \xleftarrow{{}'b} & K & \xrightarrow{b'} & B' \\ g\downarrow & & d\downarrow & & \downarrow h \\ G & \xleftarrow[{}'c]{} & D & \xrightarrow[c']{} & H \end{array}$$

The algebraic approach works with a wide variety of graphs and morphisms. But for the rest of this paper we need a special category.

Definition 1.3 (Labelled graph) *A* (directed) labelled graph *consists of*

- *a set* V *of* vertices,
- *a set* L_V *of* vertex labels,
- *a* labelling map $l_V \colon V \to L_V$,

- *a set* L_E *of* edge labels *and*
- *a set* $E \subseteq V \times V \times L_E$ *of* edges.

Definition 1.4 (Morphism) *A* morphism *between labelled graphs* $f\colon G \to H$ *with common sets of labels* $L_V = L_{V_G} = L_{V_H}$ *and* $L_E = L_{E_G} = L_{E_H}$ *is a map* $f_V\colon V_G \to V_H$*, such that*

1. $l_{V_H} \circ f_V = l_{V_G}$
2. $(\forall v, v' \in V_G)(\forall m \in L_V)((v, v', m) \in E_G \implies (f_V(v), f_V(v'), m) \in E_H)$

A monomorphism (i. e. f_V *is injective) which satisfies the stronger condition*

2'. $(\forall v, v' \in V_G)(\forall m \in L_V)((v, v', m) \in E_G \iff (f_V(v), f_V(v'), m) \in E_H)$

is called subgraph isomorphism.

Labelled graphs over fixed sets of labels L_V and L_E together with their morphisms form a category LGraph.

The maps $s, t\colon E \to V$ and $l_E\colon E \to L_E$ are obtained as projections onto the first, second and third component of a triple, respectively. For an edge $e \in E$ we call $s(e)$ the *source vertex*, , $t(e)$ the *target vertex* and $l_E(e)$ the *edge label.* A morphism f in LGraph determines a map f_E of edges by $f_E(v, v', m) = (f_V(v), f_V(v'), m)$ for all $(v, v', m) \in E$. These agreements turn LGraph into a reflective subcategory of the "category of labelled graphs" most often used (see, for example, [6]). The difference is that no multiple, parallel, equally labelled edges between two vertices are allowed here. Since the embedding functor is continuous, but not cocontinuous (i. e. pushouts are formed in a different way), this choice really matters.

Differently from [10] we do not exclude loops here, i. e. edges with coinciding source and target vertices. They are not very useful in practical applications, however, they show up when forming pushouts.

Pushouts (and all the other finite (co-)limits) always exist in LGraph.

2 Göttler's Operational Graph Grammars

In analogy to the algebraic approach, there are a lefthand side whose corresponding part is to be deleted from the host graph, and a righthand side to be added. An embedding part determines how the newly generated part is to be connected to the rest graph.

Definition 2.1 (Production) *A* production *is a labelled graph, whose set of vertices is divided into four disjoint parts: a* left side V_L*, a* right side V_R *and an embedding part. The latter consists of a* determined environment V_B *("below") and an* indetermined environment V_A *("above"). Edges between left and righthand side are forbidden.*

We do not mind if the left side and/or the determined environment are empty. [10] allows for a special treatment of edges between the determined and indetermined environments

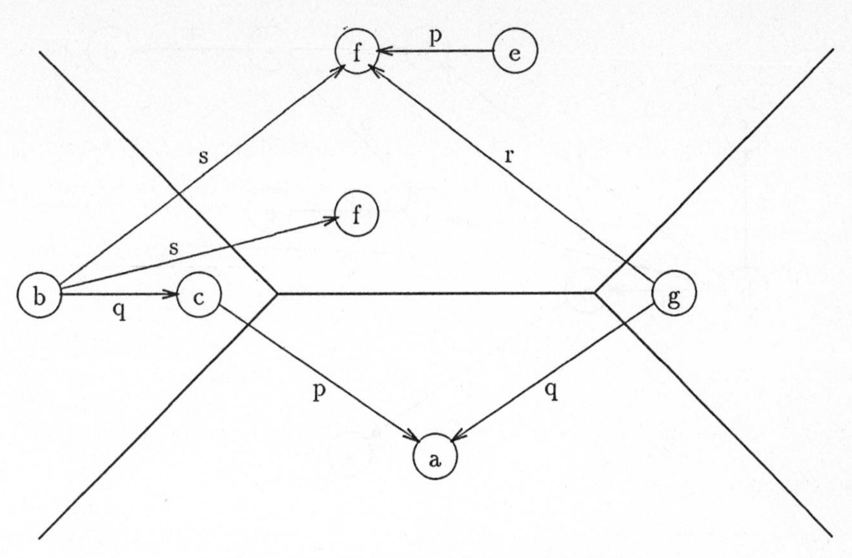

Figure 1: Example of a production

that we do not consider here. In practical examples the abilities of the algebraic approach to add and remove single edges seem better suited.

Figure 1 shows an example of a production where the vertices are spatially separated by an "X".

We may imagine each production as a set of pins (the vertices) connected by rubber bands (the edges). When applying a production to a host graph, the pins that belong to the lefthand side or to the determined environment are stitched into the host graph. The pins belonging to the righthand side point into the void; corresponding vertices are generated later. In all that, the labels of pins have to match those of the corresponding vertices of the host graph and all rubber bands lie in parallel to equally labelled edges. The embedding part determines how the generated vertices (and edges) corresponding to the righthand side are to be connected to the remaining part of the host graph after the (image of the) lefthand side was removed. Every connected component, i. e. every set of pins connected by rubber bands within the indetermined environment, is considered separately. *Each* matching occurrence of a connected component in the host graph is visited. New edges are created in correspondence to rubber bands leading to the righthand side.

The application of the production in figure 1 to the host graph in figure 2 locates the bigger of the two connected components of the indetermined environment three times (see figure 3) because one f-labelled vertex is matched two times (it is connected with two vertices labelled e). The other connected component which consists of the f-labelled vertex only, has three matching occurrences also (where two of them overlap with those already found). After removing the lefthand side and adding the righthand side together with two edges labelled r the graph in figure 4 results.

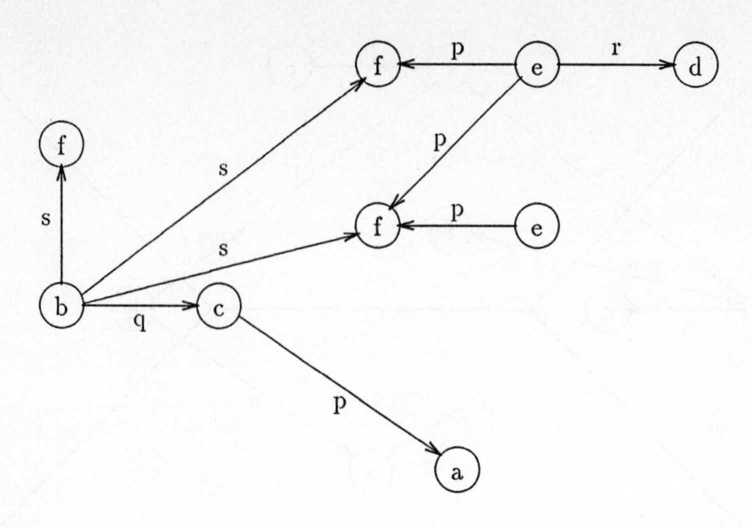

Figure 2: Example of a host graph

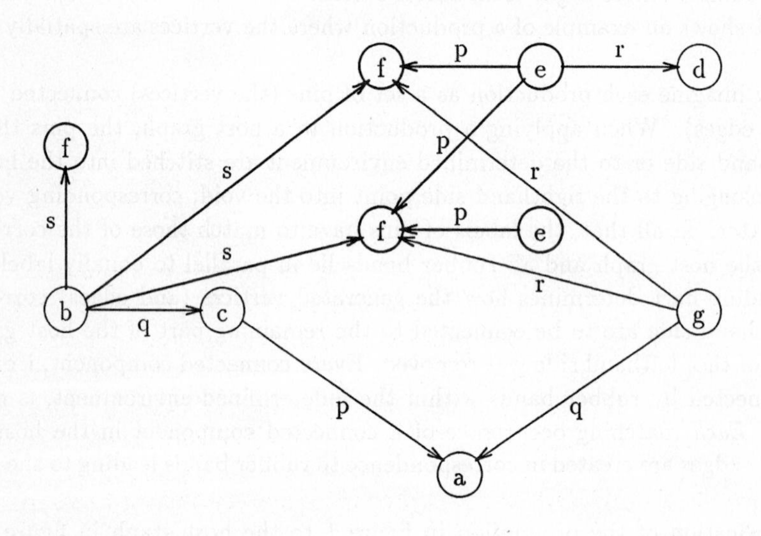

Figure 3: Correspondence between vertices and edges of the production in figure 1 and the graph in figure 2

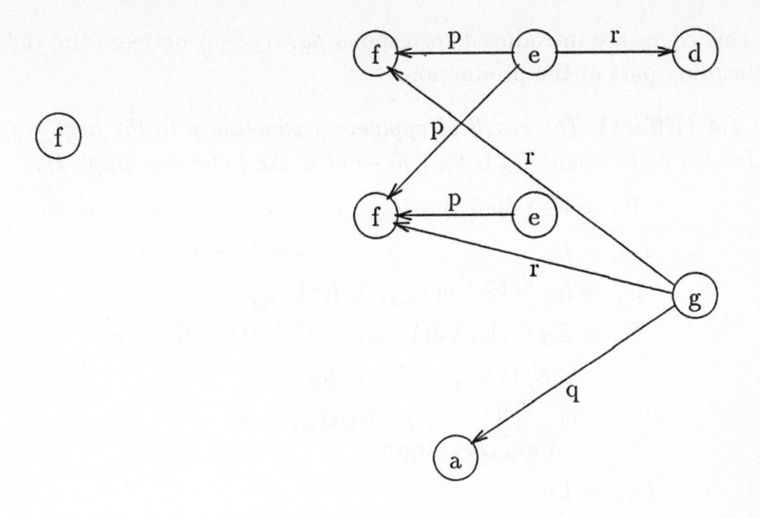

Figure 4: Derived graph

We now come to a formal definition. Some minor deviations from the definitions given in [10] are necessary to clear the way for the next section. An applicability condition is added and the indetermined environment may be embedded more freely into the host graph. We shall make some remarks accordingly.

Definition 2.2 (Spanned subgraph) *The subgraph* span(V', G) *of* G *spanned by a set of vertices* $V' \subseteq V_G$ *consists, besides* V', Lv_G *and* L_{E_G}, *of the edge set* $V' \times V' \times Lv_G \cap E_G$ *and the restricted labelling function* $l_V \restriction V'$.

The operator span corresponds to the operator con from [8].

The production p and the host graph G have to use the same sets of labels L_V and L_E.

The first step is to locate the lefthand side and the determined environment in the host graph. Let δ: span$(V_{L,p} \cup V_{B,p}, p) \rightarrow G$ be a monomorphism that satisfies additional restrictions discussed below. Then the image of the lefthand side under δ is removed and replaced by a copy of the righthand side. All extensions μ: span$(V_{L,p} \cup V_{B,p} \cup C, p) \rightarrow G$ of δ are considered, where C is any connected component of the indetermined environment.

Definition 2.3 (Connected component) *An* (undirected) path *from* v *to* v' *within the indetermined environment is a sequence of vertices* $v_0, \ldots, v_n \in V_A$ $(n \geq 0)$ *with* $v = v_0$, $v' = v_n$ *and*

$$(\forall i \in \{1, \ldots, n\})(\exists m \in L_E)((v_{i-1}, v_i, m) \in E \ \vee \ (v_i, v_{i-1}, m) \in E).$$

A connected component *of the indetermined environment of a production is the subgraph of* V_A *spanned by an equivalence class of vertices relative to the equivalence relation* $\sim$, *where* $v \sim v'$ *iff there exists an (undirected) path from* v *to* v' *within* V_A.

Finally, new edges are introduced corresponding to edges between the righthand side and the embedding part of the production.

Definition 2.4 (Effect) *The result of applying production p to the host graph G at the occurrence located by* $\delta\colon \mathrm{span}(V_{L,p} \cup V_{B,p}, p) \to G$ *is the following graph H:*

$$\begin{aligned} V_H &= V_G \setminus \delta(V_{L,p}) \uplus V_{R,p} \\ L_{V_H} &= L_V \\ l_{V_H} &= l_{V_G} \restriction (V_G \setminus \delta(V_{L,p})) \uplus l_{V_p} \restriction V_{R,p} \\ E_H &= E_G \cap (V_G \setminus \delta(V_{L,p})) \times (V_G \setminus \delta(V_{L,p})) \times L_E \\ &\quad \cup E_p \cap V_{R,p} \times V_{R,p} \times L_E \\ &\quad \cup \bigcup_{C \in V_{A,p}/\sim} \bigcup_{\mu \in M_\delta(C)} \mathrm{New}(\mu) \\ L_{E_H} &= L_E \end{aligned}$$

$\sim$ *is the equivalence relation from definition 2.3 and*

$$\begin{aligned} M_\delta(C) &= \{\mu\colon \mathrm{span}(V_{L,p} \cup V_{B,p} \cup C, p) \to G \mid \mu \restriction \mathrm{span}(V_{L,p} \cup V_{B,p}) = \delta\} \\ \mathrm{New}(\mu) &= \{(\mu(v), v', m) \mid (v, v', m) \in E_p \cap (V_{A,p} \cup V_{B,p}) \times V_{R,p} \times L_E\} \\ &\quad \cup \{(v', \mu(v), m) \mid (v', v, m) \in E_p \cap V_{R,p} \times (V_{A,p} \cup V_{B,p}) \times L_E\} \end{aligned}$$

We call $\bigcup_{C \in V_{A,p}/\sim} \bigcup_{\mu \in M_\delta(C)} \mathrm{New}(\mu)$ *the set of* new edges.

This definition is not yet symmetric. All generated edges appear in the production (of course!), because either they belong to the righthand side or are new. But edges can be deleted without naming them, simply by removing an adjacent vertex. The algebraic approach, however, requires full symmetry between the left and righthand sides. So, in order to move to the next section we have to add a kind of *gluing condition* concerning δ.

Definition 2.5 (Applicability) *A production p is* applicable *to a host graph G at the occurrence located by* $\delta\colon \mathrm{span}(V_{L,p} \cup V_{B,p}, p) \to G$ *iff the following two conditions hold (we use the notation of the preceding definition):*

$$(\forall v \in V_{L,p})(\forall (\delta(v), v', m) \in E_G)$$
$$(\delta(v), v', m) \in \delta(E_p \cap V_{L,p} \times V_{L,p} \times L_E) \cup \bigcup_{C \in V_{A,p}/\sim} \bigcup_{\mu \in M_\delta(C)} \mathrm{Old}(\mu)$$

and

$$(\forall v \in V_{L,p})(\forall (v', \delta(v), m) \in E_G)$$
$$(v', \delta(v), m) \in \delta(E_p \cap V_{L,p} \times V_{L,p} \times L_E) \cup \bigcup_{C \in V_{A,p}/\sim} \bigcup_{\mu \in M_\delta(C)} \mathrm{Old}(\mu)$$

where

$$\begin{aligned} \mathrm{Old}(\mu) &= \{(\delta(v), \mu(v'), m) \mid (v, v', m) \in E_p \cap V_{L,p} \times (V_{A,p} \cup V_{B,p}) \times L_E\} \\ &\quad \cup \{(\mu(v'), \delta(v), m) \mid (v', v, m) \in E_p \cap (V_{A,p} \cup V_{B,p}) \times V_{L,p} \times L_E\} \end{aligned}$$

We call $\bigcup_{C \in V_{A,p}/\sim} \bigcup_{\mu \in M_\delta(C)} \mathrm{Old}(\mu)$ *the set of* old edges.

Definition 2.5 requires that edges may only be removed explicitly, i. e. they have to match edges listed in the production. Therefore the restriction $\delta \upharpoonright V_{L,p}$ must be a subgraph isomorphism. In this respect we follow the older definitions given in [8] rather than [10].

Old and new edges can easily be read from the applied production (e. g. figure 1): They correspond to edges crossing the left or right side of the "X", respectively.

The applicability condition is no serious constraint if the labelling sets L_V and L_E are both finite. In this case we can add all possible neighbours to the indetermined environment:

- $V_{A,p}$ is extended by $L_V \times L_E \times \{\leftarrow, \rightarrow\}$ and we set $l_V(v, e, d) := v$ for all $(v, e, d) \in L_V \times L_E \times \{\leftarrow, \rightarrow\}$.
- E_p is extended by

$$\{((v, e, \leftarrow), v', e) \mid (v, e) \in L_V \times L_E \wedge v' \in V_{L,p}\}$$
$$\cup \{(v', (v, e, \rightarrow), e) \mid (v, e) \in L_V \times L_E \wedge v' \in V_{L,p}\}$$

The new constraint even increases the expressive power of Göttler's approach. Now it becomes possible to prevent the application of a production not only if some vertices (and edges) are missing, but also if some unwanted vertices (or edges) are present. Prohibitive vertices and edges are just not listed in the production.

The only serious digression from the semantics given in [10] therefore consists in the kind of morphisms μ permitted in definition 2.4. [10] requires that all μ are monomorphisms. But in the present framework only the condition that the restriction $\mu \upharpoonright C$ should be a monomorphism for every connected component C in the indetermined environment can be stated in a natural way. Experience shows, however, that in most practical examples the vertices in question are labelled differently and cannot be identified anyway. Furthermore, in some cases an overlapping of (the images of) determined and indetermined environments appears to be quite natural, although it does not agree with the philosophy that the determined environment can be seen as a common part of left and righthand sides.

3 Categorical Description of Operational Graph Grammars

First we construct an algebraic graph production having the same effect on a (given!) host graph as the operational one. This algebraic production is chosen from a family by having a maximal effect. Each member of the family is linked to a pattern by three pushouts. The building morphisms must preserve connected components of the indetermined environment (as part of the pattern). Furthermore, those edges crossing boundaries in the X-notation (cf. figure 1) must be handled in the right way.

3.1 Algebraic Productions Do the Work

The embedding transformation in Göttler's approach, which consists of the determined and indetermined environments, corresponds to the gluing graph in an algebraic production. Because the indetermined environment adjusts to the host graph (see again figure 3), one operational production will be equivalent to a denumerable family of algebraic productions. Each production of this family has a gluing graph which consists of the determined environment and an arbitrary collection of connected components of the indetermined environment (whereby some components may appear several times). It would be nice if exactly one member of this family would be applicable to a given host graph, but this could only be achieved by additional constraints that do not fit into the algebraic framework.

Definition 3.1 (Family of productions) *The family* Fam(p) *of algebraic productions which corresponds to a single operational production* p *consists of all elements* $\tilde{p} = ({}'B \xleftarrow{{}'b} K \xrightarrow{b'} B')$, *that satisfy the following conditions:*

- ${}'b$ *and* b' *are monomorphisms.*
- $\mathcal{C}$ *is a finite multiset of connected components of the indetermined environment of* p.
- ${}'B$ *is the colimit (the "generalized pushout") of the family of canonic monomorphisms*

$$\{\text{span}(V_{L,p} \cup V_{B,p}, p) \to \text{span}(V_{L,p} \cup V_{B,p} \cup C, p) \mid C \in \mathcal{C}\}.$$

- B', *analogously, is the generalized pushout of the family of canonic monomorphisms*

$$\{\text{span}(V_{R,p} \cup V_{B,p}, p) \to \text{span}(V_{R,p} \cup V_{B,p} \cup C, p) \mid C \in \mathcal{C}\}.$$

- *Finally,* K *is the generalized pushout of the family of canonic monomorphisms*

$$\{\text{span}(V_{B,p}, p) \to \text{span}(V_{B,p} \cup C, p) \mid C \in \mathcal{C}\}$$

- *The canonic monomorphisms defined above*

$$\begin{array}{ccc} \text{span}(V_{B,p}, p) & \longrightarrow & \text{span}(V_{L,p} \cup V_{B,p}, p) \\ \downarrow & & \downarrow \\ \text{span}(V_{B,p} \cup C, p) & \longrightarrow & \text{span}(V_{L,p} \cup V_{B,p} \cup C, p) \end{array}$$

and

$$\begin{array}{ccc} \text{span}(V_{B,p}, p) & \longrightarrow & \text{span}(V_{R,p} \cup V_{B,p}, p) \\ \downarrow & & \downarrow \\ \text{span}(V_{B,p} \cup C, p) & \longrightarrow & \text{span}(V_{R,p} \cup V_{B,p} \cup C, p) \end{array}$$

respectively, obviously form pushouts for every $C \in \mathcal{C}$, *because all unions are disjoint. Therefore we require that* ${}'b$ *and* b' *are induced by the canonic monomorphisms* $\text{span}(V_{B,p}, p) \to \text{span}(V_{L,p} \cup V_{B,p}, p)$ *and* $\text{span}(V_{B,p}, p) \to \text{span}(V_{R,p} \cup V_{B,p}, p)$, *respectively.*

Proposition 3.1 *The algebraic production* $\tilde{p} = ('B \xleftarrow{'b} K \xrightarrow{b'} B') \in \mathrm{Fam}(p)$ *has the same effect as the operational production* p *at the occurrence located by* δ *when applied to the host graph* G

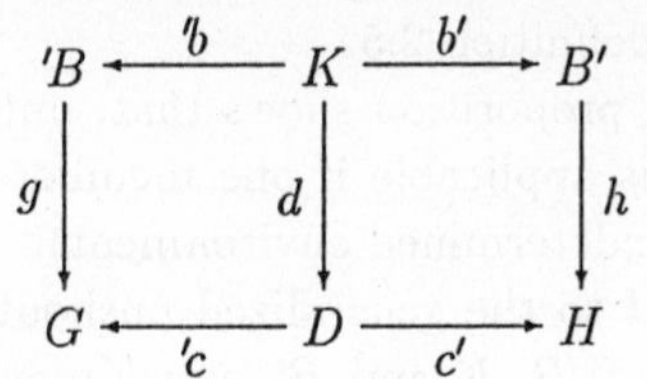

if

- *$'c$ and c' are monomorphisms.*
- *$\mathcal{C}$ is the family $\{C_\mu \mid C \in V_{A,p}/\!\sim\ \wedge\ \mu \in M_\delta(C)\}$. We adopted the notation of definition 2.4. The generalized pushouts extend over all connected components of the indetermined environment of p and all possible embeddings into G, where the lefthand side and the determined environment are fixed.*
- *$g\colon 'B \to G$ is the morphism induced by the family*

$$\{\mu\colon \mathrm{span}(V_{L,p} \cup V_{B,p} \cup C_\mu, p) \to G \mid C \in \mathcal{C}\}.$$

Proof: By the pushout property of the left square and the injectivity of $'c$, D is isomorphic to

$$\mathrm{span}(V_G \setminus (g(V_{'B}) \setminus {'b}(V_K)), G) \cong \mathrm{span}(V_G \setminus \delta(V_{L,p}), G).$$

$d := {'c}^{-1} \circ g \circ {'b}$ is defined on V_K (because $(g \circ {'b})(V_K) \subseteq V_D$) and determines a morphism from K to D. The pushout H results in

$$\begin{aligned} V_H &= c'(V_D \setminus d(V_K)) \uplus h(V_{B'}) \\ &\cong (c' \circ {'c}^{-1})(V_G \setminus \delta(V_{L,p}) \setminus (g \circ {'b})(V_K)) \uplus h(V_{R,p} \uplus b'(V_K)) \\ &\cong (V_G \setminus \delta(V_{L,p})) \uplus V_{R,p}, \end{aligned}$$

because $(c' \circ {'c}^{-1} \circ g \circ {'b})(V_K) = (c' \circ d)(V_K) = (h \circ b')(V_K)$ and $h \restriction V_{R,p}$ is injective, for $V_{R,p} \cap b'(V_K) = \emptyset$.

$$\begin{aligned} E_H &= \{(c'(v), c'(v'), m) \mid (v, v', m) \in E_D\} \\ &\quad \cup \{(h(v), h(v'), m) \mid (v, v', m) \in E_{B'}\} \\ &\cong \{(c'(v), c'(v'), m) \mid ({'c}(v), {'c}(v'), m) \in E_G \wedge {'c}(v), {'c}(v') \in V_G \setminus \delta(V_{L,p})\} \\ &\quad \cup \{((h \circ b')(v), (h \circ b')(v'), m) \mid (v, v', m) \in E_K\} \\ &\quad \cup \{(h(v), h(v'), m) \mid (v, v', m) \in E_{R,p}\} \\ &\quad \cup \bigcup_{C \in V_{A,p}/\sim} \ \bigcup_{\mu \in M_\delta(C)} \mathrm{New}(\mu) \end{aligned}$$

The other components are as expected. $\diamond$

Here we made essential use of the properties of the category LGraph. Two parallel, equally labelled edges already coincide if their source and target vertices coincide!

An analogous argument for the old edges shows that the gluing condition is exactly the applicability condition of definition 2.5.

The proof of the preceding proposition shows that, unfortunately, an infinite number of productions in the family is applicable if one member is (that contains at least one connected component in the indetermined environment). To see this, we just add a few more copies of this component to the generalized pushouts. Each additional connected component C is already part of $'B$, K and B', so we may choose one of the morphisms $\mu \in M_\delta(C)$ that makes the images coincide.

Lemma 3.2 *The applicable productions in a family constitute a directed system.*

Proof: The *amalgamation* $\bar{p} = ('\bar{B} \leftarrow \bar{K} \rightarrow \bar{B}')$ of two applicable productions $p_1 = ('B_1 \leftarrow K_1 \rightarrow B_1')$ and $p_2 = ('B_2 \leftarrow K_2 \rightarrow B_2')$ with respect to a third applicable production $r = ('R \leftarrow R \rightarrow R')$ is also applicable. That is, we have the following situation

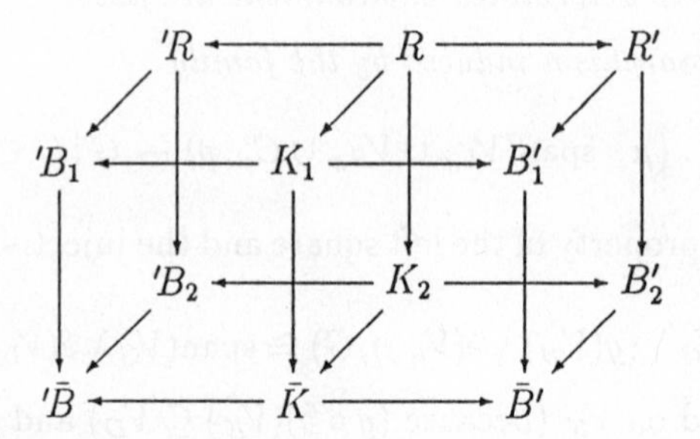

where all the squares $'R'B_1'B_2'\bar{B}$, $RK_1K_2\bar{K}$, $R'B_1'B_2'\bar{B}'$, $RK_1'R'B_1$, $RK_1R'B_1'$, $RK_2'R'B_2$ and $RK_2R'B_2'$ are pushouts and the double pushout originating from the application of r factors through p_1 as well as p_2:

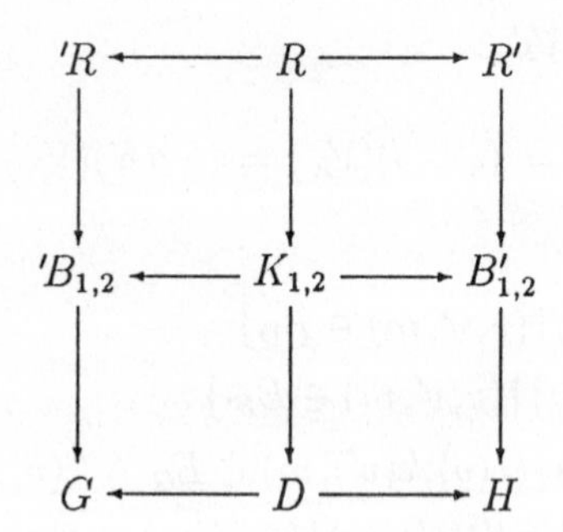

This, in turn, follows from a simple calculation with connected components like in proposition 3.1. ◇

Lemma 3.3 *No algebraic production in a family belonging to one operational production generates a bigger graph H out of G than the production in proposition 3.1.*

Proof: The righthand side of the production in proposition 3.1 creates the maximal image in H, whereas the context graph D is fixed. $\diamond$

3.2 Morphisms and Connected Components

A connected component of a graph was given by an equivalence relation on the set of vertices in definition 2.3. The set of equivalence classes is exactly the coequalizer to the maps $s, t\colon E \to V$ in the category of sets (see [17, section 6.3.2]). A graph can easily be obtained from these equivalence classes by categorical means.

Now we formalize the requirement that the image of a morphism contains no incomplete connected components.

Definition 3.2 *A morphism g* reflects connected components *iff both of the following diagrams are pullbacks*

$$\begin{array}{ccc} E & \xrightarrow{g_E} & E' \\ {\scriptstyle s}\downarrow & & \downarrow{\scriptstyle s'} \\ V & \xrightarrow[g_V]{} & V' \end{array} \qquad \begin{array}{ccc} E & \xrightarrow{g_E} & E' \\ {\scriptstyle t}\downarrow & & \downarrow{\scriptstyle t'} \\ V & \xrightarrow[g_V]{} & V' \end{array}$$

and for every connected component $f\colon C \to G$ of G the morphism $g \circ f\colon C \to G'$ is monic.

Lemma 3.4 *If the morphism $g\colon G \to G'$ reflects connected components then a multiset $\mathcal{C}$ exists, containing only connected components of G', that has G' as coproduct: $G' = \biguplus \mathcal{C}$.*

Proof: Consider an arbitrary vertex $v \in V_G$. Because g reflects connected components, for each edge $e' \in E_{G'}$ with $s_{G'}(e') = g_V(v)$ there exists a unique edge $e \in E_G$ with $s_G(e) = v$ and $g_E(e) = e'$. The target vertex $t_G(e)$ is mapped to the target vertex of e': $g_V(t_G(e)) = t_{G'}(e')$. Analogous arguments apply if the roles of source and target vertices are exchanged.

$g_V(v)$ lies in a connected component C of G'. The considerations above show that for every path in C there exists a corresponding path in G which is mapped to it. Of course, all paths in G are mapped to paths in G'. It remains to be shown that different paths in G have different images in G'. But g was supposed to be monic on each connected component. Furthermore, G is built from copies of connected components of G'. $\diamond$

3.3 Generating the Family of Algebraic Productions

The productions that are to be applied to host graphs are linked to "patterns" by reverse morphisms. Göttler's productions consist of four parts, so we have to add one part to the algebraic ones to cope with the indetermined environment.

Definition 3.3 (Production pattern) *Consider an operational production p where the vertex set is partitioned into four pieces V_L, V_R, V_B and V_A. Assume that* $\mathrm{span}(V_A, p)$ *is the* *coproduct* $\biguplus_{1\leq i\leq n} C_i$ *of connected components* C_i $(1 \leq i \leq n)$. *The finite set of* patterns *for* p *are algebraic productions* p_C *with an additional graph and an additional morphism:*

$$p_C = ('B \xleftarrow{'b} K \xrightarrow{b'} B'), \quad F \xrightarrow{k} K$$

where $'b$, b' *and* k *are subgraph isomorphisms. The graphs* F, K, $'B$ *and* B' *are constructed as follows: Fix a subgraph* $C = \biguplus_{i\in I} C_i$ *of* $\mathrm{span}(V_A)$, *where* $I \subseteq \{1,\ldots,n\}$. *Then set*

- $F = \mathrm{span}(V_C, p)$,
- $K = \mathrm{span}(V_C \cup V_B, p)$,
- $'B = \mathrm{span}(V_C \cup V_B \cup V_L, p)$ *and*
- $B' = \mathrm{span}(V_C \cup V_B \cup V_R, p)$.

The morphisms $'b$, b' *and* k *are the canonic ones.*

Definition 3.4 *Given graph morphisms* $G \xrightarrow{\alpha} G' \xrightarrow{\beta} G''$, β reflects connecting edges *with respect to* α *iff the outer rectangle of the following diagram—and the similar diagram with* t *instead of* s*—is a pullback*

$$\begin{array}{ccccc} P_s & \xrightarrow{e_s} & E' & \xrightarrow{\beta_E} & E'' \\ \downarrow{\scriptstyle \bar{s}} & & \downarrow{\scriptstyle s'} & & \downarrow{\scriptstyle s''} \\ V & \xrightarrow[\alpha_V]{} & V' & \xrightarrow[\beta_V]{} & V'' \end{array}$$

where, by definition, the lefthand square is a pullback.

This constraint means that all the edges in E'' with source or target in $(\beta_V \circ \alpha_V)(V)$ come from edges in E' (via β_E).

Theorem 3.5 *The family* $\mathrm{Fam}(p)$ *consists of elements of the form (where in definition 3.1 we "forgot" about the components* $\tilde{F}$ *and* $\tilde{k}$*)*

$$\tilde{p} = ('\tilde{B} \xleftarrow{'\tilde{b}} \tilde{K} \xrightarrow{\tilde{b}'} \tilde{B}'), \quad \tilde{F} \xrightarrow{\tilde{k}} \tilde{K}$$

where $'\tilde{b}$, $\tilde{b}'$ *and* $\tilde{k}$ *are subgraph isomorphisms.*

Each of these elements is linked to one pattern p_C by morphisms $\tilde{g}$, $\tilde{d}$, $\tilde{h}$ and f such that f reflects connected components, $\tilde{g}$, $\tilde{d}$ and $\tilde{h}$ reflect connecting edges with respect to $\tilde{'b} \circ \tilde{k}$, $\tilde{k}$ and $\tilde{b}' \circ \tilde{k}$, respectively, and the following diagram consists of three pushouts.

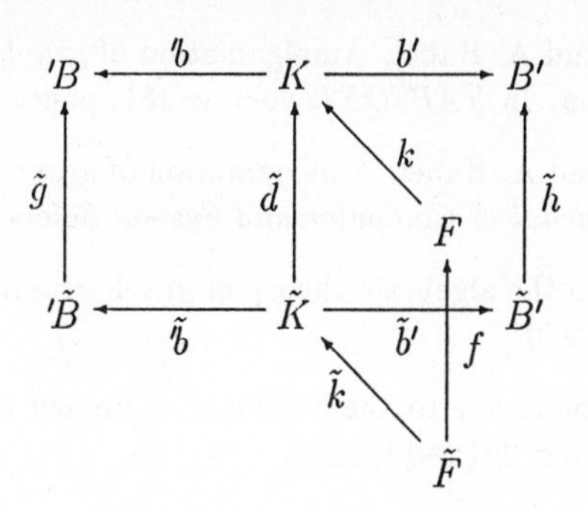

Proof: See [1]. ◇

4 Conclusion

The algebraic approach is fully contained in the formalism proposed in this paper if we allow monomorphisms instead of subgraph isomorphisms between the gluing graph and lefthand/righthand sides. Göttler's approach is contained also, within certain limits. Some important special cases, for example the NLC and NCE graph grammars, are handled in the right way.

Many interesting ideas could be carried over between these approaches. Among them are:

- The representation of productions as graphs suggests to define two-level graph grammars [8]. Productions could be manipulated by other productions before they are applied to a concrete graph.

- Having productions given in one piece instead of three as in the algebraic approach could be useful in cases where graphs are allowed to contain variables. A single substitution would be sufficient for lefthand side, gluing graph and righthand side (see, for example, [11] and this volume).

- The algebraic approach works for almost any kind of graphs. Some of the well-known theoretical results (for example the amalgamation theorem [2, 3] or the concurrency theorem [7]) might be slightly generalized, although definitely not to the full formalism.

Acknowledgements

The author is pleased to thank the referees for their valuable suggestions to improve this paper.

References

[1] K. Barthelmann. Graphgrammatikalische Hilfsmittel zur Beschreibung verteilter Systeme. Dissertation, IMMD2, Universität Erlangen–Nürnberg, 1990 (to appear).

[2] P. Böhm, H.–R. Fonio, and A. Habel. Amalgamation of graph transformations with applications to synchronization. In *TAPSOFT*, volume 185, pages 267–283, 1985.

[3] P. Böhm, H.–R. Fonio, and A. Habel. Amalgamation of graph transformations: A synchronization mechanism. *Journal of Computer and System Sciences*, 34:377–408, 1987.

[4] H. Ehrig. Introduction to the algebraic theory of graph grammars. *Lecture Notes in Computer Science*, 73:1–69, 1979.

[5] H. Ehrig. Tutorial introduction to the algebraic approach of graph grammars. *Lecture Notes in Computer Science*, 291:3–14, 1987.

[6] H. Ehrig, M. Pfender, and H.J. Schneider. Graph-grammars: An algebraic approach. In *Proc. of the IEEE Conf. on Automata and Switching Theory, Iowa City*, pages 167–180, 1973.

[7] H. Ehrig and B. K. Rosen. Parallelism and concurrency of graph manipulations. *Theoretical Computer Science*, 11:247–275, 1980.

[8] H. Göttler. Zweistufige Graphmanipulationssysteme für die Semantik von Programmiersprachen. Arbeitsbericht (Dissertation) 10, 12, IMMD2, Universität Erlangen–Nürnberg, 1977.

[9] H. Göttler. Semantical description by two-level graph-grammars for quasihierarchical graphs. In *Proc. WG78 'Graphs, Data Structures, Algorithms'*. Hanser–Verlag, 1979.

[10] H. Göttler. *Graphgrammatiken in der Softwaretechnik (Theorie und Anwendungen)*. Springer–Verlag, 1988.

[11] L. Hess and B. H. Mayoh. Graph grammars for knowledge representation. DAIMI PB – 304, Computer Science Department, Aarhus University, 1990.

[12] D. Janssens and G. Rozenberg. On the structure of node-label controlled graph languages *and* restrictions, extensions and variations of NLC grammars. *Information Sciences*, 20:191–244, 1980.

[13] D. Janssens and G. Rozenberg. Graph grammars with neighbourhood-controlled embedding. *Theoretical Computer Science*, 21:55–74, 1982.

[14] M. Nagl. Formale Sprachen von markierten Graphen. Arbeitsbericht (Dissertation) 7, 4, IMMD2, Universität Erlangen–Nürnberg, 1974.

[15] M. Nagl. *Graph-Grammatiken (Theorie, Implementierung, Anwendungen)*. Vieweg–Verlag, 1979. Habilitation.

[16] G. Rozenberg. An introduction to the NLC way of rewriting graphs. *Lecture Notes in Computer Science*, 291:55–66, 1987.

[17] D. E. Rydeheard and R. M. Burstall. *Computational Category Theory*. Prentice–Hall, 1988.

General Solution to a System of Recursive Equations on Hypergraphs

Michel Bauderon[1]

Laboratoire Bordelais de Recherche en Informatique

Unité Associée au C.N.R.S. n° 1304

Département Informatique I.U.T. 'A', Université Bordeaux I

F-33405 Talence Cedex

Abstract : A categorical framework has been described in [Ba89] to extend to systems of recursive equations on hypergraphs the classical results available for trees, such as the existence of an initial solution generalizing that of a least solution. As in the case of trees, the solution is not in general unique, but the situation is much more involved for hypergraphs. The aim of this paper is to present a classification of all the solutions of a system of recursive equations on hypergraphs.

Keywords : hypergraphs, hyperedge rewriting, systems of equations, initial solution, general solution

CONTENTS

---

[1] This work has been partially supported by the C.N.R.S. PRC "Mathématiques et Informatique" and by the ESPRIT BRA 3299 "Computing with graph transformations".

0. INTRODUCTION.

In an earlier paper ([Ba89]), we have described on its own a general categorical framework which is suitable to solve systems of recursive equations on a certain class of hypergraphs (namely, directed edge labelled hypergraphs with a set of distinguished elements that we call sources, cf [BC87] and section 1 for a precise definition) and to define a particular class of possibly infinite (but countable) hypergraphs as initial solutions of such systems. Those hypergraphs are called *equational* or *regular*. In this paper, we provide a classification of all the solutions to such a system. Although all the results stated in this paper hold for *hypergraphs*, we shall for the sake of graphical simplicity present examples of most of the issues using mere *graphs*.

Let us consider the following equation on directed edge-labelled graphs, where u is the unknown (a precise definition will be given in section 1) :

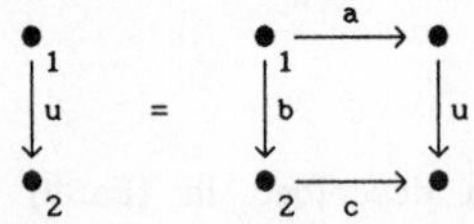

$(\Sigma 1)$

An intuitive resolution of this equation by the method of iterated substitution classically used on trees (cf [Co83]) would yield successively :

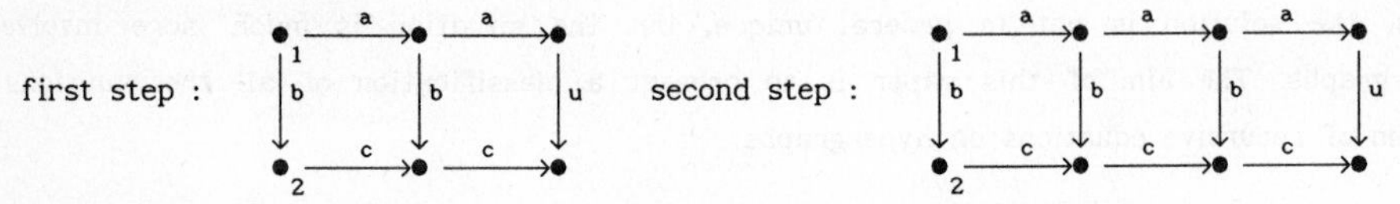

and lead finally to the following initial solution G_1 for the equation $(\Sigma 1)$:

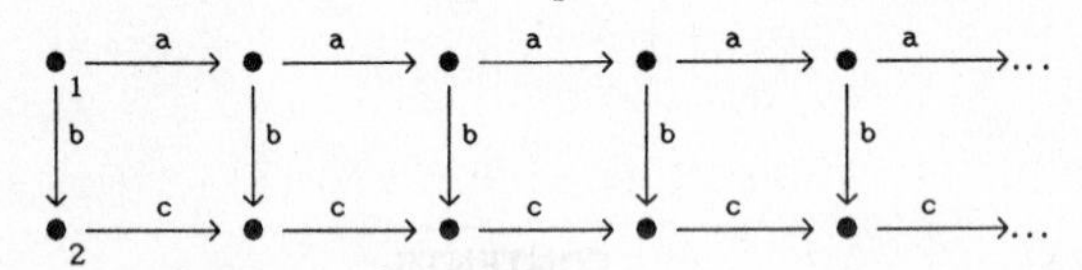

But it is pretty clear that this equation has an *infinite* number of solution. if we let indeed H be any hypergraph without sources, $G_1 \oplus H$ (the disjoint union of G_1 and H) is a solution of $(\Sigma 1)$ as well. Let us now consider the equation :

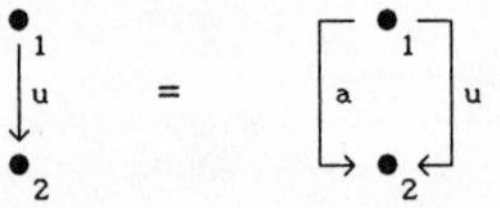

$(\Sigma 2)$

The same resolution process clearly leads to the following solution G_2 which does not have a locally finite degree :

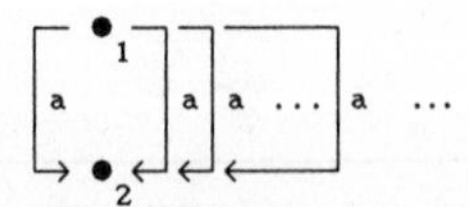

Of course, we can generate other solutions in the very same way we used for $(\Sigma 1)$. However, the situation is more complex here since it is easily checked that the following hypergraph, with an infinity of a-labelled loops is still a solution of equation ($\Sigma 2$) :

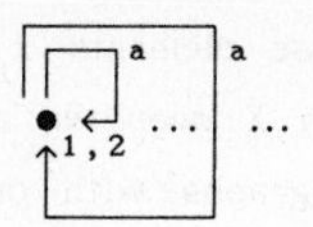

As a last example let us merely present the following equation whose resolution is not quite obvious :

1 u 2 = 1 u u 2 ($\Sigma 3$)

In this paper, we shall first (sections 1 to 3) recall from [Ba89] the basic notions and tools which enables us to solve systems of recursive equations on hypergraphs : n-hypergraphs, a notion of approximation and a notion of substitution. In the following sections, we show how to solve a system and how the most general solution to such a system can be described. The key result is the fact that any solution is generated through an iteration process. For the sake of briefness, all results will be stated without proofs, but relevant examples will be given. Complete proofs will be found in two forthcomming papers, [Ba90a,b].

We shall assume some very basic notions of category theory (objects, morphisms, category, product category and functor) to be known. More details can be found in any standard textbook ([Mc71]) or in [Ba90a]. Any other definition will be provided when necessary.

1. Hypergraphs.

1.1. A *sourced hypergraph of type n* (or an *n-hypergraph) over A*, is a sextuplet $G = \langle V_G, E_G, vert_G, lab_G, src_G, n\rangle$ where :

- V_G and E_G are countable sets (respectively of vertices and hyperedges),
- $vert_G : E_G \longrightarrow V_G^*$ assigns to each e a word on V_G representing the ordered sequence of its vertices, $vert_G(k,e)$ denoting the kth element of the sequence,
- $lab_G : E_G \longrightarrow A$ labels every hyperedge e with some letter in A such that $\tau(lab_G(e)) = |vert_G(e)|$,
- $src_G : [n] \longrightarrow V_G$ defines the sequence $src_G([n])$ of *sources* of the hypergraph G.

The set of hypergraphs of type n over A shall be denoted by $\underline{\mathfrak{CG}}^{\infty}_n(A)$. Such hypergraphs have been defined in [BC87] while a very similar notion was introduced independently in [HK87].

1.2. A sourced hypergraph is *bounded* when V_G is finite and *has locally finite degree* when every vertex has finite degree, where the degree of a vertex v is the number of times v appears as a vertex of some hyperedge of the hypergraph.

Let X be a ranked alphabet whose elements $x_1,\ldots,x_n,\ldots$ will be called *variables.* The set of *n-hypergraphs with variables in X over the alphabet A* is the set $\underline{\mathfrak{CG}}^{\infty}_{n}(A \cup X)$. Most definitions and results hold for hypergraphs with or without variables. The variables will explicitly appear only when necessary.

Numerous examples of finite sourced hypergraphs may be found in [BC87]. Let us simply recall that with each integer n, we associate the discrete n-hypergraph $\underline{n}$ having n distinct isolated vertices, no hyperedge and a source mapping sending $i \in [n]$ to the ith vertex of $\underline{n}$. This hypergraph will be drawn as :

$$\bullet_1 \quad \bullet_2 \quad \bullet_3 \quad \ldots \quad \bullet_{n-1} \quad \bullet_n$$

Similarly, with a letter $a \in A$ with rank n, we associate the n-hypergraph $\underline{a}$ with n distinct vertices, one hyperedge labelled by a, and source mapping sending $i \in [n]$ to the ith vertex of $\underline{a}$:

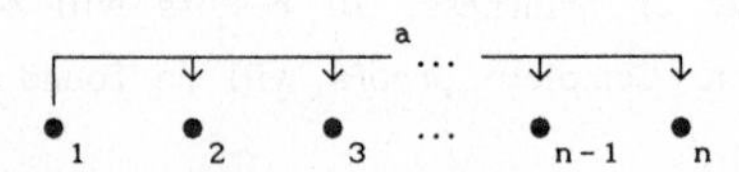

The following 2-hypergraph G_1 is infinite, but has a locally finite degree :

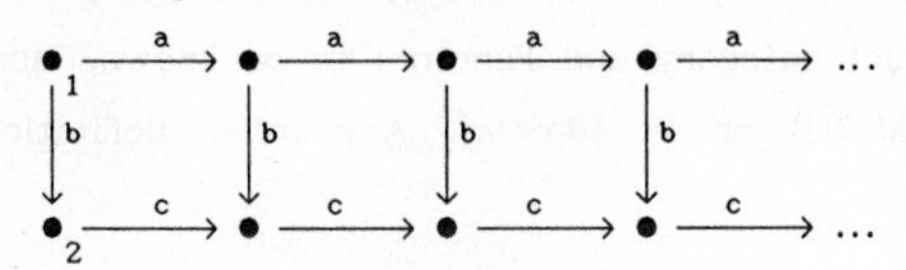

According to the definition, this 2-hypergraph may be formally described by :

- $A = \{a,b,c\}$, $V_G = E_G = \mathbb{N}$, $n = 2$, $vert_G(1) = 1.2$,
- $vert_G(i) = vert_G(1,i+1).vert_G(1,i+2)$,
- $vert_G(i+1) = vert_G(1,i).vert_G(1,i+3)$,
- $vert_G(i+2) = vert_G(2,i).vert_G(2,i+3)$,
- $lab_G(i) = b$, $lab_G(i+1) = a$, $lab_G(i+2) = c$,
- $src_G(1) = 1$, $src_G(2) = 2$.

Finally, the following 2-hypergraph G_2 is bounded but not finite, hence does not have a locally finite degree (a formal definition is left to the reader) :

2. Approximations.

2.1. Let G and G' be two concrete n-hypergraphs. A *morphism of sourced hypergraphs*

$g : G \longrightarrow G'$ is a pair $g = (Vg : V_G \longrightarrow V_{G'}, Eg : E_G \longrightarrow E_{G'})$ of arrows in **Set** such that : $lab_G = lab_{G'} \circ Eg$, $Vg^* \circ vert_G = vert_{G'} \circ Eg$ and $Vg \circ src_G = src_{G'}$ (where Vg^* denotes the canonical extension of Vg to the monoid of words on V_G).

It is easily checked that for each integer n these morphisms turn the set $\underline{\mathfrak{CG}}^{\infty}_n(A)$ of n-hypergraphs into a category. We shall let $\underline{\mathfrak{CG}}_n(A)$ denote the full subcategory of finite n-hypergraphs and the category $\underline{\mathfrak{CG}}^{\infty}(A)$ be the union of the categories $\underline{\mathfrak{CG}}^{\infty}_n(A)$ for $n \in \mathbb{N}$.

2.2. An *ω-diagram* in $\underline{\mathfrak{CG}}^{\infty}_n(A)$ is a countable sequence $(G_i, g_i : G_i \longrightarrow G_{i+1})_{i \in \mathbb{N}}$ of hypergraphs and morphisms which we shall draw as :

$$G_0 \xrightarrow{g_0} G_1 \xrightarrow{g_1} G_2 \xrightarrow{g_2} G_3 \xrightarrow{g_3} \dots \longrightarrow G_n \xrightarrow{g_n} \dots$$

An *ω-limit* for such an ω-diagram is a hypergraph $\underline{G}$ together with a countable family $(\underline{g}_i : G_i \longrightarrow \underline{G})_{i \in \mathbb{N}}$ of hypergraph morphisms such that for each $i \in \mathbb{N}$, $\underline{g}_{i+1} \circ g_i = \underline{g}_i$ and that, for any hypergraph H and any family $(h_i : G_i \longrightarrow H)_{i \in \mathbb{N}}$ of hypergraph morphisms such that for each $i \in \mathbb{N}$, $h_{i+1} \circ g_i = h_i$, there exists a (unique up to an isomorphism) hypergraph morphism h such that for each $i \in \mathbb{N}$, $h_i = h \circ \underline{g}_i$. The hypergraph H and the family $(h_i : G_i \longrightarrow H)_{i \in \mathbb{N}}$ are called an *ω-cone* over the diagram $(G_i, g_i : G_i \longrightarrow G_{i+1})_{i \in \mathbb{N}}$.

These notions clearly generalize those of an ω-chain and of a least upper bound in a partial order. With these definitions we can formulate the following result which sums up the main properties of the category of n-hypergraphs.

Theorem. The category $\underline{\mathfrak{CG}}^{\infty}_n(A)$ is *algebroidal*, which means that it has the following properties :

(i) it is ω-complete (each ω-diagram has an ω-limit) and every hypergraph G is the ω-limit of an ω-diagram of finite hypergraphs,

(ii) the discrete n-hypergraph $\underline{n}$ is an initial object of the category (i.e. for any n-hypergraph G there is a unique morphism $\bot_G : \underline{n} \longrightarrow G$).

This notion has been introduced by Plotkin and Smyth ([SP82]) in order to generalize to categories the notion of algebraic complete partial orders ("they are categories with a countable basis of finite objects"). Indeed, when dealing with algebroidal categories, one can actually rely on the intuition drawn from the practise of algebraic cpo's.

3. Substitution or hyperedge replacement.

3.1. Let G be an n-hypergraph with one occurrence of a variable x of type m i.e., one hyperedge e_i of which is labelled by x. Let then Γ be the (n+m)-hypergraph defined by (with ↓ denoting restriction) :

$V_\Gamma = V_G$, $E_\Gamma = E_G - \{ e_i \}$, $lab_\Gamma = lab_G \downarrow E_\Gamma$, $vert_\Gamma = vert_G \downarrow E_\Gamma$

and $src_\Gamma : [n+m] \longrightarrow V_\Gamma$

$k \longrightarrow src_G(k)$ for $1 \le k \le n$

$$k \longrightarrow vert_\Gamma(k-n,e_i) \text{ for } n+1 \le k \le n+m$$

The hypergraph Γ is said to be the *context* of the occurrence of x in G and we write : $G = \Gamma[x]$. The vertices of the hyperedge e_i labelled by x will be called the *occurrence vertices* of x. They are the vertices common to the context and to the occurrence of the variables.

3.2. Let $G = \Gamma[x]$ be an n-hypergraph with one occurrence of a variable x of type m and let H be an m-hypergraph. Then, *the result of the substitution* of H for x in G is the n-hypergraph $G' = \Gamma[H/x]$ defined in the following way :

$V_{G'} = (V_\Gamma \cup V_H)/_\equiv$ where $\equiv$ is the equivalence relation generated by the pairs $(src_\Gamma(k+n), src_H(k))$ for all k, $1 \le k \le m$,

$E_{G'} = E_\Gamma \cup E_{H'}$

$lab_{G'} = lab_\Gamma \cup lab_{H'}$

$vert_{G'} = vert_\Gamma \cup vert_{H'}$

$src_{G'} = src_G$

Intuitively, the hyperedge of G corresponding to the variable x has been removed and replaced by the hypergraph H, the sources of H being glued to the corresponding sources of the context of x (hence the expression *hyperedge replacement)*. The simultaneous substitution of H for all occurrences of the variable x in G is defined in the very same way, only the definiion is slightly more complicated. Similarly, the definition may be extended to simultaneous substitution for a finite number of variables. Note that this substitution may be (and has been) described in a purely categorical framework as a pushout (cf. [Eh79] and [BC87] where the algebraic and categorical approach are shown to coincide).

3.3. Let **B**, **C** and **D** be categories. A functor $\mathcal{F}$ from **C** to **D** is *ω-continuous* if it preserves the ω-limits and the related ω-cones. A bifunctor $\mathcal{F}$ from **B** × **C** into **D** is ω-continuous if its two components are. Clearly, ω-continuity is preserved by composition of functors (or bifunctors).

Let $\mathcal{F}$ be an endofunctor of a category **C**. A *fixpoint* of $\mathcal{F}$ is a pair (C,u) where C is an object of **C** and $u : \mathcal{F}C \longrightarrow C$ is an isomorphism. As usual in category theory, "equal" has to be replaced by "isomorphic" in the definition of fixpoints. The following theorem is essential to the resolution of systems of recursive equations :

Theorem : If $\mathcal{F}$ is an ω-continuous endofunctor of an algebroidal category **C**, the fixpoints of $\mathcal{F}$ form a category, with an initial object which is the ω-limit of the ω-diagram :

$$\bot \longrightarrow \mathcal{F}\bot \longrightarrow \mathcal{F}^2\bot \longrightarrow \mathcal{F}^3\bot \longrightarrow \dots \longrightarrow \mathcal{F}^n\bot \longrightarrow \dots$$

If there exists an object C and a morphism $C \longrightarrow \mathcal{F}C$, then the diagram :

$$C \longrightarrow \mathcal{F}C \longrightarrow \mathcal{F}^2C \longrightarrow \mathcal{F}^3C \longrightarrow \dots \longrightarrow \mathcal{F}^nC \longrightarrow \dots$$

has an ω-limit which is a fixpoint of the functor.

3.4. *Proposition :* Any finite or infinite n-hypergraph $G = \Gamma[x]$ with one occurrence of a variable x of type m gives rise to an ω-continous functor $\mathcal{F}_\Gamma$ from $\underline{\mathfrak{CG}}^\infty_m(A \cup X)$ to $\underline{\mathfrak{CG}}^\infty_n(A \cup X)$ whose hypergraph components maps $\Gamma[x]$ into $\Gamma[H/x]$. In other words, substitution is an ω-continuous functor whose morphism component is defined for any morphism $g : H \longrightarrow H'$ to be the identity on Γ and g on H. Moreover, if x and G have the same type n, $\mathcal{F}_\Gamma$ is an endofunctor of $\underline{\mathfrak{CG}}^\infty_n(A \cup X)$, and the previous theorem can be applied to $\mathcal{F}_\Gamma$, yielding a category of fixpoints with an initial object.

This proposition will be the key to the resolution of a system of recursive equations on hypergraph : we only need to interpret such a system as a fixpoint problem for an ω-continuous functor, and this will come straight out of the definition of such a system.

4. Systems of recursive equations on hypergraphs.

4.1. A *system of recursive equations on hypergraphs* is a finite system of the form $\Sigma = \langle x_1 = \Sigma_1, \ldots, x_n = \Sigma_n \rangle$ where $U = \{x_1, \ldots, x_n\}$ is the set of unknown, and for each integer i, $1 \le i \le n$, $\Sigma_i \in \underline{\mathfrak{CG}}_{\tau(x_i)}(A \cup U)$ is a finite hypergraph of appropriate type (i.e., both x_i and Σ_i have the same type $\tau(x_i)$). When needed, we shall denote by Σ the n-uple of sourced hypergraphs $\langle \Sigma_1, \ldots, \Sigma_n \rangle$ which are the contexts of the unknown and by $\tau(\Sigma)$ the sequence of the types, $\langle \tau(x_1), \ldots, \tau(x_n) \rangle$.

A *solution* of the system Σ is an n-uple of sourced hypergraphs $(\overline{G}_1, \ldots, \overline{G}_n)$ of appropriate types such that, for each i, $1 \le i \le n$, one has $\overline{G}_i = \Sigma_i[\overline{G}_1/x_1, \ldots, \overline{G}_n/x_n]$.

Once again, since we are using a categorical framework, we shall have to talk up to hypergraphs isomorphims, and in the sequel, = will mean "isomorphic in $\underline{\mathfrak{CG}}^\infty_k(A)$".

It follows from the previous section, that with a system of recursive equations, substitution associates an ω-continuous endofunctor $\mathcal{F}_\Sigma$ of the category $\underline{\mathfrak{CG}}^\infty_{\tau(x_1)}(A \cup U) \times \ldots \times \underline{\mathfrak{CG}}^\infty_{\tau(x_n)}(A \cup U)$. In this setting, it is clear that a solution of the system Σ is merely a *fixpoint* of the functor $\mathcal{F}_\Sigma$, and that the theorem of section 3.3. can be reformulated as :

4.2. *Theorem.* The solutions of a system Σ of regular equations form a category with an initial element. The unique (up to some isomorphism) *initial solution* of the system of equations, which is the initial object of the category of solutions), is obtained through iterated application of the functor $\mathcal{F}_\Sigma$ starting from the initial object of the category ($\underline{k}$ for an equation of type k, $\underline{[\tau(x_1)]} \times \ldots \times \underline{[\tau(x_n)]}$ for a general system of type $\tau(\Sigma)$).

An n-hypergraph is *regular* or *equational* if it is a component of the initial solution of a system of regular equations.

4.3. *Examples.* We can now check that applying the theorem to the examples of the introduction actually yields the expected solutions.

(i) Let us consider the first equation :

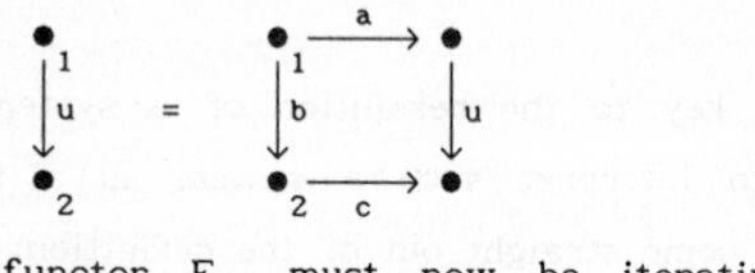

$(\Sigma 1)$

The corresponding functor F_{E1} must now be iteratively applied, starting from the initial object $\underline{2}$ in $\underline{\mathfrak{CG}}^{\infty}_{2}(A)$, yielding successively :

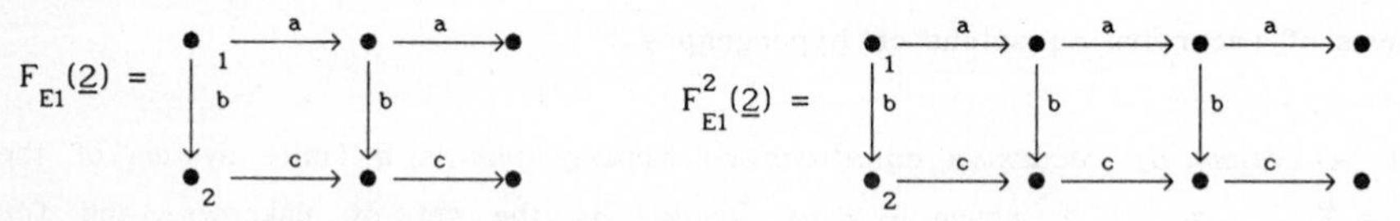

which finally leads to the following solution G_1 for the equation $(\Sigma 1)$:

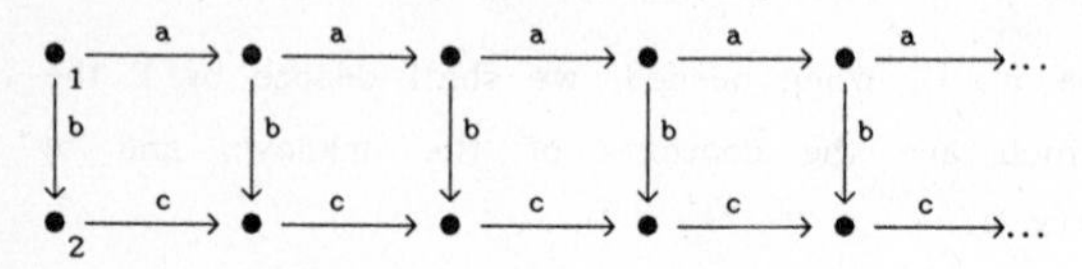

(ii) Let us now consider the equation :

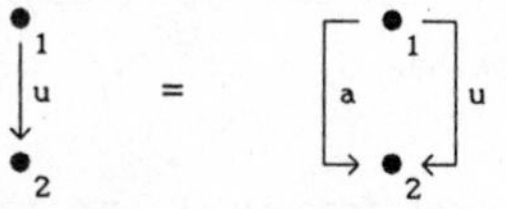

$(\Sigma 2)$

The same resolution process yields (starting once again from $\underline{2}$) :

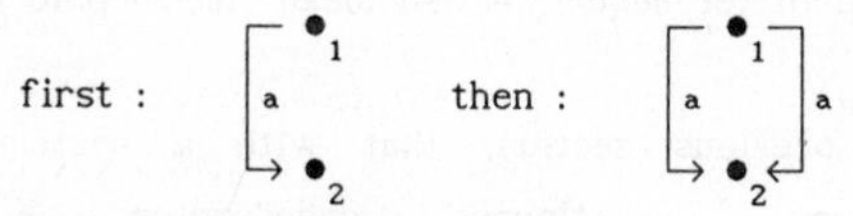

and leads finally to the following expected solution :

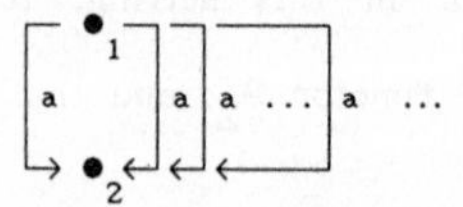

(iii) Let us now look a the last equation :

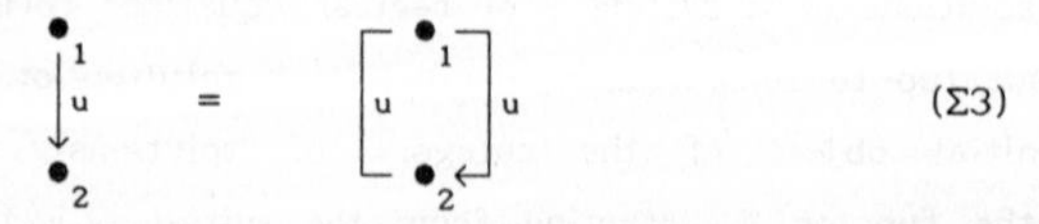

The initial solution to this equation is now quite trivially found. Starting with $\underline{2}$, one gets the ω-diagram whose terms are constantly equal to the initial 2-hypergraph $\underline{2}$. Note that the initial solution is not connected although the equation was.

4.4. A system Σ of recursive equations can also be interpreted as a rewriting system, simply by orienting its equations from left to right. As such it does not produce any terminal hypergraph, i.e., without any occurrence of the variable, and one has to add some rules producing those terminals. The following result (which needs the more general notion of colimit of a general diagram) relates the two approches :

Proposition : Let Σ be a system of recursive equations and Σ' be the rewriting system obtained by orienting from left to right the equations of Σ and adding the rule $u_i \longrightarrow \underline{p}$ for each unknown u_i of type p. Then, Σ' generates a vector of languages of terminal hypergraphs whose *colimit* is the initial solution of Σ.

5. General solution.

5.1. As was noted in the introduction, a system of equations on hypergraphs may have several distinct solutions. Indeed, from section 3.3, a system Σ of n equations may have more than one solution, since any morphism ϕ of the form $C \longrightarrow \mathcal{F}_\Sigma C$ generates a solution (for some n-tuple $C = (C_1,\ldots C_n)$ of hypergraphs with appropriate types).

Let us consider the examples of the previous sections. First for the equation ($\Sigma 2$), we can start the process from any other 2-hypergraph such as H_0 :

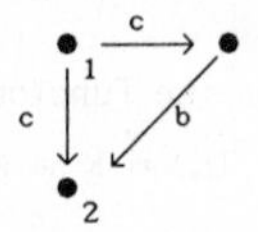

It would yield first : in which case there is an (injective) morphism $H_0 \longrightarrow \mathcal{F}_\Sigma H_0$ which allows us to iterate the application of $\mathcal{F}$, yielding then :

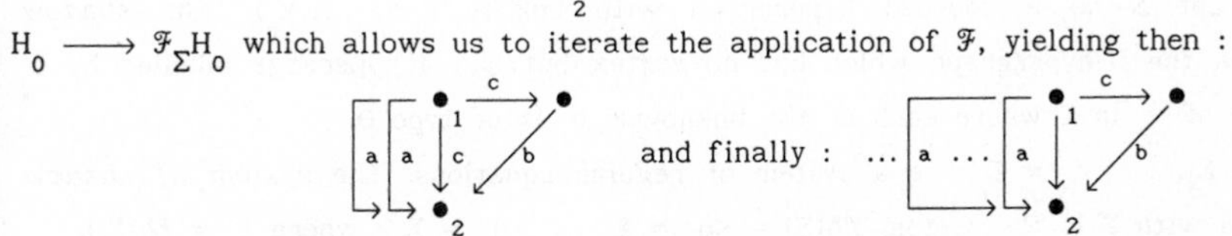

Last, starting from the hypergraph : , one will obtain through the same process the solution with an infinity of a-labelled loops already described in the introduction :

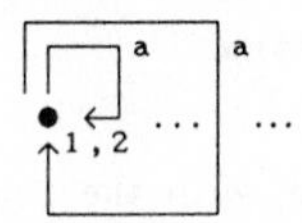

In the same way, starting with any 2-hypergraph G, the functor associated with equation

(Σ3) gives as a solution consisting of the parallel composition of an infinity of copies of G (all are glued together along their respectives sources).

The basic result of this paper is the following theorem, which allows an actual complete classification of the solutions. Its proof strongly relies on the fact that the category of n-hypergraphs is algebroidal (cf [Ba90b]) and on some structural properties of hypergraphs (we do not know whether it holds in a more general context) :

Theorem : Any solution of a system Σ of regular equations is *germ-computable,* which means that it is the ω-limit of an ω-diagram $(\mathcal{F}_\Sigma^n C, \mathcal{F}_\Sigma^n \phi)$ for some morphism $\phi : C \longrightarrow \mathcal{F}_\Sigma C$ in the appropriate product category, which we shall call the *germ* of the solution.

5.2. This means that classifying the solutions of a system is equivalent to classifying the germs. This will be done in three steps. The first will explain the first remark we made in the introduction for equation (Σ1) : one can build a solution by adding any 0-hypergraph to the initial solution. In the general case it is a bit more complex. Let us for instance consider the very simple equation $x = F \oplus x \oplus x$ and let $\underline{K}$ be its initial solution. Let K' be any 0-hypergraph such that $K = \underline{K} \oplus K'$ is a solution of the equation. One clearly has :

$$K = \mathcal{F}_F(K) = F \oplus (\underline{K} \oplus K') \oplus (\underline{K} \oplus K'))$$
$$= \mathcal{F}_F(\underline{K}) \oplus K' \oplus K' = \mathcal{F}_F(\underline{K}) \oplus K'^2 = \underline{K} \oplus K'^2$$

since $\underline{K}$ and K' are disjoint and K' has no source (with K'^n denoting n disjoint copies of K').

It is easily seen that applying m times the functor would yield :

$$K = \mathcal{F}_F^m(K) = \underline{K} \oplus K'^{2m}.$$

This remark is captured in the following definition :

Definition : Let Σ be a sourced hypergraph with unknowns $\{x_1,\dots,x_n\}$. The *shadow hypergraph* $\mathcal{S}h(\Sigma)$ is the 0-hypergraph which has no vertex but has a hyperedge labelled by u_j for each occurrence of x_j in Σ where each of the unknowns u_i is of type 0.

Let $\Sigma = \langle x_1 = \Sigma_1, \dots, x_n = \Sigma_n \rangle$ be a system of regular equations. The *system of shadow equations* associated with Σ is the system $\mathcal{S}h(\Sigma) = \langle u_1 = \Sigma_1', \dots, u_n = \Sigma_n \rangle$ where $\Sigma_i = \mathcal{S}h(\Sigma_i)$.

Proposition : If $C \longrightarrow \mathcal{F}_\Sigma C$ is a germ of a solution of Σ and $C' \longrightarrow \mathcal{F}_{\Sigma'} C'$ is a germ of a solution of the shadow equation $\mathcal{S}h(\Sigma)$, then $C \oplus C' \longrightarrow \mathcal{F}_\Sigma C \oplus \mathcal{F}_{\Sigma'} C'$ is a germ for a new solution of Σ (with ⊕ disjoint union of hypergraphs).∎

Examples. Let us try to convince the reader with the following two simple yet representative systems.

First of all, the system (Σ1) yields the shadow equation $u_0 = u_0$, of which any 0-graph is a solution, which explains the remar of the introduction.

The shadow system of the system (Σ4) :

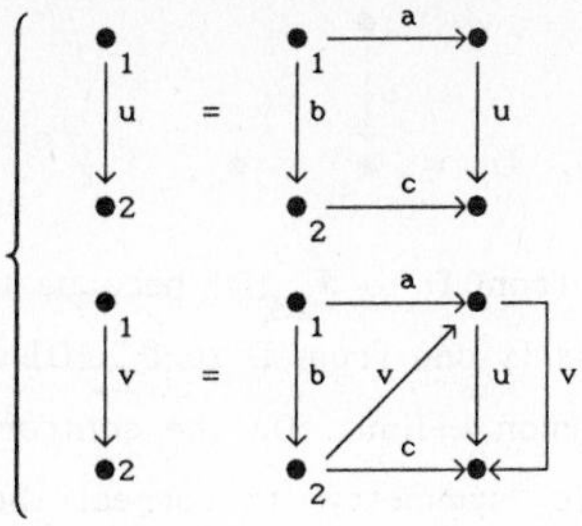

(Σ4)

is : $\begin{cases} u_0 = u_0 \\ v_0 = u_0 \oplus v_0 \oplus v_0 \end{cases}$ (Σ4')

and the 0-hypergraphs used to build new solutions shall have to respect the structure imposed by this system.

To conclude the first step, let us introduce a few more definitions. Two vertices v and v' of a hypergraph G are *connected* if there is a sequence $e_1,\ldots e_p$ of hyperedges in G such that v is a vertex of e_1, v' is a vertex of e_p and for each i, $1 \le i \le p-1$, e_i and e_{i+1} have a common vertex. Similarly, two hyperedges e and e' are connected if there is such a sequence with $e = e_1$ and $e' = e_p$, and a vertex v is connected to a hyperedge e if v belongs to e_1 and $e = e_p$. We shall use the expression : "the two items i and i' are connected ..." where item stands equally for vertex or hyperedge.

An item (or a set of items) is *source-connected* iff it is (all its elements are) connected to a source of the graph.

A solution K of a system of equations is *⊥-connected* iff all its items are connected to some item of the occurrence in K of the initial solution of the system. Of course, since a solution of a system is a vector of hypergraphs, this definition is to be understood componentwise. If follows from the previous proposition, that ⊥-connected solutions are of particular importance since any solution may be seen as the disjoint sum of a ⊥-connected solution and a 0-hypergraph.

5.2. We can now take the second step, concentrating on ⊥-connected solutions. Let us first consider some motivating examples, through the following equation :

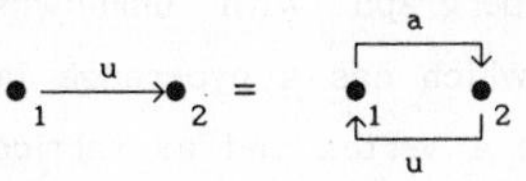

(Σ5)

It is fairly easy to find several 2-graphs which give rise to germs :

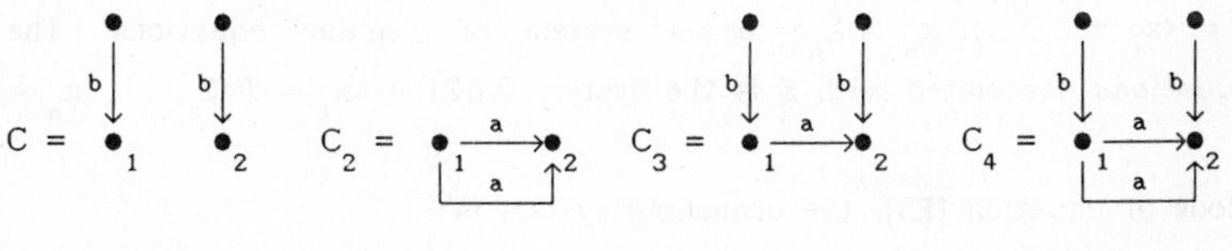

$C =$ $\quad C_2 =$ $\quad C_3 =$ $\quad C_4 =$

and graphs such as

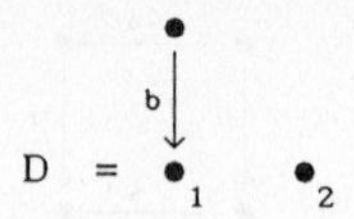

which do not : there is no morphism from D to $\mathcal{F}_{\Sigma 5}[D]$ because the unknown in the rhs of $(\Sigma 3)$ "goes the wrong way". Of course there is one from D to $\mathcal{F}^2_{\Sigma 5}[D]$ and one from $\mathcal{F}_{\Sigma 5}[D]$ to $\mathcal{F}^3_{\Sigma 5}[D]$ defining two ω-sequences with no common ω-limit. On the contrary, C_1 gives a germ and thus a *new solution* since it has the suitable "symmetry" to correct the effect of u going the wrong way.

The case of C_2 is different : C_2 defines a germ $C_2 \longrightarrow \mathcal{F}_{\Sigma 5}[C_2]$, and a solution $\underline{C}_2$, but this is *not a new solution* : it is isomorphic to the initial solution. This clearly stems from the fact that there is a morphism $C_2 \longrightarrow \mathcal{F}^2_{\Sigma 5}[\underline{2}]$ into an approximant of the initial solution.

This situation is rather general indeed. Let us consider once more the equation $(\Sigma 1)$ and the 2-hypergraph :

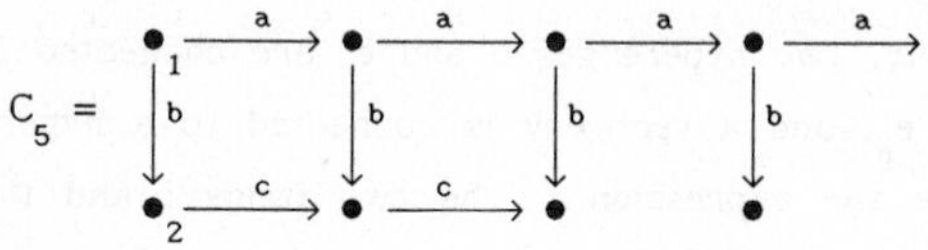

$C_5 =$

Then clearly the above graph C_5 is a germ since there is a morphism $C_5 \longrightarrow \mathcal{F}^4_{\Sigma 1}[\underline{2}]$, but is does not give rise to a new solution. The "discrepancy" introduced by C_5 can be understood as being rejected to infinity and the two solutions generated can be proven to be isomorphic. This is stated by the next proposition.

Proposition : Let $\phi : C \longrightarrow \mathcal{F}_{\Sigma}C$ be a germ and assume that there is a morphism $\varphi : C \longrightarrow G$. Then the solution K generated by ϕ is isomorphic to the initial solution G. Such a germ will be called *neutral*.

5.3. The third step will explain all the other remarks made in the previous example. Let us first define a new system of equations associated with Σ.

Definition : Let G be a n-hypergraph with unknowns $\{x_1,\ldots,x_n\}$. The *branching hypergraph* $\mathcal{Br}(G)$ is the n-hypergraph which has a hyperedge labelled by x_j for each occurrence of x_j in G which has a source as a vertex and as vertices only those of these hyperedges and the sources.

Let $\Sigma = \langle x_1 = \Sigma_1, \ldots, x_n = \Sigma_n \rangle$ be a system of regular equations. The *system of branching equations* associated with Σ is the system $\mathcal{Br}(\Sigma) = \langle x_1 = \mathcal{Br}\Sigma_1, \ldots, x_n = \mathcal{Br}\Sigma_n \rangle$.

If we look at equation $(\Sigma 5)$, the branching system is :

$\bullet_1 \xrightarrow{u} \bullet_2 \; = \; \bullet_1 \xleftarrow{u} \bullet_2$ $\quad \mathcal{B}r(\Sigma 5)$

of which C = ($\bullet \xrightarrow{b} \bullet_1$) ($\bullet \xrightarrow{b} \bullet_2$) is clearly a solution while D = ($\bullet \xrightarrow{b} \bullet_1$) $\bullet_2$ is not.

5.4. These three successive steps can be summed up in the following theorem :

Theorem : The general solution of a system Σ of recursive equations is the disjoint union of a $\perp$-connected solution and a 0-hypergraph which is a solution of the shadow system $\mathcal{S}h(\Sigma)$. Any $\perp$-connected solution is generated by a germ $\phi : C \longrightarrow \mathcal{F}_\Sigma C$ which is the parallel composition (gluing of the graphs along the sources) of a *neutral component* and of an *active component*, i.e. is a germ of the branching system $\mathcal{B}r(\Sigma)$.

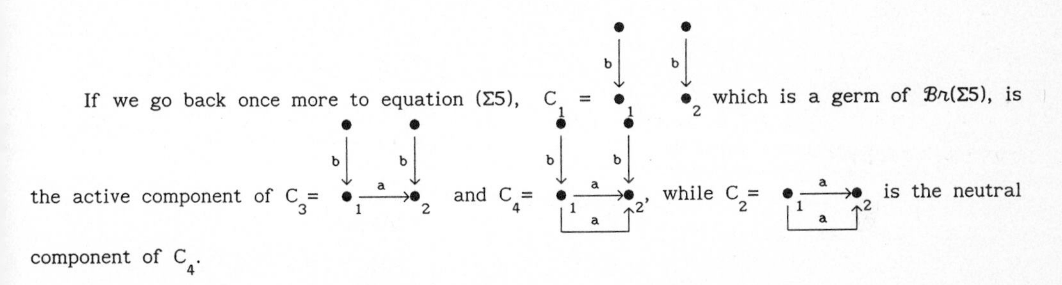

If we go back once more to equation ($\Sigma 5$), C_1 = which is a germ of $\mathcal{B}r(\Sigma 5)$, is the active component of C_3 = and C_4 = , while C_2 = is the neutral component of C_4.

As a corollary to this theorem, one may state that a *separated system* (i.e. a system whose rhs hypergraphs all have distinct sources) has a unique solution if and only if no unknown has a source as a vertex if and only if its initial solution has a finite degree (a similar result may be found in [Co88]).

6. References

[Ba89] BAUDERON M., On some properties of infinite graphs, Proc. of the WG'88, Amsterdam, Lect. Not. Comp. Sci. 344, Springer Verlag, 1989, 54-73.

[Ba90a] BAUDERON M., Infinite hypergraphs : I. Basic Properties, Report 90-20, Univeristy Bordeaux I, *to appear in* Theor. Comp. Sci,

[Ba90b] BAUDERON M., Infinite hypergraphs : II. Systems of recursive equations on hypergraphs, *revised version to appear in* Theor. Comp. Sci..

[BC87] BAUDERON M., COURCELLE B., Graph expressions and graph rewritings, Math. Systems Theory **20**, 83-127 (1987)

[Ca88] CAUCAL D., Pattern graphs, Research report 441, IRISA, 1988

[Co83] COURCELLE B., Fundamental properties of trees, Theor. Comp. Sci. 25 (1983) 95-169

[Co88] COURCELLE B. Definable properties of equational graphs, *to appear in* Annals of Pure and Applied Logic

[Co88c] COURCELLE B., Some applications of logic of universal algebra and of category theory to the theory of graph transformations, Tutorial, Bull. of the EATCS, 36, October 1988

[Eh79] EHRIG H., Introduction to the algebraic theory of graphs, Lect. Not. Comp. Sci. 73 Springer 1977 1-69

[HA89] HABEL A., Hyperedge replacement grammars and language, Thesis, Bremen 1989

[HK87] HABEL A., KREOWSKY H-J., Some structural aspects of hypergraph languages generated by hyperedge replacement, Lect. Not. Comp. Sci; 247, 207-219 (1987) and May we introduce to you : hyperedge replacement, Lect Not. Comp. Sci.,291, 15-26, 1987

[Mc71] McLANE S., Category for the working mathematician, Springer-Verlag 1971

[SP82] SMYTH M.B., PLOTKIN G., The category theoretic solution to recursive domain equations, SIAM J. Comput. Vol. 11, No 4 (1982)

CONSTRUCTION OF MAP OL-SYSTEMS FOR DEVELOPMENTAL SEQUENCES OF PLANT CELL LAYERS

Martin J. M. de Boer
Department of Computer Science
University of Regina
Regina, Saskatchewan, CANADA S4S 0A2

ABSTRACT: The construction of map OL-systems for the generation of observed developmental sequences of cell patterns in plant cell layers is discussed. The aim is to construct a map OL-system which emphasizes regularities in the patterns. The indirect construction method discussed in this paper is based on the successive specification of a developmental sequence by a cell system, a double wall map OL-system (dm-system) and an mBPMOL-system (sm-system), respectively. The final sm-system is used for growth analysis and computer simulation.

Keywords: cellular layers, development, map L-systems, construction procedure, plant morphology, computer simulation.

CONTENTS

0. INTRODUCTION

The development of cell layers in plants has been extensively studied within the framework of map OL-systems in the past decade (e.g. Lindenmayer and Rozenberg [10]; Abbott and Lindenmayer [1]; Lück and Lück [14]; Lück et al. [15]). Computer implementations of map OL-systems (De Does and Lindenmayer [6], De Boer and De Does [4]; Fracchia et al. [7]) allowed for the simulation of cell layer development in morphological studies. Having such tools available necessitates a formal approach to the construction of map OL-systems from detailed microscopic observations. Here, I discuss a method that could serve as a basis for a formalized construction procedure and for automated inference of map OL-systems for developmental sequences of plant cell layers.

1. DESCRIPTION AND ANALYSIS OF CELL LAYER DEVELOPMENT

1.1. Maps and map L-systems

Maps (Tutte [21]) consist of connected regions bounded by circular strings of edges. For the purpose of describing the topological structure of a cell layer we use maps. In maps, the edges correspond to cell walls and the regions to cells.

Map L-systems (Lindenmayer and Rozenberg [10]; Lück and Lück [11, 12]; Nakamura et al. [17, 18]) are parallel map rewriting systems which operate on maps. They are used to simulate cell division in a cell layer by binary splitting of regions in the maps. A map L-system is determined by a starting map and a set of production rules. The map sequence generated by the L-system represents the developmental sequence of a cell layer.

Map L-systems operate as follows. In the first phase of map rewriting, edges may subdivide and form branching edges inside the regions. In the second phase, matching branches in a region are connected, effecting region division.

Map L-systems can have *region label control* (Tuza and Lindenmayer [22]) or *edge label control.* The systems with edge label control can have *marker* or *cyclic control* of the positioning of division edges in the second rewriting phase. Moreover, edges can be *doubly labeled* or *singly labeled.* Doubly labeled map OL-systems with marker control or *dm-systems* were introduced by Lück and Lück [11, 12]. Singly labeled cyclic controlled BPMOL-systems (*sc-systems*) were introduced by Lindenmayer and Rozenberg [10], and singly labeled marker controlled BPMOL-systems or *sm-systems* by Nakamura et al. [17].

Each of these controls allows slightly different biological interpretations (Lindenmayer [9]). Present knowledge of biological control is not detailed enough to decide beforehand in favour of a particular type of control as the biologically most realistic one. Therefore, computational power, simplicity and ease of implementation motivates the choice for the types of map L-systems to be used for morphological studies.

Marker control is more powerful computationally than cyclic control, and cyclic controlled systems can be translated into marker controlled systems (De Boer and Lindenmayer [5]). Edge label control is at least as powerful as region label control, and region label control is translatable into edge label control (Lindenmayer [9]; Culik and Wood [2]).

Because sm-systems are simpler to implement on a computer system than dm-systems (De Does and Lindenmayer [6]; De Boer and De Does [4]; Fracchia et al. [7]) we want them as a final product of the construction procedure. In dm-systems the spatial-temporal organization of cell division can naturally be understood in terms of the behaviour of cell types, due to the integrity of cells in these systems. This property makes the dm-systems relatively easy to construct. Therefore, in the method to be discussed, a dm-system is constructed first and it is then translated into an sm-system.

1.2. Modeling approach

The problem is to construct an sm-system for a developmental sequence of a cell layer for the purpose of morphological studies and computer simulation. It is assumed that the cell lineage, and consequently the wall lineage, is known from detailed observations. In contrast to branching structures (Jürgensen and Lindenmayer [8], this requirement can easily be fulfilled for two-dimensional cell patterns. From frequent observations, the lineages can be inferred from the changes in topology arising from cell divisions. Sometimes, lineages can also be inferred over several generations from differences in wall thickness (Lück et al. [15]). Therefore, the present problem must not be confused with a form of syntactical inference in which the lineage is unknown.

Furthermore, it is assumed that there is no pre-existing knowledge about the biological control mechanisms involved in the development of the cell layer. This motivates not only the choice for sm-systems, as explained previously, but also the choice for interactionless map L-systems, termed map OL-systems. In map OL-systems edges and regions are modified independent of adjacent edges and regions.

Depending on the frequency of the observations, the relative timing of all the cell divisions may not be known exactly, and some cells may have given rise to more than two offspring cells between two observations. The first step is to modify the timing in the observed sequence, in order to obtain a topologically equivalent *binary sequence*, where in every step each cell gives rise to at most two offspring cells. The construction of a binary sequence is necessary since BPMOL-systems are binary and moreover cell division is binary. The choice for the timing of cell divisions in the binary sequence determines the map OL-system to a great extent.

An sm-system that generates a finite binary developmental sequence precisely can be constructed by simply assigning unique labels to all the edges in the sequence. The edge production rules follow directly from the edge lineage. The system can be reduced by identifying labels that have isomorphic derivation trees. Although many cell division patterns in plant meristems show regularities, an sm-system derived this way will in general have many rules, and in most cases as many rules as there are edges. This is simply because no one cell division pattern is perfectly regular.

Therefore, the approach taken here is to construct a concise map OL-system with recursive rules that captures and emphasizes the regularities in the spatial-temporal organization of the cell division pattern. In order for it to generate a perfectly regular pattern the system must be a deterministic map OL-system, or map DOL-system. Map DOL-systems are suitable for mathematical growth analysis and for simulating growth for an arbitrary number of steps by computer.

2. CONSTRUCTION OF SM-SYSTEMS

2.1. Step 1: Construction of cell systems

In Fig. 1a, a (binary) map sequence is shown that serves as a theoretical example of a developmental sequence. The corresponding cell lineage diagram is depicted in Fig. 1b. (This example was inspired by cell triplet formation in the leaves of the moss *Sphagnum palustre* (Zepf [23])).

Cell division patterns are determined by timing and orientations of cell divisions. In order to express regularities in the pattern, we start to consider timing and orientation separately, and subsequently combine these two features in a cell system.

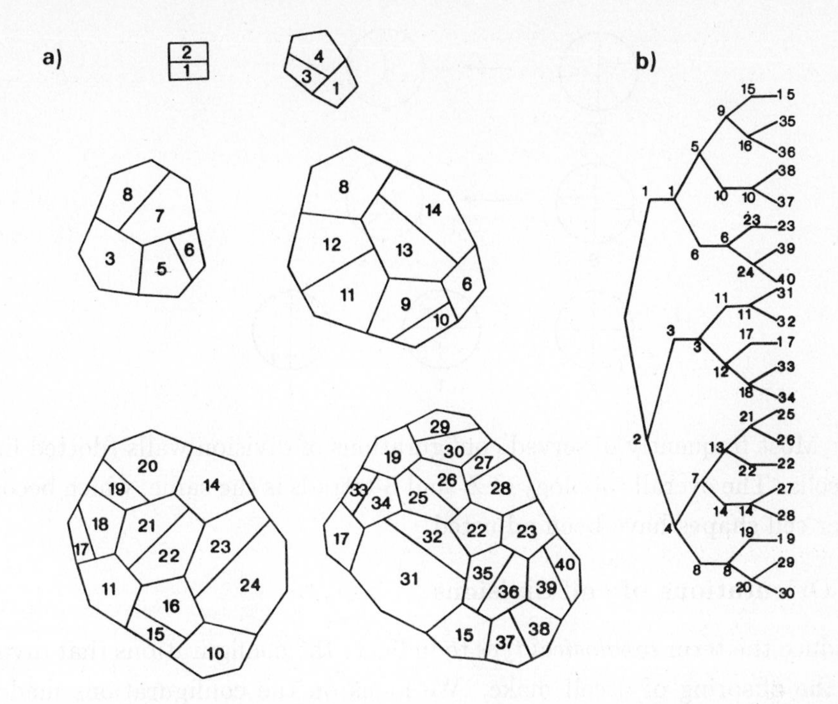

Figure 1: a. Example of a theoretical developmental sequence of a cell layer. Numbers identify individual cells. b. Diagram depicting the cell lineage in the cell layer. Horizontal lines in the lineage diagram result from cells of type B.

2.1.1. Timing of cell divisions

The precise order in which all the cell divisions have taken place in the real organism is not a concern here. The binary sequence, mentioned before, may be modified in order to obtain regularities in the timing of cell divisions. However, this must be done such that the cell growth functions in the binary sequence correspond closely to the cell growth functions in the observed sequence. In the sequence shown in Fig. 1, two *growth types*, named A and B, can be distinguished: cells that divide in the present observation (A) and cells that do not divide in the present observation, but that divide in the next observation (B). These transformations can be formally expressed with the production rules:

$$A \rightarrow A\,B$$
$$B \rightarrow A$$

The number of descendants of each cell follows a Fibonacci sequence.

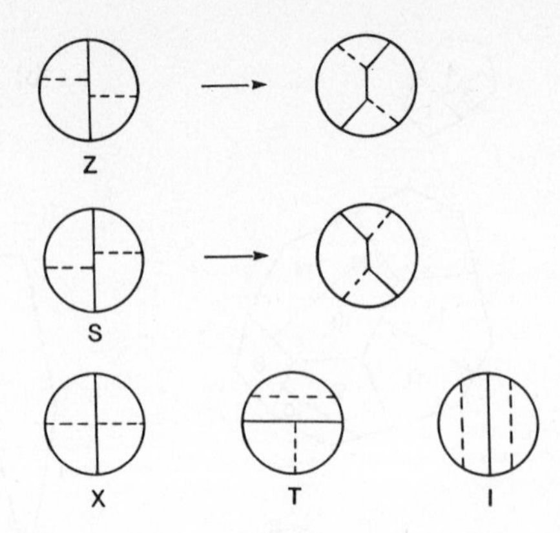

Figure 2: Most frequently observed configurations of division walls (dotted lines) of sister cells. The overall topology of Z- and S-tetrads is the same, which becomes clear after cell shapes have been adjusted.

2.1.2. Orientations of cell divisions

We introduce the term *topological type* to indicate the configurations that division walls of the offspring of a cell make. We focus on the configurations made by division walls of sister cells. Five types of such configurations (so called tetrad types) that can be found in cell division patterns are shown in Fig. 2 (Lück et al. [15]). The transitions of topological types can be expressed with a DOL-system. The production rule of the single topological type in the example sequence of Fig. 1a is

$$S \rightarrow S\,S.$$

It turns out that the pattern (Fig. 1a) is perfectly regular. If there were many growth types or topological types their number should be reduced by eliminating apparent abberations from an ideal regular pattern. In the case of very regular patterns one can also consider the configurations of division walls of cousin cells. However, it seems that in many cell division patterns, regularities are most expressed in tetrad types (Lück et al. [15]; Morris et al. [16]). Therefore, topological analysis is restricted to tetrad types in this example.

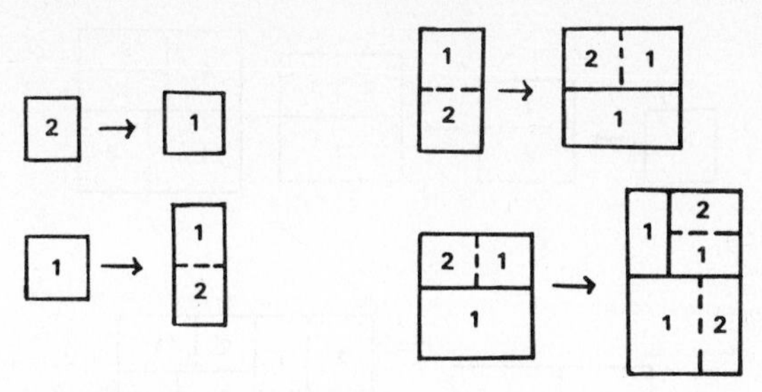

Figure 3: Production rules of the example cell system. Dotted lines indicate new division walls.

2.1.3. Cell systems

The next step is to combine the growth types and the topological types into *cell types.* The transition rules are:

$$\begin{aligned}(A,S) &\rightarrow (A,S)\,(B,S)\\(B,S) &\rightarrow (A,S)\end{aligned}$$

Replacing (A,S) and (B,S) by 1 and 2 results in the production rules:

$$\begin{aligned}1 &\rightarrow 1\,2\\2 &\rightarrow 1\end{aligned}$$

In order to specify the spatial distribution of the cell types, we need a *cell system.* The cell system is a DOL-system, with production rules that specify the cell types on both sides of a division wall. After having filled in the cell types in the cells in Fig. 1a, we find the cell system depicted in Fig. 3. By replacing cell groups in an observation (Fig. 1a), according to these production rules, the successor cell pattern in the next observation is derived. The map sequence generated by the cell system is shown in Fig. 4. Because the new division walls attach to the division wall of the mother cell, the divisions are more or less orthogonal. Therefore, cells are depicted as quadrangles, and division walls are shown to attach to opposite walls. The relative positions of division walls on the outside of the cell groups, which are not specified by the cell system, are indicated by arrows.

Note that the left-hand side of the production rule for the two-cell group does not apply to just any pair of neighboring cells with types 1 and 2, but to a pair of sister cells, which is indicated by the dotted lines. Systems like the above cell system were used by Szilard and Quinton [20] for the generation of space-filling patterns by DOL-systems.

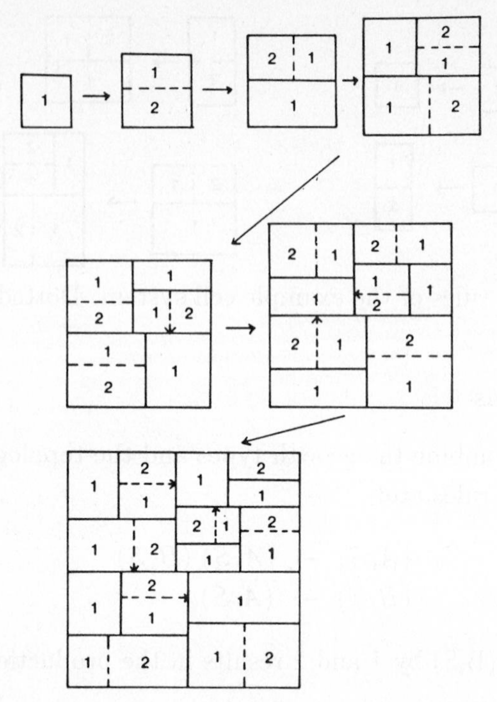

Figure 4: Map sequence of the cell system. Dotted lines indicate new division walls. Arrows indicate division wall attachment sites that are not specified by the cell system.

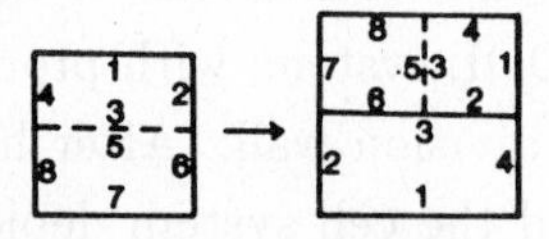

Figure 5: Cell production rule with double edge labeling.

2.2. Step 2: Construction of a dm-system from a cell system

In order to arrive at an sm-system that describes the complete topology of the cell division pattern (including the attachment sites of division walls which are indicated by arrows in Fig. 4), we construct a dm-system from the cell system (Fig. 3). This is done by translating the cell label control of the cell system into edge label control.

The edge production rules of the dm-system are found from the cell system as follows. Label the edges in the production rules of the cell system such that cells of equal type have the same clockwise sequence of edge labels, and that identical cell groups have identical edge labelings. Moreover, all edges within a cell must be given distinct labels and the same edge label may not occur in different cell types. The edge labeling in the cell groups in the right-hand side of the production rules can be found from the edge labeling in the left-hand side, by identifying left-hand side components in the right-hand side structures. The resulting edge labeled cell production rule, that captures all the edge production rules, is given in Fig. 5. The edge production rules of the dm-system follow directly from this cell production rule. They are as follows:

$$
\begin{array}{llll}
1 \to 8\,/(5,3)\,4 & & & 5 \to 3 \\
2 \to 1 & & & 6 \to 4 \\
3 \to 2\,/(3,5)\,6 & & & 7 \to 1 \\
4 \to 7 & & & 8 \to 2
\end{array}
$$

Edge labels that have isomorphic descendant trees can be replaced by one label. It follows that 7 can be replaced by 2 and that 8 can be replaced by 4, yielding the reduced system:

$$
\begin{array}{llll}
1 \to 4\,/(5,3)\,4 & & & 4 \to 2 \\
2 \to 1 & & & 5 \to 3 \\
3 \to 2\,/(3,5)\,6 & & & 6 \to 4
\end{array}
$$

The starting map is a region with boundary 1 2 3 4. In the first phase of the map rewriting step, the edge production rules are applied in parallel and the *markers* /(5,3) and /(3,5) are positioned at vertices where division walls will attach. The convention is that the production rules are applied in clockwise direction with respect to the cell in which the edge label is placed. In the second phase, division walls are inserted between matching markers (i.e. /(5,3) and /(3,5)). For a detailed definition of dm-systems see Lück and Lück [14]. The map sequence generated by the dm-system is shown in Fig. 6.

The edge production rules of the dm-system do not specify whether the offset configurations of division walls of sister cells are S or Z. Therefore, we need *alignment rules* that specify the relative positions of attachment sites of division walls on both sides of a wall (Fig. 7). The alignment rules are derived from the map sequence of the dm-system. They have been chosen to match the map topology (Fig. 6). The alignment rules determine the combinations of edge labels on both sides of the walls. For each combination that occurs, a rule is given that specifies the new combination it gives rise to. Therefore, the rules determine the complete map topology. This is possible, because in this example, the number of combinations is finite. This property of a dm-system is called *parity* (Lück and Lück [13]).

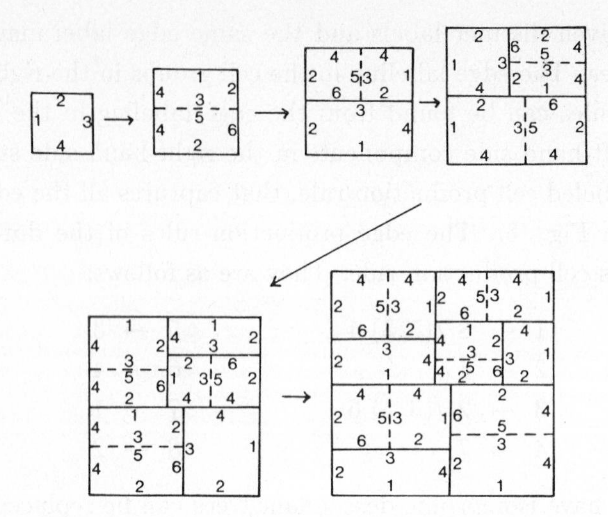

Figure 6: Map sequence of the dm-system.

The decidability question, whether a dm-system has parity is an open problem. This problem is related to the unsolved decidability question of locally catenativeness of DOL-systems (Rozenberg and Lindenmayer [19]).

2.3. Step 3: Construction of an sm-system from a dm-system

From the alignment rules of a dm-system with parity, an equivalent sm-system can be constructed (Lindenmayer [9]). The sm-system is obtained by labeling each edge in the alignment rules of the dm-system by a single label (Fig. 7). Edges with equal combinations of number labels in the dm-system, receive equal letter labels in the sm-system. The offset edges between division walls in adjacent cells do not split and are labeled x. A more complicated example of assigning letter labels to combinations of number labels is discussed in the appendix.

In contrast to the dm-system, edge directions must be specified in the sm-system (except for edges labeled x) in order to indicate the directions of application of the edge production rules to the edges. The line segments below the singly labeled graphs indicate the edge lineages (Fig. 7). The edge production rules of the sm-system, in graph notation, can be directly derived from the diagram. The operation of sm-systems is similar to the operation of dm-systems. In the second map rewriting phase, division walls inside a cell are inserted between markers that have the same label, and that are equally directed. The label and direction of the division wall is the label and direction of the markers. For a detailed definition of sm-systems see Nakamura et al. [17, 18]. The starting map

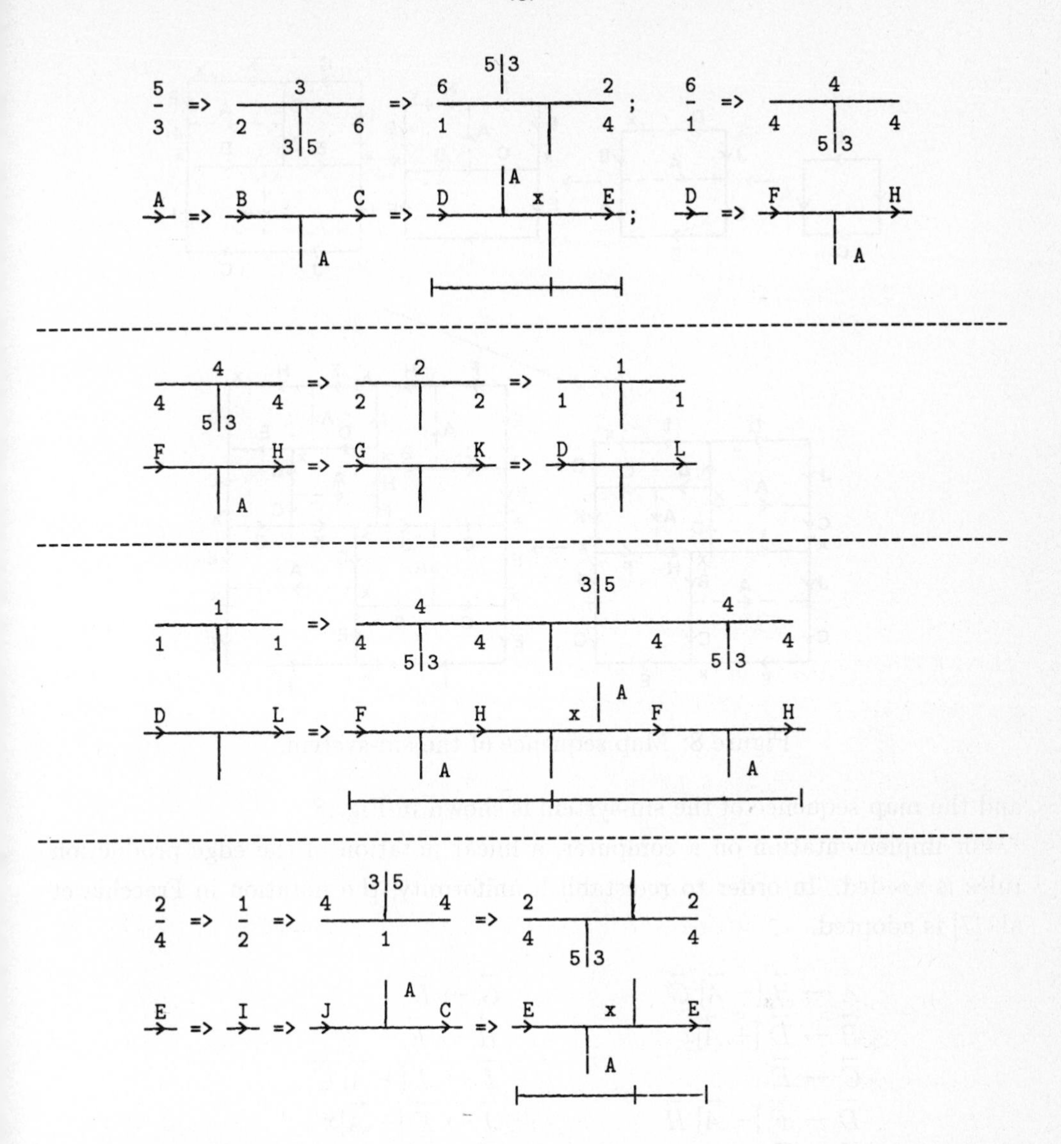

Figure 7: Graphical representation of the alignment rules of the dm-system (edges have numeric labels) and of the edge production rules of the corresponding sm-system (edges have letter labels). Edge production rules of the sm-system are shown directly below each corresponding alignment rule of the dm-system. Line segments without labels, directly below edge production rules of the sm-system indicate the edge lineage, where edge lineage is otherwise ambiguous. Dotted vertical lines indicate new division walls. Solid vertical lines indicate division walls from previous steps.

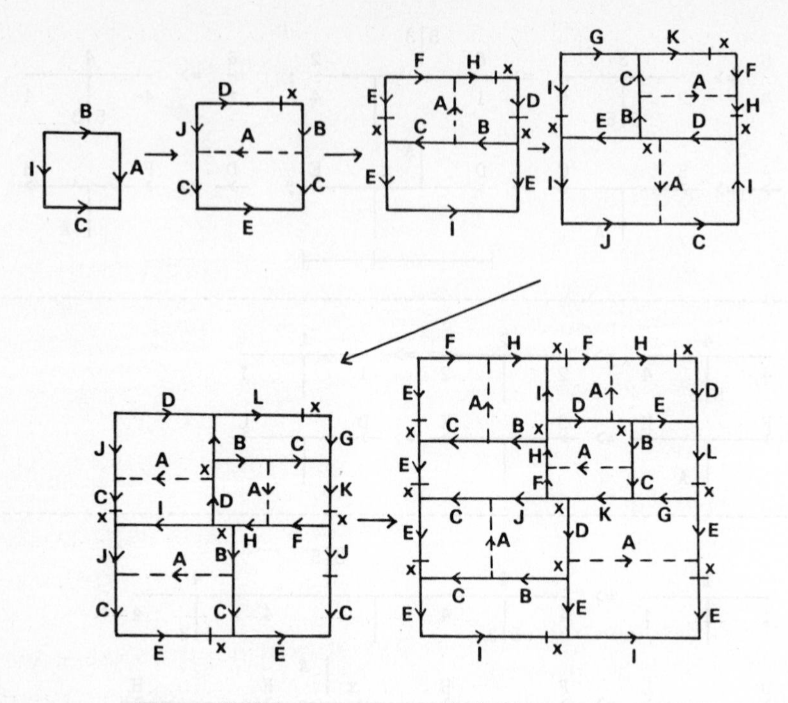

Figure 8: Map sequence of the sm-system.

and the map sequence of the sm-system is shown in Fig. 8.

For implementation on a computer, a linear notation of the edge production rules is needed. In order to re-establish uniformity, the notation in Fracchia et al. [7] is adopted.

$$
\begin{array}{ll}
\vec{A} \rightarrow \vec{B}\,[-\,\vec{A}]\,\vec{C} & \vec{G} \rightarrow \vec{D} \\
\vec{B} \rightarrow \vec{D}\,[+\,\vec{A}]x & \vec{H} \rightarrow \vec{K} \\
\vec{C} \rightarrow \vec{E} & \vec{I} \rightarrow \vec{J}\,[+\,\overleftarrow{A}]\,\vec{C} \\
\vec{D} \rightarrow \vec{F}\,[-\,\overleftarrow{A}]\,\vec{H} & \vec{J} \rightarrow \vec{E}\,[-\,\overleftarrow{A}]x \\
\vec{E} \rightarrow \vec{I} & \vec{K} \rightarrow \vec{L} \\
\vec{F} \rightarrow \vec{G} & \vec{L} \rightarrow \ x[+\,\overleftarrow{A}]\,\vec{F}\,[-\,\overleftarrow{A}]\,\vec{H} \\
x \ \rightarrow \ x &
\end{array}
$$

Edge directions are indicated by $\rightarrow$ and $\leftarrow$. Brackets specify marker positions. A marker to the right is indicated by "−" and a marker to the left by "+". The starting map (Fig. 8) is indicated by the *circular word* $\overleftarrow{C}\overleftarrow{I}\vec{B}\vec{A}$. The convention is that the sequence of edge labels is read in clockwise direction from the cell in the starting map. In circular words, $\leftarrow$ indicates that an edge is directed in counterclockwise direction, and $\rightarrow$ that an edge is directed clockwise.

Finally, it can be checked whether the above sm-system is deterministic (Nakamura et al. [17]; De Boer and Lindenmayer [5]), which is required for the generation of a regular pattern.

3. DISCUSSION

The construction method for sm-systems, described in this paper, allows one to first express regularities in a cell division pattern on the level of cells (in terms of cell systems), and to subsequently make the specification of the topology complete (in terms of sm-systems), which is required for computer simulation. Recently developed visualization methods (De Boer and De Does [4]; Fracchia et al. [7]) have made map L-systems a powerful tool for computer simulation in morphological studies.

Cell systems are not only important for the construction of sm-systems, but they are also essential for a good understanding of the developmental sequences generated by sm-systems. Cell growth functions can be derived from the cell systems and can be compared with cell growth in the real pattern in order to determine the correspondence between the simplified cell system and growth in the cell layer (De Boer [3]).

It remains to be investigated whether recursion in the cell system guarantees parity of the dm-system. In the affirmative case, the fact that dm-systems without parity cannot be translated into sm-systems, would not be of practical importance for this construction method.

From a methodological point of view, a good understanding and a standardization of the construction of mBPMOL-systems is important. The proposed method is meant to serve this purpose and will be used as a basis for automated construction of mBPMOL-systems.

ACKNOWLEDGEMENTS

Many thanks are due to the late Prof. Aristid Lindenmayer for his guidance and comments on an earlier version of this paper. I am very grateful to Prof. Helmut Jürgensen for many helpful discussions on the inference and construction of map OL-systems during my visit at the Dept. of Comp. Science, Univ. of Western Ontario, Canada, in Spring 1988. Financial support for this visit was provided by the Natural Sciences and Engineering Research Council of Canada (NSERC), Grant A0243. Travel expenses were covered by the Univ. of Utrecht and the Netherlands Organization for Scientific Research (NWO). I thank two anonymous referees for their comments on the manuscript.

APPENDIX

The following example is to illustrate a more complicated case of the construction of edge production rules of an sm-system from the edge production rules of a dm-system (step 3). Consider a doubly labeled edge

$$\frac{1}{2}$$

and the edge production rules

$$1 \rightarrow 1\,2$$
$$2 \rightarrow 1$$

The convention in dm-systems is that the production rules are applied in clockwise direction with respect to the cell in which they are located. This means that the direction is from right to left for the top label, and from left to right for the bottom label. The following doubly labeled sequence (1) is generated:

$$\frac{1}{2} \Rightarrow \frac{2\ 1}{1} \Rightarrow \frac{1\ 2\ 1}{1\ 2} \Rightarrow \frac{2\ 1\quad 1\quad 2\ 1}{1\quad 2\quad 1} \Rightarrow \frac{1\ 2\ 1\quad 2\ 1\quad 1\ 2\ 1}{1\ 2\quad 1\quad 1\ 2} \qquad (1)$$

$$a \Rightarrow b \Rightarrow c \Rightarrow b\ a\ b \Rightarrow c\ b\ c \qquad (2)$$

The second sequence (2) shows repetitions of certain combinations of top and bottom label occurences, designated by lowercase letters. Sequence (2) is described by the following alignment rules:

$$a \rightarrow b$$
$$b \rightarrow c$$
$$c \rightarrow bab$$

We designate the label combinations associated with each individual edge in the combinations a, b and c by uppercase letters A, BC, and DEF, respectively.

$$\overset{A}{\rightarrow} \Rightarrow \overset{B}{\rightarrow}\overset{C}{\rightarrow} \Rightarrow \overset{D}{\rightarrow}\overset{E}{\rightarrow}\overset{F}{\rightarrow} \Rightarrow \overset{B}{\rightarrow}\overset{C}{\rightarrow}\overset{A}{\rightarrow}\overset{B}{\rightarrow}\overset{C}{\rightarrow} \qquad (3)$$

The following edge production rules describe the occurrence of these combinations, and generate sequence (3) which is equivalent to sequence (1). They are the edge production rules of the sm-system.

$$\vec{A} \rightarrow \vec{B}\vec{C} \qquad \vec{D} \rightarrow \vec{B}\vec{C}$$
$$\vec{B} \rightarrow \vec{D} \qquad \vec{E} \rightarrow \vec{A}$$
$$\vec{C} \rightarrow \vec{E}\vec{F} \qquad \vec{F} \rightarrow \vec{B}\vec{C}$$

Note that, due to the ambiguity of the edge lineage, the rules $\vec{B} \rightarrow \vec{D}\vec{E}$ and $\vec{C} \rightarrow \vec{F}$ could have been chosen instead of $\vec{B} \rightarrow \vec{D}$ and $\vec{C} \rightarrow \vec{E}\vec{F}$, respectively. Labels D and F have productions identical to A. Substitution of A for D and F gives identical productions for B and E. From the substitution of B for E, the following reduced set of edge production rules of the sm-system results:

$$\vec{A} \rightarrow \vec{B}\vec{C}$$
$$\vec{B} \rightarrow \vec{A}$$
$$\vec{C} \rightarrow \vec{B}\vec{A}$$

Note that the two oppositely running string sequences in (1), and the sequences (2) and (3) can be described by the locally catenative formula (Rozenberg and Lindenmayer [19]): $S_n = (S_n - 2)(S_n - 3)(S_n - 2)$.

REFERENCES

[1] Abbott, L.A. and A. Lindenmayer. 1981. Models for growth of clones in hexagonal cell arrangements: applications in Drosophila wing disc epithelia and plant epidermal tissues. J. Theor. Biol. 90: 495-514.

[2] Culik II, K. and D. Wood. 1979. A mathematical investigation of propagating graph-OL systems. Information and Control 43: 50-82.

[3] De Boer, M.J.M. 1990. The relationship between cell division pattern and global shape of young fern gametophytes II. Morphogenesis of heart-shaped thalli. Bot. Gaz. 151 (In press).

[4] De Boer, M.J.M. and M. de Does. 1990. The relationship between cell division pattern and global shape of young fern gametophytes I. A model study. Bot. Gaz. 151 (In press).

[5] De Boer, M.J.M., and A. Lindenmayer. 1987. Map OL-systems with edge label control: Comparison of marker and cyclic systems. Pages 378-392 in: H. Ehrig, M. Nagl, G. Rozenberg and A. Rosenfeld eds. Graph grammars and their applications to computer science. Lecture Notes in Computer Science Vol. 291. Springer-Verlag, Berlin.

[6] De Does, M. and A. Lindenmayer. 1983. Algorithms for the generation and drawing of maps representing cell clones. Pages 39-57 in: H. Ehrig, M. Nagl, and G. Rozenberg, eds. Graph Grammars and their applications to computer science. Lecture Notes in Computer Science Vol. 153. Springer-Verlag, Berlin.

[7] Fracchia, F.D., P. Prusinkiewicz and M.J.M de Boer. 1990. Animation of the development of multicellular structures. Pages 3-18 in: N. Magnenat-Thalmann and D. Thalmann, eds. Springer-Verlag, Tokyo.

[8] Jürgensen, H. and A. Lindenmayer. 1987. Modelling development by OL-systems: inference algorithms for developmental systems with cell lineages. Bull. Math. Biol. 49: 93-123.

[9] Lindenmayer, A. 1987. An introduction to parallel map generating systems. Pages 27-40 in H. Ehrig, M. Nagl, G. Rozenberg, and A.Rosenfeld, eds. Graph grammars and their application to computer science. Lecture Notes in Computer Science. Vol. 291. Springer-Verlag, Berlin.

[10] Lindenmayer, A., and G. Rozenberg. 1979. Parallel generation of maps: developmental systems for cell layers. Pages 301-316 in V.Claus, H. Ehrig, and G. Rozenberg, eds. Graph grammars and their application to computer science. Lecture Notes in Computer Science Vol. 73. Springer-Verlag, Berlin.

[11] Lück, J. and H.B. Lück. 1978. Automata theoretical explanation of tissue growth. Pages 974-185 in Proc. Int. Symp. Math. Topics Biol., Res. Inst. Math. Sci., Kyoto.

[12] Lück, J. and H.B. Lück. 1979. Two-dimensional, differential, intercalary plant tissue growth and parallel graph generating and graph recurrence systems. Pages 301-316 in V.Claus, H. Ehrig, and G. Rozenberg, eds. Graph grammars and their application to computer science. Lecture Notes in Computer Science Vol. 73. Springer-Verlag, Berlin.

[13] Lück, J. and H.B. Lück. 1982. Sur la structure de l'organisation tissulaire et son incidance sur la morphogénèse. Pages 385-397 in H. Le Guyader, ed. Actes du 2-ème séminaire de l'Ecole de Biologie théorique du C.N.R.S. Publications de l'Université de Rouen, Rouen.

[14] Lück, J. and H.B. Lück. 1987. From OL and IL map systems to indeterminate and determinate growth in plant morphogenesis. Pages 393-410 in H. Ehrig, M. Nagl, G. Rozenberg, and A. Rosenfeld, eds. Graph grammars and their application to computer science. Lecture Notes in Computer Science. Vol. 291. Springer-Verlag, Berlin.

[15] Lück, J., A. Lindenmayer and H.B. Lück. 1988. Models of cell tetrads and clones in meristematic cell layers. Bot. Gaz. 149: 127-141.

[16] Morris, V.B., K.E. Dixon and R. Cowan. 1989. The topology of cleavage patterns with examples from embryos of *Nereis*, *Styela* and *Xenopus*. Philos. Trans. Royal Society London Ser. B. 325: 1-36.

[17] Nakamura, A., A. Lindenmayer and K. Aizawa. 1986. Some systems for map generation. Pages 323-332 in G. Rozenberg and A. Salomaa, eds. The Book of L. Springer-Verlag, Berlin.

[18] Nakamura, A., A. Lindenmayer and K. Aizawa. 1987. Map OL-systems with markers. Pages 479-495 in H. Ehrig, M. Nagl, G. Rozenberg and A. Rosenfeld, eds. Graph grammars and their application to computer science. Lecture Notes in Computer Science. Vol 291. Springer-Verlag, Berlin.

[19] Rozenberg, G. and A. Lindenmayer. 1973. Developmental systems with locally catenative formulas. Acta informatica 2: 214-248.

[20] Szilard, A.L. and R.E. Quinton. 1979. An interpretation for DOL-systems by computer graphics. The Science Terrapin, Faculty of Science, Univ. of Western Ontario, 4: 8-13.

[21] Tutte, W.T. 1982. Graph Theory. Addison-Wesley Publ. Co., Reading, Mass.

[22] Tuza, Z. and A. Lindenmayer. 1987. Locally generated colourings of hexagonal cell division patterns: application to retinal cell differentiation. Manuscript.

[23] Zepf, E. 1952. Über die Differenzierung des Sphagnumblattes. Zeitschrift für Botanik 40: 87-118.

Layout Graph Grammars: the Placement Approach

Franz J. Brandenburg

Lehrstuhl für Informatik, University of Passau

Innstr. 33, D 8390 Passau, Germany

e-mail: brandenb@unipas.fmi.uni-passau.de

Abstract

Layout graph grammars are extensions of context-free graph grammars and are introduced as a tool for syntax directed constructions of graph layouts. The constructions are based on a layout specification of the productions, which are consistently transferred to the derivations. The layout specification consists of rules for a placement of the vertices and a partial routing of the edges. It specifies minimal distances between the vertices in X- or Y-dimension. These distances can be optimized according to some formal cost measures.

There is a very intuitive visual representation of the layout specifications, which stems from an elegant graphic representation of the graph grammar productions. Alternatively, the layout specifications are expressed in graph theoretic terms, and so are completely integrated into usual graph grammars.

The computation of optimal layouts of graphs is a well-known NP-complete problem, even for binary trees. Therefore, we design layout graph grammars which guarantee polynomial time constructions of optimal layouts of graphs. This is achieved by the restriction to polynomial graph grammars and layout specifications, which can be computed efficiently by an attributation technique. Hence, layout graph grammars are a new and powerful tool for efficient solutions of graph layout problems. They help jumping accross the NP-completeness barrier.

Keywords

graphs, graph grammars, graph layout, VLSI, placement,

The support of this work by the DFG under grant Br835/1 and by the ESPRIT Basic Research Group "Computing by Graph Transformations" is gratefully acknowledged.

0. Introduction

Graphs are one of the most important tools in Computer Science. Their usefulness is based on at least three facts. On the informal side graphs are very flexible and possess a high expressive power with a natural distinction between static facts, expressed by the vertices, and dynamic relations, expressed by the edges. Secondly, they have a simple formal definition in set theoretic terms or by adjacency lists for an internal representation. Finally, and perhaps most importantly, graphs have a natural visual representation, i.e, they can be drawn in the plane. A visual representation supports human cognitive capabilities, it activates our intuition, helps finding solutions and makes complex situations understandable. However, the visual representation of a graph is an aid only if it is readable, if the graph is nicely drawn, if there is a good graph layout. Particular applications of graphs are Petri nets, entity relationship diagrams, transition graphs, schedules, circuits, VLSI, CAD models, or just drafts.

Our goal are new tools for efficient constructions of nice drawings of graphs. The need for these tools stems from the desire to automate layout processes and to improve the quality of a drawing. For our formal approach we measure the quality in terms of cost measures, which are e.g. the area and the maximal edge length of grid embeddings. However, optimal grid embeddings of graphs are NP-complete, in general. The NP-completeness holds even for binary trees, provided there are sharp bounds.

We overcome the NP-completeness by layout graph grammars. A layout graph grammar is a graph grammar and enriched by a layout component. Concerning graph grammars we follow the algorithmic or set theoretic approach, which is the most efficient and best suited for our purpose. A graph grammar is a canonical generalization of a context-free string grammar. Its core is a finite set of productions of the form (A, R, C), where A is the label of the replaced vertex, R is the graph from the right hand side, and the connection relation C specifies the embedding of the right hand side R into the local environment of the replaced vertex. A graph grammar defines a set of graphs, called a graph language, which is usually infinite.

The layout component of a graph grammar comes with an elegant graphic representation of the productions. Given (A, R, C) we draw the graph R in some standard fashion and draw the left hand side as a rectangle enclosing R. The connection relation C is represented by edges from outside the rectangle to the specified vertices of the right hand side R. This graphic representation is used to define layout rules, which come automatically with the drawing of R and A. These rules specify minimal distances and order relations in X- and Y-dimension. The rules are consistently transferred to the derivation process and induce a syntax directed layout of the generated graphs. The generated language consists of a set of terminal graphs together with a system of constraints for their layout. Alternatively, layout graph grammars can be seen as syntax directed translations converting graphs from a set theoretic representation into a visual representation.

Most importantly, for a wide class of layout graph grammars, a consistent and area minimal layout can be computed in polynomial time. Hence, layout graph grammars are well-suited to outwit the NP-completeness of optimal graph layouts. For the polynomiality we pay with restrictions on the sets of feasible graphs and on the induced layouts. The graphs must be derivable by a graph grammar, which induces a hierarchical structure and implies structural properties. The layout must be consistent with the derivation process, which needs more area than the most compact and compressed layouts. The latter, however, are not hierarchical and are intractable to compute.

In this paper we concentrate on a placement of the vertices. There is no complete routing for the edges; for a partial routing, one may determine certain points on the edges and handle these like vertices. For the computation of the layouts we employ attributation techniques. The width and the height of certain rectangles corresponding to nonterminals and the subgraphs derived from them are computed as synthesized attributes and the coordinates of the vertices are computed by inherited attributes. Clearly, the most intuitive layout comes with a top down application of the productions. However, this strategy is very area consuming, since the replaced vertex explodes, it expands such that the graph from the right hand side fits in the created space and moves all other objects to the outer boundary. Our approaches are based on local optimizations, constructing layouts of graphs that are consistent with the specifications given by the graph grammar and are computable in polynomial time. The jump from NP-completeness to polynomial time is seen as the main achievement.

1. Graphs and Embeddings

First, we recall some notions from graph theory. We consider labeled graphs, which are necessary for graph grammars. This induces some special notions. Then we consider grid embeddings of graphs as a formal framework for graph layouts.

<u>**Definition**</u>

A *directed, labeled graph* $g = (V, E, m)$ consists of a finite set of *vertices* V, a *vertex labeling function* $m : V \rightarrow \Sigma$ and a finite set $E = \{(u, a, v) \mid u, v \in V \text{ with } u \neq v \text{ and } a \in \Delta\}$ of *directed, labeled edges.* The vertex labels are taken as nonterminal or terminal symbols, and accordingly, a vertex is called a nonterminal vertex or a terminal vertex. An edge $e = (u, x, v)$ with label x is called an *x-edge* from vertex u to vertex v. Σ and Δ are the alphabets of vertex labels and edge labels, respectively.

For a graph g let $V(g)$, $E(g)$ and $m(g)$ denote its components, so that $g = (V(g), E(g), m(g))$. Notice that there may be multiple edges between two vertices, which however, are distinguished by their direction or by their edge label. Self-loops are excluded.

Special graphs are the *grids* with vertices $v = (x, y)$ in the discrete plane and edges $e = (u, a, v)$ as horizontal or vertical lines of unit length, such that $u = (x, y)$ implies $v = (x-1, y)$ or $v = (x+1, y)$ or $v = (x, y-1)$ or $v = (x, y+1)$. The coordinates of v are denoted by $v.X$ and $v.Y$.

The *size* of a graph g is defined by the number of its vertices and is denoted by $|g|$. A *directed path* p from vertex v_0 to vertex v_n is an alternating sequence of vertices and edges $p = v_0, e_1, v_1, \ldots, e_n, v_n$ with $e_i = (v_{i-1}, x_i, v_i)$ for $1 \le i \le n$. p is called an *x-path*, if each edge has the label x. The *length* of a path p is the number of its edges and is denoted by $|p|$.

Vertices u and v of a graph g are called *neighbors*, if they are connected by an edge. The neighbors of a vertex v may be distinguished by the direction of an edge and by the edge and vertex labels. The (B, x) in-neighbors of a vertex v of a given graph g are all those vertices u with label B from which there is an x-edge into v. Thus in-neigh(v, B, x) = $\{u \in V(g) \mid m(u) = B, (u, x, v) \in E(g)\}$. Accordingly define the out-neighbors. The neighbors of v are the in- or out-neighbors for any vertex label B and any edge label x.

Let $\Delta' \subseteq \Delta$ be a set of distinguished edge labels. This selection induces a canonical partition of a graph g into graphs $g_{\Delta'}$ and $g_{\Delta-\Delta'}$ with the same sets ov vertices, such that $g = g_{\Delta'} \cup g_{\Delta-\Delta'}$, where $g_{\Delta'} = (V(g), E(g) \cap \{(u, x, v) \mid x \in \Delta'\}, m(g))$ is the restriction of g to Δ', and accordingly $g_{\Delta-\Delta'} = (V(g), E(g) \cap \{(u, x, v) \mid x \in \Delta-\Delta'\}, m(g))$ is the restriction of g to $\Delta-\Delta'$. Conversely, we define the union of graphs with the same sets of vertices and disjoint sets of edges by taking the union of the edge sets.

Next, we turn to a formalization of graph drawings and define cost measures for the quality of drawings. The targets are the grid graphs, which have a natural representation in the plane. Graph drawings are now identified with grid embeddings.

Definition

Let $g = (V, E, m)$ and $g' = (V', E', m')$ be graphs, called the *guest graph* and the *host graph*, respectively. An *embedding* of g into g' is a mapping $f: g \to g'$, which maps the vertices of g one-to-one into the vertices of g' and maps the edges of g incidence preserving to the paths of g'. Thus, an edge $e = (u, a, v)$ is mapped to a path f(e) from vertex f(u) to vertex f(v). An embedding f is a *grid embedding*, if the host graph g' is a grid graph.

There is a wide spectrum of properties of embeddings and of cost measures for an evaluation of their quality. A combination of properties and cost measures tells by how much the guest graph differs from the host graph. Typical properties are direction, which means that the edges are mapped into directed paths, *vertex and edge disjointness*, which means that the paths of different edges do not cross, or do not have a common edge, cf. [3].

Definition

For a grid embedding $f: g \to g'$ let $A_f(g)$ denote the size of the smallest rectangular grid graph containing f(g). Let $T_f(g)$ denote the maximal edge length, i.e., the length of the longest path f(e) of any edge $e \in E(g)$.

Graph embeddings have been used as a formal description for placement and routing problems in VLSI and for formal descriptions of aesthetically nice drawings of graphs. There are many results in the literature showing that embedding problems are NP-complete, in general. E.g., the NP-completeness of minimal grid embeddings has been improved to trees in [1] and finally to binary trees [8, 3].

PROPOSITION

For binary trees the following problems are NP-complete:

INSTANCE: Given a binary tree g and some $K \geq 0$.

PROBLEM: Is there a vertex disjoint grid embedding $f: g \rightarrow g'$ such that

$$(1)\ A_f(g) \leq K \quad \text{or} \quad (2)\ T_f(g) \leq K.$$

From this result we can conclude that optimal grid embeddings are NP-hard for all "nontrivial" classes of graphs. Here, nontrivial means that the binary trees are (implicitly or explicitly) included in the class. Recall that for many classes of graphs there are efficiently computable grid embeddings, which are optimal up to some factor. A typical example is the H-tree layout for trees [17]. The NP-hardness for the sharp bound is a starting point for our investigations for new tools, which guarantee polynomiality.

2. Graph Grammars

We follow the set theoretic or algorithmic approach to graph grammars and consider graph grammars from the NLC family. This is motivated by the ease of handling such devices and a natural graphic representation. We do not consider graph grammars in their full generality but restrict ourselves to context-free graph grammars with a polynomial time membership problem.

Definition

A *graph grammar* is a system $GG = (N, T \cup \Delta, P, S)$, where N, T and Δ are the alphabets of nonterminal vertex labels, of terminal vertex labels and of terminal edge labels. $S \in N$ is the axiom and P is a finite set of productions. A *production* is of the form $p = (A, R, C)$, where $A \in N$ is the label of a replaced vertex, R is the right-hand side and is a labeled graph, and C is the *connection relation* of p. C consists of tuples (B, x, d, u, y, d'), where $B \in N \cup T$ is a vertex label, $x, y \in \Delta$ are edge labels, $\{d, d'\} \in \{in, out\}$ determine the direction of edges with respect to the replaced vertex and the vertex u with $u \in V(R)$.

There is an elegant graphic representation of graph grammar productions first introduced by Kaul [12]. This representation can be taken as a starting point for layout graph grammars.

Consider a production $p = (A, R, C)$ with $R = (V(R), E(R), m(R))$. Represent each vertex v of V(R) by a rectangle of unit size, which may degenerate to a point for terminal vertices, and attach the identification v and the label m(v). Place these objects such that they do not intersect. Usually, the objects are placed at grid points (with integer coordinates for the center of the rectangles). The edges

$e \in E(R)$ are drawn as straight lines or as grid polylines. Their routing may be specified by a finite set of points, e.g., where they make a bend. The left hand side of the production is drawn as a big rectangle, box(p). It surrounds all of R, and the label A may be attached e.g. at the upper left corner. box(p) has an expansion width(box(p)) in X-dimension and height(box(p)) in Y-dimension. Finally, the connection relation is represented by arrows or polylines from outside box(p) to the vertices $u \in V(R)$ inside box(p). For clarity, these arrrows or polylines should be drawn in a different colour or style. For every tuple (B, x, d) with $(B, x, d, u, y, d') \in C$ for some (u, y, d') there is a point with label B outside box(p) and a distinguished entry point q on the boundary of box(p). Then draw an edge with label x in direction d from B to q and draw an edge with label y in direction d' from q to the distinguished vertex u. Thus the line from B to u is broken at the boundary of box(p), where the edge label and the direction may change.

As our "running example" we use series parallel graphs. There are three productions for the serial and parallel compositions and for a termination. They are self-explaining and are shown in Fig. 1 resp. in Fig. 2. The right hand sides are drawn in bold style, and the boxes and connections in normal style. The first box of Fig. 1 is of size 6×3. The axiom consists of a nonterminal vertex and connected with two terminal vertices with label c and at distance 1 in X-dimension. The edge label is omitted.

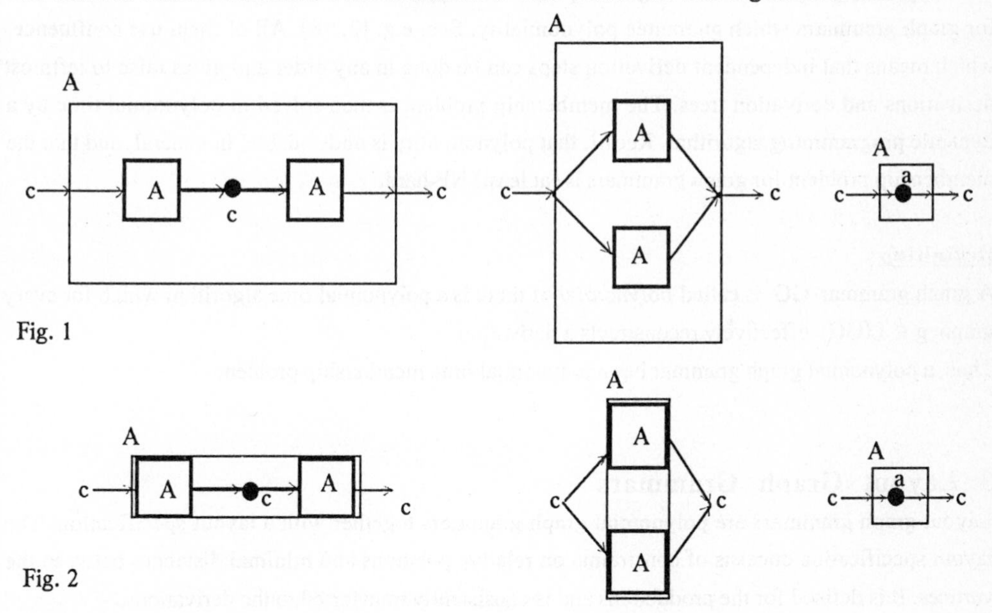

Fig. 1

Fig. 2

Definition

A direct rewriting step $g \Rightarrow_{(v,p)} g'$ is defined as follows: Select a vertex v of g with label A and a production p = (A, R, C). Replace v by (an isomorphic copy of) R. Then establish connections between the neighbors of v and vertices of R as specified by the connection relation C.
Formally, $V(g') = V(g)-\{v\} \cup V(R)$ with $V(g) \cap V(R) = \emptyset$. An edge e = (s, y, t) is in E(g') if and only if $e \in E(g)$ and $s \neq v \neq t$ or $e \in E(R)$ or e is established by the embedding. If $s \in V(g)-\{v\}$ and $t \in V(R)$, then $(s, y, t) \in E(g')$, if $s \in$ in–neigh(v, B, x) and $(B, x, in, t, y, in) \in C$ or $s \in$ out–neigh(v, B, x) and $(B, x, out, t, y, in) \in C$. The case with e = (t, y, s) is similar.

The graphic representation of the productions can be used for an illustration of derivation steps and an animation of graph rewriting. The idea is quite simple. Select the vertex v and move it (its center) to the origin. All other vertices are moved accordingly. Then expand v to the size of box(p). To this effect, every vertex w of V(g)–{v} is shifted by the vector $(c_x(v,w) \cdot \Delta_x(p), c_y(v,w) \cdot \Delta_y(p))$, where $c_x(v,w) = 1$, if w lies right of v, $c_x(v,w) = 0$, if v and w have the same X-coordinate, and $c_x(v,w) = -1$, if w lies to the left of v. $\Delta_x(p)$ is width(box(p))/2. The situation in Y-dimension is similar.

As usual, the *language* generated by a grammar consists of all terminal objects derivable from the axiom, $L(GG) = \{g \mid S \Rightarrow^* g, g = (V, E, m) \text{ and } m(v) \in T \text{ for every } v \in V\}$.

There is a rich theory on graph grammars, that has been developed particularly during the past five years. As a general reference, see this Volume and [4].

The membership problem plays a central role in our further applications of graph grammars. If the membership problem is NP-hard, then graph grammars are of no use for graph layouts, at least from the complexity point of view. Our goal is polynomiality, and there are several sufficient conditions for graph grammars which guarantee polynomiality. See, e.g. [2, 16]. All of them use confluence, which means that independent derivation steps can be done in any order and gives raise to leftmost derivations and derivation trees. The membership problem is then solved in polynomial time by a dynamic programming algorithm. Recall, that polynomiality is undecidable, in general, and that the membership problem for graph grammars is (at least) NP-hard.

Definition

A graph grammar GG is called *polynomial,* if there is a polynomial time algorithm which for every graph $g \in L(GG)$ effectively reconstructs a derivation.
Thus, a polynomial graph grammar has a polynomial time membership problem.

3. Layout Graph Grammars

Layout graph grammars are polynomial graph grammars together with a layout specification. The layout specification consists of constraints on relative positions and minimal distances between the vertices. It is defined for the productions and is consistently transferred to the derivations.

Definition

A *layout graph grammar* LGG consists of a polynomial graph grammar GG together with a layout specification LS. With each production p = (A, R, C), LS associates a finite set of layout constraints, $c_1,\ldots,c_q$. A pair (p, c_i) is called a *layout production.* Each c_i defines a finite set of relations on the vertices of R, the left hand side A and the tuples from C, which describe minimal distances between these objects in X-and Y-dimension. These relations must be consistent such that there exists a realization in terms of a grid embedding of R, which can be extended to an embedding of (A, R, C). The distance constraints are additive, which means that $u.X \geq v.X + k$ and $v.X \geq w.X+m$ implies $u.X \geq w.X + (k+m)$. Here u.X denotes the X-coordinate of the object u in some grid embedding. Moreover, the constraints are complete, i.e., each pair of vertices $u, v \in V(R)$ is related by at least one constraint.

Example

The productions in Fig. 1 are given together with a layout specificaton. E.g., in the first production there is are unit distances between the objects of the right hand side and to the walls of box(p), which is drawn dashed and has 6×3. There is a left to right relation between the vertices of the right hand side as shown by the drawing. The productions in Fig. 2 have the same layout specifications for the right hand sides, but the boxes are tight.

In a *derivation step* with a layout production (p, c), a graph g is rewritten into a graph g' by the application of the ordinary graph grammar production p, and a new grid embedding f'(g') for g' is constructed from the grid embedding f(g) of g and c.

The *language* of a layout graph grammar consists of a set of terminal graphs g together with a grid embedding f(g). Alternatively, LGG can be seen as a transformation, converting the set representation of the graphs from L(GG) into a grid drawing. This is illustrated by the system illustrated in Fig. 3.

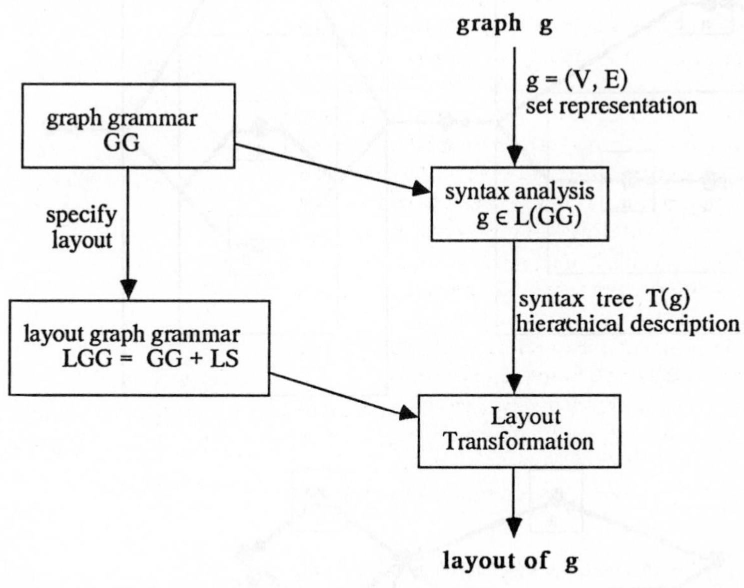

Fig 3.

Next, we introduce two concrete types of layout graph grammars. In both cases, the layout is restricted to a placement of the vertices. There is an intuitive description of layout specifications, which for the first type is directly retrieved from the graphic representation of the productions and for the second type is defined by a separate graph component.

Definition

Let $G = (N, T \cup \Delta, P, S)$ be a polynomial graph grammar. For each production $p = (A, R, C)$ let $d_i(p) = (box(p), layout(R), entry(C))$ be a particular drawing of p, $1 \le i \le q_p$ for some q_p. layout(R) is a grid embedding of R, such that the images of vertices do not overlap, box(p) is a rectangle surrounding all of layout(R) and entry(C) consists of a set of entry points for the embedding relations at the boundary of box(p). The constraints consist of the lower bounds on the distances in X-dimension and in Y-dimension between each pair of vertices of R, between the vertices and the outer walls of the surrounding rectangle box(p), and between the vertices of R and the entry points of C. Then $d(P) = \{d_i(p) \mid p \in P, 1 \le i \le q_p\}$ is the layout specification for GG and $LGG = (N, T \cup \Delta, \{(p, d_i(p)) \mid p \in P, 1 \le i \le q_p\}, S)$ is a layout graph grammar associated with GG.

Example

For an illustration, the series parallel graph in Fig. 4 and Fig 5 are constructed using the layout productions from Fig. 1 and Fig. 2. The graphs are in bold type, and the intermediate boxes from the "derivation" are drawn in normal style.

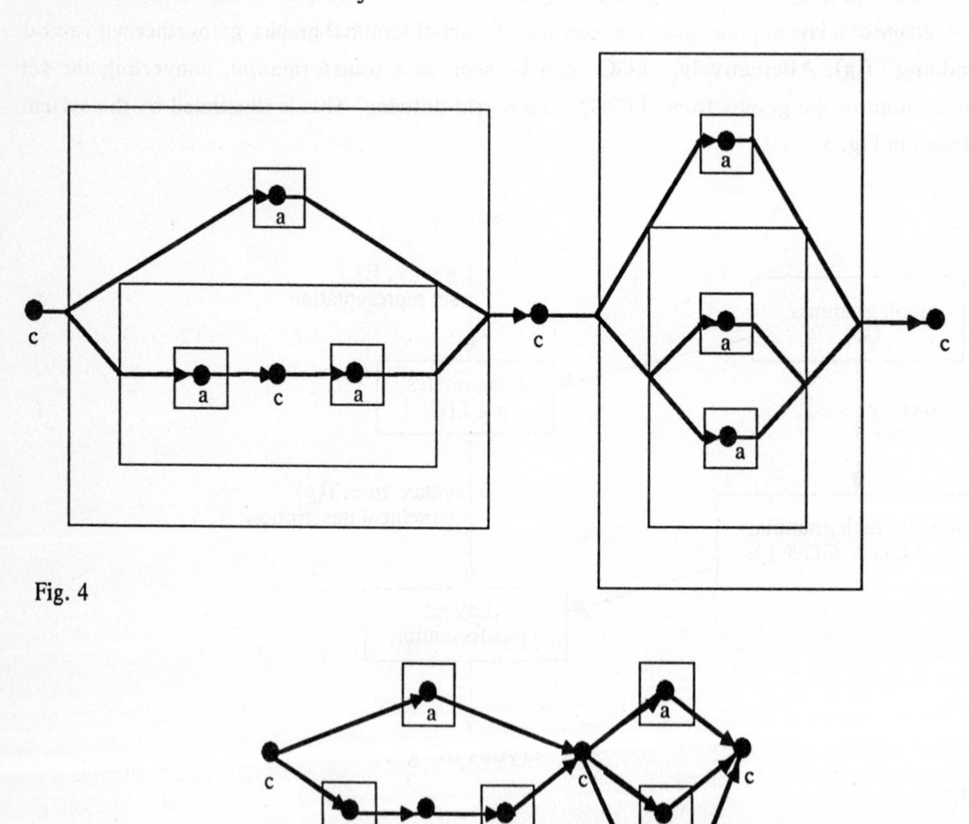

Fig. 4

Fig .5

THEOREM 1

Let GG be a polynomial graph grammar and let LGG be a layout graph grammar associated with GG. There is a polynomial time algorithm which for every graph $g \in L(GG)$ constructs a layout or grid embedding f(g) of g by LGG, such that f(g) has minimal area among all layouts of g by LGG.

Proof (Sketch).

Since GG is polynomial, g can be parsed in polynomial time. For each vertex v occurring in a reconstructed derivation for g, define attributes width(v), height(v), Xpos(v) and Ypos(v). width and height are synthesized attributes and define the minimal size of a box containing the graph generated from v. Xpos and Ypos are inherited attributes. Relative to the next enclosing box they define the X- and Y-coordinates for v. If a vertex v is replaced by a graph R using the production p = (A, R, C) and d(p) = (box(p), layout(R), entry(C)) is a layout production associated with p, then width(v) and height(v) can be computed from the width and height of the vertices of R and the distances specified by layout(R) and box(p). Conversely, given the width and the height of box(p) and the formula for its computation, for every vertex u of R and every entry point, the relative positions u.X and u.Y can be computed.

The parsing algorithm for GG is augmented by these attributes. It computes minimal values for the width and the height of each vertex. Since width and height are bounded by O(n), where n = |g|, there are at most $O(n^2)$ incomparable pairs for every vertex v occurring in the derivation. This means a quadratic increase of the running time of the syntax analysis algorithm, which is supposed to be polynomial.

Example

Consider binary trees and their generation by a layout graph grammar LGG with productions as shown in Fig. 6. LGG draws "inorder binary trees" with each vertex having its own X-coordinate. These trees have maximal width, which, however, is minimal with respect to the constraints imposed by the layout graph grammar.

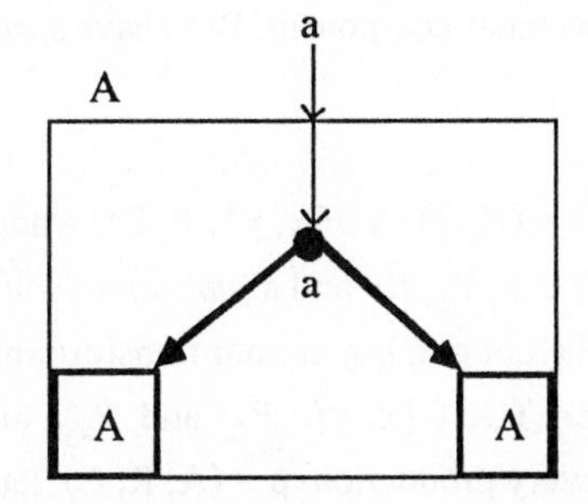

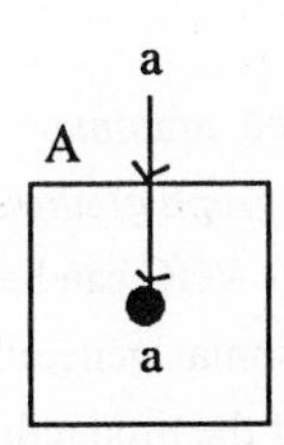

Fig. 6

Every grid embedding of a graph R specifies a "left-of" and a "below" relation between each pair u, v of vertices, where u is left-of v, if $u.X \leq v.X$, and u is below v, if $u.Y \leq v.Y$. At least one of there "≤"-relations must be strict, since distinct vertices are mapped to distinct grid points. This is an overspecification and leaves little space for optimizations. Therefore we relax the system of constraints derived from the graphic representation of the graph grammar productions.

In a uniform way, we define the layout specification by special edge labels {x, y}. An x-edge stands for "left of" and a y-edge for "below". This graph theoretic definition of the layout specification converts layout graph grammars to ordinary graph grammars. Conversely, we can consider particular graphs and graph grammars over edge labels $\Delta \cup \{x,y\}$ with $\Delta \cap \{x,y\}=\emptyset$ and partition them into a base part and a layout part, such that $g = g_\Delta \cup g_{xy}$. g_{xy} can be partitioned according to the X- and Y-coordinates into g_x and g_y. We shall omit the subscript "Δ" and write the subscript xy for $\{x, y\}$ and x for $\{x\}$. This approach has been introduced in [3]. It is recalled here for completeness and is extended to graph layouts which minimize the area or the maximal edge length; the latter needs special properties of the derived graphs.

Definition

A graph g_{xy} with edge label alphabet $\{x,y\}$ is called a *partially ordered graph,* if

(i) g_x and g_y are acyclic

(ii) for every pair of vertices (u, v) with $u \neq v$, there is an x-path from u to v, or a y-path from u to v, or conversely (a path from v to u).

A *placement graph* is a graph $h = g \cup g_{xy}$, where g and g_{xy} are graphs with the same sets of vertices and disjoint sets of edges and edge labels, and g_{xy} is a partially ordered graph. g_{xy} is called a *placement component associated* with g and is a layout specification of g.

Consider some basic properties of partially ordered graphs. An x-edge (u, x, v) is interpreted as "left of" and a y-edge is interpreted as "below". Then (i) means that the placement component is consistent with this interpretation. Hence, for any grid embedding f, $f(u).X < f(v).X$ must hold, and similarly, $f(u).Y < f(v).Y$ for a y-edge (u, y, v). Such grid embeddings are called *consistent with a placement component.* (ii) is a completeness condition. It says that for every pair of vertices there is a directed x-path from u to v or a directed y-path from u to v, or vice versa. Hence, for every pair of vertices {u,v} either "u lies left of v" or "u lies above v", or vice versa.

We are now ready to define placement graph grammars. According to the edge labels $\Delta \cup \{x, y\}$ a placement grammar splits into a base- and a placement-component. Both have special properties.

Definition

A *placement graph grammar* is a tuple $PGG = (N, T \cup \Delta \cup \{x, y\}, P, S)$, where $x, y \notin \Delta$. PGG consists of a *base graph grammar* $GG = (N, T \cup \Delta, P_\Delta, S)$ and a *placement component* $G_{xy} = (N, T \cup \{x, y\}, P_{xy}, S)$. PGG can be regarded as a layout graph grammar transforming graphs over $T \cup \Delta$ or as a graph grammar generating graphs over $T \cup \Delta \cup \{x, y\}$. P_Δ and P_{xy} are the base and the placement parts of the productions of P. For every production $p = (A, R, C)$ let $p_\Delta = (A, R_\Delta, C_\Delta)$ and $p_{xy} = (A, R_{xy}, C_{xy})$ be such that $R = R_\Delta \cup R_{xy}$ is a placement graph and $C = C_\Delta \cup C_{xy}$ is split according to the edge labels. C_Δ contains all tuples (B, a, d, u, a', d) with $a, a' \in \Delta$ and $d \in \{in, out\}$ and C_{xy} contains all such tuples with $a, a' \in \{x, y\}$. Moreover, the connection relation C_{xy} preserves the edge label and the direction, and it transfers an incoming x-edge to each x-source of the right hand side R_{xy} and an outgoing x-edge to each x-sink of R_{xy}, and similarly for the y-edges. A vertex of R_{xy} is a x-source (x-sink), if it has no incoming (outgoing) x-edge. y-sinks and y-sources are defined accordingly. Thus C_{xy} is completely specified by the graph R_{xy}.

The placement-component GG_{xy} generates partially ordered graphs. This follows from the fact that each right hand side graph R_{xy} is a partially ordered graph and the connection relation C_{xy} preserves the acyclicity of the induced x- and y- subgraphs and the completeness condition (ii).

THEOREM 2

Let PGG be a placement graph grammar with base graph grammar GG and placement component G_{xy}, and let $c > 0$.
There exists a polynomial time algorithm A (resp. A') which for every connected graph $g \in L(GG)$ of degree c constructs a placement component g_{xy} (g'_{xy}) associated with g and a grid embedding f (f') such that f (f') is consistent with g_{xy} (g'_{xy}) and for all placement components g''_{xy} with $g \cup g''_{xy} \in L(PGG)$ and all grid embeddings consistent with g''_{xy}

(1) f(g) has minimal area and

(2) f'(g) minimizes the maximal edge length.

Proof (sketch).

Let PPG = $(N, T \cup \Delta \cup \{x, y\}, P, S)$ be composed of a base graph grammar GG and a placement component G_{xy} as described above. Let $g = (V(g), E(g), m(g)) \in L(GG)$ with $|g| = n$.
First, consider only GG. Suppose that $g \in L(GG)$ is parsed e.g., by constructing a table with parsing elements of the form $PE = (A, V, p, \pi_1, \ldots, \pi_r)$, where $A \in N \cup T$ is a vertex label, $V \subseteq V(g)$ is a subset of the vertices of g such that $A \Rightarrow^* d$ is a derivation in GG with d being the subgraph of g induced by V, $p = (A, R, C)$ is a production applied to the current vertex, and $\pi_1, \ldots, \pi_r$ are parsing elements corresponding to the vertices of R with with $r = |R|$. These are used to trace the derivation, recursively.
We suppose that there are only polynomially many parsing elements, which can be constructed in polynomial time. See, e.g. [14, 15] for concrete cases.
For the construction of a grid embedding of $g \in L(GG)$, which is consistent with an associated placement graph g_{xy} and has minimal area we augment the parsing elements by a pair of integers (w, h) denoting the width and the height of a rectangle, such that a grid embedding of the subgraph d induced by V fits into this rectangle and is consistent with the placement component. Clearly, $1 \leq w, h \leq n$. For each PE choose all minimal pairs (w, h), where $(w, h) < (w', h')$ iff $w \leq w'$, $h \leq h'$ and $(w, h) \neq (w', h')$. Hence, there are at most n^2 such pairs for every PE. Given PEs $\pi_1, \ldots, \pi_r$ with disjoint sets of vertices and augmented with pairs (w_i, h_i) for the width and the height, and a production (p_Δ, p_{xy}) with $p_\Delta = (A, R_\Delta, C_\Delta)$, then the new augmented parsing element $(A, V, (p, p_{xy}), \pi_1, \ldots, \pi_r, w, h)$ can be constructed in O(n) steps with (w, h) minimized. Hence, the algorithm is polynomial in the number of augmented parsing elements, and thus polynomial, in total.

Concerning the maximal edge length we proceed similarly. Fix a bound b for the maximal edge length and determine the best bound by binary search. The parsing elements $PE = (A, V, p, \pi_1, \ldots, \pi_r)$ are now augmented by integers w and h, and by a set of triples $\{(v_i, x_i, y_i) \mid 1 \leq i \leq K\}$ for some constant K. K bounds the number of connection vertices and depends only on the degree of g and on the graph grammar GG, see [15]. A connection vertex connects an induced subgraph with the rest graph, i.e. there is an edge (v, a, u) in g with $v \in V$ and $u \in V(g)-V$. For every induced subgraph, the number of connection vertices is bounded and independent of the size of the induced subgraph or g. The integers w and h denote the width and the height of a tight rectangle such that the induced subgraph fits into the rectangle with edge length at most b. Now $b \le w$, $h \le K \cdot b \cdot n$. Each pair (x_i, y_i) represents the coordinates of a connection vertex v_i of V in the rectangle of size $w \times h$.

Since K is a constant, the total number of augmented parsing elements still remains polynomial, the construction of new parsing elements can be done in polynomial time.
The correctness and complexity of this procedure can be proved along the lines of the proofs of the polynomiality of the membership problem, cf. [2, 3, 14, 15].
For $g \in L(GG)$ and an augmented parsing element $(S, V(g), p, \pi_1,..., \pi_r, w, h, \emptyset)$, a grid embedding $f: g \rightarrow g'$ can effectively be constructed in linear time by tracing back the derivation via p and $\pi_1,..., \pi_r$, such that $A_f(g)$ resp. $T_f(g)$ are minimized.

Acknowledgement

I wish to thank the anonymous referees for their careful reading and helpful comments.

References

[1] S.N. Bhatt, S. S. Cosmadakis, "The complexity of minimizing wire lenghts in VLSI layouts", Inform. Proc. Letters 25 (1987), 263-267
[2] F.J. Brandenburg, "On polynomial time graph grammars" Lecture Notes in Computer Science 294 (1988), 227-236
[3] F.J. Brandenburg, "On the complexity of optimal drawings of graphs" Lecture Notes in Computer Science 411 (1990), 166-180
[4] H. Ehrig, M. Nagl, G. Rozenberg, A. Rosenfeld (Eds.), "Graph Grammars and Their Application to Computer Science", Lecture Notes in Computer Science 291 (1987)
[5] S. Even, "Graph Algorithms", Computer Science Press (1979)
[6] M.J. Fischer, M.S. Paterson, "Optimal tree layout", Proc. 12 ACM STOC (1980), 177-189
[7] M.R. Garey, D.S. Johnson, "Computers and Intractability: a Guide to the Theory of NP-Completeness", Freeman and Company, San Francisco (1979)
[8] A. Gregori, "Unit-Length embedding of binary trees on a square grid" Inform. Proc. Letters 31 (1989), 167-173
[9] D. Janssens, G. Rozenberg, "On the structure of node label controlled graph languages", Inform. Sci. 20 (1980), 191-216
[10] D. S. Johnson, "The NP-completeness column: An ongoing guide" J. Algorithms 3 (1982), 89-99
[11] M. Kaul, "Practical Applications of Precedence Graph Grammars" Lecture Notes in Computer Science 291 (1987), 326-342
[12] M. Kaul, "Syntaxanalyse von Graphen bei Präzedenz-Graph-Grammatiken" Ph. D. Thesis and MIP-Bericht Universität Passau (1986).
[13] Z. Miller, J.B. Orlin, "NP-completeness for minimizing maximum edge length in grid embeddings", J. Algorithms 6 (1985), 10-16
[14] G. Rozenberg, E. Welzl, "Boundary NLC graph grammars - basic definitions, nornal forms and complexity", Inform. Contr. 69 (1986), 136-137
[15] R. Schuster, "Graph Grammatiken und Graph Einbettungen: Algorithmen und Komplexität", Ph.D Thesis and MIP Bericht, Universität Passau (1987)
[16] A.O. Slisenko, "Context-free grammars as a tool for describing polynomial-time subclasses of hard problems", Inform. Proc. Letters 14 (1982), 52-56
[17] J. D. Ullman, "Computational Aspects of VLSI", Computer Science Press (1984)

CYCLE CHAIN CODE PICTURE LANGUAGES

Franz J. Brandenburg +
Fakultät für Mathematik und Informatik
Universität Passau, Innstr. 33
D 8390 Passau

Michal P. Chytil +
Charles University Prague, Malostranské nam. 25
CS 118 00 Prague

Abstract

A picture word is a string over the alphabet {u, d, l, r}. These symbols mean drawing of a unit line in direction up, down, left, and right, respectively. A picture word describes a walk in the plane; its trace is the picture it describes. A set of picture words describes a (chain code) picture language.

A cycle means a closed curve in the discrete Cartesian plane. It is elementary, if the curve is simple and has no crossings. Cycles are among the most important features for chain code pictures. They are used as fundamental objects to build more complex objects, and occur, e.g., in the pictures of the digits "0", "6", "8" and "9", in icons for snowflakes, houses or trees and in complex kolam patterns. In general, cycles are non-context-free constructs and cannot be captured by context-free picture grammars. Therefore, we extend the concept of context-free picture grammars attaching cycles by the requirement that certain subpictures must be cycles or elementary cycles. We investigate basic properties of context-free cycle and elementary cycle grammars with emphasis on the complexity of the recognition problem. In particular, it is shown that the description complexity is polynomial for cycle languages and for unambiguous elementary cycle languages, and is NP-complete for ambiguous elementary cycle languages.

Keywords

chain code pictures, picture words, context-free picture languages, cycles, cycle grammars, recognition and membership problems, syntax directed transformations

---

+ This work was supported by the cooperation of the University of Passau and the Charles University of Prague and was done during mutual visits in Prague and Passau.

0. Introduction

String encodings of pictures have been used successfully for automatic pattern recogniton problems [6]. Important contributions to this field are Freeman´s "chain codes" on the descriptive side [5] and syntactic methods as formal and analytic tools [6, 13]. These approaches have been combined by Maurer et al. [12] initiating an intensive study of picture languages.
Originally, chain codes have been defined over an eight-letter alphabet to denote unit movements on grid points in direction north, south, east, west, northeast, southeast, southwest and northwest. More recently, these movements have been simplified to the four-letter alphabet {u, d, l, r} denoting movements up, down, left, and right, and have been augmented by the letters $\{\uparrow, \downarrow\}$, denoting to lift and sink a plotter pen [9]. A string over such sets of symbols can be seen as a straight-line program for a plotter pen, a cursor or a laser or an electron ray. Each letter means an elementary command describing a simple action. This approach of chain code descriptions of pictures is suitable especially to line drawings, e.g., maps, charts, characters, electrocardiograms, or arbitrary contour images.

Closed curves are one of the most important features of line drawings. They are often used as basic objects to build more complex objects, e.g. kolam patterns [16]. In graphics this has lead to special primitives like polygon, ellipse, or rectangle [4]. However, for picture descriptions, closed curves are non-context-free and cannot be captured by context-free picture grammars [12]. This holds, e.g., for special cycles such as the rectangles or the squares, which are described by the sets of picture words $\{u^n r^m d^n l^m \mid n,m \geq 1\}$ and $\{u^n r^n d^n l^n \mid n \geq 1\}$, respectively. Furthermore, cycles are important for the picture languages themselves and have a strong influence on the properties of the picture description languages. Cycles appear in technical constructions of several proofs, they imply the gap between the size of a picture and the size of its description [12, 17], and there are several approaches to eliminate special cycles from the picture descriptions [2, 7, 8, 12].

In this paper we introduce cycles and elementary cycles and integrate them as basic features into context-free picture grammars. Formally, we distinguish cycle, elementary cycle and non-cycle nonterminals. If a derivation starts from a cycle nonterminal, then is is legal and contributes to the generated language, only if it generates a picture word representing a cycle. Similarly, starting from an elementary cycle nonterminal means selecting only those derivations generating picture words which describe elementary cycles, when all elementary cycle nonterminals occurring in the derivation are ignored. This mechanism leads to a controlled nesting of elementary cycles. Notice that every cycle nonterminal filters appropriate picture words from its language. In this way, the squeezing is integrated continuously into the derivation processes and is not static at the end.

Clearly, the extension to cycle and elementary cycle grammars increases the power of regular, linear and context-free grammars in generating picture languages. For example, the set of rectangles is a regular cycle language and the set of squares is a linear cycle language, and both are non-context-free picture languages. Cycles and elementary cycles are also used as basic features for transformations of pictures by means of syntax directed translations on picture words. Examples show that this extension is natural.

1. Preliminaries

We shall assume that the reader is familiar with the basic concepts from formal language theory and from graph theory. For basic notions on picture words we refer to [12].

Let Σ be a finite alphabet and let Σ^* denote the set of words over Σ. A word x is a *subword* of a word w if w = uxv for words u, v. x is a *subsequence*, if $x = x_1...x_r$ and $w = y_0x_1y_1...x_ry_r$ for words x_i, y_i. The *length* of a word w is denoted by |w|.

Let T = {u, d, l, r} denote the picture description alphabet corresponding to the directions up, down, left, and right, respectively. Let $u = d^{-1}$ and $l = r^{-1}$. A word $w \in \{u, d, l, r\}^*$ is called a *picture word.*

We use *context-free grammars* as tuples G = (N, T, P, S), where N and T are the sets of nonterminals and terminals, P is the set of productions, and $S \in N$ is the axiom. G is a linear (regular) grammar, if the productions are of the corresponding type. The *language* of words generated by G is denoted by L(G). If T = {u, d, l, r}, then G is called a *context-free picture grammar* and L(G) is a *context-free picture description language.*

Next, we turn to notions from the ´picture part´. The universe is the infinite grid consisting of the two-fold Cartesian product of the integers as the universal point set and of horizontal and vertical lines of unit length between any two neighbored grid points (x, y) and (x', y') with $|x-x'| + |y-y'| = 1$. The infinite grid is also regarded as an infinite graph.
A (chain code) *picture* p is a finite, connected subgraph of the infinite grid.

A picture word $w \in \{u, d, l, r\}^*$ is regarded as a sequence of commands for a cursor and defines a walk on the grid. For a grid point (x, y) define the *up-successor* by pos(u, (x,y)) = (x+1, y). Its trace is the line or the (directed) edge from (x,y) to (x+1, y). Accordingly, define the down, left, and right-successors, and their traces. Cursor positions and traces are extended to picture words by pos(λ, (x, y)) = (x, y) and pos(vw, (x, y)) = pos(w, pos(v, (x, y))). Thus, picture words are processed from left to right. They define a walk on the grid, which is the picture described by w.

For every picture word w, initialize the cursor position at the origin (0, 0) and let sh(w) = pos(w, (0,0)) denote the *shift* of w and of the picture described by w. This normalization defines standard representatives in the sense of [12]. With each picture word $w = a_1...a_n$, $a_i \in \{u, d, l, r\}$ for $1 \le i \le n$, associate an undirected grid graph g(w) = (V, E) with vertices $V = \{sh(a_1..a_i) \mid 0 \le i \le n\}$ and edges $E = \{\{sh(a_1...a_{i-1}), sh(a_1...a_i)\} \mid 1 \le i \le n\}$. Define the *standard drawn picture* of a picture word w by q(w) = (g(w), (0,0), sh(w)). Define the *drawn picture* of w by the equivalence class of pictures under the set of translations $t_{m,n}$. A translation $t_{m,n}$ with integers m,n means a shift of vertices and edges by the vector (m, n). Similarly, the *standard basic picture* is defined by the graph g(w) and the *basic picture* p(w) of w is the class of graphs which are translation equivalent to g(w).

To simplify the terminology we shall address a translation equivalence class by its standard representative and for a picture word w we refer to q(w) and p(w) as the *drawn picture* and the *basic picture* of w. p(w) is the *picture described* by w. It is the trace of a cursor started in the origin and driven by w, and q(w) is p(w) together with the final position of the cursor.
Forthcoming we shall concentrate on basic pictures and sets of basic pictures. Similar notions and results can be obtained for drawn pictures.

A set of (basic) pictures Π is called a *picture language* and a set L of picture words is called a *picture description language*, describing the picture language $p(L) = \{p(w) \mid w \in L\}$. A set of pictures Π is a *context-free picture language*, if $\Pi = p(L(G))$ for some context-free grammar G.

A picture word w describes a *cycle* or a *closed curve*, if sh(w) = (0,0), i.e. the cursor returns to the start point. In [12] the term normal is used. A picture word w describes an *elementary cycle*, if it describes a cycle and none of its proper subwords describes a cycle. Then p(w) is a simple closed curve. Note that cycles may be degenerated curves. E.g., the picture word ud describes a cycle although its picture looks like a line. Also, uurdld and uldrrurdll are picture words desrcibing a cycle.

Let $C = \{w \in \{u, d, l, r\}^* \mid w$ describes a cycle$\}$ and $E = \{w \in \{u, d, l, r\}^* \mid w$ describes an elementary cycle$\}$ denote the set of cycle and elementary cycle picture words. If $\#_a(w)$ denotes the number of occurences of the symbol a in a word w, then $C = \{w \in \{u, d, l, r\}^* \mid \#_u(w) = \#_d(w)$ and $\#_l(w) = \#_r(w)\}$ and $E = \{w \in C \mid w = uxv$ and $uv \neq \lambda$ implies $x \notin C\}$.

<u>**Lemma 1**</u>

Let $w \in \{u, d, l, r\}^*$ be a picture word.

(1) $w \in C$ is decidable in real time by a 2-counter machine.

(2) $w \in E$ is decidable in real time by a Turing machine with a 2-dimensional work tape and is decidable in real time on a unit cost RAM.

Proof.

(1) is the origin crossing problem studied in [3]. For (2) the real time solution on a 2-dimensional work tape is obvious and the real time solution on a RAM follows from [15].

<u>**Lemma 2**</u>

The sets of cycles p(C) and elementatry cycles p(E) are non-context-free picture languages.

Proof.

Consider $R = C \cap u^+r^+d^+l^+ = E \cap u^+r^+d^+l^+$.

$R = \{u^ir^jd^il^j \mid i, j \geq 1\}$ describes the set of rectangles and this language is known as non-context-free.

2. Context-free cycle grammars

In this section we introduce cycle picture grammars and establish some basic properties. The cycles are related to distinguished cycle nonterminals, whose cycle languages consist of picture words describing cycles. Thus, we squeeze cycles from the distinguished cycle nonterminals by filtering their languages of picture words for cycle descriptions. Simple nonterminals are left untouched. In the next section we introduce elementary cycle grammars in a similar way.

Definition

Let $N = N_S \cup N_C$ be a set of nonterminals with $N_S \cap N_C = \emptyset$. N_S consists of the *simple* nonterminals and N_C of the *cycle* nonterminals. A tuple $G = (N_S, N_C, T, P, S)$ is called a *cycle grammar*, if $A(G) = (N_S \cup N_C, T, P, S)$ is a picture grammar. A(G) is the picture grammar associated with G.

If A is a cycle nonterminal, then a derivation $A \Rightarrow^* y$ with $y \in T^*$ is *cycle respecting* for A, if y describes a cycle. A derivation $x \Rightarrow^* y$ is cycle respecting, if it is cycle respecting for each cycle nonterminal occurring in the derivation. The *cycle description language* of a cycle grammar G is $L_C(G) = \{w \in T^* |$ there is a cycle respecting derivation $S \Rightarrow^* w\}$ and the set of pictures $p(L_C(G))$ is the *cycle language* of G. A set of pictures Π is called a *context-free cycle language*, if $\Pi = p(L_C(G))$ for some context-free grammar G. Linear and regular cycle languages are defined accordingly.

A cycle grammar G partitions its nonterminals into two classes, the simple nonterminals and the cycle nonterminals. The special restriction is that derivations starting from a cycle nonterminal yield a cycle; otherwise they are illegal and are useless for the cycle description language. The selection of special derivations implies $L_C(G) \subseteq L(A(G))$ for every cycle grammar G and its associated picture grammar A(G). If the axiom is a cycle nonterminal, then the cycle language consists only of cycles, i.e. $w \in L_C(G)$ implies $sh(w) = (0, 0)$ and w is normal in the terminology of [12].

Example 1

Consider a regular picture grammar G as a picture or as a cycle grammar with nonterminals N = {A, B, C, D, S}, productions $P = \{S \rightarrow A, A \rightarrow uA, A \rightarrow uB, B \rightarrow rB, B \rightarrow rC, C \rightarrow dC, C \rightarrow dD, D \rightarrow lD, D \rightarrow l\}$ and axiom S.

If $N = N_S$ and $N_C = \emptyset$ then $L(G) = L_C(G) = u^+r^+d^+l^+$ describes all pictures made by four strokes, which may or may not intersect.

If S is a cycle nonterminal, $N_C = \{S\}$ and $N_S = \{A, B, C, D\}$, then $L_C(G)$ describes all rectangles and if B, C, or D is taken as a cycle nonterminal, then $L_C(G) = \emptyset$.

Example 2

Consider the cycle grammar $G = (N_S, N_C, T, P, S)$ with $N_C = \{S, R\}$, $N_S = \{A, B, C, D, A', B', C', D'\}$, and $P = \{S \rightarrow A', A' \rightarrow uA', A' \rightarrow uB', B' \rightarrow rB', B' \rightarrow rC', C' \rightarrow dC', C' \rightarrow dD', D' \rightarrow lD', D' \rightarrow l, A' \rightarrow RuA', B' \rightarrow RrB', C' \rightarrow RdC', D' \rightarrow RlD', R \rightarrow uA, A \rightarrow uA, A \rightarrow uB, B \rightarrow rB, B \rightarrow rC, C \rightarrow dC, C \rightarrow dD, D \rightarrow lD, D \rightarrow l\}$.

Then the pictures described by $L_C(G)$ consist of two levels of rectangles. The lower level rectangle is generated from S using the primed nonterminals. The rectangles on the second level are generated whenever the cycle nonterminal R occurs. The rectangles themselves are generated using the unprimed nonterminals.

Example 3

Let $G = (\{A, B, C\}, \{S\}, T, P, S)$ with $P = \{S \rightarrow A,\ A \rightarrow uA,\ A \rightarrow B,\ B \rightarrow rB,\ B \rightarrow C,\ C \rightarrow dCl,\ C \rightarrow \lambda\}$ be a linear cycle grammar G. Then $L_C(G)$ describes the set of all squares.

It is evident that there is a proper inclusion between the regular, the linear and the context-free cycle languages. This follows from standard examples and cycle-free picture description languages, e.g. $L = \{l^i d l u^i l \mid i \geq 1\}$ and $L = \{r d^i r u^i r r d^j r u^j r \mid i,j \geq 1\}$, cf. [12]. It is easy to see that the results also holds for the restriction to cycles, i.e., if the axiom is a cycle nonterminal.

Theorem 1

The class of regular cycle languages is properly contained in the class of linear cycle languages, which is properly contained in the class of context-free cycle languages.

The class of regular (linear, context-free) picture languages is properly contained in the class of regular (linear, context-free) cycle languages.

The classes of regular cycle languages and of context-free picture languages are incomparable.

By standard techniques from formal language theory it can be shown that for every cycle grammar G there is a cycle grammar G' in Chomsky normal form with $L(G) = L(G')$ and $L_C(G) = L_C(G')$.

Theorem 2

The recognition problem for context-free cycle grammars is solvable in polynomial time, i.e., given a picture word $w \in \{u, d, l, r\}^*$ and a cycle grammar G the membership problem $w \in L_C(G)$ is solvable in time $O(n^3)$.

Proof.

The recognition algorithm is an extension of the Cocke-Younger-Kasami (CYK) algorithm for the membership problem of context-free string grammars, see [10]. For a string $w = a_1 \ldots a_n$ construct a table of sets of nonterminals $V_{i,j}$ such that $A \in V_{i,j}$ iff $A \Rightarrow^* a_i \ldots a_{i+j-1}$. In the case of cycle grammars attach the shift $sh(a_i \ldots a_{i+j-1})$ to every $V_{i,j}$ and add a cycle nonterminal A to $V_{i,j}$ only if $sh(a_i \ldots a_{i+j-1}) = 0$. Then $S \in V_{1,n}$ iff $S \Rightarrow^* w$ iff $w \in L(G)$.

The recognition problem is directed towards the description of pictures and uses picture words as problem instances. The membership problem is more abstract. Its problem instance is a picture p and a cycle grammar, and the question is whether or not there is an appropriate description for p. The step from the description to the object itself increases the complexity from tractable to intractable.

Theorem 3

The membership problem for regular and context-free cycle grammars is NP-complete, i.e., given a picture p and a cycle grammar G, is there a cycle picture word $w \in L_C(G)$ such that $p = p(w)$.

Proof.

The NP-hardness of the membership problem for regular picture grammars has been shown by Sudborough and Welzl [17]. The membership in NP follows along the lines of Kim and Sudborough [11] adding tests for cycles.

The usefulness of cycles is demonstrated by simple transformations on picture words, which are defined in terms of syntax-directed translations, cf. [1].

Example 4

Let S, A, C, D, E, F be simple nonterminals and X_C, and Y_C simple nonterminals. Define transformation rules:

S	→	rS, rS	S	→	AS, AS	S	→	r, r
A	→	uX_Cd, uX_Cd	A	→	uY_Cd, uY_Cd			
C	→	rCu, ruCul	C	→	ru, rul			
D	→	lDd, dlDrd	D	→	ld, d			
E	→	ruEul, rEu	E	→	rul, ru			
F	→	dlFrd, lFd	F	→	d, ld			
X_C	→	CD, CD	Y_C	→	EF, EF			

This syntax-directed translation scheme performs transformations of pictures as indicated in Fig. 1. It operates only on pictures consisting of three "flags", where the left and right flags are rectangles and the middle is a diamond. It translates a rectanglar flag into a diamond, and conversely and respects the size of the flags.

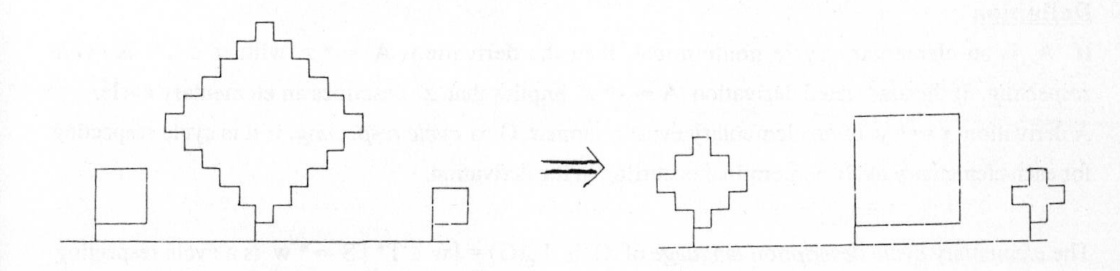

Fig. 1

3. Elementary Cycle Grammars

In this section we investigate elementary cycles and introduce elementary cycle grammars and their languages. An elementary cycle describes a simple closed curve. Elementary cycles can be nested, when they are integrated into context-free concepts. Technically, we proceed as for cycle grammars, and distinguish simple and elementary cycle nonterminals and squeeze the derivation processes with elementary cycle nonterminals involved.

Definition

Let $N = N_s \cup N_e$ be a set of nonterminals with $N_s \cap N_e = \emptyset$. N_s contains the simple nonterminals and N_e the *elementary* nonterminals. A tuple $G = (N_s, N_e, T, P, S)$ is called an *elementary cycle grammar*, if $A(G) = (N_s \cup N_e, T, P, S)$ is a picture grammar. $A(G)$ is the picture grammar associated with G.

Definition

Let $G = (N_s, N_e, T, P, S)$ be an elementary cycle grammar.

Define a homomorphism h over $N_s \cup N_e \cup T$ erasing the elementary cycle nonterminals by $h(b) = b$ for $b \in N_s \cup T$ and $h(b) = \lambda$ for $b \in N_e$. The homomorphism h induces the subgrammar G_s and for every elementary cycle nonterminal A, the subgrammar G_A. $G_s = (N_s, T, h(P), S)$ with $h(P) = \{A \rightarrow h(\alpha) \mid A \rightarrow \alpha \in P, A \in N_s\}$ and $G_A = (N_s \cup \{A\}, T, h(P) \cup \{A \rightarrow h(\alpha) \mid A \rightarrow \alpha \in P\}, A)$. Thus G_s has only simple nonterminals and is the associated reduction of G. In G_A the cycle nonterminal A is added as axiom, and this A appears only on the left hand side of the productions.

The reductions of G to G_s and G_A induce a reduction on the derivations, such that all elementary cycle nonterminals are erased, except in the very first step.

If $A \Rightarrow^* z$ is a derivation in G starting from the nonterminal A, then $A \Rightarrow_A^* z'$ denotes the *associated derivation* in G_A. $A \Rightarrow_A^* z'$ results from $A \Rightarrow^* z$ by using the same productions or an empty step, if the corresponding left hand side does not exist any more. For an illustration, consider the derivation tree of $A \Rightarrow^* z$. Then $A \Rightarrow_A^* z'$ is obtained by erasing all proper subtrees with an elementary cycle nonterminal as a root. Observe, that z' is a subsequence of z and is obtained by erasing nested elementary cycles.

Definition

If A is an elementary cycle nonterminal, then the derivation $A \Rightarrow^* z$ with $z \in T^*$ is *cycle respecting*, if the associated derivation $A \Rightarrow_A^* z'$ implies that z' describes an elementary cycle.

A derivation $x \Rightarrow^* y$ of an elementary cycle grammar G is *cycle respecting*, if it is cycle respecting for each elementary cycle nonterminal occurring in the derivation.

The *elementary cycle description language* of G is $L_e(G) = \{w \in T^* \mid S \Rightarrow^* w$ is a cycle respecting derivation$\}$ and the set of pictures $p(L_e(G))$ is the elementary cycle language of G. A set of pictures Π is called a *context-free elementary cycle language*, if $\Pi = p(L_c(G))$ for a context-free grammar G. Linear and regular elementary cycle languages are defined accordingly.

Example 5

Consider example 1 and let $N_e = \{S\}$. Then $L_e(G)$ describes the set of all rectangles.

Using cycle-free languages and the sets of rectangles and squares as counter-examples we obtain strict inclusions among the basic classes of elementary cycle languages.

Theorem 4

The class of regular elementary cycle languages is properly contained in the class of linear elementary cycle languages, which is properly contained in the class of context-free elementary cycle languages. The class of regular (linear, context-free) picture languages is properly contained in the class of regular (linear, context-free) elementary cycle languages.
The classes of regular elementary cycle languages and of context-free picture languages are incomparable.

Lemma 3

For every elementary cycle grammar G there is an elementary cycle grammar G' with $L(G) = L(G')$ and $L_c(G) = L_c(G')$ (modulo the empty word) such that the productions are of the form $A \rightarrow BC$, $A \rightarrow B$, or $A \rightarrow a$ with elementary or simple nonterminals A, B, C and $a \in T$.

Proof.

The elimination of erasing productions can be done by standard techniques.Then introduce simple nonterminals as copies of the terminals so that $A \rightarrow \alpha$ implies $\alpha \in T$ or $\alpha \in (N_s \cup N_e)^+$.
For $|\alpha| \geq 2$, decompose it as usual, if $A \in N_s$. If $A \in N_e$ and $\alpha = \alpha_o\beta_1 \ldots \beta_k \alpha_k$ with $\alpha_i \in N_s^*$ and $\beta_i \in N_e$, then introduce new simple nonterminals $A_1,\ldots,A_k$ and decompose α by $A \Rightarrow^* A_o \ldots A_k$, $A_o \Rightarrow^* \alpha_o$, and $A_i \Rightarrow^* \beta_i \alpha_i$ for $i = 1 \ldots k$.

Lemma 4

Let $G = (N_s, N_e, T, P, S)$ be an elementary cycle grammar with $S \in N_e$.
If $w \in L_e(G)$ and $w = xyz$ has a cycle y, then there are derivations $S \Rightarrow^* xAz$ and $A \Rightarrow^* z$ with $A \in N_e$.

Proof.

Assume the contrary. For a cycle respecting derivation $S \Rightarrow^* w$ consider the derivation tree and let A be the "lowest" cycle nonterminal such that $A \Rightarrow^* y'yy''$ and $S \Rightarrow^* x'Az'$ with $w = xy'yy''z$. Lowest means the distance from the leaves. Then $A \Rightarrow^* y'yy''$ is not cycle respecting, if $y'y'' \neq \lambda$.

Definition

An elementary cycle grammar G is *unambiguous*, if so is the associated picture grammar $A(G)$. Thus, every word w generated by $A(G)$ has a unique leftmost derivation.

Theorem 5

The recognition problem for unambiguous elementary cycle grammars is solvable in polynomial time.

Proof.

As for cycle grammars we extend the CYK algorithm. Let the grammar be in binary normal form. For a string $w = a_1 \ldots a_n$ construct a table with sets $V_{i,j}$ with entries (A, v) where $A \in N_s \cup N_e$

and v is a subsequence of $a_i...a_{i+j-1}$ such that $A \Rightarrow^* a_i... a_{i+j-1}$ is a cycle respecting derivation and $A \Rightarrow_A^* v$ is the associated derivation in G_A, or v = \$. \$ marks the fact that the associated derivation contains a cycle, which would be illegal for an elementary cycle nonterminal A.
For i = 1,...,n and j = 1 initialize $V_{i,1} = \{(A, a_i) \mid A \in N_s, A \in a_i, A \rightarrow a_i \in P\}$.
For j > 1, suppose there exists a k < j such that $A \rightarrow BC \in P$, $(B, u) \in V_{i,k}$ and $(C, v) \in V_{i+k, j-k}$.
If $A \in N_s$, $u \neq \$ \neq v$ and uv does not contain a substring describing a cycle, then add (A, uv) to $V_{i,j}$.
If $A \in N_s$ and u = \$ or v = \$ or if uv contains a cycle, then add (A, \$) to $V_{i,j}$. If $A \in N_e$, $u \neq \$ \neq v$, $A \in N_e$, sh(uv) = (0, 0) and uv describes an elementary cycle, then add (A, λ) to $V_{i, j}$.
For productions of the form $A \rightarrow B$ proceed similarly with $(B, u) \in V_{i,j}$.
Then $(S, v) \in V_{1,n}$ for some subsequence v of w iff $S \Rightarrow^* w$ and there is a cycle respecting derivation $S \Rightarrow^* w$ such that $S \Rightarrow^*_S v$ is the associated derivation in G_S. If v = \$, then there are illegal cycles generated from elementary nonterminals.

Concerning the complexity, since the grammar is unambiguous, every $V_{i,j}$ contains at most two entries with first component A, (A, v) or (A, \$), where v is the unique subsequence of $a_i ... a_{i+j-1}$ with all cycles deleted. The computation of each new entry of $V_{i,j}$ costs at most 0(j), so that the total time complexity is $O(n^4)$.

By the same arguments as for Theorem 3 we obtain for elementary cycle grammars:

Theorem 6
The membership problem for elementary regular cycle grammars and for context-free cycle grammars is NP-complete, i.e., given a picture p and a cycle grammar G, is there a cycle picture word $w \in L_c(G)$ such that p = p(w).

4. Ambiguous grammars

The following theorem shows that unambiguity is essential for the tractability of the recognition problem; ambiguity causes a jump to NP-completeness. The crucial idea behind the proof is the fact that cycle respecting derivations are powerful enough to make the syntactic analysis sensitive to the "geometrical context" of the pictures determined by the analysed picture words. This is exploited by the ambiguity.

Theorem 7
The recognition problem for elementary cycle grammars is NP-complete.

Proof.
It is easy to verify that the problem is in NP. For a given elementary cycle grammar G and a terminal string w nondeterministically construct a derivation tree for w and check whether it describes a cycle respecting derivation.
To prove the NP-completeness we shall describe a picture language T of "templates" and an algorithm which for any nondeterministic Turing machine M, any string w, and any positive integer t constructs a template u and an elementary cycle grammar G such that $u \in L_e(G) \Leftrightarrow$ M accepts

w in t steps. The grammar is ambiguous, so that there are several different derivations witnessing that $u \in L(G)$. However, at most one of them is cycle respecting. The algorithm constructs u and G in time bounded by $c(|w| + t)^2$, for some constant c depending only on M. (In fact, u depends only on t and |w| and G only on M and w, and the dependence on w appears only in one production).
The underlying idea of the proof is as follows. A computation of a Turing machine can be encoded into a picture word describing a picture schematically represented by Fig. 2. Each of the horizontal sections in the picture corresponds to a configuration of the computation. The vertical ordering reflects the time ordering of the computation. Passing information correctly from one horizontal section to its upper neigbour is guaranteed by the condition that the whole picture is an elementary cycle. If there were an error in the information passing, two of the adjacent horizontal sections would intersect and the derivation of the picture word would not be cycle respecting.
We start with the definition of the set of T of templates. To simplify things we introduce several abbreviations of picture words.

$<u\rightarrow> = (urdlr)^6$, $<u\leftarrow> = (uldrl)^6$, $<d\rightarrow> = (drulr)^6$, and $<d\leftarrow> = (dlurl)^6$

and for all $i, j \geq 0$ define

$<i,j\text{-}template> = r\,[r(<u\rightarrow><d\rightarrow>)^i\, ruul(<u\leftarrow><d\leftarrow>)^i\, luu]^j\, ld^{4j}$.

Fig. 3 shows the string <3,2-template>.

The templates can be derived by a grammar we shall design in an ambiguous way. For instance, the grammar will contain rules $A_c \rightarrow urdl$ and $C_c \rightarrow lr$ and will be able to derive substrings $urdC_c$ and A_cr (cf. Fig. 4 a, b). The two rules turn the two strings in the same string urdlr (cf. Fig. 4c), i.e., the basic square from which the fraction $<u\rightarrow>$ of templates is built. For a while we will regard context-free grammars as analytic tools for string reduction. Thus urdlr can be reduced to $urdC_c$ by the rule $C_c \rightarrow lr$ or to A_cr by the rule $A_c \rightarrow urdl$. In what follows we shall draw the curve from Fig. 4d and 4e instead of 4a and 4b, respectively, and call them their *terminal projections*.
Hence, the templates will serve as the raw material from which the curves encoding Turing machine computations can be "carved". The tool for carving the curves is the elementary cycle grammar G_e, with all nonterminals being elementary cycle nonterminals and consisting of the following rules:

$A_c \rightarrow urdl$, $B_c \rightarrow drul$, $C_c \rightarrow lr$, $D_c \rightarrow uC_c d$,

$A'_c \rightarrow uldr$, $B'_c \rightarrow dlur$, $C'_c \rightarrow rl$, $D'_c \rightarrow uC'_c d$.

The productions of the grammar G_e reduce the "horizontal" segments of templates to different shapes. Fig. 5a demonstrates the result of a possible reduction of the string $(<u\rightarrow><d\rightarrow>)^2$ and Fig. 5b gives an example of reducing the string $(<u\leftarrow><d\leftarrow>)^2$.

Note that the grammar G_e is asymmetric on $<u\rightarrow>$ and $<d\rightarrow>$ (or $<u\leftarrow>$ and $<d\leftarrow>$). The substring $urdC_curdC_c$ cannot be further reduced, but the symmetric string $druC_cdruC_c$ can be reduced to drD_cruC_c.

Now we come to an encoding of Turing machine computations by picture words.

Definition

A binary relation R over an alphabet Σ is a *regular relation* iff there is a regular language L such that for any two strings $u = a_1...a_m$, $v = b_1...b_n$ $(a_1,...,a_m,b_1...b_n \in \Sigma)$ the following holds:

$u \, R \, v$ iff $n = m$ and $a_1b_1...a_nb_n \in L$.

We shall then call L the regular representation of R and denote $L = L(R)$.

Claim 1

Let M be an arbitrary Turing machine with input alphabet Σ. Then there exists a regular relation R on $\{0,1\}$, a constant $c > 0$, a string $v \in \{0,1\}^*$ and a homomorphism $h: \Sigma^* \to \{0,1\}^*$ such that for any $w \in \Sigma^*$ and any $t > 0$ the following equivalence holds:

M accepts w in t steps iff there are strings $u_1,...,u_q \in \{0,1\}^*$ such that

(a) q is even and equals t or $t+1$,

(b) $u_0 = 0^{ct} \, v \, h(w) \, 0^{ct}$,

(c) $|u_0| = |u_1| = ... = |u_q|$,

(d) $u_i \, R \, u_{i+1}$ for $0 \le i < q$,

(e) $u_q \in \{0\}^*$.

Claim 2

If R is a regular relation over $\{0,1\}^*$ then there exist

(1) a right linear grammar $G_R^{\rightarrow}$ with rules of the form $X \to abY$ or $X \to \Lambda$, where X, Y are nonterminals and $a,b \in \{0,1\}$, with the initial nonterminal $S^{\rightarrow}$ and a nonterminal $F^{\rightarrow}$ such that $S^{\rightarrow} \Rightarrow^* wF^{\rightarrow}$ iff $w \in L(R)$.

(2) a right linear grammar $G_R^{\leftarrow}$ with rules of the form $X \to abY$ or $X \to \Lambda$, where X, Y are nonterminals and $a,b \in \{0,1\}$, with the initial nonterminal $S^{\leftarrow}$ and nonterminals $F_1^{\leftarrow}, F_2^{\leftarrow}$ such that

(a) $S^{\leftarrow} \Rightarrow^* a_1b_1...a_nb_nF_1^{\leftarrow}$ iff $b_na_n...b_1a_1 \in L(R)$ and $a_1 = ... = a_n = 0$,

(b) $S^{\leftarrow} \Rightarrow^* a_1b_1...a_nb_nF_2^{\leftarrow}$ iff $b_na_n...b_1a_1 \in L(R)$ and $a_i = 1$ for some $1 \le i \le n$.

We shall construct a grammar generating picture words which encode sequences $u_1,...,u_n$ as described in Claim 1. The symbols 0 and 1 will be encoded by strings defined as follows.

Definition

We define four encodings $U_{\rightarrow}$, $D_{\rightarrow}$, $U_{\leftarrow}$, $D_{\leftarrow}$ of the symbols $0,1$.

$U_{\rightarrow}(0) = A_c rurdC_c(A_c r)^4$ | $U_{\leftarrow}(0) = (A'_c l)^4 uldC'_c A'_c l$

$U_{\rightarrow}(1) = (A_c r)^4 \, urdC_c(A_c r)^4$ | $U_{\leftarrow}(1) = A'_c luldC'_c(A'_c l)^4$

$D_{\rightarrow}(0) = (B_c r)^3 drD_c rD_c ruC_c$ | $D_{\leftarrow}(0) = dlD'_c lD'_c luC'_c(B'_c l)^3$

$D_{\rightarrow}(1) = drD_c rD_c ruC_c(B_c r)^3$ | $D_{\leftarrow}(1) = (B'_c l)^3 \, dlD'_c lD'_c luC'_c$.

Fig. 6 shows the terminal projections of the codes.

The codes are designed such that, e.g., if the picture corresponding to $U_{\rightarrow}(0)$ is drawn at distance 2 below $D_{\leftarrow}(0)$, the two pictures do not interfere. But if $U_{\rightarrow}(1)$ is drawn at distance 2 below $D_{\leftarrow}$ (0), the pictures intersect. Henceforth the following two claims hold.

Claim 3

Let $n \geq 1$ and $a_i, b_i, c_i, d_i \in \{0,1\}$ for $1 \leq i \leq n$. Then the picture word

$$U_{\rightarrow}(a_1)D_{\rightarrow}(b_1)\ldots U_{\rightarrow}(a_n)D_{\rightarrow}(b_n)\mathrm{ruul}U_{\leftarrow}(d_n)D_{\leftarrow}(c_n)\ldots U_{\leftarrow}(d_1)D_{\leftarrow}(c_1)$$

is cycle-free iff $a_1 \ldots a_n = c_1 \ldots c_n$.

Fig. 7 illustrates Claim 3 for $a_1a_2 = 01, b_1b_2 = 11, c_1c_2 = 01, d_1d_2 = 10$.

Claim 4

Let $n \geq 1$ and $a_i, b_i, c_i, d_i \in \{0,1\}$ for $1 \leq i \leq n$. Then the picture word

$$U_{\leftarrow}(b_n)D_{\leftarrow}(a_n)\ldots U_{\leftarrow}(b_1)D_{\leftarrow}(a_1)\mathrm{luur}U_{\rightarrow}(c_1)D_{\rightarrow}(d_1)\ldots U_{\rightarrow}(c_n)D_{\rightarrow}(d_n)$$

is cycle-free iff $b_1 \ldots b_n = d_1 \ldots d_n$.

The last basic essential observation is that the grammar G_e turns all codes from the above definition into the uniform four type of words comprising the templates. The observation is expressed by the following fact.

Claim 5

Let G_e be the grammar from above. Then there are following derivations.

$U_{\rightarrow}(0) \Rightarrow^* <u\rightarrow>$ $\quad$ $U_{\rightarrow}(0) \Rightarrow^* <u\rightarrow>$

$D_{\rightarrow}(0) \Rightarrow^* <d\rightarrow>$ $\quad$ $D_{\rightarrow}(1) \Rightarrow^* <d\rightarrow>$

$U_{\leftarrow}(0) \Rightarrow^* <u\leftarrow>$ $\quad$ $U_{\leftarrow}(1) \Rightarrow^* <u\leftarrow>$

$D_{\leftarrow}(1) \Rightarrow^* <d\leftarrow>$ $\quad$ $D_{\leftarrow}(1) \Rightarrow^* <d\leftarrow>$

We still have to add another feature ensuring that the endpoints of horizontal sections of picture words encoding computations will match. That is, we have to exclude picture words of the form schematically illustrated by Fig. 8.

Therefore we introduce an auxiliary symbol # ("lock") and define its codes as follows.

Definition

$U_{\rightarrow}(\#) = (A_c r)^2 \mathrm{urd} C_c (A_c r)^3$ $\quad$ $D_{\rightarrow}(\#) = \mathrm{dru} C_c (A_c r)^3 \mathrm{dru} C_c A_c r$

$U_{\leftarrow}(\#) = (A'_c l)^3 \mathrm{uld} C_c (A'_c l)^2$ $\quad$ $D_{\leftarrow}(\#) = A'_c \mathrm{ldlu} C'_c (A'_c l)^3 \mathrm{dlu} C'_c$.

Fig. 9 shows the picture words corresponding to the codes of #.

It is now easy to see that claims as above hold.

Claim 6

Let G_0 be the grammar from above. Then the following derivations are possible by G_0.

$U_{\rightarrow}(\#) \Rightarrow^* <u\rightarrow>$ $\quad$ $D_{\rightarrow}(\#) \Rightarrow^* <d\rightarrow>$

$U_{\leftarrow}(\#) \Rightarrow^* <u\leftarrow>$ $\quad$ $D_{\leftarrow}(\#) \Rightarrow^* <d\leftarrow>$

Claim 7

Let $m, n \geq 1$ and $a_i, b_i, c_i, d_i \in \{0,1\}$ for $1 \leq i \leq n$. Define picture words $W_1 = W'_1 \mathrm{ruul} W''_1$

and $W_2 = W'_2$ ruulW''_2 with $W'_1 = U_{\rightarrow}(\#)D_{\rightarrow}(\#)U_{\rightarrow}(a_1)D_{\rightarrow}(b_1)..U_{\rightarrow}(a_n)D_{\rightarrow}(b_n)U_{\rightarrow}(\#)D_{\rightarrow}(\#)$,
$W''_1 = U_{\rightarrow}(\#)D_{\rightarrow}(\#)U_{\leftarrow}(d_n)D_{\leftarrow}(c_n)...U_{\leftarrow}(d_1)D_{\leftarrow}(c_1)U_{\rightarrow}(\#)D_{\rightarrow}(\#)$,
$W'_2 = U_{\leftarrow}(\#)D_{\leftarrow}(\#)U_{\leftarrow}(b_n)D_{\leftarrow}(a_n)...U_{\leftarrow}(b_1)D_{\leftarrow}(a_1)U_{\leftarrow}(\#)D_{\leftarrow}(\#)$
and $W''_2 = U_{\leftarrow}(\#)D_{\leftarrow}(\#)U_{\rightarrow}(c_1)D_{\rightarrow}(d_1)...U_{\rightarrow}(c_n)D_{\rightarrow}(d_n)U_{\leftarrow}(\#)D_{\rightarrow}(\#)$.
Then W_1 is cycle-free iff $m = n$ and $a_1...a_n = c_1...c_n$
and W_2 is cycle-free iff $m = n$ and $b_1...b_n = d_1...d_n$.

Now we are ready to describe the elementary cycle grammar G meeting the conditions summarized in the paragraph following Theorem 7. The terminal alphabet of the grammar is {u,d,r,l}, the initial nonterminal is S_c and its system of productions is given below. Assume that $vh(w) = a_1...a_n$, where the string v and the homomorphism h are as in Claim 1.

The description of the grammar G contains two subgrammars $H_R^{\rightarrow}$ and $H_R^{\leftarrow}$, which are grammars obtained from grammars $G_R^{\rightarrow}$ and $G_R^{\leftarrow}$ by replacing the symbols 0, 1 by their picture codes.
Thus $H_R^{\rightarrow}$ contains the rule $X \rightarrow U_{\rightarrow}(a)D_{\rightarrow}(b)Y$ iff the grammar G contains the rule $X \rightarrow abY$, etc. The encoded grammars have the same nonterminals as the original ones. Furthermore G contains the grammar G_e defined above and the following productions.

S_c	$\rightarrow rrU_{\rightarrow}(\#)D_{\rightarrow}(\#)S_1$,	S_1	$\rightarrow S_2$,
S_1	$\rightarrow U_{\rightarrow}(0)D_{\rightarrow}(0)S_1$	S_2	$\rightarrow U_{\rightarrow}(a_1)D_{\rightarrow}(0)...U_{\rightarrow}(a_n)D_{\rightarrow}(0)S_3$,
S_3	$\rightarrow U_{\rightarrow}(0)D_{\rightarrow}(0)S_3$	S_3	$\rightarrow$ ruulS_4,
S_4	$\rightarrow U_{\rightarrow}(\#)D_{\rightarrow}(\#)S^{\leftarrow}$,	S_5	$\rightarrow U_{\rightarrow}(\#)D_{\rightarrow}(\#)S^{\rightarrow}$,
S_6	$\rightarrow ddS_6 \mid dd$,	$F^{\rightarrow}$	$\rightarrow U_{\rightarrow}(\#)D_{\rightarrow}(\#)$tuul$S_4$
$F_2^{\rightarrow}$	$\rightarrow U_{\leftarrow}(\#)D_{\leftarrow}(\#)$luur$S_5$,	$F_1^{\leftarrow}$	$\rightarrow U_{\leftarrow}(\#)D_{\leftarrow}(\#)llS_6$,

It is not difficult to check that grammar G meets all the specified properties.

First, let us assume that the machine M accepts the word w in t steps. Let the homomorphism h, the string v and the constant c be as in Claim 1. Define $q = t/2$ and $m = |h(w)| + |v| + 2ct$. It is easy to see that grammar G can generate a string α (using only the above rules) such that
(a) $\alpha \Rightarrow^*_{G_e}$ <m,p - template> and
(b) the terminal projection of α is cycle-free.
Hence $S_c \Rightarrow^*_G$ <m,p-template> by a cycle respecting derivation.

Conversely, let $S_c \Rightarrow^*_G$ <m,p-template> by a cycle respecting derivation, for some m, p. Let us consider the derivation tree of this derivation. The rules from G_e can be applied only at the lowest levels of the tree. Reducing the <m,p-template> in the bottom-up fashion using only these rules, yields a string β. Since $S_c \Rightarrow^*_G \beta$ and all the cycle nonterminals of the grammar G with the exception of S_c belong to G_e, the terminal projection of β is a cycle-free string. The string can be reduced to S_c using only productions listed above. These facts imply that the string β is the encoding of a sequence of strings $u_0,...,u_q$ meeting the conditions of Claim 1. Hence w is accepted by a computation of M in q steps.

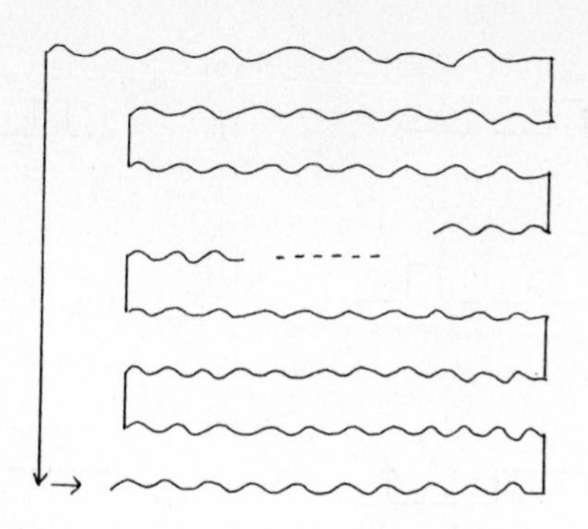

Fig. 2

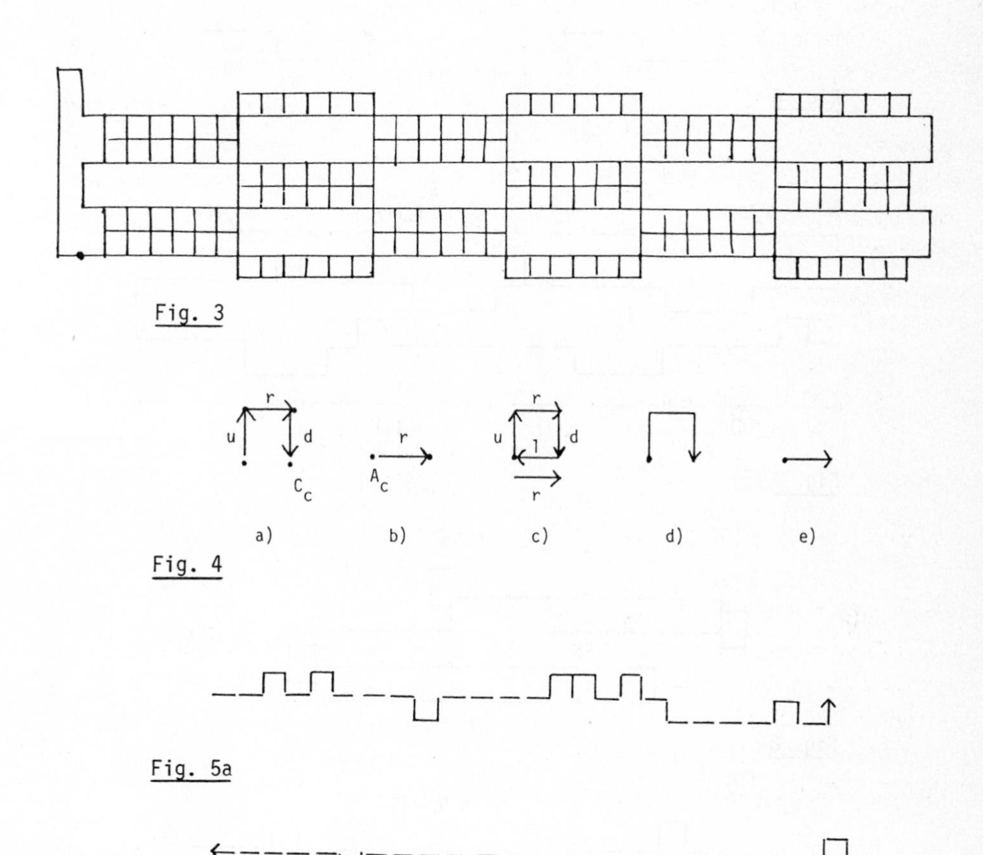

Fig. 3

Fig. 4

Fig. 5a

Fig. 5b

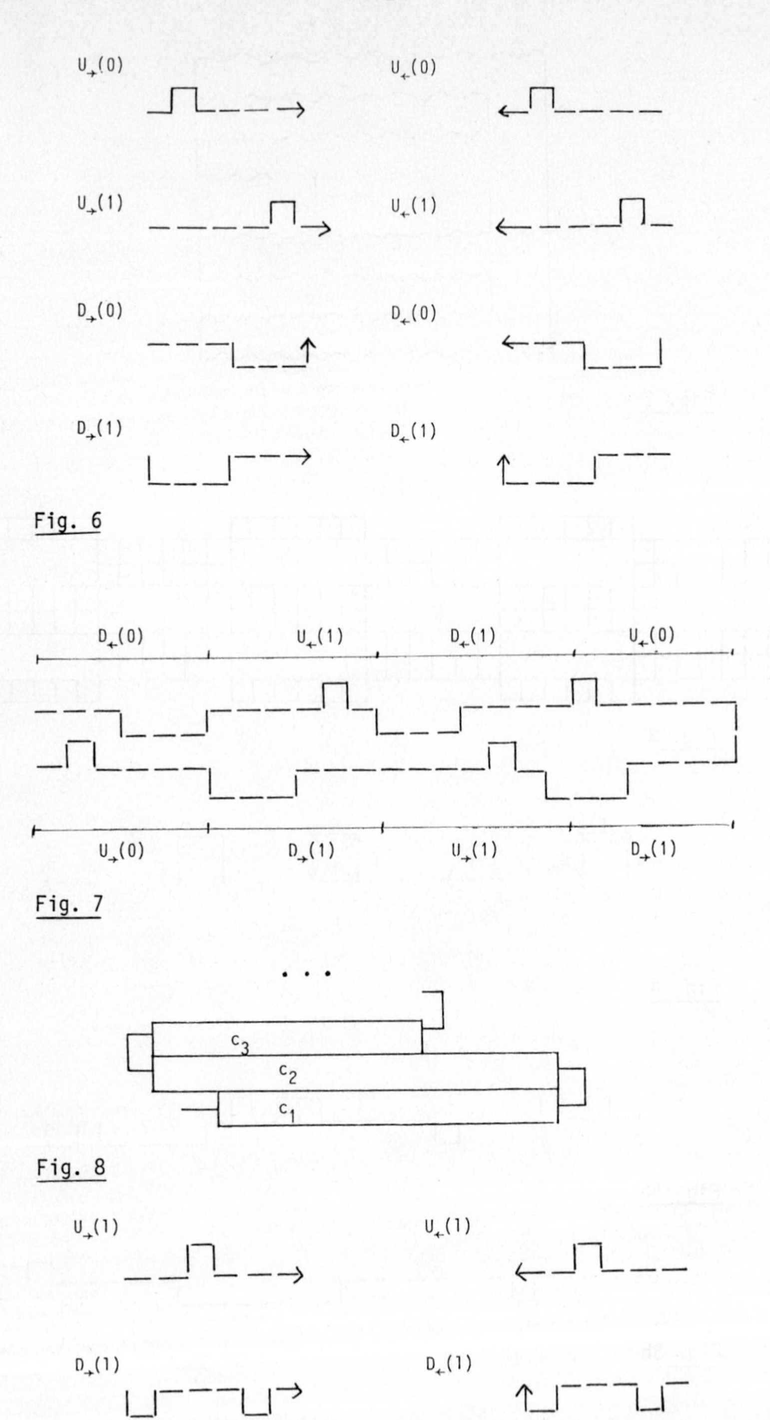

Fig. 6

Fig. 7

Fig. 8

Fig. 9

5. Conclusion

The notion of cycle grammars introduced in this paper enables to grasp an important feature of pictures as a syntactic pattern. It makes it possible generating and parsing a wider class of picture languages without an essential increase of computational complexity. Another possibility offered by this approach is to employ the technique of syntax directed translation to transform only selected subparts of pictures as briefly indicated in Example 4.

The last theorem shows that further amplification of this approach grasping also the natural hierarchical structure of subcycles goes beyond the limits of practical tractability.

We feel that this is only a first step towards the integration of cycles into picture descriptions and that cycle grammars and elementary cycle grammars and particularly syntax directed transductions of picture description languages deserve further investigations.

References

[1] A.V. Aho, J.D. Ullman,
"The Theory of Parsing, Translation, and Compiling", Vol 1, Prentice Hall (1972)

[2] F.J. Brandenburg, J. Dassow,
"Reductions of Picture Words", MIP 8905, Universität Passau (1989)

[3] M.J. Fischer, A.L. Rosenberg
"Real Time Solutions of the Origin-Crossing Problem", Math. Sys. Theory 3 (1967), 257- 263

[4] J.D. Foley, A. van Dam,
"Fundamentals of Interactive Computer Graphics", Addison Wesley (1982)

[5] H. Freeman,
"On the encoding of arbitrary geometric configurations", IRE Trans. EC-10 (1961), 260-268

[6] K.S. Fu,
"Syntactic Pattern Recognition and Applications", Prentice-Hall (1982)

[7] R. Gutbrod,
"A transformation system for generating description languages of chain code pictures" Theoret. Comput. Sci. 68 (1989), 239-252

[8] F. Hinz,
"Classes of picture languages that cannot be distinguished in the chain code concept and the deletion of redundant retreats", Lecture Notes in Computer Science 349 (1989), 132-143

[9] F. Hinz, E. Welzl,
"Regular chain code picture languages with invisible lines",
Report 252, IIG, Techn. Univ. Graz (1988)

[10] J.E. Hopcroft, J.D. Ullman,
"Introduction to Automata Theory, Languages and Computation", Addison-Wesley (1979)

[11] C. Kim, I.H. Sudborough,
"The membership and equivalence problems for picture languages"
Theor. Comput. Sci. 52 (1987), 177-191

[12] H.A. Maurer, G. Rozenberg, E. Welzl,
"Using string languages to describe picture languages",
Inform. and Control 54 (1982), 155-185

[14] A. Rosenfeld,
"Picture Languages - Formal Models for Picture Recognition", Academic Press (1979)

[15] G. Rozenberg, E. Welzl,
"Graph Theoretic Closure Properties of the Family of Boundary NLC Graph Languages"
Acta Informatica 23 (1986), 289-309

[16] A. Schönhage,
"Storage modification machines", SIAM J. Computing 9 (1980), 490-508

[17] G. Siromoney, R. Siromoney,
"Rosenfeld's Cycle Grammars and Kolam",
Lecture Notes in Computer Science 291 (1987), 564-579

[18] I.H. Sudborough, E. Welzl,
"Complexity and decidability for chain code picture languages"
Theoret. Comput. Sci. 36 (1985), 173-202

An efficient implementation of graph grammars based on the RETE matching algorithm

H. Bunke, T. Glauser, T.-H. Tran
University of Bern,
Länggassstr. 51, CH-3012 Bern,Switzerland

ABSTRACT: This paper is concerned with the efficient determination of the set of productions of a graph grammar that are applicable in one rewriting step. We propose a new algorithm that is a generalization of a similar algorithm originally developed for forward chaining production systems. The time complexity of the proposed method is not better than that of a naive solution, in the worst case. In the best case, however, a significant speedup can be achieved. Some experiments supporting the results of a theoretical complexity analysis are described.

Keywords: graph grammars, forward chaining, conflict set, RETE-matching algorithm, computational complexity

CONTENTS

0 Introduction

Graph grammars have been a subject of research for about twenty years now and significant progress in both theory and applications can be observed over the past years [1, 2, 3]. A problem that is inherent to graph grammars, however, is the high computational complexity of many operations as compared to string grammars. In this paper, we consider the problem of running graph grammars in a forward directed fashion. That is, we start with the initial graph of a grammar and successively apply productions until a final graph has been derived. The basic algorithm for this task is shown in Fig. 1. The conflict set is determined by checking each production if its left-hand side occurs as a subgraph in the actual graph g. If the left-hand side of a production is contained in g in a number of different ways, we say there are several instances, or occurrences, of p in g. In the second step of the body of the loop in Fig. 1 we select one instance p of an applicable production. This selection can be made at random or according to some predefined rules, for example, the control diagram of a programmed graph grammar [4]. Finally, in the third step we apply the production that has been chosen in step 2. This is done by deleting the left-hand side of p in g, inserting the right-hand side of p into g, and applying the embedding transformation procedure in order to ensure the proper connections between the newly inserted right-hand side and that part of g that remains unchanged

while termination condition is not true **do**
determine the conflict set (i.e., the set of applicable productions);
select one applicable production p from the conflict set;
apply p;

Figure 1: Basic algorithm for running a graph grammar in a forward directed way

after deleting the left-hand side of p. After a particular production has been applied, the derivation is continued, i.e. the body of the loop is executed again, until a termination condition becomes true. The derivation is usually stopped if a graph with only terminal labels has been derived or if there isn't any production applicable to the actual graph.

A naive implementation of the forward derivation of a graph grammar that exactly follows the algorithm shown in Fig. 1 is conceptually fairly simple but computationally expensive, particularly in determining the conflict set. If we want to determine the conflict set in a straightforward way, we go in a loop over all productions. For each production p we check the actual graph g in order to find all occurrences of p in g. This is done by taking one after the other node and comparing it to the nodes of the actual graph. As a result, the naive determination of the conflict set requires of the order $O(P \cdot G^L)$ operations for an actual graph of size G and a grammar with P productions where each left-hand side is of size L. Note that the factor P comes from the outer loop over all productions while G^L is the time complexity for finding all occurrences of a particular production in the actual graph.

In this paper, we present a new method for the efficient determination of the set of all applicable productions in one rewriting step of a graph grammar. As it will be discussed later, the application of this method can lead to a significant speed-up of the execution of productions. Our new algorithm is mainly based on the observation that only local changes occur in the actual graph if a production is applied. In other words, most of the underlying graph will remain unaffected by a production application. It follows that a number of productions that have been found applicable in the last cycle are still applicable in the actual cycle (see Fig. 1). So in the course of a derivation sequence there are only incremental changes to the conflict set. For the determination of these incremental changes it is sufficient to do a local analysis that is based on the left- and right-hand side of the production actually applied. So it is not necessary to inspect all grammar productions and the complete underlying graph.

The algorithm shown in Fig. 1 is not only applicable to graph grammars but also to forward chaining production systems. A forward chaining production system consists of a database, which is the counterpart to the underlying graph in a graph grammar, a finite set of productions, and an interpreter that follows the algorithm in Fig. 1 in order to derive conclusions from an initial database. For more details about production systems see [5]. A particular example of a forward chaining production system is the OPS5 software tool [5]. A production system differs from a graph grammar mainly in the data structures, which are string-oriented rather than being graphs. Nevertheless, the determination of the conflict set in a forward chaining production system is a problem of exponential time complexity [6]. This is caused by the fact that there may be variables in the productions that require consistent bindings. An algorithm for the efficient determination of the conflict set in forward chaining production systems has been proposed in [6]. In fact, the RETE algorithm according to [6] has been incorporated in the interpreter of the OPS5 system [5].The algorithm presented in this paper is a generalization of the RETE algorithm to graph grammars. As it will be seen in the following sections, most of the concepts can be transferred from the original algorithm, i.e. from string oriented structures, to graphs in a straightforward way.

There are applications of graph grammars where it is not necessary to determine the complete conflict set in each rewriting step. Instead, it could be sufficient to find any occurence of any applicable production. In such a situation the selection of an applicable production could be based on

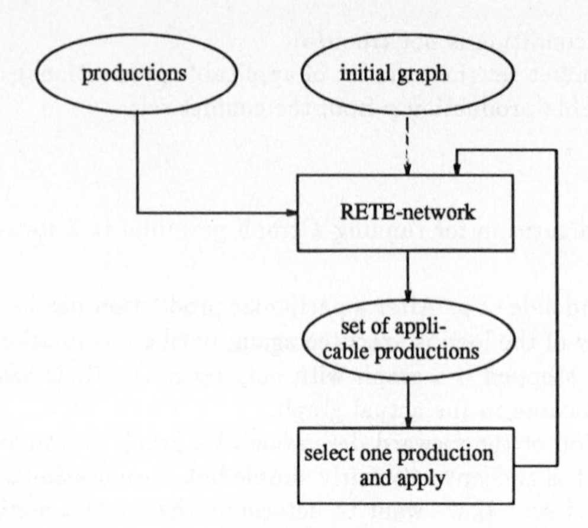

Figure 2: Overview of RETE-based graph grammar implementation

a simpler procedure than the one proposed in this paper. Unfortunately, however, any such simpler procedure, which is based on a sequential search for an applicable production in the underlying graph, requires still of the order $O(P \cdot G^L)$ operations in the worst case. Our proposed graph grammar implementation provides a very general basis that can be combined with any particular control strategy for searching and selecting applicable productions. This is useful, for example, when one needs to experiment with different control strategies during the development of a graph grammar application. If one considers graph grammars as a tool with a similar range of potential applications as rule based systems, it is particularly advisable to provide an implementation that is compatible with a variety of different conflict resolution strategies.

1 Informal Overview and Example

An overview of our RETE-based graph grammar implementation is shown in Fig. 2. First, we compile the given set of grammar productions into a network, the so-called RETE-network. This network is a compact representation of the set of productions such that any subgraph that occurs in different productions is stored only once. In this way, each node in the RETE-network represents a particular graph that is contained as a subgraph in the left-hand side of one or more productions. The crucial idea of our algorithm for efficient conflict set determination is to store with each node of the RETE-network, which represents a certain subgraph g' of the left-hand side of one or more productions, all occurrences of g' in the underlying graph. In this way, the RETE-network is a dynamic data structure that not only represents the set of grammar productions but also implicitly contains parts of the underlying graph at any time during a derivation sequence leading from an initial to a final graph. It is important to note, furthermore, that the network isn't a passive structure. Instead, attached to its nodes are active procedures that receive and transmit data to and from other nodes, respectively.

After the RETE-network has been compiled from the productions the grammar can be run. For this purpose, we input the initial graph into the RETE-network, i.e. we propagate its nodes and edges through the network. After the propagation of the initial graph we get, at the output of the network, the conflict set, i.e. all instances of productions that are applicable to the initial graph. Next, one production is selected from this set. The particular selection strategy is not important in the context of this paper; any procedure may be used. The application of the selected production p proceeds in two steps. First, we input the left-hand side of p into the RETE-network and delete all occurrences of all nodes and edges of the left-hand side in the network. Then the right-hand side of p is propagated through the network. This results in an update of the information that is stored

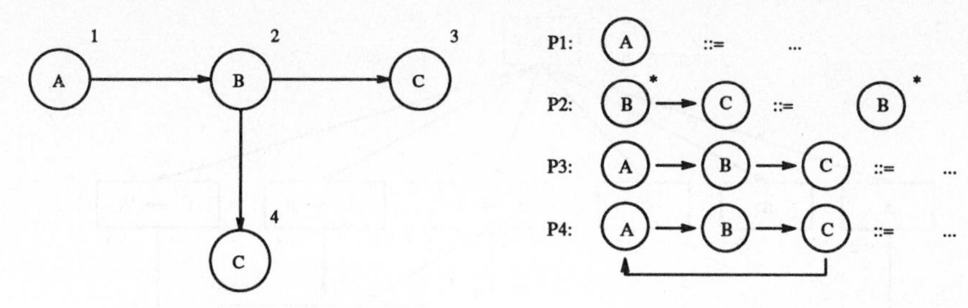

Figure 3: Start graph

Figure 4: Productions

with some network nodes. More precisely, whenever the right-hand side of p either completely or partially matches the subgraph represented by a network node, the matching part gets stored at that network node. The embedding edges are determined immediately after the production has been selected from the conflict set, and are treated in the same way as the edges of the left-hand and right-hand side, respectively.

After processing the left-hand side, right-hand side, and embedding of the selected production p as sketched above, the RETE-network is in an updated state that exactly reflects the changes of the underlying graph that are due to the application of p. At the output of the network, the set of applicable production has been properly updated, too. Now the derivation proceeds with the selection of the next production from the conflict set.

Next we illustrate the ideas sketched above by means of an example. Generally, we deal with directed graphs with labeled nodes and edges. (In the present example, however, the edges are unlabeled.) Consider the initial graph in Fig. 3 and the graph grammar productions in Fig. 4. Since the right-hand sides and the embeddings of the productions p_1, p_3, and p_4 are not needed in this example, they have been omitted. Suppose that the embedding of the production p_2 is such that the B-node in the right-hand side inherits all edges from the B-node in the left-hand side.

The network that is compiled from the productions in Fig. 4 is given in Fig. 5. (For details see section 2.) The root serves as the entrance for all nodes and edges input to the network. There are four output- or production-nodes, each representing a particular grammar production. At an output labeled with p_i all occurrences of p_i in the underlying graph will be stored when running the grammar. Thus these occurrences stored at the output nodes represent the conflict set.

For the following discussion it is crucial to distinguish between nodes and edges of the RETE-network and nodes and edges of the underlying graph. Therefore, we use the notations n-node and n-edge for elements of the network, and g-node and g-edge for elements of the underlying graph. For running the productions p_1 to p_4 we first input the initial graph shown in Fig. 3 into the RETE-network. Generally, any g-node and g-edge that is input to the network is propagated from one n-node to another n-node as long as certain tests performed by a n-node on the g-node or g-edge are successful. Some g-nodes and g-edges that pass such a test get stored at certain locations in the network. First we input the g-nodes 1 to 4 of the initial graph shown in Fig. 3 into the network. Since the label A matches the n-node 2 and because the n-node 2 is connected to the output labeled p_1, the g-node 1 is reported as an applicable instance of the production p_1. The g-nodes 2–4 match the n-nodes 3 and 4 respectively. However, as these n-nodes don't have sucessors, the g-nodes 2–4 are discarded.

We continue with inputting the g-edges of Fig. 3 into the RETE-network. The g-edge (1,2) matches the n-node 6 and gets stored there while both (2,3) and (2,4) get stored at n-node 5. Furthermore, two combinations consisting of two g-edges each, namely ((1,2),(2,3)) and ((1,2),(2,4)), match the n-node 8 and get stored there. Since the n-node 5 is directly connected to the p_2-output both edges (2,3) and (2,4) are reported as instances of the production p_2. Similarly ((1,2),(2,3)) and ((1,2),(2,4)) are reported as instances of p_3. We conclude that the conflict set consists of one instances of p_1, two instances of p_2, and two instances of p_3 after the initial graph shown in Fig. 3

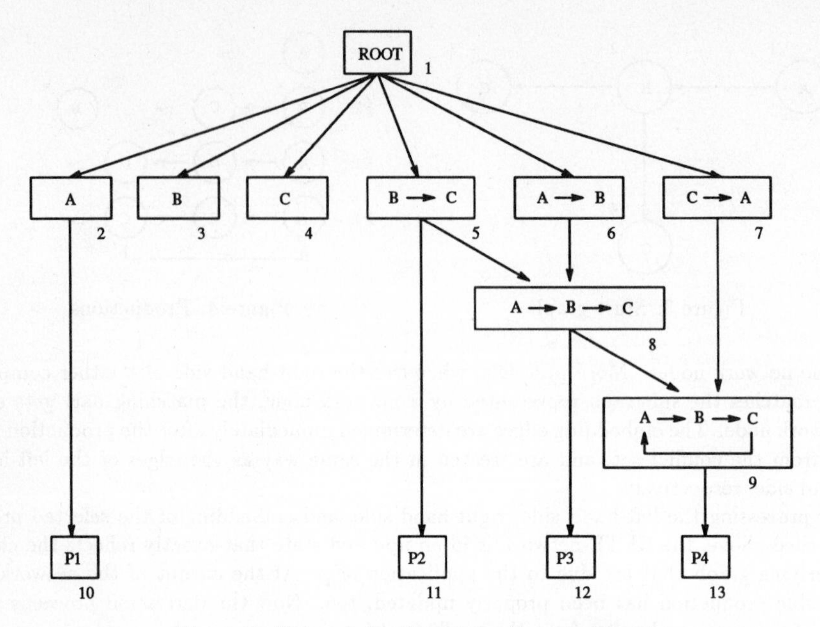

Figure 5: Network for the productions in Fig. 4

has been completely input to the network shown in Fig. 5. A snapshot of this situation is depicted in Fig. 6.

Now suppose that the instance (2,3) of p_2 is selected for application. This means that any item stored at a n-node is deleted if it contains either one of the g-nodes 2 or 3. In Fig. 6, all g-edges stored at the n-nodes 5, 6 and 8 are deleted. Consequently all elements from the conflict set are removed except for the instance of p_1, consisting of the g-node 1. Next the node in the right-hand side of p_2 is input as new g-node, say 5, to the RETE-network. Aditionally, we have to input the embedding edges (1,5) and (5,4) to the net. They get stored at the n-nodes 6 and 5, respectively. The g-edge (5,4) is reported as an instance of p_2. Furthermore, the g-edge combination ((1,5),(5,4)) gets stored at the n-node 8 and is reported as an instance of p_3. At this moment, the network is in an updated state that correctly represents the situation after the production p_2 has been applied. A graphical illustration is given in Fig. 7. We notice that there are three elements in the conflict set.

This simple example shows the fundamental ideas underlying our algorithm for efficient determination of the conflict set in graph grammars. A more detailed description of the steps involved is given in the following sections.

2 The RETE-network

The grammars considered in this paper work on directed labeled graphs. Formally, such a graph is defined by a 4-tuple

$g = (V, E, \alpha, \beta)$, where

- V is a finite set of nodes, or vertices,
- $E \subseteq V \times V$ is the set of directed edges,
- $\alpha : V \rightarrow L_V$ is the node labeling function, and
- $\beta : E \rightarrow L_E$ is the edge labeling function.

L_V and L_E are finite alphabets of node and edge labels, respectively. To simplify the following considerations we assume that any graph g is represented by a node list

$$n_g = ((v_1, \alpha(v_1)), (v_2, \alpha(v_2)), \ldots, (v_n, \alpha(v_n))), \text{ where } V = \{v_1, \ldots, v_n\} \quad (3.1)$$

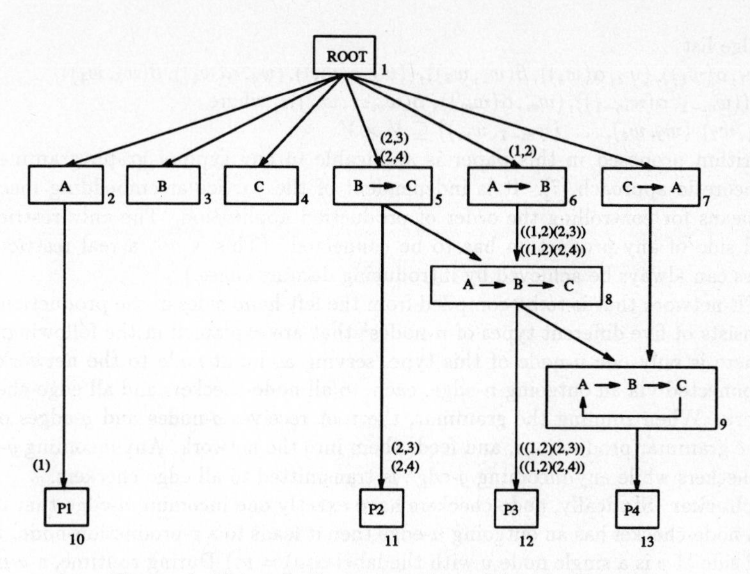

Figure 6: Network after processing of the start graph

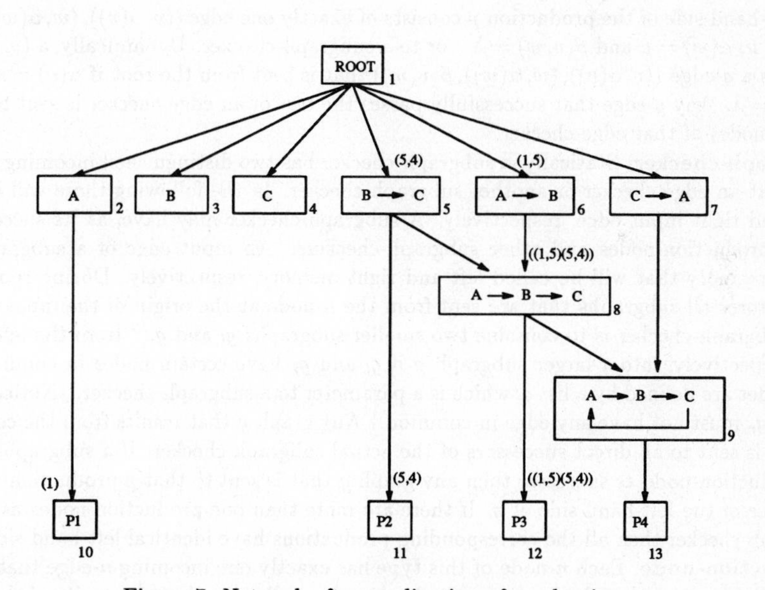

Figure 7: Network after application of production p_2

and an edge list

$$e_g = (((w_1, \alpha(w_1)), (w_2, \alpha(w_2)), \beta(w_1, w_2)), ((w_3, \alpha(w_3)), (w_4, \alpha(w_4)), \beta(w_3, w_4)), \ldots, ((w_{m-1}, \alpha(w_{m-1})), (w_m, \alpha(w_m)), \beta(w_{m-1}, w_m))), \text{ where} \quad (3.2)$$

$$E = \{(w_1, w_2), (w_3, w_4), \ldots, (w_{m-1}, w_m)\} \subseteq V \times V.$$

The algorithm proposed in this paper is applicable to any type of graph grammar belonging to the set theoretic approach [7]. It is independent of the particular embedding mechanism and allows any means for controlling the order of production application. The only restriction is that the left-hand side of any production has to be connected. (This is not a real restriction because connectedness can always be achieved by introducing dummy edges.)

The RETE-network that is to be compiled from the left-hand sides of the productions of a graph grammar consists of five different types of n-nodes[1] that are explained in the following.

root: There is only one n-node of this type, serving as input-node to the network. Statically the root is connected via an outgoing n-edge, each, to all node-checkers and all edge-checkers in the RETE-network. When running the grammar, the root receives g-nodes and g-edges of the initial graph and the grammar productions, and feeds them into the network. Any incoming g-node is sent to all node-checkers while any incoming g-edge is transmitted to all edge checkers.

ν-node checker: Statically, node-checkers have exactly one incoming n-edge that originates at the root. If a node-checker has an outgoing n-edge then it leads to a p-production-node. (In this case the left-hand side of p is a single node v with the label $\alpha(v) = \nu$.) During runtime, a ν-node checker checks a g-node $(v, \alpha(v))$ that is sent from the root whether $\alpha(v) = \nu$. Any g-node that successfully passes the test $\alpha(v) = \nu$ is transmitted from the node-checker to each p-production-node that is a direct successor.

(ν, μ, λ)-edge-checker: Edge-checkers have, similarly to node-checkers, exactly one incoming n-edge that originates at the root. In contrast with node-checkers, however, an edge-checker can have any number of outgoing n-edges, each of which may either lead to a p-production-node – in this case the left-hand side of the production p consists of exactly one edge $((v, \alpha(v)), (w, \alpha(w)), \beta(v, w))$ with $\alpha(v) = \nu$, $\alpha(w) = \mu$ and $\beta(v, w) = \lambda$ – or to a subgraph-checker. Dynamically, a (ν, μ, λ)-edge-ckecker tests a g-edge $((v, \alpha(v)), (w, \alpha(w)), \beta(v, w))$ that is sent from the root if $\alpha(v) = \nu \wedge \alpha(w) = \mu \wedge \beta(v, w) = \lambda$. Any g-edge that successfully passes the test of an edge-checker is sent to all direct successor n-nodes of that edge-checker.

c-subgraph-checker: Statically, a subgraph-checker has two distinguished incoming edges each originating at an edge-checker or another subgraph-checker. In the following these will be referred to as left and right input edge, respectively. A subgraph-checker may have, as its successors, any number of production-nodes and other subgraph-checkers. An input edge of a subgraph-checker has a local memory that will be called left and right memory, respectively. During runtime, such a memory stores all subgraphs that are sent from the n-node at the origin of the input edge. The task of a subgraph-checker is to combine two smaller subgraphs g_l and g_r [2] from the left and right memory, respectively, into a larger subgraph g if g_l and g_r have certain nodes in common. These common nodes are defined by a list c, which is a parameter to a subgraph-checker. (Notice, however, that g_l and g_r must not have any edge in common.) Any graph g that results from the combination of g_l and g_r is sent to all direct successors of the actual subgraph-checker. If a subgraph-checker sc has a p-production-node as successor then any graph g that is sent to that p-production-node by sc is an instance of the left-hand side of p. If there are more than one production-nodes as successors of a subgraph-checker then all the corresponding productions have identical left-hand sides.

p-production-node: Each n-node of this type has exactly one incoming n-edge that originates at a subgraph-checker, edge-checker, or node-checker, depending on the complexity of the left-hand side of the production p. During runtime, a p-production-node gets transmitted, from one of its predecessors in the RETE-network, an instance of the left-hand side of p. This instance is contained

[1] Again, we use the terminology n-node, n-edge, g-node and g-edge in order to distinguish between elements of the RETE-network and the underlying graph

[2] From now on we consider g-edges as particular instances of subgraphs.

in the underlying graph. It will be stored in the local memory of the production-node. Thus the union of the local memories of all production-nodes represent the conflict set.

Generally, ν, (ν, μ, λ) and c are parameters that determine the precise type of test performed by a node-, edge-, and subgraph-checker, respectively. Examples of the different types of network nodes can be found in Fig. 5: 1 is the root; 2,3,4 are node-checkers; 5,6,7 are edge-checkers; 8 and 9 are subgraph-checkers; finally, 10 to 13 are production nodes. As it was explained above, there are two different types of local memories in the RETE-network. First, left and right input edges to a subgraph-checker have a local memory, each. Secondly, each production node has a local memory. In the graphical representation in Fig. 6 and 7, the contents of the local memories of the left and right input edges to a subgraph-checker are not drawn next to the respective input edge but next to the node at which the input edge originates. In Fig. 6, for example, the contents (2,3) and (2,4) drawn next to the n-node 5 actually represent the contents of the memory of the left input edge to the n-node 8, and (1,2) drawn next to 6 represents the contents of the memory of the right input edge to 8. Similarly ((1,2),(2,3)) and ((1,2) (2,4)) drawn next to 8 represent the contents of the left memory of 9. Notice that the right memory of 9 is empty.

The edges of a RETE-network are directed and unlabeled. They are used as one-way channels for transmitting g-nodes, g-edges, and subgraphs. The procedures attached to the different types of RETE-network nodes will be explained in section 4.

3 Compilation of the RETE-network

The compilation of a set of productions into a RETE-network is a preprocessing step that has to be executed exactly once for any set of productions. Its aim is to create for a given set of productions the corresponding RETE-network, i.e. the ν-node-, (ν, μ, λ)-edge- and c-subgraph-checkers and the respective production nodes. Here follows the algorithm for this task:

create-RETE-network
generate the ROOT-n-node;
for every production p
  for every g-node in the left-hand side with node label ν
    if ν-node-checker for this ν not yet in network
    then generate node-checker for this ν; (1)
    insert a n-edge from the root node to it;
  for every g-edge (ν, μ, λ) in the left-hand side
    if (ν, μ, λ)-edge-checker for actual combination (ν, μ, λ) not yet in network
    or such an edge-checker exists but is already used in the actual production
    then generate edge-checker for actual (ν, μ, λ) combination, insert a n-edge
    from the root node to it and put reference to edge-checker on *open-list*; (2)
    else put reference to existing edge-checker on *open-list*;
  if no element on *open-list* (3)
  /* production-LHS consists of a single g-node */
  then generate a production-node as successor of the actual ν-node-checker;
  else /* generate c-subgraph-checkers */
    while *open-list* has more than 1 element **do**
      set *changes* to false;
      for every n-node referenced in *open-list* (4)
        for every successor in the network of the actual n-node considered (5)
          if the other antecedent of the actual successor is on *open-list* too
          and the subgraph-checker at the actual successor n-node checks for
          a valid subgraph of the actual production (6)
          then remove references to actual n-nodes from *open-list*;
          put reference to successor n-node on *open-list*;

```
            set changes true;
        if no changes
        then remove from open-list the references to
            - the subgraph-checker of the largest subgraph g of the actual
              production's left-hand side                                  (7)
            - some n-node g' that checks an edge or a subgraph
              connected to the subgraph g;
        generate a new n-node as successor of the two n-nodes, i.e.
        a subgraph-checker checking for the combination of g and g';
        put a reference to the new n-node on open-list;
    /* what is left on the list is a reference to a subgraph equal to the
    left-hand side of the actual production */
    generate a p-production-node as successor of the only n-node left on open-list;   (8)
end create-RETE-network
```

In step (7) we use g and g' to denote both a subgraph of the left-hand side of a production and the n-node representing this subgraph in the RETE-network. The largest subgraph is formally defined by having the greatest number of edges, or equivalently, the greatest distance from the root-node in the RETE-network. Generally, the largest subgraph will not be unique. In the extreme case, i.e. at the beginning of the creation of the n-nodes for a production, it can consist of only a single edge. The combination of g and g' is obtained by taking the union of the set of edges of g and g'.

As an example for the algorithm *create-RETE-network* we consider the left-hand sides of Fig. 4 and the resulting network in Fig. 5. The production p_1 generates node 2 in (1) and node 10 in (3). Next, the production p_2 generates node 5 in (2) and node 11 in (8) not going through (4) - (7) as there is only one edge in p_2. Now the production p_3 finds node 5 and generates node 6 in (2). There are no successors of nodes 5 and 6 in (6), so in (7) node 8 is generated. As this completes the left-hand side of p_3, node 8 gets a successor, node 12, in (8). Next, the production p_4 finds nodes 5 and 6 in (2) and generates node 7. *open-list* contains now references to nodes 5,6 and 7. In (6) node 8 is found as a successor of nodes 5 and 6, thus the new *open-list* contains references to nodes 7 and 8. Now, nodes 8 and 7 are selected in (7) as largest subgraph-checker and remaining edge-checker, respectively, leading to the generation of node 9, and finally node 13 in (8).

As can be seen, the algorithm creates a network, reusing checkers that have already been generated. There are two cases where checkers are not reused, namely in (2) where as many similar edge-checkers are needed as there are similar g-edges in a production, and another more complex case: Consider two productions with left-hand sides G_1 and G_2 such that G_2 is a subgraph of G_1, and G_1 has already been compiled into the network. If the subgraph-checkers using edges not contained in G_2 come first in the network, then the following subgraph-checkers (although checking for elements in G_2,) cannot be used.

For complexity considerations let E and N be the maximum number of edges and nodes of the left-hand side of a production, respectively and P the number of productions. For any production there are at most E edge-checkers and E subgraph-checkers, thus the space complexity is $O(PE)$. For any production, there are at most E loops at while, E loops at (4) and PE loops at (5). Checking at (6) can be done in time N as we know what parts of the left-hand side are represented by the different n-nodes. Thus the time-complexity for the compilation amounts to $O(P^2E^3N)$.

4 Dynamic Behaviour of the RETE-network

The different types of n-nodes and their role in receiving, testing, storing, and transmitting of subgraphs have already been informally discussed in section 3. In the present section, we give a more formal description of the procedures that are attached to the n-nodes of the RETE-network. As already done in in section 3, we call any data element that is stored in a local memory or that is transmitted from one n-node to another a subgraph, no matter if it consists of only a single node, a single edge, or a collection of nodes and edges. Any edges belonging to the embedding are to be handled as if they belonged to the left-hand or the right-hand side of a production.

When running the procedures in the RETE-network we must distinguish if we are processing a subgraph that has its origin in the initial graph, or in the right-hand side of a production, or in the left-hand side of a production. In the first two cases we eventually have to add new elements to the conflict set while the third case usually causes the deletion of elements from the conflict set. It is not necessary to further distinguish between the first and the second case. Here we input the nodes and edges of a subgraph - i.e., the initial graph or a right-hand side that has to be added to the underlying graph - according to the format given by (3.1) and (3.2) element by element to the root-node of the network. More precisely, for each input element we call the procedure *root (input-element)*. The procedure *root* calls other procedures that will be given below.

procedure *root (input-element)*
determine type of input-element;
if type = node **then**
  for all ν-node-checkers that are direct successors of the root-node **do**
    call *ν-node-checker (input-element)*;
else /* type = edge */
  for all (ν, μ, λ)-edge-checkers that are direct successors of the root-node **do**
    call *(ν, μ, λ)-edge-checker (input-element)*;
end *root*

procedure *ν-node-checker (input-node)*
/* input-node = $(v, \alpha(v))$ * /
if $\alpha(v) = \nu$ **then**
  for all p-production-nodes that are direct successors of the actual ν-node-checker **do**
    call *p-production-node (input-node)*;
    /* such a production node represents a left-hand side consisting of only
    one node labeled with ν * /
end *ν-node-checker*

procedure *(ν, μ, λ)-edge-checker (input-edge)*
/* input-edge = $((v, \alpha(v)), (w, \alpha(w)), \beta(v, w))$ * /
if $\alpha(v) = \nu \wedge \alpha(w) = \mu \wedge \beta(v, w) = \lambda$ **then**
  for all c -subgraph-checkers that are direct successors
  of the actual (ν, μ, λ)-edge-checker **do**
    call *c-subgraph-checker (input-edge)*;
  for all p-production nodes that are direct successors
  of the actual (ν, μ, λ)-edge-checker **do**
    call *p-production-node (input-edge)*;
end *(ν, μ, λ)-edge-checker*

procedure *p-production-node (input-subgraph)*
store input-subgraph in local memory;
/* since the conflict set is the set of all such local memories, this operation adds a new instance of the production p to the conflict set */
end *p-production-node*

procedure *c-subgraph-checker (input-subgraph)*
/ * c is a list of pairs of nodes, as described in the text below */

if call of procedure was from left predecessor **then**
store input-subgraph in left memory;
for all subgraphs g_R in right memory **do**
call *test (input-subgraph,* g_R, c);
if *test* successful **then**
call *construct (input-subgraph,* g_R, g);
for all c'-subgraph-checkers that are direct successors
of the actual c-subgraph-checker **do**
call *c'-subgraph-checker* (g);
for all p-production-rules that are direct successors
of the actual c-subgraph-checker **do**
call *p-production-node* (g);
else /* call was from right prodecessor */
⋮
this part is in exact analogy to the then-part
⋮
end *c-subgraph-checker*

The procedure *c-subgraph-checker* calls two other procedures, *test* and *construct*, that were already mentioned in section 3. Their formal description is given below. Let the input-subgraph g_I be given by its node list n_I and its edge list e_I:

$n_I = ((v_1, \alpha(v_1)), (v_2, \alpha(v_2)), ..., (v_n, \alpha(v_n))),$

$e_I = (((v_{i1}, \alpha(v_{i1})), (v_{i2}, \alpha(v_{i2})), \beta(v_{i1}, v_{i2})), ((v_{i3}, \alpha(v_{i3})), (v_{i4}, \alpha(v_{i4})), \beta(v_{i3}, v_{i4}))$
$, \ldots, ((v_{im-1}, \alpha(v_{im-1})), (v_{im}, \alpha(v_{im})), \beta(v_{im-1}, v_{im}))).$

Similarly, we have for the graph g_R

$n_R = ((w_1, \alpha(w_1)), (w_2, \alpha(w_2)), ..., (w_r, \alpha(w_r))),$

$e_R = (((w_{j1}, \alpha(w_{j1})), (w_{j2}, \alpha(w_{j2})), \beta(w_{j1}, w_{j2})), ((w_{j3}, \alpha(w_{j3})), (w_{j4}, \alpha(w_{j4})), \beta(w_{j3}, w_{j4}))$
$, \ldots, ((w_{js-1}, \alpha(w_{js-1})), (w_{js}, \alpha(w_{js})), \beta(w_{js-1}, w_{js}))).$

These graphs are stored in the local memories of the left and right input, respectively, of the subgraph-checker that calls the procedure *test.* The task of this procedure is to check if there exists a graph g_R in the right memory of the actual subgraph-ckecker such that the combination of g_I and g_R, i.e. the union of their edges, is isomorphic to the subgraph represented by the actual subgraph-checker. (If the call to *test* is from the else-part of the procedure *c-subgraph-checker* then the left memory will be searched for g_R.) As a consequence of the static structure of the RETE-network, the only task that has to be actually solved by *test* is to check if some nodes of g_I are identical with some nodes of g_R. Such a check can be completely specified by giving a list of pairs of nodes $c = ((v_{k1}, w_{k1}), ..., (v_{kt}, w_{kt}))$. This list plays the role of a parameter for the actual subgraph-checker, similar to the node and edge labels for the node- and subgraph-checkers. The first node of a pair in c is from g_I while the second node is from g_R. In detail, the procedure *test* consists of two parts. First it is checked if $v_{k1} = w_{k1} \wedge v_{k2} = w_{k2} \wedge \ldots \wedge v_{kt} = w_{kt}$. Secondly, it is checked if $v \neq w$ for any pair (v, w) of nodes not contained in c, where v is from g_I and w is from g_R. As an example, the instance of *test* that runs in the n-node 8 in Fig. 6 checks if the first node of an edge in the left memory is identical to the second node of an edge in the right memory. Similarly, in node 9 it is tested whether (1) the first node of a pair of edges stored in the left memory is identical to the second node of an edge in the right memory, and (2) the last node of the same pair of edges in the left memory is identical to the first node of the same edge in the right memory.

The procedure *construct* that is called from a subgraph-checker constructs a graph g – the third element in the parameter list of the procedure *construct* – from two other graphs g_I and g_R – the first and second element in the parameter list – by "gluing together" g_I and g_R at their common nodes. More formally, g is given by a node list n and an edge list e:

$n = n_I \circ ((w\prime_1, \alpha(w\prime_1)), ..., (w\prime_l, \alpha(w\prime_l)))$, [3]
$e = e_I \circ e_R$.
The list $((w\prime_1, \alpha(w\prime_1)), ..., (w_{l}\prime, \alpha(w_{l}\prime)))$ is obtained from n_R by deleting the elements $(w_{k1}, \alpha (w_{k1})), \ldots, (w_{kt}, \alpha (w_{kt}))$ which are already contained in n_I.

After the procedure *root* has been called for all nodes and edges of either the initial graph or the right-hand side of a production, the conflict set will be properly updated. The corresponding procedures that have to be called for the nodes and edges of a left-hand side, which has to be deleted from the conflict set, are exactly the same. The only exception is that, instead of storing a subgraph into a local memory of either a subgraph-checker or a production-node, we have to delete it from that memory. As an example, the reader may verify the procedures given above by going step by step through the example presented in section 1.

5 Computational Complexity Analysis

In this section we shall discuss the computational complexity of a naive implementation and the RETE-approach for the determination of the conflict set. For these considerations we use G as the number of edges of the underlying graph and L (resp. R) as the number of edges of a production's left-hand side (resp. right-hand side). As the numbers of nodes in these graphs are at most of the same order as the edges, it is not necessary to define them separately for this discussion. Let P be the number of productions for a given grammar.

First we shall discuss the complexity of the naive version. The best case is given when all node and edge labels in all productions and in the underlying graph are pairwise different. By application of a production, a right-hand side of size R is joined to the underlying graph. This right-hand side happens to be the left-hand side of a production that wasn't in the conflict set before. The naive algorithm has to compare all edges, resulting in a time complexity of $O(PL(G + R))$. The worst case is given, when there are no labels whatsoever and the time complexity in this case is $O(PL^2(G + R)^L)$. Notice that $O(L^2)$ time is needed to check the adjacencies of a graph of size L.

For the RETE-approach the best case is the same case as the best case for the naive approach. Notice that every subgraph-checker in the network has a left and a right memory list, storing the subgraphs that passed the test at the predecessing n-node. Every p-production-node has one memory-list containing all applicable instances of the corresponding left-hand side. In the best case now there are $P \cdot L$ edge-checkers for all types of edges, and all memory lists, where edges and subgraphs of the new right-hand side will be stored, are empty before the right-hand side is processed by the network. As all labels are distinct there is at most one edge found by any edge-checker and all respective memory lists will contain at most one subgraph. Therefore, the time-complexity in the best case is $O(PLR)$ for insertion of a right-hand side. In order to delete a left-hand side we have to compare all edges with the edges stored at the edge-checkers and to process the memory lists of all successor n-nodes when a deletion actually took place in the memory list of a n-node. As all labels are distinct there is at most one element in any memory list and the deletion of a left-hand side has a time complexity of $O(PL^2)$.

In the worst case all edges and nodes have the same label. If a right-hand side is processed by the network, any edge-checker delivers all edges to all successor memory lists and every subgraph-checker can combine every element of the left memory list with every element of the right memory list to pass them on as subgraphs to its successors. Without loss of generality, we assume that all subgraph-checkers have edge-checkers as right predecessors. So, for a production of length L there are $L - 1$ subgraph-checkers. Two edge-checkers transmit R new and G old edges to the first subgraph-checker sc_1, each; sc_1 transmits $(R+G)^2$ subgraphs to sc_2, a.s.o.This results in $(R+G)^L$ applicable instances in the p-production-node's memory-list. There are $P \cdot L \cdot R$ edge-checks and $(R+G)^2 + (R+G)^3 + \ldots + (R+G)^L$ subgraph checks for every production resulting in a time-complexity of $O(PL^2(R+G)^L)$.

[3] $\circ$ is the concatenation operator

For the deletion of a left-hand side all elements of all memory lists have to be processed. The number of memory elements is $(L+G) + (L+G) + (L+G)^2 + (L+G) + (L+G)^3 + \ldots +(L+G)^{L-1} + (L+G) + (L+G)^L$, for every production. The time complexity in the worst case amounts thus to $O(PL^2(L+G)^L)$.

In summary we consider the case where L and R are roughly of the same size, resulting in a best case performance of $O(PL^2)$ for the RETE approach (compared to $O(PL(G+L))$ in the naive version) and a worst case performance of $O(PL^2(L+G)^L)$ (compared to $O(PL^2(L+G)^L)$ in the naive version). So in the worst case the RETE algorithm performs not better than the naive version. However, as it can be seen from the best case performance we do not always have to reconsider all edges of the underlying graph for the determination of the conflict set after the application of a production, and can therefore expect that the RETE algorithm performs better than the naive version. As a matter of fact the inital processing of the start graph has the same time complexity as the naive version.

A problem left to future research is an analysis of the computational complexity of the average-case, as it was done for the original RETE-algorithm in [8].

6 Experimental Examples

A graph grammar implementation according to Fig. 2 based on the RETE algorithm has been done on an IBM-RT computer using the programming language C. The resulting program is portable and includes a comfortable interface to the user for defining a grammar and monitoring its execution. We run a number of experiments in order to empirically assess the efficiency of the proposed algorithm. For the purpose of comparison we also implemented the naive algorithm according to Fig. 1.

In the first experiment, we used a grammar consisting of three productions. Two of these productions contract an edge with a particular label to a single node while the third production just changes a node label. The start graphs and the productions were designed such that there existed in total only six occurrences of the productions in any start graph, independent of the size of the start graph. Furthermore, no production application created a new instance in the conflict set. The curve in Fig. 8 shows the CPU-time in seconds used for running the productions on a number of start graphs of varying size until the termination condition became true. In this experiment a derivation sequence was terminated if there were no more applicable productions. The selection of a production from the conflict set was based on a production priority order. This order was the same for both the RETE based and the naive version. The time behaviour shown in Fig. 8 is easy to explain. Raising the number of nodes in the initial graph causes a growing effort for the naive version when searching for the applicable productions in *each* run through the while-loop in Fig. 1. Formally, the factor G in the expression $O(PL^2(G+R)^L)$, see section 5, is increased. In the RETE based version, however, an increased effort is required only in the processing of the initial graph. Subsequently, the time behaviour is completely independent of the initial graph because any application of a production affects the RETE network only locally.

In a second experiment, we were concerned with circuit diagrams representing serial and parallel compositions of resistors. The grammar describes how certain configurations of resistors can be simplified to a canonical standard form. Again, we run the grammar productions until the conflict set became empty, using initial graphs of varying size. The CPU-times measured in this experiment are shown in Fig. 9. The two curves have an explanation that is similar to the first experiment.

Further experiments included the interpretation of circuit diagrams using the grammar described in [9], and the analysis of graphs representing simple scenes of three-dimensional objects. The grammars that were used comprised up to thirty productions, and the initial graphs consisted of up to 200 nodes. In all these experiments we noticed that the RETE based graph grammar implementation is significantly faster than the naive version. Generally, we recognized that the time behaviour observed in the experiments is in good agreement with the observations made in section 5.

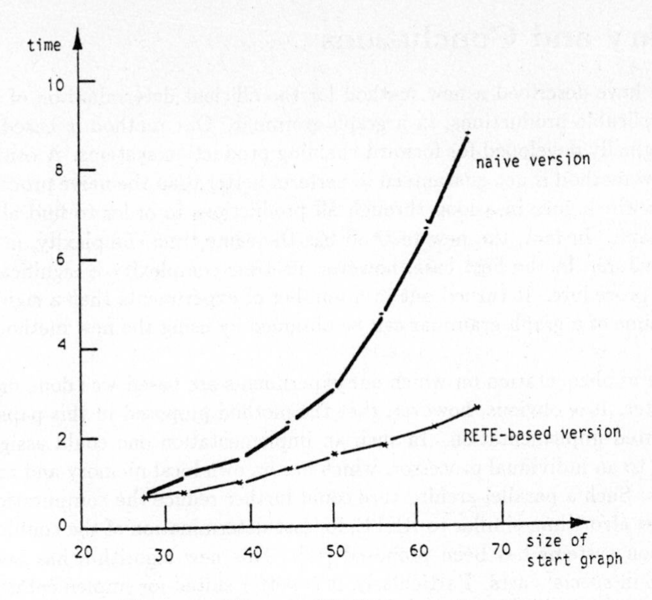

Figure 8: Time used in first experiment

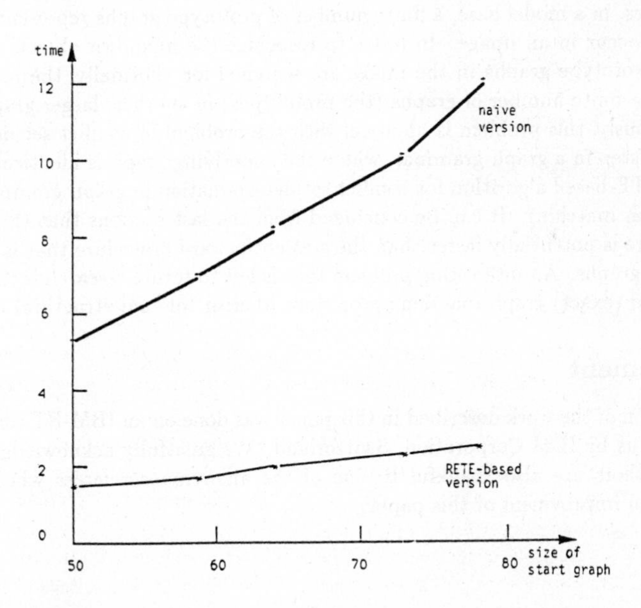

Figure 9: Time used in experiment with circuit diagrams

7 Summary and Conclusions

In this paper, we have described a new method for the efficient determination of the conflict set, i.e. the set of applicable productions, in a graph grammar. Our method is based on a technique that has been originally developed for forward chaining production systems. A complexity analysis shows that the new method is not guaranteed to perform better than the naive procedure for conflict set determination which goes in a loop through all productions in order to find all occurrences in the underlying graph. In fact, the new method has the same time complexity, in the worst case, as the naive procedure. In the best case, however, its time complexity is significantly lower than that of the naive procedure. It turned out in a number of experiments that a significant speed-up in the execution time of a graph grammar can be obtained by using the new method for conflict set determination.

The prototype implementation on which our experiments are based was done on a conventional sequential computer. It is obvious, however, that the method proposed in this paper lends itself to a parallel distributed implementation. In such an implementation one could assign, for example, one network node to an individual processor, which has its own local memory and can communicate with its neighbors. Such a parallel architecture could further reduce the computational complexity.

Recently, a new algorithm, similar to RETE, for fast determination of the conflict set in forward chaining production systems has been proposed [10]. This new algorithm has been shown to be superior to RETE in special cases. Particularly, it is better suited for implementation on a parallel tree-structured architecture. It would be interesting to investigate if this new algorithm could be generalized to graph grammars, similar to the RETE algorithm, and to compare it to the approach proposed in this paper.

It is known that graph grammars can be applied to the recognition of structures in images [9]. Another useful method for this problem is prototype graph matching. In prototype graph matching one stores, in a model base, a finite number of prototype graphs representing prototypical objects that may occur in an image. In order to recognize the unknown objects in an image, all instances of the prototype graphs in the image are searched for. Formally, the problem is to find all occurrences of a finite number of graphs (the prototypes) in another, larger graph (representing the image). Obviously, this problem is identical with the problem of conflict set determination for the first rewriting step in a graph grammar, where the underlying graph is identical with the initial graph. So the RETE based algorithm for conflict set determination in graph grammars can be used for prototype graph matching. It can be concluded from the last sections that this new prototype matching procedure is potentially faster than the straightforward procedure that is based on a loop over all prototype graphs. An interesting problem that is left to future research is the generalization of the RETE based (exact) graph matching procedure to error tolerant structural matching [11].

Acknowledgement

The implementation of the work described in this paper was done on an IBM-RT computer that was made available to us by IBM Corporation, Switzerland. We greatfully acknowledge their generous support. The authors are also thankful to one of the anonymous referees whose very detailed comments led to an improvment of this paper.

References

[1] Claus, V./ Ehrig, H./ Rozenberg, G. (ed.): *Graph-Grammars and their application to computer science and biology*, Springer Lecture Notes 73, 1979

[2] Ehrig, H./ Nagl, M./ Rozenberg, G. (ed.): *Graph-Grammars and their application to computer science*, Springer Lecture Notes 153, 1983

[3] Ehrig, H./ Nagl, M./ Rozenberg, G./ Rosenfeld, A. (ed.): *Graph-Grammars and their application to computer science*, Springer Lecture Notes 291, 1987

[4] Bunke, H.: *On the generative power of sequential and parallel programmed graph grammars*, Computing Vol. 29, pp 89–112, 1982

[5] Brownston, L./ Farell, R./ Kant, E./ Martin, N.: *Programming expert systems in OPS5. An introduction to rule-base programming*, Addison-Wesley, 1986

[6] Forgy, C.L.: *RETE, a fast algorithm for the many pattern / many object pattern match problem*, Artificial Intelligence Vol. 19, pp 17–37,1982

[7] Nagl, M.: *Set theoretic approaches to graph-grammars*, in Ehrig, H./ Nagl, M./ Rozenberg, G./ Rosenfeld, A. (ed.): Graph-Grammars and their application to computer science, Springer Lecture Notes 291, pp 41–54, 1987

[8] Albert, L. / Fayes, F.: *Average case complexity analysis of the RETE pattern-match algorithm*, Springer Lecture Notes 317, pp. 18-37, 1988

[9] Bunke, H.: *Graph grammars as a generative tool in image understanding*, in Ehrig, H./ Nagl, M./ Rozenberg, G. (ed.): Graph-Grammars and their application to computer science, Springer Lecture Notes 153, pp 8–19, 1983

[10] Miranker, D.P.: *TREAT: A new and efficient match algorithm for AI production systems*, Research Notes in AI, Morgan Kaufmann Publishers, 1990

[11] Bunke, H./ Allermann, G.: *Inexact graph matching for structural pattern recognition*, Pattern Recognition Letters Vol. 1, pp 245–253, 1983

AN APPLICATION OF GRAPH GRAMMARS

TO THE ELIMINATION OF REDUNDANCY

FROM FUNCTIONS DEFINED BY SCHEMES

Didier CAUCAL

IRISA , Campus de Beaulieu , F35042 Rennes Cedex , France

email : caucal@irisa.fr

ABSTRACT: The infinite tree obtained classically by unfolding the definition of a recursive scheme, contains several identical subtrees. When they are identified, the resulting graph is generated by a deterministic graph grammar, if the scheme is monadic. We show how to extract one such a grammar from the scheme.

Keywords: recursive program scheme, context-free grammar, deterministic graph grammar.

CONTENTS

0. INTRODUCTION

A good strategy for evaluating functional programs should (1) avoid computing useless values, i.e. values that are never used, and (2) avoid computing twice the same value. Neither point can be detected statically, but point (2) can be done syntactically by a shared representation of the program. This shared representation will be deduced from the text of the program and has to be finite. The method that we use for this construction is to consider a free interpretation of the recursive scheme obtained classically by unfolding the definitions infinitely. The resulting infinite tree is then contracted as much as possible: identical subtrees, corresponding to the same computation are shared. The result is an infinite graph. It turns out that this infinite graph is generated by a finite system of patterns, which will be the shared representation of the program. To clarify matters, let us introduce the classical example of the Fibonacci function :

$$F(n) = \text{if } n \leq 1 \text{ then } 1 \text{ else } F(n-1) + F(n-2) \text{ endif}.$$

An associated recursive program scheme [Ni 75], [Gu 81], in short RPS, can be the following equation :

$$F(n) = f(p(n), g(F(h(n)), F(h(h(n)))))$$

where f,g,p,h are base function symbols with respective arities 2,2,1,1. The base function f is interpreted as the conditional which gives 1 when its condition is realized and its second argument otherwise ; g is the addition ; p is the predicate $n \leq 1$, and h is the decrement function.

Every RPS can be seen as a term rewriting system : the defining equation of F can be oriented to give the following rule :

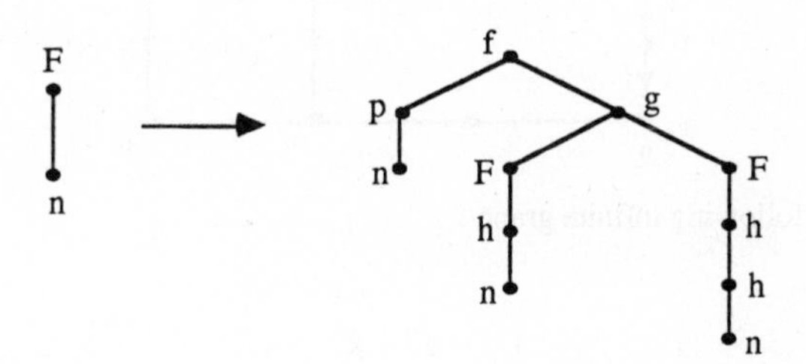

To the defined function F, we associate the infinite tree below, obtained by unfolding the right-hand side of the rule, starting from F(n).

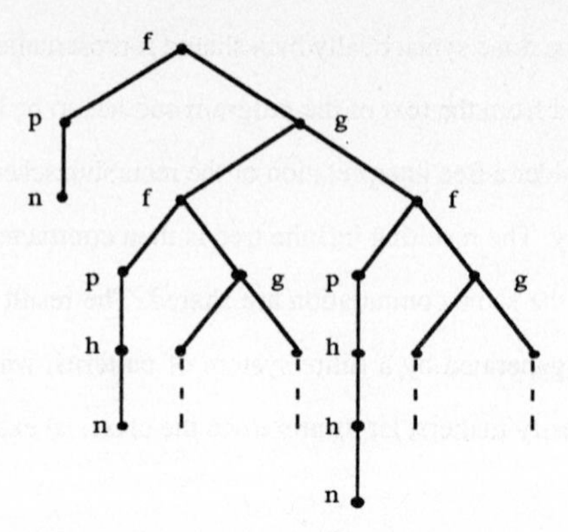

This unfolded tree describes all possible computations of the defined function, but it does not impose any evaluation strategy. Nevertheless, because of the two recursive calls of F in its definition, every evaluation of F by unfolding its tree, is of exponential complexity.

As in [St 80], [Pa 82], [Ba et al. 87], [Ho-Pl 88], we can identify the occurrences of identical subtrees in the definition. We then obtain the following graph grammar where the edge labels indicate the rank of the argument :

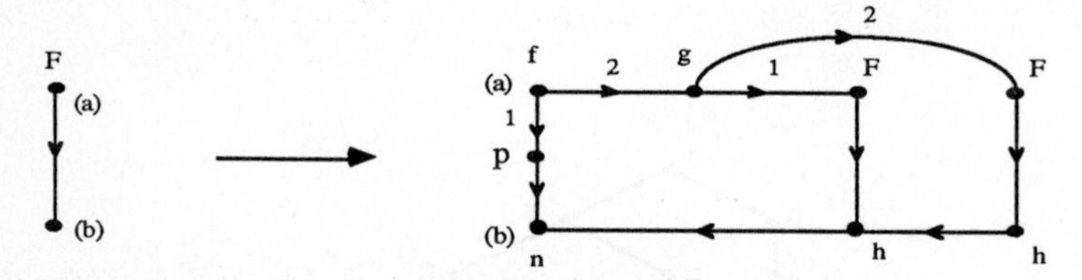

The unfolding of this rule gives the following infinite graph :

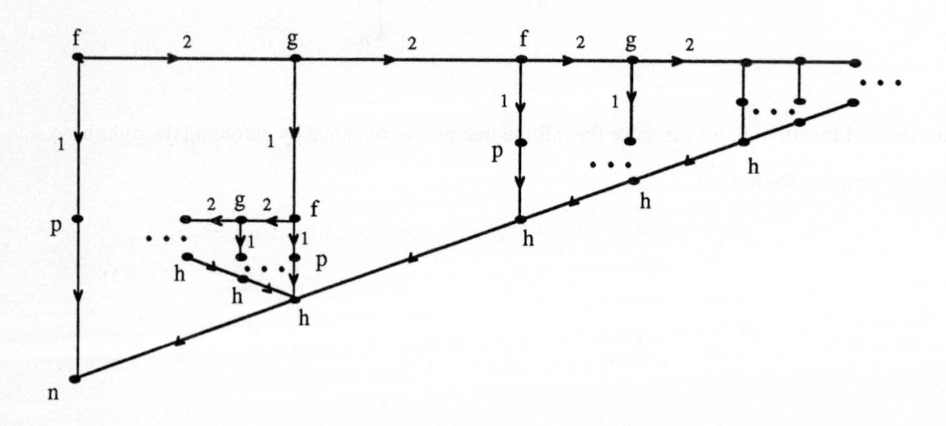

Here again, because of the two recursive calls of F in the right-hand side of the grammar rule, every evaluation of F by unfolding the above graph is of exponential complexity.

There exists a better quotient obtained by identifying all the identical subtrees in the unfolded tree. We obtain the following 'canonical' graph :

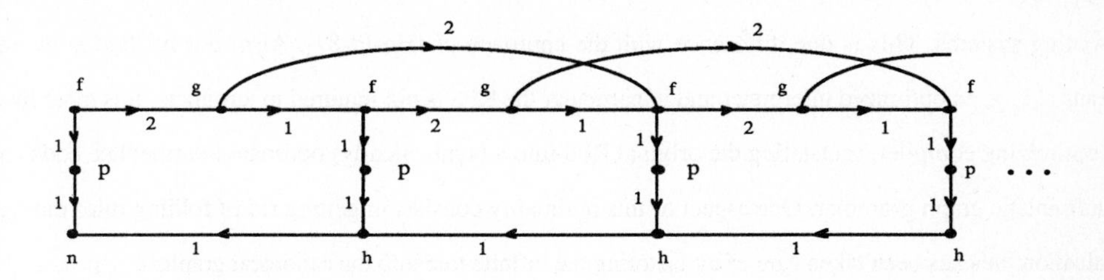

Such a graph is a 'pattern' graph, or an equational graph in the sense of Bauderon [Ba 89] and Courcelle [Co 89 a] [Co 89 b], i.e. this graph is obtained by iterated rewritings of a deterministic (see Section 3) graph grammar, having the following unique rule :

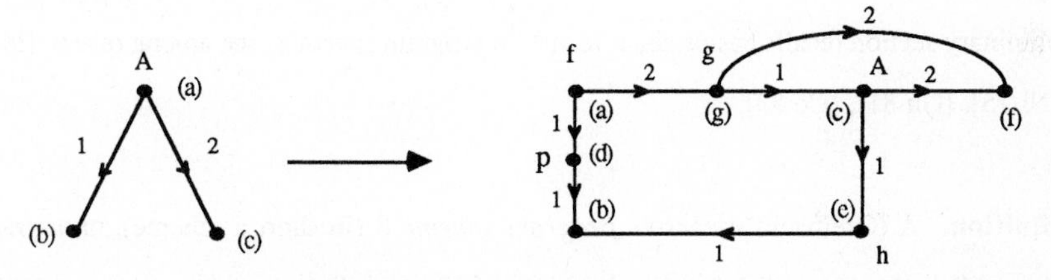

The right-hand side of that rule is the basic pattern of the Fibonacci function. Taking into account the interpretation of the base functions, this grammar can be transformed easily into the following evaluation program of the Fibonacci function :

```
procedure A(a,b,c) ;
    d ← b ≤ 1 ;
    if d then a ← 1 endif ;
    e ← b - 1 ;
    call   A(!c,e,!f) ;    {the ! signs mean that the call returns immediately}
    g ← c + f ;                                          {if c and f are evaluated}
    a ← g ;
endprocedure
```

and the execution of A(a,b,c) gives a = Fib(b).

This evaluation program of the Fibonacci function is of complexity linear in time and space.

Of course, the difficulty of this method is to construct the pattern(s) of the canonical graph from the RPS. In this paper, we solve this problem for 'monadic' RPS (meaning that only the base functions may have several arguments). Since there exists [Ca-Mo 90] (or Example 7 in Section 3) canonical graphs of polyadic RPS which are not pattern graphs, this method cannot be applied to all RPS, a fortiori to all term rewriting systems. This is one difference with the approach of [Ho-Pl 87]. Also, our method is by no means akin to an optimized interpreter and in particular the RPS is not required to terminate. It is more like an optimizing compiler, translating the original RPS into a (syntactically) optimate intermediate code : a deterministic graph grammar. One aspect of this optimality consists in getting rid of folding rules during evaluation: this has been taken care of by factoring the infinite tree into the canonical graph.

1. SCHEME

This preliminary section recalls basics about recursive program schemes, see among others [Ia 60], [Ga-Lu 73], [Ni 75], [Gu 81], [Co 83].

Definition. A (Greibach) *recursive program scheme* S (in short a scheme), on a graded alphabet F and on a set $X = \{x_1,\ldots,x_n,\ldots\}$ of variables, is a finite set of equations of the form $fx_1\ldots x_n = t$ (where $f \in F_n$, $t \in T(F,\{x_1,\ldots,x_n\})$), satisfying the following conditions :

(i) the scheme is functional : $((fx_1\ldots x_n = s) \in S \wedge (fx_1\ldots x_n = t) \in S) \Rightarrow s = t$

(ii) the scheme is in Greibach form : $(t = gx_1\ldots x_m) \in S \Rightarrow g \notin \Phi(S)$ where

$\Phi(S)$ denotes the set $\{ f \mid (fx_1\ldots x_n = t) \in S \}$ of *defined functions* by S.

Example 1. The recursive function A defined by

A(n) = if n = 0 then 0 else n - A(B(n-1)) endif

B(n) = if n = 0 then 0 else n - B(A(n-1)) endif

can be represented by the following scheme :

$A(n) = f(p(n),g(n,A(B(h(n)))))$

$B(n) = f(p(n),g(n,B(A(h(n)))))$.

The equations of a scheme S may be used as rewriting rules, and allow to unfold a given term t into a (usually infinite) derivation tree S(t).

Definition. Given a scheme S and a term $t \in T(F,X)$, the *unfolded tree* S(t) is defined as

(i) $S(t) = f(S(t_1),\ldots,S(t_n))$ if $t = ft_1\ldots t_n$ and $f \notin \Phi(S)$

(ii) $S(t) = S(t'[x_1 \leftarrow t_1, \ldots, x_n \leftarrow t_n])$ if $t = ft_1\ldots t_n$ and $(fx_1\ldots x_n = t') \in S$.

The existence (resp. unicity) of such a tree results from condition (ii) (resp. (i)) of the scheme definition.

Example 2. From the scheme of Example 1, the unfolded tree from An is equal to

$$f(pn,g(n,f(p(f(phn,g(hn,\ldots)),\ldots))) .$$

Unfolded trees generally contain isomorphic subtrees, describing redundant computations. Avoiding redundancy via systematic sharing of isomorphic subtrees leads to 'canonical' graphs. We show in the sequel that canonical graphs turn out to have a regular structure for schemes, satisfying the following three conditions :

(i) the scheme is *monadic* : $\Phi(S) \subseteq F_1$, i.e. every defined function is of arity 1

(ii) the scheme is *without constant* : $F_0 = \emptyset$

(iii) the scheme is *reduced* : $(fx_1\ldots x_n = t) \in S \Rightarrow S(fx_1\ldots x_n)$ has a leaf.

From now on, a scheme is to be understood as a monadic and reduced scheme without constant (we let the reader check that this is the case with Example 1) and the set X of variables is restricted to $\{x\}$.

For any graded alphabet F, the 'split' W(F) of F is defined as the set of 'split' symbols :

$$W(F) = \{ (f,i) \mid f \in F_n \wedge 1 \leq i \leq n \} .$$

Given a scheme S, let $(fx = t)$ be an equation in S. All leaves of the unlabelled tree S(fx) are then labeled by x (since S is monadic and uses no constant). In the sequel, we allow ourselves to identify S(fx) with the set of labeled arcs $u \xrightarrow{(g,i)} ui$ where u and ui are nodes of the unfolded tree, and g is the label of u.

Example 3. The tree of Example 2 is identified with the following set of labeled arcs :

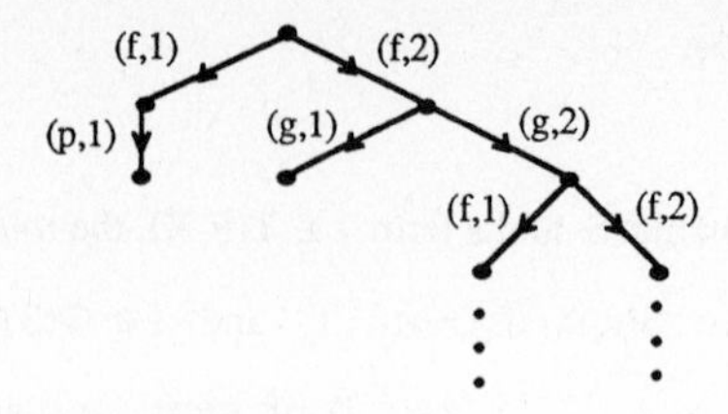

To eliminate syntactic redundancy, we identify isomorphic subtrees.

Definition. The *canonical graph* Can(S(fx)) of every unfolded tree S(fx) is defined by

$$\mathrm{Can}(S(fx)) = \{ [u]_\equiv \xrightarrow{a} [v]_\equiv \mid (u \xrightarrow{a} v) \in S(fx) \}$$

where $u \equiv v$ if and only if the subtrees of S(fx) on u and v are isomorphic.

To decide the isomorphism of subtrees of an unfolded tree T, we associate to every vertex u the *path-language* L(T,u) of the path labels from u to a *terminal vertex* (a vertex which is source of no arc), i.e. L(T,u) is the smallest language over W(F) such that

$$L(T,u) = \cup\{ a.L(T,v) \mid (u \xrightarrow{a} v) \in T \}.$$

In particular $L(T,u) = \emptyset$ if u is a terminal vertex. So, the isomorphism of subtrees of T is equivalent to the equality of their path-languages.

Proposition 1.1 *The subtrees of an unfolded tree* S(fx) *on* u *and* v *are isomorphic if and only if* u *and* v *have the same path-languages :* L(S(fx),u) = L(S(fx),v).

Proof.

Clearly, isomorphic subtrees have the same path-language. Conversely, the unfolded tree S(fx) is *deterministic* : two arcs with the same source have different labels. Furthermore S being reduced, from every vertex u of S(fx), there exists a path from u to a terminal vertex. Consequently, the subtree of S(fx) on u is isomorphic to the deterministic graph associated with L(T,u), i.e. the following graph :

$$\{ v^{-1}.L(T,u) \xrightarrow{a} (va)^{-1}.L(T,u) \mid v \in W(F)^* \wedge a \in W(F) \wedge \exists w,\ vaw \in L(T,u) \}$$

where $v^{-1}.L(T,u) = \{ w \mid vw \in L(T,u) \}$ is the left quotient of L(T,u) by v.

So, the subtrees of S(fx) are isomorphic if they have the same path-languages. ♦

Now, we need an effective construction of the canonical graph Can(S(fx)) from S and fx.

2. PREFIX TRANSITION GRAPH

In order to construct the canonical graph Can(S(fx)) of S(fx), we want to show that such a graph is isomorphic to the prefix transition graph [Ca 90] of a reduced simple grammar constructible from S.

Let us recall that a *simple grammar* is an ε-free cf-grammar in Greibach normal form which is LL(1), i.e. for all rules $A \to au$ and $A \to av$ of G where a is a terminal, we have $u = v$. A cf-grammar is *reduced* if G generates a terminal word from any non-terminal.

Given an ε-free cf-grammar in Greibach normal form, and a non-terminal A, the *prefix transition graph* P(G,A) of G accessible from a non-terminal A is the following graph :

$$P(G,A) = \{ u \xrightarrow{a} v \mid A \underset{G}{\longmapsto}^{*} u \wedge u \overset{a}{\underset{G}{\longmapsto}} v \}$$

where $\overset{a}{\underset{G}{\longmapsto}}$ is a *prefix rewriting step* on the non-terminal words, labelled by a terminal a, and defined by $u \overset{a}{\underset{G}{\longmapsto}} v$ if there exist a rule $X \to ax$ of G and a word y such that $u = Xy$ and $v = xy$,

and $\underset{G}{\longmapsto}^{*}$ is an arbitrary sequence of such unlabelled steps.

Theorem 2.1 *Any pair* (S,f) *of a scheme* S *and an axiom* f *may be effectively transformed into a reduced simple grammar* G *such that the graphs* Can(S(fx)) *and* P(G,f) *are isomorphic.*

The sketch of the proof is given in 3 steps.

Step 1 : we put the scheme S into *Greibach normal form* : each rule has the form $fx = gt_1...t_n$ where the t_i 's are in $T(\Phi(S),\{x\})$ which corresponds to $\Phi(S)^*.x$ because S is monadic.

To do this, it suffices to replace iteratively each proper tree $t \in (F\text{-}\Phi(S)).(\Phi(S)^*.x)^*$ in a right-hand side of a rule of S by hx where h is a new symbol in F_1, and to add in S the equation hx = t. So, we obtain a scheme S' in Greibach form which is equivalent to S : S'(fx) = S(fx) for every $f \in \Phi(S)$.

Example 4. From Example 1, we obtain the following scheme in Greibach normal form :

Ax = f(Px,Cx)

Bx = f(Px,Dx)

Cx = g(x,ABHx)

Dx = g(x,BAHx)

$Px = px$

$Hx = hx$.

Step 2 : As [Co-Vu 76], we convert the scheme S in Greibach normal form into a simple grammar G. This elementary transformation amounts to replace each equation $fx = gu_1x\ldots u_nx$ to a grammar rule $f = (g,1)u_1 + \ldots + (g,n)u_n$ (here, terminals are in the split alphabet $W(F\text{-}\Phi(S))$, and non-terminals are in $\Phi(S)$). So, the language of terminal words generated by G from f is equal to the path-language $L(S(fx),f)$ of the unfolded tree $S(fx)$ (see Proposition 1.1), hence G is reduced.

Example 5. From Example 4, we obtain the following simple grammar

$A = (f,1)P + (f,2)C$

$B = (f,1)P + (f,2)D$

$C = (g,1) + (g,2)ABH$

$D = (g,1) + (g,2)BAH$

$P = (p,1)$

$H = (h,1)$.

Step 3 : Finally, we transform the grammar to an equivalent simple one, and ensures that non-terminal words define always different languages [Co 74]. Such a transformation may be done by a polynomial algorithm [Ca 89]. By Proposition 1.1, the prefix transition graph of the simplified grammar (from the axiom f) is isomorphic to the canonical graph of the tree $S(fx)$.

Example 6. From Example 5, we obtain the following canonical simple grammar :

$A = (f,1)P + (f,2)C$

$C = (g,1) + (g,2)AAH$

$P = (p,1)$

$H = (h,1)$

which corresponds to the scheme

$Ax = f(Px,Cx)$

$Cx = g(x,AAHx)$

$Px = px$

$Hx = hx$.

In the next and final section, we show that every prefix transition graph of an ε-free and reduced cf-grammar in Greibach normal form may be effectively generated by finite 'patterns'.

3. DETERMINISTIC GRAPH GRAMMAR

In this section, we indicate an algorithm which, given a prefix transition graph of a reduced cf-grammar in Greibach normal form, produces its building blocks, called patterns.

Graphs considered henceforth are representation of (possibly infinite) terms : each node s has a finite outdegree, corresponding to some function symbol $f \in F_n$, and s has exactly n outgoing arcs, respectively labeled (f,1) , ... , (f,n). Let us specialize for the above 'term graphs' the usual notion of a graph grammar.

Definition. A *graph grammar* is a finite set of production rules of the form

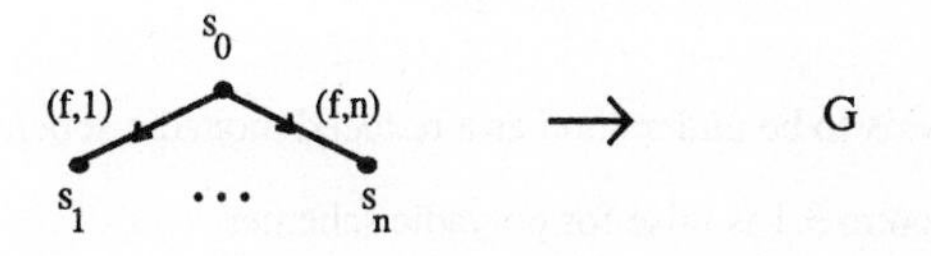

where $f \in F_n$, $s_i \neq s_j$ if $i \neq j$ and all the s_i are vertices of the finite (term) graph G.

A *non-terminal* of a graph-grammar G is a label of a left-hand side of a rule of G ; a *terminal* of G is a label of a right-hand side of a rule of G which is not a non-terminal.
Graph grammars induce graph rewriting as usual [Ra 84], [Ha-Kr 87], [Ke 89].

Definition. Applying a rule $L \rightarrow R$ to a graph G consists in finding a total mapping g from the set V_R of vertices of R to the set of vertices such that

(i) $g(L) \subseteq G$ where $g(L) = \{ (g(s) \xrightarrow{a} g(t)) \mid (s \xrightarrow{a} t) \in L \}$

(ii) g restricted to $V_R - V_L$ is one-to-one and $g(V_R - V_L) \cap V_G = \emptyset$.

The obtained graph is $(G - g(L)) \cup g(R)$.

The condition (i) amounts to finding the pattern L in G, possibly collapsed. The second condition is an alpha-conversion ensuring that the vertices of R which do not occur in L are also not in G. Beware that the rewriting relation is not a function.

In analogy to Kleene's substitutions on scheme, we introduce parallel rewriting for graphs : G rewrites in parallel into H if H results from the simultaneous application of the rules at all the occurences of the left members (in G).

We focus now on *deterministic* graph grammars, meaning that there is one rule per non-terminal. Given a graph H represented as a set of arcs and a deterministic graph grammar G, we define $G^{\omega}(H)$ as the unique (up to isomorphism) graph represented by the set of terminal arcs of a sequence $G^{n}(H)$ where $G^{0}(H) = H$ and $G^{n}(H)$ rewrites in parallel into $G^{n+1}(H)$. We use the term *pattern graphs* to denote the class of term graphs $G^{\omega}(H)$ for finite H ; the right members of the rules in G are called the patterns of $G^{\omega}(H)$.

In order to construct the canonical graph Can(S(fx)) of the unfolded tree S(fx) of a scheme S from fx, we want to show that Can(S(fx)) may be effectively generated by a deterministic graph grammar.

Theorem 3.1 *Every canonical graph of a scheme is effectively a pattern graph.*

Let us recall that a scheme is to be understood as a reduced monadic scheme without constant. As shown in Example 7 below, Theorem 3.1 is false for polyadic schemes.

Example 7. (given by G. Sénizergues). Let us consider the following reduced and polyadic scheme S without constant :

$$Axy = f(x,y,A(Bx,BBy))$$

$$Bx = gx$$

The canonical graph Can(S(Axx)) of the unlabelled tree of S from Axx is represented below :

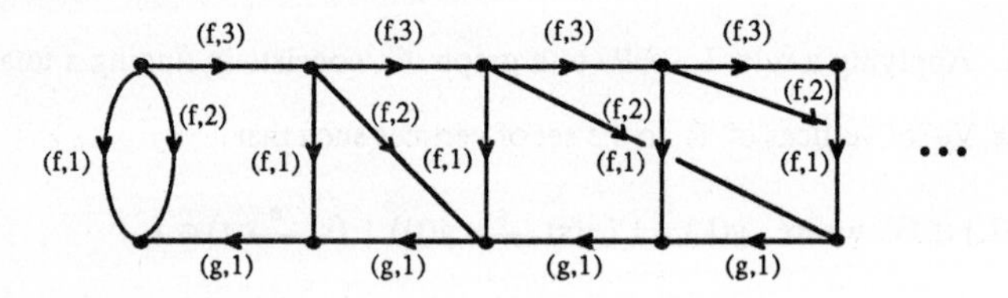

and is not a pattern graph.

To prove Theorem 3.1 and from Theorem 2.1, it suffices to establish constructively the result of [Mu-Sc 85] restricted to reduced simple grammars : every accessible prefix transition graph of a reduced simple grammar is a pattern graph.

Theorem 3.2 *There exist a procedure which from a given reduced ε-free context-free grammar* R *in Greibach normal form, and a letter* r, *produces a deterministic graph-grammar* G *and a finite graph* H *such that the pattern graph generated by* G *from axiom* H *is equal to the prefix transition graph of* R *from* r.

Proof.

The grammar G to be constructed, generates progressively the prefix transition graph

$$P(R,r) = P(R,r)_1 \cup \ldots \cup P(R,r)_n \cup \ldots$$

by slices

$$P(R,r)_n = \{ (s \xrightarrow{a} t) \in P(R,r) \mid |s| = n \}$$

of arcs whose sources are of growing lengths.

We recall [Bü 64] that $\underset{R}{\overset{*}{\longmapsto}} = \underset{R}{\longmapsto}^*$ is decidable and that the language $\{ u \mid r \underset{R}{\overset{*}{\longmapsto}} u \}$ of accessible words by prefix rewriting from a given word r is a rational language recognized by an automaton constructible from (R,r). In particular, the relation

$$\underset{R}{\overset{*}{\longmapsto}} \circ \underset{R}{\overset{*}{\longleftarrow\!\!\dashv}} = \{ (u,v) \mid \exists\, w ,\ u \underset{R}{\overset{*}{\longmapsto}} w \wedge v \underset{R}{\overset{*}{\longmapsto}} w \}$$

is decidable.

i) From now on, s is a vertex of P(R,r) different from ε . From s , the grammar G will generate the graph

$$P_s = P(R,r)_s \cup \{ u \xrightarrow{a} v \mid u \in V_{P(R,r)_s} \wedge u \underset{R}{\overset{a}{\longmapsto}} v \wedge |v| < |u| = |s| \}$$

where $P(R,r)_s$ is the connected component of the set

$$\{ (u \xrightarrow{a} v) \in P(R,r) \mid |u| \geq |s| \wedge |v| \geq |s| \} .$$

containing s.

To establish finiteness of the P_s (up to isomorphism), let us express $P(R,r)_s$ by the rewriting $\underset{R}{\overset{*}{\longmapsto}}_n = (\underset{R}{\longmapsto}_n)^*$ by vertices of length greater or equal to $n \geq 0$, i.e.

$$u \underset{R}{\longmapsto}_n v \quad \text{if} \quad u \underset{R}{\longmapsto} v \wedge |u| \geq n \wedge |v| \geq n .$$

As $|r| = 1$, we have the following property (1) :

$$P(R,r)_s = \{ u \xrightarrow{a} v \mid \exists\, t \in V(s) ,\ t \underset{R}{\overset{*}{\longmapsto}}_{|s|} u \underset{R}{\overset{a}{\longmapsto}}_{|s|} v \} \tag{1}$$

where V(s) is the subset of vertices of $P(R,r)_s$ which are targets of arcs in P(R,r) whose sources have of length $< |s|$, i.e.

$$V(s) = (\{r\} \cup \{ v \mid \exists u ,\ r \underset{R}{\overset{*}{\longmapsto}} u \underset{R}{\longmapsto} v \wedge |u| < |s| \}) \cap V_{P(R,r)_s} .$$

We establish in (iii) that all the vertices of $P(R,r)_s$ have a same suffix $Suff(s)$ of length $|s| - 1$ and that P_s is isomorphic to $P(R,V(s).Suff(s)^{-1})$ where

$$V(s).Suff(s)^{-1} = \{ u \mid u.Suff(s) \in V(s) \} \text{ is the right quotient of } V(s) \text{ by } Suff(s)$$

and $\qquad P(R,E) = \bigcup \{ P(R,e) \mid e \in E\}$

is the graph of the prefix transitions accessible from a set E of words.

ii) First, we show that $V(s)$ is constructible. As $\underset{R}{\overset{*}{\longmapsto}}$ is decidable, we can construct the set

$$W(s) = \{r\} \cup \{ v \mid \exists u ,\ r \underset{R}{\overset{*}{\longmapsto}} u \underset{R}{\longmapsto} v \wedge |u| < |s| \}$$

of the vertices of $P(R,r)$ accessible by arcs whose sources have of length $< |s|$.

To extract the subset $V(s)$ of $W(s)$, we establish that $V(s)$ is the smallest part of $W(s)$ closed by $\underset{R}{\overset{*}{\longmapsto}}_{|s|} \circ {}_{|s|}\underset{R}{\overset{*}{\longleftarrow\!\!\dashv}}$ and reaching s by $\underset{R}{\overset{*}{\longmapsto}}_{|s|}$.

Indeed, if $u \underset{R}{\overset{*}{\longmapsto}}_{|s|} \circ {}_{|s|}\underset{R}{\overset{*}{\longleftarrow\!\!\dashv}} v$ with $u \in V(s)$ and $v \in W(s)$ then v is a vertex of $P(R,r)_s$ so $v \in V(s)$. Hence $V(s)$ is a subset of $W(s)$ closed by $\underset{R}{\overset{*}{\longmapsto}}_{|s|} \circ {}_{|s|}\underset{R}{\overset{*}{\longleftarrow\!\!\dashv}}$. Furthermore s is accessible by $\underset{R}{\overset{*}{\longmapsto}}_{|s|}$ from an element of $V(s)$.

From (1) and for every proper subset P of $V(s)$, there are two paths in $P(R,r)_s$ with the same target, such that the source of one of them is in P and the other is in $V(s) - P$, i.e.

$$\forall P ,\ \emptyset \neq P \subset V(s) ,\ \exists u \in P , \exists v \in V(s) - P ,\ u \underset{R}{\overset{*}{\longmapsto}}_{|s|} \circ {}_{|s|}\underset{R}{\overset{*}{\longleftarrow\!\!\dashv}} v .$$

So, $V(s)$ is included in the smallest part of $W(s)$ containing an element t of $V(s)$ and closed by $\underset{R}{\overset{*}{\longmapsto}}_{|s|} \circ {}_{|s|}\underset{R}{\overset{*}{\longleftarrow\!\!\dashv}}$. Furthermore, if $t \in W(s)$ and $t \underset{R}{\overset{*}{\longmapsto}}_{|s|} s$ then $t \in V(s)$.

Finally, $V(s)$ is the smallest part of $W(s)$ closed by $\underset{R}{\overset{*}{\longmapsto}}_{|s|} \circ {}_{|s|}\underset{R}{\overset{*}{\longleftarrow\!\!\dashv}}$ whose an element is accessible by ${}_{|s|}\underset{R}{\overset{*}{\longleftarrow\!\!\dashv}}$ from s.

To construct $V(s)$, it remains to decide on $\underset{R}{\overset{*}{\longmapsto}}_{|s|} \circ {}_{|s|}\underset{R}{\overset{*}{\longleftarrow\!\!\dashv}}$ restricted to $W(s)$.

But for every word $u = u'u''$ and $v = v'v''$ of length $\geq |s|$ where $|u''| = |s| - 1 = |v''|$, we have

$$u \underset{R}{\overset{*}{\longmapsto}}_{|s|} \circ {}_{|s|}\underset{R}{\overset{*}{\longleftarrow\!\!\dashv}} v \quad \text{iff} \quad u'' = v'' \wedge \exists w \neq \varepsilon ,\ u' \underset{R}{\overset{*}{\longmapsto}} w \underset{R}{\overset{*}{\longleftarrow\!\!\dashv}} v' .$$

Then $\underset{R}{\overset{*}{\longmapsto}}_{|s|} \circ {}_{|s|}\underset{R}{\overset{*}{\longleftarrow\!\!\dashv}}$ is decidable and $V(s)$ is constructible.

iii) From (ii), the equivalence relation $\equiv$ on the set of the vertices of $P(R,r)$ different from ε, defined by

$$s \equiv t \quad \text{iff} \quad V(s).Suff(s)^{-1} = V(t).Suff(t)^{-1}$$

is decidable.

Furthermore $\equiv$ is of finite index. Indeed from (ii), $Suff(s)$ is a common suffix to all elements of $V(s)$, and every element of $V(s)$ is of length at most $|s| + K - 1$ where K is the maximal length of the right-hand side of the rules of R.

Finally $\equiv$ is finer than the equivalence relation of the couples (s,t) such that P_s is isomorphic to P_t. Indeed, from (1) and as $Suff(s)$ is a common suffix to all elements of $V(s)$, $Suff(s)$ is a common suffix to all vertices of $P(R,r)_s$, then

$$P(R,r)_s = \{ u \xrightarrow{a} v \mid \exists t \in V(s).Suff(s)^{-1} ,\ t \overset{*}{\underset{R}{\longmapsto}}_1 u \overset{a}{\underset{R}{\longmapsto}}_1 v \}.Suff(s) ,$$

so $\qquad P_s = P(R,V(s).Suff(s)^{-1}).Suff(s)$.

iv) We can now construct a graph grammar G generating $P(R,r)$ by vertices of growing length.

As $\equiv$ is an equivalence of finite index, a set A of representatives is constructible. It suffices to construct A by vertices of growing length. In fact, if $s \equiv t$ then $P(R,r)_t = (P(R,r)_s.Suff(s)^{-1}).Suff(t)$. So, if $s \overset{a}{\underset{R}{\longmapsto}} s'$ and $|s'| \geq |s|$ then $t \overset{a}{\underset{R}{\longmapsto}} t'$ and $s' \equiv t'$ for $t' = (s'.Suff(s)^{-1}).Suff(t)$.

Take an injection j of A into the subset of the functions of F which are not in R, and such that $j(s)$ is of arity $\#V(s)$ for all $s \in A$. Then to every s in A, we associate the following elementary graph

$$L(s) = \{ Suff(s) \xrightarrow{(f,i)} s_i \mid 1 \leq i \leq p \} \text{ where } \{s_1,\ldots,s_p\} = V(s) .$$

Such a graph $L(s)$ will be the left-hand side of a rule of G where the right-hand side has as terminal arcs set, the graph

$$T_s = \{ u \xrightarrow{a} v \mid u \overset{a}{\underset{R}{\longmapsto}} v \wedge u \in V(s) \}$$

of the arcs of $P(R,r)$ of sources in $V(s)$, and as non-terminal arcs set the graph $N(s)$ to be defined. For this and as R is reduced, to every vertex s of $P(R,r)$ different from ε, we take a set R_s of representatives of length $|s| + 1$ of the classes $V(t)$ where t is a vertex of $P(R,r)_s$ of length $|s| + 1$, i.e. R_s is a minimal set of vertices of $P(R,r)_s$ of length $|s| + 1$, such that

$$\bigcup\{ V(t) \mid t \in R_s \} = \{ u \in V_{T(s)} \cup V(s) \mid |u| > |s| \} .$$

This allows us to construct the deterministic graph grammar G:

$$G = \{ (K(s),T(s) \cup N(s)) \mid s \in A \}$$

where $\quad N(s) = \{ (L(u).Suff(u)^{-1}).Suff(t) \mid t \in R_s \wedge u \in A \wedge u \equiv t \}$.

By construction and for every $s \in A$, $G^\omega(K(s))$ is isomorphic to P_s. In particular $G^\omega(K(r_0))$ is isomorphic to P_{r_0} where $r_0 \in A$ and $r_0 \equiv r$; so $G^\omega(K(r_0))$ is isomorphic to $P_r = P(R,r)$. ♦

A similar result has been given by Baeten, Bergstra and Klop [Ba-Be-Kl 87]. But, our construction is slightly more complex, and is different of [Ca 90], because in our case, the progressive decomposition of the transition graph proceeds by layers of term-graphs, not by layers of nodes.

Example 8. From Example 6, we obtain the following graph grammar :

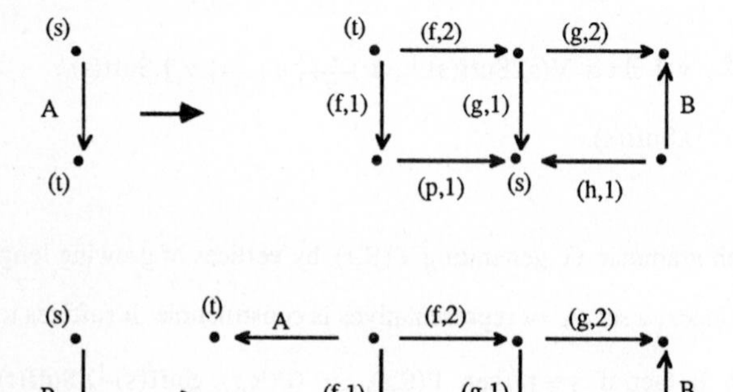

which generates from A the canonical graph of the unfolded tree of Example 3 (or of the scheme of Example 1).

Even taking into account a such optimization, the evaluation complexity of a recursive function depends on the structure of its definition.

Example 9. From Example 8, we obtain a program of exponential complexity to evaluate the recursive function A of Example 1. But this function can also be defined as follows :

$$\begin{aligned} A(n) &= B(n,0,i(T)) \\ B(n,m,T) &= \text{if } m = n \text{ then } T(m) \text{ else } B(n,m+1,j(m+1,T)) \text{ endif} \\ i(T) &= T \text{ where } T(0) \leftarrow 0 \\ j(m,T) &= T \text{ where } T(m) \leftarrow m - T(T(m-1)) \end{aligned}$$

by using a table T. The complexity becomes linear.

This points out that our construction need be carried over also for a class of polyadic schemes.

CONCLUSION

Given a reduced monadic scheme without constant, we have presented a method to construct a deterministic graph grammar generating the canonical graph of the scheme. This construction has just been extended to a general subclass of polyadic schemes which includes the class of reduced monadic schemes.

ACKNOWLEDGEMENTS

Let me thank P. Darondeau , R. Monfort and J.-C. Raoult for their help in the drafting of this paper. I thank also an anonymous referee for his detailed remarks.

REFERENCES

Ba-Be Kl 87 J.C.M. Baeten, J.A. Bergstra, J.W. Klop *Decidability of bisimulation equivalence for processes generating context-free languages*, LNCS 259, p. 94-111, 1987.

Ba-Ee-Gl Ke-Pl-Sl 87 H.P. Barendregt, M.C.J.D. van Eekelen, J.R.W. Glauert, J.R. Kennaway, M.J. Plesmeijer, M.R. Sleep *Term graph rewriting*, LNCS 259, p. 141-158, 1987.

Ba 89 M. Bauderon *On systems of equations defining infinite graphs*, LNCS 344, p. 54-73, 1989.

Bü 64 R. Büchi *Regular canonical systems*, Archiv für Mathematische Logik und Grundlagenforschung 6, p. 91-111, 1964.

Ca 89 D. Caucal *A fast algorithm to decide on simple grammars equivalence*, LNCS 401, p. 66-85, 1989.

Ca 90 D. Caucal *On the regular structure of prefix rewritings*, LNCS 431, p. 87-102, 1990.

Ca-Mo 90 D. Caucal, R. Monfort *On the transition graphs of automata and grammars*, WG 90, to appear in LNCS, 1990.

Co 74 B. Courcelle *Une forme canonique pour les grammaires simples déterministes*, Rairo 1, p. 19-36, 1974.

Co 83 B. Courcelle *Fundamental properties of infinite trees*, TCS 25, p. 95-169, 1983.

Co 89 a B. Courcelle *The monadic second-order logic of graphs, II : infinite graphs of bounded width*, Math. Syst. Theory 21, p. 187-222, 1989.

Co 89 b B. Courcelle *The definability of equational graphs in monadic second order logic*, LNCS 372, p. 207-221, 1989.

Co-Vu 76 B. Courcelle, J. Vuillemin *Completeness result for the equivalence of recursive schemes*, JCSS 12, p. 179-197, 1976.

Ga-Lu 73 S. Garland, D. Luckam *Program schemes, recursion schemes, and formal languages*, JACM 7, p. 119-160, 1973.

Gu 81 I. Guessarian *Algebraic semantics*, LNCS 79, 1981.

Ha-Kr 87 A. Habel, H.J. Kreowski *Some structural aspects of hypergraph languages generated by hyperedge replacement*, LNCS 247, p. 207-219, 1987.

Ho-Pl 88 B. Hoffmann, D. Plump *Jungle evaluation for efficient term rewriting*, LNCS 343, p. 191-203, 1988.

Ia 60 Ianov *The logical schemes of algorithms*, Problems of cybernetic, USSR, p. 82-140, 1960.

Ke 88 R. Kennaway *On 'on graph rewritings'*, TCS 52, p. 37-58, 1988.

Mu-Sc 85 D. Muller, P. Schupp *The theory of ends, pushdown automata, and second order logic*, TCS 37, p. 51-75, 1985.

Ni 75 M. Nivaţ *On the interpretation of polyadic recursive schemes*, Symposia Mathematica 15, Academic Press, 1975.

Pa 82 P. Padawitz *Graph grammars and operational semantics*, TCS 19, p. 117-141, 1982.

Ra 84 J.-C. Raoult *On graph rewritings*, TCS 32, p. 1-24, 1984.

St 80 J. Staples *Computation on graph-like expressions*, TCS 10, p. 171-185, 1980.

Graphic Equivalence and Computer Optimization

Tien Chi Chen
Department of Computer Science
The Chinese University of Hong Kong, Shatin, HONG KONG

ABSTRACT. Procedure graphs are directed graphs with arc labels denoting precedence, useful in modelling computer algorithms. Equivalent procedure graphs yield the same data outcome under identical inputs, yet imply different performance, cost and practicability; they offer opportunities for optimization through judicious selection. Examples are given together with observations on computer architecture.

Keywords: Procedure graphs, graphic equivalence, computer optimization, computer system design, internal forwarding, graphic modelling, edge-labelling.

CONTENTS

1. Procedure graphs

A procedure graph is a directed graph with arc labels denoting invocation precedence. It can specify the dynamic use of static resources as found in executing computer instructions. With multiple labelling, and using the freedom to specify sequential use of resources in parallel, one could model the specification of entire algorithms; and with suitable representation of computing resources, their computer processing.

A given procedure graph can be mapped into alternative equivalent graphs delivering the same data outcome when subjected to identical inputs. Optimization can result from choices based on economic criteria. A preliminary study has been made by Chen and King [Chen 89]; we shall now put the theory on a firmer footing, generalize it and discuss its practical implications.

The procedure graph resembles the precedence graph which uses pairs of node labels to denote precedence bounds [Dewilde85]. But arc labels are more direct and more intuitively appealing than node labels, as algorithmic alternatives are seen in terms of explicit data movement along explicit paths, rather than just the consequence of such movement. There is also more room to specify multiple labels on arcs than on nodes.

1.1. Nodes and arcs

A directed arc is used to transport data, which are assumed to be words of equal standard length. A node can receive input data from entering arcs; can store, copy, and alter the data; and can forward them as output through leaving arcs. A storage node stores and distributes received data without change, and comes in two types: a memory node can deliver output repeatedly until overwritten; a temporary node, on the other hand, becomes empty immediately after the first delivery.

An arithmetic node has exactly two ("left" and "right") entering arcs for data input, and is "fired" automatically when both input arcs contain valid data. After firing, the inputs are immediately annihilated to make room for new ones, and after some specific delay, an output is produced; the output remains available until replacement. Detailed arithmetic may call for a third input for control information, but such is not our immediate concern. This node could be viewed as a composite of an arithmetic operator preceded by two temporary nodes and followed by a memory node. An arithmetic node with only one input can be taken as a two-input arithmetic node with one input always present.

Other nodes can be defined as composites of storage and arithmetic nodes. These include multiplexers, decoders, and memory modules.

1.2. Precedence labels

Data moving along an arc marks an event in space-time. The precedence order of each event on arc α is represented by a pseudo-time label $T(\alpha)$. We shall simply use the term time where no misunderstanding arises. Anywhere in the graph, if $T(\alpha) < T(\beta)$, then arc α is invoked ahead of arc β. When $T(\alpha) = T(\beta)$, the invocations are simultaneous for practical purposes. $T(\alpha)$ could be a real number, a variable within a real interval, or a vector of such variables denoting repeated transmissions along the same arc. To avoid ambiguity, arcs joining at a storage node are not permitted to have equal time labels.

1.3. Tracks

A (space-time) track is either an event or a sequence of events along a (directed) path dealing with the input data or its transformation: if data d is supplied at the beginning of a track, then at the end F(d) is received. A transmission track traverses storage nodes exclusively, with $F(d) \equiv d$. An arithmetic track passes through at least one arithmetic node. The tracks as defined need not be completely disjoint; different tracks could share arcs or even subtracks.

A track with arc α entering node P can include a leaving arc β as part of the track, subject to the following data delivery conditions:

a. $T(\beta) > T(\alpha)$, allowing a discernible time lapse to ensure data transmission and/or transformation;

b. If P is a storage node, and there is no other entering arc σ such that $T(\beta) > T(\sigma) \geq T(\alpha)$. Such an intervening event would transmit its data to β, making the latter a part of its own track extension. For arithmetic nodes events, The rules still applies with a minor change: there is no intervening event occurring on the same input arc.

In a procedure graph one can apply the above procedure to extend a given track both ways until one can proceed no more; the resultant track is said to be maximal (in length). Repeating by starting with an event not covered by the known maximal tracks, one ends up eventually with a collection of maximal tracks the union of which spans the entire graph. Their starting nodes together form the set of data sources in the entire graph. We further assume that true outputs are needed only at or beyond the ending nodes of the maximal tracks; then all ending nodes together form the set of (gateways to the) data sinks.

Often arcs are incompletely labelled. Unlabelled arcs which could be labelled to form a track are understood to do so. Two arcs incident on a node but could not be combined into a track are assumed to be traversed in any relative order if either or both are unlabelled.

1.4. Practical analogs

The present work was motivated initially by a study of computer CPU architecture, where the analogs for storage nodes are memory cells (memory node) and registers (memory node and temporary node), and arithmetic nodes are autonomous arithmetic mechanisms. Movement among the nodes follow causality-preserving sequencing rules with quantization of time into CPU cycles.

The theory is believed relevant to the description, processing and optimization of all algorithms. For algorithms written in procedural languages storage nodes may be named operands, and arithmetic nodes are arithmetic operators.

2. Computational equivalence and transformations

When the outcomes of a given procedure graph G for any given input can be obtained via an alternate graph G', then G and G' are said to be (computationally) equivalent. Optimization of a procedure then becomes making the best selection from the equivalent graphs, or more explicitly, by a series of equivalence transformations on an initially given procedure. We shall use

G <==> G' to denote equivalence between G and G'
G ===> G' to denote an equivalent transformation from G to G'

Labels above or below the signs refer respectively to the rightward and leftward tranformations.

Sections 2.1 and 2.2 describe basic equivalence properties, and section 2.3 discusses properties derived from the basic ones. The use of each equivalence statement, say from left to right, consititutes a rule to be invoked when economic conditions, described in Section 4. The application of a rule within a complex procedure graph, as discussed in Section 3, constitutes a derivation.

Relabelling

From the previous section we can already derive an important equivalence property: all time labels can be changed without affecting the global data delivery, as long as <u>every</u> maximal track receives and delivers data at the end-points at prescribed time-label values.

These end-points provide fixed boundaray values for the track which could be redefined internally with a high degree of freedom. And if the final delivery time is unimportant, the ending arcs need not even be fixed in pseudo-time. We shall consider only integer time labels in the present study.

2.1. Data transmission: <u>Serial-Parallel(SP) / Parallel-Serial(PS)</u>

A graph with a serial track (PQR) is transformed into one with parallel (i.e., fork) transmission (PQ; PR) or <u>vice versa</u>. PQR on the left-hand side must form a transmission track. The data source P can be either a storage node or an arithmetic node. Note that PS is not unique without further qualification, as the result could as well be (PRQ).

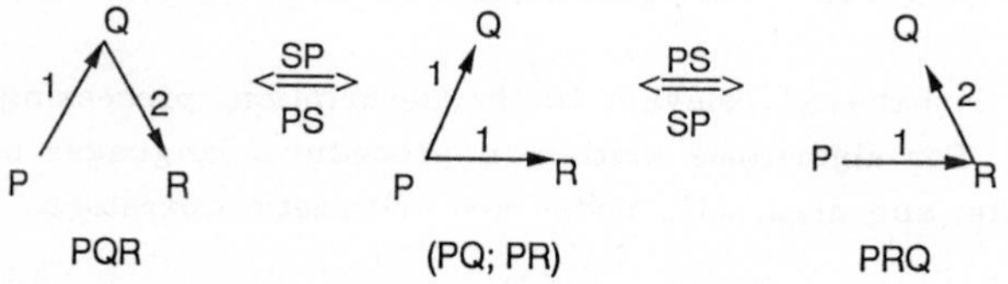

PQR $\underset{PS}{\overset{SP}{\Longleftrightarrow}}$ (PQ; PR) $\underset{SP}{\overset{PS}{\Longleftrightarrow}}$ PRQ

2.2. Removal and insertion of resources

Tracks and nodes can be removed (expunged) if they serve no useful purpose, or if their purpose could be assigned to others. Conversely they can be inserted. An arbitrary arithmetic node is designated by Ω.

2.2.1. Overwriting: <u>Overwritten input eXpunction(OX) / EXtra Overwritten input(XO)</u>

When two input tracks join at the node Q, without any intervening leaving arc, the earlier track can be deleted. The reverse (XO) inserts an input to be overwritten. Note P, R can be the very same node.

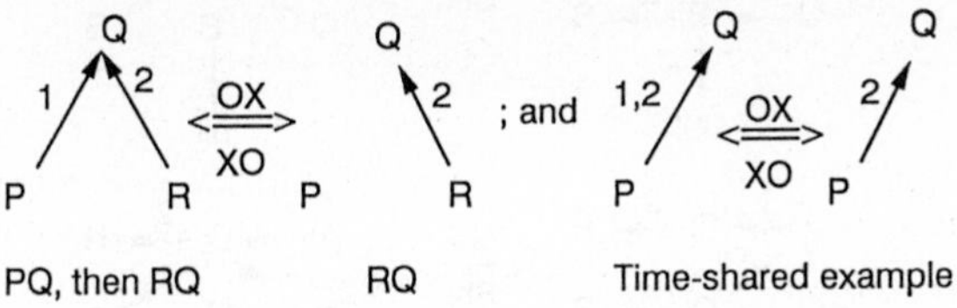

PQ, then RQ <=OX/XO=> RQ ; and Time-shared example

2.2.2. Redundant input: <u>Input eXpunction(IX) / EXtra Input (XI)</u>

We shall use <u>nil</u> for a nonexistent output. If a node is shown with some explicit outputs, also an output to nil, this means all possible outputs are as shown. Thus a node with nil as sole ouput has no true output at all.

If a (storage or arithmetic) node has no output, and if its contents is not part of the final result, it can be disengaged or engaged. The disengaged node can be eliminated if desired.

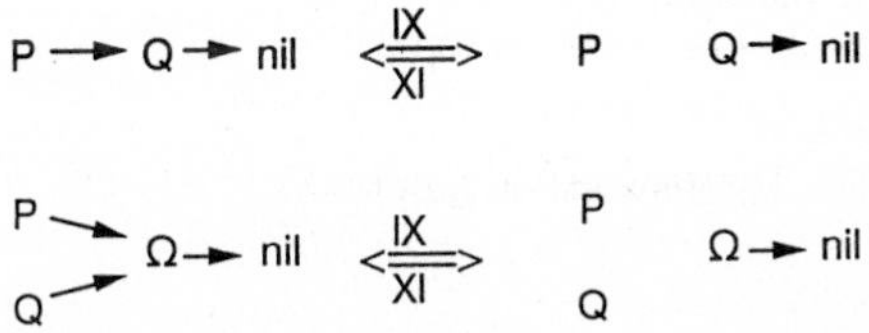

P → Q → nil <=IX/XI=> P Q → nil

P, Q → Ω → nil <=IX/XI=> P, Q Ω → nil

2.2.3. Redundant node: <u>Node eXpunction(NX) / EXtra Node(XN)</u>

A storage node which leads to no output is deleted (through SP, IX) from, or inserted (through XI, PS) into, a transmission track.

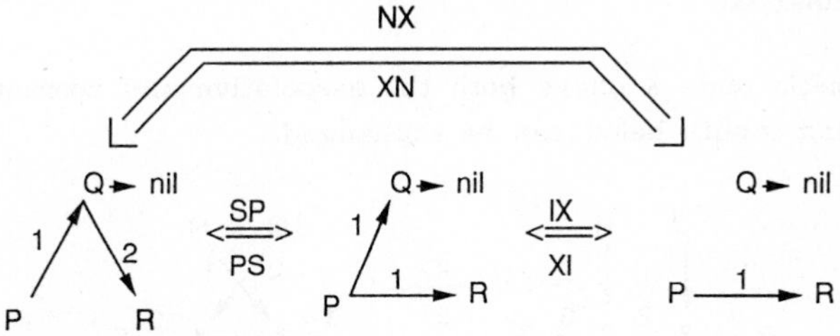

NX / XN

Q → nil <=SP/PS=> Q → nil <=IX/XI=> Q → nil; P —1→ R

2.2.4. Time-sharing: <u>Node Shared(NS) / Split Node(SN)</u>

Two similar nodes are merged through time-sharing; or conversely a time-shared node is split into two nodes.

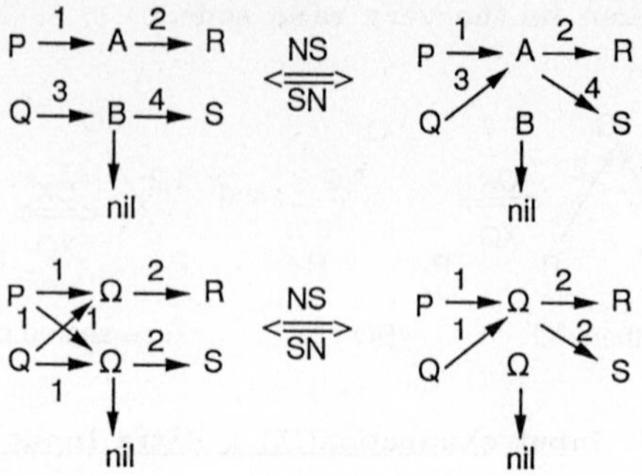

2.3. Arithmetic equivalence

Arithmetic equivalence involves arithmetic with changed data, exploiting invariance laws.

Direct application of the arithmetic laws

Detailed figures are omitted.

P ¢ Q <==> Q ¢ P (CML / CML) (CoMmutative Law)

(P @ Q) @ R <==> P @ (Q @ R) (ASL / ASL) (ASsociative Law)

P δ (Q Ω R) <==> (P δ Q) Ω (P δ R) (DIL / DIL) (DIstributive Law by operator δ over the operation Ω)

Input-Input eXchange (II)

If the arithmetic node & obeys both the associative and commutative laws, the second and the third inputs below can be exchanged.

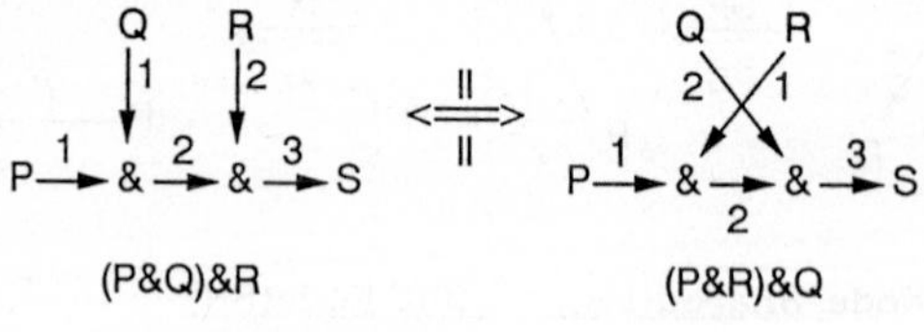

(P&Q)&R <==> (P&R)&Q (II / II)

2.3. Derived equivalence properties

2.3.1. Tracks and forks: Transmission subtrack Transplantation (TT)

Part of a transmission track can be transplanted to become a branch of another node in the same track, or vice versa, in an extension of SP/PS equivalence. For example, if A, B, ..., P, are storage nodes below, then

A $\xrightarrow{1}$ B ... $\xrightarrow{n}$ P $\xrightarrow{n+1}$ Q → R <==TT==> A $\xrightarrow{1}$ B ... $\xrightarrow{n}$ P   Q $\xrightarrow{3}$ R (P $\xrightarrow{2}$ Q)

A,B, Q and R could be arithmetic nodes.

2.3.2. Loops and forks: The loop-fork /fork-loop (LF/FL)

When a transmission track loops upon itself, say re-entering at node B, the graph can be replaced by a fork configuration with early exit. The re-entry input to B is omitted, as are all second inputs in the overlapped arcs. This could apply to arithmetic tracks, but the loop itself should not traverse arithmetic nodes. Note that this works in a direction opposite to that in OX. The reverse operation is FL.

P ←n-2— ... ← E; P ↓n-1 B; A —1→ B —2,n→ C —3→ D; D → E; C ↓n+1 Q   <==LF / FL==>   P ←n-2— ... ← E; A —1→ B —2→ C —3→ D; D → E; C ↓3 Q

2.3.3. The dL / Ld equivalence

This equivalence property can be memorized easily using the letters d and L. The requirement is that (QR) is the only one to read from Q before the second writing into Q; the parts connected by dotted lines need not even be related in reality. The reverse transformation is Ld. (Note that if d represents a pure transmission track, the FL transformation would also apply.)

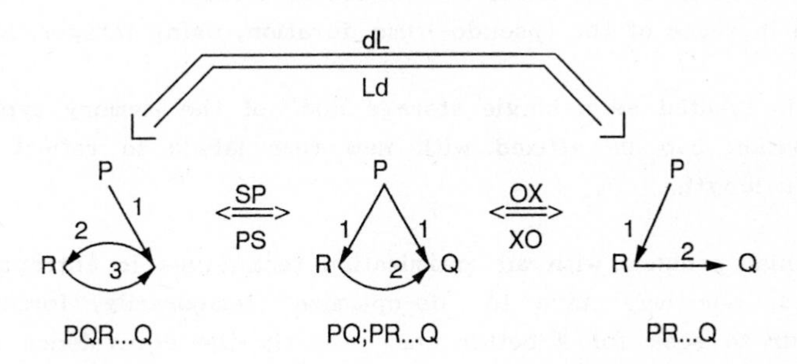

3. Complex graphs

Equivalence transformations need not be confined to simple procedure graphs. They apply to domains within complex procedure graphs, provided the global data delivery remain unchanged.

3.1. Local remapping

Specifically, one can choose one of the equivalence relationships, and define a production rule (left-to-right transformation, or vice versa. The choice defines new left- and right hand sides. The vertices on the new left-hand side are taken as the gluing points in the production. See Ehrig [Ehrig 79] [Ehrig 87].

The direct derivation scheme in Ehrig does not seem to apply literally. There is a need to adjust the production rules to fit the current situation, particularly the matching of the boundary labels and the exploitation of the freedom to relabel the internal arcs. One way to achieve this is to relabel the graphs in the production rules to match the new environment, then apply the direct derivation normally.

After gluing, the areas surrounding the previously boundaries can be examined for further remapping, if needed; this is often highly desirable. In practice a sequence of remappings can be carried out more or less intuitively under human supervision. The remapping derivations can continue as long as needed.

4. Economic factors

The economic factors which might favor one graph over another equivalent graphs (or, running the production rules forwards or backwards) include performance enhancement, cost reduction, and easy realizability of the final procedure (but not necessarily the intermediate ones).

In the absence of detailed information, the following factors have been chosen:

- δM: The increase in the memory access involvement
- δA: The increase in the number of arcs involved
- δN: The increase of the number of nodes involved
- δT: The increase of the (pseudo-)time duration, using integer labels

A memory cell is treated as a single storage node of the memory type. The arcs after transformation can be affixed with new time labels to reflect the possible changes in track length.

A perennial problem with all optimization techniques is entrapment by local optima. Indeed, one may have to "de-optimize" temporarily, forsaking a local optimum in oredr to seek for a better one. Clearly the equivalence mappings are applicable in either directions, yet the detailed prescriptions are elusive.

5. Examples

5.1. Classical internal forwarding

There are three classical optimization tranformation techniques, formulated mainly to reduce memory operations, are grouped under the name "(internal) forwarding." See Chen [Chen 80]. In the following table, M is a memory cell, and P, Q, R are registers.

Subgraph	Map	Result	Classical name	δM	δA	δN	δT
PMR	SP	PM,PR	SFF (Store-Fetch Forwarding)	-1	0	0	-1
MQ, MR	PS	MQR	FFF (Fetch-Fetch Forwarding)	-1	0	0	+1
PM, RM	OX	RM	SSC (Store-Store Cancellation)	-1	-1	-1	-1

The first two are essentially reverses of each other. Vector chaining in the Cray 1 is essentially SP, using vector operands [Russell 83], M then becomes a composite node capable of housing an entire vector.

5.2. A Gaussian elimination innerloop

A striking use of the dL transformation is in the handling of the following FORTRAN innerloop for a version of the Gaussian elimination algorithm.

```
            Do 300 K =J, N
300         M(I,K) = M(I,K) * Q - M(J,K)
```

where Q = M(J,J)/M(I,J). It is compiled in the IBM System/360 and related machines into a sequence of two-address instructions

```
M(I,K)                --> R(0)
R(0) *    Q           --> R(0)
R(0) -    M(J,K)      --> R(0)
R(0)                  --> M(I,K)
```

as represented by the first graph below with two loops along the same arithmetic track. One of the inputs to the arithmetic nodes have been omitted for clarity. The busiest node there is R(0), with 3 entering tracks and 3 leaving tracks. After three successive dL transformations, R(0) is completely bypassed by M(I,K). The loop is handled by a direct fetch into the arithmetic input stations, and a direct store from the adder output. The next loop traversal cancels the need to store into R(0).

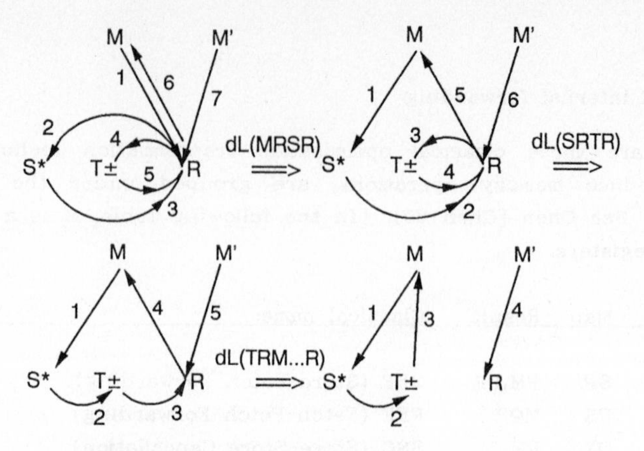

Remapping of the Gaussian elimination loop

The effect is the recovery of the original innerloop, free of register involvements. R(0) is loaded exactly once, namely during the last innerloop traversal. We have $\delta A = -3$ for each innerloop traversal except the very last, with $\delta A = -2$.

This optimization was actually done in the IBM System 360 Model 91 and related machines, using automatic internal forwarding operations encoded by backward pointers in realtime. See Chen [Chen80], pp. 471-474.

5.3. Instruction pairs involving Fetch-and-Add

In a node of an interconnection network linking processors with memory units, the instruction pairs involving the FETCH-AND-ADD synchronism primitive can be replaced by alternative instruction sequences offering the convenience of decentralized processing, short turnaround, and often simplification [Gottlieb 83]. These instruction pairs also allows nontrivial tests of our rules on equivalence and optimization, here discussed as operations with a computer CPU.

Let the contents of memory location M be m, and that of register A by a. The Fetch-and-add instruction (FA A,M) is an atomic instruction which loads A with m, and stores (m+a) in M. We shall discuss a few salient cases below.

5.3.1. FA with Fetch

The combination (FA A,M; L B,M) is mapped using SP (on PMB) then PS (on MA, MP), to reduce 2 memory accesses.

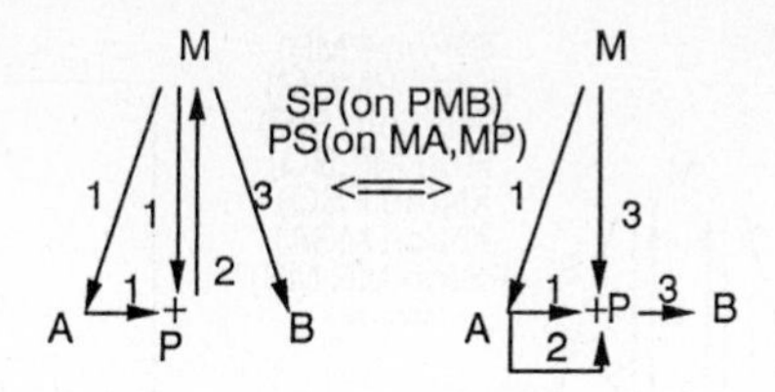

5.3.2. FA with store

The sequence (FA A,M; ST B,M) leads to sweeping simplification with the decommission of an adder.

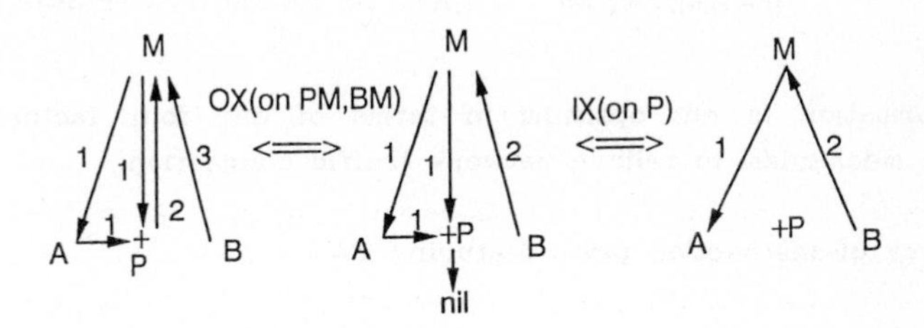

5.3.3. FA followed by FA

The combination (FA A,M; FA B,M) is interesting because of its role in network forwarding using local adders in the Ultra computer.

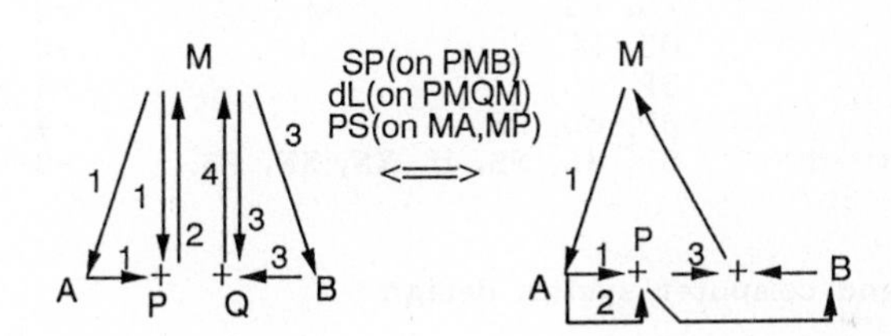

In Gottlieb's network forwarding case the register operands are added; one addend stays in the network node, while the sum (indicated by S below) goes on to perform FA. The returned operand replaces the waiting operand; but is also added to the latter to form another result. The mapping can be expressed as the result of seven successive transformations:

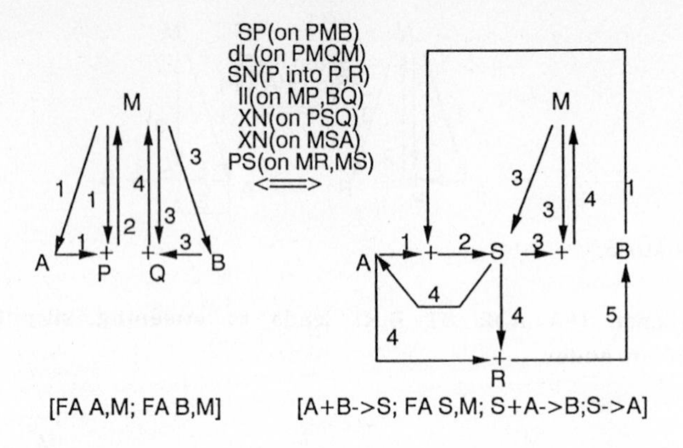

Their transformation is not optimum in terms of our four factors, but offers a programmable mechanism to relieve network traffic congestion.

5.3.4. Summary of instruction pairs featuring FA

A summary of the instruction pairs involving FA is shown in the table below. All instructions refer to the same memory address.

Instructions	Transformations	δM	δA	δN	δT
FA, Load	SP, PS	-2	0	0	0
Load, FA	PS, PS	-2	0	0	0
FA, Store	OX, IX	-2	-3	-1	-1
Store, FA	SP, dL	-3	-1	0	-1
FA, FA	SP, dL, PS	-4	-1	0	0
FA, FA (Gottlieb)	SP, dL, NS, II, XN, XN, PS	-3	3	2	1

6. Optimization and computer system design

Computer optimization occurs in many forms, involving hardware, software, the programmer, and the use of program packages. The examples above have revealed some weaknesses in conventional processing, and also suggest remedies. Some general comments may be in order.

6.1. Granularity matching

The examples shown would seem to be too fine-grained, hence meaningful only for hardware use. Indeed, the forwarding operations were initially installed as hardware features in realtime. It is all the more surprising, therefore, that the outcome of machine self-optimization in Section 5.2 essentially recreates the original multi-operator statement by the user. Thus the Model 91 machine hardware appears to be closer to higher level languages than expected.

The fact is, many machines have a fine-grained operation repertoire which can accommodate the coarse-grained procedural statement exactly, yet the "medium grain" instructions creates much mismatch with higher-level procedural statements, and simulates the latter inadequately. The direct addressing of output registers from arithmetic mechanisms would eliminate many redundant register operations.

6.2. Arithmetic registers

Most instruction sets today prescribe the mandatory use of arithmetic registers, yet there is no register specification in a procedural language statement. In other words, to the higher-language programmer virtually all floating-point registers are temporary, to be invoked exactly once but never again because it bears no explicit name addressable by the user. A temporary register which cannot be used again can surely be ignored after it has outlived its purpose; it need not be saved after the completion of the statement.

Nevertheless, in conventional computing the user has no good mechanism to declare disinterest in a register, and consequently every register has to be saved on program switching before completion. Nor has the user any good way to specify the multiple reuse of arithmetic outcome using procedure languages.

6.3. Atomicity

A synchronization primitive is assumed to be atomic, allowing no intervention until completion. Nevertheless, the cases studied using the Fetch-and-add primitive suggests that, if another instruction referring to the same memory address is present, the combined processing with may be faster than the sum of the two individual execution times, indeed, perhaps faster than the processing of the primitive alone.

7. Summary and conclusions

We have described the use of procedure graphs in description of computer algorithms.

Techniques have been provided to transform among data-equivalent graphs, the choice among them offer a means for optimization. An entire graph can be handled through processing and gluing of subdomains. The consideration of economic factors is decoupled from the equivalence question, to achieve full generality.

The use of four heuristic economy factors in optimization have been studied through examples, and their implications on computer designs of their use have been discussed. It appears that the machines can accommodate efficiently the coarse-grained structures in procedural language programs, but instructions as we know them today cannot.

The theory, however, should be applicable to characterizations and optimizations in computer algorithms, and their handling by hardware and software. It is hoped that future studies will add mathematical rigor and will lead to optimization techniques more transparent and more amenable to automatic implementation.

REFERENCES

[Chen 80] Tien Chi Chen, "Overlap and pipeline processing," Chapter 9 in **Introduction to Computer Architecture** (H. Stone, Ed.), 2nd. Ed., Science Research Associates, Chicago, (1980).

[Chen 89] Tien Chi Chen and Willis K. King, "Computational invariance and generalized internal forwarding," Proc. Int. Symp. on Comput. Arch. and Digital Signal Processing (CA-DSP '89), Hong Kong, (1989),212-216.

[Dewilde 85] P. Dewilde, E. Deprettere and R. Nouta, "Parallel and pipelined VLSI implementation od signal processing algotithms," Chapter 15 in S.Y. Kung, H.J.Whitehouse and T. Kailath (Eds.), **VLSI and Modern Signal Processing,** Prentice-Hall, Englewood Cliffs, NJ, 1984, 257-276.

[Ehrig 79] H. Ehrig, "Introduction to the algebraic theory of graph grammars (a survey)," Lecture Notes in Computer Science 73, **Graph-Grammars and their Application to Computer Science and Biology**, Springer-Verlag, Berlin, 1979.

[Ehrig 87] H. Ehrig, "Tutorial introduction to the algebraic approach of graph grammars,"Lecture Notes in Computer Science 291, **Graph Grammars and their Application to Computer Science**. Springer-Verlag, Berlin, 1987.

[Gottlieb 83] A. Gottlieb et. al., "The NYU Ultracomputer - Designing an MIMD Shared Memory Parallel Computer," IEEE Trans. Comput. **C-32,** (1983), 175-189.

[Russell 78] R. M. Russell, "The Cray-1 Computer System," Comm. ACM **21**, (1978), 63-72.

Graph Grammars and Logic Programming*

Andrea Corradini, Ugo Montanari, Francesca Rossi
Università di Pisa
Dipartimento di Informatica
Corso Italia 40
56125 Pisa - ITALY

Hartmut Ehrig, Michael Löwe
Fachbereich Informatik 20
Technische University Berlin
D-1000 Berlin 10 - GERMANY

ABSTRACT: In this paper we investigate the relationship between the algebraic definition of graph grammars and logic programming. In particular, we show that the operational semantics of any logic program can be faithfully simulated by a particular context-free hypergraph grammar. In the process of doing that, we consider the issue of representing terms, formulas, and clauses as particular graphs or graph productions, by first evaluating the approaches already proposed for Term Rewriting Systems (TRS), and then by giving an original extension of those approaches, to be able to deal with the unique features of logic programming. Actually, not only does our representation of definite clauses by graph productions allow us to deal correctly with logical unification, but also it overcomes some of the problems encountered by other approaches for representing TRS's as graph grammars. The main result of the paper states the soundness and completeness of the representation of clauses by productions, and this correspondence is extended to entire computations, showing how a context-free grammar (over a suitable category of graphs) can be associated with a logic program. The converse holds as well, i.e. given any context-free graph grammar (over that category), a logic program can be extracted from it.

Keywords: Logic Programming, Graph Grammars, Term Rewriting Systems, Hypergraphs, Jungles, Dags.

CONTENTS

* Research supported by the GRAGRA Basic Research Esprit Working Group n. 3299.

0. Introduction

Many papers in the literature tackle the issue of representing Term Rewriting Systems (TRS's) using graph rewriting, either exploiting the algebraic theory of graph grammars (like [PEM87, HKP88], which follow the so called 'double-pushout approach', and [Rao84, Ken87, Ken91], which use instead a single pushout construction), or using a more operational presentation of graph rewriting (e.g. [BvEGKPS87]). In general, rewriting rules are translated into graph productions, while terms are represented by (hyper-)graphs of a suitable category: usually many non isomorphic graphs can represent the same term. All the mentioned papers prove the soundness of their graph representation of rewriting rules, in the sense that if a production can be applied to a graph, then the corresponding rule can be applied to the term represented by the graph. However, in all the referred papers that translation is not complete, i.e. it is possible that a graph cannot be rewritten with a production, although the term it represents can be reduced by the corresponding rewriting rule. Usually, completeness just holds for left-linear rules, i.e. rules which do not have two occurrences of the same variable in the left hand side (cf. [BvEGKPS87]).

In this paper we apply essentially the same idea to Logic Programming ([Ll87]), showing how a program can be translated into a graph grammar. A related work is reported in [Cou88], where context-free graph grammars are used to represent a proper subset of logic programs, i.e. recursive queries in relational databases.

With respect to TRS's, the representation of logic programs with graph grammars has to take into account two peculiar aspects:

1) the 'terms' appearing in a program are built over a two-sorted signature, including function and predicate symbols, and
2) logic program goals can include variables: the matching condition between the head of a clause and the goal to be reduced is unification.

The first aspect can be easily handled, by defining a suitable category of *enriched directed acyclic graphs* whose objects represent collections of both terms and atomic formulas. We also show that the two preferred ways of representing terms by graphs (i.e. *directed acyclic graphs*, in short *dags*, and *jungles*), as used in the papers mentioned above, are, as expected, completely equivalent.

The second point is more complex to deal with, since the presence of variables in a goal is a new potential source of incompleteness. We solve this problem by representing a definite clause as graph production only after transforming it into a *canonical form*. Incidentally, it turns out that the resulting production is *context-free* (in the sense of [BC87], but not of [Ha89]). The main result of the paper states that our representation is sound and complete: more precisely, completeness means that if there is a resolution step from a goal G_1 to a goal G_2 using the definite clause C, then the production associated with C can be applied *to every graph D_1 which represents G_1*, yielding a graph D_2 which represents G_2. Therefore the use of the canonical form (which is actually a left-linear clause equivalent to the given one) not only allows us to deal correctly with unification, but also overcomes the other causes of incompleteness already present in the approaches mentioned above. Indeed, although we stick to logic programming, it should be possible to apply a similar solution to Term Rewriting Systems as well.

The main result is then extended to entire computations, by representing any logic program (together with a goal) by a context-free grammar whose initial graph represents the goal, and whose productions represent the definite clauses. We show that the soundness and completeness result holds also for sequences of resolution steps, and thus also for entire refutations.

We also front the opposite problem of associating a logic program to an arbitrary context-free grammar in the category of enriched dags. Clearly this translation causes a loss of information, but as far as one is interested just in the *term substitutions associated with the derivations*, the resulting logic program is equivalent to the grammar, and can be used as an executable implementation of it.

In this paper we discuss the representation of logic programs by graph grammars, focusing on the soundness and completeness results. The advantages of this translation are many, and can be topics for future research. For example,

1) logic programming can provide an efficiently executable representation of a subclass of graph-grammars;
2) well known graph grammar techniques can be applied to logic programs, providing new tools for program transformation;
3) a rich collection of results about parallelism and concurrency in the graph-grammar theory could be exploited in the logic programming framework, in order to analyze and prove properties of parallel execution frameworks in a formal manner.

The paper is organized as follows. In Section 1 we present a brief overview of the algebraic approach to graph grammars and of the main concepts of logic programming. Section 2 fronts the problems involved in representing terms and formulas as graphs, studies some already proposed formalisms (for terms only) and proposes a representation for terms and formulas as enriched dags. Section 3 uses the category of dags introduced in Section 2 to represent clauses as suitable context-free hypergraph productions. Section 4 contains the main result of the paper, i.e. the soundness and completeness of graph representation of clauses, and its extension to entire computations. Finally, Section 5 concludes the paper summarizing its achievements and suggesting some other issues to be studied in the future.

1. Background

This section has the purpose of recalling the basic concepts of the algebraic approach to graph grammars (as summarized for example in [EPS73, Eh87]) and of logic programming ([Ll87]). Nevertheless, in order to fully understand the subsequent treatment the reader is required to have some familiarity with both fields, and also with some basic notions of category theory ([ML71] is a standard reference).

1.1. Graph grammars

We remind the reader of the very basic concepts of the algebraic theory of graph grammars.

- A *graph* is a tuple G = (V, A, s, t), where V is the set of *vertices* (or *nodes*), A is the set of *arcs* (or *edges*), and s and t are functions from A to V which respectively return the *source* and the *target* node of an arc. If C = (C_V, C_A) is a fixed pair of sets of *colors*, a *colored graph* (over C) is a graph equipped with two mappings $l: V \to C_V$ and $m: A \to C_A$, which give the colors of nodes and arcs respectively.
- A *graph morphism* f from G to H, written f: G → H, consists of a pair of functions between arcs and nodes respectively which are compatible with source and target mappings, and which, in the case of colored graphs, are assumed to be color preserving. The category of graphs and graph morphisms over C is denoted by $\mathbf{Graph}_C$.
- A *graph production* p is a pair of graph morphisms $p = (B_1 \xleftarrow{l} K \xrightarrow{r} B_2)$, such that l is injective. The graphs B_1, B_2, and K are called *left side*, *right side*, and *interface* of p, respectively.
- Given a production $p = (B_1 \xleftarrow{l} K \xrightarrow{r} B_2)$, a graph G, and a graph morphism $g: B_1 \to G$ (called an *occurrence* of the left side of p in G), there exists a *direct derivation from G to G' via p and g* (written $G \underset{p,g}{\Rightarrow} G'$) iff the diagram below can be constructed, where the left and the right squares are *pushouts* in $\mathbf{Graph}_C$. In this case the pair of arrows ‹h, l'› is called the *pushout complement* of ‹l, g› (cf. [Eh87], where necessary and sufficient conditions for the existence of the pushout complement in category $\mathbf{Graph}_C$ are reported).

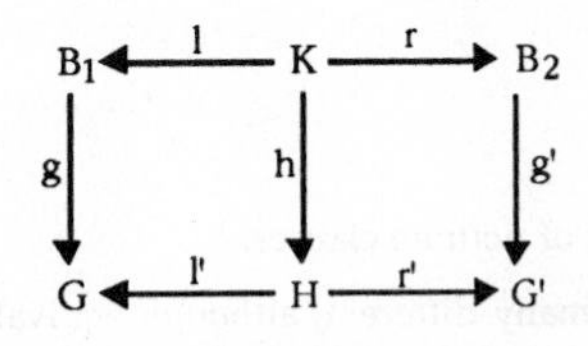

1.1 Diagram

- A *graph grammar* is a tuple GG = (S, PROD, C, T), where S is the initial graph, PROD is a finite set of productions, C is a pair of sets of colors, and T = (T_V, T_A) is a pair of sets of *terminal colors*, included in C.

- Given a graph grammar GG = (S, PROD, C, T), a *derivation* is a sequence of direct derivation steps starting from S and ending with a graph colored with terminal colors only.

The definitions of *production, direct derivation, grammar*, and *derivation* that we recalled here for (colored) graphs are essentially independent from the specific underlying category. Actually, the same notions have been applied to a variety of categories (including various kinds of graphs, relational structures, software modules, pictures, etc.: cf. [CER79, ENR83, ENRR87, EKR91], the proceedings of the first four international workshops on graph-grammars). Recently some efforts can be reported towards the generalization of the algebraic theory of graph grammars to arbitrary underlying categories [EHKP91, Ken91].

In this paper we will use the above definitions in a suitable category (defined in Section 2) having a special kind of *hypergraphs* as objects, which will be used to represent the terms and formulas of a logic program. Hypergraphs are a simple generalization of graphs.

1.2 Definition *(category $HGraph_C$)*

A *directed, colored hypergraph over* C is a tuple H = (V, A, s, t, l, m), where A is the set of *hyperarcs* (or *hyperedges*), V, l, and m are as for graphs above, and s, t: $A \to V^*$ associate with each hyperedge a *tuple* of source and target nodes, instead of a single node. A *hypergraph morphism* f: G → H is as a graph morphism, where the condition that sources (resp. targets) of an edge must be preserved becomes $s_H \circ f_A = f_V^* \circ s_G$ (and similarly for t), where f^* is the obvious extension of f to tuples. The category of directed, colored hypergraphs over C is denoted by $\mathbf{HGraph_C}$. ♦

Graphs can be considered as special hypergraphs where for each edge e, $|s(e)| = |t(e)| = 1$.[1] The prefix *hyper-* will be often omitted in the sequel.

1.2. Logic programming

An introductory and complete treatment of logic programs can be found in [Ll87]. We recall here just the concepts to be used in the rest of the paper.

Let $\Sigma = \cup_n \Sigma_n$ be a ranked set of function symbols, where Σ_n is the set of functions of rank n (in particular, Σ_0 is the set of constants). Also, let $\Pi = \cup_n \Pi_n$ be a ranked set of predicate symbols (where Π_n is the set of predicates with n arguments), and let X be a set of variables. A *term over* Σ is an element of $T_\Sigma(X)$, the free Σ-algebra generated by X, that is

- a variable in X, or
- a constant in Σ_0, or
- $f(t_1, \ldots, t_n)$, if $t_1, \ldots, t_n$ are terms over Σ, and $f \in \Sigma_n$.

If $t_1, \ldots, t_n$ are terms over Σ, and $p \in \Pi_n$, then $p(t_1, \ldots, t_n)$ is an *(atomic) formula over* (Σ, Π). The set of formulas over (Σ, Π) with variables in X will be sometimes indicated by $T_{\Sigma,\Pi}(X)_\pi$, which denotes the carrier of sort π of the free (Σ, Π)-algebra generated by X.

A *definite clause* C is an expression of the form

$$H :\!- B_1, \ldots, B_n. \qquad (n \geq 0)$$

where H, $B_1, \ldots, B_n$ are formulas, ':-' means logic implication (right to left), and ',' means logical conjunction.

A *goal* G is an expression of the form

$$:\!- G_1, \ldots, G_n \qquad (n > 0)$$

where $G_1, \ldots, G_n$ are formulas.

A *logic* (or *HCL*) *program P* is a finite set of definite clauses.

A logic program can be interpreted in many different, although equivalent, ways. As a first order theory, its semantics is defined in a model-theoretic way as its *least Herbrand model*. Under the operational reading, instead, a *resolution rule* states how to transform a goal into another. The operational semantics is then defined

[1] If T is a tuple, then |T| denotes its length. Similarly, if S is a set then |S| is its cardinality.

as the set of all (ground) atomic formulas which can be transformed into the empty goal through a sequence of resolution steps. The equivalence between the operational and model-theoretic semantics of a program is proved by showing that the resolution inference rule is both sound and complete for definite clauses.

However, it has been stressed in [FLMP89] that the classic semantics does not capture the operational behaviour of a program, because the *computed answer substitutions* (i.e. the actual results produced by refutations) cannot be recovered from the least model. It is shown that this problem can be solved by allowing also *non ground* atomic formulas to appear in the models of a program, and by slightly changing the definition of models. In this paper, since we are interested in showing that the operational behaviour of a program can be faithfully simulated in the graph grammar context, we will focus on the operational behaviour of a program in the sense of [FLMP89], i.e. on the answer substitutions computed by the program for a given goal, as defined in the rest of this section.

Given a set of variables X and a ranked set of function symbols Σ, a *substitution* is a function $\theta: X \rightarrow T_\Sigma(X)$ which is the identity on all but a finite number of variables. Thus, θ can be written as $\theta = \{x_1/t_1, \ldots, x_n/t_n\}$, where $\theta(x_i) = t_i$ for each $1 \leq i \leq n$, and $\theta(x) = x$ otherwise. The *application* of θ to a term t produces a new term (written $t\theta$) obtained from t by replacing each occurrence of x_i with t_i, for $1 \leq i \leq n$. This definition, and the corresponding notation, is also used for the application of substitutions to formulas.

Given two substitutions θ and δ, their *composition* is a substitution $\theta \bullet \delta$ such that for all terms t, $t(\theta \bullet \delta) = (t\theta)\delta$. Given two substitutions θ and δ, θ is said to be *more general* than δ if there exists a substitution ω such that $\theta \bullet \omega = \delta$. Two atomic formulas A and B *unify* if there exists a substitution θ such that $A\theta = B\theta$. In this case θ is called a *unifier* of A and B. The set of unifiers of any two atomic formulas is either empty, or it has a most general element (up to variable renaming) called the *most general unifier* (*mgu*).

Given a clause $C = (H \text{ :- } B_1, \ldots, B_n)$ and a goal $G = (\text{:- } G_1, \ldots, G_n)$, a *resolution step* involves the selection of an atomic goal G_i and the construction of the most general unifier (if any) θ between H and G_i. The result of such a step is the new goal $G' = (\text{:- } G_1, \ldots, G_{i-1}, B_1, \ldots, B_n, G_{i+1}, \ldots, G_n)\theta$. In this case we will say that there is a *resolution step from G to G' via C and θ.*

A *refutation* of a goal G is a finite sequence of resolution steps which starts with G and ends with the empty goal. If the refutation has length n, where step i uses clause C_i and the mgu θ_i, then the substitution $\theta = (\theta_1 \bullet \ldots \bullet \theta_n)\mid_{Var(G)}$ (i.e. the restriction of $\theta_1 \bullet \ldots \bullet \theta_n$ to the variables appearing in G), is called a *computed answer substitution* for G. In this case we say that there is a *refutation of G via $C_1,\ldots,C_n$ and $\theta = (\theta_1 \bullet \ldots \bullet \theta_n)_{Var(G)}$*.

2. Representing formulas by graphs

In this section we show how the goals of a logic program (i.e. conjunctions of atomic formulas) can be represented by suitable graphs. Goals will be regarded as *unordered* collections of formulas, possibly sharing variables.

Many proposals have been presented in the literature for representing terms built over a given signature by graphs. For example in [Rao84, Ken87, PEM87] *directed acyclic graphs* (*dags*) are used, while in [HKP88] *jungles* (i.e. hypergraphs satisfying some additional requirements) are considered. In Section 2.1 we briefly introduce these two approaches, showing their equivalence: this fact permits us to exploit well known results about both dags and jungles. Next, in Section 2.2, we define a suitable category of enriched dags in order to represent not only terms, but also formulas.

2.1. Directed acyclic graphs vs. jungles

Let Σ be a (one sorted) signature, as defined in Section 1.2, and $\Sigma^\top$ be the flat partial ordering over $\Sigma \cup \{\top\}$, defined as $x \leq y$ iff $x = y$ or $y = \top$. As shown for example in [Rao84, PEM87], terms over Σ (possibly including variables from a set X) can be represented as *directed acyclic graphs* having $\Sigma \cup X$ as the set of node colors. In the following definition we essentially follow [PEM87], but we label all nodes representing variables with the same color '$\top$'.

2.1 Definition *(category DAG_Σ)*

A *directed acyclic graph (dag) D over signature Σ* is a colored graph over $(\Sigma^\top, N)$, $D = (V_D, A_D, s_D, t_D, l_D$:

$V_D \to \Sigma^T$, m_D: $A_D \to N$) (where N denotes the set of natural numbers) such that

- $l_D(v) \in \Sigma_n$ iff there exist arcs $a_1, \ldots, a_n \in A_D$ such that $m_D(a_i) = i$ and $s_D(a_i) = v$, and for all $a \in A_D$ s.t. $s_D(a) = v$, $a = a_i$ for some $i \in \{1, \ldots, n\}$.
- if $l_D(v) = T$, then there exists no arc a such that $s_D(a) = v$.
- D is acyclic.

A *dag morphism* h: $D \to D'$ is a graph morphism (h_A: $A_D \to A_{D'}$, h_V: $V_D \to V_{D'}$) which preserves the edge colors (i.e. $m_{D'} \circ h_A = m_D$), and which is decreasing w.r.t. node colors, i.e. $l_{D'}(h_V(v)) \leq l_D(v)$ for each $v \in V_D$. DAG_Σ is the category having directed acyclic graphs over Σ as objects, and dag morphisms as arrows. ♦

A dag D represents a collection of terms over Σ, possibly sharing subterms.

2.2 Definition *(variables and terms represented by a dag)*

Let D be a dag over Σ. Then the *set of variables of D*, denoted by Var_D, is defined as the set of nodes of D labeled by T:

$$Var_D \equiv \{v \in V_D \mid l_D(v) = T\}.$$

Every node of D is the root of a subgraph which represents a term in the free Σ-algebra generated by Var_D, $T_\Sigma(Var_D)$. The function $term_D$: $V_D \to T_\Sigma(Var_D)$ is defined inductively as follows:

- $term_D(v) = v$ if $l_D(v) = T$
- $term_D(v) = c$ if $l_D(v) = c \in \Sigma_0$
- $term_D(v) = f(term_D(v_1), \ldots, term_D(v_n))$ if $l_D(v) = f \in \Sigma_n$, $\forall i \in \{1, \ldots, n\}\ \exists a_i \in A_D$ s.t. $s_D(a_i) = v$, $m_D(a_i) = i$, and $t_D(a_i) = v_i$.

By the definition of dags the function $term_D$ is well defined. Finally, let us denote by $Terms_D \subseteq T_\Sigma(Var_D)$ the set of terms represented by a dag D, i.e.

$$Terms_D = \{term_D(v) \mid v \in V_D\}$$ ♦

2.3 Remarks

1) *Rooted* dags (i.e. dags with a distinguished node) are used in [PEM87] to represent a single term: for our purposes it is better to regard a dag as representing a *collection* of terms, possibly sharing subterms.
2) Two isomorphic dags represent the same set of terms, up to variable renaming. If Φ and Φ' are two sets of terms over Σ, in the rest of the paper we will write $\Phi \equiv \Phi'$ to say that they are equal up to variable renaming. ♦

2.4 Example *(a directed acyclic graph)*

Let D be the dag in the following picture: it has variables $Var_D = \{v_2, v_5\}$, and represents the terms $Terms_D = \{f(v_2, h(a, v_5)), g(h(a, v_5), v_5), h(a, v_5), v_2, a, v_5\}$.

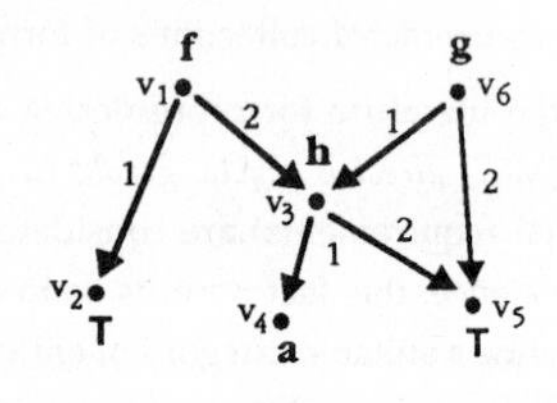

♦

Since by Definition 2.1 the node colors need not be preserved by morphisms, it is easy to see that every enriched dag morphism f: $G \to H$ induces a substitution θ_f on the variables in Var_G. In fact a variable node of G (labeled with T) can be mapped by f to a node of H with label in Σ, i.e. to the root of a sub-dag representing a term, thus modeling an instantiation of the variable.

2.5 Definition *(the substitution induced by a morphism).*

Every enriched dag morphism f: $G \to H$ induces:

1) a substitution θ_f: $Var_G \to Terms_H$, defined as $\theta_f(x) = term_H(f_V(x))$.
2) a mapping ϕ_f: $Terms_G \to Terms_H$ defined as $\phi_f(t) = t\theta_f$. ♦

2.6 Example *(dag morphism)*

Let D' be the following dag, and $k: D' \to D$ be the unique dag morphism mapping D' to dag D of Example 2.4. Clearly, $Vars_{D'} = \{n_2, n_3\}$, and the substitution associated with k is $\theta_k = \{n_2/v_2, n_3/h(a, v_5)\}$.

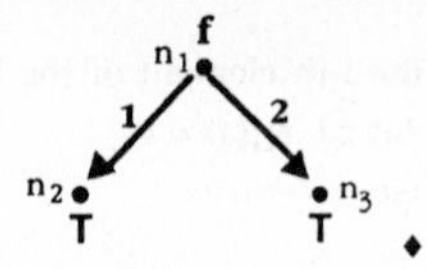

♦

A different way of representing terms by graphs has been used in [HKP88], where a special kind of hypergraphs, *jungles*, are considered. In a jungle over a signature Σ the operator symbols of Σ label the edges, instead of the nodes as for dags; in addition the number of connections of an edge must be consistent with the rank of its label.

2.7 Definition *(the category of jungles over Σ)*

An edge-colored hypergraph $G = (V_G, A_G, s_G, t_G, m_G: A_G \to \Sigma)$ (cf. Definition 1.2) is a *jungle over Σ* iff

- for each $a \in A_G$, if $m_G(a) \in \Sigma_n$ then $|s_G(a)| = 1$ and $|t_G(a)| = n$.
- for each $v \in V_G$, $outdegree_G(v) \leq 1$.[2]
- G is acyclic.[3]

A *jungle morphism* $h: G \to G'$ is simply a hypergraph morphism. We denote by $\mathbf{Jungle}_\Sigma$ the category including all jungles over Σ as objects, and jungle morphisms as arrows. ♦

In a jungle G over Σ the set of variables is defined as the set of nodes with outdegree 0. Notions similar to those introduced in Definitions 2.2 and 2.5 can be defined also for jungles.

2.8 Example *(a jungle)*

The following jungle represents exactly the same terms and variables as the dag of Example 2.4. Small natural numbers indicate the ordering of the outgoing tentacles of a hyperedge.

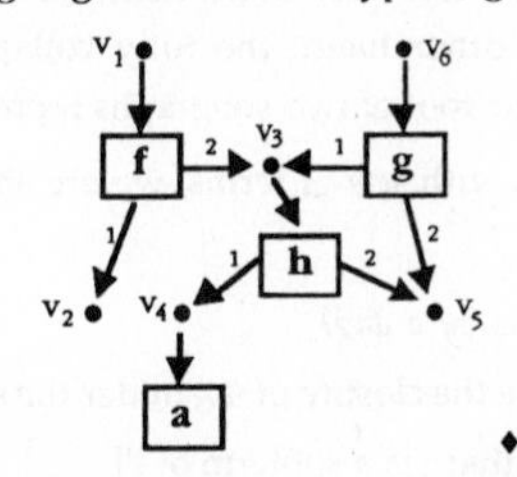

♦

The next proposition states that the representations of terms with dags or jungles are equivalent.

2.9 Proposition *($\mathbf{Jungles}_\Sigma$ and $\mathbf{DAG}_\Sigma$ are equivalent).*

Let Σ be a signature. Then the categories $\mathbf{DAG}_\Sigma$ and $\mathbf{Jungle}_\Sigma$ are equivalent, i.e. there exist two functors $\mathbf{D}: \mathbf{Jungle}_\Sigma \to \mathbf{DAG}_\Sigma$, and $\mathbf{J}: \mathbf{DAG}_\Sigma \to \mathbf{Jungle}_\Sigma$ such that $D(J(D)) \cong D$ and $J(D(G)) \cong G$ for each dag D and jungle G.

Proof Let $D = (V_D, A_D, s_D, t_D, l_D, m_D)$ be a dag over Σ. Define $J(D) = (V_J, A_J, s_J, t_J, m_J)$ by putting

- $V_J = V_D$ (the nodes are the same)
- $A_J = \{\langle v, f\rangle \mid v \in V_D \text{ and } l_D(v) = f \neq T\}$ (one hyperarc for each Σ-colored node of D)
- $s_J(\langle v, f\rangle) = v$
- $t_J(\langle v, f\rangle) = v_1 \cdot \ldots \cdot v_n$ if $f \in \Sigma_n$, $\forall i \in \{1, \ldots, n\}$ $\exists a_i \in A_D$ s.t. $s_D(a_i) = v$, $m_D(a_i) = i$, and $t_D(a_i) = v_i$.
- $m_J(\langle v, f\rangle) = f$.

---

[2] Informally, the *outdegree* of a node is the number of tentacles outgoing from it. For a jungle G, $outdegree_G(v) = |\{a \in A_G \mid v \in s_G(a)\}|$.

[3] For a hypergraph G, its *underlying bipartite graph* U(G) includes all nodes and hyperarcs of G as nodes, and the tentacles of G as arcs. Then G is acyclic iff U(G) is acyclic.

Conversely, let $G = (V_G, A_G, s_G, t_G, m_G)$ be a jungle over Σ. Define $D(G) = (V_D, A_D, s_D, t_D, l_D, m_D)$ as

- $V_D = V_G$
- $A_D = \{\langle a, i\rangle \mid a \in A_G, m_G(a) \in \Sigma_n, \text{ and } 1 \le i \le n\}$
- $s_D(\langle a, i\rangle) = s_G(a)$
- $t_D(\langle a, i\rangle) = t_G(a)\mid_i$ that is the i-th element of the list $t_G(a)$
- $l_D(v) = m_G(a)$ if $\exists a \in A_G$ s.t. $s_G(a) = v$
  $l_D(v) = T$ otherwise
- $m_D(\langle a, i\rangle) = i$

It is straightforward to check that **J** and **D** are well defined, and that they are indeed functors between the considered categories, since they can be easily extended to morphisms. Moreover, they are the inverse of each other up to isomorphisms. ♦

The last result states that jungles and dags are completely equivalent as far as we are concerned with results or constructions expressible in categorical terms. It can also be shown that the set of terms represented by a dag or jungle is preserved by functors **J** and **D**. This implies that, for example, all the properties and results reported in [PEM87] about dags and their relationships with terms also hold for jungles.

A single term can be represented by many (non isomorphic) graphs, since identical subterms can either be collapsed in a single subgraph or not (cf. [Rao84, PEM87, HKP88]). This freedom in the representation of terms by graphs can increase the efficiency of a Term Rewriting System modeled by a set of graph productions: the application of a production to a graph corresponds to n applications of the corresponding rewrite rule, if the rewritten subgraph represents n occurrences of the same subterm [HKP88].

Among the many possible representations of a (single) term t by a (rooted) graph, there are two which are special: in [HKP88] these are called the *variable-collapsed tree*, and the *fully collapsed tree* of t, respectively, and they are the initial and the final representation of t (in the categorical sense). The variable collapsed tree of t is characterized by the fact that a node can have more than one ingoing arc (i.e. it is the root of a *shared* subterm) only if it is a variable; on the other hand, the fully collapsed tree is such that function *term* is injective, i.e. two distinct nodes cannot be the root of two subgraphs representing the same term.

In our framework, since we are concerned with *sets* of terms, we are able to characterize a final representation, but not an initial one.

2.10 Definition *(representing a set of terms by a dag)*

Let Φ_σ be a set of terms over Σ, and let $\overline{\Phi}_\sigma$ be the closure of Φ_σ under the subterm relation, i.e.

$$\overline{\Phi}_\sigma = \{t \mid \text{there is a term } t' \in \Phi_\sigma \text{ such that } t \text{ is a subterm of } t'\}$$

Then a dag $D \in |\mathbf{DAG}_\Sigma|$ *represents* Φ_σ iff $Terms_G \cong \overline{\Phi}_\sigma$. ♦

2.11 Proposition *(the fully collapsed dag representing a set of terms)*

Let Φ_σ and $\overline{\Phi}_\sigma$ be as in the previous definition. Then the set of dags representing Φ_σ has a final element $\Delta(\Phi_\sigma)$, the *fully collapsed dag of* Φ_σ. That is, for all $D \in |\mathbf{DAG}_\Sigma|$ which represent Φ_σ, there exists a unique dag morphism $!D: D \to \Delta(\Phi_\sigma)$.

Proof $\Delta(\Phi_\sigma) = (V_\Phi, A_\Phi, s_\Phi, t_\Phi, l_\Phi, m_\Phi)$ is defined as follows:

- $V_\Phi = \overline{\Phi}_\sigma$
- $A_\Phi = \{\langle t, i\rangle \mid t \in \overline{\Phi}_\sigma$, the outermost operator of t has rank n, and $1 \le i \le n\}$
- $s_\Phi(\langle t, i\rangle) = t$
- $t_\Phi(\langle t, i\rangle) = t_i$ if $t \equiv f(t_1, \ldots, t_m)$, $1 \le i \le m$
- $l_\Phi(t) = f$ if $t \equiv f(t_1, \ldots, t_m)$
  $l_\Phi(t) = T$ if t is a variable
- $m_\Phi(\langle t, i\rangle) = i$

By construction, obviously $Terms_{\Delta(\Phi_\sigma)} \equiv \overline{\Phi}_\sigma$. Now, let D be a dag representing Φ_σ. Then the morphism $!D: D \to \Delta(\Phi_\sigma)$ is uniquely determined by:

- $!D_V(v) = term_D(v)$
- $!D_A(a) = \langle term_D(s_D(a)), m_D(a)\rangle$ ♦

It is easy to check that for a given set of terms Φ_σ, there is no initial representation as dag. For example, let $\Phi_\sigma = \{c\}$, with $c \in \Sigma_0$: then every discrete dag with all nodes labeled by 'c' represents Φ_σ.

2.2. From terms to formulas

In the previous section we considered the representation of terms built over a one sorted signature Σ. However, the 'terms' appearing in a logic program are two-sorted, since they are constructed from two disjoint sets of operators, including predicate and function symbols respectively. Although both dags and jungles over *many sorted* signatures can be easily defined (and the equivalence result reported above still holds), a different approach can be taken thanks to the observation that a predicate symbol must always appear as the outermost operator in a formula. Informally, this constraint permits us to construct a hypergraph representing a collection Φ_π of atomic formulas in two steps: first we take a dag representing all the terms which appear as arguments of the formulas in Φ_π; and then we add, for each atom $p(t_1, ..., t_n) \in \Phi_\pi$, a new hyperedge labeled 'p', with n outgoing tentacles connecting it to the roots of the dags representing the terms $t_1, ..., t_n$. This choice allows us to reflect the two-sorted nature of the operators into their representation as hypergraph components: while function symbols appear as labels for nodes, predicate symbols will color the hyperedges.

In the rest of the paper, we will use a fixed signature (Σ, Π), as introduced in Section 1.2. Σ^T is the flat partial ordering defined in Section 2.1. Let us now introduce the category of enriched dags used to represent collections of terms and formulas.

2.12 Definition *(category Π–DAG_Σ)*

An *enriched dag G over (Σ, Π)* is a hypergraph

$$G = (V_G, A_G, s_G, t_G, l_G: V_G \to \Sigma^T, m_G: A_G \to \Pi \cup N)$$

such that:

- erasing from G all the hyperarcs with label in Π, the resulting graph is a dag over Σ. More formally, let $A'_G = \{a \in A_G \mid m_G(a) \in N\}$; then $U(G) \equiv (V_G, A'_G, s_G|_{A'_G}, t_G|_{A'_G}, l_G: V_G \to \Sigma^T, m_G|_{A'_G}: A'_G \to N)$ is an object of $\mathbf{DAG}_\Sigma$.[4] The graph U(G) is called *the underlying dag of G*.
- if $m_A(a) \in \Pi_n$, then $|s_G(a)| = 0$ and $|t_G(a)| = n$.

An *enriched dag morphism* $h: G \to G'$ is a hypergraph morphism $(h_A: A_G \to A_{G'}, h_V: V_G \to V_{G'})$ such that

- its restriction to the underlying dags $U(h): U(G) \to U(G')$ is a dag morphism
- the edge colors in Π are preserved, i.e. if $m_G(a) \in \Pi$, then $m_{G'}(h_A(a)) = m_G(a)$.

Π-$\mathbf{DAG}_\Sigma$ is the category having enriched dags over (Σ, Π) as objects, and enriched dag morphisms as arrows. ♦

Obviously, Definitions 2.2 and 2.5 also apply to enriched dags, considering their underlying dags. Besides a set of variables and terms, an enriched dag also represents a multiset of atomic formulas. Conversely, a given collection of terms and formulas can be represented by many enriched dags. Let us now define these notions formally. In the sequel, by *collection* we mean *set* of terms, but *multiset* of atomic formulas.

2.13 Definition *(the formulas represented by an enriched dag)*

Let $G = (V_G, A_G, s_G, t_G, l_G, m_G)$ be an enriched dag over (Σ, Π), and let A''_G be the set of Π-colored edges, i.e. $A''_G = \{a \in A_G \mid m_G(a) \in \Pi\}$. Moreover, let Var_G be the set of variables of G, as in Definition 2.2. Then every edge in A''_G is the root of a subgraph which represents an atomic formula in $T_{\Sigma,\Pi}(Var_G)_\pi$. The function $form_G$: $A''_G \to T_{\Sigma,\Pi}(Var_G)_\pi$ is defined as

$$form_G(a) = p(term_G(v_1), ..., term_G(v_n)) \qquad \text{if } m_G(a) = p \in \Pi_n, \text{ and } t_G(a) = v_1 \cdot ... \cdot v_n.$$

It can be easily shown that function $form_G$ is well defined. We also denote by $Forms_G$ the *multiset* of formulas in $T_{\Sigma,\Pi}(Var_G)_\pi$ represented by an enriched dag G, i.e.

[4] For $f: A \to B$ and $A' \subseteq A$, $f|_{A'}: A' \to B$ denotes the obvious restriction.

$Forms_G = \{\langle form_G(a), a\rangle \mid a \in A''_G\}$[5].

Furthermore, let Φ_π be a multiset of atomic formulas over (Σ, Π), Φ_σ be a set of terms over Σ, and $\Phi = \Phi_\pi \cup \Phi_\sigma$. In addition, let $\overline{\Phi}_\sigma$ be the set of all terms which occur in Φ. In general $\Phi_\sigma \subset \overline{\Phi}_\sigma$ because $\overline{\Phi}_\sigma$ contains also all the arguments of formulas in Φ_π, and is closed under the subterm relation. Then an enriched dag G *represents* Φ iff $Forms_G \cong \Phi_\pi$,[6] and $Terms_G \cong \overline{\Phi}_\sigma$. Finally, the *collection of terms and formulas associated with* G, Φ_G, is defined as $\Phi_G = Forms_G \cup Terms_G$, while *the goal associated with* G is simply $Forms_G$. ♦

2.14 Example *(an enriched dag)*

Let B be the enriched dag in the following picture. Then its set of variables is $Var_D = \{v_1, v_2, v_3, v_6\}$, and it represents the set of formulas $Forms_D = \{\langle$**reverse**$(v_1, v_2), h_1\rangle, \langle$**append**$(v_2,$ **cons**$(v_3,$ **nil**$), v_6), h_2\rangle\}$ and the set of terms $Terms_D = \{v_1, v_2,$ **cons**$(v_1, v_3), v_3,$ **cons**$(v_3,$ **nil**$),$ **nil**$, v_6\}$.

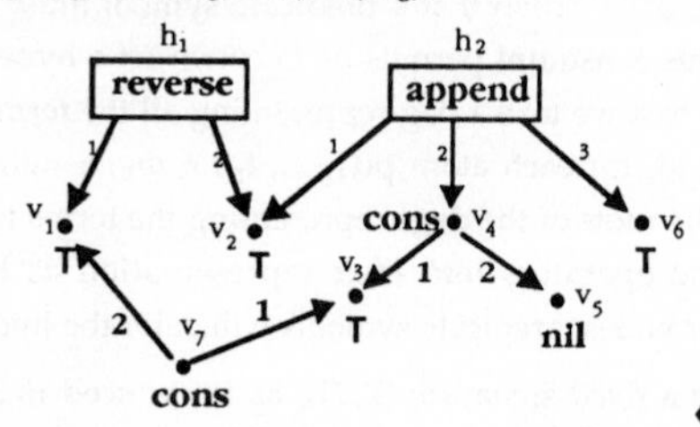

♦

We decided to associate a multiset of formulas with an enriched graph, instead of a set, in order to be able to capture the fact that in logic programming, for example, the two goals ":- p(x)" and ":- p(x), p(x)" can behave quite differently, although they are logically equivalent. However, it must be noticed that since the elements of a multiset are not ordered, we cannot represent the fact that a logic programming goal is an *ordered list* of atomic formulas. Thus it is not possible to define a selection rule based on such ordering in the graph grammar framework we are defining. Nevertheless, we believe that this is mainly a matter of implementation, because the operational semantics of a logic program does not depend on the selection rule (cf. [Ll87]).

For a given collection of terms and formulas, a final representation as enriched dag (in the sense of Proposition 2.11) does not always exist. Nevertheless a unique enriched dag (up to isomorphisms) can be characterized by a slightly weaker condition, if we restrict ourselves to a suitable subset of the possible morphisms.

2.15 Proposition *(the fully collapsed enriched dag representing a collection of terms and formulas)*

Let Φ be a collection of terms and formulas, as in Definition 2.13. Then the set of enriched dags representing Φ includes an element $\Delta(\Phi)$ (determined up to isomorphisms), such that for all G which represent Φ, there exists an enriched dag morphism $h_G: G \to \Delta(\Phi)$ which is injective on Π-colored edges. This morphism is unique up to an arbitrary permutation of the edges representing distinct occurrences of the same atomic formula.

Proof. Let Φ_π, Φ_σ and $\overline{\Phi}_\sigma$ be as in Definition 2.13, and let $\Delta(\overline{\Phi}_\sigma)$ be the fully collapsed dag of $\overline{\Phi}_\sigma$ as defined in Proposition 2.11. Then add to the set of edges A_Φ the set Φ_π. For each $\langle p(t_1, ..., t_n), x\rangle \in \Phi_\pi$ put $s_\Phi(\langle p(t_1, ..., t_n), x\rangle) = \varepsilon$ (the empty list); $t_\Phi(\langle p(t_1, ..., t_n), x\rangle = t_1 \cdot ... \cdot t_n$, and $m_\Phi(\langle p(t_1, ..., t_n), x\rangle) = p$. Clearly the resulting enriched dag $\Delta(\Phi)$ represents Φ. For any other G which represents Φ, a morphism $h_G: G \to \Delta(\Phi)$ is uniquely determined on nodes and N-colored arcs as in Proposition 2.11, while on Π-colored edges it can be forced to be a bijection, because the number of such edges in G and $\Delta(\Phi)$ is identical. ♦

To conclude this section, let us present a lemma which characterizes the existence of certain pushouts in category

---

[5] We represent a multiset M of elements of S as a set of pairs $M = \{\langle s, x\rangle\}$, where $s \in S$ and x belongs to a set of tags used to distinguish among different occurrences of the same element. The usual representation of M as a function $f_M: S \to N$ can easily be obtained by putting $f_M(s) = |\{\langle x,y\rangle \in M \mid x = s\}|$. All the operations on multisets will not use the identity of tags.

[6] For two multisets of formulas M and M', we write $M \approx M'$ if $f_M = f_{M'}$, and $M \cong M'$ if $f_M = f_{M'}$ modulo variable renaming.

Π-$\mathbf{DAG}_\Sigma$ in terms of the existence of the most general unifier of two tuples of terms. This result will be used in the proof of the main theorem in Section 4.

2.16 Lemma *(existence of certain pushouts in Π-DAG_Σ)*
Let K be a discrete enriched dag where all nodes are variables (i.e. $l_K(k) = T$ for each $k \in K_V$), and let $d: K \to D$, $g: K \to G$ be two arrows in Π-$\mathbf{DAG}_\Sigma$. Without loss of generality, we can suppose that Var_G and Var_D are disjoint. Then the pushout of $\langle d, g\rangle$ exists in Π-$\mathbf{DAG}_\Sigma$ iff for any enumeration $\langle k_1, \ldots, k_n\rangle$ of the nodes of K, the tuples $\tau_d \equiv \langle term_D(d(k_1)), \ldots, term_D(d(k_n))\rangle$ and $\tau_g \equiv \langle term_G(g(k_1)), \ldots, term_G(g(k_n))\rangle$ have a most general unifier. Moreover, if $\langle f, h\rangle$ is the pushout of $\langle d, g\rangle$ (as in the following diagram), then the enriched dag H represents the collection of terms and formulas $\Phi_G\theta_f \cup \Phi_D\theta_h$, where θ_f and θ_h are as in Definition 2.5.

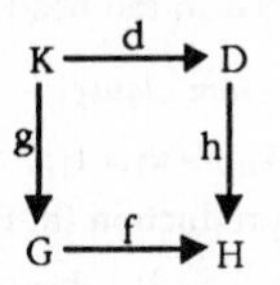

Proof We need the following fact, which is easy to check: if $r: A \to B$ and $r': B \to C$ are morphisms in Π-$\mathbf{DAG}_\Sigma$, then for every $v \in A_V$, $term_C(r' \circ r(v)) = term_B(r(v))\theta_{r'}$.
Only if part. Let $\langle h: D \to H, f: G \to H\rangle$ be a pushout of $\langle d, g\rangle$. Then, by commutativity of pushouts and the above fact, for each $1 \le i \le n$ we have $term_D(d(k_i))\theta_h = term_H(h \circ d(k_i)) = term_H(f \circ g(k_i)) = term_G(g(k_i))\theta_f$. Therefore the substitution $\theta = \theta_h \cup \theta_f$ unifies τ_d and τ_g (remember that Var_G and Var_D are disjoint by hypothesis). The fact that θ is actually the mgu of τ_d and τ_g easily follows from the universal property of pushouts.

If part. In category $\mathbf{Jungle}_\Sigma$ there exist all colimits over *consistent* diagrams (i.e. diagrams for which a cocone exists: cf. [Ken91]). By Proposition 2.9 this also holds for category $\mathbf{DAG}_\Sigma$; it could be easily shown that the same is true in Π-$\mathbf{DAG}_\Sigma$. Therefore it suffices to check that if the above tuples of terms unify with substitutions θ (i.e. for each $1 \le i \le n$ $term_D(d(k_i))\theta = term_G(g(k_i))\theta)$, then there exist arrows $\langle h: D \to H, f: G \to H\rangle$ such that the resulting square commutes. But this can easily be obtained by putting $H = \Delta(\Phi_D\theta \cup \Phi_G\theta)$, and forcing h, f in such a way that $\theta_h = \theta \mid_{Var_D}$ and $\theta_f = \theta \mid_{Var_G}$. ♦

The relationship between most general unifiers and categorical universal constructions has been stressed in many places. For example, mgu's are characterized in [RB85] (resp. [Go89]) in terms of coequalizers (resp. equalizers), and also as pullbacks in [AM89]; in these papers terms and substitutions are represented as arrows of a category. On the contrary, our characterization is much closer to the one in [PEM87], where terms are objects and mgu's are pushouts.

3. Representing clauses by graph productions

In this section we show how to represent a logic program clause by a graph production, i.e. by a pair of morphisms in category Π-$\mathbf{DAG}_\Sigma$ of the following shape: $H \leftarrow h - K - b \rightarrow B$.

We introduce this representation in two steps: first a clause is transformed in an equivalent *canonical form*, and then the canonical form is straightforwardly represented by a diagram like the one above, exploiting the representation of collections of formulas and terms by enriched dags introduced in Section 2.

3.1 Definition *(the canonical form of a clause)*
Let $C \equiv p(t_1, \ldots, t_m) \text{ :- } B_1, \ldots, B_n$. be a logic program clause, and let $\{x_1, \ldots, x_m\}$ be a set of variables not occurring in C. Then the *canonical form of* C is the clause

$$can(C) \equiv p(x_1, \ldots, x_m) \text{ :- } x_1 = t_1, \ldots, x_m = t_m, B_1, \ldots, B_n.$$

where the equality predicate in the body stands for term unification. ♦

Obviously *can*(C) and C are equivalent from the logic programming viewpoint, in the sense that for a given goal G, either the application to G of both clauses fail, or both succeed producing the same substitution for the variables in G. This canonical form of clauses has been used for example in [Le88], for comparing different logic languages. In our case, the transformation of definite clauses into a canonical form is fully justified by the results presented in the next section.

3.2 Example *(the canonical form of a clause)*

Consider the following definite clause which defines the *reverse* predicate:

$$C \equiv \text{reverse}(\text{cons}(x, y), z) :- \text{reverse}(y, w), \text{append}(w, \text{cons}(x, \text{nil}), z).$$

The corresponding canonical form is

$$can(C) \equiv \text{reverse}(x_1, x_2) :- x_1 = \text{cons}(x, y), x_2 = z, \text{reverse}(y, w), \text{append}(w, \text{cons}(x, \text{nil}), z). \blacklozenge$$

A clause in canonical form can easily be represented by a graph production, taking as interface graph the discrete graph containing all the variables included in the head.

3.3 Definition *(the dag representation of a program clause)*

Let C be a program clause, and $can(C) \equiv p(x_1, \ldots, x_m) :- x_1 = t_1, \ldots, x_m = t_m, B_1, \ldots, B_n$. be its canonical form. The *dag representation* of C, $\Delta(C)$, is the enriched dag production (in the sense of Section 1.1)

$$H_C \equiv \Delta(\{p(x_1, \ldots, x_m)\} \xleftarrow{h_C} K_C \equiv \Delta(\{x_1, \ldots, x_m\}) \xrightarrow{b_C} B_C \equiv \Delta(\{t_1, \ldots, t_m, B_1, \ldots, B_n\})$$

where $\Delta(\Phi)$ is as in Proposition 2.15, h_C is the obvious inclusion, and b_C maps each x_i to the root of the dag representing t_i. ♦

3.4 Example *(representing a definite clause with an enriched dag production)*

The enriched dag production of the following picture represents the clause of Example 3.2. The dotted arrows show how the nodes of K are mapped by $h: K \to H$ and $b: K \to B$, respectively.

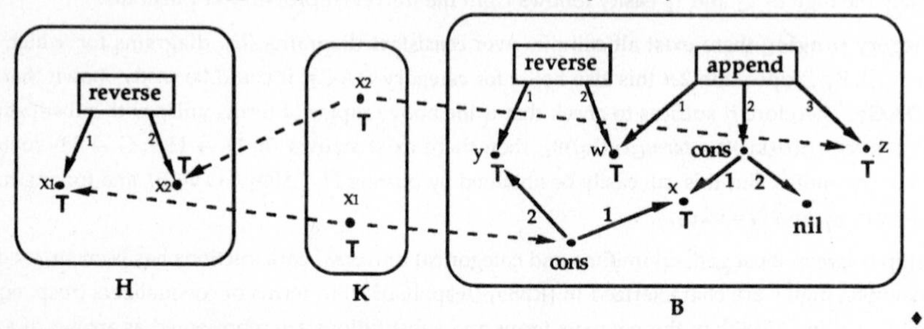

♦

By Definitions 3.1 and 3.3, the dag representation of a definite clause satisfies some properties which actually characterize a *context-free production*, in the following sense.

3.5 Definition *(context-free productions)*

A production $p = (H \xleftarrow{h} K \xrightarrow{b} B)$ over Π-**DAG**$_\Sigma$ is *context-free* iff:

- the left side H contains a single Π-colored hyperedge connected to m distinct variable nodes, where m is the rank of its label,
- the interface K is a discrete graph including exactly m variable nodes, and
- the arrow $h: K \to H$ is a bijection on nodes. ♦

This characterization implies that not only can a definite clause be represented by a context free production (via its canonical form), but also that from every context free production over category Π-**DAG**$_\Sigma$ a definite clause can be extracted. It should be noticed that our notion of context-freedom coincides with the one in [BC87]. On the other hand, the notion proposed in [Ha89] is more restrictive, since it also requires that no pair of nodes can be identified by the production, that is, morphism $b: K \to B$ must be injective.

3.6 Definition *(the clause associated with a context free hypergraph production)*

Let $p = (H \xleftarrow{h} K \xrightarrow{b} B)$ be a context-free production of category Π-**DAG**$_\Sigma$. Then the *clause associated with p* is

$q(t_1, \ldots, t_m)$:- $\{B_1, \ldots, B_n\}$[7]

where q is the label of the unique hyperarc *a* of H, $\{B_1, \ldots, B_n\} = Forms_B$, and for each $i \in \{1, \ldots, m\}$ $t_i = term_D(b_V(v_i))$, where $\langle v_1, \ldots, v_m \rangle$ is the unique ordering of the nodes of K such that $t_H(a) = h_V(v_1) \cdot \ldots \cdot h_V(v_m)$. ♦

It must be stressed that passing from a production to the associated definite clause, some information might get lost. More precisely, it is easy to show that if there is a node $v \in B_V$ such that $term_B(v)$ does not occur as a subterm in $Forms_B$, nor in $term_B(b_V(v'))$ for any node v' in K_V, then $term_B(v)$ does not appear in the associated definite clause.

In the classic presentation of the double-pushout approach to graph grammars (briefly recalled in Section 1.1), since the underlying category (usually a category of colored graphs) has all pushouts, much emphasis has been placed on the *gluing analysis*, i.e. the analysis of conditions which ensure the existence of the pushout complement of $\langle l, g \rangle$ in Diagram 1.1. In fact if the pushout complement H can be determined, the direct derivation step can be successfully completed, because the right pushout is sure to exist. Moving to different categories (like the ones introduced in Section 2) where pushouts do not always exist, the gluing analysis loses its predominance, in favor to the analysis of conditions for the existence of the second pushout. In fact the next lemma states that if there is an occurrence of the left hand side of a context free production p in a dag D, then the pushout complement can always be constructed in a unique way.

3.7 Lemma *(the pushout complement of an occurrence of a context-free production always exists)*

Let $p \equiv H_C \xleftarrow{h_C} K_C \xrightarrow{b_C} B_C$ be a context-free production over **Π-DAG**$_\Sigma$. Moreover, let $d: H_C \to D$ be an occurrence of H_C in D. Then the pushout complement of $\langle h_C, d \rangle$ exists and is unique. ♦

3.8 Remark

In the hypotheses of the Lemma, the pushout complement is a graph Q (cf. Diagram 4.2 below) obtained from D by removing the image of the unique edge in H_C. Arrow $l: Q \to D$ is the inclusion, and $q: K_C \to Q$ is the unique arrow making the square commuting. Notice that, in order to make the square commuting, q maps the i-th node of K_C to the root of the i-th argument of the atomic formula selected in D by arrow d. ♦

Intuitively, if p is the dag representation of a clause C, the existence of an arrow $d: H_C \to D$ just means that in the goal represented by D there is an atomic goal with the same predicate symbol which appears in the head of clause C. In this case the pushout complement Q can always be constructed, simply removing that predicate symbol from the goal.

4. Logic programs as graph grammars and *viceversa*

We present here the main result of the paper, which states that the representation of a program clause C by a dag production Δ(C), introduced in the last section, is sound and complete. Next the result is extended to entire programs, showing the existence of a tight relation between logic programs and context-free graph grammars over **Π-DAG**$_\Sigma$.

Informally, by soundness we mean that if there is a direct derivation from an enriched dag D to D' via production Δ(C), then there is a resolution step from goal $Form_D$ to goal $Form_{D'}$ via clause C. Conversely, by completeness we mean that if there is a resolution step from G to G' via C, then *for any enriched dag* D *which represents* G the production Δ(C) can be applied to D, and the derived dag D' represents G'. Although we do not consider here Term Rewriting Systems, this result should be compared with [Rao84, Ken87, BvEGKPS87, PEM87, HKP88], where in representing term rewriting rules by graph productions just soundness is ensured. We also discuss two possible causes of incompleteness, showing that both are avoided by the use of the canonical form. This suggests that similar completeness results could be obtained for TRS as well, using a canonical form of rewriting rules.

---

[7] The clause is defined up to the ordering of the atomic formulas in the body.

4.1 Theorem *(soundness and completeness of dag representation of program clauses)*

Let C be a definite clause and $\Delta(C)$ be its dag representation, as described in Definition 3.3. Then:

Soundness. If D is an enriched dag and d: $H_C \to D$ is an occurrence of H_C in D such that $D \overset{\Rightarrow}{}_{\Delta(C),d} D'$ (i.e. there exists a direct derivation from D to D' via $\Delta(C)$ and d), then there exists a resolution step from goal $Forms_D$ to goal $Forms_{D'}$ via clause C and substitution $\theta \cong \theta_r \cup \theta_{d'}$ (cf. Definition 2.5 and Diagram 4.2).

Completeness. If there exists a resolution step from goal G to goal G' via C and θ (cf. Section 1.2), then *for every enriched dag* D *which represents* G, there exists an occurrence d: $H_C \to D$ such that $D \overset{\Rightarrow}{}_{\Delta(C),d} D'$, D' represents G', and $\theta \cong \theta_r \cup \theta_{d'}$.

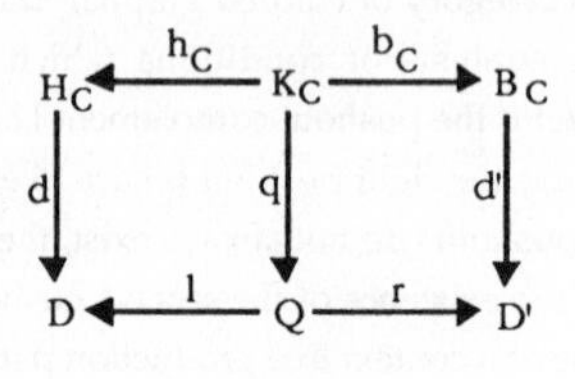

4.2 Diagram

Proof *Soundness.* Suppose that $D \overset{\Rightarrow}{}_{\Delta(C),d} D'$, and $form_D(a) = p(s_1, ..., s_m)$, where *a* is the image through d of the unique hyperarc of H_C. Then by Remark 3.8, the dag Q represents the collection of formulas and terms $\Phi_D \backslash \{\langle p(s_1, ..., s_m), a \rangle\} \cup \{s_1, ..., s_m\}$. By Lemma 2.16, since the pushout of $\langle q, b_C \rangle$ exists in Π-DAG_Σ by hypothesis, the tuples of terms $\langle t_1, ..., t_m \rangle$ and $\langle s_1, ..., s_m \rangle$ unify, where by Definition 3.3 $\langle t_1, ..., t_m \rangle$ are the arguments of the atomic goal in the head of C. Therefore C can be applied to $Forms_D$, because its head unifies with $p(s_1, ..., s_m)$. By Lemma 2.16, $Forms_{D'} = Forms_{B_C}\, \theta_{d'} \cup Forms_Q\, \theta_r = \{B_1, ..., B_n\}\theta_{d'} \cup Forms_D \backslash \{\langle p(s_1, ..., s_m), a \rangle\}\theta_r = (\{B_1, ..., B_n\} \cup Forms_D \backslash \{\langle p(s_1, ..., s_m), a \rangle\}) \circ \theta$ where θ is the most general unifier of $p(s_1, ..., s_m)$ and $p(t_1, ..., t_m)$.

Completeness. If there exists a resolution step from goal G to goal G' via C and θ, D is a dag which represents G, and $p(s_1, ..., s_m)$ is the atomic goal in G rewritten by the application of C, let d: $H_C \to D$ be defined in such a way that $form_D(d(a)) = p(s_1, ..., s_m)$, for the unique hyperarc a of H_C. Notice that morphism d can be defined because all the nodes of H_C are non shared variables. Then by Lemma 3.7 and Remark 3.8 the pushout complement of $\langle h_C, d \rangle$ exists, and is a graph Q which represents $G \backslash \{\langle p(s_1, ..., s_m), d_A(a) \rangle\} \cup \{s_1, ..., s_m\}$. By Lemma 2.16 and the hypotheses, the pushout of $\langle b_C, q \rangle$ exists, and is a graph D' which represents the goal $(G \backslash \{\langle p(s_1, ..., s_m), d_A(a) \rangle\} \cup \{B_1, ..., B_n\})\theta$ which is exactly G' (up to variable renaming). ♦

The just proven completeness result is a direct consequence of the use of the canonical form, and does not hold for the representations of term rewriting rules by graph production proposed for example in [PEM87] and [HKP88]. In those papers, the left side of a production is a graph which represents the whole left hand side of the rewrite rule, and therefore it includes in general more than a single hyperedge. In our framework, if $C \equiv p(t_1, ..., t_m)$:- $B_1, ..., B_n$ is a clause, this corresponds to defining a production $\Delta'(C)$ representing C, having as left side a graph H'_C which represents the entire head $\{p(t_1, ..., t_m)\}$. In this case there are two possible sources of incompleteness, i.e. of situations where given a dag D, clause C can be applied to the goal $Forms_D$, but production $\Delta'(C)$ cannot be applied to D because there are no morphisms from H'_C to D: we discuss them briefly.

1) *Incompleteness due to non ground goals.* Let C ≡ "p(a) :- $B_1, ..., B_n$", and D be the unique representation of the goal "p(x)". Clearly C can be applied to p(x) (producing the substitution {x/a}), but there is no morphism from the first to the second of the following graphs, since the constraint of Definition 2.1 cannot be satisfied.

It must be stressed that this situation is peculiar to logic programming, where the goals can contain variables and the matching condition between the head of a clause and a goal is unification. In a term rewriting system, on the contrary, variables can appear only in the rewrite rules, since all the terms in a rewriting sequence are ground; therefore a situation like that one can never happen. Clearly, the choice of the canonical form avoids these situations, since the left side of $\Delta(C)$ does not include function symbols.

2) *Incompleteness due to non left-linearity.* Let C ≡ "p(x, x) :- B_1, ..., B_n", $H'_C = \Delta(\{p(x, x)\})$, and D be the representation of the goal "p(a, a)" containing two distinct nodes labeled by a. Again clause C can be applied to p(a, a) (producing the substitution {x/a}), but obviously there is no morphism from H'_C to D.

H'_C p • T   D p • a • a

Using the canonical form, this situation is avoided, without losing generality, since no variable can appear twice in the left side of a production. This solution could be applied also for term rewriting systems, for which the same problem holds (cf. [Rao84, Ken87, BvEGKPS87, PEM87, HKP88]). In [HKP88] a different solution is proposed: a set of *folding rules* is added to the productions representing the rules of a TRS. In the above situation a folding rule can be applied to D, causing the fusion of the two nodes labeled with *a*: afterwards a morphism from H'_C to the resulting graph can be found, allowing the application of production $\Delta'(C)$. An analogous solution is suggested in [Ken87], where the folding of the graph to be rewritten is forced by a more general definition of 'occurrence'.

4.2. Logic programs as context-free hypergraph grammars and *viceversa*

Let us analyze now the correspondence between logic programs (over the fixed signature (Σ, Π)) and a specific class of graph grammars, namely context-free hypergraph grammars over category Π-$\mathbf{DAG}_\Sigma$. More precisely, for a given logic program **P** and a fixed goal G we define a context free graph grammar GG(**P**, G) such that each derivation in GG(**P**, G) exactly corresponds to a refutation for G and viceversa. On the other hand, with each context free grammar GG we associate a logic program **P**(GG) (which is defined up to the ordering of the formulas in the body of the clauses) and a goal G such that, again, refutations for G in **P**(GG) and derivations of GG are in bijective correspondence.

To do this we first need to extend the correspondence between a resolution step and a direct derivation, stated in Theorem 4.1, to entire computations. The point is that, from the computational point of view, the relevant aspect of a refutation is that it returns a *computed answer substitution*, i.e. the substitution (defined on the variables of the initial goal) obtained as the composition of the substitutions computed at each resolution step. We can define a corresponding notion of *substitution associated with a graph grammar derivation*, showing first how a substitution can be associated with a single direct derivation step.

4.3 Definition *(the substitutions associated with direct derivations and with derivations)*

By Lemma 3.7 and Remark 3.8, in a direct derivation like the one depicted in Diagram 4.2, arrow $l: Q \to D$ is a bijection on nodes. This allows us to consider the function $\sigma \equiv r_V \circ l_V^{-1}: D_V \to D'_V$, and its associated substitution $\theta_\sigma: Var_D \to T_\Sigma(Var_{D'})$. We will call it *the substitution associated with the direct derivation* $D \overset{\Rightarrow}{_{\Delta(C),d}} D'$.

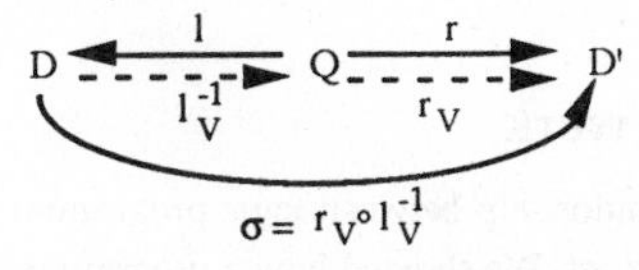

With this definition, the substitutions associated with the direct derivation steps of a derivation can be composed easily. If $\mathbf{Der} = (D_0 \Rightarrow D_1 \Rightarrow \ldots \Rightarrow D_n)$ is a derivation, and θ_{σ_i} is the substitution associated with the direct derivation $D_{i-1} \Rightarrow D_i$, then the substitution $\theta: Var_{D_0} \to T_\Sigma(Var_{D_n})$, defined as $\theta = \theta_{(\sigma_n \circ \ldots \circ \sigma_1)} = \theta_{\sigma_1} \bullet \ldots \bullet \theta_{\sigma_n}$ is called *the substitution associated with the derivation Der.* ♦

Let us now define the graph grammar associated with a logic program and a goal. In order to do that, by the definition in Section 1.1 we still have to choose a set of terminal colors. In the case of logic programming a successful derivation is clearly a refutation, i.e. a sequence of resolution steps ending with the empty goal. Therefore, the successful final states of the corresponding grammar should be all the possible representations of the empty goal as graph. It is quite obvious that these are all the enriched dags which contain no hyperedge labeled with a predicate symbol: thus we take as terminal colors the pair of sets $T = (\Sigma^T, N)$.

4.4 Definition *(the graph grammar associated with a logic program and a goal)*

Let P be a logic program (over (Σ, Π)), that is a set $\mathbf{P} = \{C_1, \ldots, C_k\}$ of definite clauses, and let G be a goal. Then the *graph grammar associated with* **P**, G is

$$GG(\mathbf{P}, G) = (\Delta(G), PROD(\mathbf{P}), C = (\Sigma^T, N \cup \Pi), T = (\Sigma^T, N))$$

where $\Delta(G)$ is the fully collapsed dag of G, PROD(**P**) is the set $\{\Delta(C_1), \ldots, \Delta(C_k)\}$ of productions representing the clauses of **P**, C is the pair of sets of colors, and T is the pair of sets of terminal colors. ♦

Clearly in the last definition all the involved graphs and morphisms are considered as elements of category Π-**DAG**$_\Sigma$. The next theorem states that there is a bijection between the refutations of a goal G in a program P and the derivations in the associated grammar.

4.5 Theorem *(the refutations for a goal are exactly the derivations in the associated grammar)*

Let P be a logic program, G a goal, and GG(**P**, G) the associated graph grammar. Then there exists a refutation of G in **P** producing the computed answer substitution θ, if and only if there exists a derivation in GG(**P**, G) with an associated substitution $\theta' \cong \theta$.

Proof Since by Theorem 4.1 there is a perfect correspondence between resolution steps and direct derivations, it is sufficient to notice that if there is a refutation for G via $C_1, \ldots, C_n$, then we can build a derivation for $\Delta(G)$ via $\Delta(C_1), \ldots, \Delta(C_n)$, and viceversa. Clearly, the associated substitutions are the same (up to variable renaming). ♦

In the opposite direction, let us show that from every context-free grammar GG over Π-**DAG**$_\Sigma$ a logic program P(GG) with a goal G can be extracted, in such a way that the bijective correspondence between refutations for G in P(GG) and derivations of GG still holds.

4.6 Theorem *(from context-free grammars to logic programs)*

Given a context-free grammar over Π-**DAG**$_\Sigma$, GG = (S, PROD, C, T), where PROD = $\{p_1, \ldots, p_n\}$ and C and T are as in Definition 4.4, let **P**(GG) be the program $\{C_1, \ldots, C_n\}$, where C_i is the clause associated with p_i (cf. Definition 3.6), and let G be the goal $Forms_S$. Then for each derivation of GG with associated substitution θ there is a refutation for G in **P**(GG) with computed answer substitution $\theta' \cong \theta|_X$, where X is the subset of variables of S appearing in $Forms_S$.

Proof The same soundness and completeness statements of Theorem 4.1 also hold if one starts with an arbitrary context-free production over Π-**DAG**$_\Sigma$ and considers its associated definite clause, as in Definition 3.6, provided that one takes into account that some of the variables of an enriched dag D can disappear when considering its associated goal $Forms_D$ (cf. the remark after Definition 3.6). Then the extension of this result to entire derivations is straightforward. ♦

The last result implies that a context-free grammar over Π-**DAG**$_\Sigma$ can be directly executed as a logic program, as far as one is interested just in the term substitutions associated with the derivations, and not in the language generated by the grammar.

5. Conclusions and future work

In this paper we analyzed the deep relationship between logic programming and the class of context-free graph grammars over a category of enriched dags. We showed how a program (equipped with a goal) can be translated into a grammar, and, *viceversa*, how to extract a logic program from a given context-free grammar. The main result states that our translation of definite clauses as graph productions is sound and complete, while previous proposals (of representing term rewriting systems as grammars) could not achieve completeness.

This result has been obtained by first studying an appropriate representation for terms, formulas, and clauses as graphs and graph productions. In particular, terms and formulas have been represented as enriched dags, while clauses have been represented as truly context-free productions (by first considering a canonical form for definite clauses, and then by faithfully representing such a canonical form as a production). In this way we could model in a correct way the unification mechanism, which is a fundamental technique peculiar to logic programming.

The results of this paper can be considered as a starting point for future research towards a deeper understanding of the relationship between logic programming and graph grammars. Some well known constructions on graph productions could be applied to the clauses of a logic program. As an example, the *parallelism theorem* (cf. [Eh83]) could suggest how to combine two clauses into a single, non definite clause (as formalized in [FLP84]). On the other hand the *concurrency theorem* could provide a formal ground for a theory of unfolding clauses (cf. [Le88]).

6. References

[AM89] A. Asperti, S. Martini, *Projections instead of variables, A category theoretic interpretation of logic programs*, Proc. 6th Int. Conf. on Logic Programming, MIT Press, 1989, pp. 337-352.

[BC87] M. Bauderon, B. Courcelle, *Graph Expressions and Graph Rewritings*, Mathematical System Theory **20**, 1987, pp. 83-127.

[BvEGKPS87] H.P. Barendregt, M.C.J.D. van Eekelen, J.R.W. Glauert, J.R. Kennaway, M.J. Plasmeijer, M.R. Sleep, *Term graph reduction*, in Proc. PARLE, LNCS 259, 1987, pp. 141-158.

[CER79] V. Claus, H. Ehrig, G. Rozenberg, (Eds.) *Proceedings of the 1st International Workshop on Graph-Grammars and Their Application to Computer Science and Biology*, LNCS 73, 1979.

[Cou88] B. Courcelle, *On using context-free graph grammars for analyzing recursive definitions*, in *Programming of Future Generation Computers II*, K. Fuchi, L. Kott (Eds.), Elsevier-North-Holland, 1988, pp. 83-122.

[Eh83] H. Ehrig, *Aspects of concurrency in graph grammars*, in [ENR83], pp. 58-81.

[Eh87] H. Ehrig, *Tutorial introduction to the algebraic approach of graph-grammars*, in [ENRR87] pp. 3-14.

[EHKP91] H. Ehrig, A. Habel, H.-J. Kreowski, F. Parisi-Presicce, *High-Level Replacement Systems*, in [EKR91].

[EKR91] H. Ehrig, H.-J. Kreowski, G. Rozenberg, (Eds.) *Proceedings of the 4th International Workshop on Graph-Grammars and Their Application to Computer Science*, LNCS, 1991, this volume.

[ENR83] H. Ehrig, M. Nagl, G. Rozenberg, (Eds.) *Proceedings of the 2nd International Workshop on Graph-Grammars and Their Application to Computer Science*, LNCS 153, 1983.

[ENRR87] H. Ehrig, M. Nagl, G. Rozenberg, A. Rosenfeld, (Eds.) *Proceedings of the 3rd International Workshop on Graph-Grammars and Their Application to Computer Science*, LNCS 291, 1987.

[EPS73] H. Ehrig, M. Pfender, H.J. Schneider, *Graph-grammars: an algebraic approach*, Proc, IEEE Conf. on Automata and Switching Theory, 1973, pp. 167-180.

[FLMP89] M. Falaschi, G. Levi, M. Martelli, C. Palamidessi, *Declarative Modeling of the Operational Behaviour of Logic Languages*, Theoretical Computer Science, **69**(3), 1989, pp. 289-318.

[FLP84] M. Falaschi, G. Levi, C. Palamidessi, *A Synchronization Logic: Axiomatics and Formal Semantics of Generalized Horn Clauses*, in Information and Control, **60**(1-3), Academic Press, 1984.

[Go89] J.A. Goguen, *What is Unification? A Categorical View of Substitution, Equation and Solution*, in M. Nivat and H. Aït-Kaci (Eds.), *Resolution of Equations in Algebraic Structures*, Academic Press, 1989.

[Ha89] A. Habel, *Hyperedge Replacement: Grammars and Languages*, Ph.D. Thesis, University of Bremen, 1989.

[HKP88] A. Habel, H-J. Kreowski, D. Plump, *Jungle evaluation*, in Proc. Fifth Workshop on Specification of Abstract Data Types, LNCS 332, 1988, pp. 92-112.

[Ken87] J.R. Kennaway, *On 'On Graph Rewritings'*, Theoretical Computer Science, **52**, 1987, pp. 37-58.

[Ken91] J.R. Kennaway, *Graph rewriting in some categories of partial morphisms*, in [EKR91].

[Le88] G. Levi, *Models, Unfolding Rules and Fixpoint Semantics*, in Proc. 5th Int. Conf. Symp. on Logic Programming, Seattle, MIT Press, pp. 1649-1665, 1988.

[Ll87] J.W. Lloyd, *Foundations of Logic Programming*, Springer Verlag, 1984, (Second Edition 1987).

[ML71] S. Mac Lane, *Categories for the Working Mathematician*, Springer Verlag, New York, 1971.

[PEM87] F. Parisi-Presicce, H. Ehrig and U. Montanari, *Graph Rewriting with Unification and Composition*, in [ENRR87], pp. 496-514.

[Rao84] J.C. Raoult, *On Graph Rewritings*, Theoretical Computer Science, **32**, 1984, pp. 1-24.

[RB85] D.E. Rydeheard, R.M. Burstall, *A Categorical Unification Algorithm*, Proc. of the Workshop on Category Theory and Computer Programming, LNCS 240, 1985.

GRAPHS AS RELATIONAL STRUCTURES :
An algebraic and logical approach

Bruno COURCELLE
Université Bordeaux-I
Laboratoire d'Informatique[(+)]
351, Cours de la Libération
33405 TALENCE - Cedex -
France

Abstract : Relational structures form a unique framework in which various types of graphs and hypergraphs can be formalized and studied. We define operations on structures that are compatible with monadic second-order logic, and that are powerful enough to represent context-free graph- and hypergraph-grammars of various types, namely, hyperedge replacement, C-edNCE, and separated handle replacement ones. Several results on monadic second-order properties of the generated sets are obtained in a uniform way.

Keywords : Context-free graph-grammar, C-edNCE graph-grammar, Graphs, Graph operation, Hypergraph, Hyperedge replacement, Monadic second-order logic, Monadic second-order definable graph transformation, Relational structure.

CONTENTS

INTRODUCTION

There are many notions of graphs. A graph can be oriented or not, colored (i.e., labeled) or not. with or without multiple edges. It may have distinguished vertices called sources (or terminals). Hypergraphs are not essentially different

---

Notes :(+) Laboratoire associé au CNRS. Email : courcell@geocub.greco-prog.fr.This work has been supported by the ESPRIT-Basic Research Action 3299 ("Computing by Graph Transformation").

from graphs, and graph grammars deal heavily with them. This variety of definitions is actually motivated by a variety of uses of the concept. Accordingly, many notions of graph grammars have been introduced. The field of graph grammars definitely needs some unification. This paper aims to contribute to it. Rather than choosing one type of graph, and declaring dogmatically that this type is the "right" one, in terms of which everything should be defined, we prefer to consider that all definitions are useful and deserve consideration, preferably in a unified framework. For this purpose, several concepts already exist that play the role of unifying tools. These tools are graph morphisms, graph operations and relational structures.

(1) Graph morphisms

A morphism $h : G \longrightarrow H$ is useful to specify precisely an occurrence of G in H, subject to substitution by another graph. This idea is the starting point of the formalization of graph rewriting rules in terms of push-out diagrams developed by Ehrig and others. It applies virtually to any kind of graphs, provided the notion of a morphism is well-chosen.

(2) Graph operations

By a graph operation, we mean a mapping that constructs a graph from k given graphs $G_1,...,G_k$, for example by gluing them together in some fixed way. (Series-composition and parallel-composition are typical examples of graph operations). These graph operations are necessarily deterministic : adding a new edge "anywhere" in a graph is not a graph operation. They allow to build graphs and to describe them by "graph expressions" of various types.

Let us assume that we fix a class of graphs and a set F of graph operations. Every system of recursive set equations written with F and set union has a least solution for inclusion. Context-free graph grammars of several types (hyperedge-replacement, C-edNCE) can be expressed as systems of equations over appropriate sets of graph operations, and many of their properties can be described at the abstract level of systems of equations.

(3) Relational structures

All the existing notions of graphs we know can be formalized in the uniform framework of relational structures. Logical formulas, to be evaluated in the relational structure $|G|$ describing a graph G, define properties of this graph. Hierarchies of graph properties follow from hierarchies of logical languages (first-order, monadic second-order, second-order, etc...), previously introduced in the context of complexity theory.

We shall be essentially interested by monadic-second order logic, for which general decidability results exist (see below). There are two variants of this language, that we denote here by MS_2 (when quantifications can be done on sets of vertices and sets of edges) and by MS_1 (when quantifications are restricted to sets of vertices).

These two variants correspond actually to two distinct representations of graphs by relational structures. In the first case, edges belong to the domain (that contains also vertices). In the second, edges do not belong to the domain, and are treated as given relations on the domain of vertices. This shows that the single notion of a relational structure can handle the two cases.

In the present paper we shall focus our attention on relational structures and graph operations, and on the links between them. The rewriting of structures, that is, the link between (1) and (3) above has been considered by Ehrig et al. [13] . We aim to present in a unified way the following two known results :

Result 1 [7] : - One can decide whether all graphs of a set L generated by a hyperedge replacement grammar satisfy a given MS_2-formula.

Result 2 [5,12] : - One can decide whether all graphs of a set L generated by a C-edNCE (or a B-NLC) graph grammar satisfy a given MS_1-formula.

These two results can be established essentially in the same way: Appropriate graph operations are defined that make possible to represent these graph grammars as systems of equations. A certain compatibility property between formulas (of either MS_2 or MS_1 according to the case) and the graph operations is established, and yields the results .(We say that a graph property is compatible with a graph operation if its truth value for a graph composed from other graphs by means of this operation, depends only, via a Boolean function, of the truth values of finitely many auxiliary graph properties for the composing graphs; this notion extends to logical formulas in an obvious way since they define graph properties.) We generalize these two results in the following way.

We let $\mathfrak{T}$ be the class of all finite relational structures over some set of relation symbols. We define on $\mathfrak{T}$ several operations. We let S//S' be a certain gluing of two structures S and S'. We also consider operations of the form S'=f(S), where S' is defined in terms of S as follows : its domain is a subset of that of S ; its relations are defined by quantifier-free first-order formulas in terms of those of S.

Monadic second-order logic is compatible with these operations.(See Lemmas (3.2) and (3.3) below. This yields the two above results as special cases. It is not yet clear whether the sets of graphs defined by systems of equations over these operations are generable by known grammars or not, or whether new grammars should be designed to generate them.

1 - RELATIONAL STRUCTURES

(1.1) Definitions - We let $\mathcal{C}$ be a countable set of symbols called constants. We let $\mathcal{R}$ be a set of relation symbols. Each symbol r in $\mathcal{R}$ has an arity $\rho(r)$ in $\mathbb{N}_+$, and $\mathcal{R}$ has countably many symbols of each arity.

Let R and C be finite subsets of $\mathcal{R}$ and $\mathcal{C}$. An (R,C)-structure is an object of the form $S=<\mathbf{D}_S, (r_S)_{r\in R}, (c_S)_{c\in C}>$ where $\mathbf{D}_S$ is a set, called the domain of S, each c_S belongs to $\mathbf{D}_S$, and each r_S is a subset of $\mathbf{D}_S^{\rho(r)}$, i.e., is a $\rho(r)$-ary relation on $\mathbf{D}_S$. One may have $c_S = c'_S$ with $c \neq c'$. We denote by $\mathcal{S}(R,C)$ the set of (R,C)-structures having a finite domain. By a structure we shall mean in this paper an element of $\mathcal{S}(R,C)$ for some R and C.

(1.2) Example - Sourced hypergraphs.

Let A be a set of ranked edge labels, let $\mathbf{R}_A := \{\mathbf{edg}_a \mid a \in A\}$, let $\mathbf{C}_n := \{\mathbf{s}_1,\dots,\mathbf{s}_n\}$.We let m+1 be the rank of $\mathbf{edg}_a$ where a is of rank m. A hyperedge with label a of rank m will have a sequence of m vertices, and will be said to be of type m. (We shall say an edge rather than a hyperedge for simplicity.) With a hypergraph $G=<\mathbf{V}_G, \mathbf{E}_G, \mathbf{lab}_G, \mathbf{vert}_G, \mathbf{src}_G>$ as defined in [1,3, 6], we associate the structure $|G|$ in $\mathcal{S}(\mathbf{R}_A,\mathbf{C}_n)$ such that $\mathbf{D}_{|G|} := \mathbf{V}_G \cup \mathbf{E}_G$ (we assume that $\mathbf{V}_G$ is the set of vertices, that $\mathbf{E}_G$ is the set of edges, and that $\mathbf{V}_G \cap \mathbf{E}_G = \emptyset$),

$$\mathbf{edg}_{aG}(x,y_1,\dots,y_n) :\Leftrightarrow \mathbf{true} \quad \text{iff } x \in \mathbf{E}_G,\ y_1,\dots,y_m \in \mathbf{V}_G,\ \mathbf{lab}_G(x) = a \text{ and } \mathbf{vert}_G(x) = (y_1,\dots,y_m)$$

$$\mathbf{s}_{iG} := \mathbf{src}_G(i) \text{ for each } i=1,\dots,n.$$

Such a hypergraph is called an n-hypergraph over A. The set of n-hypergraphs over A is denoted by $\mathbf{HG}(A)_n$. It is clear that two structures $|G|$ and $|G'|$ are isomorphic iff G and G' are isomorphic. Note that some structures may not be associated with any hypergraph. This is the case if, for example, some constant $\mathbf{s}_i$ denotes an edge.

Let us say that a hypergraph is simple if it has no edge of type 0 and no pair of edges with the same sequence of vertices and the same label. Let $\|G\|$ be the structure $<\mathbf{V}_G, (\mathbf{edg}'_{aG})_{a \in A}, (\mathbf{s}_{iG})_{i \in [n]}>$ associated with G as above with :

$$\mathbf{edg}'_{aG}(x_1,\ldots,x_n) : \Leftrightarrow \mathbf{true} \text{ iff there is an edge y having the label a , and the sequence of vertices } (x_1,\ldots,x_n)$$

For any two simple hypergraphs G and G', the structures $\|G\|$ and $\|G'\|$ are isomorphic iff G and G' are isomorphic.

2 - OPERATIONS ON RELATIONAL STRUCTURES

(2.1) Definition - The gluing operation.

Let S ∈ $\mathcal{S}$(R,C) and S' ∈ $\mathcal{S}$(R',C') be disjoint structures, i.e., be structures such that $\mathbf{D}_S \cap \mathbf{D}_{S'} = \emptyset$. We let S//S' be the structure T defined as follows. We let D := $\mathbf{D}_S \cup \mathbf{D}_{S'}$, we let ~ be the least equivalence relation on D such that $c_S \sim c_{S'}$ for every c ∈ C ∩ C'. We let $\mathbf{D}_T$:= D/~ and we denote by [d] the equivalence class of an element d in D. We complete the definition of T as follows :

$c_T := [c_S]$ if $c \in C$

$c_T := [c_{S'}]$ if $c \in C'$ (we have $[c_S] = [c_{S'}]$ if $c \in C \cap C'$)

$r_T([d_1],\ldots,[d_n])$ holds if $r_S(d'_1,\ldots,d'_n)$ holds or if

$r_{S'}(d'_1,\ldots,d'_n)$ holds for some $d'_1 \in [d_1],\ldots,d'_n \in [d_n]$.

If S and S' are not disjoint, we replace S' be an isomorphic copy, disjoint from S. Note that S//S' ∈ $\mathcal{S}$(R∪R', C ∪ C'). If C∩C'=∅, then $\mathbf{D}_{S//S'}=\mathbf{D}_S \cup \mathbf{D}_{S'}$ and S//S' is the disjoint sum of S and S'.

(2.2) Definition - Quantifier-free definable operations.

We denote by QF(R,C,X) the set of quantifier-free formulas written with R, C, and the variables of X. Our purpose is to specify by quantifier-free formulas total mappings : $\mathcal{S}(R,C) \longrightarrow \mathcal{S}(R',C')$. We let Δ be a tuple of formulas of the form $<\delta,(\theta_r)_{r\in R'},(\tau_c)_{c\in C'}>$ such that :

- $\delta \in QF(R,C,\{x_1\})$ and is of the form $\delta' \vee W\{x_1=\tau_c \mid c \in C'\}$, for some formula δ',(we indicate by W the disjunction of a set of formulas.)

- $\theta_r \in QF(R,C,\{x_1,\ldots,x_n\})$ where $n= \rho(r)$,

- $\tau_c \in C$ for each $c \in C'$.

The set of such tuples will be denoted by $\mathcal{D}((R,C),(R',C'))$. With every Δ in $\mathcal{D}((R,C),(R',C'))$, we associate the total mapping $\mathbf{def}_\Delta : \mathcal{S}(R,C) \longrightarrow \mathcal{S}(R',C')$ such that, for every S in $\mathcal{S}(R,C)$, we have $S' = \mathbf{def}_\Delta(S)$ iff S' is the structure in $\mathcal{S}(R',C')$ such that :

$\mathbf{D}_{S'} := \{x \in \mathbf{D}_S \mid S \models \delta(x)\}$,
$c_{S'} := (\tau_c)_S$ for each $c \in C'$,
$r_{S'}(x_1,\ldots,x_n)$ holds iff $x_1,\ldots,x_n \in \mathbf{D}_S$
and $S \models \delta(x_1)\wedge\ldots\wedge\delta(x_n)\wedge\theta_{r'}(x_1,\ldots,x_n)$.
(i.e., $x_1,\ldots,x_n \in \mathbf{D}_{S'}$ and $S \models \theta_{r'}(x_1,\ldots,x_n)$).

A mapping : $\mathcal{S}(R,C) \longrightarrow \mathcal{S}(R',C')$ is quantifier-free definable (qfd) iff it is of the form $\mathbf{def}_\Delta$ for some Δ in $\mathcal{D}((R,C),(R',C'))$.

If two formulas φ and φ' can be transformed into each other by the laws of propositional calculus ,like De Morgan's law, then they are equivalent in every structure.(this will be written $\varphi \sim \varphi'$). Since we have no function symbols, there are only finitely many ~-equivalence classes of quantifier-free formulas. We shall assume that each formula is everywhere replaced by some canonical representative of its class, chosen in a way we need not specify. It follows from this convention that each set QF(R,C,X) is finite if X is finite, and that $\mathcal{D}((R,C),(R',C'))$ is finite for each pair ((R,C),(R',C')).

(2.3) Definition - A signature of operations on structures.

Any pair (R,C) consisting of finite subsets R and C of $\mathcal{R}$ and $\mathcal{C}$ will be called a sort. We let $\mathcal{O}$ be the (countable) set of all sorts.

For every s and s' in $\mathcal{O}$, we let $//_{s,s'}$ be an operation symbol of profile : $s\times s' \longrightarrow s''$ where $s''=(R \cup R', C \cup C')$ if $s=(R,C)$, $s'=(R',C')$. The associated operation will be the restriction of // to $\mathcal{S}(R,C)\times \mathcal{S}(R',C')$. For every $s=(R,C)$ and $s'=(R',C')$ in $\mathcal{O}$, for every Δ in $\mathcal{D}(s,s')$, we introduce an operation symbol $\mathbf{def}_{\Delta,s,s'}$ of profile: $s \longrightarrow s'$. The associated mapping is $\mathbf{def}_\Delta : \mathcal{S}(R,C) \longrightarrow \mathcal{S}(R',C')$.

We let $F= \{//_{s,s'}, \mathbf{def}_{\Delta,s,s'} \mid s,s' \in \mathcal{O}, \Delta \in \mathcal{D}(s,s')\}$. We obtain in this way a many sorted F-algebra

$$\mathcal{S} =<(\mathcal{S}(R,C))_{(R,C)\in\mathcal{O}}, (//_{s,s'})_{s,s' \in \mathcal{O}}, (\mathbf{def}_{\Delta,s,s'})_{s,s' \in \mathcal{O}, \Delta\in\mathcal{D}(s,s')}>$$

In most concrete cases, we shall consider restrictions of $\mathcal{S}$ to finitely many sorts (whence to finitely many operations, since the sets $\mathcal{D}(s,s')$ are finite).We shall also use constants denoting fixed finite structures. Every such structure can be taken as the value of a constant. We also put in F all these constants, extending in this way the above definition.

Any variable-free term t written with the symbols of F , and that is well-formed w.r.t. sorts and profiles denotes a structure $\mathbf{val}(t)$ in $\mathcal{S}(R,C)$, where (R,C) is the sort of t.

We shall also use <u>derived operations</u>, that is, operations : $\mathcal{S}(R_1,C_1) \times \ldots \times \mathcal{S}(R_n,C_n) \longrightarrow \mathcal{S}(R',C')$ defined by finite terms written with the operations and constants of F and with variables $x_1,\ldots,x_n$ of respective sorts $(R_1,C_1),\ldots,(R_n,C_n)$. We now give a few useful examples.

(2.4) Examples : Operations on sourced hypergraphs (Bauderon, Courcelle [1, 3, 6]).

Let $s_{n,p} : \mathcal{S}(\mathbf{R}_A,\mathbf{C}_p) \longrightarrow \mathcal{S}(\mathbf{R}_A,\mathbf{C}_{n+p} - \mathbf{C}_n)$ be the qfd operation that transforms S into S' by making the value of $\mathbf{s}_i$ in S be the value of $\mathbf{s}_{i+n}$ in S' for all i=1,...,p. The disjoint union of sourced hypergraphs $\oplus_{n,p}$ can be expressed in terms of structures by $|G\oplus_{n,p}G'|=|G|//s_{n,p}|G'|$, for every n-hypergraph G and every p-hypergraph G', both over A.

Let $i,j \in [n]$. The operation $\theta_{i,j}$ that fuses the ith and the jth source of a hypergraph can be expressed by : $|\theta_{i,j}(G)| := |G| // \mathbf{B}_{i,j}$,where $\mathbf{B}_{i,j}$ is the structure in $\mathcal{S}(\emptyset, \{\mathbf{s}_i, \mathbf{s}_j\})$ consisting of a single element that is the value of both $\mathbf{s}_i$ and $\mathbf{s}_j$.

The operation σ_α , that renames sources is clearly a qfd operation.

(2.5) Example : Graphs with ports (Courcelle, Engelfriet, Rozenberg [12]).

Let A be a finite set of edge labels. We let **GR**(A) be the set of finite simple oriented graphs, the edges of which are labeled in A. These graphs can be defined as structures of the form $G=<\mathbf{V}_G, (\mathbf{edg'}_{aG})_{a\in A}>$, the domain of which is the set of vertices $\mathbf{V}_G$ of the graph G. The binary relations $\mathbf{edg'}_{aG}$ express the existence of a-labelled edges linking two vertices. A graph with ports is a graph as above, equipped with sets of distinguished vertices. Formally, a graph with ports of type n is a structure of the form :

$$G = <\mathbf{V}_G, (\mathbf{edg}_{aG})_{a\in A}, (\mathbf{pt}_i)_{i\in[n]}>,$$

where each $\mathbf{pt}_i$ is a unary relation on $\mathbf{V}_G$. We say that x is an i-port iff $\mathbf{pt}_i$ (x) holds.

Ports are useful in the following operation $\eta_{a,\,i,j}$ that augments a graph G by adding to it all possible edges with label a, directed from an i-port to a j-port. Hence, $\eta_{a,\,i,j}$ is the qfd operation defined by the tuple $\Delta=<\delta,(\theta_{\mathbf{edg}_a})_{a\in A}, (\theta_{\mathbf{pt}_i})_{i\in[n]}>$ where

δ is **true**,

$\theta_{\mathbf{edg}_a}(x,y)$ is $\mathbf{edg}_a(x,y) \vee [\mathbf{pt}_i(x) \wedge \mathbf{pt}_j(y)]$,

$\theta_{\mathbf{edg}_b}(x,y)$ is $\mathbf{edg}_b(x,y)$ (where $b \in A$, $b \neq a$),),

$\theta_{\mathbf{pt}_i}(x)$ is $\mathbf{pt}_i(x)$.

The disjoint union of graphs is also used in [12] and is nothing but the operation //. (Since there are no constants, S//S' is here the disjoint union of the structures S and S').

A third operation is used in [12], the renaming of ports : if z is a finite subset of $\mathbb{N}_+\times \mathbb{N}_+$, then $\pi_z(G)$ is the graph $<G,(\mathbf{edg}_{aG})_{a\in A},(\mathbf{pt}_i)_{i\in[n']}>$, where $\mathbf{pt}_i$ (x) : $\Leftrightarrow$ $\mathbf{pt}_j$ (x) for some $(i,j) \in z$ and $n'=\mathbf{Max}\{i/(i,j)\in z$ for some $j\}$. It is easy to construct Δ such that $\pi_z=\mathbf{def}_\Delta$. Hence the operation π_z is qfd. □

(2.6) Definition : Graph-grammars.

It is shown in Courcelle [5] that with every context-free graph grammar Γ, one can associate a system $\mathbf{S}_\Gamma$ of equations in sets of graphs, the least solution of which is the tuple of sets of graphs generated by the nonterminals of the grammar Γ.The system $\mathbf{S}_\Gamma$ is built with set union and graph operations corresponding to the right-hand sides of the production rules of the grammar. (Some of these operations are actually denotations for fixed graphs corresponding to the terminal productions of the grammar).

In the present paper, we define a grammar as a system of equations, that does not necessarily correspond to any concrete rewriting mechanism on structures.A system of equations is a tuple of the form $S=\langle u_1=p_1,\ldots,u_n=p_n\rangle$ where :

• u_i is an unknown, i.e., a symbol having a sort (R_i, C_i) ; this unknown is intended to denote a subset of $\mathcal{S}(R_i,C_i)$,

• p_i is a polynomial of sort (R_i,C_i), i.e., a sum of the form $t_1 \cup t_2 \cup \ldots \cup t_m$ where each t_j is a finite term of sort (R_i,C_i) written with the unknowns, the operations and constants of F. ($\mathbf{B}_{i,j}$ is an example of a constant; it is used in Example (2.4)). Each term t_i is called a monomial.

Such systems are investigated in general by Courcelle [2]. A grammar is a pair $\Gamma=(S,u_i)$ consisting of a system S and an unknown u_i of S playing the role of the initial nonterminal in standard grammars. Every system S as above has a least solution in $\mathcal{P}(\mathcal{S}(R_1,C_1))\times\ldots\times\mathcal{P}(\mathcal{S}(R_n,C_n))$ (see [2]), denoted by $(L(S,u_1),\ldots,L(S,u_n))$. The set defined by $\Gamma=(S,u_i)$ is $L(\Gamma):=L(S,u_i)$. Sets of this form are called equational sets of structures.

It is explained in [2,5] that derivation trees can be defined, that they represent the way objects are generated by polynomial systems. This applies in particular to the present case.

(2.7) Example : Context-free HR sets of graphs.

It is known from Bauderon and Courcelle [1,3,6] that a set of k-hypergraphs, $L \subseteq \mathbf{HG}(A)_k$ (see Example (1.2)) can be defined by a hyperedge-replacement graph grammar (also Habel and Kreowski [16], or Engelfriet and Rozenberg [15]) iff it is defined by a system of equations built with the operations $\oplus_{n,p},\theta_{i,j},\sigma_\alpha$ reviewed in Example (2.4). Since these operations are derived operations of F, defined by terms where each variable ocurs at most once, it follows that the sets of structures representing HR sets of hypergraphs, i.e., sets defined by hyperedge replacement grammars, are equational.

(2.8) Example : C-edNCE sets of graphs.

It is proved in Courcelle et al. [12] that every C-edNCE set of graphs is equational with respect to the algebraic structure associated with the operations defined in Example (2.5). Hence, every such set is equational in the sense of Definition (2.6). □

3 - MONADIC SECOND-ORDER LOGIC

(3.1) Definitions.

Let $(R,C) \in \mathcal{O}$. Let W be a set of variables. This set consists of object (lowercase) and set (uppercase) variables. We denote by W_0 the set of object variables. We denote by $\mathcal{L}(R,C,W)$ the set of monadic second-order logical formulas written with R, C, and that have their free variables in W. We omit the mention of W when it is empty. We denote by $\mathcal{L}^h(R,C)$ the set of formulas in $\mathcal{L}(R,C)$ having at most h levels of nested quantifications. We refer the reader to Courcelle [5,7] for formal definitions in specific cases concerning graphs and hypergraphs.

(3.2) Lemma : Let $s=(R,C)$ and $s'=(R',C')$ belong to $\mathcal{O}$. Let f be a qfd operation : $s \longrightarrow s'$. Then for every formula φ in $\mathcal{L}^h(R',C')$, one can construct a formula ψ in $\mathcal{L}^h(R,C)$, such that, for every S in $\mathcal{S}(R,C)$:

$$f(S) \models \varphi \text{ iff } S \models \psi.$$

Proof : Let $\Delta=\langle\delta,(\theta_r)_{r\in R'},(\tau_c)_{c\in C'}\rangle$. For every $t \in W_0 \cup C'$, we let $\overline{t}$ denote t if $t \in W_0$ and τ_c if $c \in C'$. We let $\overline{\varphi}$ be defined as follows by induction on the structure of φ, where $\varphi\in \mathcal{L}(R',C',W)$:

If φ is $\exists X.\varphi_1$, then $\overline{\varphi}$ is $\exists X.\ \overline{\varphi_1}$.
If φ is $\exists x.\varphi_1$, then $\overline{\varphi}$ is $\exists x.[\delta(x) \wedge \overline{\varphi_1}]$.
If φ is $\neg\varphi_1$, then $\overline{\varphi}$ is $\neg\ \overline{\varphi_1}$.
If φ is $\varphi_1\vee \varphi_2$, then $\overline{\varphi}$ is $\overline{\varphi_1}\vee \overline{\varphi_2}$.
If φ is $t=t'$, or $t \in X$, or $r(t_1,...,t_n)$, then $\overline{\varphi}$ is $\overline{t}=\overline{t}'$, or $\overline{t}\in X$ or $\theta_r[\overline{t_1}/x_1,..., \overline{t_n}/x_n]$ respectively..(We denote in this way the result of the substitution of ti , for each i = 1,...,n.)

We let then ψ be the formula $\overline{\varphi}$ □

In the following lemma, we denote by θ_S the truth value of a closed formula θ, in a structure S of the appropriate type.

(3.3) Lemma : Let (R,C) and (R',C') ∈ $\mathcal{O}$. For every formula φ in $\mathcal{L}^h(R \cup R', C \cup C')$, one can construct two sequences of formulas, $\psi_1,\ldots,\psi_n$ in $\mathcal{L}^h(R,C)$, and $\psi'_1,\ldots,\psi'_n$ in $\mathcal{L}^h(R',C')$, of the same length, such that, for every S in $\mathcal{S}(R,C)$ and S' in $\mathcal{S}(R',C')$:

$$\varphi_{S//S'} = W \{\psi_{i_S} \wedge \psi'_{i_{S'}} \mid 1 \leq i \leq n\}.$$

Proof : In the special case where $C \cap C' = \emptyset$, the structure S//S' is the disjoint union of S and S'. The result follows then as an easy adaptation of the proof of Lemma 4.5 of Courcelle [7]. In the general case, one can express S//S' by means of $\oplus$ (disjoint union as above), renamings of sources, and fusions of sources. The general result follows then from those of Courcelle [7]. □

As in Courcelle [7], one obtains the following results, where $\mathbf{M}(F')_s$ denotes the set of terms of sort s, constructed with symbols from F'.

(3.4) Theorem : Let F' be a finite subset of F . For every sort s=(R,C) in $\mathcal{O}$, for every formula φ in $\mathcal{L}(R,C)$, one can construct a linear time algorithm, that decides, for every term t in $\mathbf{M}(F')_s$ whether $\mathbf{val}(t) \models \varphi$.

(3.5) Theorem : For every grammar Γ defining a subset of $\mathcal{S}(R,C)$, for every formula φ in $\mathcal{L}(R,C)$ one can construct a grammar Γ' that defines $\{G \in L(\Gamma) \mid G \models \varphi\}$. One can decide whether $G \models \varphi$ for some (or for each) graph G in $L(\Gamma)$.

The key idea for proving these results is the following. Let h be the quantification level of φ. Let us say that two structures are equivalent if they satisfy the same formulas of level at most h. Since there are finitely many formulas of each level, this equivalence has finitely many classes. Lemmas (3.2) and (3.3) show that this equivalence is a congruence w.r.t. the operations of F. It follows that a tree automaton recognizing the set of terms which define structures satisfying φ can be constructed. This automaton yields the algorithm of Theorem (3.4). From a grammar Γ, one can construct a grammar Γ' , the derivation trees are those of Γ that are recognized by the tree-automaton.This gives a proof of Theorem (3.5).

These theorems have the following special cases.

Case 1 : (R,C) = ($\mathbf{R_A}$,$\mathbf{C_n}$) introduced in Example (1.2), and the operations are those used in Definition (2.3). Then the corresponding grammars generate all sets of structures of the form {|G| / G ∈ L} where L is a HR set of hypergraphs. Theorems (3.4) and (3.5) entail then Proposition (4.14) and Corollary (4.8) of [7] respectively (because the construction of Theorem (3.5) does not modify the derived operations used in the right handsides of productions, i.e., that if Γ is a HR grammar, then so is Γ'). Let us precise that monadic second-order formulas use quantifications on sets of edges, by contrast with the case of the following example.

Case 2 : (R,C) is as in Example (2.5) and the operations on structures are those defined also in (2.5). One obtains by grammars all sets of the form {‖G‖ / G ∈ L}, where L is C-edNCE. Note that, in terms of graphs, the monadic-second order formulas considered here use quantifications on sets of vertices, and <u>not</u> on sets of edges. The sets of hypergraphs with ports defined by separated handle replacement grammars (Courcelle et al. [12]) can also be handled with essentially the same operations, (and a few others making possible to create hyperedges), and the same restriction of quantifications to sets of vertices. Thus, Theorem (3.4) yields Theorem (6.7) of [12].

4 - FURTHER RESULTS

Lacking of space, we only indicate a few additional results and the main research directions. Formal definitions can be found in Courcelle [8,9]. Full proofs will be given in Courcelle [10]. Let us only indicate that a monadic second-order definable graph transformation (a MSDGT for short), is a mapping from graphs to graphs defined as follows. Given G, one first construct G' by augmenting G with a fixed number of disjoint copies of it. Then, one defines H inside G' , more or less like we did in Definition (2.2), except that here , we can use MS formulas as opposed to quantifier-free ones.

(4.1) Proposition : <u>For every finite subset</u> F' <u>of</u> F, <u>for every sort</u> s=(R,C) <u>in</u> $\mathcal{O}$, <u>the transformation</u> **val** : $\mathbf{M}(F)_s \longrightarrow \mathcal{S}(R,C)$ <u>is monadic second-order definable</u>.

This proposition is a straightforwoard adaptation of Lemma (4.3) of [8]. The following corollary is immediate.

(4.2) Corollary : <u>If</u> L ⊆ $\mathcal{S}$(R,C) <u>is an equational set of structures, then</u> L = **f**(K) <u>for some recognizable set</u> K <u>of finite trees and some monadic second-order definable graph transformation</u>.

We now consider the converse.

(4.3) Theorem : If L ⊆ $\mathcal{S}$(R,C) is a set of structures of the form **f**(K) for some recognizable set of finite trees K, and some MSDGT **f**, then it is equational.

A related result by Engelfriet [14] shows that, if L is a set of finite, oriented, edge labelled simple loop-free graphs, then it is generated by a C-edNCE grammar iff it is monadic-second order definable in a recognizable set of finite trees K (i.e., iff the set of structures representing these graphs, the domain of which consists only of vertices , is of the form **f**(K) for some MSDGT **f**). It gives a grammar-independent characterization of C-edNCE, proving that this class is in some sense maximal. We obtain it as a Corollary of our proof of Theorem (4.3), because we know by the results of Courcelle, Engelfriet and Rozenberg [12] that the subsignature of F introduced in Example (2.8) corresponds to C-edNCE sets of graphs.

Research directions

It remains to interpret the operations used in the proof of Theorem (4.3) in terms of concrete hypergraph rewritings. This would yield a class of context-free graph grammars generating the class of sets of structures of the form **f**(K) for some recognizable set of trees K, and some MSDGT **f**.

(4.4) Open problem : Find a set F' of "natural" and "simple" qfd operations such that, every qfd operation is a derived operation over F'.

(4.5) Theorem [11] : If L is a set of sourced hypergraphs and {|G|/G∈L} is equational, then L is HR (i.e., can be generated by a hyperedge-replacement grammar.)

From this result, one obtains, by using the main theorem of Courcelle [9] the following result :

(4.6) Theorem : If L is a C-edNCE set of graphs of bounded tree-width, then it is HR.

Acknowledgement : I thank J.Engelfriet and the referees for many helpful comments.

REFERENCES

[1] BAUDERON M., COURCELLE B., Graph expressions and graph rewritings, Mathematical System Theory 20 (1987) 83-127.

[2] COURCELLE B. , Equivalences and transformations of regular systems. Applications to recursive program schemes and grammars, Theor. Comp. Sci. 42 (1986), 1-122.

[3] COURCELLE B., A representation of graphs by algebraic expressions and its use for graph rewriting systems, Proceedings of the 3rd International Workshop on Graph Grammars, L.N.C.S. 291, Springer, 1987, pp. 112-132.

[4] COURCELLE B., On context-free sets of graphs and their monadic second-order theory, same volume as [3], pp. 133-146.

[5] COURCELLE B., An axiomatic definition of context-free rewriting and its application to NLC graph grammars, Theoretical Computer Science 55 (1987) 141-181.

[6] COURCELLE B., Graph rewriting : An algebraic and logic approach, in "Handbook of Theoretical Computer Science,Volume B", J. Van Leeuwen ed., Elsevier,1990, pp.193-242

[7] COURCELLE B., The monadic second-order logic of graphs I, recognizable sets of finite graphs. Information and Computation 85 (1990) 12-75.

[8] COURCELLE B., The monadic second-order logic of graphs V : On closing the gap between definability and recognizability, Research Report 89-91, Bordeaux I University, to appear in Theor. Comput. Sci.

[9] COURCELLE B., The monadic second order logic of graphs VI : On several representations of graphs by relational structures, Report 89-99, (see also Logic in Computer Science 1990, Philadelphia).

[10] COURCELLE B., The monadic second-order logic of graphs VII: Graphs as relational structures, in preparation.

[11] COURCELLE B., ENGELFRIET J., A logical characterization of hypergraph languages generated by hyperedge replacement grammars, in preparation.

[12] COURCELLE B., ENGELFRIET J., ROZENBERG G., Handle-rewriting hypergraph grammars, this volume.(Long version as research report 90-84, Bordeaux-I University, or reports 90-08 and 90-09 of the University of Leiden).

[13] EHRIG H. et al., Transformations of structures, an algebraic approach, Math. Systems Theory 14 (1981) 305-334.

[14] ENGELFRIET J., A characterization of context-free NCE graph languages by monadic-second order logic on trees, preprint, 1990.

[15] ENGELFRIET J., ROZENBERG G., A comparison of boundary graph grammars and context-free hypergraph grammars,Information and Computation 84 (1990) 163-206.

[16] HABEL A., KREOWSKI H.J., May we introduce to you : Hyperedge replacement, same volume as [3], pp. 15-26.

Context-free Handle-rewriting Hypergraph Grammars

Bruno Courcelle
LaBRI, Université de Bordeaux 1
351, Cours de la Libération, 33405 Talence, France

Joost Engelfriet
Grzegorz Rozenberg
Department of Computer Science, Leiden University
P.O.Box 9512, 2300 RA Leiden, The Netherlands

ABSTRACT. Separated handle-rewriting hypergraph grammars (S-HH grammars) are introduced, where separated means that the nonterminal handles are disjoint. S-HH grammars have the same graph generating power as the vertex rewriting context-free NCE graph grammars, and as recursive systems of equations with four types of simple operations on graphs.

Keywords: graph grammar, hypergraph, recursive system of equations.

CONTENTS

1. Introduction

Roughly speaking one may distinguish three types of context-free graph grammars, according to which unit of a given graph is rewritten into another graph: a vertex, an edge, or a handle (i.e., an edge together with its two incident vertices, cf. [EhrRoz, MaiRoz87/90]). In general one can say that handle-rewriting grammars are more powerful than vertex-rewriting grammars, which on their turn have more power than edge-rewriting grammars.

To increase the power of edge-rewriting grammars, hyperedge-rewriting grammars have been introduced in [HabKre87a/87b, BauCou] (and studied in, e.g., [Hab, Cou88b/90, Lau, MonRos, EngRoz, EngHey89/91a]). Such grammars generate (directed) hypergraphs, of which the hyperedges are rewritten into hypergraphs. Still, these grammars are not as powerful as vertex-rewriting grammars, with the set of all complete graphs as a counter-example.

In this paper we introduce the so-called separated handle-rewriting hypergraph grammar (S-HH grammar). In such a grammar, a handle, i.e., a hyperedge together with all its incident vertices, is rewritten into a hypergraph. The adjective "separated" signifies that the nonterminal handles (i.e., the handles that are rewritten) do not overlap, which means that they have no common incident vertices. This ensures that the rewriting is context-free.

Our first main result is that S-HH grammars have the same (graph generating) power as the largest known class of context-free graph grammars: the vertex-rewriting confluent NCE grammars (or C-edNCE grammars) studied in, e.g., [Kau, Bra, Sch, Eng89]. C-edNCE grammars are a special case of the grammars of [Nag] and a generalization of the NLC grammars of [JanRoz80/83]; subclasses of C-edNCE were considered in, e.g., [EngLeiWel, EngLei, EngLeiRoz87/88, Cou87].

A completely different type of graph grammar is the (algebraic) term grammar ([MezWri, Cou86, BauCou]). In general, a term grammar defines a set of objects to be the (first component of the) least fixed point of a system of equations, where the equations consist of terms (or expressions). This assumes that the set of all objects is made into an algebra by specifying certain operations on the objects, to be used in the equations. Our second main result is that, just as C-edNCE grammars, term grammars have the same power as S-HH grammars, for some appropriate choice of operations on graphs.

The full version of this paper is presented in [CouEngRoz].

2. Hypergraphs and graphs

We will use $\mathbb{N}$ to denote $\{0,1,2,\ldots\}$ and $\mathbb{N}_+$ to denote $\{1,2,\ldots\}$. For $m,n \in \mathbb{N}$, $[m,n] = \{i \in \mathbb{N} \mid m \leq i \leq n\}$. For a set A, $\mathcal{P}(A)$ denotes its powerset. A (positively) ranked alphabet is a finite set A of symbols together with a mapping rank: $A \rightarrow \mathbb{N}_+$; it is binary if $\text{rank}(a) \in \{1,2\}$ for every $a \in A$.

Let A be a ranked alphabet (of edge labels). A (directed, edge labeled) hypergraph over A is a tuple $H = (V,E)$ where V is the finite set of vertices and E is the finite set of hyperedges (or edges). Each hyperedge is a tuple $(a,v_1,\ldots,v_k)$ with $a \in A$, $k = rank(a)$, and $v_i \in V$ for $i \in [1,k]$. Note that H may have multiple edges, but not with the same label. For an edge $e = (a,v_1,\ldots,v_k)$ in E, we write $lab(e) = a$, $vert(e) = (v_1,\ldots,v_k)$, $vert(e,i) = v_i$, and $vset(e) = \{v_1,\ldots,v_k\}$. Each v_i is said to be a vertex of e, or a vertex incident with e; the number i is said to be a tentacle of e. The integer k is called the rank of e, denoted rank(e) (i.e., rank(e) = rank(lab(e))). As usual, we will add a subscript H to indicate that we deal with the hypergraph H; thus, V_H stands for V, E_H for E, and, e.g., $vert_H$ for vert.

HG(A) denotes the set of all hypergraphs over A. We do not distinguish between isomorphic hypergraphs. But, as usual, when defining properties of, or operations on, hypergraphs we use concrete representatives of the involved isomorphism classes. In particular, when defining binary operations we assume the two hypergraphs to be disjoint (i.e., to have disjoint vertex sets).

Let A be a binary ranked alphabet (of edge and vertex labels). A (directed, edge labeled, vertex labeled) _graph_ over A is a hypergraph H = (V,E) over A such that for every $v \in V$ there is exactly one hyperedge $e \in E$ such that $e = (a,v)$ for some $a \in A$. The hyperedges of rank 2 are the edges of the graph, whereas the hyperedges of rank 1 are the vertex labels. Note that loops, i.e., edges (a,u,u), are allowed.

GR(A) denotes the set of all graphs over A. Thus, $GR(A) \subseteq HG(A)$.

3. Separated handle-rewriting hypergraph grammars

Before we explain the working of separated handle-rewriting hypergraph grammars (S-HH grammars), we introduce some more terminology. In the sequel N and A are (disjoint) ranked alphabets of nonterminals and terminals, respectively. The hypergraphs in $HG(N \cup A)$ will be the sentential forms of the S-HH grammar. As we are interested in the graph generating power of S-HH grammars, we always assume that A is binary. Also, to simplify the presentation, we assume (without always mentioning it) that every hypergraph $H \in HG(N \cup A)$ is in a normal form that consists of two properties: (1) H is _parasite-free_, i.e., no nonterminal edge $e \in E_H$ has a parasite: a terminal edge $e' \in E_H$ such that $vset(e') \subseteq vset(e)$, and (2) H is _loop-free_, i.e., no nonterminal edge $e \in E_H$ has two distinct tentacles i and j such that $vert_H(e,i) = vert_H(e,j)$.

A _handle_ is an edge together with its incident vertices. To obtain a context-free way of rewriting handles, we will demand that they are disjoint:

a hypergraph $H \in HG(N \cup A)$ is separated if $vset(e) \cap vset(e') = \emptyset$, for every two distinct nonterminal edges $e, e' \in E_H$.

Since we want S-HH grammars to generate graphs, we will restrict attention to "graph-generating" hypergraphs: $H \in HG(N \cup A)$ is graph-generating if, for every $v \in V_H$ that is not incident with a nonterminal edge, there is exactly one edge of rank 1 incident with v. Thus every vertex that is not part of a nonterminal handle has a (unique) vertex label (and note that, since H is parasite-free, vertices of a nonterminal handle are not incident with edges of rank 1).

As an example, consider the hypergraph given in Fig.1 (with $N = \{X,Y\}$, $A = \{a,b,c,d,q\}$, $rank(X) = 3$, $rank(Y) = rank(q) = 1$, and $rank(x) = 2$ for $x \in \{a,b,c,d\}$). Fat dots represent vertices. A nonterminal edge e is represented as a lab(e)-labeled box, with lines (representing tentacles) connecting it to its incident vertices. The integers at these lines numbered from 1 to rank(e) indicate the order of vert(e). A terminal edge e of rank 2 is represented as a lab(e)-labeled arrow from vert(e,1) to vert(e,2), and a terminal edge e of rank 1 is represented by labeling vert(e,1) with lab(e). Note that this hypergraph is separated and graph-generating (and also parasite-free and loop-free).

A hypergraph $H \in HG(N \cup A)$ together with a partial function port: $V_H \longrightarrow \mathbb{N}_+$ is called a hypergraph with ports. It is denoted (H,port). The function 'port' indicates which vertices of H serve as gluing points in a rewriting step of an S-HH grammar (cf. the "external vertices" or "sources" in hyperedge-rewriting grammars). If $port(v) = i$, then v is also called an i-port of H. In fact, in [CouEngRoz] 'port' is taken to be a finite subset of $\mathbb{N}_+ \times V_H$. However, in [EngHey91b] it is shown that this relation may be assumed to be "non-overlapping", i.e., every $v \in V_H$ occurs at most once as second component of a tuple of 'port'.

$HG^P(N \cup A)$ denotes the set of all hypergraphs with ports, over N and A. Thus, identifying $(H,\emptyset)$ with H, $HG(N \cup A) \subseteq HG^P(N \cup A)$. $GR^P(A)$ denotes the set of all graphs with ports, over A.

A production of an S-HH grammar is of the form $X \longrightarrow (H,port)$, where $X \in N$ and $(H,port) \in HG^P(N \cup A)$; H is separated and graph-generating (and also parasite-free and loop-free), and port: $V_H \longrightarrow [1,rank(X)]$. Applying this production to an X-labeled edge e of a hypergraph K consists of: (1) removing e and its incident vertices (i.e., the handle) from K, together with all (terminal!) edges incident with those vertices, (2) adding H to the remainder, and (3) embedding H in this remainder as follows. If the source (target) of a terminal edge was the i-th vertex of e in K, and for vertex $v \in V_H$, $port(v) = i$, then a terminal edge (with the same label) is established with source (target) v and the same target (source) in K as the original edge. Note that this embedding involves both duplication (in case there are several i-ports in H) and deletion (in case there are none) of terminal edges, as in NLC and NCE grammars. Restricting the (partial) function port to be injective and surjective (i.e., there is exactly one vertex v such that port(v) = i for each $i \in [1,\text{rank}(X)]$), one is back in the case of hyperedge-rewriting grammars. Thus, the embedding process of S-HH grammars is a natural generalization of the one used in hyperedge-rewriting grammars, using a feature of the embedding process of NLC and NCE grammars.

Consider the production $X \longrightarrow (H,\text{port})$ pictured in Fig.2, where rank(X) = 3 and a vertex v is labeled i if port(v) = i. Thus, there are three 1-ports, no 2-ports, and two 3-ports. (To avoid confusion with vertex labels from A we take $A \cap \mathbb{N}_+ = \emptyset$ in examples.) If we apply this production to e, the left-most nonterminal edge labeled X, of the hypergraph K of Fig.1, we obtain the hypergraph given in Fig.3. Note that the a-labeled edge with target vert(e,1) is tripled, the edges incident with vert(e,2) are deleted, and the c-labeled edge with source vert(e,3) is doubled.

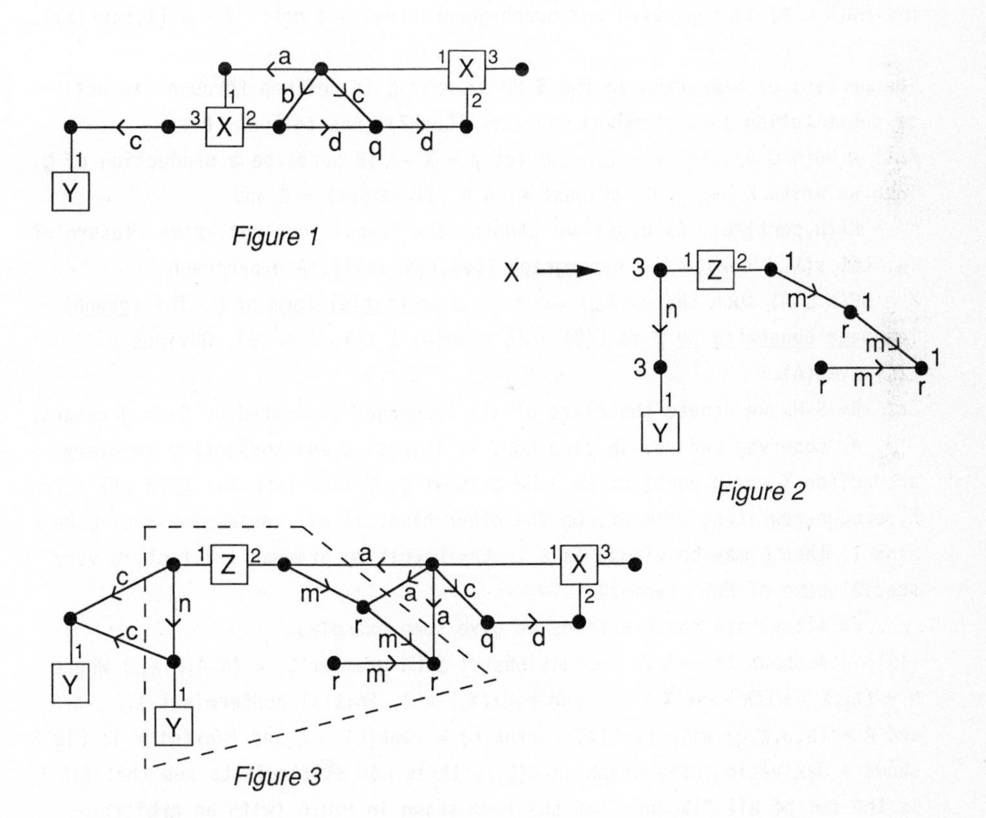

Figure 1

Figure 2

Figure 3

Formally, the <u>substitution</u> of a hypergraph with ports $(H,port) \in HG^p(N \cup A)$ for an edge e of a hypergraph $K \in HG(N \cup A)$ (disjoint with H), denoted $K[(H,port)/e]$, is the hypergraph $(V,E) \in HG(N \cup A)$, where $V = (V_K - vset(e)) \cup V_H$, and

$$E = (E_K - \{(a,u,v) \in E_K \mid u \in vset(e) \text{ or } v \in vset(e)\}) \cup E_H$$
$$\cup \{(a,v,w) \mid \exists u \in V_K\colon u = vert(e,port(v)) \text{ and } (a,u,w) \in E_K\}$$
$$\cup \{(a,w,v) \mid \exists u \in V_K\colon u = vert(e,port(v)) \text{ and } (a,w,u) \in E_K\}.$$

It is easy to see that if H and K are separated and graph-generating (and parasite-free and loop-free), then so is $K[(H,port)/e]$.

We now define our main notion.

<u>Definition</u>. A <u>separated</u> <u>handle-rewriting</u> <u>hypergraph</u> <u>grammar</u> (S-HH grammar) is a tuple $\mathcal{G} = (N,A,P,X_{in})$, where N is the ranked nonterminal alphabet, A is the binary ranked terminal alphabet (disjoint with N), P is the finite set of productions, and $X_{in} \in N$ is the initial nonterminal, with $rank(X_{in}) = 1$. A production in P is of the form $X \longrightarrow (H,port)$, where $X \in N$, $H \in HG(N \cup A)$ is separated and graph-generating, and port: $V_H \longrightarrow [1,rank(X)]$.

The process of rewriting in the S-HH grammar $\mathcal{G}$ is defined through the notion of substitution in a standard way (cf. [Cou87]), as follows. Let $K,K' \in HG(N \cup A)$, let $e \in E_K$, and let $p = X \longrightarrow (H,port)$ be a production of $\mathcal{G}$. Then we write $K \Rightarrow_{(e,p)} K'$ or just $K \Rightarrow K'$ if $lab_K(e) = X$ and $K' = K[(H,port)/e]$. As usual $\Rightarrow^*$ denotes the transitive, reflexive closure of $\Rightarrow$. Let $s(X_{in})$ denote the hypergraph $(\{v\},\{(X_{in},v)\})$. A hypergraph $K \in HG(N \cup A)$ such that $s(X_{in}) \Rightarrow^* K$ is a <u>sentential</u> <u>form</u> of $\mathcal{G}$. The (<u>graph</u>) <u>language</u> <u>generated</u> <u>by</u> $\mathcal{G}$ is $L(\mathcal{G}) = \{G \in HG(A) \mid s(X_{in}) \Rightarrow^* G\}$. Obviously, $L(\mathcal{G}) \subseteq GR(A)$.

By S-HH we denote the class of all languages generated by S-HH grammars.

As observed before, in case port is injective and surjective in every production $X \longrightarrow (H,port)$ of an S-HH grammar $\mathcal{G}$, $\mathcal{G}$ is a (special type of) hyperedge-rewriting grammar. On the other hand, if all nonterminals of $\mathcal{G}$ have rank 1, then $\mathcal{G}$ may be viewed as a vertex-rewriting grammar (in fact, a very special case of NCE grammar).

To illustrate these notions we give some examples.

(1) Fig.4 shows the three productions of S-HH grammar $\mathcal{G}_1 = (N,A,P,X_{in})$ where $N = \{X_1,X_2\}$ with $rank(X_1) = 1$ and $rank(X_2) = 2$, initial nonterminal $X_{in} = X_1$, and $A = \{a,b,c,q\}$ with $rank(a) = rank(b) = rank(c) = 2$ and $rank(q) = 1$. Fig.5 shows a derivation of a graph in $L(\mathcal{G}_1)$. It is not difficult to see that $L(\mathcal{G}_1)$ is the set of all "ladders" of the form shown in Fig.6 (with an arbitrary number of squares rather than four). Note that there is one 1-port and one 2-port in each of the productions for X_2. Thus, there is no duplication or deletion of vertices (and their incident edges). In fact, $\mathcal{G}_1$ is also a

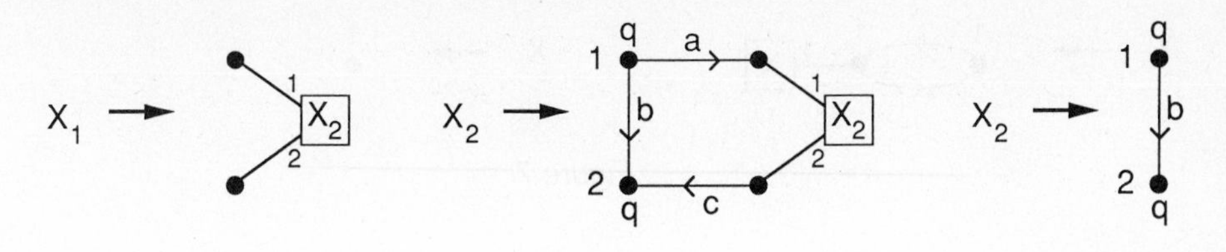

Figure 4

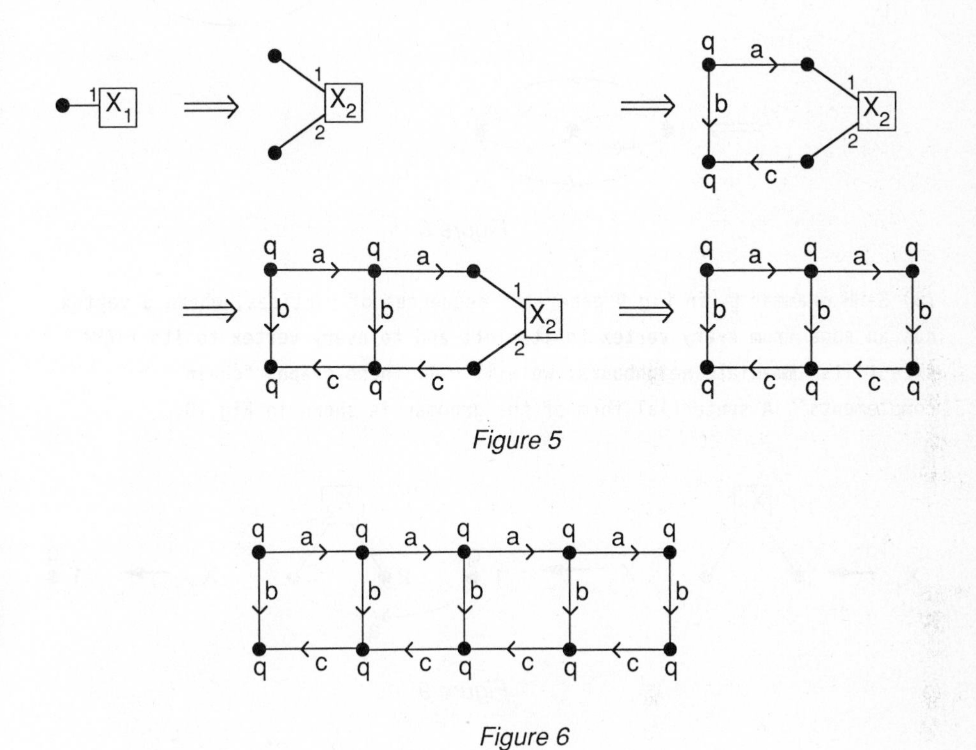

Figure 5

Figure 6

hyperedge-rewriting grammar (and even, since $\text{rank}(X_2) = 2$, an edge-rewriting grammar).

(2) The productions of grammar $\mathcal{G}_2$ generating all complete directed graphs (without loops) are shown in Fig.7. We have dropped the unique terminal labels. Application of the first production to a nonterminal hyperedge (X,v) results in the duplication of vertex v and all its incident edges. Fig.8 shows a derivation of the complete graph with three vertices. Note that $\mathcal{G}_2$ is a vertex-rewriting grammar. The language $L(\mathcal{G}_2)$ cannot be generated by a hyperedge-rewriting grammar (see Propositions 3.17 and 4.17 of [BauCou]; see also [HabKre87a, Hab]).

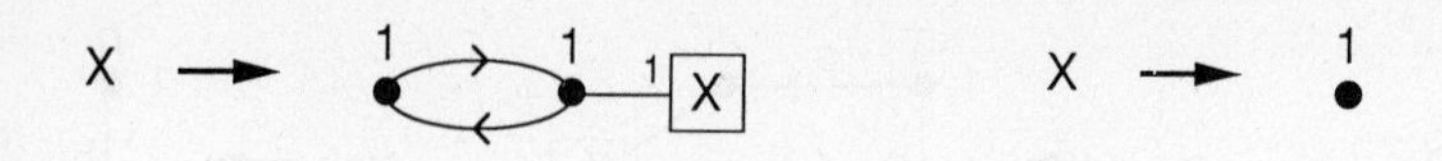

Figure 7

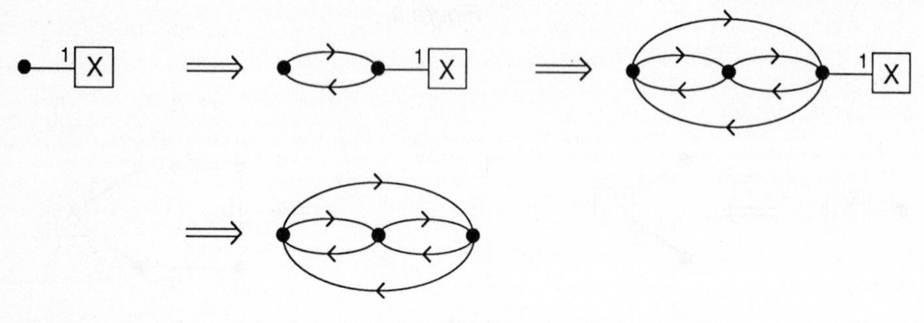

Figure 8

(3) S-HH grammar $\mathcal{G}_3$ in Fig.9 generates sequences of vertices, where a vertex has an edge from every vertex to its left and to every vertex to its right, except its immediate neighbours; we will call these graphs "chain complements". A sentential form of the grammar is shown in Fig.10.

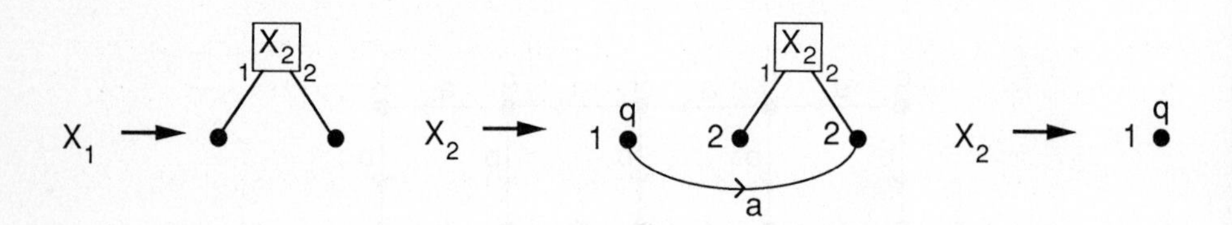

Figure 9

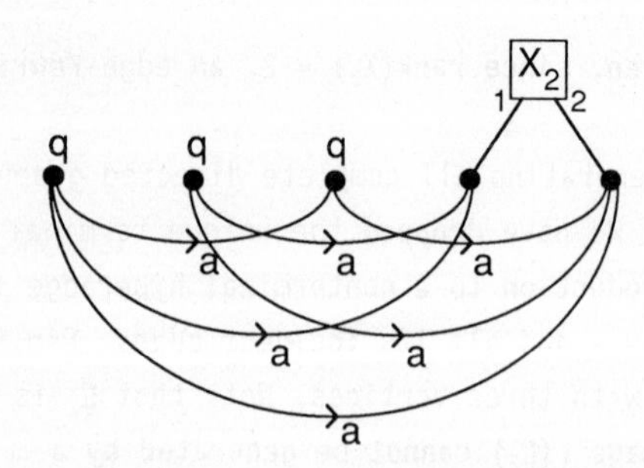

Figure 10

4. Confluent NCE grammars

The separation property guarantees that every S-HH grammar is confluent, i.e., that the order of applying productions in a derivation does not influence the derived graph. Confluence is a property that graph grammars should have to deserve the adjective context-free (see [Cou87]). The vertex-rewriting NCE grammars are not confluent in general, which is the reason to restrict attention to confluent NCE grammars: C-edNCE grammars (cf. [Eng89]). In "edNCE" the e stands for "edge- (and vertex-) labeled", d for "directed", and NCE for "neighbourhood controlled embedding".

By C-edNCE we denote the class of all languages generated by C-edNCE grammars. This is the largest known class of context-free graph languages. It contains, e.g., the classes of all (graph) languages generated by confluent NLC grammars [Cou87], boundary NLC and NCE grammars [RozWel, EngLeiWel], and hyperedge-rewriting grammars [BauCou, HabKre87b].

Our first main result is the following.

Theorem 1. S-HH = C-edNCE.

Thus, S-HH grammars can be used instead of C-edNCE grammars. One advantage is that they are defined by easy structural properties, rather than by the operational property of confluence in the case of C-edNCE grammars. Another is that they have a simpler embedding mechanism (almost as simple as that of hyperedge-rewriting grammars).

Several important subclasses of C-edNCE have their counterpart as natural subclasses of S-HH grammars. For instance, linear edNCE grammars (LIN-edNCE [EngLei]) correspond to linear S-HH grammars (where linear means that there is at most one nonterminal occurrence in every right-hand side of a production). As another example, boundary edNCE grammars (B-edNCE [EngLeiWel]) correspond to S-HH grammars such that every terminal edge has at least one vertex that is not incident with any nonterminal edge.

5. The algebra of graphs with ports

In this section, the second main result of the paper is stated: the equivalence of S-HH grammars and systems of equations with four types of operations on graphs with ports, viz., disjoint union, edge creation, port relabeling, and port deletion. These operations make it possible to build "large" graphs from "smaller" ones, and to denote graphs by algebraic expressions formed with the operations, together with constants denoting "elementary" graphs. Let A be a binary ranked alphabet. Recall that $GR^p(A)$ denotes the set of all graphs with ports over A.

Let G and G′ be two (disjoint) graphs with ports. The <u>disjoint</u> <u>union</u> of G and G′ is $G \oplus G' = (V_G \cup V_{G'}, E_G \cup E_{G'}, port)$ with $port(v) = port_G(v)$ for all $v \in V_G$, and $port(v) = port_{G'}(v)$ for all $v \in V_{G'}$.

The second operation, <u>edge</u> <u>creation</u>, creates new edges between ports. For every $a \in A$ with $rank(a) = 2$, $i,j \in \mathbb{N}_+$, and for every graph $G \in GR^p(A)$, we define $\eta_{a,i,j}(G) = (V_G, E, port_G)$ with $E = E_G \cup \{(a,v,w) \mid port(v) = i, port(w) = j\}$.

The operation <u>port</u> <u>relabeling</u> is needed to be able to redefine the port numbers of vertices. Let $i,j \in \mathbb{N}_+$, and let G be a graph with ports. Then $\rho_{i,j}(G) = (V_G, E_G, port)$ with $port(v) = [\text{if } port_G(v) = i \text{ then } j \text{ else } port_G(v)]$.

The last operation is <u>port</u> <u>deletion</u>, needed to remove port numbers. For every $i \in \mathbb{N}_+$, and for every graph with ports G, we define $\delta_i(G) = (V_G, E_G, port)$ with $port(v) = [\text{if } port_G(v) = i \text{ then undefined else } port_G(v)]$.

Finally we introduce the elementary graphs; they will be used as basic objects in expressing graphs by algebraic expressions. We denote by ε the empty graph (i.e., $V_\varepsilon = \emptyset$). And, for every $q \in A$ with $rank(q) = 1$, we denote by σ_q the graph with ports consisting of a single 1-port labeled q, i.e., $(\{v\},\{(q,v)\},port)$ with $port(v) = 1$.

For example, if G is the graph σ_q, then $G \oplus G$, $\eta_{a,1,1}(G \oplus G)$, and $\delta_1(\eta_{a,1,1}(G \oplus G))$ are given in Fig.11 a, b, and c, respectively.

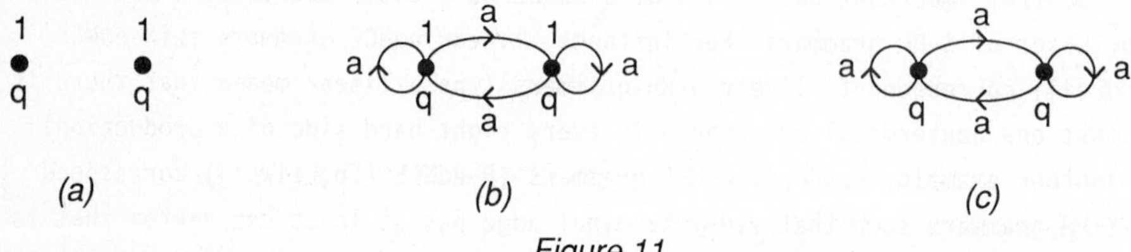

Figure 11

It is easy to see that every graph with ports G = (V,E,port) can be obtained in at least one way from the elementary graphs by the above operations. In fact, let $V = \{v_1,\ldots,v_n\}$ and let q_i be the label of v_i, i.e., $(q_i,v_i) \in E$. Let G_1 be the graph $\sigma_{q_1} \oplus \rho_{1,2}(\sigma_{q_2}) \oplus \cdots \oplus \rho_{1,n}(\sigma_{q_n})$; it is the discrete graph with vertices $v_1,\ldots,v_n$, where v_i has label q_i and port number i. Let G_2 be the graph obtained from G_1 by applying all $\eta_{a,i,j}$ with $(a,v_i,v_j) \in E$, in some order. Then $G = \phi_1(\cdots\phi_n(G_2)\cdots)$, where $f_i = \delta_i$ if $port(v_i)$ is undefined and $f_i = \rho_{i,j}$ if $port(v_i) = j$.

The reader will have noticed that we have not defined four operations, but infinitely many. This makes $GR^p(A)$ into an F_A-algebra, where F_A is the (one-sort) infinite signature consisting of

- the binary symbol $\oplus$,
- the unary symbols $\eta_{a,i,j}$, $\rho_{i,j}$, and δ_i (for all $a \in A$ with rank(a) = 2, and $i,j \in \mathbb{N}_+$), and
- the nullary symbols (i.e., constants) ε and σ_q (for all $q \in A$ with rank(q) = 1).

We denote by $T(F_A)$ the set of well-formed terms over F_A. Each of these terms, say t, is called a graph expression, and denotes a graph val(t) in $GR^p(A)$: the value of t, in the usual way. For example, $val(\eta_{a,1,2}(\sigma_q \oplus \sigma_q \oplus \rho_{1,2}(\sigma_q \oplus \sigma_q)))$ is the graph with ports given in Fig.12. As discussed above, every graph with ports is the value of at least one graph expression.

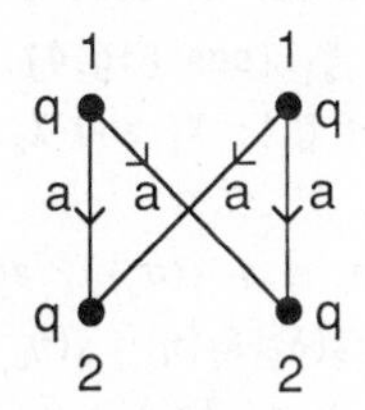

Figure 12

Since we now have expressions to denote graphs, we can use systems of equations (or term grammars) to define sets of graphs (cf. [MezWri]). This closely corresponds to the intuitive way in which recursive definitions of graph properties are usually given, in contrast to the ordinary graph grammars that have a more operational flavour. See [Cou86] for a thorough study of recursive systems of equations and their least solutions in arbitrary algebras.

Let $\mathcal{X}$ be the set of variables $\{X_1,\dots,X_n\}$. $T(F_A,\mathcal{X})$ denotes the set of all well-formed terms over $F_A \cup \mathcal{X}$ (where every variable is a nullary symbol). For every $t \in T(F_A,\mathcal{X})$, the derived operation $\|t\|$ of t in the subset algebra $\mathcal{P}(GR^p(A))$ is defined in the usual way, as follows. For $L_1, \dots, L_n \subseteq GR^p(A)$: $\|X_i\|(L_1,\dots,L_n) = L_i$, $\|t' \oplus t''\|(L_1,\dots,L_n) = \{G' \oplus G'' \mid G' \in \|t'\|(L_1,\dots,L_n), G'' \in \|t''\|(L_1,\dots,L_n)\}$, if c is a constant then $\|c\|(L_1,\dots,L_n) = \{c\}$, and if f is unary then $\|f(t')\|(L_1,\dots,L_n) = \{f(G) \mid G \in \|t'\|(L_1,\dots,L_n)\}$.

Definition. By a term grammar S over F_A we will mean a polynomial system of equations over the signature F_A, in the sense of [Cou86]. That is: $S = \langle X_1 = p_1, \dots, X_n = p_n \rangle$ with $n \geq 1$, where each p_i is a polynomial of the form $p_i = m_{i,1} \cup \dots \cup m_{i,r(i)}$, and each $m_{i,j}$ is a monomial, i.e., an expression in $T(F_A,\mathcal{X})$, where $\mathcal{X} = \{X_1,\dots,X_n\}$.

A <u>solution</u> of S in $\mathcal{P}(GR^p(A))$ is an n-tuple $(L_1,\ldots,L_n)$ of subsets of $GR^p(A)$ such that $L_i = \|p_i\|(L_1,\ldots,L_n)$ for every $i \in [1,n]$, where $\|p_i\|(L_1,\ldots,L_n) = \cup\{\|m_{i,j}\|(L_1,\ldots,L_n) \mid j \in [1,r(i)]\}$. Every term grammar has a least solution in $\mathcal{P}(GR^p(A))$, with respect to componentwise inclusion.

The <u>language</u> <u>defined</u> <u>by</u> S, denoted L(S), is the first component of the least solution of S in $\mathcal{P}(GR^p(A))$. □

Note that $L(S) \subseteq GR^p(A)$; we will only be interested in the case that S defines a graph language, i.e., $L(S) \subseteq GR(A)$.

We explain the above definition with some examples of term grammars that define graph languages. To understand the explanation, note that an alternative definition of "solution", producing the same least solution, is to require that $\|p_i\|(L_1,\ldots,L_n) \subseteq L_i$ for every $i \in [1,n]$.

(1) We first consider a term grammar S_1 that defines the set of "ladders", generated by S-HH grammar $\mathcal{G}_1$ (see Fig.4). S_1 has two unknowns (corresponding to the nonterminals of $\mathcal{G}_1$): X_1 and X_2. It has two equations:

$$X_1 = \delta_1(\delta_2(X_2)), \text{ and}$$

$$X_2 = m_1 \cup m_2, \text{ where } m_2 = \eta_{b,1,2}(\sigma_q \oplus \rho_{1,2}(\sigma_q)), \text{ and}$$

$$m_1 = \rho_{3,1}(\rho_{4,2}(\delta_1(\delta_2(\eta_{c,2,4}(\eta_{b,3,4}(\eta_{a,3,1}(t)))))))$$

$$\text{with } t = \rho_{1,3}(\sigma_q) \oplus \rho_{1,4}(\sigma_q) \oplus X_2.$$

Note that the three monomials of S_1 closely correspond to the three productions of $\mathcal{G}_1$. Namely, m_2 expresses that the graph with two vertices (both ports) and one b-labeled edge is a "ladder with ports". m_1 expresses that if G is a ladder with ports then so is the one that is obtained from G by adding one square to the left (and moving the port numbers appropriately). And the monomial of the first equation expresses that if G is a ladder with ports, then one obtains a ladder by dropping the port numbers from G. Thus, if (L_1,L_2) is the least solution of S_1, then L_1 is the set of all ladders, as in Fig.6, whereas L_2 is the set of all ladders with ports, i.e., ladders of which the two "left-most" vertices are ports: the "upper" one is a 1-port and the "lower" one a 2-port. Hence, $L(S_1) = L_1 = L(\mathcal{G}_1)$.

(2) The set of all complete graphs (without loops) is defined by the term grammar $S_2 = \langle X_1 = \delta_1(X_2),\ X_2 = \sigma_q \cup \rho_{2,1}(\eta_{a,2,1}(\eta_{a,1,2}(\rho_{1,2}(\sigma_q) \oplus X_2)))\rangle$ (cf. S-HH grammar $\mathcal{G}_2$ in Fig.7). Intuitively, X_2 defines all complete graphs of which all vertices are 1-ports. The second monomial of the second equation expresses that if G is a complete graph, then another complete graph can be obtained from G by adding one vertex v and adding all edges from the vertices of G to v, and vice versa.

(3) The set of all chain complements is defined by the following term grammar S_3 (cf. S-HH grammar $\mathcal{G}_3$ in Fig.9):

$$X_1 = \delta_1(\delta_2(X_2)), \text{ and}$$

$$X_2 = \sigma_q \cup \rho_{3,2}(\eta_{a,1,2}(\sigma_q \oplus \rho_{1,3}(X_2))).$$

Intuitively, X_2 defines all chain complements of which the "head" vertex is a 1-port and all other vertices are 2-ports. The last monomial expresses that if G is a chain complement with ports, then another one can be obtained by turning the "head" vertex of G into a 3-port, adding one vertex v (the "new head") and adding all edges from v to the 2-ports of G, followed by turning the "old head" (i.e., the 3-port) into a 2-port.

It turns out that a set of graphs is defined by a term grammar if and only if it is generated by an S-HH grammar. This can be called an algebraic fixed point characterization of S-HH (because it is based on systems of equations in the algebra $GR^p(A)$, and because a solution of such a system S is also said to be a fixed point of S).

<u>Theorem 2</u>. Let $L \subseteq GR(A)$. $L \in$ S-HH iff L is defined by a term grammar over F_A.

<u>Proof</u> (sketch). A term grammar $S = \langle X_1 = p_1,\ldots,X_n = p_n\rangle$ over F_A and an S-HH grammar $\mathcal{G} = (N,A,P,X_{in})$ correspond if $N = \{X_1,\ldots,X_n\}$, $X_{in} = X_1$, and $P = \{X_i \to val(m) \mid m \text{ is a monomial of } p_i\}$, where val is extended to a mapping from $T(F_A,N)$ to $HG^p(N \cup A)$ by defining, for $X \in N$, $val(X) = ([1,k], \{(X,1,\ldots,k)\}, port)$ with $k = rank(X)$ and $port(i) = i$ for all $i \in [1,k]$. In the direction from S to $\mathcal{G}$ the problem is how to find the ranks of the X_i. In the direction from $\mathcal{G}$ to S one should, if necessary, take $X_{in} = X_0$ and add an additional equation $X_0 = \delta_1(\ldots\delta_k(X_1)\ldots)$ where $k = rank(X_1)$, cf. example (2) above. □

Combined with our first result, this theorem provides an algebraic fixed point characterization of C-edNCE languages. Such a characterization did not yet exist for vertex-rewriting graph grammars (for hyperedge-rewriting grammars, see [BauCou]).

Theorem 2 allows us to adapt to S-HH languages the results of [Cou87/88a/89/90] on graph properties expressible in monadic second-order logic. We assume the reader to be familiar with [Cou87/90].

A graph with ports G = (V,E,port) can be completely described by the relational structure $|G| = (V,(edg_a)_{a\in A},(pt_i)_{i\in\mathbb{N}_+})$ where (for rank(a) = 1) $edg_a(v)$ is true iff $(a,v) \in E$, (for rank(a) = 2) $edg_a(v,w)$ is true iff $(a,v,w) \in E$, and $pt_i(v)$ is true iff $port(v) = i$. Monadic second-order formulas can thus be written, with quantifications over vertices and sets of vertices, and properties of graphs can be expressed. Typical examples of such monadic second-order properties are vertex k-colourability, or properties concerning paths such as connectivity. We refer the reader to [Cou88a/89/90] for more details. The adaptation to the present situation is straightforward. Let us observe that we do not allow quantification over edges and sets of edges. It follows that the existence of a Hamiltonian path is not expressible in the language we consider here, whereas it is in those that are considered in [Cou88a/89/90]; see Section 3 of [Cou88b].

Theorem 3. Let $\mathcal{G}$ be an S-HH (C-edNCE, term) grammar, and let φ be a closed monadic second-order formula.

(1) One can construct from $\mathcal{G}$ and φ an S-HH (C-edNCE, term) grammar generating $\{G \in L(\mathcal{G}) \mid G \text{ satisfies } \varphi\}$.

(2) One can decide for $\mathcal{G}$ and φ whether or not all $G \in L(\mathcal{G})$ satisfy φ.

6. Conclusion

In [Eng91] the class of C-edNCE languages is completely characterized in terms of monadic second-order logic on trees. In [Eng89] another characterization of C-edNCE is given in terms of regular tree languages and regular string languages (generalizing the one of B-edNCE in [EngLeiWel]). Thus, altogether there are five different descriptions of this class of context-free graph languages: through handle-rewriting S-HH grammars, vertex-rewriting C-edNCE grammars, algebraic term grammars (or systems of equations), monadic second order logic on trees, and regular string/tree languages. This demonstrates that it is a natural class of graph languages.

In [CouEngRoz], S-HH grammars actually generate hypergraph languages (we have considered here a restriction of them). It is shown in [CouEngRoz] that the class of S-HH hypergraph languages is incomparable with the class of hypergraph languages generated by hyperedge-rewriting grammars. It would be nice to have a natural class of context-free hypergraph grammars that could generate both kinds of hypergraph languages.

Acknowledgment. The authors are grateful to Linda Heyker for her contribution to the writing (and drawing) of this paper.

References

[BauCou] M.Bauderon, B.Courcelle; Graph expressions and graph rewritings, Math. Systems Theory 20 (1987), 83-127

[Bra] F.J.Brandenburg; On polynomial time graph grammars, Proc. STACS 88, LNCS 294, 227-236

[Cou86] B.Courcelle; Equivalences and transformations of regular systems - applications to recursive program schemes and grammars, Theor. Comput. Sci. 42 (1986), 1-122
[Cou87] B.Courcelle; An axiomatic definition of context-free rewriting and its application to NLC graph grammars, Theor. Comput. Sci. 55 (1987), 141-181
[Cou88a] B.Courcelle; The monadic second-order logic of graphs. III: Tree-width, forbidden minors and complexity issues, Report I-8852, Bordeaux
[Cou88b] B.Courcelle; Some applications of logic, of universal algebra and of category theory to the theory of graph transformations, Bulletin of the EATCS 36 (1988), 161-218
[Cou89] B.Courcelle; The monadic second-order logic of graphs: Definable sets of finite graphs, LNCS 344 (1989), 30-53
[Cou90] B.Courcelle; The monadic second-order logic of graphs. I: Recognizable sets of finite graphs, Inform. Comput. 85 (1990), 12-75. See also [EhrNagRozRos], 112-133
[CouEngRoz] B.Courcelle, J.Engelfriet, G.Rozenberg; Handle-rewriting hypergraph grammars, parts I and II, Reports 90-08 and 90-09, Leiden, or Report 90-84, Bordeaux
[EhrNagRozRos] H.Ehrig, M.Nagl, G.Rozenberg, A.Rosenfeld (eds.); Graph-Grammars and their Application to Computer Science, LNCS 291, 1987
[EhrRoz] H.Ehrig, G.Rozenberg; Some definitional suggestions for parallel graph grammars, in "Automata, Languages, and Development" (A.Lindenmayer, G.Rozenberg, eds.), North-Holland Pub. Co., Amsterdam, 1976, 443-468
[Eng89] J.Engelfriet; Context-free NCE graph grammars, Proc. FCT '89, LNCS 380, 148-161
[Eng91] J.Engelfriet; A characterization of context-free NCE graph languages by monadic second-order logic on trees, this Volume
[EngHey89] J.Engelfriet, L.M.Heyker; The string generating power of context-free hypergraph grammars, Report 89-05, Leiden, to appear in JCSS
[EngHey91a] J.Engelfriet, L.M.Heyker; The term generating power of context-free hypergraph grammars, this Volume
[EngHey91b] J.Engelfriet, L.M.Heyker; Hypergraph languages of bounded degree, 1991, in preparation
[EngLei] J.Engelfriet, G.Leih; Linear graph grammars: power and complexity, Inform. Comput. 81 (1989), 88-121
[EngLeiRoz87] J.Engelfriet, G.Leih, G.Rozenberg; Apex graph grammars, in [EhrNagRozRos], 167-185
[EngLeiRoz88] J.Engelfriet, G.Leih, G.Rozenberg; Nonterminal separation in graph grammars, Report 88-29, Leiden, to appear in TCS
[EngLeiWel] J.Engelfriet, G.Leih, E.Welzl; Boundary graph grammars with dynamic edge relabeling, JCSS 40 (1990), 307-345
[EngRoz] J.Engelfriet, G.Rozenberg; A comparison of boundary graph grammars and context-free hypergraph grammars, Inform. Comput. 84 (1990), 163-206
[Hab] A.Habel; Hyperedge replacement: grammars and languages, Ph.D.Thesis, Bremen, 1989
[HabKre87a] A.Habel, H.-J.Kreowski; Some structural aspects of hypergraph languages generated by hyperedge replacement, Proc. STACS '87, LNCS 247, 207-219
[HabKre87b] A.Habel, H.-J.Kreowski; May we introduce to you: hyperedge replacement, in [EhrNagRozRos], 15-26
[JanRoz80] D.Janssens, G.Rozenberg; On the structure of node-label-controlled graph languages, Inform. Sci. 20 (1980), 191-216
[JanRoz83] D.Janssens, G.Rozenberg; A survey of NLC grammars, Proc. CAAP '83, LNCS 159, 114-128
[Kau] M.Kaul; Syntaxanalyse von Graphen bei Präzedenz-Graph-Grammatiken, Ph.D.Thesis, Osnabrück, 1985
[Lau] C.Lautemann; Efficient algorithms on context-free graph languages, Proc. ICALP '88, LNCS 317, 1988, 362-378
[MaiRoz87] M.G.Main, G.Rozenberg; Handle NLC grammars and r.e. languages, JCSS 35 (1987), 192-205
[MaiRoz90] M.G.Main, G.Rozenberg; Edge-label controlled graph grammars, JCSS 40 (1990), 188-228. See also [EhrNagRozRos], 411-426

[MezWri] J.Mezei, J.B.Wright; Algebraic automata and context-free sets, Inform. Contr. 11 (1967), 3-29

[MonRos] U.Montanari, F.Rossi; An efficient algorithm for the solution of hierarchical networks of constraints, in [EhrNagRozRos], 440-457

[Nag] M.Nagl, "Graph-grammatiken", Vieweg, Braunschweig, 1979

[RozWel] G.Rozenberg, E.Welzl; Boundary NLC graph grammars - basic definitions, normal forms, and complexity, Inform. Contr. 69 (1986), 136-167

[Sch] R.Schuster; Graphgrammatiken und Grapheinbettungen: Algorithmen und Komplexität, Ph.D.Thesis, Report MIP-8711, Passau, 1987

FROM GRAPH GRAMMARS TO HIGH LEVEL REPLACEMENT SYSTEMS

Hartmut Ehrig
Technical University Berlin
Franklinstraße 28/29
W-1000 Berlin 10 (Germany)

Annegret Habel
Hans-Jörg Kreowski
University of Bremen
Postfach 33 04 40
W-2800 Bremen 33 (Germany)

Francesco Parisi-Presicce
Università degli Studi Aquila
Via Vetoio
I-67100 L'Aquila (Italia)

Abstract: The algebraic approach to graph grammars - well-known in the literature for several types of graphs and structures - is extended to include several new types of replacement systems, especially the replacement of algebraic specifications which were recently introduced for a rule-based approach to modular system design.
This leads to the new concept of high level replacement systems which is formulated in an axiomatic algebraic framework based on categories and double-pushouts. In this paper only basic notions like productions, derivations, parallel and sequential independence are introduced for high-level replacement systems leading to Church-Rosser and Parallelism Theorems previously shown in the literature for special cases only.

Keywords: graph grammars, high level replacement systems, category theory, independent derivations, Church-Rosser Theorem, parallelism theorem

CONTENTS

1. INTRODUCTION

The algebraic approach to graph grammars introduced in [EPS 73] and in the detailed survey [Ehr 79] has been applied to several fields in Computer Science and related areas. In the basic approach (see [Ehr 79]) directed graphs with colored nodes and edges and color-preserving graph morphisms are used. Since color-preserving graph morphisms turned out to be too restrictive for some applications grammars based on partial graphs (see [SE 76]), on structures ([EKMRW 81]), and on graphs with partially ordered color sets ([PEM 87], [CMREL 90]) have been introduced. In each of these generalizations the rewriting process is formulated in terms of double-pushouts in suitable categories of graphs or structures, written:

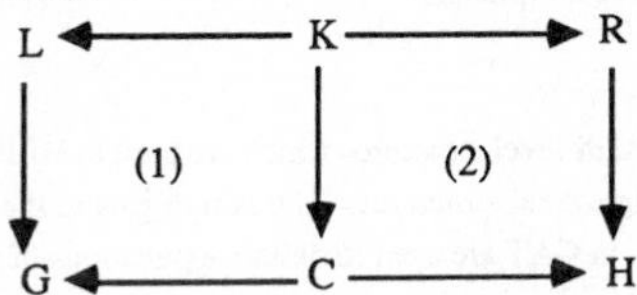

The top row $L \leftarrow K \rightarrow R$ is a production p with left hand side L, right hand side R, and interface K. Given a context C and morphism $K \rightarrow C$ the gluing of L with C along K leads to G, the pushout object in the left diagram (1). The gluing

of R with C along K leads to H, the pushout object in the right diagram (2). Both diagrams together, i.e. the double pushout, represent a direct derivation from G to H using production p written in short notation:

$$p: G \Rightarrow H$$

This basic idea of direct derivations for graphs and structures has recently been applied replacing graphs and structures by algebraic specifications leading to a rule based approach to modular system design ([PP 89], [PP 90]). This was the starting point for "algebraic specification grammars" ([PE 90]), where the theory of graph grammars is going to be extended to algebraic specifications, and also for "high level replacement systems", where graphs and graph morphisms are replaced by objects and morphisms of a suitable category.
In this paper we introduce this new notion of "high level replacement systems" which on one hand unifies the theories of grammars and replacement systems based on different types of graphs and structures. On the other hand it opens up new kinds of applications, including algebraic specifications, hypergraphs, graphics, figures, collages, place-transition nets, colored nets, and different kinds of structured transition systems (see [Ha 89], [MM 88], [Hu 89] and different papers in [EKR 91]).

In section 2 of this paper we introduce the basic concepts of high level replacement systems including productions, direct derivations, applicability of productions, derivation sequences, HLR-systems and languages (HLR = High-Level-Replacement). In fact, this can be considered as step 0 of the theory of HLR-systems, where we only need a category with a distinguished class M of morphisms which are used in the productions.

In section 3 we introduce four basic examples of HLR0-systems: Graph Grammars ([Ehr 79]), graph grammars with partially ordered label sets ([PEM 87]), structure grammars ([EKMRW 81]), and algebraic specification grammars ([PE 90]). In each case we define the corresponding category, the distinguished class of morphisms, the construction of pushouts and of pushout complements which are useful to apply productions, and a sample derivation step.
In section 4 we introduce HLR-concepts for independence and parallelism leading to Church-Rosser and Parallelism Theorems for HLR-systems under suitable conditions.

These conditions, called HLR1-conditions, are formulated in section 5 and the main ideas of the proof of the Church-Rosser and Parallelism Theorem based on these conditions is given. Moreover we show how far these conditions are satisfied for the examples studied in section 3. In fact, for graphs with partially ordered color sets we need rather strong additional requirements to satisfy the conditions, while there are no problems with the other examples.

In section 6 we summarize the results of this paper and discuss further developments including additional results, like the Concurrency Theorem (see [Ehr 79]), and further examples of HLR-systems mentioned above.

In the appendix basic concepts and results from category theory are introduced (see [AM 75], [HS 73], [ML 72]) which are used to prove the results in sections 4 and 5.
Finally we would like to thank the referees for several useful comments and Helga Barnewitz for excellent typing.

2. BASIC HLR-CONCEPTS

In this section we introduce the basic concepts of high-level-replacement systems, short HLR-systems, including productions, derivations, systems and languages. For some basic notions of category theory, like object, morphism, isomorphism and pushout we refer to the appendix, while the interpretation for HLR-systems are given in this section.

2.1 General Assumption for HLR-systems

Let CAT be a category with a distinguished class M of morphisms.

Interpretation:
The objects of the category CAT can be regarded as high-level structures which are used in HLR-systems, the morphisms in CAT as structure preserving functions between high-level structures. The morphisms in the class M are those which are used in the productions, while general morphisms in CAT are used to define applications of productions to high-level structures.

2.2 Definition (Productions and Derivations)

1. A production $p = (L \leftarrow K \rightarrow R)$ in **CAT** consists of a pair of objects (L, R), called left and right hand side respectively, an object K, called gluing object or interface, and two morphisms $K \rightarrow L$ and $K \rightarrow R$ belonging to the class M.

2. Given a production p as above and an object C, called context object, together with a morphism $K \rightarrow C$ a direct derivation $G \Rightarrow H$ via p, short $p: G \Rightarrow H$ or $G \Rightarrow PH$, from an object G to an object H via p is given by two pushout diagrams (1) and (2) in the category CAT of the following shape:

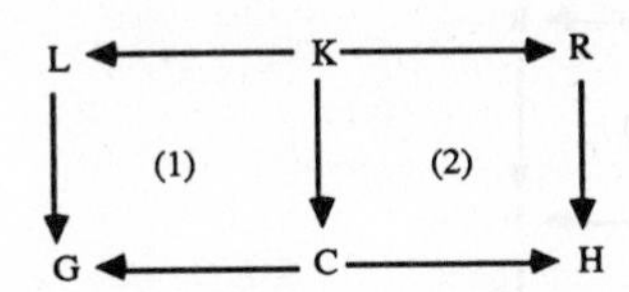

The morphisms $L \rightarrow G$ resp. $R \rightarrow H$ are called occurrence of L in G resp. R in H.

3. A derivation sequence $G \Rightarrow^* H$, short derivation, between objects G and H means $G \cong H$ (isomorphism) or a sequence of $n \geq 1$ direct derivations

$$G = G0 \Rightarrow^{p1} G1 \Rightarrow^{p2} \ldots \Rightarrow^{pn} Gn = H$$

For this sequence we may also write $G \Rightarrow^* H$ via (p1,...,pn).

Interpretation

1. The idea of a production is that the left hand side L is replaced by the right hand side R, where the gluing object K together with morphisms $K \rightarrow L$ and $K \rightarrow R$ in M designates corresponding gluing items in L and R. If M is a specific class of monomorphisms the gluing object K can be considered as common subobject of L and R which remains unchanged if the production is applied.
2. The pushouts (1) resp. (2) can be interpreted as gluing of objects L and C along K leading to object G resp. gluing of R and C along K leading to object H. This supports a symmetric view of direct derivations where in addition to the production p the context object C together with $K \rightarrow C$ is given and G and H are constructed both by pushout. For application of a production p to an object G with occurrence $L \rightarrow G$ (occurrence view) we refer to 2.4 below. In the symmetric view the objects G and H are uniquely determined up to isomorphism by the pushout properties of (1) and (2) (see appendix).
3. A derivation sequence is a sequence of $n \geq 0$ direct derivation steps where each step is at most unique up to isomorphism. For this reason we require for $n = 0$ only $G \cong H$ instead of $G = H$.

2.3 Fact (Symmetry and Induced Productions)

1. Given a production $p = (L \leftarrow K \rightarrow R)$ and a direct derivation $p: G \Rightarrow H$, the inverse production $p^{-1} = (R \leftarrow K \rightarrow L)$ leads to an inverse direct derivation $p^{-1}: H \Rightarrow G$.
2. Each direct derivation $p : G \Rightarrow H$ as given in 2.2.2 leads to an induced production $p^* = (G \leftarrow C \rightarrow H)$ provided that the class M is closed under pushouts, i.e. $(K \rightarrow L) \in M$ implies $(C \rightarrow G) \in M$ in (1) and $(K \rightarrow R) \in M$ implies $(C \rightarrow H) \in M$ in (2) of 2.2.2.
3. Given an induced production p^* of p as above for each direct derivation $p^*: G^* \Rightarrow H^*$ there is also a direct derivation $p: G^* \Rightarrow H^*$.

Proof

1. The inverse direct derivation $p^{-1}: G \Rightarrow H$ is given by

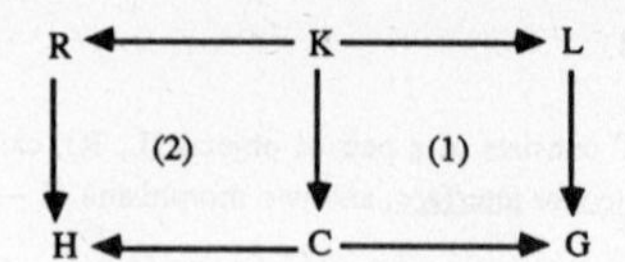

using the same pushout diagrams (1) and (2) as in 2.2.2 but in opposite order.

2. If M is closed under pushouts the morphisms $C \to G$ and $C \to H$ belong to M such that p^* is a production.
3. Given a direct derivation $p: G \Rightarrow H$ we have pushouts (1) and (2) while a direct derivation $p^*: G^* \Rightarrow H^*$ leads to pushouts (3) and (4)

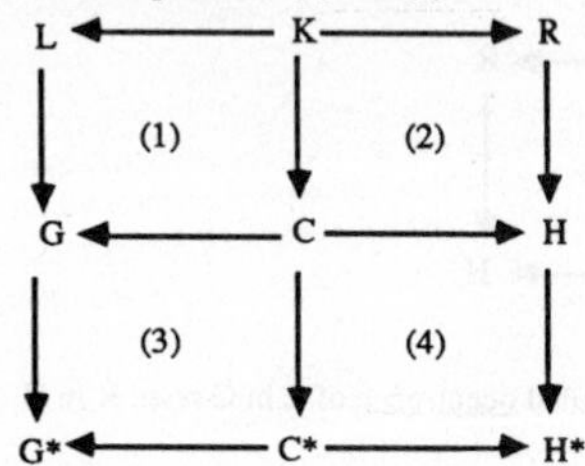

using the composition property of pushouts (see appendix) also (1) + (3) and (2) + (4) are pushouts leading to a direct derivation $p: G^* \Rightarrow H^*$.

2.4 Definition (Applicability of Productions)

Given a production $p = (L \leftarrow K \to R)$, an object G, and an occurrence $L \to G$ of L in G the production p is called applicable to G via $L \to G$ if the following two conditions are satisfied:

1. There are an object C and morphisms $K \to C$ and $C \to G$ such that (1) in 2.2.2 becomes a pushout (in this case C is called pushout-complement of L w.r.t. G and K in (1)).
2. There is an object H which is the pushout of $K \to R$ and $K \to C$ in (2) of 2.2.2 (pushout construction).

Remarks and Interpretations

1. Even if the category **CAT** has pushouts the pushout complement will only exist under suitable conditions. In the case of graph grammars these conditions are summarized in the gluing condition which requires that all boundary points in L w.r.t. $L \to G$ are gluing points w.r.t.
$K \to L$ (see [Ehr 79] and example 3.1 below). For general HLR-systems conditions for the existence of pushout complements will be called gluing condition.
Roughly spoken the pushout complement C of L w.r.t. G and K is a subobject C of L such that the gluing of C with L via K is isomorphic to G. If the gluing condition is satisfied there may be more than one pushout complement C of L w.r.t. G and K. We say that the pushout complement construction is unique if there is only one pushout complement up to isomorphism:
2. If the category **CAT** has pushouts the pushout construction in the second step always exists. Otherwise we have to require the existence of this pushout construction and it is useful to provide necessary and / or sufficient conditions for the existence of pushouts (see example 3.2 below).
3. If both conditions are satisfied a direct derivation $p: G \Rightarrow H$ can be constructed in two steps. It is unique up to isomorphism if and only if the pushout complement construction is unique.

2.5 Definition (HLR-Systems and Languages)

1. An HLR-system

$$HLRS = (S, P, T)$$

in a category **CAT** is given by an object S, called start object, a set **P** of productions (see 2.2.1), and a class T of objects in **CAT**, called terminal colored objects.

2. The language L(HLRS) of an HLR-system HLRS = (S, **P**,T) is given by the set of all terminal colored objects in **CAT** derivable from S by **P**

$$L(HLRS) = \{G \in T / S \Rightarrow^* G\}$$

Remarks and Interpretations

1. The main part of an HLR-system is the set of productions **P** together with the notion of derivations as defined in 2.2. For a suitable notion of language it is important to distinguish a start object S and a class of "terminal colored" objects **T** which may be equal to the class of all objects in **CAT**. We speak of "terminal colored" objects to avoid confusion with the notion of a "terminal object" T in a category, which would mean that for each other object A there is a unique morphism from A to T.
2. The language L(HLRS) is given by all those objects G in the class **T** of terminal colored objects which are derivable in $n \geq 0$ steps from the start object S (see 2.2.3). If **T** is closed under isomorphisms in **CAT** also the language is closed under isomorphisms, i.e. for each $G \in L(HLRS)$ and $G' \cong G$ we also have $G' \in L(HLRS)$ and $G' \cong G$ we have $G' \in L(HLRS)$.

3. EXAMPLES OF HLR-SYSTEMS

In this section we present four basic examples of HLR-systems. In each case we will give the definition of the HLR-category underlying this replacement system, the construction of the pushout objects and diagrams in this category and the applicability conditions for productions, especially existence and uniqueness of pushout complements given by a gluing condition (see 2.4).

3.1 Graph Grammars

The algebraic theory of Graph Grammars provides the first example of a High Level Replacement System. It is the first not only because it is probably the most intuitive and the easiest to visualize, but also because it predates the other examples, having been introduced in [EPS 73].

The category **GRAPHS** underlying the graph grammars depends on a fixed color or label alphabet CO consisting of two sets, a label set CO_A for arcs and a label set CO_N for nodes. Taking M to be the class of all injective graph morphisms it satisfies the general assumptions 2.1 for HLR-systems.

3.1.1 Definition (Category GRAPHS)

1) A (labelled) graph $G = (G_A, G_N\ s, t, m_N, m_A)$ consists of a set of nodes G_N, a set of arcs G_A, two functions s, t: $G_A \to G_N$ (selecting the source and target node of each arc, respectively) and two labelling maps m_N: $G_N \to CO_N$ and m_A: $G_A \to CO_A$.
2) A graph morphism $f{:}G \to G'$ is a pair (f_A, f_N) of functions f_A: $G_A \to G_{A'}$ and f_N: $G_N \to G'_N$ such that
   a) $m'_N \circ f_N = m_N$, $m'_A \circ f_A = m_A$ (color preserving)
   b) $f_N \circ s = s' \circ f_A$, $f_N \circ t = t' \circ f_A$ (structure preserving)
3) The category GRAPHS is obtained by taking as objects the labelled graphs, as arrows the graph morphisms and by defining the composition of morphisms $h = g \circ f = G \to G' \to G''$ componentwise with $h_N = g_N \circ f_N$ and $h_A = g_A \circ f_A$.
4) The distinguished class M of morphisms in **GRAPHS** is the class of all injective graph morphisms, where a graph morphism f: $G \to G'$ possesses properties like injectivity, surjectivity and bijectivity if both its components f_N and f_A do.

3.1.2 Lemma (Existence and Construction of Pushouts in GRAPHS)

The category **GRAPHS** has pushouts. For graph morphisms f1: $G0 \to G1$ and f2:$G0 \to G2$, with f2 injective, the pushout object G3 of f1 and f2 is given by:

$$G3_N = G1_N + (G2_N - f2_N(G0_N))$$
$$G3_A = G1_A + (G2_A - f2_A(G0_A))$$

$s_{G3}(a) = \quad s_{G1}(a) \quad$ if $a \in G1_A$

$\quad s_{G2}(a) \quad$ if $a \notin G1_A$, $s_{G2}(a) \in G2_N - f2_N(G0_N)$

$\quad f1_N(n) \quad$ if $a \notin G1_A$, $s_{G2}(a) = f2_N(n)$ for $n \in G0$

t_{G3} is similar to s_{G3}, and with m_G short for m_A resp. m_N of graph G we have

$m_{G3}(x) = \quad m_{G1}(x) \quad$ if $x \in G1$

$\quad m_{G2}(x) \quad$ if $x \in G2 - f2(G0)$.

The morphism g1: G1 → G3 is just an inclusion while g2: G2 → G3 is defined by

$f2(x) = \quad x \quad$ if $x \in G2 - f2(G0)$

$\quad f1(y) \quad$ if $x = f2(y)$ for $y \in G0$.

A similar construction can be given, using quotient sets, for cases when f2 is not injective.

A direct derivation is based on the context object and on two pushout constructions to obtain the starting object and the result of applying the production. Not all occurrences L → G of L in G allow the production p = (L ← K → R) to be applied to G because not all occurrences guarantee the existence of the context C. The context exists if the following Gluing Condition is satisfied.

3.1.3 Definition (Gluing Condition for GRAPHS)

Given a production $p = (L \xleftarrow{l} K \xrightarrow{r} R)$, a morphism g: L → G satisfies the <u>Gluing Condition</u> with respect to p if $DANG(g) \cup ID(g) \subseteq l(K)$

where

$ID(g) = \{x \in L \mid g(x) = g(y) \text{ for some } y \in L, y \neq x\}$

$DANG(g) = \{x \in L_N \mid g(x) = s_G(a) \text{ or } g(x) = t_G(a) \text{ for some } a \in G_A - g_A(L_A)\}$

The inclusion $DANG \subseteq l(K)$ is strictly related to the structure of a graph: no node can be removed without removing all arcs incident to the node.

3.1.4 Lemma (PO-Complement in GRAPHS)

Given morphisms g: L → G and l: K → L, then

a) (Existence) g satisfies the Gluing Condition if and only if there exists a graph C = G - g(L) + g(l(K)) such that G is the pushout object of l: K → L and c: K → C where c is the restriction of $g \circ l$ to C.

b) (Uniqueness) The pushout complement is unique if bl is injective.

3.1.5 Example (Derivation in GRAPHS)

The production $p = (L \xleftarrow{l} K \xrightarrow{r} R)$ below (where the letter for each node is the label, while the numbers indicate the images of the nodes of K under l and r and the arcs are uniquely labelled with the empty word λ) can be applied to the graph G below because the Gluing Condition is satisfied with

$ID = \emptyset$ and $DANG = L_N$. The context graph is the graph C below which, glued to R via K, gives the result of the direct derivation H. The direct derivation $p: G \Rightarrow H$ is given by the following double-pushout:

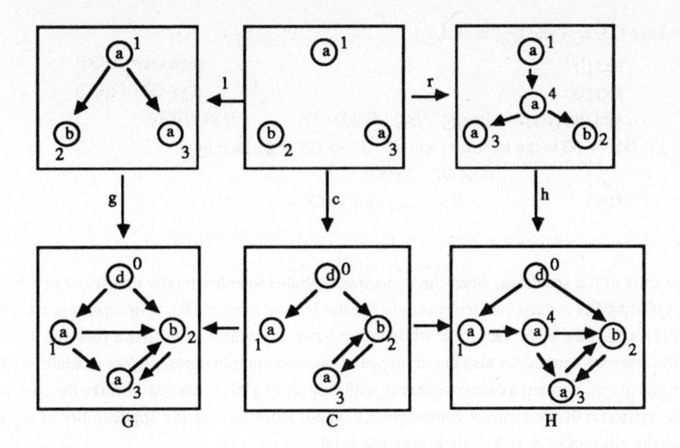

3.2 Graph Grammars with Preordered Label Sets

Another example of a High Level Replacement System is given by [PEM 87] in which labelled graphs are considered where a structure is imposed on the set of labels. The structure consists of a preorder which determines which of the color-changing graph morphisms are part of our framework. The minimal structure used on the set of labels includes, as special cases, the graphs with color-preserving morphisms and graphs with morphisms which either preserve the labelling or label items previously unlabeled.

The category **POGRAPHS** underlying this type of graph grammars is based on a structured color alphabet SC = (SC_A, SC_N), consisting of two sets SC_A and SC_N, each equipped with a preorder (i.e. a reflexive and transitive) binary relation, called $<_A$ and $<_N$, respectively.

3.2.1 Definition

1) A (labelled) SC-graph G is a graph (G_A, G_N, s, t, m_N, m_A) as in the category **GRAPHS** where m_N: $G_N \to sc_N$ and m_A: $G_A \to sc_A$ for the given structured alphabet SC.
2) An SC-graph morphism f: G → G' is a pair f_A: $G_A \to G'_A$ and f_N: $G_N \to G'_N$ of functions such that
   a) $f_N \circ s = s' \circ f_A$, $f_N \circ t = t' \circ f_A$ (structure preserving)
   b) $m'_A(f_A(a)) <_A m_A(a)$, $m'_N(f_N(x)) <_N m_N(x)$ for all $a \in G_A$ and $x \in G_N$ (label compatible)

   A sc-graph morphism is called color preserving if $<_A$ and $<_N$ in b) are replaced by = (equality).
3) The category **POGRAPHS** consists of the SC-graphs as objects, the SC-graph morphisms as arrows and the composition of morphisms is taken componentwise.
4) As class M of morphisms in **POGRAPHS** we can consider the class of all injective SC-graph morphisms or the subclass of all injective color preserving SC-graph morphisms.

Remark

Notice that the transitivity of the binary relation < guarantees composability of morphisms. For $B \subseteq SC_A$, we denote by glb(B) (greatest lower bound of B) an element $b \in SC_A$ such that $b <_A x$ for all $x \in B$ and if $y <_A x$ for all $x \in B$, than $y <_A b$. Similarly for $B \subseteq SC_N$. Note that glb(B) need not exist and, if it does, it need not be unique (unless $<_A$ is a partial order). If $<_A$ and $<_N$ are partial orders and glb(b) and dually lub(B) (least upper bound of B) for each $B \subseteq SC_A$, SC_N the category **POGRAPHS** is denoted by **CPOGRAPHS**, graphs with completely partially ordered color alphabet.

3.2.2 Lemma (Existence and Construction of Pushouts)

For sc-morphisms f1: G0 → G1 and f2: G0 → G2, with f2 injective, the pushout object G3 of f1 and f2 exists iff for each $y \in f1(G0)$, glb($\{m_{G1}(y)\} \cup \{m_{G2}(f2(x)):f1(x) = y\}$) exists.

In this case we have G3, s_{G3} and t_{G3} as in 3.1.1

$$m_{G3}(y) = \begin{cases} m_{G1}(y) & \text{if } y \in G1 - f1(G0) \\ m_{G2}(y) & \text{if } y \in G2 - f2(G0) \\ glb(\{m_{G1}(y)\} \cup \{m_{G2}(f2(x)) : f1(x) = y\}) & \text{if } y \in f1(G0) \end{cases}$$

The morphism g1: G1 → G3 is the inclusion and g2: G2 → G3 is defined by

$$g2(x) = \begin{cases} x & \text{if } x \in G2 - f2(G0) \\ f1(y) & \text{if } x = f2(y) \text{ for } y \in G0 \end{cases}$$

□

At the opposite ends of the spectrum, when the structured alphabet is either totally unordered or a boolean algebra, then the category **POGRAPHS** is closed under pushouts. In the former case, in fact, sc-morphisms must be color preserving and **POGRAPHS** coincides with **GRAPHS** while in the latter any subset of SC has a (unique) glb.
The nature of the preorder determines also the existence of pushout complements. In fact in addition to the property about the structure of the graph, existence of complements with respect to glb's is needed. Unlike the case for color preserving morphisms, the existence of the pushout complement C is not sufficient for the applicability of the production L ← K → R to G since the pushout of R and C via K may not exist.

3.2.3 Lemma (PO-complement in POGRAPHS)

For SC-morphisms g: L → G and l: K→ L,

a) (Existence) there exists an SC-graph C = G - g(L) + g(l(K)) such that G is the pushout object of l: K → L and c : K → C (where c is the image restriction of g ∘ l to C) if and only if g satisfies the Gluing Condition for **GRAPHS** (see 3.1.3).

b) (Uniqueness) the pushout complement C is unique if l is injective and color preserving and for each $x \in L$, the set $\{a \in sc: glb(m_L(x),a) = m_G(g(x))\}$ has cardinality 1.

3.2.4 Example (Derivation in POGRAPHS)

Let $SC_N = \{a, b, c, x\}$ with $c <_N$ and $y <_N y$ for all $y \in SC_N$ and $SC_A = \{\lambda\}$ with $\lambda <_A \lambda$.
The production p = (L ← K → R) below is applicable to the graph G below with g: L→ G as indicated by the numbers, because g is color preserving except for node 4 where we have $m_G(g_N(4)) = c <_N x = m_L(4)$. This yields a direct derivation p: G ⇒ H given by the following double-pushout:

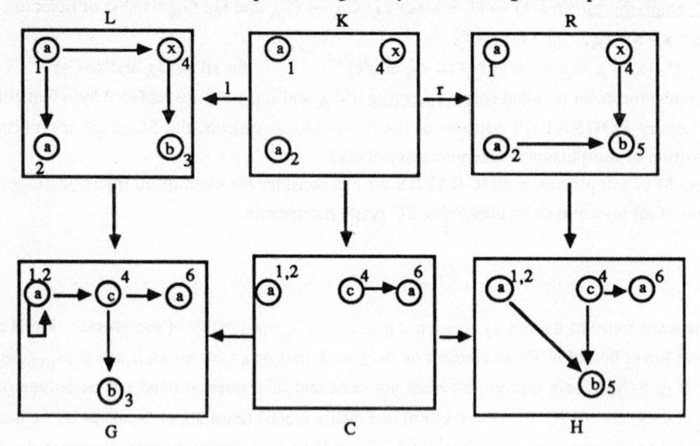

3.3 Structure Grammars

A generalization of Graph Grammars leads to the notion of transformation of structures and to structure grammars (see [EKMRW 81]). If we consider each node label as a unary predicate to be applied to a node, an arc label as a unary predicate over arcs and the functions selecting the source and target node of each arc as a binary predicate, then each

labelled graph can be seen as a set of unary and binary predicates over a set of atoms.
The category **STRUCT** of structures is based on a fixed set $\mathbf{P}$, called set of predicates, where each element has an associated arity $n \geq 0$. For any set X, n-ary predicate Q and $x_1,...,x_n \in X$ the formal expression $Q(x_1,...,x_n)$ is an atomic formula over X. The set of such atomic formulas is denoted by X^*.

3.3.1 Definition (Category STRUCT)

1) A <u>structure</u> G over $\mathbf{P}$ is a pair (G_A, G_F) of sets where G_A is the set of <u>atoms</u> and G_F is a subset of the set G_A^* of all the atomic formulas over G_A.

2) A <u>structure morphism</u> f: $G \to H$ is a pair of functions $f_A: G_A \to H_A$ and $f_F: G_F \to H_F$ such that for all atomic formulas $Q(x_1,...,x_n) \in G_F$,
$f_F(Q(x_1,...,x_n)) = Q(f_A(x_1),...,f_A(x_n)) \in H_F$.
(Note that f_F is completely determined by f_A).
The morphism $f = (f_A, f_F)$ is called injective if f_A is injective.
It is called <u>strict</u> if $Q(f(x_1),...,f(x_n)) \in H_F$ implies $Q(x_1,...x_n) \in G_F$.

3) The category **STRUCT** is obtained by taking as objects the structures over $\mathbf{P}$, as morphisms the structure morphisms and by defining composition of morphisms componentwise.

4) The class M can be taken to be the class of all injective or all injective strict **STRUCT**-morphism.

3.3.2 Lemma (Existence and Construction of Pushouts in STRUCT)

The category **STRUCT** has pushouts.
For the structure morphisms f1: $G0 \to G1$ and f2: $G0 \to G2$, with f2 injective, the pushout object G3 of f1 and f2 is given by

$G3_A = G1_A + (G2_A - f2_A(G0_A))$

$G3_F = G1_F \cup (G2_F - f2_F(G0_F))$

The morphism g1: $G1 \to G3$ is the inclusion and g2: $G2 \to G3$ is defined by

$g2(x) = f1(y)$ if $x = f2(y), y \in G0$

$\quad\quad\quad x$ if $x \in G2 - f2(G0)$.

The pushout is called <u>natural</u> if f1 or f2 is injective and we have $g1_F(G1_F) \cap g2_F(G2_F) \subseteq g1_F\, f1_F(G0_F)$.

A similar construction, using pushouts in the category **SETS** of sets, can be given when f2 is not injective.
The applicability of a production $p = (L \leftarrow K \to R)$ of structures to the structure G requires an occurrence (a structure morphism $L \to G$) of L in G and the existence of a pushout complement of L in G. Two conditions must be satisfied for the pushout complement: one depends on the equivalent problem in the category **SETS**, the other one on the fact that formulas in a structure can use only atoms of that structure.

3.3.3 Lemma (PO Complement in STRUCT)

Given structure morphisms g: $L \to G$ and l: $K \to L$, then

a) (Existence) there is a structure C and morphisms $K \to C$ and $C \to G$ such that the diagram

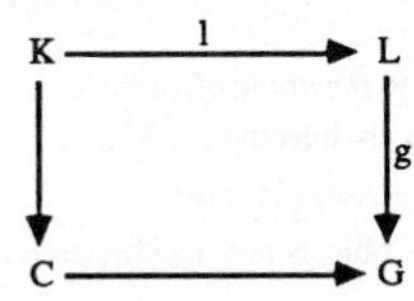

is a pushout in **STRUCT** if and only if
$ID(g) \cup DANG(g) \subseteq l_A(K_A)$
where
$ID(g) = \{y \in L_A : g(y) = g(y') \text{ for some } y' \neq y, y' \in L_A\}$ and
$DANG(g) = \{y \in L_A : g(y) \text{ is an atom in some } Q \in G_F - g(L_F)\}$.

b) (Uniqueness) Natural pushout complements are unique (up to isomorphism).

3.3.4 Example (Derivation in STRUCT)

A derivation in **STRUCT** is given by the following double-pushout where a,b,c,d,l,a' are atoms with l(a') = a, A, B unary and M, N binary predicates and C a natural pushout complement

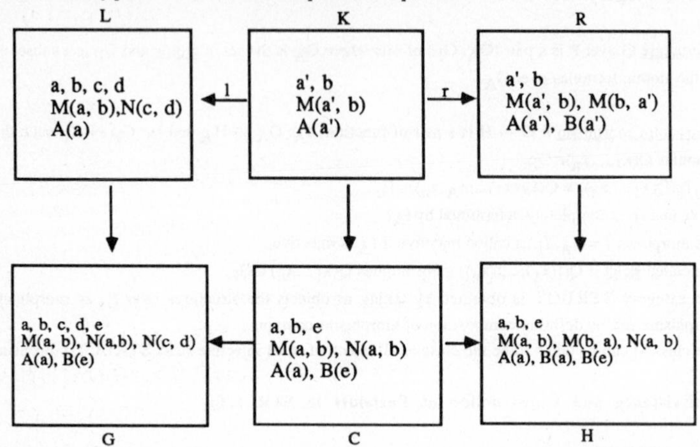

Note, that in this example all structures can be interpreted as graphs with arcs colored by M, N and nodes a,...,e,a' partially colored by A, B.

3.4 Algebraic Specification Grammars

Algebraic specification grammars have been introduced most recently as a device to generate algebraic specifications in the sense of [EM 85] using productions (see [PP 89], [PP 90], [PE 90]). More than a grammar, it is in fact a replacement system in that there are no non-terminal specifications or part of specifications. The motivation for this type of replacement system comes from the design of modular software systems (see [PP 89], [PP 90]). A production in an algebraic specification grammar corresponds exactly to the interface of a module specification and if each production is realized by the body of a module specification, then the transformation of a specification A into a specification B using these productions can be translated into the design of a modular system with A and B as import and export interfaces ([PP 89]).

3.4.1 Definition (Category SPEC)

1) An algebraic specification SPEC is a triple (S, OP, E) consisting of a set S of sorts, a set OP of operator symbols and a set E of equations over the signature SIG = (S, OP).

2) A specification morphism f: (S1, OP1, E1) → (S2, OP2, E2) consists of a pair (f_S: S1 → S2, f_{OP}: OP1 → OP2) of functions such that

   a) for each operator symbol N∈OP1 with signature N: s1...sn → s, the operator symbol $f_{OP}(N)$ has signature $f_{OP}(N)$: $f_S(s1)...f_S(sn) \to f_S(s)$

   b) for each e∈E1, the translated equation $f^{\#}(e)$ is in E2 up to renaming of variables.

   The morphism f = (f_S, f_{OP}) is called injective if f_S and f_{OP} are injective.

   It is called strict if $f^{\#}(e) \in$ E2 implies e∈E1 for all equations e over (S1, OP1).

3) The category **SPEC** consists of algebraic specifications as objects and specification morphisms as arrows, with componentwise composition of morphisms.

4) The class M can be taken to be the class of all injective or all injective strict specification morphisms.

3.4.2 Lemma (Existence and Construction of Pushouts in SPEC)

The category **SPEC** has pushouts.

Given specification morphisms f1: SPEC0 → SPEC1 and f2: SPEC0 → SPEC2, the pushout object SPEC3 = (S3, OP3, E3) is defined by:

$S3 = S1 +_{S0} S2$ (the pushout object of $f1_S$ and $f2_S$ in the category **SETS** of sets)

$OP3 = OP1 +_{OP0} OP2$ (the pushout object of $f1_{OP}$ and $f2_{OP}$ in **SETS**)

$E3 = g1^{\#}(E1) \cup g2^{\#}(E2)$

where g1: (S1, OP1) → (S3, OP3) and g2: (S2, OP2) → (S3, OP3) are the signature morphisms induced by (S3, OP3). The pushout is called natural if f1 or f2 is injective and we have $g1^{\#}(E1) \cap g2^{\#}(E2) \subseteq g1^{\#} f1^{\#}(E0)$.

Remark

For simplicity we only consider a restricted kind of specification morphisms in 3.4.1. In the general case the translated equation $f^{\#}(e)$ is only required to be derivable from E2. In our case with $f^{\#}(e)$, in E2 an alternative way would be to consider specification morphisms as triples $f = (f_S, f_{OP}, f_E)$ with f_S and f_{OP} as above and f_E:E1 → E2 such that for $e \in E1$ with e = (X, L, R) we have $f_E(e) = (X^{\#}, f^{\#}(L), f^{\#}(R))$. This suggests to take pushout constructions separately in each component. But then we would have to consider disjoint copies of the same equation which contradicts the usual idea and of sets of equations (e.g. the pushout **nat*** of f1 and f2 with f1 = f2 = f:**sig(nat)** → **nat**, where **sig(nat)** is the signature of the specification **nat** of natural numbers (see [EM 85]), would have disjoint copies of each equation in **nat** if the equations are constructed as separate pushout). Our notion of strict injective morphisms allows to construct pushouts separately in each component avoiding disjoint copies of the same equation (e.g. f:**sig(nat)** → **nat** is not strict, the strict version would be f1:**nat** → **nat** and the pushout of f1 and f1 is **nat** without disjoint copies of equations) (see proof of 7.7.6 and 7.7.7 below).

In order for a production of specifications p = (L ← K → R) to be applied to G, there must be an occurrence L → G of L in G for which there is a pushout complement C of L in G. Such a C must be well-defined and in particular its operator symbols must have the proper sorts and its equations composed of the appropriate operator symbols.

3.4.3 Definition (Gluing Condition in SPEC)

For a specification production $p = (L \xleftarrow{l} K \xrightarrow{r} R)$ and a specification morphism g: L → G, define

$ID(g)_S = \{s \in S_L : g_S(s) = g_S(s')$ for some $s \neq s', s' \in S_L\}$

$ID(g)_{OP} = \{N \in OP_L : g_{OP}(N) = g_{OP}(N')$ for some $N \neq N', N' \in OP_L\}$

$DANG(g) = \{s \in S_L : \exists N \in OP_G - g_{OP}(OP_L)$ which contains $g_S(s)$ as one of the sorts in its signature$\}$

then g satisfies the Gluing Condition with respect to p if

$ID(g)_S \cup DANG(g) \subseteq l_S(S_K)$ and

$ID(g)_{OP} \subseteq l_{OP}(OP_K)$.

3.4.4 Lemma (PO Complement in SPEC)

Given a specification production $p = (L \xleftarrow{l} K \xrightarrow{r} R)$ and a specification morphism g:L → G

a) (Existence) There exists a pushout complement $C = (S_C, OP_C, E_C)$ such that $G = L +_K C$ if and only if

(1) g satisfies the Gluing Condition with respect to p

(2) the set $E_C = E_G - g^{\#}(E_L - l^{\#}(E_K))$ is a set of equations over the signature (S_C, OP_C) where

$S_C = S_G - g_S(S_L - l_S(S_K))$

$OP_C = OP_G - g_{OP}(OP_L - l_{OP}(OP_K))$.

b) (Uniqueness) Natural pushout complements are unique (up to isomorphism).

3.4.5 Example (Derivation in SPEC)

Let **nat**, **int**, **data**, and **string(data)** be algebraic specifications of natural numbers, integers, data elements, and strings over data elements as given in [EM 85] we obtain a direct derivation

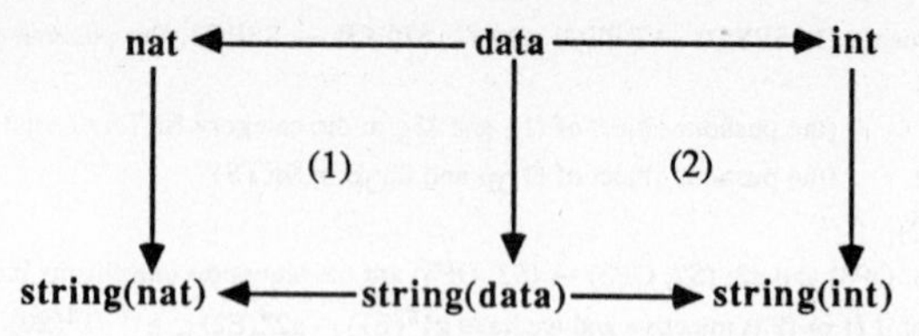

from **string(nat)** to **string(int)** which are the results of (standard) parameter passing in pushout (1) and (2) respectively.

4. HLR-CONCEPTS FOR INDEPENDENCE AND PARALLELISM

In this section we formulate the notions of sequential and parallel independence of derivations for HLR-systems, and we present two Church-Rosser Theorems and the Parallelism Theorem for HLR-systems, which are well-known in the case of graph grammars (see [Kr 77], [Ehr 79]). For most of the proofs, however, we need additional conditions for the HLR-category **CAT** (see 2.1) which will be formulated together with the corresponding proof ideas in section 5. Categories satisfying these conditions are called HLR1-categories (see 5.1). As a general assumption for all results in this section we assume to have a HLR1-category.

4.1 Definition (Independence)

1. Given two productions $p = (L \leftarrow K \rightarrow R)$ and $p' = (L' \leftarrow K' \rightarrow R')$, a derivation sequence $G \Rightarrow H \Rightarrow X$ via (p, p') given by the following pair of (non-dotted arrow) double-pushouts

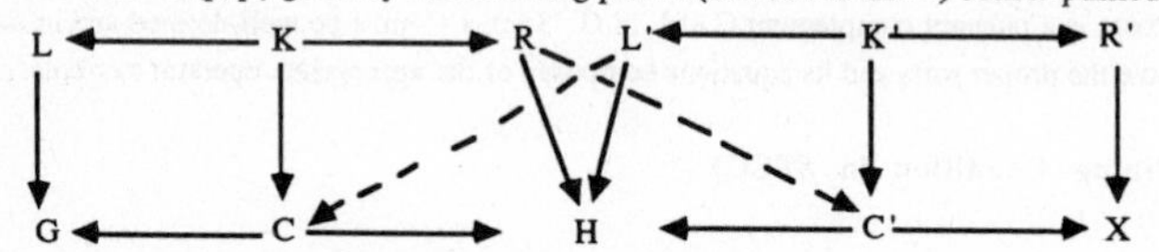

is called <u>sequentially independent</u>, if there are morphisms $L' \rightarrow C$ and $R \rightarrow C'$ (dotted arrows) such that $L' \rightarrow C \rightarrow H = L' \rightarrow H$ and $R \rightarrow C' \rightarrow H = L' \rightarrow H$.

2. Dually, two direct derivations $G \Rightarrow H$ via p and $G \Rightarrow H'$ via p' given by the following pair of (non-dotted arrow) double-pushouts

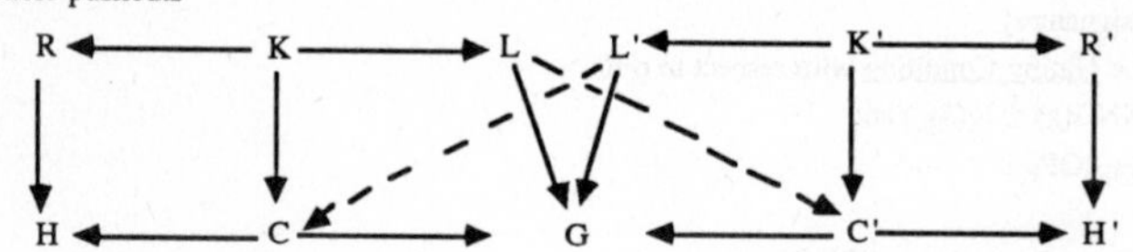

is called <u>parallel independent</u>, if there are morphisms $L' \rightarrow C$ and $L \rightarrow C'$ (dotted arrows) such that $L' \rightarrow C \rightarrow G = L' \rightarrow G$ and $L \rightarrow C' \rightarrow G = L \rightarrow G$.

Remarks and Interpretation

1. In the case of graph grammars (see 3.1) these conditions for sequential (resp. parallel) independence are equivalent to the fact that the intersection of the images of $R \rightarrow H$ and $L' \rightarrow H$ in H (resp. $L \rightarrow G$ and $L' \rightarrow G$ in G) consists of common gluing points only, i.e. is included in the intersection of the images of $K \rightarrow R \rightarrow H$ and $K' \rightarrow R' \rightarrow H$ in H (resp.$K \rightarrow L \rightarrow G$ and
$K' \rightarrow L' \rightarrow G$ in G). This equivalence is shown in [Ehr 79] in the graph case. For HLR-systems it is easier to define independence directly by conditions 1 and 2 above, because this avoids to require general pullback constructions generalizing intersections.
2. Using inverse productions and derivations (see 2.3.1) it is easy to see that sequential independence of $G \Rightarrow H \Rightarrow X$ via (p, p') is equivalent to parallel independence of $H \Rightarrow G$ via p^{-1} and $H \Rightarrow X$ via p'.
3. A consequence of sequential and parallel independence are the following Church-Rosser properties which allow to apply the productions p and p' in opposite order where the composite morphisms $L' \rightarrow C \rightarrow G$ and $R \rightarrow C' \rightarrow X$

(resp. $L' \to C \to H$ and $L \to C' \to H'$) are occurrence morphisms (see 2.2.2). In fact, these properties are local Church-Rosser properties including only one derivation step in each direction. Of course this can be extended to finite derivation sequences.

4.2 Theorem (Church-Rosser-Property I)

Given parallel independent direct derivations $G \Rightarrow H$ via p and $G \Rightarrow H'$ via p' there is an object X and direct derivations $H \Rightarrow X$ via p' and $H' \Rightarrow X$ via p such that the derivation sequences $G \Rightarrow H \Rightarrow X$ via (p, p') and $G \Rightarrow H' \Rightarrow X$ via (p', p) become sequentially independent.

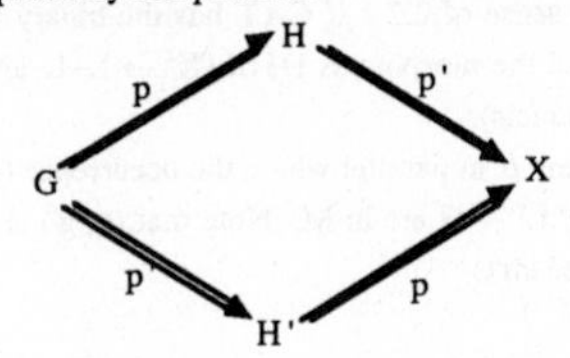

Remark

The proof of this result is given in 5.2 and requires the HLR1-conditions stated in 5.1.

4.3 Theorem (Church-Rosser-Property II)

Given sequentially independent direct derivations $G \Rightarrow H \Rightarrow X$ via (p, p') there exists also a sequentially independent sequence $G \Rightarrow H' \Rightarrow X$ via (p', p) using the same objects G and X and the same productions in opposite order. Moreover, $G \Rightarrow H$ via p and $G \Rightarrow H'$ via p' become parallel independent.

Remark

The following proof is based on 4.2 and hence on the HLR1-conditions stated in 5.1.

Proof

Let $G \Rightarrow H \Rightarrow X$ via (p, p') be sequentially independent. By remark 2 in 4.1 this means that $H \Rightarrow G$ via p^{-1} and $H \Rightarrow X$ via p' are parallel independent. Using Church-Rosser-Property I (see 4.2) there is an object H' and direct derivations $G \Rightarrow H'$ via p' and $X \Rightarrow H'$ via p^{-1} such that the sequence $H \Rightarrow G \Rightarrow H'$ via (p^{-1}, p') is sequentially independent. But this means that $G \Rightarrow H$ via p and $G \Rightarrow H'$ via p' are parallel independent.

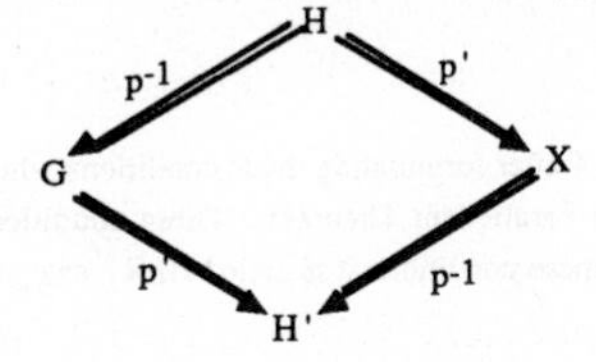

If the category **CAT** (see 2.1) has binary coproducts denoted by + (see appendix) we are able to formulate parallel productions and derivations (see 4.4) leading to the Parallelism Theorem in 4.5.

4.4 Definition (Parallel Productions and Derivations)

1. Given productions $p = (L \leftarrow K \to R)$ and $p' = (L' \leftarrow K' \to R')$ the production
   $$p + p' = (L{+}L' \leftarrow K{+}K' \to R{+}R')$$
   is called <u>parallel production</u> of p and p'.
2. A direct derivation $G \Rightarrow X$ via p + p' is called parallel derivation:

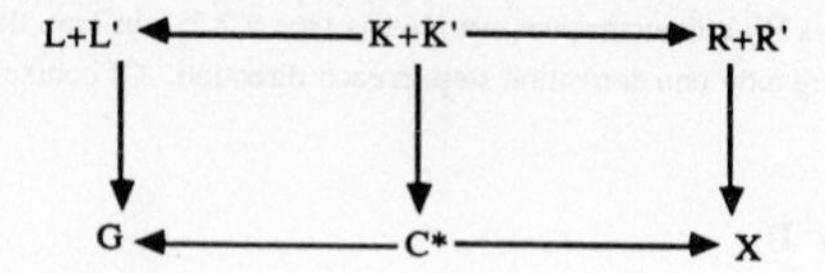

Remarks and Interpretation

1. The parallel production p + p' is a production in the sense of 2.2.1 if **CAT** has the binary coproducts L+L' of L and L', K+K' of K and K', and R+R' of R and R', and the morphisms $l+l':K+K' \to L+L'$ and $r+r':K+K' \to R+R'$ are in the class M (i.e. M is closed under binary coproducts).
2. The idea of a parallel production p + p' is to apply p and p' in parallel where the occurrence $(g,g'):L+L' \to G$ needs not to be in M (e.g. injective) even if $g:L \to G$ and $g':L' \to G$ are in M. Note that (g, g') is the unique morphism out of the coproduct L+L' induced by g and g' (see appendix).

4.5 Theorem (Parallelism Theorem)

1. <u>Synthesis</u>: Given a sequentially independent derivation sequence $G \Rightarrow H \Rightarrow X$ via (p, p') there is a synthesis construction leading to a parallel derivation $G \Rightarrow X$ via p+p'.
2. <u>Analysis:</u> Given a parallel derivation $G \Rightarrow X$ via p+p' there is a analysis construction leading to two sequentially independent derivation sequences $G \Rightarrow H \Rightarrow X$ via (p, p') and $G \Rightarrow H' \Rightarrow X$ via (p', p).
3. <u>Bijective Correspondence:</u> The constructions "synthesis" and "analysis" are inverse to each other in the following sense: Given p, p', and p+p' there is a bijective correspondence (up to isomorphism) between sequentially independent derivation sequences $G \Rightarrow H \Rightarrow X$ via (p, p') and parallel derivations $G \Rightarrow X$ via p+p'.

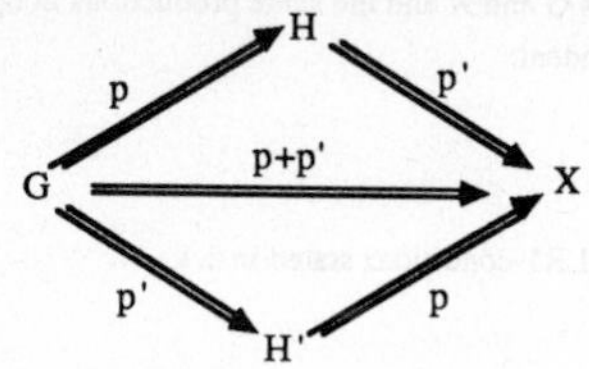

Remark

The proof of this result is given in 5.3 and is based on the HLR1-conditions stated in 5.1.

5. HLR1-CATEGORIES

In this section we present the remaining proofs for section 4 after formulating those conditions which are required in the proofs of the Church-Rosser-Properties I and II and the Parallelism Theorem. These conditions are called HLR1-conditions. If our basic category **CAT** (see 2.1) satisfies these conditions it is called HLR1-category. Finally we study the examples of section 3 concerning the HLR1-properties.
In addition to the general assumptions 2.1 we require in this section:

5.1 Definition (HLR1-Conditions and -Categories)

The following conditions 1 - 4 are called HLR1-Conditions and each category **CAT** with designated class M satisfying these conditions is called <u>HLR1-category:</u>

1. <u>Existence of Pushouts and Pullbacks for M-Morphisms</u>
   (a) For objects A, B, C and morphisms $A \to B$, $A \to C$ where at least one is in M there exists a pushout diagram (1):

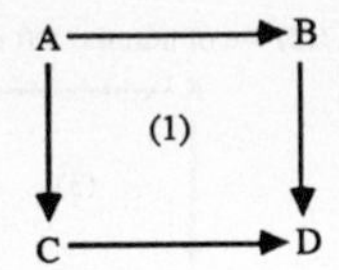

(b) For objects B, C, D and morphisms B $\to$ D, C $\to$ D which are both in M there exists a pullback diagram (1).

2. Inheritance of M-Morphisms under Pushouts and Pullbacks
   (a) For each pushout diagram (1) as above A $\to$ B in M implies C $\to$ D in M.
   (b) For each pullback diagram (1) with B $\to$ D and C $\to$ D in M as above also the morphisms A $\to$ B and A $\to$ C are in M.
3. Triple-Pushout-Condition
   Given the following diagram such that A $\to$ B, C $\to$ D, D' $\to$ E, A' $\to$ B', C $\to$ D', and D $\to$ E are in M then we have:
   (a) If (1) + (2) and (1') + (2) are pushouts and (2) pullback then (1), (1'), and (2) are pushouts.
   (b) If (2) is a pushout then it is also a pullback

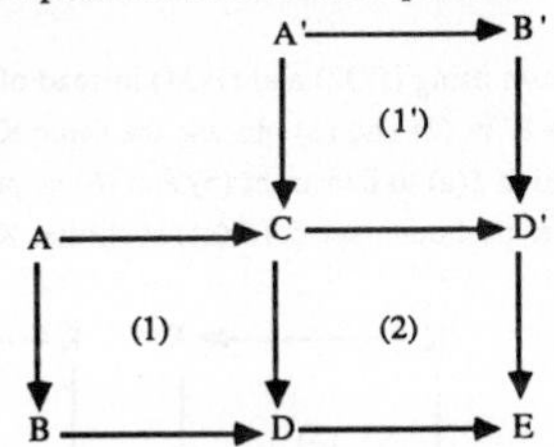

4. Existence of Binary Coproducts Compatible with M
   (a) For each pair of objects, A, B there is a coproduct A+B with universal morphisms A $\to$ A+B and B $\to$ A+B.
   (b) For each pair of morphisms f:A $\to$ A' and g:B $\to$ B' in M also the coproduct morphism f+g:A+B $\to$ A'+B' is in M.

Remarks and Interpretation

1. The HLR1-conditions 1, 2 and 3(a) will be used in the proof of the Church-Rosser-Theorem I (see 5.2) and conditions 3(b) and (4) in the proof of the Parallelism Theorem (see 5.3).
2. It is easy to check that all the HLR1-conditions are satisfied in the category **SETS** of sets with M being the class of all injective morphisms. For the examples of section 3 they will be analysed in theorem 5.4 below. However, none of these conditions is true for all categories **CAT** and classes M.
3. Conditions (a) and (b) of the triple pushout condition are equivalent to the following conditions (a') resp. (b'), provided that we have inheritance of M-morphisms under pushouts (condition 2):
   (a') If (1) + (2) is pushout and (2) pullback, then (1) is pushout.
   (b') A pushout of a pair of M-morphisms is a pullback.

5.2 Proof of Church-Rosser-Property I

Theorem 4.2 can be proved in any HLR1-category (see 5.1) as follows:

Given parallel independent direct derivations G $\Rightarrow$ H via p and G $\Rightarrow$ H' via p' we have the following pushouts where all the horizontal morphisms are in M by HLR1-condition 2(a):

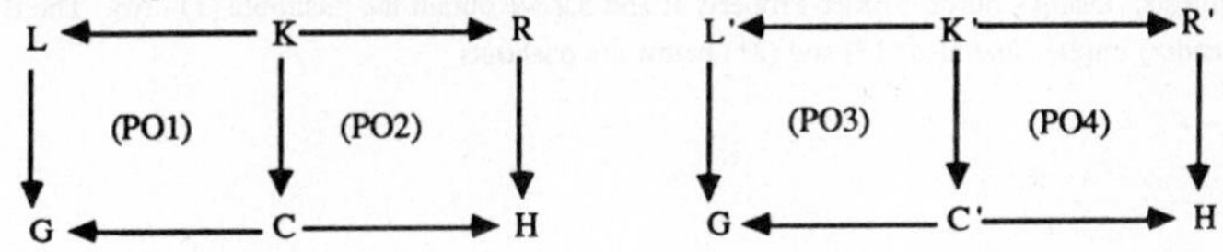

Combining (PO1) and (PO3) and using parallel independence (see 4.1.2) we obtain the left of the following diagrams:

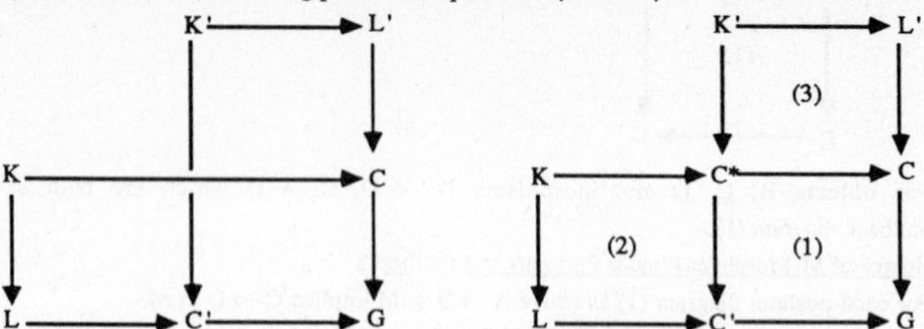

The right diagram above is obtained from the left one by constructing the pullback (1) of $C \to G$ and $C' \to G$ which are both in M (see above) such that the pullback exists (see HLR1-condition 1(b)). Moreover $C^* \to C$ and $C^* \to C'$ are in M by HLR1-condition 2(b) and the universal property of pullbacks implies that there are unique morphisms $K \to C^*$ and $K' \to C^*$ such that (2) and (3) commute and $K \to C = K \to C^* \to C$, $K' \to C' = K' \to C^* \to C'$. Now the pushouts (PO1) and (PO3) are decomposed by the diagrams (1) and (2) resp. (1) and (3) where all the morphisms required in the Triple-Pushout-Condition (see 5.1.3) are in M. This implies by 5.1.3a that (1), (2), and (3) are pushouts separately.

Next we will construct a similar triple-pushout-diagram using (PO2) and (PO4) instead of (PO1) and (PO3). That means we replace $K \to L$ by $K \to R$ and $K' \to L'$ by $K' \to R'$ in (2) and (3) but use the same $K \to C^*$ and $K' \to C^*$. Since $K \to R$ and $K' \to R'$ are in M we can use HLR1-condition 1(a) to construct (5) and (6) as pushouts leading to $C^* \to D$ and $C^* \to D'$ in M by inheritance of M-morphisms under pushouts (see 5.1.2(a)) such that X can be constructed as pushout in (4) using again 5.1.1(a).

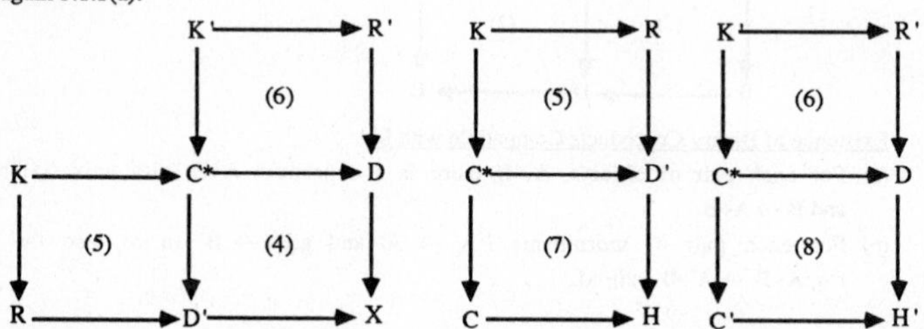

Then we observe that the pushouts (PO2) and (PO4) can be decomposed by pushouts (5) resp. (6) and the decomposition of $K \to C$ and $K' \to C'$ shown above. The universal pushout property of (5) (resp. (6)) implies that there is $D' \to H$ (resp. $D \to H'$) such that $R \to H = R \to D' \to H$ and diagram (7) above (resp. $R' \to H' = R' \to D \to H'$ and diagram (8)) commutes. But also (7) and (8) become pushouts by the Decomposition Lemma for pushouts (see appendix). Now we have all pushout pieces to construct the direct derivations

$H \Rightarrow X$ via p' using (3) + (7) and (6) + (4), and

$H' \Rightarrow X$ via p using (2) + (8) and (5) + (4)

where all the combined diagrams are pushouts by the Composition Lemma for pushouts (see appendix).

Finally note that there are morphisms $R \to D'$ and $L' \to C$ such that $R \to H = R \to D' \to H$ and $L' \to H = L' \to C \to H$ which implies that the derivation $G \Rightarrow H \Rightarrow X$ via (p, p') - and for similar reasons also $G \Rightarrow H' \Rightarrow X$ via (p', p) - becomes sequentially independent.

□

5.3 Proof of the Parallelism Theorem

Theorem 4.5 can be proved in any HLR1-category (see 5.1) as follows:

1. Synthesis: Using Church-Rosser-Property II and 5.2 we obtain the pushouts (1) - (6). The Butterfly-Lemma (see appendix) implies that also (1*) and (2*) below are pushouts

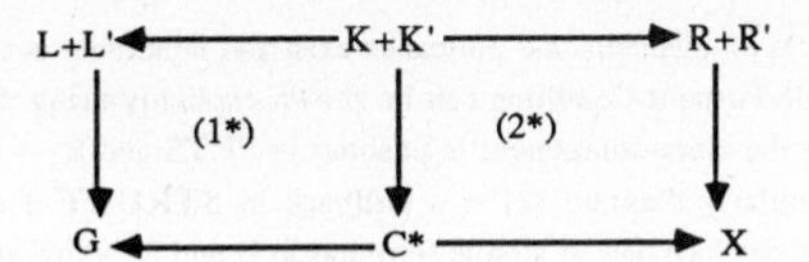

Hence we have a parallel derivation p+p':G ⇒ X where p+p' is a production by HLR1-property 4.

2. Analysis: Given a parallel derivation p+p':G ⇒ X by pushouts (1*) and (2*) above the Butterfly-Lemma leads to pushouts (1) - (6) as given in 5.3. Now H and H' can be constructed as pushouts in (7) and (8) of 5.2 using HLR1-condition (1(a) because C* → C and C* → D are in M by inheritance of M under pushouts (3) and (6). Using the last part of the proof in 5.2 we obtain sequentially independent derivation sequences G ⇒ H ⇒ X via (p, p') and G ⇒ H' ⇒ X via (p', p).

3. Bijective Correspondence: All pushout and pullback constructions are unique up to isomorphism (see appendix). Using first the construction "synthesis" we obtain pushouts (1) - (8), (1*), and (2*). Applying "analysis" to (1*) and (2*) we reconstruct pushouts (1) - (8) and hence (PO1) -

(PO4). Vice versa applying first "analysis" to (1*) and (2*) we obtain pushouts (1) - (8) and hence (PO1) - (PO4). Applying now "synthesis" to (PO1) - (PO4) we reconstruct pushouts (1) - (8), (1*) and (2*) because (1) was already pullback after "analysis" using Triple-Pushout-Condition 3(b) (see 5.3).

5.4 Theorem (HLR1-Categories)

The following categories **CAT** with morphism-classes M are HLR1-categories (see 5.1):

1. The category **SETS** of sets with M = class of all injective functions.
2. The category **GRAPHS** of graphs with M = class of all injective graph morphisms (see 3.1.1).
3. The category **CPOGRAPHS** of graphs with completely partially ordered color alphabets with M = class of all injective color preserving graph morphisms (see remark of 3.2.1).
4. The category **STRUCT** of structures with M = class of all injective structure morphisms (see 3.3.1).
5. The category **SPEC** of specifications with M = class of all injective specification morphisms (see 3.4.1).

In the following cases the HLR1-properties are not satisfied:

6. The category **TOP** of topological spaces with M = class of all injective continuous functions.
7. The category **POGRAPHS** of graphs with preordered color alphabets and the category **CPOGRAPHS** (see 3.) both with M = class of all injective graph morphisms (see 3.2.1).

Proof

1. The category **SETS** has pushouts, pullbacks and coproducts which preserve injective functions. This implies HLR1-conditions 1, 2, and 4. The Triple-Pushout-Condition is satisfied because each pushout with injective functions is also a pullback. Moreover the pullback (2) in 5.1.3 is also a pushout because D → E and D' → E are jointly surjective which follows from pushout properties of (1) + (2) or (1') + (2). Moreover (1) and (1') can be shown to be pushouts using injectivity of C → D and C → D' and the pullback property of (2).
2. In the category **GRAPHS** of graphs a commutative diagram is a pushout (resp. pullback) if and only if the arc- and the node-components are pushouts (resp. pullbacks) in **SETS**. This means that the HLR1-properties for **GRAPHS** follow from those for **SETS** (see example 1 above).
3. In **CPOGRAPHS** pushouts (resp. pullbacks) exist using greatest lower bounds (resp. least upper bounds) for the label constructions (see 3.2.2), especially if one or both morphisms are injective and color preserving. Moreover injective color preserving graph morphisms are closed under pushout, pullbacks and coproducts. This implies HLR1-conditions 1, 2, and 4(which are also valid for M = class of all injective graph morphisms). In the Triple-Pushout-Condition (1), (1'), and (2) become pushouts because all graph morphisms in (2) are color preserving. Vice versa the pushout (2) is also a pullback for the same reason. For counterexample with non-color preserving morphisms in M see example 7 below.

4. In the category **STRUCT** of structures all pushouts and pullbacks exist and injectivity is closed under pushouts, pullbacks and coproducts. The Triple-Pushout-Condition can be shown explicitly using that e.g. diagram (1) is pushout in **STRUCT** if and only if the atom-component is pushout in **SETS** and $B \to D$, $C \to D$ are jointly surjective on atomic formulas. Similarly diagram (2) is a pullback in **STRUCT** if and only if the atom-component is a pullback in **SETS** and for each pair of atomic formulas in D and D' which are identified in E there is a common preimage in C.
5. The verification of HLR1-conditions for **SPEC** is completely similar to that of **STRUCT** above where atomic formulas have to be replaced by equations.
6. In the category **TOP** the Triple-Pushout-Condition is violated for M = class of all injective continuous functions. Let A and B be the topological spaces with underlying set {x, y} where in A the sets {x} and {y} are open, while in B only {x} is open. Then the following diagram with identity functions is a pushout but not a pullback because the pullback object is B.

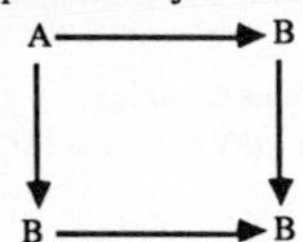

This violates HLR1-condition 3(b).
However, it is possible to restrict the class M such that the Triple-Pushout-Condition is satisfied.
7. In the category **POGRAPHS** we don't have pushouts (resp. pullbacks) unless we have greatest lower bounds (resp. least upper bounds). Even if both of them exist the Triple-Pushout-Condition is not satisfied for M = class of all injective graph morphisms:
For colors c1, c2 with $c1 \neq c2$ and $c1 > c2$ the following diagram is a pushout with injective morphisms but not a pullback, because $c1 \neq lub(c2, c2)$.

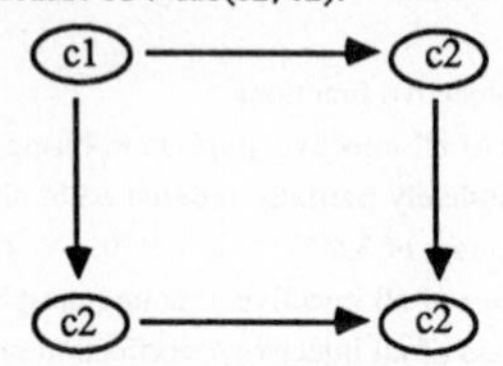

This violates HLR1-condition 3(b). But also 3(a) is violated as the following diagrams (1), (1') and (2) demonstrate using the partial order given by the left part of the figure:

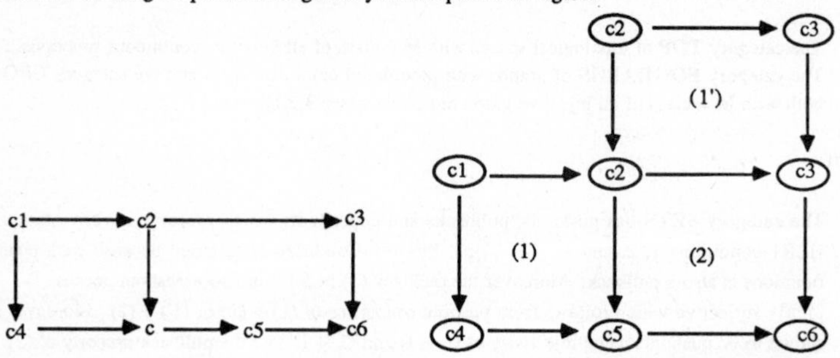

It remains open, whether there is a class M not restricted to color preserving graph morphisms such that the Triple-Pushout-Lemma is satisfied.

6. CONCLUSION

In this paper we have introduced high-level-replacement systems which are based on a category **CAT** together with a distinguished class M of morphisms. The main idea is to replace in the algebraic approach to graph grammars, using double-pushouts in the category **GRAPHS** graphs, graph morphisms, and injective graph morphisms by objects in **CAT**, morphisms in **CAT**, and morphisms in M respectivley. This allows already to formulate productions, derivations, HLR-systems and languages. In order to generalize the Church-Rosser-Theorem and the Parallelism Theorem to HLR-systems we need certain additional properties, called HLR1-conditions, which are shown to be satisfied for the categories **SET, GRAPHS, CPOGRAPHS, STRUCT,** and **SPEC** with suitable classes M of injective morphisms.

In an extended version of this paper (see [EHKP 90] and subsequent papers we will give further examples of HLR-categories including hypergraphs, place-transition nets, and other kinds of transition systems and nets. Moreover, we will formulate HLR2-conditions which are sufficient to prove the Concurrency Theorem, and HLRn-Conditions for $n \geq 3$ to generalize other important results of the algebraic theory of graph grammars to HLR-systems.

7. REFERENCES

[AM 75] Arbib, M.A.; Manes, E.G.: Arrows, Structures and Functors, Academic Press, New York - San Francisco - London, 1975

[CMREL 90] Corradini, U. Montanari, F. Rossi, H. Ehrig, M. Löwe: Graph Grammars and Logic Programming, in [EKR 91]

[Ehr 79] Ehrig, H.: Introduction to the Algebraic Theory of Graph Grammars, LNCS 73 (1979), p. 1-69

[EHR 86] Ehrig, H.; Habel, A.; Rosen, B.K.: Concurrent Transformations of Relational Structures, Fundamenta Informaticae IX (1986), 13-50

[EHKP 90] Ehrig, H.; Habel, A.; Kreowski, H.-J.; Parisi-Presicce, F.: Parallelism and Concurrency in High Level Replacement Systems, Techn. Report, TU Berlin, FB 20, No. 90/35

[EK 75] Ehrig, H.; Kreowski, H.-J.: Categorical Theory of Graph Grammars, Techn. Report TU Berlin, FB 20, Bericht Nr. 75-08 (1975)

[EKR 91] Ehrig, H.; Kreowski, H.-J.; Rozenberg, G.: Graph Grammars and Their Applications to Computer Science, to appear in LNCS 1991

[EKMRW 81] Ehrig, H.; Kreowski, H.-J.; Maggiolo-Schettini, A.; Rosen, B.; Winkowski, J.: Transformation of Structures: An Algebraic Approach, Math. Syst. Theory 14 (1981), p. 305-334

[EM 85] Ehrig, H.; Mahr, B.: Fundamentals of Algebraic Specification 1. Equations and Initial Semantics. EATCS Monographs on Theoretical Computer Science, Vol. 6, Springer (1985)

[ER 80] Ehrig, H.; Rosen, B.K.: Parallelism and Concurrency of Graph Manipulations. Theor. Comp. Sci. 11 (1980), 247-275

[EPS 73] Ehrig, H.; Pfender, M.; Schneider, H.J.: Graph Grammars: An Algebraic Approach, Proc. IEEE Conf. SWAT'73, Iowa City (1973), p. 167-180

[Ha 89] Habel, A.: Hyperedge Replacement: Grammars and Languages; Ph.D. Thesis, Univ. Bremen, 1989

[Hu 89] Hummert, U.: Algebraische Theorie von High-Level-Netzen, Ph.D.Thesis, TU Berlin, 1989

[HS 73] Herrlich, H.; Strecker, G.E.: Category Theory. Allyn and Bacon, Boston 1973

[Kr 77] Kreowski, H.-J.: Manipulation von Graphmanipulationen, Ph.D. Thesis, TU Berlin, 1977

[ML 72] MacLane, S.: Categories for the working mathematician. Springer New York - Heidelberg - Berlin 1972

[MM 88] Meseguer, P.; Montanari, U.;: Petri Nets are Monoids: A New Algebraic Foundation for Neth Theory; Proc. Logics in Comp. Sci., 1988

[PE 90] Parisi-Presicce, F.; Ehrig, H.: Algebraic Specification Grammars, in [EKR 91]

[PEM 87] Parisi-Presicce, F.; Ehrig, H.; Montanari, U.: Graph Rewriting with Unification and Composition, Proc. 3rd Int. Workshop on Graph Grammars, Springer LNCS 291 (1987), p. 496-511

[PP 89] Parisi-Presicce, F.: Modular System Design Applying Graph Grammar Techniques, Proc. 16th ICALP, Springer LNCS 372 (1989), p. 621-636

[PP 90] Parisi-Presicce, F.: A Rule Based Approach to Modular System Design, Proc. 12th Inf. Conf. Software Engineering, March 1990

[SE 76] Schneider, H. J.; Ehrig, H.: Grammars on Partial Graphs, Acta Informatica 6 (1976), p. 297-316

APPENDIX: Basic Concepts From Category Theory

In this appendix, the basic concepts from category theory are recalled as far as they are needed in our paper. All notions and results in sections 2 - 8 are defined w.r.t. an arbitrary category **CAT** and examples are given in the category **SETS** of sets as defined in section 1. (See, e.g. [AM 75, HS 73, ML 72] for more details and for proofs.)

1. CATEGORIES

A category comprises a class of objects like sets, topological spaces, monoids, groups, etc. and a set of morphism for each two objects like functions, partial functions or binary relation for sets, continuous functions for topological spaces, homomorphisms for algebraic structures, etc. Moreover, an associative composition is provided for morphisms like functional composition, and each object corresponds to an identity morphism like the identity function. More formally, we get the following notion:

Definition: A category **CAT** consists of a class **obj(CAT)** of objects together with a set **CAT(A, B)** of morphisms for each pair A, B of objects subject to the following conditions:

(i) For any three objects A, B, C (that may not be distinct) there is a mapping

$$\circ : \mathbf{CAT}(A, B) \times \mathbf{CAT}(B, C) \to \mathbf{CAT}(A, C)$$

called composition, where the image of $f \in \mathbf{CAT}(A, B)$ and $g \in \mathbf{CAT}(B, C)$ is denoted by $g \circ f$.

(ii) The composition is associative, i.e for any four objects A, B, C, D and any morphisms $f \in \mathbf{CAT}(A, B)$, $g \in \mathbf{CAT}(B, C)$ and $h \in \mathbf{CAT}(C, D)$ we have $h \circ (g \circ f) = (h \circ g) \circ f$.

(iii) For any object A, there is a morphism $1_A \in \mathbf{CAT}(A, A)$, called the identity of A.

(iv) The identity is neutral, i.e. for any object B and any morphisms $f \in \mathbf{CAT}(A, B)$ and $g \in \mathbf{CAT}(B, A)$ we have $(f \circ 1_A) = f$ and $(1_A \circ g) = g$.

Remarks: We may write $f: A \to B$ or $A \to^f B$ for $f \in \mathbf{CAT}(A, B)$. A is called the domain of f and B the codomain. Moreover, we may write $A \to^f B \to^g C$ for $A \to^{g \circ f} B$. Note that the associativity law guarantees that longer arrow chains like $A \to^f B \to^g C \to^h D$ can be interpreted without ambiguity as $h \circ g \circ f = h \circ (g \circ f) = (h \circ g) \circ f$. Sometimes, morphisms are called arrows, and we may write $A \to B$ for $A \to^f B$ if the name of the morphism does not matter.

Examples: Sets as objects and functions as morphisms together with functional composition and the identity functions form the category **SETS**. Further examples of categories can be found in section 3.

2. DIAGRAMS

Various notions and constructions in categories are based on certain patterns of objects and morphisms, called diagrams.

Definition: A diagram (in a category **CAT**) is a directed graph the nodes of which are labelled by objects and the arcs by morphisms subject to the following condition:
If e is an arc with source v and target v' and label $f \in \mathbf{CAT}(A, B)$, then v is labelled by A and v' by B.

Remarks: Note that each path from a node labelled by an object A to a node labelled by an object B represents a morphism in **CAT(A, B)** (obtained by composition of the label morphisms along the path). In drawings of diagrams, nodes are given by the location of their labels while arcs are drawn as usually as arrows with their labels next to them. We skip the arc labels if their notation is not important:

Definition: A diagram is commutative if any two paths with the same start node and the same end node represent the same morphism.

Examples: Each of the following graphs is a diagram:

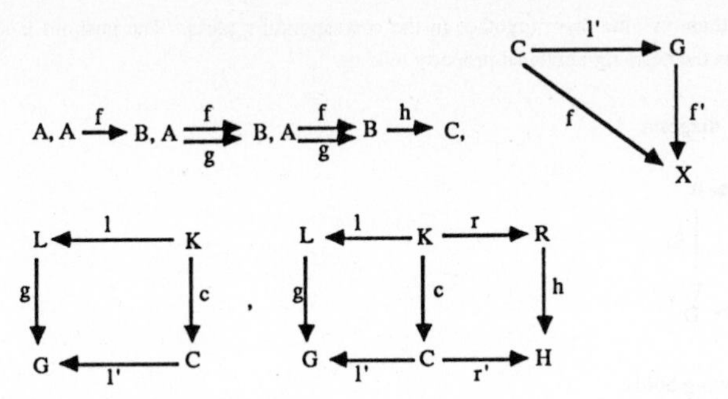

Commutativity means for the third diagram f = g, for the fifth one $f' \circ l' = f$, for the sixth one $g \circ l = l' \circ c$ and for the seventh one: $g \circ l = l' \circ c$ and $h \circ r = r' \circ c$.

3. ISOMORPHISMS

Isomorphisms are morphisms which have unique inverse morphisms. They pick up those objects that are essentially equal to each other. The notion of an isomorphism is extremely important because many categorical constructions are only unique up to isomorphism.

Definition: If the two diagrams

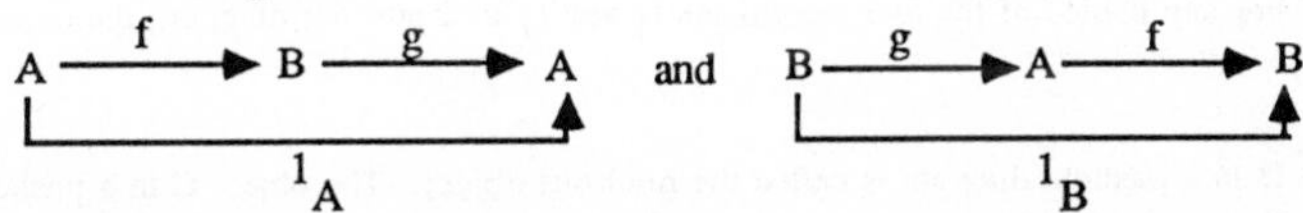

are commutative, f is called an isomorphism and the objects A, B are isomorphic, denoted by $A \cong B$.

Remarks: If f is an isomorphism, the corresponding morphism g is unique and may be denoted by f^{-1}, called inverse morphism of f. Clearly, f^{-1} is an isomorphism too because of the symmetry of the situation.
Examples: In **SETS**, a function is an isomorphism if and only if it is bijective.

4. COPRODUCTS

The coproduct is a construction that joins two objects (or any number of objects) in such a way that one keeps full access to the components and can tell them apart.

Definition: Given two objects A and B, an object C together with two morphisms $in_A: A \to C$ and $in_B: B \to C$ is a (binary) coproduct of A and B if the following condition holds:
For any object X and any two morphisms $f_A: A \to X$ and $f_B: B \to X$, there is a unique morphism $f: C \to X$ with $f \circ in_A = f_A$ and $f \circ in_B = f_B$.

Uniqueness Lemma: The coproduct object is unique up to isomorphism.

Remarks: The coproduct object of two objects A and B is denoted by A + B. The unique morphism out of the coproduct is denoted by (f_A, f_B). If X = A' + B' with corresponding injection $in_{A'}$, and $in_{B'}$, then morphisms $g_A: A \to A'$ and $g_B: B \to B'$ induce a morphism $g_A + g_B: A + B \to A' + B'$ by $g_A + g_B = (in_{A'} \circ g_A, in_{B'} \circ g_B)$.

Example: In **SETS**, the disjoint union of two sets A, B together with the obvious injections form a coproduct of A and B.

5. PUSHOUTS

Given two objects B and C with corresponding parts specified by two morphisms $f_1: A \to B$ and $f_2: A \to C$ for some object A, one may like to join them or glue them together in the corresponding parts. The pushout is such a gluing construction - the best possible as the defining universal property tells us.

Definition: The commutative diagram

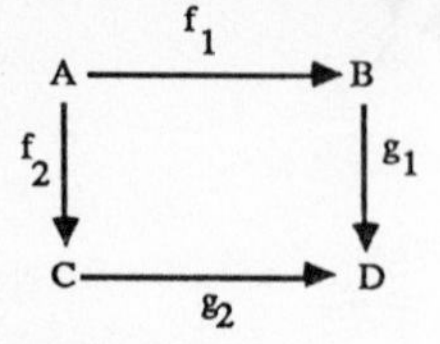

is a pushout diagram if the following holds:
For any commutative diagram

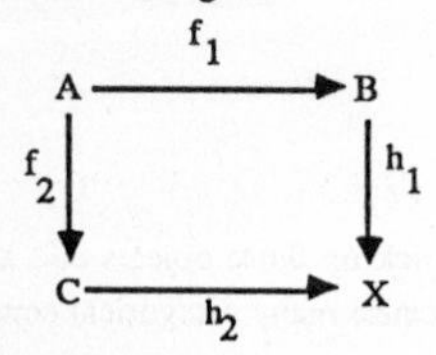

there is a unique morphism $k:D \to X$ with $k \circ g_1 = h_1$ and $k \circ g_2 = h_2$.

Uniqueness Lemma: For any choice of the two morphisms f_1 and f_2 in a pushout diagram, the object D is unique up to isomorphism.

Remarks: The object D in a pushout diagram is called the **pushout** object. The object C in a pushout diagram is called a pushout complement (of the morphisms f_1 and g_1). Note that f_1 and g_1 do not establish a unique pushout complement in general.

Example: Given two functions $f_1A \to B$ and $f_2:A \to C$ for some sets A, B, C the pushout object D can be constructed as follows:

(i) Take the disjoint union B + C with the injections $in_1:B \to B + C$ and $in_2:C \to B + C$.

(ii) Factor it through the equivalence relation $\equiv$ induced by the relation ~ on B + C defined by $in_1(f_1(x)) \sim in_2(f_2(x))$ for all $x \in A$.

Completing the pushout diagram, we define $g_1:B \to D$ and $g_2:C \to D$ by $g_1(y) = [in_1(y)]$ and $g_2(z) = [in_2(z)]$ for all $y \in B$ and $z \in C$ where $[x]$ denotes the equivalence class of $x \in B + C$.
If B and C are subsets of a set S with $A = B \cap C$ and f_i are the inclusions of A into B and C resp., then the pushout object may be chosen as $B \cup C$.

6. PUTTING PUSHOUTS TOGETHER

If various pushout diagrams are put together in a certain way, they behave quite nice yielding new pushout diagrams. The proofs of the paper are based on these "inference rules".

Composition- and Decomposition Lemma: Let

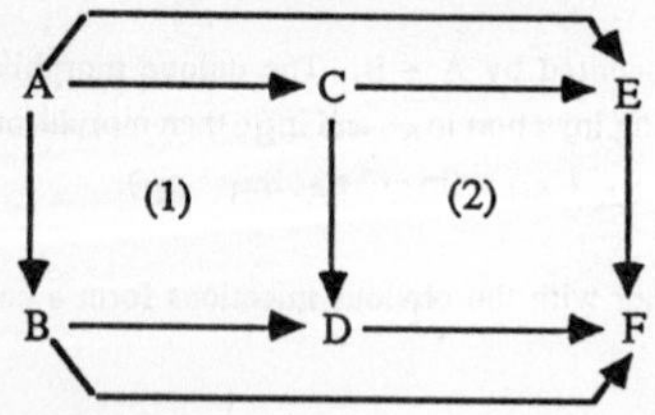

be a commutative diagram.

If diagrams (1) and (2) are pushout diagrams, then the composite diagram (1) + (2) is a pushout diagram.
If (1) and (2) is a pushout diagram and (1) is a pushout diagram then also (2) is a pushout diagram.

Butterfly-Lemma: Consider the commutative diagrams

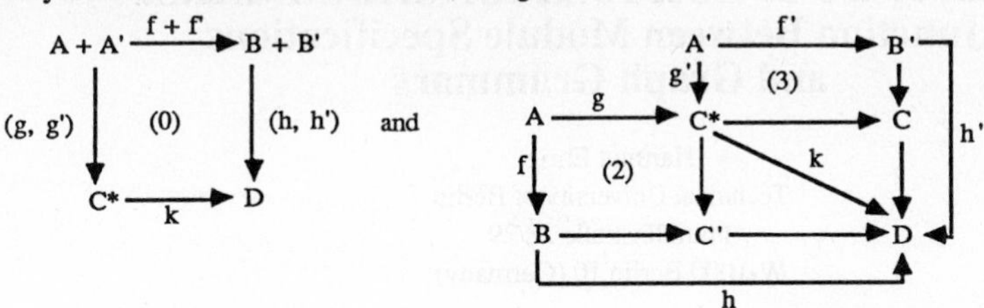

where (1) refers to the subdiagram

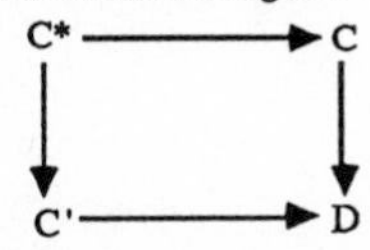

Then the diagram (0) is a pushout diagram if and only if the diagrams (1), (2) and (3) can be chosen as pushout diagrams.

7. PULLBACKS

If two objects B and C are "mapped" into a third object A by morphisms f_1:B → A and f_2:C → A. Then, intuitively, their images in A may overlap. The pullback construction describes this idea in an arbitrary category.

Definition: If all arcs in the definition of a pushout diagram (see section 5) are reversed, we obtain a pullback diagram.

Remarks: All properties of pushout diagrams are transferred in properties of pullback diagrams, if all arrows are reversed. Particularly, the pullback object is unique up to isomorphism for given morphisms f_1:B → A and f_2:C → A.

Example: Given two functions f_1:B → A and f_2:C → A, the pullback object in **SETS** can be chosen as the set D = $\{(b, c) \in B \times C \mid f_1(b) = f_2(c)\}$. The pullback diagram is completed by the projections g_1:D → B and g_2:D → C given by $g_1(b, c) = b$ and $g_2(b, c) = c$ for all $(b, c) \in D$.
If f_1 and f_2 are both inclusions, D can be chosen as the intersection of B and C in A.

8. INITIAL OBJECTS

If an object of a category CAT can be "mapped" by a unique morphism into any other object, it is called initial.

Definition: An object I is initial if for any object A there is a unique morphism f_A:I → A.

Uniqueness Lemma: The initial object of a category is unique up to isomorphism.

Example: In **SETS**, the empty set Ø is initial because the empty function is the only function into a set A.

ALGEBRAIC SPECIFICATION GRAMMARS: A Junction Between Module Specifications and Graph Grammars

Hartmut Ehrig
Technical University of Berlin
Franklinstraße 28/29
W-1000 Berlin 10 (Germany)

Francesco Parisi-Presicce
Università degli Studi L'Aquila
Via Vetoio
I-67100 L'Aquila (Italia)

Abstract: Algebraic specification grammars have recently been introduced implicitly by the second author as a new kind of graph grammars in order to generate algebraic specifications using productions and derivations. In fact, in the well-known algebraic approach to graph grammars, also known as "Berlin-approach", we mainly have to replace the category of graphs by the category of algebraic specifications to obtain the basic definitions, constructions and results for this new kind of grammars. Since a production in an algebraic specification grammars corresponds exactly to an interface of an algebraic module specification for modular software systems this new kind of grammars can be used for modular system design. For this purpose we give an overview how notions in the theory of grammars and that of module specifications correspond to each other and discuss how both theories can benefit from each other. Concerning full technical detail and proofs we refer to other published or forthcoming papers.

Keywords: algebraic specification, graph grammars, algebraic module specifications, algebraic specification grammars, match theorem, rule based design, modular systems

CONTENTS

1. Introduction

The algebraic approach to graph grammars introduced in [EPS 73] and [Ehr 79] is based on the notion of directed graphs with colored arcs and nodes and of morphisms between graphs which preserve both the structure of the graphs and their coloring. The central idea of this approach is to view the productions of a grammar to generate graphs as a triple of colored graphs $p = (L \leftarrow K \rightarrow R)$ related by graph morphisms, and to define a direct derivation $p{:}G \Rightarrow H$ as a pair of pushout diagrams

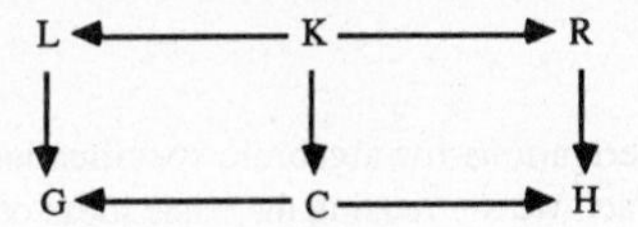

This approach has been extended to structures [EKMRW 81] as pairs of sets of atoms and sets of formulas, where morphisms transform atoms and preserve formulas. The derivation is still a pair of pushouts in the category of structures. In both cases, K represents the part of the production not changed by the replacement of L by R and is used to "glue" the new part R to the context C to obtain H.

Quite independently, initiated by [Zi 74], the algebraic theory of data type specifications has been developing, from data type specifications having as semantics a unique (initial or final) algebra ([Gu 75], [GTW 76], [Wa 79]) to parameterized data type specifications with a functorial semantics ([TWW 78], [EM 85]).

Parameterized specifications and their structuring mechanisms of parameter passing have been extended, to better model the specification of large software systems, to algebraic module specifications ([BEPP 87], [EW 85]) which consist of four specifications (IMP, PAR, EXP, BOD) where IMP and EXP represent the import (what is needed) and the export (what is provided) interfaces, PAR a parameter part shared by the interfaces, and BOD is the specification of the realization (implementation) of the export interface using the import interface. Interconnection mechanisms, viewed as operations on module specifications, have been defined, which preserve the correctness and the semantics of the operands. The basic operations are composition, where the import of a module is matched with the export of another module, union, defined componentwise as pushouts, and actualization where the parameter part is replaced by an actual specification [EM 90]. The visible part of a module specification consists of the import and export interfaces and their shared parameter part IMP $\leftarrow$ PAR $\rightarrow$ EXP and the functorial semantics consists of a transformation from algebras (models) of the import to algebras of the export in such a way that the parameter part of the import algebra is left unchanged. This correspondence between the form of a production of graph grammars and that of the visible part of a module specification has led to the following problem addressed in [PP 89]:

> Given a library LIB of module specifications and two specifications SPEC0 and SPEC1, is it possible to design a modular system using only the modules in LIB and such that SPEC0 and SPEC1 are the overall import and export interfaces, respectively!

The solution proposed in [PP 89, PP 90] is to represent the modules in SET only by their interfaces, thus defining a set LIB of productions on specifications and to view the existence of such a modular system as the existence of a derivation from SPEC0 to SPEC1 which uses as productions those in LIB. The correspondence between the interfaces of a module specification and the production of a graph grammar allows us to use the results of graph grammar theory, adapting them to algebraic specifications. In Section 2, we introduce explicitly the notions of algebraic specification grammar and of derivation, and the gluing conditions and illustrate the equivalent of the Parallelism and Concurrency Theorems of Graph Grammars. In Section 3, we review the basic notions of module specifications emphasizing the match operation. In Section 4, we conclude by illustrating the correspondence between the results of Specification Grammars and the operations on module specifications, showing how a direct derivation can be translated into modular system design.

For proofs and technical details, we refer the reader to other published or forthcoming papers.

2. Algebraic Specification Grammars

In this section we introduce productions and derivations for algebraic specifications leading to the notion of algebraic specification grammars. In fact, we are reusing the basic ideas of graph grammars in [EPS 73] and [Ehr 79]. In the case of graphs direct derivations are defined by double-pushouts in the category **GRAPH** of graphs and in the case of algebraic specifications we use double-pushouts in the category **SPEC** of algebraic specifications. We discuss how notions of independence and R-relations of direct derivations in [Ehr 79] can be reformulated for algebraic specification grammars and state explicitly the corresponding parallelism and R-concurrency theorems.

2.1 Definition (Category **SPEC** of Algebraic Specifications)

1. An <u>algebraic specification</u> SPEC = (S, OP, E) consists of a set S of sorts, a set OP of operation symbols, and a set E of equations over the signature SIG = (S, OP) as defined in [EM 85].
2. A <u>morphism</u> f:SPEC1 → SPEC2 of <u>algebraic specifications</u> SPECi = (Si, OPi, Ei) for i = 1, 2 consists of a pair $f = (f_S, f_{OP})$ of functions f_S:S1 → S2 and f_{OP}:OP1 → OP2 such that
   a) for each operation symbol N1 ∈ OP1 with signature N1:s1...sn → s the corresponding operation symbol $f_{OP}(N1)$ ∈ OP2 has signature $f_{OP}(N1)$:$f_S(s1)...f_S(sn) \to f_S(s)$.
   b) For each equation e1 ∈ E1 the translated equation f#(e1) belongs to E2 (strict case) or can be derived from E2 (general case) where f# is the extension of f_{OP} to (pairs of) terms (see [EM 85] for more details).
3. The category **SPEC** of algebraic specifications has objects and morphisms as defined in part 1 and 2 above respectively. The composition of morphisms is defined componentwise in the category **Sets** of sets.

Remark

The category **SPEC** of algebraic specifications has pushouts (see [EM 85] or [EM 90]): Given algebraic specifications SPECi = (Si, OPi, Ei) for i = 0, 1, 2 and morphisms f1:SPEC0 → SPEC1, and f2:SPEC0 → SPEC2 the pushout

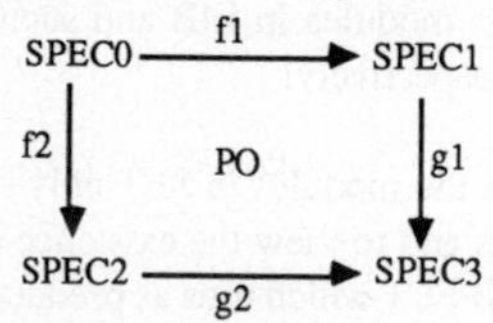

of f1 and f2 in the category **SPEC** is defined by

SPEC3 = (S3, OP3, E3)

where $S3 = S1 +_{S0} S2$ (pushout of $f1_S$ and $f2_S$ in **Sets**)

$OP3 = OP1 +_{OP0} OP2$ (pushout of $f1_{OP}$ and $f2_{OP}$ in **Sets**)

defining signature morphisms gj:(Sj, OPj) → (S3, OP3) for j = 1, 2, and

E3 = g1#(E1) ∪ g2#(E2)

leading to specification morphisms gj:SPECj → SPEC3 for j = 1,2. Intuitively SPEC3 is the result of gluing of SPEC1 and SPEC2 along SPEC0, i.e. along corresponding items f1(x) in SPEC1 and f2(x) in SPEC2 for each item x in SPEC0.

2.2 Definition (Production and Direct Derivation)

1. An algebraic specification production, short **SPEC**-production,

$$p = (LSPEC \xleftarrow{l} GSPEC \xrightarrow{r} RSPEC)$$

consists of algebraic specifications
- LSPEC (left hand side specification L(p) of p)
- GSPEC (gluing specification G(p) of p)
- RSPEC (right hand side specification R(p) of p)

and two injective specification morphisms
- l:GSPEC → LSPEC (left gluing morphism)
- r:GSPEC → RSPEC (right gluing morphism).

2. A direct derivation of algebraic specifications, short direct **SPEC**-derivation, for a **SPEC**-production p (as given in part 1) from specification SPEC1 to SPEC2, written

p:SPEC1 ⇒ SPEC2,

consists of a context specification CSPEC and two pushout diagrams (PO1) and (PO2) including p, SPEC1, SPEC2, and CSPEC in the following way:

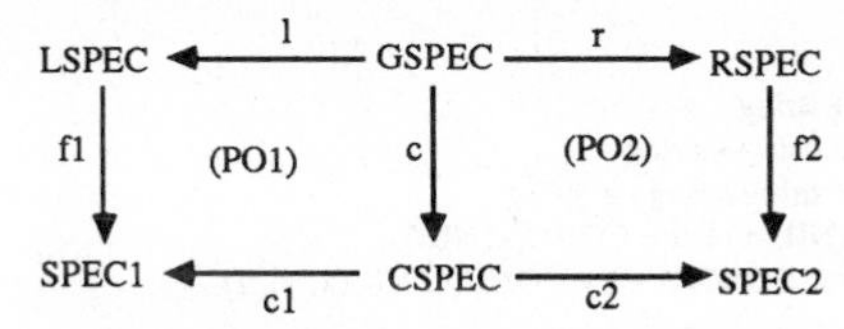

Remarks and Interpretation

1. In the direct derivation p:SPEC1 ⇒ SPEC2 given by the pushouts (PO1) and (PO2) SPEC1 (resp. SPEC2) is the gluing of the left hand side LSPEC (resp. right hand side RSPEC) of the production p with the context graph CSPEC along the gluing specification GSPEC. Given a production p we can just take a context specification CSPEC together with a specification morphism c:GSPEC → CSPEC to construct SPEC1 and SPEC2 as pushouts in (PO1) and (PO2) respectively leading to a direct derivation p:SPEC1 ⇒ SPEC2.

2. If we want to apply a production p to a specification SPEC1 we have to find an occurrence of LSPEC in SPEC1, i.e. a specification morphism f1:LSPEC → SPEC1. In a second step we have to construct the context specification CSPEC such that the gluing of LSPEC with CSPEC along GSPEC is equal to SPEC1, i.e. (PO1) becomes a pushout in **SPEC**. In fact, we need a gluing condition (see remark 3) in order to be able to construct such a context specification CSPEC. Intuitively CSPEC is SPEC1 minus the left hand side LSPEC but still including the gluing terms in GSPEC. In a second step we glue together the context specification CSPEC with the right hand side RSPEC of the production p leading to the pushout (PO2) including the derived specification SPEC2.

3. Similar to the graph case the gluing condition for specifications mentioned in remark 2 states that dangling and identification point must be gluing points. In addition we must have in the case of strict specification morphisms f1 and l that the set of equations

$$E_{CSPEC} = E_{SPEC1} - f1^{\#}(E_{LSPEC} - l^{\#}(E_{GSPEC}))$$

are equations over the signature (S_{CSPEC}, OP_{CSPEC}) where

$$S_{CSPEC} = S_{SPEC1} - f1_S(S_{LSPEC} - l_S(S_{GSPEC})), \text{ and}$$

$$OP_{CSPEC} = OP_{SPEC1} - f1_{OP}(OP_{LSPEC} - l_{OP}(S_{LSPEC})).$$

In fact, the condition that dangling and identification points must be gluing points is equivalent to the fact that (S_{CSPEC}, OP_{CSPEC}) as constructed above becomes a well-defined signature, the signature of the context specification CSPEC. If the gluing condition is satisfied the context specification CSPEC, also called pushout complement of SPEC1 w.r.t. LSPEC, GSPEC, f1 and l, is given by

(CSPEC = (S_{CSPEC}, OP_{CSPEC}, E_{CSPEC}).

In the case of general specification morphisms f1 and l the construction of E_{CSPEC} is slightly more difficult because we have to consider closed sets of equations w.r.t. equational deduction or derivability (see [EM 85]). In [PP 89] it is shown for the case of general specification morphisms that the gluing condition is necessary and sufficient for the existence of pushout-complements, and hence for the existence of the context specification CSPEC in step1 of remark 2.

2.3 Example

Define

```
GSPEC = data =   sort data
                 opns CONST: → data

LSPEC = GSPEC +      sort string
                     opns   NIL: → string
                            MAKE: data → string
                            CONC: string string → string
                     eqns   CONC(NIL, x) = x = CONC(x, NIL)
                            CONC(x, CONC(y, z)) = CONC(CONC(x, y), z)

RSPEC = GSPEC +sorts string, set
                     opns   CREATE: → set
                            INSERT: data set → set
                            NIL: → string
                            LADD: data string → string
                            RADD: string data → string
                     eqns   INSERT(d1, INSERT(d2, x)) = INSERT(d2, INSERT(d1, x))
```

with inclusion morphisms l:GSPEC → LSPEC and r:GSPEC → RSPEC

The production p = (LSPEC ← GSPEC → RSPEC) is applicable to the specification

```
SPEC1 =   sorts nat, string, bintree
          opns  ZERO: → nat
                SUCC: nat → nat
                NIL: → string
                MAKE: nat → string
                CONC: string string → string
                LEAF: nat → bintree
                LEFT: bintree nat → bintree
                RIGHT: nat bintree → bintree
                BOTH: bintree nat bintree → bintree
          eqns  CONC(NIL, x) = x = CONC(x, NIL)
                CONC(x, CONC(y, z)) = CONC(CONC(x, y), z)
```

via the occurrence morphism f1:LSPEC → SPEC1 with $f1_S$(data) = nat, $f1_{OP}$(CONST) = ZERO and the other obvious associations. This morphism satisfies the Gluing Condition and we obtain the context specification

```
CSPEC =  sorts nat, bintree
         opns  ZERO: → nat
               SUCC: nat → nat
               LEAF: nat → bintree
               LEFT: bintree nat → bintree
               RIGHT: nat bintree → bintree
               BOTH: bintree nat bintree → bintree
```

which can be "glued" to RSPEC via GSPEC to obtain the specification

```
SPEC2 =  sorts nat, bintree, set, string
         opns  ZERO: → nat
               SUCC: nat → nat
               LEAF: nat → bintree
               LEFT: bintree nat → bintree
               RIGHT: nat bintree → bintree
               BOTH: bintree nat bintree → bintree
               CREATE: → set
               INSERT: nat set → set
               NIL: → string
               LADD: nat string → string
               RADD: string nat → string
         eqns  INSERT(n, INSERT(m, s)) = INSERT(m, INSERT(n, s))
```

which is the result of applying the SPEC-production p to SPEC1.

2.4 Definition (Algebraic Specification Grammars)

Given a set PROD of **SPEC**-productions p in the sense of 2.2.1 we define:

1. A derivation of algebraic specifications, short **SPEC**-derivation, written

$$SPEC0 \overset{*}{\underset{PROD}{\Longrightarrow}} SPEC$$

is given by a sequence of $n \geq 0$ direct **SPEC**-derivations

$$SPEC0 \overset{p1}{\Longrightarrow} SPEC1 \overset{p2}{\Longrightarrow} \ldots \overset{pn}{\Longrightarrow} SPECn = SPEC$$

with $pi \in PROD$ for $i = 1,\ldots,n$.

2. Given in addition a distinguished start specification SPEC0 we obtain an algebraic specification grammar

$$ALG\text{-}SPEC\text{-}GRA = (SPEC0, PROD, \overset{*}{\underset{PROD}{\Longrightarrow}})$$

and the language L(ALG-SPEC-GRA) is given by

$$L(ALG\text{-}SPEC\text{-}GRA) = \{SPEC \,/\, SPEC0 \overset{*}{\underset{PROD}{\Longrightarrow}} SPEC\},$$

i.e. the set of all specifications derivable from SPEC0 using productions in PROD.

Remark

Our notion of algebraic specification grammars contains no nonterminals.

In fact, the pair $(PROD, \overset{*}{\underset{PROD}{\Longrightarrow}})$ can be considered as a replacement system for algebraic specifications.

We could also distinguish between terminals and nonterminals concerning sorts and operation symbols, which are both preserved by specification morphisms. In this case only specifications with terminal sorts and operation symbols would be considered to belong to the language.

2.5 Discussion (Translation of Graph Grammar Theory to Theory of Algebraic Specification Grammars)

There are two simple steps allowing to translate concepts and results from graph grammars as given in [Ehr 79] to algebraic specification grammars as defined in 2.2 and 2.4 above.
Step 1 (Generalization): Formulate each concept and result for graph grammars only in terms of graphs, graph morphisms and suitable properties in the category **GRAPH** of graphs.
Step 2 (Replacement): Replace graphs, graph morphisms, and each property in the category **GRAPH** by algebraic specifications, specification morphisms and the corresponding property in the category **SPEC** of algebraic specifications respectively, in order to obtain the translated concept and result for algebraic specification grammars.

Examples for this translation are the following:

1. A graph grammar production p = (B1 ← K → B2) as given in [Ehr 79] becomes a **SPEC**-production p = (LSPEC ← GSPEC → RSPEC) as in 2.2.1.
2. A direct derivation of graph grammars defined by a pair of pushouts in **GRAPH** in [Ehr 79] becomes a direct **SPEC**-derivation as defined in 2.2.2.
3. A graph derivation (see [Ehr 79]) becomes a **SPEC**-derivation (see 2.4.1).
4. A graph grammar (see [Ehr 79]) becomes an algebraic specification grammar (see 2.4.2).
5. Parallel and sequential independence of graph derivations is defined by the fact that the corresponding occurrences of graphs only overlap in gluing items. But there is also a categorical characterization of this property by existence of suitable graph morphisms (see [Ehr 79]) which can be translated directly into the category **SPEC** of algebraic specifications. Moreover, this property can be reinterpreted for **SPEC**-derivations by the fact that the corresponding occurrences of algebraic specifications only overlap in gluing items corresponding to the gluing specification GSPEC in the productions (see 2.2.1). Using the concepts of parallel and sequential independence we are able to translate the Church-Rosser and Parallelism Theorem from graph grammars to algebraic specification grammars (see 2.6).
6. The notion of R-related productions and derivation sequences for graph grammars (see [Ehr 79]) can be formulated by commutativity of diagrams and by the existence of certain pushout complements in the category **GRAPH**. This allows a direct translation to the category **SPEC** (see remark 3 in 2.2) and a translation of the Concurrency-Theorem (see [Ehr 79] and 2.7).

2.6 Church-Rosser and Parallelism Theorem

1. Given parallel independent direct **SPEC**-derivations p:SPEC1 ⇒ SPEC2 and p':SPEC1 ⇒ SPEC3 (see 2.2.2 and 2.5.5) there is an algebraic specification SPEC4 and direct **SPEC**-derivations p':SPEC2 ⇒ SPEC4 and p:SPEC3 ⇒ SPEC4 such that the following derivation sequences are sequential independent

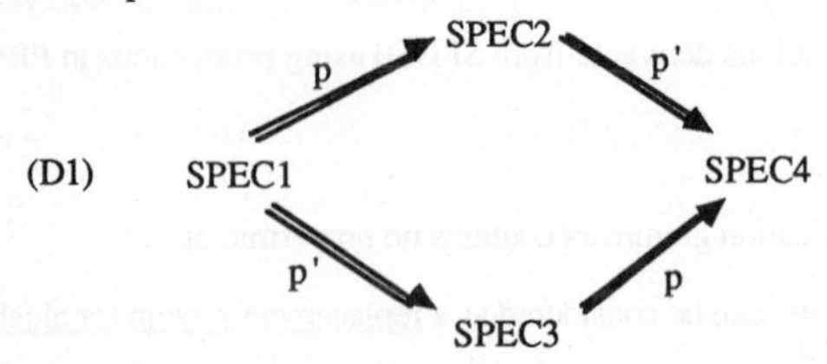

2. Under the assumptions of part1 there is also a direct **SPEC**-derivation

(D2) p + p': SPEC1 ⇒ SPEC4

using the parallel production defined by

$$p + p' = (LSPEC + LSPEC' \xleftarrow{l+l'} GSPEC + GSPEC' \xrightarrow{r+r'} RSPEC + RSPEC')$$

where + is the disjoint union (coproduct) of specifications resp. specification morphisms.

3. Vice versa each direct **SPEC**-derivation (D2) using the parallel production p+p' leads to parallel and sequential independent **SPEC**-derivations as given in diagram (D1).

Remark

In part 1 we can also start with a sequential independent sequence SPEC1 $\Rightarrow^{p}$ SPEC2 $\Rightarrow^{p'}$ SPEC4 or SPEC1 $\Rightarrow^{p'}$ SPEC3 $\Rightarrow^{p}$ SPEC4 leading to parallel and sequential independent sequences as given in diagram (D1).

2.7 Construction of MATCH-Productions

Given a sequence of direct **SPEC**-derivations

$$(S1) \qquad SPEC1 \xrightarrow{p} SPEC2 \xrightarrow{p'} SPEC3$$

using the **SPEC**-productions

$$p = (LSPEC \leftarrow GSPEC \rightarrow RSPEC) \text{ and } p' = (LSPEC' \leftarrow GSPEC' \rightarrow RSPEC')$$

we construct a new **SPEC**-production

$$p *_M p' = (LSPEC* \xleftarrow{l*} GSPEC* \xrightarrow{r*} RSPEC*)$$

called <u>match-production</u> of p and p' as follows:

1. Let M = (MATCH, m, m') be the intersection of RSPEC and LSPEC' in SPEC2, i.e. MATCH with m:MATCH → RSPEC and m':MATCH → LSPEC' is the pullback (PB) of the occurrences f2:RSPEC → SPEC2 and f1':LSPEC' → SPEC2 of the right (resp. left) hand side specification of p (resp. p') in the given sequence (S1).

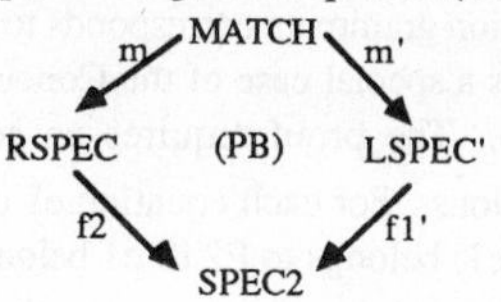

2. Let MSPEC be the pushout object of m and m' in the diagram (PO) below, G and G' pushout-complements in (PO2) resp. (PO1'), LSPEC* resp. RSPEC* pushouts in (PO1) resp. (PO2'), GSPEC* pullback in (PB*), and l* = h ° g resp. r* = h' ° g' in the diagram below:

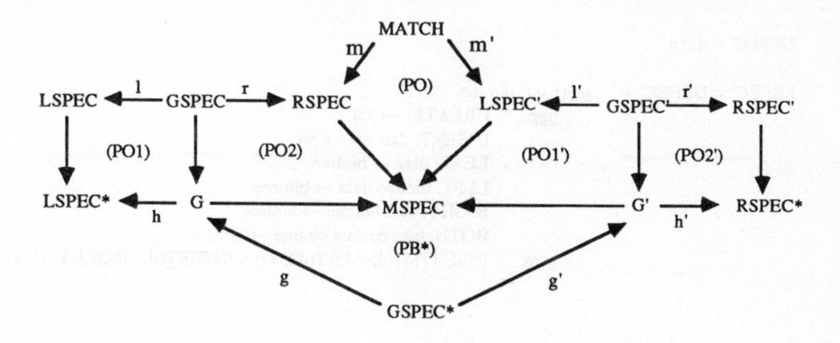

Remarks

1. The existence of the pushout complements G resp. G' in (PO2) resp. (PO1') can be shown using the pushouts corresponding to the given SPEC-derivations in (S1).
2. A matching fork between productions p and p' is given by a triple M = (MATCH, m, m') consisting of a specification MATCH and morphisms m:MATCH → R(p) and m':MATCH → L(p') into the right (resp. left) hand side of p (resp. p') if there are pushout complements G resp. G' in (PO2) and (PO1') above where (PO) is again the pushout of m1 and m2.
3. The match-production corresponds to the R-concurrent production in the case of graph grammars (see [Ehr 79]).
4. Pullbacks in the category **SPEC** are constructed by the pullbacks of the corresponding sets of sorts and operation symbols in each component and the equations are constructed as the intersection of the corresponding preimages of the closures of sets of equations in the given specifications.

2.8 Match-Theorem

Given a derivation sequence (S1) as in 2.7 there is a direct **SPEC**-derivation

$$(S2) \qquad p *_M p'\text{:SPEC1} \Rightarrow \text{SPEC3}$$

called matched derivation of the sequence (S1), where $p *_M p'$ is the match-production of p and p' (see 2.7). Vice versa each direct **SPEC**-derivation (S2) using $p *_M p'$ leads to a derivation sequence (S1) as in 2.7 using p and p'.

Remarks

1. We don't have to assume that the given sequence (S1) is sequential independent as in the case of theParallelism Theorem (see remark of 2.6).
2. The Match-Theorem for algebraic specification grammars corresponds to the Concurrency-Theorem for graph grammars (see [Ehr 79]). Hence it is a special case of the Concurrency-Theorem for high-level replacement systems (see [EHKP 91]). The proof requires an additional property for all morphisms f:SPEC1 → SPEC2 in the productions: For each equation e1 over the signature SIG1 of SPEC1 we have that the translated equation f#(e1) belongs to E2 iff e1 belongs to E1.

2.9 Example

The production p' = (LSPEC' ← GSPEC' → RSPEC') defined by

GSPEC' = **data**

```
LSPEC' = GSPEC' +   sorts set, bintree
                    opns  CREATE: → set
                          INSERT: data set → set
                          LEAF: data → bintree
                          LEFT: bintree data → bintree
                          RIGHT: data bintree → bintree
                          BOTH: bintree data bintree → bintree
                    eqns  INSERT(d1, INSERT(d2, x)) = INSERT(d2, INSERT(d1, x))
```

RSPEC' = GSPEC' + sorts set, bintree
opns LEAF: data → bintree
LEFT: bintree data → bintree
RIGHT: data bintree → bintree
BOTH: bintree data bintree → bintree
NODES: bintree → set

can be applied to the specification SPEC2 of example 2.3 with a morphism f1':LSPEC' → SPEC2 mapping data into nat, CONST into ZERO and the identity on the rest. The result of this direct derivation is

SPEC3 = sorts nat, bintree, set, string
opns ZERO: → nat
SUCC: nat → nat
LEAF: nat → bintree
LEFT: bintree data → bintree
RIGHT: data bintree → bintree
BOTH: bintree data bintree → bintree
NODES: bintree → set
NIL: → string
LADD: nat string → string
RADD: string nat → string

The "intersection" of RSPEC and LSPEC' in SPEC2 (as in 2.7) is the specification

MATCH = **data** + sort set
opns CREATE: → set
INSERT: data set → set
eqns INSERT(d1, INSERT(d2, x)) = INSERT(d2, INSERT(d1,x))

with the inclusion morphisms m:MATCH → RSPEC and m':MATCH → LSPEC'

The pushout object of m and m' is the specification

MSPEC = LSPEC' + sort string
opns NIL: → string
LADD: data string → string
RADD: string data → string

The matched production p $*_M$ p' as in 2.7 consists of the specification

GSPEC* = **data**
LSPEC* = GSPEC* + sorts string, bintree
opns NIL: → string
MAKE: data → string
CONC: string string → string
LEAF: data → bintree
LEFT: bintree data → bintree
RIGHT: data bintree → bintree
BOTH: bintree data bintree → bintree
eqns CONC(x, NIL) = x = CONC(NIL, x)
CONC(x, CONC(y, z)) = CONC(CONC(x, y),z)

RSPEC* = GSPEC* + sorts string, bintree, set
opns NIL: → string
LADD: data string → string
RADD: string data → string
LEAF: data → bintree
LEFT: bintree data → bintree
RIGHT: data bintree → bintree
BOTH: bintree data bintree → bintree
NODES: bintree → set

We can summarize the construction of the matched production in the following diagram, where we use keywords to represent parts of the specifications

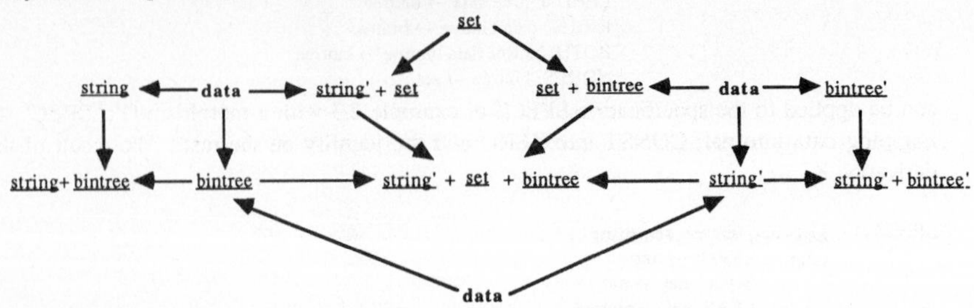

This production can be applied to SPEC1, via the specification morphism which associates data to nat and the remaining sorts and operators to those bearing the same name, to produce in one direct derivation the specification SPEC3 above.

3. Algebraic Module Specifications

In this section we start with an informal introduction to module specifications including the basic operations composition, union and actualization on module specifications as given in [BEPP 87] and chapter 3 of [EM 90].
In addition we study a new operation, called match operation on module specifications, which plays an important role for rule based modular system design. The match operation on module specifications extends the construction of match productions in section 2. Using these constructions a derivation sequence of specifications can be extended to the construction of compound modular systems (see [EPP 90] and section 4).

3.1 Informal Introduction to Algebraic Module Specifications

The importance of decomposing large software systems into smaller units, called modules, to improve their clarity, facilitate proofs of correctness, and support reusability has been widely recognized within the programming and software engineering community. For all stages within the software development process modules resp. module specifications are seen as completely self-contained units which can be developed independently and interconnected with each other.

An algebraic module specification MOD as developed in [EW 85], [BEPP 87] and [EM 90] consists of four components

MOD:

PAR	EXP
IMP	BOD

which are given by four algebraic specifications in the sense of [EM 85]. The export EXP and the import IMP represent the interfaces of a module while the parameter PAR is a part common to both import and export and represents a part of the parameter of the whole system. These interface specifications PAR, EXP, and IMP are allowed to be algebraic specifications with constraints in order to be able to express requirements for operations and domains in the interfaces by suitable logical formalisms. The body BOD, which makes use of the resources provided by the import and offers the resources provided by the export, represents the constructive part of a module.

The semantics of a module specification MOD as above is given by the loose semantics with constraints of the interface specifications PAR, EXP, and IMP, a "free construction" from import to body algebras, and a "behavior construction" from import to export algebras given by restriction of the free construction to the export part.

A module specification is called (internally) correct if the free construction "protects" import algebras and the behavior construction transforms import algebras satisfying the import constraints into export algebras satisfying the export constraints.

Basic interconnection mechanisms to built up module specifications are composition, union, and actualization:

Composition

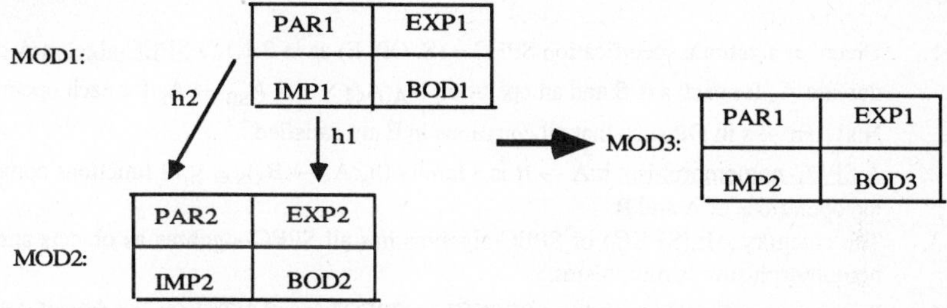

where h1 and h2 are passing morphisms and $BOD3 = BOD1 +_{IMP1} BOD2$ is a pushout construction. The result of the composition is denoted by $MOD3 = MOD1 \circ_{(h1,\, h2)} MOD2$.

Union

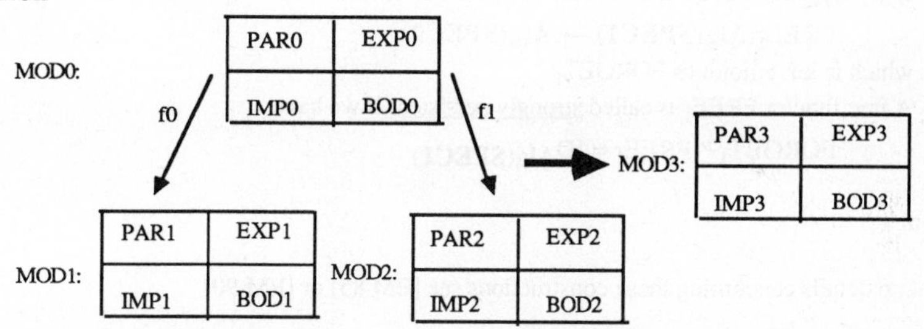

where f0 and f1 are "module specification morphisms" and $SPEC3 = SPEC1 +_{SPEC0} SPEC2$ for SPEC = PAR, EXP, IMP, BOD are pushout constructions. The result of the union is denoted by $MOD_3 = MOD1 +_{MOD0} MOD2$.

Actualization

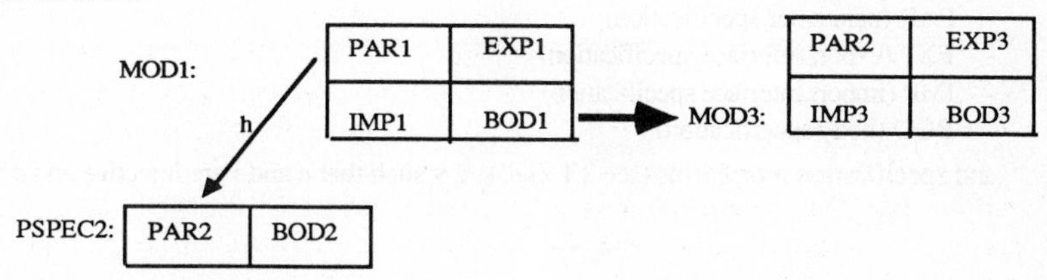

where h is a parameter passing morphism h:PAR1 → BOD2 and SPEC3 = SPEC1 $+_{PAR1}$ BOD2 for SPEC = EXP, IMP, BOD are pushout constructions. The result of the actualization is denoted by MOD3 = act_h(MOD1).

As main results for module specifications we have shown in [BEP 87] that the basic interconnection mechanisms are operations on module specifications which are preserving correctness and which are compositional w.r.t. the semantics. This means that correctness of modular system specification can be deduced from correctness of its parts and its semantics can be composed from that of its components. Moreover, there are nice compatibility results between these operations which can be expressed by associativity, commutativity and distributivity results (see [EM 90]).

In addition to the syntactical notions given in 2.1 we need the following semantical notions for algebraic module specifications in 3.3 below.

3.2 Definition (SPEC-algebras and Functors)

1. Given an algebraic specification SPEC = (S, OP, E) as in 2.1.1, a SPEC-algebra A consists of a domain A_s for each $s \in S$ and an operation $N_A:A_{s1} \times ... \times A_{sn} \to A_s$ for each operation symbol N:s1...sn → s in OP such that all equations in E are satisfied.
2. A SPEC-homomorphism h:A → B is a family $(h_s:A_s \to B_s)_{s \in S}$ of functions compatible with the operations of A and B.
3. The category **Alg(SPEC)** of SPEC-algebras has all SPEC-algebras as objects and all SPEC-homomorphisms as morphisms.
4. For each specification morphism f:SPEC1 → SPEC2 (see 2.1.2) there is a forgetful functor

   $\mathbf{FORGET_f:Alg(SPEC2) \to Alg(SPEC1)}$

   defined for each SPEC2-algebra A2 by $FORGET_f(A2)$ = A1 with $A1_s = A2_{f_S(s)}$ and $N_{A1} = f_{OP}(N)_{A2}$ for $s \in S1$ and $N \in OP1$, and a free functor

   $\mathbf{FREE_f:Alg(SPEC1) \to Alg(SPEC2)}$

   which is left adjoint to $FORGET_f$.
5. A free functor $FREE_f$ is called strongly persistent if we have

   $FORGET_f \circ FREE_f = ID_{\mathbf{Alg(SPEC1)}}$

Remark

For more details concerning these constructions see [EM 85] or [EM 90].

3.3 Definition (Algebraic Module Specification)

1. A algebraic module specification

   MOD = (PAR, EXP, IMP, BOD, e, s, i, v)

   consists of algebraic specifications (see 2.1.1)
   - PAR (parameter specification)
   - EXP (export interface specification)
   - IMP (import interface specification)
   - BOD (body specification)

   and specification morphisms (see 2.1.2) e, s, i, v such that e and i are injective and $s \circ i = v \circ e$

$$\begin{array}{ccc} PAR & \xrightarrow{e} & EXP \\ {\scriptstyle i}\downarrow & & \downarrow{\scriptstyle v} \\ IMP & \xrightarrow{s} & BOD \end{array}$$

2. The functorial semantics SEM of MOD is the functor

$$SEM = FORGET_V \circ FREE_S : Alg(IMP) \rightarrow Alg(EXP)$$

where $FREE_S$ (resp. $FORGET_V$) is the free (resp. forgetful) functor w.r.t. s (resp. v) (see 3.2.4).

3. The module specification MOD is called (internally) correct if $FREE_S$ is strongly persistent (see 3.2.5).

4. The interface production p(MOD) of MOD is an algebraic specification production (see 2.2.1), defined by

$$p(MOD) = (IMP \xleftarrow{i} PAR \xrightarrow{e} EXP)$$

Remark

In contrast to [BEP 87] and [EM 90] e and i are assumed to be injective as required for productions in 2.2.1.

3.4 Example

As an example of module specification we take the realization of the production p of example 2.3.

MOD2 = (PAR2, EXP2, IMP2, BOD2) where

```
PAR2 = GSPEC =  sort data + opns CONST: → data
EXP2 = RSPEC = PAR2 +  sorts string, set
                       opns   CREATE: → set
                              INSERT: data set → set
                              NIL: → string
                              LADD: data string → string
                              RADD: string data → string
                       eqns   INSERT(d1, INSERT(d2, x) = INSERT(d2, INSERT(d1, x))

IMP2 = LSPEC = PAR2 +  sort string
                       opns   NIL: → string
                              MAKE: data → string
                              CONC: string string → string
                       eqns   CONC(NIL, x) = x = CONC(x, NIL)
                              CONC(CONC(x, y), z) = CONC(x, CONC(y, z))

BOD2 = EXP2 ∪ IMP2 +   eqns   INSERT(d, INSERT(d, x)) = INSERT(d, x)
                              LADD(d, x) = CONC(MAKE(d), x)
                              RADD(x, d) = CONC(x, MAKE(d))
```

The functorial semantics SEM2 of MOD2 transforms IMP2-algebras where strings can be concatenated with the operation CONC into EXP2-algebras where strings can be incremented only one element at the time, either on the left (LADD) or on the right (RADD).

The following match operation defines the combination of two module specifications as an extension of the construction of the match production (see 2.7) for the corresponding interface productions.

3.5 Definition (Match-Operation)

Given algebraic module specifications

MODj = (PARj, EXPj, IMPj, BODj, ej, sj, ij, vj) for j = 1,2,

and a matching fork (see remark 2 in 2.7)

M = (MATCH, mE:MATCH $\rightarrow$ EXP2, mI:MATCH $\rightarrow$ IMP1)

between the interface productions p(MOD2) and p(MOD1) then the <u>match composition</u>, short <u>match</u>, MOD3 of MOD1 with MOD2 via M, written

MOD3 = MOD1 $\circ_M$ MOD2,

is given by the following algebraic module specification

MOD3 = (PAR3, EXP3, IMP3, BOD3, e3, s3, i3, v3)

due to the following construction steps

1. Let p(MOD2) $*_M$ p(MOD1) be the match production of p(MOD2) and p(MOD1) given by

$$p(MOD3) = p(MOD2) *_M p(MOD1) = (IMP3 \xleftarrow{i3} PAR3 \xrightarrow{e3} EXP3)$$

2. Let BOD3 together with b1 and b2 be the pushout of the following diagram

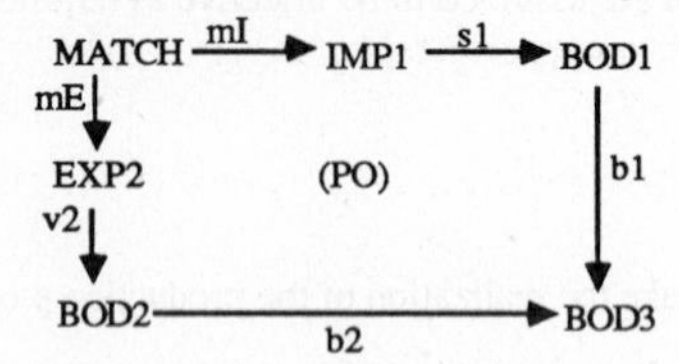

and s3:IMP3 $\rightarrow$ BOD3 and v3:EXP3 $\rightarrow$ BOD3 the unique specification morphisms induced by the pushout properties of IMP3 and EXP3 respectively.

Remark

The construction can be summarized in the following diagram

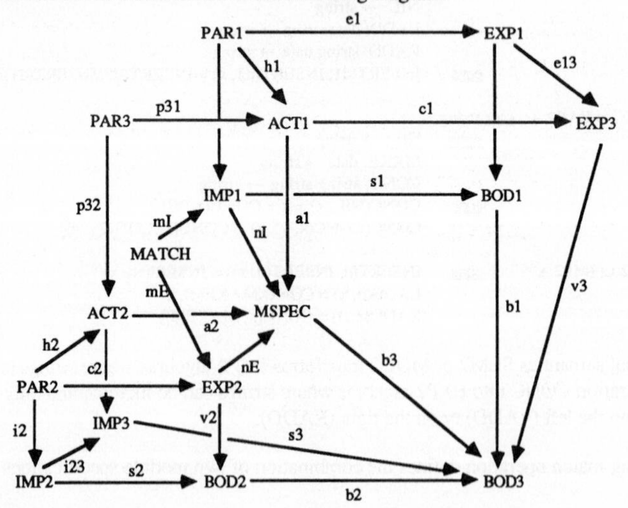

where in step1 MSPEC is constructed as pushout of (m_I, m_E), ACT1 and ACT2 as pushout complements of $n_I \circ i1$ resp. $n_E \circ i2$, and EXP3 and IMP3 as pushouts of (h1, e1) and (h2, i2) respectively, and in step2 BOD3 as pushout of ($s1 \circ m_I$, $v2 \circ m_E$), and b3, s3, v3 as induced morphisms of $b1 \circ s1$ and $b2 \circ v2$, $b3 \circ a2$ and $b2 \circ s2$, and $b3 \circ a1$ and $b1 \circ v1$ using the pushout properties of MSPEC, IMP3 and EXP3 respectively.

3.6 Example

Define the module specification MOD1 = (PAR1, EXP1, IMP1, BOD1) by

PAR1 = GSPEC' = sort data + opns CONST: → data
EXP1 = RSPEC' = PAR1 + sorts set, bintree
opns LEAF: data → bintree
LEFT: bintree data → bintree
RIGHT: data bintree → bintree
BOTH: bintree data bintree → bintree
NODES: bintree → sets

IMP1 = LSPEC' = PAR1 + sorts set, bintree
opns CREATE: → set
INSERT: data set → set
LEAF: data → bintree
LEFT: bintree data → bintree
RIGHT: data bintree → bintree
BOTH: bintree data bintree → bintree
eqns INSERT(d1, INSERT(d2, x)) = INSERT(d2, INSERT(d1,x))

BOD1 = EXP1 ∪ IMP1 + opns ADD: set set → set
eqns ADD(CREATE, s) = s
ADD(INSERT(d, x),s) = INSERT(d, ADD(x, s))
NODES(LEAF(d)) = INSERT(d, CREATE)
NODES(LEFT(b, d)) = INSERT(d, NODES(b))
NODES(RIGHT(d, b)) = INSERT(d, NODES(b))
NODES(BOTH(b1, d, b2)) = INSERT(d, ADD(NODES(b1), NODES(b2)))

The functorial semantics SEM1 of MOD1 transforms IMP1-algebras of sets and binary trees into EXP1-algebras of binary trees with an operation with return for each tree the set of the data in its nodes.
The specification MATCH of example 2.9, along with inclusion morphisms, is a matching fork for the interfaces of the module specifications MOD1 and MOD2 (see 3.4).
The match MOD3 = MOD1 $*_M$ MOD2 of MOD1 and MOD2 via M as defined above is composed of

PAR3 = GSPEC* = sort data + opns CONST: → data

EXP3 = RSPEC* = PAR3 + sorts string, bintree, set
opns NIL: → string
LADD: data string → string
RADD: string data → string
LEAF: data → bintree
LEFT: bintree data → bintree
RIGHT: data bintree → bintree
BOTH: bintree data bintree → bintree
NODES: bintree → set

IMP3 = LSPEC* = PAR3 + <u>sorts</u> string, bintree

<u>opns</u> NIL: → string
MAKE: data → string
CONC: string string → string
LEAF: data → bintree
LEFT: bintree data → bintree
RIGHT: data bintree → bintree
BOTH: bintree data bintree → bintree

<u>eqns</u> CONC(x, NIL) = x = CONC(NIL, x)
CONC(x, CONC(y, z) = CONC(CONC(x, y), z))

BOD3 = BOD1 $+_{MATCH}$ BOD2 = IMP3 ∪ EXP3 +

<u>opns</u> CREATE: → set
ADD: set set → set
INSERT: data set → set

<u>eqns</u> INSERT(d, INSERT(d, x)) = INSERT(d, x)
LADD(d, x) = CONC(MAKE(d), x)
RADD(x, d) = CONC(x, MAKE(d))
ADD(CREATE, s) = s
ADD(INSERT(d, x), s) = INSERT(d, ADD(x, s))
NODES(LEAF(d)) = INSERT(d, CREATE)
NODES(LEFT(b, d)) = INSERT(d, NODES(b))
NODES(RIGHT(d, b)) =INSERT(d, NODES(b))
NODES(BOTH(b1, d, b2)) = INSERT(d, ADD(NODES(b1), NODES(b2)))

3.7 Theorem (Properties of the Match-Operation)

1. The match operation is a derived operation of actualization and composition. Using the notation of 3.5 we have

   $$\text{MOD1} \circ_M \text{MOD2} = act_{h1}(\text{MOD1}) \circ act_{h2}(\text{MOD2})$$

   where act_{hi}(MODi) is the actualization of MODi by ACTi using hi:PARi → ACTi for i = 1, 2 and ° corresponds to the composition of act_{h1}(MOD1) with import interface MSPEC and act_{h2}(MOD2) with export interface MSPEC.
2. The match-operation is correctness preserving and compositional, i.e. correctness of MOD1 and MOD2 implies that of MOD3 = MOD1 $\circ_M$ MOD2 and in this case we have:

   $$\text{SEM3} = (\text{SEM1} +_{ID_{PAR1}} ID_{ACT1}) \circ (\text{SEM2} +_{ID_{PAR2}} ID_{ACT2})$$

   where SEMi is the functorial semantics of MODi for i = 1, 2, 3 and ID_{SPEC} denotes the identity functor on the category Alg(SPEC).
3. The interface productions are in the following relations

   $$p(\text{MOD1} \circ_M \text{MOD2}) = p(\text{MOD2}) *_M p(\text{MOD1})$$

Proofidea (see [EPP 90] for more detail)

1. Follows directly from the definition of actualization and union concerning the interface parts and from general colimit results concerning the body constructions.

2. Follows from correctness and compositionality of actualization and composition.
3. Follows directly from construction step1 in 3.5.

4. Concluding Remarks

We have seen how the basic notions of graph grammars can be converted into grammars of algebraic specifications and how a production in such grammars corresponds exactly to the visible part (import and export interfaces and shared parameter) of a module specification. The problem of designing a modular system with overall import (resp. export) interfaces IMP (resp. EXP) is reduced to the problem of deriving a specification EXP in the algebraic specification grammar

$$SPEGRA = (IMP, LIB, \xrightarrow[LIB]{*}).$$

Once the membership of EXP in the language defined by SPEGRA is verified, it is possible to construct from each derivation sequence IMP $\Rightarrow^*$EXP a corresponding modular system with overall import (resp. export) interfaces IMP (resp. EXP) using the following facts (see [PP 89] and 3.7 above):

- a direct derivation p:SPEC1 $\Rightarrow$ SPEC2 via a production p = (LSPEC $\leftarrow$ GSPEC $\rightarrow$ RSPEC) corresponds to the actualization of MOD = (LSPEC, GSPEC, RSPEC, BOD) by the context specification CSPEC of the direct derivation
- a sequentially independent derivation sequence p1:SPEC1 $\Rightarrow$ SPEC2, p2:SPEC2 $\Rightarrow$ SPEC3 corresponds to the union MOD1 $+_{MOD0}$ MOD2 of the realizations of the interfaces p1 and p2, actualized by the part common to the two contexts
- a sequentially dependent derivation sequence p1:SPEC1 $\Rightarrow$ SPEC2, p2:SPEC2 $\Rightarrow$ SPEC3 can be replaced by a direct derivation via the interfaces (IMP3 $\leftarrow$ PAR3 $\rightarrow$ EXP3) of the match MOD1 $\circ_M$ MOD2 of the modules realizing p1 and p2.

The correspondence between the productions of an algebraic specification grammar and the interfaces of a module specification can be extended to results other than the composability of modules and productions, by translating other developments in one field to the other. For example, the approach to the vertical development of modular systems ([EM 90]) can be restricted to the interfaces of the module specifications to obtain a theory for the vertical ("over time") development of a specification grammar, viewed as a step-by-step evolution of a rule-based system. Conversely, results in the theory of Graph Grammars such as the Amalgamation Theorem can be translated into specification grammars and then used in the theory of module specifications.
Both Graph Grammars and Algebraic Specification Grammars are examples of replacement systems which share much of their theory. The common features of these and other systems on hypergraphs, structures, transition systems and nets are investigated under the common label of High Level Replacement Systems [EHKP 90/91], in which general conditions on the underlying category are characterized with respect to results such as Church-Rosser properties, Parallelism and Concurrency Theorems. It is than sufficient to check the properties of the category underlying a specific application to be able to use these results.

5. References

[BEPP 87] E.K. Blum, H. Ehrig, F. Parisi-Presicce: Algebraic Specification of Modules and their Basic Interconnections, J. Comp. System Sci. 34, 2/3 (1987), 239-339

[COMP 89] B. Krieg-Brückner, ed.: A Comprehensive Algebraic Approach to System Specification and Development, ESPRIT BRWG 3264, Univ. Bremen, Bericht 6/89

[Ehr 79] H. Ehrig: Introduction to the Algebraic Theory of Graph Grammars, Lect. Notes in Comp. Sci. 73 (1979), 1-69

[EFPB 86] H. Ehrig, W. Fey, F. Parisi-Presicce, E.K. Blum: Algebraic Theory of Module Specifications with Constraints, invited, Proc MFCS, Lect. Notes Comp. Sci. 233 (1986), 59-77

[EHKP 90] H. Ehrig, A. Habel, H.-J. Kreowski, F. Parisi-Presicce: From Graph Grammars to High Level Replacement Systems, this volume

[EHKP 91] H. Ehrig, A. Habel, H.-J. Kreowski, F. Parisi-Presicce: Parallelism and Concurrency in High-Level-Replacement Systems, to appear

[EKMRW 81] H. Ehrig, H.-J. Kreowski, A. Maggiolo-Schettini, B. Rosen, J. Winkowski: Transformation of Structures: an Algebraic Approach, Math. Syst. Theory 14 (1981), 305-334

[EM 85] H. Ehrig, B. Mahr: Fundamentals of Algebraic Specifications 1: Equations and Initial Semantics, EATCS Monographs on Theoret. Comp. Sci., vol 6, Springer-Verlag 1985

[EM 90] H. Ehrig, B. Mahr: Fundamentals of Algebraic Specifications 2: Module Specifications and Constraints, EATCS Monographs on Theoret. Comp. Sci., vol 21, Springer-Verlag 1990

[EPS 73] H. Ehrig, M. Pfender, H.J. Schneider: GRAPH GRAMMARS: An Algebraic Approach, Proc IEEE Conf. SWAT 73, Iowa City (1973), 167-180

[EPP 90] H. Ehrig, F. Parisi-Presicce: A Match Operation for Rule Based Modular System Design, to appear in Proc 7th ADT Workshop, Wusterhausen 1990, Springer LNCS

[EW 85] H. Ehrig, H. Weber: Algebraic Specifications of Modules, in "Formal Models in Programming" (E.J. Neuhold, G. Chronist, eds.), North-Holland 1985

[GTW 76] J.A. Goguen, J.W. Thatcher, E.G. Wagner: An initial algebra approach to the specification, correctness and implementation of abstract data types. IBM Research Report RC 6487, 1976. Also: Current Trends in Programming Methodology IV: Data Structuring (R. Yeh, ed.), Prentice Hall (1978), 80-144

[Gu 75] J.V. Guttag: The specification and application to programming of abstract data types. Ph.D. Thesis, University of Toronto, 1975

[PP 89] F. Parisi-Presicce: Modular System Design applying Graph Grammar Techniques, Proc. 16 ICALP, LNCS 372 (1989), 621-636

[PP 90] F. Parisi-Presicce: A Rule-Based Approach to Modular System Design, Proc. 12th Int. Conf. Soft. Eng., Nice (France) 1990, 202-211

[TWW 78] J.W. Thatcher, E.G. Wagner, J.B. Wright: Data type specification: parameterization and the power of specification techniques. 10th Symp. Theory of Computing (1978), 119-132. Trans. Prog. Lang. and Syst. 4 (1982), 711-732

[Wa 79] M. Wand: Final algebra semantics and data type extensions. JCSS 19 (1979), 27-44

[Zi 74] S.N. Zilles: Algebraic specification of data types. Project MAC Progress Report 11, MIT 1974, 28-52

A characterization of context-free NCE graph languages by monadic second-order logic on trees

Joost Engelfriet

Dept. of Computer Science, Leiden University,
P.O.Box 9512, 2300 RA Leiden, The Netherlands

ABSTRACT. A graph language L is in the class C-edNCE of context-free NCE graph languages if and only if L = f(T) where f is a function on graphs that can be defined in monadic second-order logic and T is the set of all trees over some ranked alphabet. This logical characterization implies a large number of closure and decidability properties of the class C-edNCE.

Keywords: context-free graph grammar, monadic second-order logic, tree

CONTENTS

1. Introduction

Context-free graph grammars are a general formalism to define sets of graphs in a recursive fashion, just as context-free grammars are used to recursively define sets of strings. Since many interesting graph properties are recursive in one way or another, context-free graph grammars provide a means to study such properties in general. Unfortunately there are many types of context-free graph grammars, and there is no agreement on which is the "correct" one. Here we consider the class C-edNCE of graph languages generated by the context-free

NCE graph grammars, studied in [Kau85, Bra88, Sch87, Eng89a]. One advantage of the class C-edNCE is that it is the largest known class of context-free graph languages (it includes, e.g., the hyperedge replacement languages of [HabKre 87, BauCou87] and the B-NLC languages of [RozWel86/87]). Thus, results on C-edNCE apply to a quite large class of recursive graph properties. Another advantage of C-edNCE is that it seems to be stable in the sense that it can be characterized in several different ways. It is shown in [CouEngRoz91] that C-edNCE is also generated by a specific type of handle-rewriting hypergraph grammars (S-HH grammars), generalizing hyperedge replacement systems, and that C-edNCE has a least fixed point characterization in terms of very simple graph operations. In [Eng89a] it is shown that C-edNCE languages can be described in a grammar independent way by regular tree and string languages. In this paper we present a characterization of C-edNCE in terms of monadic second-order logic on trees. More precisely, we first define the class MSOF of monadic second-order definable functions; they are unary functions that transform graphs into graphs. Then the result is that a graph language L is in C-edNCE if and only if $L = \{f(t) \mid t \in T_\Delta\}$ where f is in MSOF and T_Δ is the set of all trees over a (ranked) alphabet Δ. Intuitively, the recursive (context-free) aspect of a graph in L is captured by the tree t, whereas the actual construction of the graph from t is given through a monadic second-order description f. In what follows we abbreviate "monadic second-order" by MSO.

The use of MSO logic to characterize classes of languages generated by grammars dates back to [Büc60, Elg61], where it is shown that the class of regular string languages equals the class of MSO definable string languages. This was generalized to trees in [ThaWri68, Don70] (cf. also, e.g., [Eng89b]): a tree language is regular if and only if it is MSO definable. Note that these characterizations differ from the one for C-edNCE: a graph language L is MSO definable if there exists an MSO formula ϕ such that L consists of all graphs that satisfy ϕ. The class of MSO definable graph languages is incomparable to C-edNCE (cf. [Cou90a]). Recently the above MSO characterization of the regular tree languages was used by Courcelle in some very elegant results showing that several classes of context-free graph languages are closed under intersection with MSO definable graph languages (see [Cou87/90], and see [CouEngRoz91] for C-edNCE). These results provide meta-theorems for closure (and decidability) properties of these classes. Other such meta-theorems (not using MSO logic) are shown in, e.g., [HabKreVog89a/b, LenWan88]; the advantage of MSO logic is that it is grammar independent, well known, and easy to use. Our MSO characterization of C-edNCE is a generalization of Courcelle's result. It implies several new MSO meta-theorems for closure and decidability properties of C-edNCE, in particular generalizing the results of [HabKreVog89a] to C-edNCE. It was first presented in [Eng88, Oos89].

2. Graphs and graph grammars

$\mathbb{N} = \{0,1,2,\ldots\}$, and for $n,m \in \mathbb{N}$, $[m,n] = \{i \in \mathbb{N} \mid m \leq i \leq n\}$.

Let Σ be an alphabet of node labels, and Γ an alphabet of edge labels. A _graph_ over Σ and Γ is a tuple $g = (V,E,\lambda)$, where V is the finite set of nodes, $\lambda: V \to \Sigma$ is the node labeling function, and $E \subseteq \{(u,\gamma,v) \in V \times \Gamma \times V \mid u \neq v\}$ is the set of (labeled) edges. Thus, we consider directed graphs with labeled nodes and edges. There are no loops; there may be multiple edges, but not with the same label. The set of all graphs over Σ and Γ is denoted $GR(\Sigma,\Gamma)$. The components of a given graph g will be denoted by V_g, E_g, and λ_g. A (directed) _path_ from u to v is a sequence $v_1,\ldots,v_n$ of nodes ($n \geq 1$) such that $v_1 = u$, $v_n = v$, and there is an edge from v_i to v_{i+1} for every $i \in [1,n-1]$.

The edNCE grammars are a special case of Nagl's grammars [Nag79]. They belong to the family of NLC-like graph grammars [Roz87]. The main advantage of these grammars over other NLC-like grammars is that the edge labels can be changed dynamically. In "edNCE", e stands for "edge- (and node-) labeled", d for "directed", and NCE for "neighbourhood controlled embedding". The edNCE graph grammars are studied in, e.g., [Kau85, Bra88, Sch87, EngLeiRoz87, EngLei89, EngLeiWel90, Eng89a, EngRoz90].

Definition 1. An _edNCE_ _graph_ _grammar_ is a system $G = (\Sigma,\Delta,\Gamma,\Omega,P,S)$, where Σ is the alphabet of node labels, $\Delta \subseteq \Sigma$ is the alphabet of terminal node labels (the elements of Σ-Δ are nonterminal node labels), Γ is the alphabet of edge labels, $\Omega \subseteq \Gamma$ is the alphabet of terminal edge labels (the elements of Γ-Ω are nonterminal edge labels), P is the finite set of productions, and $S \in \Sigma$-Δ is the initial nonterminal. A production $\pi \in P$ is of the form $\pi: X \to (g,B)$ where $X \in \Sigma$-Δ is the left-hand side of π, $g \in GR(\Sigma,\Gamma)$ is the right-hand side of π, and $B \subseteq V_g \times \Gamma \times \Gamma \times \Sigma \times \{in,out\}$ is the embedding relation of π, satisfying the following condition: if $(x,\beta,\gamma,\sigma,d) \in B$, $\lambda_g(x) \in \Delta$, and $\sigma \in \Delta$, then $\gamma \in \Omega$. □

For convenience, by the above condition, we restrict attention to "non-blocking" edNCE grammars (cf. [EngLeiWel90]).

The productions of G are applied to elements of $GR(\Sigma,\Gamma)$. For a graph $h \in GR(\Sigma,\Gamma)$, the production $\pi: X \to (g,B)$ is applicable to a nonterminal node $v \in V_h$ if $\lambda_h(v) = X$. Application of π to v consists of the following steps. First v is removed from h, together with all edges incident with v. Then g (or, more precisely, a fresh isomorphic copy of g) is added to the remainder of h. Finally, edges are established between V_g and V_h-$\{v\}$ according to the embedding relation B: for $x \in V_g$ and $y \in V_h$-$\{v\}$, an edge (x,γ,y) is added (an edge (y,γ,x) is added) if and only if there was an edge (v,β,y) (an edge (y,β,v), respectively) in E_h and the tuple $(x,\beta,\gamma,\lambda_h(y),out)$ is in B (the tuple $(x,\beta,\gamma,\lambda_h(y),in)$ is in B, respectively). This results in a graph h'; notation: $h \Rightarrow_{(v,\pi)} h'$ or just $h \Rightarrow h'$. As usual, the <u>language generated by G</u> is $L(G) = \{h \in GR(\Delta,\Omega) \mid S \stackrel{*}{\Rightarrow} h\}$, where S denotes a graph consisting of one node, labeled S. A graph $h \in GR(\Sigma,\Gamma)$ such that $S \stackrel{*}{\Rightarrow} h$ will be called a <u>sentential form</u> of G.

In other words, a tuple $(x,\beta,\gamma,\sigma,d)$ in B means that if the rewritten nonterminal node v is connected to a σ-labeled node by a β-labeled edge in direction d, then, after application of the production, x will be connected to the same σ-labeled node by a γ-labeled edge, in the same direction d.

We will need one easy result on (non-blocking) edNCE grammars: it is decidable whether such a grammar G generates a finite language, where it is understood that isomorphic graphs in L(G) are identified.

<u>Theorem</u> 2. It is decidable, for a (non-blocking) edNCE grammar G, whether or not L(G) is finite.

<u>Proof</u> (cf. [EngLeiWel90, Lemma 16]). Transform the edNCE grammar $G = (\Sigma,\Delta,\Gamma,\Omega,P,S)$ into an ordinary context-free grammar G', with Σ-Δ as set of nonterminals, Δ as set of terminals, S as initial nonterminal, and the following productions. If P contains a production $X \rightarrow (g,B)$ and $V_g = \{v_1,\ldots,v_n\}$, then G' has the production $X \rightarrow Y_1 \cdots Y_n$, where $Y_i = \lambda_g(v_i)$ for all $i \in [1,n]$. It should be clear that L(G) is finite iff L(G') is finite. □

The edNCE grammars are context-free in the sense that one node is rewritten in each derivation step, and hence their derivations can be modeled by derivation trees. However, the grammar may still be context-sensitive in the sense that the (edges of the) graph generated according to such a derivation tree may depend on the order in which the productions are applied. An edNCE grammar that does not suffer from this context-sensitivity is said to be <u>context-free</u> or <u>confluent</u> (abbreviated C-edNCE).

<u>Definition</u> 3. An edNCE grammar G is <u>confluent</u> (C-edNCE) if the following holds for every sentential form h of G. Let v_1 and v_2 be distinct nonterminal nodes of h, and let π_1 and π_2 be productions applicable to v_1 and v_2, resp. If $h \Rightarrow_{(v_1,\pi_1)} h_1 \Rightarrow_{(v_2,\pi_2)} h_{12}$ and $h \Rightarrow_{(v_2,\pi_2)} h_2 \Rightarrow_{(v_1,\pi_1)} h_{21}$, then $h_{12} = h_{21}$ (where we assume that in both cases the same isomorphic copies of the right-hand sides of π_1 and π_2 are used). □

Confluence is decidable ([Kau85]). The class of all graph languages generated by C-edNCE grammars is denoted C-edNCE. It is shown in [Sch87, Bra88] that the language $GR(\Sigma,\Gamma)$ is not in C-edNCE. It is not difficult to show that C-edNCE contains the class of all context-free NLC graph languages, as defined in [Cou87]. Two types of C-edNCE grammars are of special interest. An edNCE grammar is boundary (B-edNCE) if there are no edges between nonterminal nodes in any right-hand side of a production [EngLeiWel90], and it is linear (LIN-edNCE) if there is at most one nonterminal in any right-hand side. Obviously, LIN-edNCE $\subseteq$ B-edNCE $\subseteq$ C-edNCE; the inclusions are shown to be proper in [EngLei89] and [Eng89a], respectively (and the counter-examples are the set of all binary trees and the set of all edge complements of binary trees, respectively). B-edNCE also contains the B-NLC languages of [RozWel86/87], and, as shown in [EngRoz90], the graph languages generated by the hyperedge replacement grammars of [BauCou87, HabKre87]; both these inclusions are proper (and counter-examples are the set of all edge complements of chains, see [EngLeiWel90], and the set of all complete graphs, respectively). Thus, C-edNCE seems to contain all known classes of context-free graph languages.

We give two examples of C-edNCE grammars.

Examples 4. (1) Consider the LIN-edNCE grammar $G_1 = (\Sigma,\Delta,\Gamma,\Omega,P,S)$ with $\Sigma = \{S,Y,q,p\}$, $\Delta = \{q,p\}$, $\Gamma = \{\alpha,\beta,\gamma\}$, $\Omega = \{\gamma\}$, and P consists of the three productions π_a, π_b, π_c drawn in Fig.1. To draw a production $X \rightarrow (g,B)$ of this grammar, we have added B to g as follows: a tuple $(x,\lambda,\mu,\sigma,in)$ of B is represented by a broken line $\sigma \overset{\lambda/\mu}{- - - - \rightarrow} \bullet$ where the dot represents node x of g. There are no out-tuples in B. Thus, production π_a is $X \rightarrow (g,B)$ with $V_g = \{x,y\}$, $E_g = \{(x,\beta,y)\}$, $\lambda_g(x) = q$, $\lambda_g(y) = Y$, and $B = \{(x,\beta,\gamma,q,in), (y,\alpha,\alpha,q,in)\}$. $L(G_1)$ consists of all "ladders" of the form given in Fig.2.

(2) A co-graph is an undirected, unlabeled graph, recursively defined as follows [CorLerSte81]. A graph consisting of one node is a co-graph. If g and h are co-graphs, then so are g+h and g*h, where g+h is the disjoint union of g and h and g*h is obtained from g+h by adding all edges between a node of g and a node of h. To generate the set of all co-graphs by a C-edNCE grammar we represent an undirected unlabeled edge $\{x,y\}$ by two edges (x,γ,y) and (y,γ,x), and we label all nodes by q. Consider the C-edNCE grammar $G_2 = (\{S,q\},\{q\},\{\gamma\},\{\gamma\},P,S)$ with the productions π_*, π_+, π_q as drawn in Fig.3; the embedding for each of these productions $X \rightarrow (g,B)$ is $B = \{(x,\gamma,\gamma,\sigma,d) \mid x \in V_g, \sigma \in \{S,q\}, d \in \{in,out\}\}$, which means that "all edges are taken over by all nodes". It should be clear that G_2 generates all co-graphs. Note that G_2 is not boundary. □

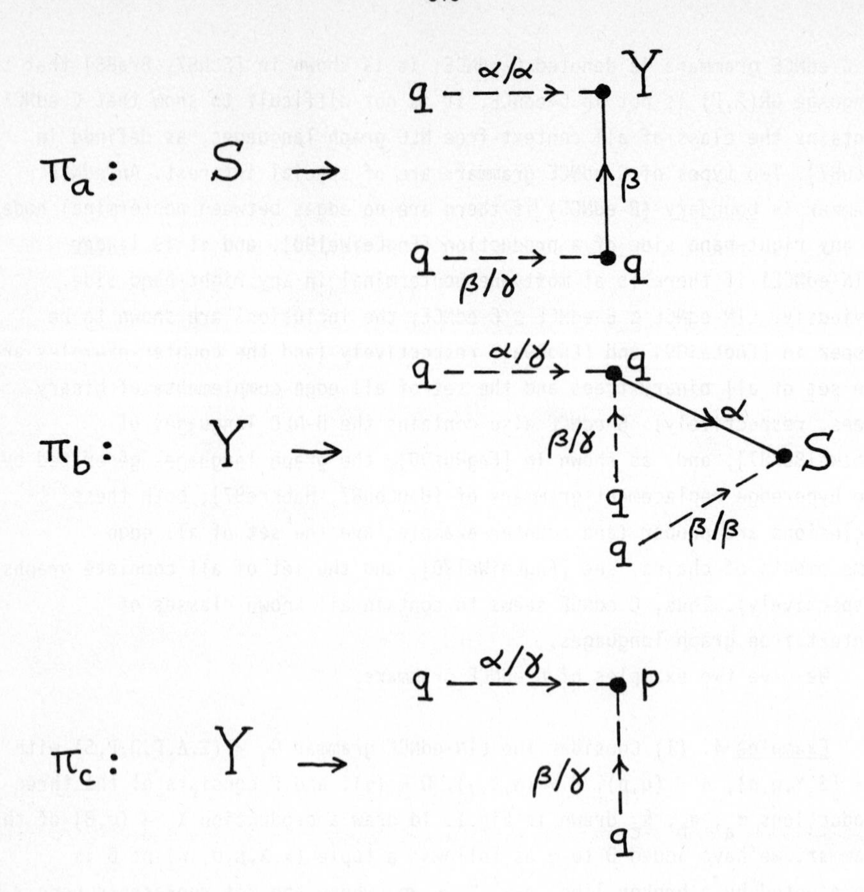

Figure 1

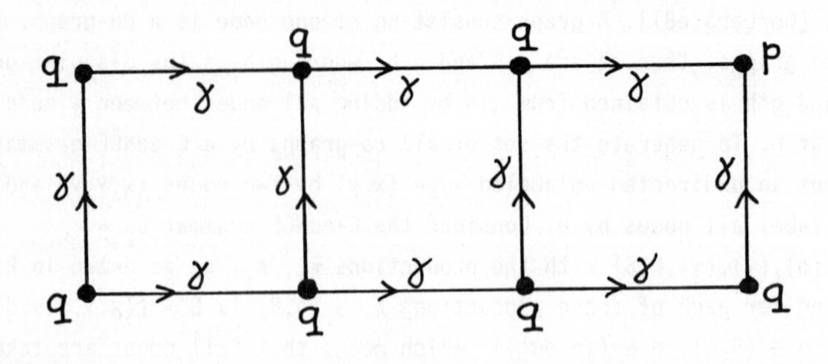

Figure 2

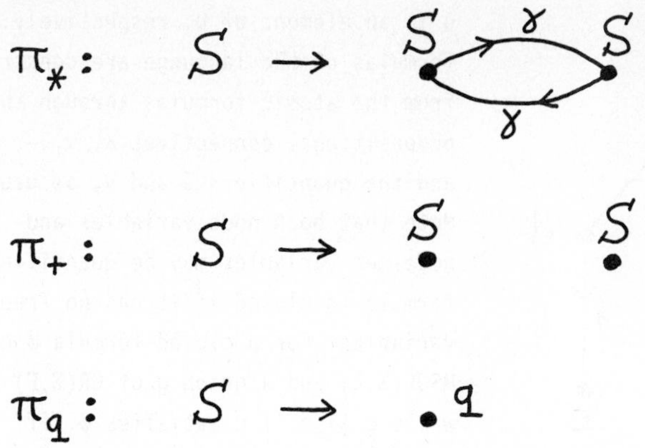

π_* : S → S γ S γ

π_+ : S → S S

π_q : S → q

Figure 3

3. Graphs, trees, and monadic second-order logic

A ranked alphabet is an alphabet Δ together with a mapping rank: Δ ⟶ ℕ. By Γ(Δ) we denote the alphabet [1,m], where m = max{rank(δ) | δ ∈ Δ}. A (ordered, directed) <u>tree</u> over Δ is a graph t ∈ GR(Δ,Γ(Δ)), defined in the usual way, where we use the edge labels to indicate the order of the sons of a node; thus, (1) there is a node r of t (its root) such that for every node v of t there is a unique path from r to v, and (2) every node of t labeled δ has exactly n outgoing edges, where n = rank(δ), and each i ∈ [1,n] occurs as label of one of these edges. The set of all trees over Δ is denoted T_Δ. Figure 4 shows a tree over the ranked alphabet Δ = {a,b,c,d,e} with rank(a) = 3, rank(b) = 2, rank(d) = 1, and rank(c) = rank(e) = 0.

For alphabets Σ and Γ, we use a monadic second-order logical language MSOL(Σ,Γ), each closed formula of which expresses a property of the graphs in GR(Σ,Γ). MSOL(Σ,Γ) has node variables, denoted u,v,..., and node-set variables, denoted U,V,.... For a given graph g, each node variable ranges over the elements of V_g, and each node-set variable ranges over the subsets of V_g. There are four types of atomic formulas in MSOL(Σ,Γ): $lab_\sigma(u)$, for every σ ∈ Σ, $edge_\gamma(u,v)$, for every γ ∈ Γ, u = v, and u ∈ U (where u and v are node variables and U is a node-set variable). Their meaning should be clear: u has label σ, there is an edge from u to v with label γ, u and v are the same, and

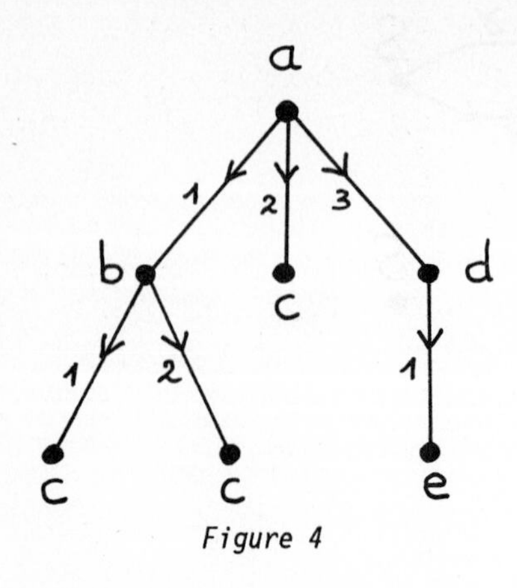

Figure 4

u is an element of U, respectively. The formulas of the language are constructed from the atomic formulas through the propositional connectives $\wedge$, $\vee$, $\neg$, $\rightarrow$, and the quantifiers $\exists$ and $\forall$, as usual. Note that both node variables and node-set variables may be quantified. A formula is closed if it has no free variables. For a closed formula ϕ of MSOL(Σ,Γ) and a graph g of GR(Σ,Γ) we write $g \models \phi$ if g satisfies ϕ. If formula ϕ has free variables, say u,v, and U (and no others), then we also write the formula as $\phi(u,v,U)$; if graph g has nodes $x,y \in V_g$ and a set of nodes $X \subseteq V_g$, then we write $g \models \phi(x,y,X)$ to mean that the formula ϕ holds in g when the values x,y, and X are given to the variables u,v, and U, respectively. A graph language L is <u>MSO definable</u> if there is a closed formula ϕ of MSOL(Σ,Γ) such that $L = \{g \in GR(\Sigma,\Gamma) \mid g \models \phi\}$.

It is well known that there exists an MSOL formula path(u,v), with free variables u and v, that expresses that there is a path from u to v. Thus, as a very simple example, the formula $\phi \equiv \forall u\ \forall v$: path(u,v) defines the language of all strongly connected graphs.

The following proposition generalizes a result of [Büc60, Elg61] for string languages. For the notion of a regular tree language see, e.g., [GecSte84].

<u>Proposition</u> 5 [ThaWri68, Don70]. A tree language is MSO definable if and only if it is regular.

4. Monadic second-order definable functions

The main concept in this paper is that of an MSO definable function f on graphs, introduced in [Eng88] and [Cou89/90b/91]. It is inspired by the notion of interpretation in [ArnLagSee88]. For a given input graph g, f describes the nodes, edges, and labels of the output graph g′ in terms of MSOL formulas on g. For each node label σ of g′ there is a formula $\phi_\sigma(u)$ expressing that u will be a node of g′ with label σ. Thus, the nodes of g′ are a subset of the nodes of g. For each edge label γ of g′ there is a formula $\phi_\gamma(u,v)$

expressing that there will be an edge from u to v in g′, with label γ. Finally, to allow partial functions, there is a closed formula ϕ_{dom} that describes the domain of f. For a partial function f, we will denote its domain by dom(f).

Definition 6. Let Σ_i and Γ_i be alphabets, for i = 1,2. An MSO definable function (or just MSO function) f: $GR(\Sigma_1,\Gamma_1) \to GR(\Sigma_2,\Gamma_2)$ is specified by formulas in $MSOL(\Sigma_1,\Gamma_1)$, as follows:

- a closed formula ϕ_{dom}, the domain formula,
- a formula $\phi_\sigma(u)$ with one free variable u, for every $\sigma \in \Sigma_2$, the node formulas,
- a formula $\phi_\gamma(u,v)$ with two free variables u and v, for every $\gamma \in \Gamma_2$, the edge formulas.

The domain of f is $\{g \in GR(\Sigma_1,\Gamma_1) \mid g \models \phi_{dom}\}$. For every g ∈ dom(f), f(g) is defined to be the graph $(V,E,\lambda) \in GR(\Sigma_2,\Gamma_2)$ such that
$V = \{u \in V_g \mid$ there is exactly one $\sigma \in \Sigma$ such that $g \models \phi_\sigma(u)\}$,
$E = \{(u,\gamma,v) \mid u,v \in V,\ u \neq v$, and $g \models \phi_\gamma(u,v)\}$, and
for $u \in V$, $\lambda(u) = \sigma$ where $g \models \phi_\sigma(u)$. □

The class of all MSO definable functions is denoted MSOF.

Note that the domain of an MSO definable function is an MSO definable language. Note that a node u of g may not be a node of f(g) for two reasons: either there is no σ such that $g \models \phi_\sigma(u)$ or there are more than one such σ. However, it is easy to see that we may always assume the $\phi_\sigma(u)$ to be mutually exclusive, in which case only the first reason remains.

Examples 7. (1) As first example we consider an MSO function f that is defined on every acyclic graph in $GR(\Sigma,\Gamma)$ and computes its transitive closure. Let τ be a new edge label, that stands for "transitive edge". Then f: $GR(\Sigma,\Gamma) \to GR(\Sigma,\Gamma\cup\{\tau\})$. First of all, $\phi_{dom} \equiv \neg(\exists u\ \exists v: \neg(u = v) \wedge path(u,v) \wedge path(v,u))$ expresses that the input graph g should be acyclic. For every $\sigma \in \Sigma$, $\phi_\sigma(u) \equiv lab_\sigma(u)$, i.e., every node of g is a node of f(g), with the same label. Finally, $\phi_\gamma(u,v) \equiv edge_\gamma(u,v)$ for every $\gamma \in \Gamma$, and $\phi_\tau(u,v) \equiv path(u,v)$. Thus, the edges of g remain in f(g), but f(g) also contains all transitive edges.

(2) We define an MSO function f that translates certain string-like trees into "ladders" (cf. Example 4(1)). Let $\Delta = \{a,b,c\}$ be the ranked alphabet with rank(a) = rank(b) = 1 and rank(c) = 0. In Fig.5 it is shown how f translates the tree abababac into the ladder of Fig.2. In general, f translates the tree $(ab)^n ac$ into a ladder with n+1 steps, $n \geq 0$. Thus, f: $GR(\Delta,\Gamma(\Delta)) \to GR(\{q,p\},\{\gamma\})$ where $\Gamma(\Delta) = \{1\}$. The formula ϕ_{dom} defines the trees

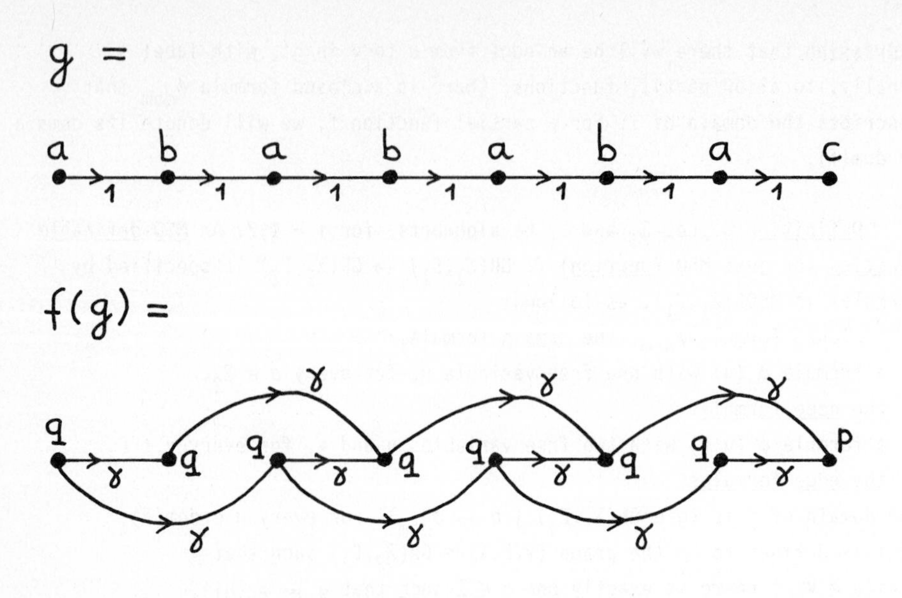

Figure 5

corresponding to the regular string language $(ab)^*ac$, cf. Proposition 5. The formula $\phi_p(u)$ is $lab_c(u)$, and the formula $\phi_q(u)$ is $\neg lab_c(u)$. The formula $\phi_\gamma(u,v)$ is $(lab_a(u) \wedge edge_1(u,v)) \vee (lab_a(u) \wedge t(u,v)) \vee (lab_b(u) \wedge t(u,v))$ where $t(u,v) \equiv \exists w: edge_1(u,w) \wedge edge_1(w,v)$.

For the sake of another example, assume that we also wish to label the "lower left-most" node of each ladder by p rather than q. Then $\phi_p(u)$ should express that either u has label c or u is the root of the tree (i.e., u has no incoming edge). Thus, $\phi_p(u) \equiv lab_c(u) \vee \neg\exists v{:}edge_1(v,u)$.

(3) For the notion of <u>co-graph</u> see Example 4(2). A <u>co-tree</u> is a tree in T_Δ where Δ is the ranked alphabet $\{+,*,q\}$ with rank(+) = rank(*) = 2 and rank(q) = 0. It is well known that the co-graph g corresponding to a co-tree $t \in T_\Delta$ (in the sense that t represents the recursive definition of g) can be defined as follows: the nodes of g are the leaves of t, and there is an edge between u and v in g if and only if the least common ancestor of u and v in t has label *. From this it easily follows that the translation from co-tree to co-graph is MSO definable. For ϕ_{dom} one can take the formula that expresses that the input graph is a tree over Δ (see (1) and (2) in the definition of tree). Taking q as the unique label of the nodes of the co-graph, $\phi_q(u) \equiv lab_q(u)$; note that this implies that all tree nodes with label * or + are

dropped. Taking γ as the unique label of its edges, $\phi_\gamma(u,v)$ expresses that the least common ancestor of u and v has label *; note that the following formula expresses that w is the least common ancestor of u and v: $path(w,u) \wedge path(w,v) \wedge \forall w': (path(w',u) \wedge path(w',v)) \rightarrow path(w',w)$. □

We now state the main result of this paper. Since the proof is quite involved, it is omitted. A full proof is given in [Oos], and will be available in two technical reports.

Theorem 8. For a graph language $L \subseteq GR(\Sigma,\Gamma)$, $L \in$ C-edNCE if and only if there exist a ranked alphabet Δ and an MSO definable function $f: GR(\Delta,\Gamma(\Delta)) \rightarrow GR(\Sigma,\Gamma)$ such that $L = f(T_\Delta)$, i.e., $L = \{f(t) \mid t \in T_\Delta, t \in dom(f)\}$.

This theorem can be written as C-edNCE = MSOF(TREES), where TREES = $\{T_\Delta \mid \Delta$ is a ranked alphabet$\}$ and MSOF(TREES) = $\{f(T) \mid f \in$ MSOF and $T \in$ TREES$\}$. The result is effective, i.e., given a C-edNCE grammar one can construct the ranked alphabet Δ and the MSOL formulas specifying f, and vice versa. A similar result holds for LIN-edNCE: Δ should be restricted to be monadic (i.e., $rank(\delta) \leq 1$ for all $\delta \in \Delta$). A generalization of Theorem 8 to hypergraphs (and structures) is presented in [Cou90b/91].

The hardest part of the proof of Theorem 8 is to show that MSOF(TREES) $\subseteq$ C-edNCE. The direction C-edNCE $\subseteq$ MSOF(TREES) is easier; to give an idea of the proof, let $G = (\Sigma,\Delta,\Gamma,\Omega,P,S)$ be a C-edNCE grammar such that each right-hand side of a production contains at most one terminal node (it is easy to see that this may always be assumed). Let $R \subseteq T_\Delta$ be the (regular) set of all derivation trees of G, where each node of a derivation tree is labeled by a production, i.e., Δ is P with the obvious ranks. We now define f in such a way that its domain is R (see Proposition 5). Moreover, the node formula $\phi_\sigma(u)$ is the disjunction of all $lab_\pi(u)$ where π is a production with a terminal node labeled σ. The main technical difficulty is to define the edge formulas of f.

The above construction is illustrated by Example 4(1) and Example 7(2), and also by Example 4(2) and Example 7(3). Note, e.g., that the derivation trees of G_2 are exactly the co-trees, and that the co-graph corresponding to a co-tree t is exactly the graph generated by G_2 according to the derivation tree t.

In the remainder of this paper we show that Theorem 8 implies some quite strong closure and decidability properties of C-edNCE.

5. Closure properties

The main result of this section is that C-edNCE is closed under MSO definable functions, i.e., if $L \in$ C-edNCE and $f \in$ MSOF, then $f(L) = \{f(g) \mid g \in L, g \in \text{dom}(f)\}$ is in C-edNCE. This follows immediately from Theorem 8 and the fact that MSOF is closed under composition (cf. Lemma 4.4 of [ArnLagSee88]).

Lemma 9. MSOF is closed under composition.

Proof. Let f: $GR(\Sigma_1,\Gamma_1) \to GR(\Sigma_2,\Gamma_2)$ and f': $GR(\Sigma_2,\Gamma_2) \to GR(\Sigma_3,\Gamma_3)$ be MSO functions. The formulas of f will be indicated by ϕ and those of f' by ψ. We first show how to turn any formula ψ of $MSOL(\Sigma_2,\Gamma_2)$ into a formula $f^{-1}(\psi)$ of $MSOL(\Sigma_1,\Gamma_1)$ such that, for every g in the domain of f, $f(g) \models \psi$ iff $g \models f^{-1}(\psi)$, and similarly for the case that ψ has free variables. The formula $f^{-1}(\psi)$ is obtained from ψ by making the following changes: change every $lab_\sigma(u)$ into $\phi_\sigma(u)$, every $edge_\gamma(u,v)$ into $\phi_\gamma(u,v)$, every $\exists u{:}(\cdots)$ into $\exists u{:}\ (node(u) \wedge \cdots)$, where node(u) is the formula expressing that u will be a node of the output graph of f (an obvious boolean combination of the node formulas of f), and similarly for $\forall u$, $\exists U$, and $\forall U$.

The composition h of f and f' is now specified by formulas η as follows: $\eta_{dom} \equiv \phi_{dom} \wedge f^{-1}(\psi_{dom})$; for every $\sigma \in \Sigma_3$, $\eta_\sigma(u) \equiv node(u) \wedge f^{-1}(\psi_\sigma(u))$ where node(u) is the same formula as above; for every $\gamma \in \Gamma_3$, $\eta_\gamma(u,v) \equiv f^{-1}(\psi_\gamma(u,v))$. □

Theorem 10. C-edNCE is closed under MSO definable functions.

The same result holds for LIN-edNCE (but not for B-edNCE).

As a corollary of Theorem 10 we reobtain Courcelle's closure result, for C-edNCE (see Theorem 3 of [CouEngRoz90]): C-edNCE is closed under intersection with MSO definable languages. In fact, if L is an MSO definable language, then the identity on L is an MSO function (and applying it to a language has the effect of intersection): if L is defined by the closed formula ϕ, then define f by taking this ϕ as domain formula, and taking all $lab_\sigma(u)$ and all $edge_\gamma(u,v)$ as node and edge formulas, respectively. We also obtain the known result [Eng89a] that C-edNCE is closed under taking edge complements: assuming that there is just one edge label γ, define f by taking $\neg edge_\gamma(u,v)$ as edge formula (and all $lab_\sigma(u)$ as node formulas, and true as domain formula). Example 7(1) shows that C-edNCE is closed under taking the transitive closure of each graph, and similarly it is easy to see that one could also throw away all existing transitive edges, i.e., turn an acyclic graph into its Hasse diagram. In general one can add or remove edges that satisfy certain MSO properties (or rather their incident nodes satisfy them); and similarly one can remove nodes that satisfy MSO properties (e.g., one can remove all isolated nodes from all graphs of the language). In [Cou89/90b/91] MSO definable functions are considered that can also add nodes; our results hold for these functions too.

As another consequence of Theorem 10 we show that if L is a C-edNCE language, then so is the language of all induced subgraphs (of graphs of L) that satisfy a given MSO property. For a graph g and $U \subseteq V_g$ we denote by g[U] the subgraph of g induced by U. Let $\phi(U)$ be an MSOL formula with one free variable U. We say that g[U] is a ϕ-subgraph of g if $g \models \phi(U)$. We show that the set of all ϕ-subgraphs of graphs in L is in C-edNCE.

Theorem 11. Let $\phi(U)$ be a formula in MSOL(Σ,Γ) with one free variable U, and let $L \subseteq GR(\Sigma,\Gamma)$.
If $L \in$ C-edNCE, then $\{g[U] \mid g \in L, U \subseteq V_g, g \models \phi(U)\} \in$ C-edNCE.
Proof. Let $\Omega = \Sigma \times \{0,1\}$, and let rel(L) be the language of all graphs in GR(Ω,Γ) that are obtained from the graphs in L by changing the node labels σ into $(\sigma,1)$ or $(\sigma,0)$ in all possible ways. Then rel(L) is in C-edNCE: if $L = f(T_\Delta)$ for some $f \in$ MSOF and some ranked alphabet Δ, then $rel(L) = f'(T_{\Delta\times\{0,1\}})$ where the formulas of f′ are obtained from those of f by changing every $lab_\delta(u)$ into $lab_{(\delta,0)}(u) \vee lab_{(\delta,1)}(u)$, and then defining (for i = 1,2) the node formula $\phi_{(\sigma,i)}(u)$ to be the conjunction of the (changed) node formula $\phi_\sigma(u)$ of f and the disjunction of all $lab_{(\delta,i)}(u)$ for $\delta \in \Delta$. It is easy to see that there is an MSO definable function h that translates rel(L) into the required language (and then the result follows from Theorem 10): the domain formula of h expresses the fact that the set U of all nodes that have bit 1 in their label satisfies $\phi(U)$, and h has node formulas $\phi_\sigma(u) \equiv lab_{(\sigma,1)}(u)$, and edge formulas $edge_\gamma(u,v)$. □

As an example, if L is in C-edNCE, then the set of all strongly connected components of graphs in L is also in C-edNCE (and the same is true for connected components).

6. Decidability properties

By Theorem 2 it is decidable whether a C-edNCE grammar generates a finite graph language. Thus, it follows from Theorem 10 that it is decidable for a C-edNCE language L and an MSO function f whether f(L) is finite. This implies that certain boundedness problems are decidable for C-edNCE (cf. [HabKreVog 89a] where boundedness problems are investigated for hyperedge replacement systems). As an example, let f be the MSO function that transforms every graph into the discrete graph consisting of all its isolated nodes. Then the above result shows that it is decidable for a C-edNCE language L whether there is a bound on the number of isolated nodes in the graphs of L. We now show two general decidability results on boundedness. For a formula $\phi(U)$ and a graph g, let $size_\phi(g)$ denote the maximal number of nodes of a ϕ-subgraph of g, and let $num_\phi(g)$ denote the number of ϕ-subgraphs of g (where isomorphic ϕ-subgraphs are not identified). We will show that it is decidable for a C-edNCE language L whether $size_\phi(g)$ is bounded on L, and similarly for $num_\phi(g)$.

Theorem 12. Let $\phi(U)$ be an MSOL formula with one free variable U. It is decidable for a graph language $L \in$ C-edNCE whether or not there exists $b \in \mathbb{N}$ such that $size_\phi(g) \leq b$ for all $g \in L$.

Proof. Let $L_\phi = \{g[U] \mid g \in L, U \subseteq V_g, g \models \phi(U)\}$. It should be clear that there is a bound on $size_\phi(g)$ if and only if L_ϕ is finite. Since Theorem 11 is effective, the language L_ϕ can be obtained from L and tested on finiteness (by Theorem 2). □

Theorem 13. Let $\phi(U)$ be an MSOL formula with one free variable U. It is decidable for a graph language $L \in$ C-edNCE whether or not there exists $b \in \mathbb{N}$ such that $num_\phi(g) \leq b$ for all $g \in L$.

Proof. For a graph g and two nodes u,v of g, define u and v to be ϕ-equivalent (notation: $u \equiv_\phi v$) if for all $U \subseteq V_g$ with $g \models \phi(U)$: $u \in U$ iff $v \in U$. Thus, u and v are in the same ϕ-subgraphs of g. Let $eq_\phi(g)$ be the number of equivalence classes of $\equiv_\phi$ in g. Clearly, $eq_\phi(g) \leq 2^{num_\phi(g)}$ and $num_\phi(g) \leq 2^{eq_\phi(g)}$. Hence, $num_\phi(g)$ is bounded on L iff $eq_\phi(g)$ is bounded on L. To decide boundedness of $eq_\phi(g)$, we consider representatives. For a graph g, define a ϕ-representative set to be a subset of V_g that contains exactly one node from each equivalence class of $\equiv_\phi$. Let $\psi(V)$ be the MSOL formula that expresses that V is a ϕ-representative set; this formula is easy to find, note that $u \equiv_\phi v$ is expressed by $\forall U: \phi(U) \rightarrow (u \in U \leftrightarrow v \in U)$. Then, clearly, $eq_\phi(g)$ is bounded on L iff $size_\psi(g)$ is bounded on L. The latter is decidable by Theorem 12. □

It follows from these theorems that (almost) all concrete decidability results proved for hyperedge replacement systems in [HabKreVog89a] are also decidable for C-edNCE grammars. The meta-results of [HabKreVog89a] can of course not be compared with the above meta-results, because they are of another nature. However, the above results seem to be easier to use than those in

[HabKreVog89a]: one just has to express a certain property in MSOL to obtain decidability. Let us give some examples.

It is decidable whether a C-edNCE language is of bounded degree. Let us say that a set U of nodes of a graph g is a "neighbourhood" if it consists of all neighbours of some node of g. Obviously, there is an MSOL formula $\phi(U)$ expressing that U is a neighbourhood. Thus, bounded degree is decidable by Theorem 12.

It is decidable whether there is a bound on the number of strongly connected components of the graphs in L. This follows from Theorem 13 because there is an MSOL formula expressing that U is a strongly connected component. Theorem 12 shows that it is decidable whether there is a bound on the size of strongly connected components. The same results hold for connected components.

The one result that cannot be generalized directly is the following. It is proved in [HabKreVog89a] that it is decidable for hyperedge replacement systems whether there is a bound on the length (or number) of simple paths between certain nodes (where a path is simple if it contains no repetitions of nodes). Since simplicity is not expressible in our MSOL (just as Hamiltonicity, cf. [Cou90a]), we will consider minimal paths instead: a path $v_1,\dots,v_n$ is minimal if there are no "shortcuts", i.e., no edges (v_i,γ,v_j) with $i+1 < j$. Let σ and τ be node labels. Let $\phi(U)$ express that U consists of the nodes of a minimal path from some node labeled σ to some node labeled τ: $\phi(U) \equiv \exists u,v\colon lab_\sigma(u) \wedge lab_\tau(v) \wedge minpath(u,v,U)$, where $minpath(u,v,U) \equiv path(u,v,U) \wedge \forall V\colon ((V \subseteq U \wedge path(u,v,V)) \rightarrow V = U)$, and $path(u,v,U)$ expresses that there is a path from u to v of which all nodes are in U (this last formula can easily be obtained by relativizing the formula $path(u,v)$ to U). Then Theorem 12 implies that it is decidable for $L \in$ C-edNCE whether there is a bound on the length of minimal paths from σ-nodes to τ-nodes in the graphs of L. Theorem 13 implies the same result for the number of such minimal paths. In fact, from these results those for hyperedge replacement languages can be reobtained: it is not difficult to see that every such language L can be transformed into a language L′ such that the size (or number) of simple paths of L is bounded iff the size (or number) of minimal paths of L′ is bounded (divide each edge into two edges, by introducing a new node for each edge).

Let $\phi(U)$ be the property that U is a (maximal) clique. Theorem 12 shows that it is decidable for $L \in$ C-edNCE whether there is a bound on the size of cliques in the graphs of L. This was shown for B-NLC in [RozWel87]; hyperedge replacement languages always have bounded clique size. Theorem 13 shows that it is decidable whether the number of cliques is bounded for the graphs of L.

Note that, of course, all results of this section hold for classes of languages that are effectively contained in C-edNCE, such as the hyperedge replacement languages [EngRoz90], the B-NLC and B-edNCE languages [EngLeiWel90], and the S-HH languages [CouEngRoz91].

Acknowledgment. I am grateful to my student Vincent van Oostrom for his help with the proof of Theorem 8.

References

[ArnLagSee88] S.Arnborg, J.Lagergren, D.Seese; Problems easy for tree-decomposable graphs, Proc. ICALP '88, Lecture Notes in Computer Science 317, Springer-Verlag, Berlin, 1988, 38-51

[BauCou87] M.Bauderon, B.Courcelle; Graph expressions and graph rewritings, Math. Systems Theory 20 (1987), 83-127

[Bra88] F.J.Brandenburg; On polynomial time graph grammars, Proc. STACS 88, Lecture Notes in Computer Science 294, Springer-Verlag, Berlin, 227-236

[Büc60] J.R.Büchi; Weak second-order arithmetic and finite automata, Zeitschr. f. Math. Logik und Grundlagen d. Math. 6 (1960), 66-92

[CorLerSte81] D.G.Corneil, H.Lerchs, L.Stewart Burlingham; Complement reducible graphs, Discr. Appl. Math. 3 (1981), 163-174

[Cou87] B.Courcelle; An axiomatic definition of context-free rewriting and its application to NLC graph grammars, Theor. Comput. Sci. 55 (1987), 141-181

[Cou89] B.Courcelle; The monadic second-order logic of graphs. V: On closing the gap between definability and recognizability, Report 89-91, LaBRI, University of Bordeaux, 1989, to appear in TCS

[Cou90a] B.Courcelle; The monadic second-order logic of graphs. I: Recognizable sets of finite graphs, Inform. Comput. 85 (1990), 12-75. See also [EhrNagRozRos87], 112-133

[Cou90b] B.Courcelle; The monadic second-order logic of graphs. VII: Graphs as relational structures, Manuscript, University of Bordeaux, 1990

[Cou91] B.Courcelle; Graphs as relational structures: an algebraic and logical approach, this Volume

[CouEngRoz91] B.Courcelle, J.Engelfriet, G.Rozenberg; Context-free handle-rewriting hypergraph grammars, this Volume

[Don70] J.Doner; Tree acceptors and some of their applications, J. Comput. System Sci. 4 (1970), 406-451

[EhrNagRozRos87] H.Ehrig, M.Nagl, G.Rozenberg, A.Rosenfeld (eds.); Graph-Grammars and their Application to Computer Science, Lecture Notes in Computer Science 291, Springer-Verlag, Berlin, 1987

[Elg61] C.C.Elgot; Decision problems of finite automata and related arithmetics, Trans. Amer. Math. Soc. 98 (1961), 21-51

[Eng88] J.Engelfriet; Monadic second-order logic for graphs, trees, and strings, Copies of transparencies, November 1988

[Eng89a] J.Engelfriet; Context-free NCE graph grammars, Proc. FCT '89, Lecture Notes in Computer Science 380, Springer-Verlag, Berlin, 1989, 148-161

[Eng89b] J.Engelfriet; A regular characterization of graph languages definable in monadic second-order logic, Report 89-03, Leiden University, 1989, to appear in TCS

[EngLei89] J.Engelfriet, G.Leih; Linear graph grammars: power and complexity, Inform. Comput. 81 (1989), 88-121

[EngLeiRoz87] J.Engelfriet, G.Leih, G.Rozenberg; Apex graph grammars, in [EhrNagRozRos87], 167-185

[EngLeiWel90] J.Engelfriet, G.Leih, E.Welzl; Boundary graph grammars with dynamic edge relabeling, J. Comput. System Sci. 40 (1990), 307-345

[EngRoz90] J.Engelfriet, G.Rozenberg; A comparison of boundary graph grammars and context-free hypergraph grammars, Inform. Comput. 84 (1990), 163-206

[GecSte84] F.Gécseg, M.Steinby; "Tree automata", Akadémiai Kiadó, Budapest, 1984

[HabKre87] A.Habel, H.-J.Kreowski; May we introduce to you: hyperedge replacement, in [EhrNagRozRos87], 15-26

[HabKreVog89a] A.Habel, H.-J.Kreowski, W.Vogler; Decidable boundedness problems for hyperedge-replacement graph grammars, in Proc. TAPSOFT '89, Lecture Notes in Computer Science 351, Springer-Verlag, Berlin, 1989, 275-289

[HabKreVog89b] A.Habel, H.-J.Kreowski, W.Vogler; Metatheorems for decision problems on hyperedge replacement graph languages, Acta Informatica 26 (1989), 657-677

[Kau85] M.Kaul; Syntaxanalyse von Graphen bei Präzedenz-Graph-Grammatiken, Ph.D.Thesis, Osnabrück, 1985

[LenWan88] T.Lengauer, E.Wanke; Efficient analysis of graph properties on context-free graph languages, Proc. ICALP '88, Lecture Notes in Computer Science 317, Springer-Verlag, Berlin, 1988, 379-393

[Nag79] M.Nagl; "Graph-grammatiken", Vieweg, Braunschweig, 1979.

[Oos89] V.van Oostrom; Graph grammars and 2nd order logic (in Dutch), M. Sc. Thesis, 1989

[Roz87] G.Rozenberg; An introduction to the NLC way of rewriting graphs, in [EhrNagRozRos87], 55-66

[RozWel86] G.Rozenberg, E.Welzl; Boundary NLC graph grammars - basic definitions, normal forms, and complexity, Inform. Contr. 69 (1986), 136-167

[RozWel87] G.Rozenberg, E.Welzl; Combinatorial properties of boundary NLC graph languages, Discr. Appl. Math. 16 (1987), 59-73

[Sch87] R.Schuster; Graphgrammatiken und Grapheinbettungen: Algorithmen und Komplexität, Ph.D.Thesis, Report MIP-8711, Passau, 1987

[ThaWri68] J.W.Thatcher, J.B.Wright; Generalized finite automata theory with an application to a decision problem of second-order logic, Math. Syst. Theory 2 (1968), 57-81

The term generating power of context-free hypergraph grammars

Joost Engelfriet & Linda Heyker

Dept. of Computer Science, Leiden University
P.O. Box 9512, 2300 RA Leiden, The Netherlands

ABSTRACT. Context-free hypergraph grammars and attribute grammars generate the same class of term languages. Extending the context-free hypergraph grammar with a context-free grammar and a semantic domain, a syntax-directed translation device is obtained that is equivalent to the attribute grammar.

Keywords: graph grammar, hypergraph, term, attribute grammar, syntax-directed translation.

CONTENTS

1. Introduction

An attractive feature of (hyper)graphs is that they can be used to represent other structures such as strings, and terms (trees, expressions). Thus a graph grammar ([EhrNagRosRoz]) can be used as a term generating device (or tree grammar), generating a set of terms. We investigate the power of context-free hypergraph grammars (cfhg's), introduced by Habel & Kreowski [HabKre, Hab] and Bauderon & Courcelle [BauCou], to generate terms, with sharing allowed (see [HabKrePlu] for the connection between hypergraph grammars and term rewriting systems).

In particular, cfhg's are compared to attribute grammars (AG's, [Knu,

DerJouLor]). An AG can be viewed as a device that translates strings into terms (i.e., expressions to be evaluated in the semantic domain of the AG). The range of this translation is a set of terms, the term language generated by the AG. Our first main result is that cfhg's have the same term generating power as attribute grammars. Thus, results on the class of term languages generated by AG's can be transferred to cfhg's.

By coupling a context-free grammar (cfg) and a semantic domain to the cfhg, a syntax-directed translation device is obtained that translates the strings generated by the cfg into the terms generated by the cfhg, and hence into values from the semantic domain. Our second main result is that such a cfhg-based translation device has the same power as the attribute grammar. In some sense, this result "explains" AG's to be syntax-directed string-to-graph translators. On the other hand, it shows that cfhg's can be used to describe the semantics of programming languages (cf. the "push-down processor" of [AhoUll]).

The relationship between attribute grammars and graph grammars has been studied before. For example, Courcelle has shown that the set of dependency graphs of an attribute grammar can be generated by a hypergraph grammar ([Cou1]), and a similar result for NLC-like grammars is shown in [EngLeiRoz]. We shall use cfhg's not only to simulate the dependencies between attributes but also to simulate (formal) attribute evaluation, in a natural way. In [Hof] a more involved kind of graph grammar is used to model parsing and evaluation together.

The full version of this paper is presented in [EngHey2].

2. Terms and hypergraphs

A ranked alphabet Γ is a finite set of (function) symbols together with a mapping rank: $\Gamma \rightarrow \mathbb{N}$, where $\mathbb{N} = \{0,1,2,...\}$. A term over a ranked alphabet Γ is, as usual, a well-formed expression written with Γ, commas, and parentheses in prefix notation. A term language is a subset of $T(\Gamma)$, the set of all terms over Γ. A term can be represented by a hypergraph in a natural way (see [Cou2, HabKrePlu]), because the order of its subterms can be expressed using the order on the incident nodes of each edge. A hypergraph (over a ranked alphabet Σ) is here a tuple $H = (V,E,nod,lab,ext)$, where V is the finite set of nodes, E is the finite set of hyperedges (or just edges), nod: $E \rightarrow V^*$ is the incidence function, lab: $E \rightarrow \Sigma$ is the labeling function, and ext $\in V^*$ is the sequence of external nodes (the "gluing points" that are used in the edge rewriting mechanism of the hypergraph grammars) [1]. It is required that for every $e \in E$, $rank(lab(e)) = \#nod(e)$ [2]; $\#nod(e)$ is called the rank of e, denoted rank(e). The i-th element of a sequence nod(e) is denoted nod(e,i), and we say that e and nod(e,i) are incident. Similarly, the i-th element of ext is denoted ext(i); #ext is called the rank of H, denoted rank(H). For a given hypergraph H, its components are denoted by V_H, E_H, nod_H, lab_H, and ext_H, respectively. The set of all hypergraphs over a ranked alphabet Σ is denoted $HGR(\Sigma)$.

To give a formal definition of hypergraphs that represent terms (the "jungles" in [HabKrePlu]) we need some more terminology. The "direction" of a hyperedge e with $nod(e) = (v_1,v_2,\ldots,v_r)$, $r \geq 1$, is as follows. The nodes $v_1,v_2,\ldots,v_{r-1}$ are the "source" nodes of e, the nodes that represent the arguments of the function symbol lab(e), and we say that e is an _outgoing_ edge of these nodes. The node v_r is the "target" node of e, the node representing the result of lab(e) applied to the arguments, and e is called an _incoming_ edge of v_r. In the sense of [HabKre, Hab] e is a hyperedge of type (r-1,1). With this notion of direction, a _hyperpath_ in a hypergraph H from v to w (of length n) is a sequence $(v_0,v_1,\ldots,v_n) \in V_H^{n+1}$ with $n \geq 0$, $v_0 = v$, and $v_n = w$ such that for every $i \in [1,n]$ [3] there exists an edge $e \in E_H$ that is outgoing for v_{i-1} and incoming for v_i.

Definition. Let H be a hypergraph. H is a _hyperterm_ if
(1) H has no hyperedges of rank 0,
(2) rank(H) = 1,
(3) each $v \in V_H$ has exactly one incoming hyperedge (i.e., every node represents the result of one function), and
(4) H is acyclic, i.e., for no $v \in V_H$ there is a hyperpath from v to v of positive length. □

With each node v of a hyperterm H a term is associated: term(v,H) is the term in $T(dec(\Sigma))$ [4] recursively defined as $lab_H(e)$ if the incoming hyperedge e of v has rank 1, and $lab_H(e)(term(nod_H(e,1),H),\ldots,term(nod_H(e,r-1),H))$ if the incoming hyperedge e of v has rank $r \geq 2$. We abbreviate $term(ext_H(1),H)$ to term(H): the _term associated with_ H.

---

1) For a set A, A^* denotes the set of all finite sequences of elements of A. In case A is an alphabet, a sequence $(a_1,a_2,\ldots,a_n)$ of A^* will be written as a string $a_1a_2\ldots a_n$.

2) By #A we denote the cardinality of a set A or the length of a sequence A.

3) For $n,m \in \mathbb{N}$, $[n,m]$ denotes the set $\{i \in \mathbb{N} \mid n \leq i \leq m\}$.

4) For a ranked alphabet Γ, $dec(\Gamma)$ denotes the alphabet $\{\gamma \in \Gamma \mid rank(\gamma) \geq 1\}$ of which for every symbol the rank is decreased by one.

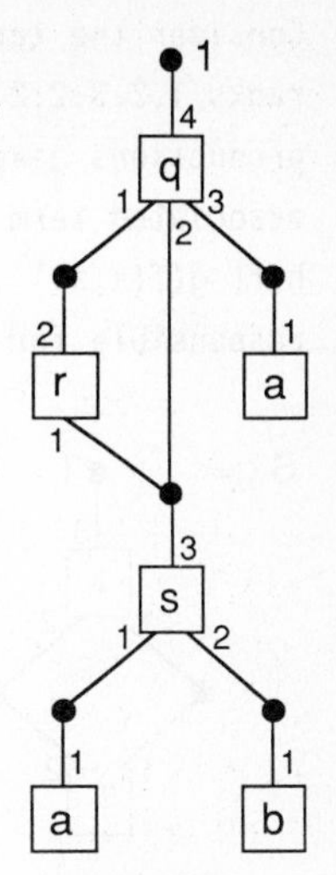

Figure 1

Notice that sharing of subterms is allowed, but not obliged. Thus a hyperterm has a unique associated term, but a term can be associated with more than one hyperterm. Observe also that sharing cannot be "seen" anymore in the terms associated with hyperterms, just like in the unfolding of DOAG's (directed ordered acyclic graphs).

The hypergraph H, over $\Sigma = \{q,r,s,a,b\}$ with ranks 4,2,3,1,1, respectively, given in Fig.1 satisfies the four demands. The lines connecting the edges (boxes) with their incident nodes (fat dots) are called tentacles. The integers at these tentacles numbered from 1 to the rank of the edge indicate the order of the incident nodes; node $ext_H(i)$ is labeled i (in this case i = 1 only). It should be clear that term(H) = q(r(s(a,b)),s(a,b),a).

We now turn to context-free hypergraph grammars (cfhg's), see, e.g., [BauCou, HabKre, Hab]. A context-free hypergraph grammar is a tuple $G = (\Sigma,\Delta,P,S)$, where Σ is a ranked alphabet, $\Delta \subseteq \Sigma$ is the terminal alphabet (and Σ-Δ is the nonterminal alphabet), P is the finite set of productions, and $S \in \Sigma$-Δ is the initial nonterminal. Every production $\pi \in P$ is of the form (X,H) with left-hand side $lhs(\pi) = X \in \Sigma$-Δ, right-hand side $rhs(\pi) = H \in HGR(\Sigma)$, and rank(X) = rank(H). Application of a production (X,H) to a nonterminal hyperedge e of a hypergraph $K \in HGR(\Sigma)$ can only take place if $lab_K(e) = X$, and consists of substituting H for e in K, identifying $nod_K(e,i)$ with $ext_H(i)$ for all $i \in [1,rank(X)]$. If H and K are not disjoint, an isomorphic copy of H should be taken. The result is indicated by K[e/H] and we write $K \Longrightarrow K'$, if K' is (isomorphic to) K[e/H]. A sentential form of G is a hypergraph $K \in HGR(\Sigma)$ that can be derived from sing(S), the hypergraph ([1,r],{e},nod,lab,ext) with nod(e) = ext = (1,2,...,r), and lab(e) = S, where rank(S) = r. The hypergraph language generated by G is $L(G) = \{H \in HGR(\Delta) \mid sing(S) \Longrightarrow^* H\}$. Note that rank(H) = rank(S) for all $H \in L(G)$. The notion of a derivation tree of a cfhg can be defined in a way similar to that of an ordinary context-free grammar (cf., e.g., [Lau, EngHey2]); for a derivation tree t we denote by yield(t) the hypergraph generated according to any derivation corresponding to t.

Definition. A cfhg $G = (\Sigma,\Delta,P,S)$ is term-generating if every hypergraph in L(G) is a hyperterm. The term language generated by G is $TERM(G) = \{term(H) \mid H \in L(G)\} \subseteq T(dec(\Delta))$. □

Consider the term-generating cfhg $G = (\Sigma,\Delta,P,S)$, where $\Sigma = \{S,T,f,g,h,a\}$ with ranks 1,2,3,2,2,1, respectively, $\Delta = \{f,g,h,a\}$, and P consists of the three productions given in Fig.2. For the hyperterm $H \in L(G)$ of Fig.3, the associated term in $T(dec(\Delta))$ is term(H) = f(g(f(g(f(a,a)) , h(f(a,a)))) , h(f(g(f(a,a)) , h(f(a,a))))). Application of the middle production is responsible for sharing of the subterms f(a,a) and f(g(f(a,a)) , h(f(a,a))).

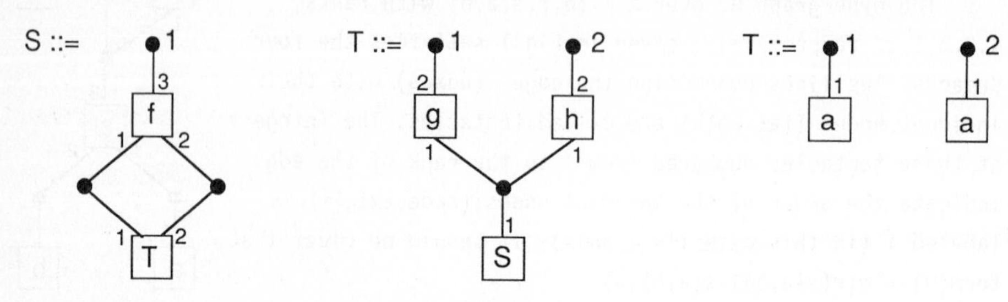

Figure 2

The term language generated by this grammar can be described inductively by (i) $f(a,a) \in TERM(G)$, and (ii) if $t \in TERM(G)$ then $f(g(t),h(t)) \in TERM(G)$. Notice that the right-hand sides of the productions of G are not hyperterms: the first one violates demand (3), and the others demand (2); also, the sentential forms of G in which T occurs are not hyperterms, because they violate demand (3).

The class of all term languages generated by cfhg's will be denoted as TERM(CFHG), i.e., TERM(CFHG) = {TERM(G) | G is a term-generating cfhg}. The definition of TERM(CFHG) is chosen to be suitable for the comparison of cfhg's with attribute grammars as far as terms are concerned. Another definition could be obtained, for example, by restricting attention to a more natural way of representing a term by a hypergraph: a <u>clean</u> <u>hyperterm</u> is a hyperterm in which every hyperedge contributes to the term associated with it, i.e., from every node there exists a hyperpath to ext(1). In Section 3 it will be shown that the class {TERM(G) | G is a clean term-generating cfhg} equals TERM(CFHG).

Moreover we could just as well allow all cfhg's (generating hypergraphs of rank 1), and consider the terms associated with the (clean) hyperterms they generate. This is stated in the following theorem. At the same time we reassure the reader who is worried about the decidability of "(clean) term-generating".

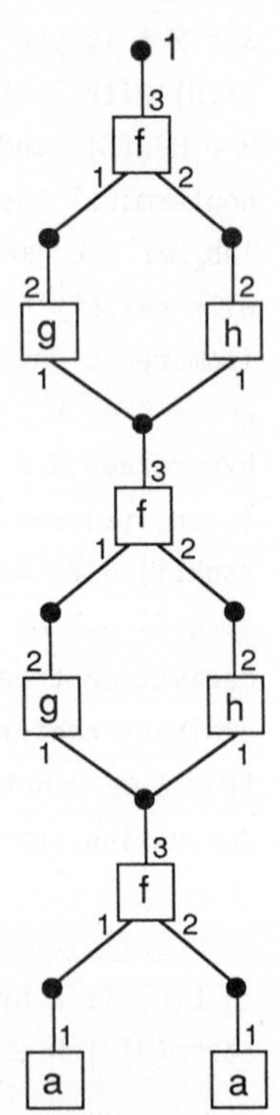

Figure 3

Theorem 1. Let G be a cfhg.
(1) It is decidable whether G is (clean) term-generating.
(2) {term(H) | H $\in$ L(G), H is a (clean) hyperterm} $\in$ TERM(CFHG).

Proof. Follows from results of [Cou3], because the property of a hypergraph to be a hyperterm can easily be expressed in monadic second order logic. □

3. Term generation: context-free hypergraph grammars versus attribute grammars

Comparing cfhg's to attribute grammars (AG's), we recall the notion of attribute grammar, with which the reader is assumed to be familiar (see, e.g., [Knu, DerJouLor]).

An attribute grammar G consists of (1)-(4) as follows.
(1) A context-free grammar (cfg) [5] $G_0 = (N_0,T_0,P_0,S_0)$ that is called the underlying grammar of G.
(2) A semantic domain $D = (V,\Gamma)$, where V is a set of values, Γ is a ranked alphabet, and each $\gamma \in \Gamma$ denotes a mapping γ_D: $V^n \longrightarrow V$ with n = rank(γ). Thus, D is a Γ-algebra where Γ is a one-sorted signature. For a term $t \in T(\Gamma)$, we denote by t_D the value of t in V, as usual.
(3) An attribute description (A,Syn,Inh,Att) where A is a finite set of attributes, and Syn, Inh, and Att are mappings from N_0 to 2^A. For each nonterminal $X \in N_0$, Syn(X) is the set of its synthesized attributes, and Inh(X) is the set of its inherited attributes, Syn(X) $\cap$ Inh(X) = $\emptyset$. Att(X) is the set of all its attributes, Att(X) = Syn(X) $\cup$ Inh(X). It is required that Inh(S_0) = $\emptyset$ and #Syn(S_0) = 1. The only (synthesized) attribute of S_0 is called the designated attribute of G and is denoted α_d.
(4) For each production $p = X_0 \longrightarrow w_0X_1w_1X_2w_2...X_nw_n$ a set of semantic rules r_p. For each $X_j\cdot\alpha \in$ ins(p), r_p contains one semantic rule of the form $X_j\cdot\alpha$ = t with $t \in T(\Gamma$,outs(p)) [6], where ins(p) = $\{X_i\cdot\beta \mid (\beta \in Inh(X_i)$ and $i \in [1,n])$ or $(\beta \in Syn(X_i)$ and $i = 0)\}$ is the set of inside attributes of p and outs(p) = $\{X_i\cdot\beta \mid (\beta \in Syn(X_i)$ and $i \in [1,n])$ or $(\beta \in Inh(X_i)$ and $i = 0)\}$ is the set of outside attributes of p.

---

5) For a cfg G = (N,T,P,S), N is the nonterminal alphabet, T is the terminal alphabet (disjoint with N), P is the finite set of productions of the form $p = X_0 \longrightarrow w_0X_1w_1X_2w_2...X_nw_n$, $n \geq 0$, all $X_j \in N$ and all $w_j \in T^*$, and $S \in N$ is the initial nonterminal. For a production p, its left-hand side X_0 is denoted lhs(p), and its right-hand side $w_0X_1w_1X_2w_2...X_nw_n$ is denoted rhs(p).

6) For a ranked alphabet Γ and a set Y with $\Gamma \cap Y = \emptyset$, $T(\Gamma,Y)$ denotes $T(\Gamma \cup Y)$ where rank(y) = 0 for all $y \in Y$.

As an example, consider the AG Gbin defined in (1.5) in [Knu]. In our notation this AG is defined as follows.

$G_0 = (N_0,T_0,P_0,S_0)$ where $N_0 = \{N,L,B\}$, $T_0 = \{0,1,\bullet\}$, $S_0 = N$, and $P_0 = \{p_1,p_2,\dots,p_6\}$ with $p_1 = N \longrightarrow L\bullet L$, $p_2 = N \longrightarrow L$, $p_3 = L \longrightarrow LB$, $p_4 = L \longrightarrow B$, $p_5 = B \longrightarrow 1$, and $p_6 = B \longrightarrow 0$.

$D = (V,\Gamma)$ where V is the set of rational numbers and $\Gamma = \{+,2\uparrow,-,0,1\}$ with ranks 2,1,1,0,0, respectively. For every $\gamma \in \Gamma$, the function γ_D is as expected.

$A = \{s,\ell,v\}$, where v represents the "value" of the nonterminals N, L, and B, ℓ represents the "length" of a list of bits L, and s represents the "scale" of a bit B, or of the last bit in a list L. $Inh(N) = \emptyset$, $Syn(N) = \{v\}$, $Inh(L) = \{s\}$, $Syn(L) = \{\ell,v\}$, $Inh(B) = \{s\}$, and $Syn(B) = \{v\}$. The designated attribute of Gbin is $\alpha_d = v$.

For the production $p_1 = N \longrightarrow L_1\bullet L_2$, $ins(p_1) = \{N\cdot v,\ L_1\cdot s,\ L_2\cdot s\}$, $outs(p_1) = \{L_1\cdot\ell,\ L_1\cdot v,\ L_2\cdot\ell,\ L_2\cdot v\}$, and its semantic rules are $N\cdot v = +(L_1\cdot v,\ L_2\cdot v)$, $L_1\cdot s = 0$, and $L_2\cdot s = -(L_2\cdot\ell)$. The other productions have the following semantic rules.

$p_2 = N \longrightarrow L$:	$N\cdot v\ = L\cdot v$,	$L\cdot s\ = 0$,
$p_3 = L_1 \longrightarrow L_2B$:	$L_1\cdot\ell\ = +(L_2\cdot\ell,\ 1)$,	$L_2\cdot s\ = +(L_1\cdot s,\ 1)$,
	$L_1\cdot v\ = +(L_2\cdot v,\ B\cdot v)$,	$B\cdot s\ = L_1\cdot s$,
$p_4 = L \longrightarrow B$:	$L\cdot\ell\ = 1$,	$B\cdot s\ = L\cdot s$,
	$L\cdot v\ = B\cdot v$,	
$p_5 = B \longrightarrow 1$:	$B\cdot v\ = 2\uparrow(B\cdot s)$, and	
$p_6 = B \longrightarrow 0$:	$B\cdot v\ = 0$.	

On the basis of this AG Gbin, we shall explain the (usual) way in which an AG translates strings generated by its underlying grammar into values of its semantic domain. Gbin translates binary numbers into their decimal values as follows. For example consider the string $1101\bullet01 \in L(G_0)$. The dependency graph of the derivation tree t of G_0 that yields this string is shown in Fig.4 (see also (3.1) in [Knu]). The encircled symbols represent nodes of t. For the nonterminals, the fat dots on their left and right indicate their inherited and synthesized attributes, respectively. For instance, the semantic rule $N\cdot v = +(L_1\cdot v,\ L_2\cdot v)$ of the production p_1 that is applied at the root of t gives rise to two edges to the attribute v of this root x, labeled N, one from the attribute v of its first nonterminal son x_1, labeled L, and one from the attribute v of its second nonterminal son x_2, labeled L too. This semantic rule expresses that the function + should be applied to the value of v of x_1, denoted $x_1\cdot v$, and the value of $x_2\cdot v$ to obtain the value of $x\cdot v$. A dashed edge from some $y\cdot\beta$ to some $x\cdot\alpha$ indicates that the semantic rule that is used to compute the value of $x\cdot\alpha$ is of the form $X_j\cdot\alpha = X_i\cdot\beta$. A short incoming arrow represents a semantic rule with a right-hand side in $T(\Gamma)$.

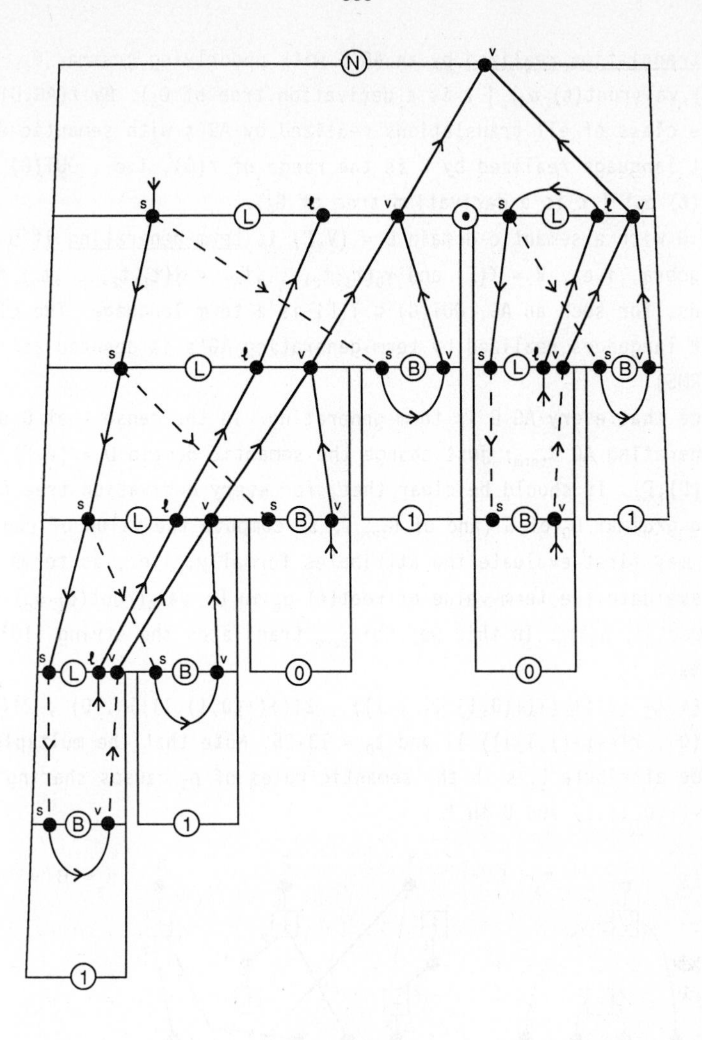

Figure 4

If we evaluate the values of all attributes of t, we find the value of $\mathrm{root}(t)\cdot\alpha_d = \mathrm{root}(t)\cdot v$: $\mathrm{val}(\mathrm{root}(t)\cdot\alpha_d)$ is the number 13•25, the "meaning" of 1101•01. Since Gbin is non-circular, i.e., no derivation tree of G_0 contains a directed cycle in its dependency graph, the value of every $x\cdot\alpha$, denoted $\mathrm{val}(x\cdot\alpha)$, is determined uniquely. From now on, we assume every AG to be non-circular.

If an AG is reduced, i.e., for every dependency graph of a derivation tree t of its underlying grammar there is a directed path from every $x\cdot\alpha$ to $\mathrm{root}(t)\cdot\alpha_d$, see [Fil], then every $\mathrm{val}(x\cdot\alpha)$ is used in the computation of $\mathrm{val}(\mathrm{root}(t)\cdot\alpha_d)$. Obviously, Gbin is not reduced (see Fig.4).

The <u>translation realized by</u> an AG G with underlying grammar G_0, is $\tau(G) = \{(yield(t), val(root(t)\cdot\alpha_d) \mid t$ is a derivation tree of $G_0\}$. By $\tau(AG,D)$ we denote the class of all translations realized by AG's with semantic domain D. The output language realized by G is the range of $\tau(G)$, i.e., $OUT(G) = \{val(root(t)\cdot\alpha_d) \mid t$ is a derivation tree of $G_0\}$.

An AG G with a semantic domain $D = (V,\Gamma)$ is <u>term-generating</u> if D is the free Γ-algebra, i.e., $V = T(\Gamma)$ and $\gamma_D(t_1,t_2,\ldots,t_n) = \gamma(t_1,t_2,\ldots,t_n)$ for all $\gamma \in \Gamma$. Thus, for such an AG, $OUT(G) \subseteq T(\Gamma)$ is a term language. The class of all output languages realized by term-generating AG's is denoted as OUT(AG,TERMS).

Notice that every AG G is term-generating, in the sense that G determines a term-generating AG G_{term}: just change the semantic domain $D = (V,\Gamma)$ into $D_{term} = (T(\Gamma),\Gamma)$. It should be clear that, for every derivation tree t of the underlying grammar G_0 of G (and of G_{term}), to compute the value of $root(t)\cdot\alpha_d$ in V, one may first evaluate the attributes formally, i.e., as terms in T(Γ), and then evaluate the term-value of $root(t)\cdot\alpha_d$ in D: $val_G(root(t)\cdot\alpha_d) = (val_{G_{term}}(root(t)\cdot\alpha_d))_D$. In this way $Gbin_{term}$ translates the string 1101•01 into the term

t = + (+ (+ (+ (2↑(+ (+(+(0,1),1) , 1)) , 2↑(+(+(0,1),1))) , 0) , 2↑(0)) , + (0 , 2↑(-(+(1,1))))), and t_D = 13•25. Note that the multiple use of the outside attribute $L_1\cdot s$ in the semantic rules of p_3 causes sharing of the subterms +(+(0,1),1) and 0 in t.

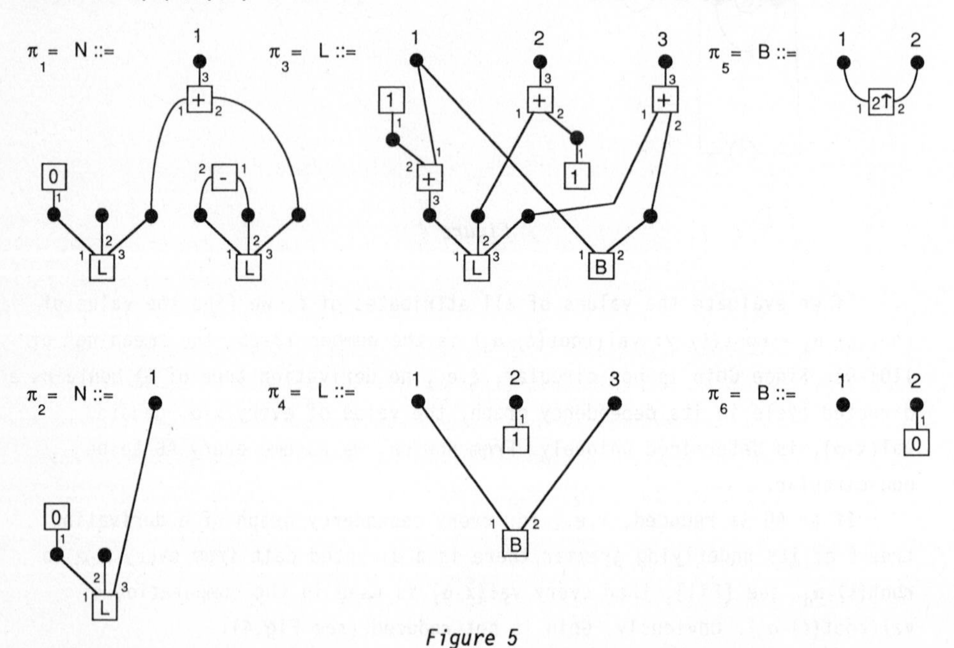

Figure 5

Now we turn to our principal aim: the comparison of the term generating power of cfhg's with that of AG's. In particular, we prove that TERM(CFHG) = OUT(AG,TERMS). First the, intuitively straightforward, inclusion OUT(AG,TERMS) $\subseteq$ TERM(CFHG) is considered.

<u>Lemma 2</u>. For every term-generating AG G there exists a term-generating cfhg G′ such that TERM(G′) = OUT(G). Moreover, if G is reduced then G′ is clean term-generating.

<u>Proof</u>. (Sketch, illustrated by Fig.5 in which the productions of G′ are given for G = $Gbin_{term}$.) Let G be an AG with semantic domain $(T(\Gamma),\Gamma)$ for some ranked alphabet Γ, and let $G_0 = (N_0,T_0,P_0,S_0)$ be its underlying grammar. The terminal alphabet of G′ must be inc(Γ) [7]. Every $X \in N_0$ is a nonterminal of G′ of rank #Att(X); in particular, S_0 is the initial nonterminal of G′ (of rank 1). These nonterminals are used to simulate the derivation trees of G_0. The terminals of G_0 need not be simulated by G′, because they have no attributes and consequently do not influence OUT(G).

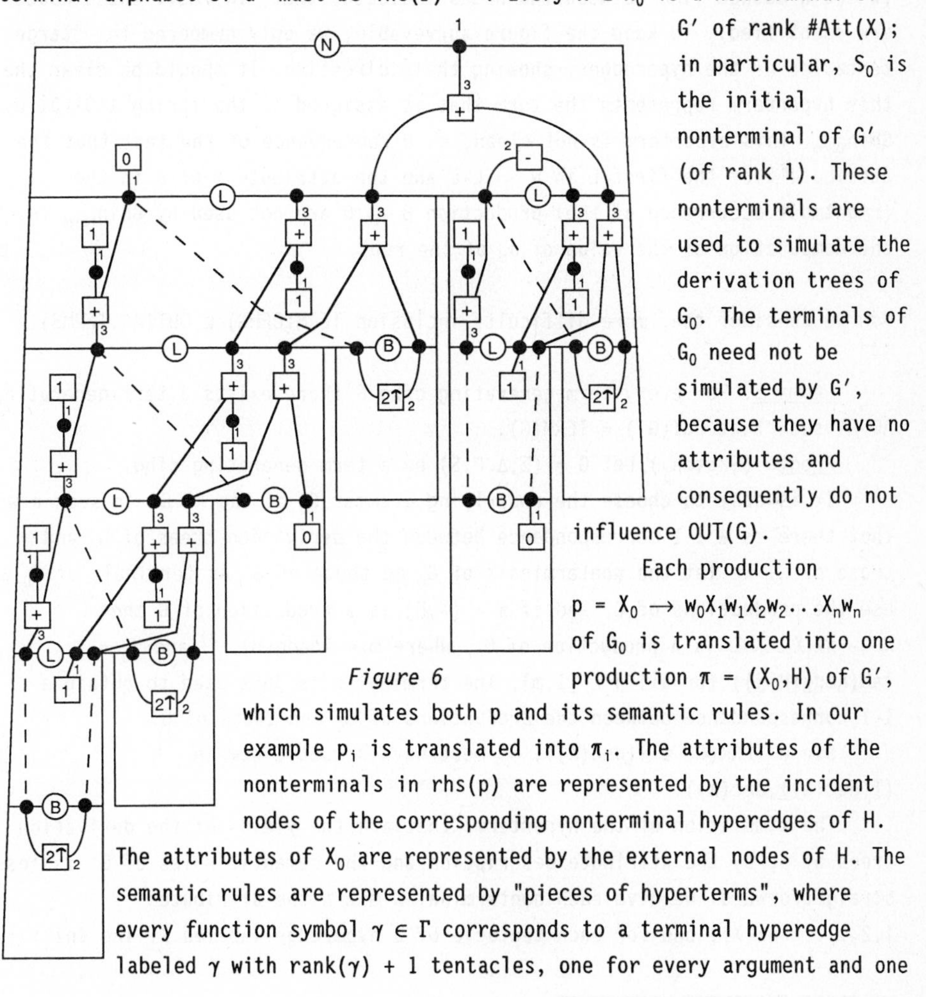

Figure 6

Each production $p = X_0 \longrightarrow w_0X_1w_1X_2w_2\ldots X_nw_n$ of G_0 is translated into one production $\pi = (X_0,H)$ of G′, which simulates both p and its semantic rules. In our example p_i is translated into π_i. The attributes of the nonterminals in rhs(p) are represented by the incident nodes of the corresponding nonterminal hyperedges of H. The attributes of X_0 are represented by the external nodes of H. The semantic rules are represented by "pieces of hyperterms", where every function symbol $\gamma \in \Gamma$ corresponds to a terminal hyperedge labeled γ with rank(γ) + 1 tentacles, one for every argument and one

7) For a ranked alphabet Γ, the ranked alphabet that contains every symbol of Γ with the rank increased by one, is denoted inc(Γ).

"extra" for the result.

Every term $t \in T(\Gamma)$ in OUT(G) is the value of the designated attribute of the root of some derivation tree t of G_0. The cfhg G′ is constructed in such a way that it has a derivation tree t' corresponding to t, such that term(yield(t')) = t. In fact, each attribute α of a node x in t is represented by a node v in yield(t') such that term(v,yield(t')) = val($x \cdot \alpha$). This representation is not necessarily injective, due to copy rules, i.e., semantic rules of the form $X_j \cdot \alpha = X_i \cdot \beta$. For instance, the attributes s of the L in lhs(p_3) and the B in rhs(p_3) are represented by the same node in π_3.

As an example, for the derivation tree t of $Gbin_{term}$ in Fig.4 the corresponding derivation tree t', together with its yield is shown in Fig.6 (where a dashed line between two nodes indicates that, in fact, these nodes are identified). To keep the figure surveyable, we only numbered the "target" tentacles of the hyperedges, showing their direction. It should be clear that this hyperterm represents the term that is assigned to the string 1101•01 by $Gbin_{term}$. This hyperterm is not clean, as a consequence of the fact that the attribute ℓ of the first L in N ⟶ L•L and the attribute s of B in the (rightmost occurrence of the) production B ⟶ 0 are not used by $Gbin_{term}$ in the computation of the value of α_d of the root. □

Now we consider the, more difficult, inclusion TERM(CFHG) ⊆ OUT(AG,TERMS).

<u>Lemma 3</u>. For every term-generating cfhg G there exists a term-generating AG G′ such that OUT(G′) = TERM(G).

<u>Proof</u>. (Sketch.) Let G = (Σ,Δ,P,S) be a term-generating cfhg.

It is easy to choose the underlying grammar G_0 of the AG G′ in such a way that there is a 1-1 correspondence between the derivation trees of G_0 and those of G. We let the nonterminals of G_0 be those of G. As terminals of G_0 we use the productions of G, and if π = (X,H) is a production of G then $X \longrightarrow \pi X_1 X_2 \ldots X_m$ is a production of G_0, where m = #nedg(H) [8] and X_j = lab(nedg(H,j)) for all $j \in [1,m]$. The terminal π is just used to obtain a 1-1 correspondence between the productions of G_0 and those of G.

Since TERM(G) ⊆ T(dec(Δ)), G′ must have semantic domain (T(dec(Δ)),dec(Δ)).

The simulation of the hyperterms that are the yields of the derivation trees of G, by the attribute description and the semantic rules of G′ is less straightforward. We give each nonterminal $X \in \Sigma$-Δ the attributes 1,2,...,rank(X), one for each tentacle of a hyperedge labeled X. The initial

8) The set of all nonterminal hyperedges of a hypergraph H is denoted nedg(H). We give this set a fixed but arbitrary order of which the j-th element is denoted nedg(H,j).

nonterminal has no inherited attributes, and the designated attribute α_d of G′ is 1. Notice that the choice of the attributes is quite obvious. Their division into inherited and synthesized, as well as the definition of the semantic rules, depends on the fact that G generates hyperterms. We roughly explain this with an example. Consider the cfhg G = (Σ,Δ,P,S), where Σ = {S,R,a,b,c,d} with ranks 1,3,2,2,2,1, respectively, Δ = {a,b,c,d}, and P ={π_1,π_2,π_3} as given in Fig.7.

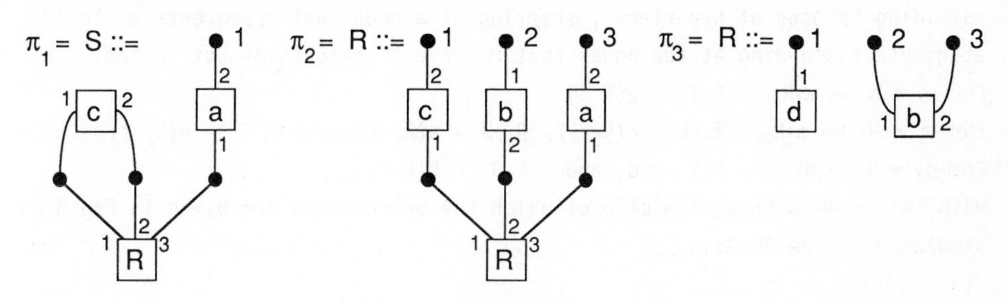

Figure 7

G generates hyperterms of the form given in Fig.8. So, TERM(G) is the string language $\{a^nb^nc^nd \mid n \geq 1\}$, if we omit the parentheses. This illustrates that non-context-free string languages can be generated in a context-free way (see [Hab, EngHey1], where the string generating power of cfhg's is studied). We define an AG G′ such that OUT(G′) = TERM(G) as follows.

The underlying grammar is $G_0 = (\Sigma\text{-}\Delta,P,P_0,S)$ with $P_0 = \{p_1 = S \longrightarrow \pi_1R$, $p_2 = R \longrightarrow \pi_2R$, $p_3 = R \longrightarrow \pi_3\}$. The semantic domain is (T(dec(Δ)),dec(Δ)). The attribute of the initial nonterminal S is 1, and the attributes of R are 1, 2, and 3. Thus, for every $p_i \in P_0$, the attribute $X_j{\cdot}\alpha$ (with $j \geq 1$) represents nod(nedg(rhs(π_i),j),α), and the attribute $X_0{\cdot}\alpha$ represents ext(α) of rhs(π_i).

We define the semantic rules of G′ in such a way that the value of an attribute $x'{\cdot}\alpha$ in a derivation tree t' of G_0 equals the term that is associated with the node that is represented by $x'{\cdot}\alpha$ in the yield of the corresponding derivation tree t of the cfhg G. In particular, val(root(t')$\cdot\alpha_d$) = term(yield(t)). In the hypergraph H of Fig.8, any three nodes v_1,v_2,v_3 that lie on an imaginary horizontal line "belong" to a node x of t (labeled R) and are represented by the attributes 1,2 and 3 of the corresponding node x′ of t', respectively. The subhypergraph H_x consisting of these three nodes and all nodes and edges below the line, is the yield of the derivation subtree of t with root x. We now define attribute j to be synthesized if the unique incoming hyperedge of v_j belongs to H_x, and

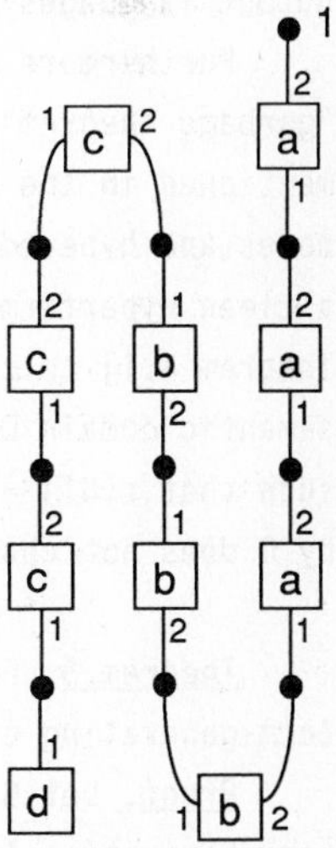

Figure 8

inherited otherwise (clearly, this definition does not depend on x, but on its label only). Thus, in our example, 1 and 3 are synthesized and 2 is inherited. As another example consider the cfhg of which the productions are given in Fig.5; looking at any derivation subtree with root L in Fig.6, one sees that 1 should be an inherited attribute of L, and 2 and 3 synthesized. Knowing which attributes are synthesized and which inherited, it is not very hard to "read off" the semantic rules from the right-hand sides of the productions of G, by unfolding "pieces of hyperterm", starting at a node that represents an inside attribute and ending at the nodes that represent the outside attributes.

For $p_1 = S \longrightarrow \pi_1 R$, $\quad S \cdot 1 = a(R \cdot 3)$.

For $p_2 = R_1 \longrightarrow \pi_2 R_2$, $\quad R_1 \cdot 1 = c(R_2 \cdot 1)$, $R_1 \cdot 3 = a(R_2 \cdot 3)$, and $R_2 \cdot 2 = b(R_1 \cdot 2)$.

For $p_3 = R \longrightarrow \pi_3$, $\quad R \cdot 1 = d$, and $\quad R \cdot 3 = b(R \cdot 2)$.

With this same method, the cfhg of which the productions are given in Fig.5 is simulated by the AG $Gbin_{term}$. □

From Lemma's 2 and 3 we conclude that cfhg's and AG's have the same term generating power.

<u>Theorem 4</u>. TERM(CFHG) = OUT(AG,TERMS).

This theorem allows us to compare TERM(CFHG) with some known classes of term (or tree) languages, by using results on OUT(AG,TERMS). For example, we conclude that TERM(CFHG) contains the IO context-free tree languages ([DusParSedSpe, EngFil]) and the output languages of deterministic top-down tree transducers ([CouFra]), whereas it is itself contained in the class of output languages of macro tree transducers (cf. [EngVog, CouFra]).

Furthermore Lemma's 2 and 3 can be used to prove the following "garbage theorem" that provides an alternative way of defining TERM(CFHG), as mentioned in the discussion preceding Theorem 1 ("garbage" is that, i.e., nodes and hyperedges, which has to be deleted from a hyperterm to turn it into a clean hyperterm). In fact the "garbage theorem" is the application of [Fil, Theorem 4.1] to cfhg's. Theorem 4.1 of [Fil] says that for every AG G with semantic domain D there is a reduced AG G′ with the same semantic domain D such that $\tau(G') = \tau(G)$. Observe that in that case the output language realized by G does not change either.

<u>Theorem 5</u>. For every term-generating cfhg G there exists a clean term-generating cfhg G′ such that TERM(G′) = TERM(G).

<u>Proof</u>. Let G be a term-generating cfhg. Then by Lemma 3 there exists a term-generating AG G_1 such that $OUT(G_1) = TERM(G)$. By [Fil, Theorem 4.1] there exists a reduced AG G_2 with the same semantic domain as G_1 such that $OUT(G_2)$ =

$OUT(G_1)$. Hence G_2 is term-generating. Thus, by Lemma 2 there exists a clean term-generating cfhg G' such that $TERM(G') = OUT(G_2) = OUT(G_1) = TERM(G)$. □

4. The translation power of context-free hypergraph grammars

Recall that every AG G determines a term-generating AG G_{term}. By the proof of Lemma 2, for an AG G with an arbitrary semantic domain D, and with underlying grammar G_0, the translation realized by G can also be defined as $\tau(G) = \{(yield_{G_0}(t),(term(yield_{G_{term}'}(\varphi(t)))_D) \mid t \text{ is a derivation tree of } G_0\}$, where $\varphi(t)$ is obtained from t by replacing every p (of G_0) by the corresponding π of the cfhg G_{term}'. This suggests the following attribute evaluation method for G. Given a derivation tree t of G_0, turn it into a derivation tree $\varphi(t)$ of G_{term}'. Compute the hyperterm H that is the yield of $\varphi(t)$ in a bottom-up fashion. Compute the value of H in D, i.e., $(term(H))_D$. An evaluation method of this type, called the DAG-evaluator, is described in [Mad] and implemented in the NEATS System (see also [DerJouLor]). The advantage of the method is that copy rules do not have to be executed, because the corresponding nodes are already identified in the right-hand sides of the productions of G_{term}'. The disadvantage of the method is that the hyperterm H may take a lot of space (which can be improved by evaluating "sub-hyperterms" as soon as possible).

Of course, any term-generating cfhg G, together with a semantic domain $D = (V,\Gamma)$, can be viewed as a syntax-directed translation device realizing the tree-to-value translation $\{(t,(term(yield_G(t)))_D) \mid t \text{ is a derivation tree of } G\}$. So we have shown that the translation of every AG can be viewed as a parsing phase followed by such a translation realized by a cfhg with the same semantic domain. Now we shall define a "cfhg-based" syntax-directed translation scheme that translates strings (rather than trees) to values of the semantic domain. It may be viewed as the generalization to graphs of the syntax-directed translation scheme of [AhoUll] (which is an AG with synthesized attributes only !).

Definition. A cfhg-based syntax-directed translation scheme (cts) is a 4-tuple $T = (D,GL,GR,\varphi)$, where $D = (V,\Gamma)$ is a semantic domain, $GL = (N,T,P_{GL},S)$ is a cfg, the left grammar, $GR = (\Sigma,inc(\Gamma),P_{GR},S)$ is a term-generating cfhg, the right grammar, with $\Sigma\text{-}inc(\Gamma) = N$, and φ is a mapping from P_{GL} to P_{GR} such that if $p = X_0 \longrightarrow w_0X_1w_1X_2w_2 \ldots X_mw_m \in P_{GL}$ and $\varphi(p) = (X,H) \in P_{GR}$, then $X_0 = X$, $m = \#nedg(H)$, and $X_j = lab_H(nedg(H,j))$, for all $j \in [1,m]$.

The translation realized by T is the relation $\tau(T) = \{(yield_{GL}(t), (term(yield_{GR}(\varphi(t))))_D) \mid t$ is a derivation tree of GL$\}$, where φ is extended to derivation trees in the obvious way. □

Notice that a cts is "simple" in the sense of [AhoUll], i.e., nonterminals of p cannot be deleted or duplicated in $\varphi(p)$.

As an example consider the cts $T = (D,Gl,GR,\varphi)$ where $D = (V,\Gamma)$ is the semantic domain of the AG Gbin, $GL = (N_0,T_0,P_0,S_0)$ is the underlying grammar G_0 of Gbin, $GR = (N_0 \cup inc(\Gamma),inc(\Gamma),P,S_0)$ is the term-generating cfhg of which the productions are given in Fig.5, and $\varphi: P_0 \longrightarrow P$ is defined by $\varphi(p_i) = \pi_i$, $1 \leq i \leq 6$, as in the discussion in the beginning of this section, where GR is called G_{term}'. The translation realized by T is $\tau(T) = \{(yield_{G_0}(t),(term(yield_{G_{term}'}(\varphi(t))))_D) \mid t$ is a derivation tree of $G_0\}$. Thus, to each binary number in L(GL) its rational value (in V) is assigned by T, just like the translation realized by the AG Gbin. Hence $\tau(T) = \tau(Gbin)$.

The set of all translations realized by cts's with a semantic domain D is denoted $\tau(CTS,D)$. As for TERM(CFHG) we can give three other possible definitions of $\tau(CTS,D)$, by allowing all cfhg's as right grammar of a cts, and/or assuming that the right grammar is clean term-generating. Moreover we can extend Lemma's 2 and 3: for every AG there is a cts (with the same semantic domain) that realizes the same translation, and vice versa.

Theorem 6. $\tau(CTS,D) = \tau(AG,D)$, for every semantic domain D.

Theorems 4 and 6 express the close relationship that exists between term-generating cfhg's and AG's.

References

[AhoUll] A.V.Aho, J.D.Ullman; "The Theory of Parsing, Translation, and Compiling", Prentice-Hall Inc., Englewood Cliffs, New Jersey, 1972.

[BauCou] M.Bauderon, B.Courcelle; Graph expressions and graph rewritings, Mathematical Systems Theory 20 (1987), 83-127.

[Cou1] B.Courcelle; Equivalences and transformations of regular systems, applications to recursive program schemes and grammars, Theoretical Computer Science 42 (1986), 1-122.

[Cou2] B.Courcelle; On using context-free graph grammars for analyzing recursive definitions, in "Programming of future generation computers, II" (K.Fuchi, L.Kott, eds.), Elsevier Pub.Co., 1988, 83-122.

[Cou3] B.Courcelle; The monadic second-order logic of graphs, I: recognizable sets of finite graphs, Information and Computation 85 (1990), 12-75. See also [EhrNagRosRoz], 133-146.

[CouFra] B.Courcelle, P.Franchi-Zannettacci; Attribute grammars and recursive program schemes I and II, Theoretical Computer Science 17 (1982), 163-191, 235-257.

[DerJouLor] P.Deransart, M.Jourdan, B.Lorho, "Attribute grammars; Definitions, Systems and Bibliography", Lecture Notes in Computer Science 323, Springer-Verlag, Berlin, 1988.

[DusParSedSpe] J.Duske, R.Parchmann, M.Sedello, J.Specht; IO-macrolanguages and attributed translations, Information and Control 35 (1977), 87-105.

[EhrNagRosRoz] H.Ehrig, M.Nagl, G.Rozenberg, A.Rosenfeld (eds.); "Graph-Grammars and their Application to Computer Science", Lecture Notes in Computer Science 291, Springer-Verlag, Berlin, 1987.

[EngFil] J.Engelfriet, G.Filè; The formal power of one-visit attribute grammars, Acta Informatica 16 (1981), 275-302.

[EngHey1] J.Engelfriet, L.M.Heyker; The string generating power of context-free hypergraph grammars, Report 89-05, Leiden University, 1989, to appear in Journal of Computer and System Sciences.

[EngHey2] J.Engelfriet, L.M.Heyker; The term-generating power of context-free hypergraph grammars and attribute grammars, Report 89-17, Leiden University, 1989.

[EngLeiRoz] J.Engelfriet, G.Leih, G.Rozenberg; Apex graph grammars and attribute grammars, Acta Informatica 25 (1988), 537-571.

[EngVog] J.Engelfriet, H.Vogler; Macro tree transducers, Journal of Computer and System Sciences 31 (1985), 71-146.

[Fil] G.Filè; Interpretation and reduction of attribute grammars, Acta Informatica 19 (1983), 115-150.

[Hab] A.Habel; Hyperedge replacement: grammars and languages, Ph.D.Thesis, Bremen, 1989.

[HabKre] A.Habel, H.-J.Kreowski; May we introduce to you: hyperedge replacement, in [EhrNagRosRoz], 15-26.

[HabKrePlu] A.Habel, H.-J.Kreowski, D.Plump; Jungle evaluation, in: "Recent Trends in Data Type Specification" (D.Sanella, A.Tarlecki, eds.), Lecture Notes in Computer Science 332, Springer-Verlag, Berlin, 1987, 92-112.

[Hof] B.Hoffman; Modelling compiler generation by graph grammars, in: "Graph-Grammars and their Application to Computer Science" (H.Ehrig, M.Nagl, G.Rozenberg, eds.), Lecture Notes in Computer Science 153, Springer-Verlag, Berlin, 1983, 159-171.

[Knu] D.E.Knuth; Semantics of context-free languages, Mathematical Systems Theory 2 (1968), 127-145. Correction: Mathematical Systems Theory 5 (1971), 95-96.

[Lau] C.Lautemann; Decomposition trees: structured graph representation and efficient algorithms, in: CAAP '88 Proceedings (M.Dauchet, M.Nivat, eds.), Lecture Notes in Computer Science 299, Springer-Verlag, Berlin, 1988, 28-39.

[Mad] O.L.Madsen, On defining semantics by means of extended attribute grammars, in: "Semantics-directed compiler generation" (N.D.Jones, ed.), Lecture Notes in Computer Science 94, Springer-Verlag, Berlin, 1980, 259-299.

ELEMENTARY ACTIONS ON AN EXTENDED ENTITY-RELATIONSHIP DATABASE

Gregor Engels
TU Braunschweig, Informatik
Postfach 33 29, D-3300 Braunschweig

ABSTRACT: Semantic data models have been widely studied for the conceptual specification of databases. However, most of these data models are restricted to the description of the static structure of a database. They do not provide means to specify the dynamic behaviour of a database.

This paper sketches a language for the specification of actions on databases which have been specified by an Extended Entity-Relationship (EER) schema. These actions are based on so-called elementary actions, which are automatically be derived from the EER schema. So, it can always be guaranteed that these schema dependent elementary actions preserve all inherent integrity constraints.

The semantics of the elementary actions is given in two steps: First, it is shown how the semantics of a database schema, i.e., a current database state, can be represented by an attributed graph. Then, the semantics of elementary actions is given by programmed graph replacements.

Keywords: conceptual database specification, operational semantics of database actions, programmed graph replacements

CONTENTS

0. INTRODUCTION

Semantic data models have been studied for the conceptual specification of database systems for certain application areas (cf. [HK 87]). However, most of these data models are restricted to the description of the static structure of a database. They do not provide means to specify the dynamic behaviour and especially database modifications.

Meanwhile, it has been widely accepted in the database community that the specification of databases should comprise the description of the static structure and the corresponding dynamic behaviour. In this sense, database specifications are comparable to abstract data type specifications.

Nevertheless, the classical, successfully used approach to specify database systems is to start with the specification of the static structure. According to the chosen data model, types of database objects and possible interrelations between instances of these object types are fixed in the first step of the database specification process. Afterwards, database actions are specified which model actions of the application area. Of course, these database actions should respect the constraints which arise from the specification of the static structure. So, database specifications can be regarded as *constructive data type specifications*, consisting of a data type construction part and a type dependent action specification part.

In the literature there exist several approaches to specify modifications of (system) states. The scope ranges from descriptive, non-deterministic specifications, for instance, by pre-/postconditions, up to procedural, deterministic ones. A great advantage of the last ones is that they are executable and, therefore, well-suited for a rapid prototyping of database specifications.

The issue of this paper is to present a language for the specification of actions on a database, the structure of which has been specified by an Extended Entity-Relationship (EER) schema. This data model together with a well-defined semantics have been developed at Braunschweig Technical University during the last years ([HNSE 87], [HoGo 88]). It is our objective to start with this definition of syntax and semantics of an EER schema, given in [HoGo 88], and to define syntax and semantics of actions on such an EER database.

The developed language has a procedural, operational style. This is motivated by the intended use of the action specification language within the database design environment CADDY ([EHHLE 89]). The environment CADDY offers an integrated set of tools to a database designer to support him/her during the specification and testing by rapid prototyping of a conceptual database schema. So, this executable action specification language enables an immediate interpretation of actions and, therefore, supports the rapid prototyping facility of CADDY.

The main ideas, presented in this paper, are the following: We define how database states of an EER database can be represented by attributed graphs. Actions on a database can then be regarded as graph transformations. Because of the constraints given by the description of the static structure of a database, sequences of such graph transformations often have to be applied to a database state graph to yield a transition between correct database states. Therefore, the approach of *programmed graph grammars* is suited to describe such sequences of graph transformations. We show that such programmed graph replacements can be derived automatically from the description of the static structure of a database. These programmed graph replacements are called *elementary actions*. They describe the modification of a database object together with all update propagation operations (cf. [SSW 80]) to yield a consistent, correct database state after a modification.

There exist some approaches in the literature which propose graph grammars for database specifications; see for example [EK 80], [FV 83], or [Na 79], who gives a survey on some early approaches in the field. Programmed graph grammars have successfully been applied in the more general case of software specifications (e.g. [Na 87], [Gö 88]). In this case, programmed graph grammars are used to specify graph classes for a specific application by fixing all allowed graph modifications. Thereby, the structure of a graph is implicitly be determined. In contrast to this, in the approach in this paper, we start with a description of the structure of a graph and show how allowed modifications of such graphs can be derived automatically from this structure description.

The paper is organized as follows: In section 1, we summarize the definition of [HoGo 88] of syntax and semantics of an EER schema. In section 2, we describe how an EER schema and a corresponding database state can be represented as attributed graph. Section 3 shows how actions on an EER database can be automatically derived from the EER schema. The semantics of these actions is given by programmed graph replacements. Section 4 summarizes the ideas and results of this paper.

1. THE EXTENDED ENTITY-RELATIONSHIP MODEL

The extended Entity-Relationship model (EER model) was developed by the database group at Braunschweig Technical University during the last years [HNSE 87, HoGo 88]. This conceptual data model is an extension of the classical Entity-Relationship model [Ch 76]. Similar to other ER extensions [EWH 85, MMR 86, TYF 86, PRYS 89], it combines concepts known from semantic data models like TAXIS

[MyW 80], SDM [HaM 81], IFO [AbH 87], or IRIS [LyK 86] to a uniform conceptual data model.

Thereby, this model and the corresponding specification language offer features to a database designer which allow a problem-oriented, natural modelling of the information structure of a certain application task. The main important features of the EER model are

- the possibility to extend the set of allowed data types for attribute domains by new, application-dependent data types,
- components, i.e., object-valued attributes, to model structured entity types,
- the concept of type construction in order to support specialization and generalization, and
- several structural restrictions like the specification of key attributes.

Let us illustrate the EER model by a very small example (cf. Fig. 1.1). It models the world of persons who might be married or not, and who live at a certain town. This is expressed by the entity types **PERSON** and **TOWN**, and the relationship **lives-at**. The partition of objects of type **PERSON** into married and unmarried persons is described by the type construction **part**, which has **PERSON** as input type and **MARRIED** resp. **NOT_MARRIED** as output types. Each object of a certain entity type is described by some attribute values. In case of key attributes (e.g. **Name** at **PERSON**, expressed by a dot in the diagram), the value of this attribute uniquely identifies an object in the set of all objects of this type. Attributes may also be object-valued, as it is in case of the attribute **Info** at **PERSON**. This enables the modelling of complex structured objects which contain subobjects, also called components.

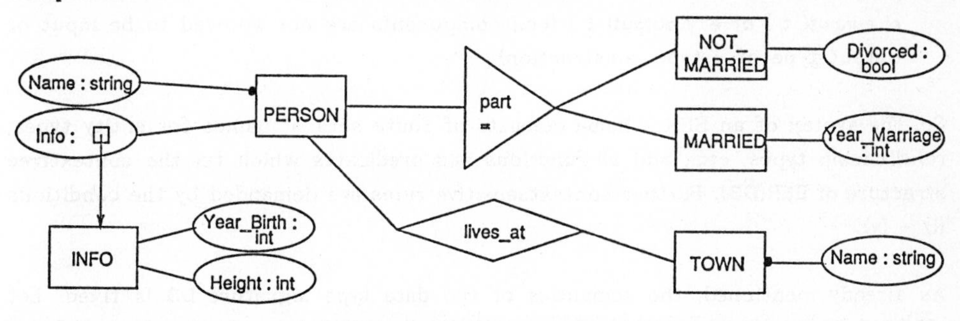

Figure 1.1: Example of an EER schema

An EER schema bases upon a data type signature **DS = (DATA, OPNS)**, which consists of a set **DATA** of data type names and appropriate operations **OPNS**, and which has a fixed semantics. This data type signature **DS** has to be specified by a database designer in a previous step so that the data types can be used as attribute domains. In the example of Fig. 1.1, **DS** contains the specifications of **int**, **string**, and **bool**.

Due to [HoGo 88], syntax and semantics of an EER schema can then be defined as follows:

<u>Def. 1.1:</u>
The <u>syntax of an extended Entity-Relationship schema</u> **EER(DS)** over **DS** is given by

- the finite sets **ENTITY, RELSHIP, ATTRIB, COMP, T_CONSTR,** and
- the partial functions

  attr : (ENTITY ∪ RELSHIP) × ATTRIB -o→ DATA

  key : ENTITY -o→ ATTRIB

  comp : ENTITY × COMP -o→ ENTITY
- the total functions

  particip : RELSHIP → ENTITY⁺

  input, output : T_CONSTR → F **(ENTITY)** (F ≙ powerset of <u>f</u>inite subsets)
- the predicate

  is-partition ⊂ T_CONSTR

The following conditions must hold:

(i) for all **e ∊** <u>**domain**</u>**(key): key (e) = a** implies **(e, a) ∊** <u>**domain**</u> **(attr)**

(ii) **comp** is injective
(i.e., that an entity type is component of at most one (other) entity type)

(iii) **output (** c_1 **) ∩ output (** c_2 **) = ∅** for two distinct c_1, c_2 **∊ T_CONSTR**
(i.e., each constructed entity type is uniquely constructed)

(iv) It is not allowed that **connection⁺ (e, e)** holds for some **e ∊ ENTITY,** where **connection⁺** is the transitive closure of the relation **connection** defined by: if e_{in} **∊ input(t)** and e_{out} **∊ output(t)** for some **t ∊ T_CONSTR,** then **connection(** e_{in}, e_{out} **)** holds (i.e., type construction is cycle free).

(v) **comp(e, c) = e'** with **e, e' ∊ ENTITY, c ∊ COMP** implies **¬ ∃ t ∊ T_CONSTR** with **e' ∊ input(t)** or **e' ∊ output(t)** (i.e., components are not allowed to be input or output type of a type construction)

So, the syntax of an EER schema consists of finite sets of names for entity types, relationship types, etc., and of functions and predicates which fix the contextfree structure of **EER(DS)**. Further contextsensitive rules are demanded by the conditions (i) - (v).

As already mentioned, the semantics of the data type signature **DS** is fixed. Let **μ[DATA]** denote the semantical domain of **DATA**. Let **|FISET|** denote the class of finite sets, and **|FUN|** the class of total functions.

<u>Def. 1.2:</u>
The <u>semantics of an extended Entity-Relationship schema</u> **EER(DS)** over DS is given by

- a function **μ[ENTITY] : ENTITY → |FISET|**

- a function $\mu[\text{RELSHIP}]$: RELSHIP $\rightarrow$ IFISETI
  such that $\text{particip}(r) = \langle e_1, ..., e_n \rangle$ for $r \in$ RELSHIP implies
  $$\mu[\text{RELSHIP}](r) \subset \mu[\text{ENTITY}](e_1) \times ... \times \mu[\text{ENTITY}](e_n)$$
- a function $\mu[\text{ATTRIB}]$: ATTRIB $\rightarrow$ IFUNI such that
  $\text{attr}(e, a) = d$ implies $\mu[\text{ATTRIB}](a) : \mu[\text{ENTITY}](e) \rightarrow \mu[\text{DATA}](d)$ resp.
  $\text{attr}(r, a) = d$ implies $\mu[\text{ATTRIB}](a) : \mu[\text{RELSHIP}](r) \rightarrow \mu[\text{DATA}](d)$
- a function $\mu[\text{COMP}]$: COMP $\rightarrow$ IFUNI such that
  $\text{comp}(e, c) = e'$ implies bijective functions
  $$\mu[\text{COMP}](c) : \mu[\text{ENTITY}](e) \rightarrow \mu[\text{ENTITY}](e')$$
- a function $\mu[\text{T\_CONSTR}]$: T_CONSTR $\rightarrow$ IFUNI such that
  $\text{input}(t) = \{ i_1, ..., i_n \}$, $\text{output}(t) = \{ o_1, ..., o_m \}$ implies injective functions
  $$\mu[\text{T\_CONSTR}](t) : \bigcup_{j=1}^{m} \mu[\text{ENTITY}](o_j) \rightarrow \bigcup_{k=1}^{n} \mu[\text{ENTITY}](i_k)$$

The following conditions must hold:

(i) ("disjoint sets of instances")
for two distinct $e, e' \in$ ENTITY: $\mu[\text{ENTITY}](e) \cap \mu[\text{ENTITY}](e') = \emptyset$

(ii) ("key attributes")
for all $e \in$ ENTITY, $a \in$ ATTRIB with $\text{key}(e) = a$ and $\text{attr}(e, a) = d$
$\mu[\text{ATTRIB}](a) : \mu[\text{ENTITY}](e) \rightarrow \mu[\text{DATA}](d)$ is injective

(iii) ("partition")
for all $t \in$ is_partition: $| \text{input}(t) | = 1$ and $\mu[\text{T\_CONSTR}](t)$ is bijective

The definition of the semantics of an EER schema assigns to the schema a fixed database state. This database state consists of sets of instances of entity resp. relationship types. The association of current attribute or component values to entity resp. relationship instances is described by functions. Analogously, the rearrangement of instances of entity types by a type construction is fixed by corresponding functions.

The definition contains several restricting conditions, which have to be fulfilled in a current database state. These restrictive conditions are termed "schema inherent integrity constraints" in the database literature, because they are expressed by syntactical means within the schema. For example, type constructions can be used as partition operator (condition (iii)). This means that, in contrast to the general case, <u>all</u> instances of the input entity types have to be contained in the set of instances of output types.

2. GRAPH REPRESENTATION OF DATABASE STATES

Section 1 summarized the definitions of syntax and semantics of an EER schema as they were given in [HoGo 88]. The main idea was to represent an EER schema and a corresponding database state by a set of functions and predicates. As it is our intention to describe state transitions by graph replacements, at first we need a graph representation of a database state. Therefore, we define how syntax and semantics of an EER schema can equivalently be represented by an attributed graph.

Def. 2.1:

A directed, labelled graph G = (Nodes, Nodelabels, nodelab, Edgelabels, Edges) is defined by

- a finite set of Nodes,
- a finite set of Nodelabels,
- a labelling function nodelab : Nodes → Nodelabels,
- a finite set of Edgelabels,
- a set of labelled Edges ⊂ Nodes × Edgelabels × Nodes.

The syntax of an EER schema can then be represented by a directed, labelled graph, where all identifiers for entity or relationship types, attributes, etc. occur as node labels. Labelled edges represent the functions and predicates of the schema and describe the contextfree and contextsensitive interrelations within an EER schema. Representation 1 defines how the constituents of the Schema_Graph are constructed for a given EER schema.

Representation 1:

Schema_Graph := (S_Nodes, S_Nodelabels, S_nodelab, S_Edgelabels, S_Edges)
with
S_Nodelabels := ENTITY ∪ RELSHIP ∪ ATTRIB ∪ COMP ∪ T_CONSTR ∪ DATA
S_nodelab : S_Nodes → S_Nodelabels bijective function
S_Edgelabels := { E_Key, E_Attrib, E_Comp, E_Input, E_Output, E_Attr_Type, E_Comp_Type, E_Partition } ∪ { E_Part_1, ..., E_Part_n },
where n := max { | particip(r) | | r ∈ RELSHIP }

The set S_Edges is constructed as follows:

For all e ∈ ENTITY ∪ RELSHIP, e' ∈ ENTITY, r ∈ RELSHIP, a ∈ ATTRIB, c ∈ COMP, t ∈ T_CONSTR, d ∈ DATA holds

- (e, E_Key, a) ∈ S_Edges *iff.* (e, a) ∈ domain(attr) and key(e) = a
- (e, E_Attrib, a) ∈ S_Edges *iff.* (e, a) ∈ domain(attr) and key(e) ≠ a
- (a, E_Attr_Type, d) ∈ S_Edges *iff.* ((e, a) ∈ domain(attr) and attr(e, a) = d)
- (e, E_Comp, c) ∈ S_Edges *iff.* (e, c) ∈ domain(comp)
- (c, E_Comp_Type, e') ∈ S_Edges *iff.* ((e, c) ∈ domain(comp) and comp(e, c) = e')
- for i = 1, ..., | particip(r) | :

  (r, E_Part_i, e) ∈ S_Edges *iff.* (particip(r) = $\langle e_1, ..., e_n \rangle$ and e_i = e)
- (t, E_Partition, e) ∈ S_Edges *iff.* e ∈ input(t) and t ∈ is_partition
- (t, E_Input, e) ∈ S_Edges *iff.* e ∈ input(t) and not (t ∈ is_partition)
- (t, E_Output, e) ∈ S_Edges *iff.* e ∈ output(t)

Figure 2.1 shows the representation of the example of an EER schema of Fig. 1.1 as **Schema_Graph**.

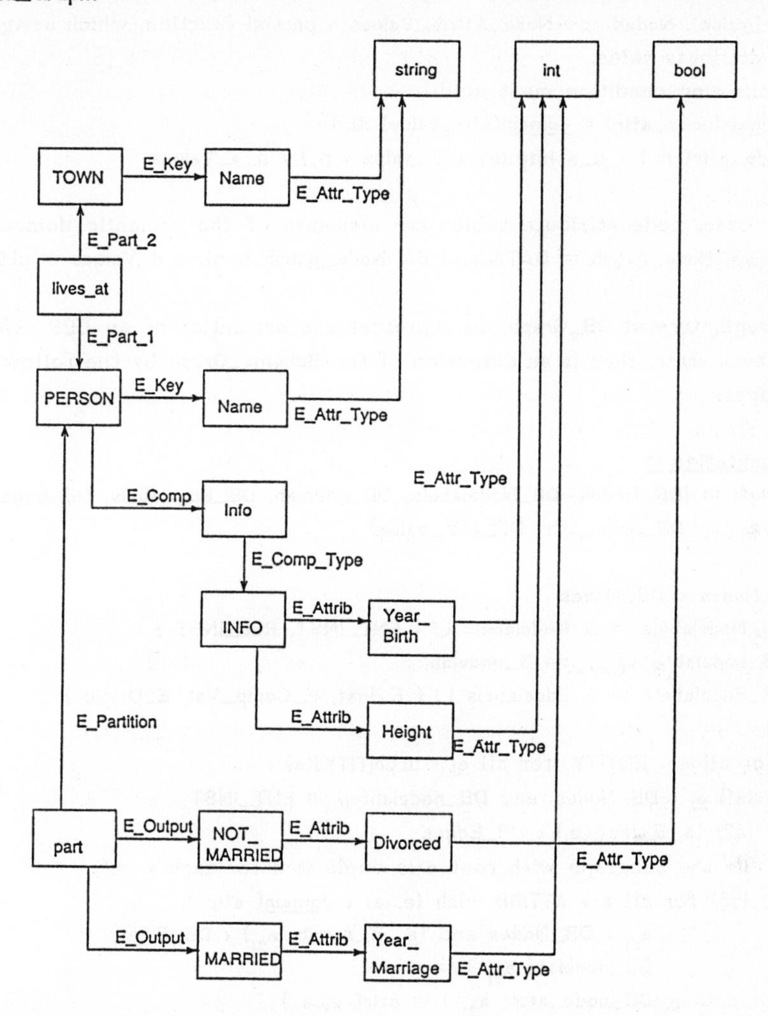

Figure 2.1: Example of a **Schema_Graph**

For the representation of a database state, we need node attributes additionally: Let **Node_Attrib** be a finite set of node attributes. Each node attribute **n_a** $\in$ **Node_Attrib** has a domain of values, termed **n_a_Values**. Let **Node_Attrib_Values** be $\bigcup_{\text{n\_a} \in \text{Node\_Attrib}}$ **n_a_Values**.

Def. 2.2:

An attributed graph AG = (**Nodes, Nodelabels, nodelab, Edgelabels, Edges, node_attr, attr_value**) over **Node_Attrib** and **Node_Attrib_Values** is defined by

- (Nodes, Nodelabels, nodelab, Edgelabels, Edges) a directed, labelled graph
- node_attr: Nodes -o→ Node_Attrib a partial function, which assigns a node an attribute
- attr_value : Nodes -o→ Node_Attrib_Values a partial function, which assigns a node an attribute value.

The following condition must hold:

domain(node_attr) = domain(attr_value) and

node_attr(n) = n_a implies attr_value (n) $\in$ n_a_Values

In our case, node attribute values are elements of the semantic domain of data types. So, Node_Attrib := DATA and d $\in$ Node_Attrib implies d_Values := μ[DATA](d)

The graph, termed DB_Graph, to represent the semantics of an EER schema, i.e., a database state, then is an extension of the Schema_Graph by the following nodes and edges:

Representation 2:

DB_Graph := (DB_Nodes, DB_Nodelabels, DB_nodelab, DB_Edgelabels, DB_Edges, DB_node_attr, DB_attr_value)

with

- S_Nodes $\subset$ DB_Nodes
- DB_Nodelabels := S_Nodelabels $\cup$ { ENT_INST, REL_INST }
- $\text{DB\_nodelab}|_{\text{S\_Nodes}}$:= S_nodelab
- DB_Edgelabels := S_Edgelabels $\cup$ { E_Inst, E_Comp_Val, E_Origin }

and

(a) for all e $\in$ ENTITY, for all $\underline{e}_i$ $\in$ μ[ENTITY](e) :

(a1) $\underline{e}_i$ $\in$ DB_Nodes and DB_nodelab($\underline{e}_i$) := ENT_INST

(a2) (e, E_Inst, $\underline{e}_i$) $\in$ DB_Edges

/* the subgraph with root e is duplicated for each $\underline{e}_i$ */

(a3) for all a $\in$ ATTRIB with (e, a) $\in$ domain(attr):

$a_{\underline{e}_i}$ $\in$ DB_Nodes and ($\underline{e}_i$, E_Attrib, $a_{\underline{e}_i}$) $\in$ DB_Edges

DB_nodelab($a_{\underline{e}_i}$) := a

DB_node_attr($a_{\underline{e}_i}$) := attr(e, a)

DB_attr_value($a_{\underline{e}_i}$) := μ[ATTRIB](a)($\underline{e}_i$)

(a4) for all c $\in$ COMP with (e, c) $\in$ domain(comp)

$c_{\underline{e}_i}$ $\in$ DB_Nodes and ($\underline{e}_i$, E_Comp, $c_{\underline{e}_i}$) $\in$ DB_Edges

DB_nodelab($c_{\underline{e}_i}$) := c

($c_{\underline{e}_i}$, E_Comp_Val, μ[COMP](c)($\underline{e}_i$)) $\in$ DB_Edges

(b) for all r $\in$ RELSHIP, for all $\underline{r}_i$ $\in$ μ[RELSHIP](r) with $\underline{r}_i$ = $\langle \underline{e}_{i_1}, \ldots, \underline{e}_{i_n} \rangle$

(b1) $\underline{r}_i$ $\in$ DB_Nodes, and DB_nodelab($\underline{r}_i$) := REL_INST

(b2) (r, E_Inst, $\underline{r}_i$) $\in$ DB_Edges

(b3) for all a $\in$ ATTRIB with (r, a) $\in$ domain(attr):

$a_{\underline{r}_i}$ $\in$ DB_Nodes and ($\underline{r}_i$, E_Attrib, $a_{\underline{r}_i}$) $\in$ DB_Edges

DB_nodelab($a_{\underline{r}_i}$) := a
DB_node_attr($a_{\underline{r}_i}$) := attr(r, a)
DB_attr_value($a_{\underline{r}_i}$) := μ[ATTRIB](a)($\underline{r}_i$)

(b4) ($\underline{r}_i$, E_Part_j, $\underline{e}_{i_j}$) ∈ DB_Edges for j = 1, ..., n

(c) for all t ∈ T_CONSTR, for all $\underline{e}$ ∈ domain(μ[T_CONSTR](t)):
($\underline{e}$, E_Origin, μ[T_CONSTR](t)($\underline{e}$)) ∈ DB_Edges

Figure 2.2 gives an example for a DB_Graph, which represents a (very small) database state.

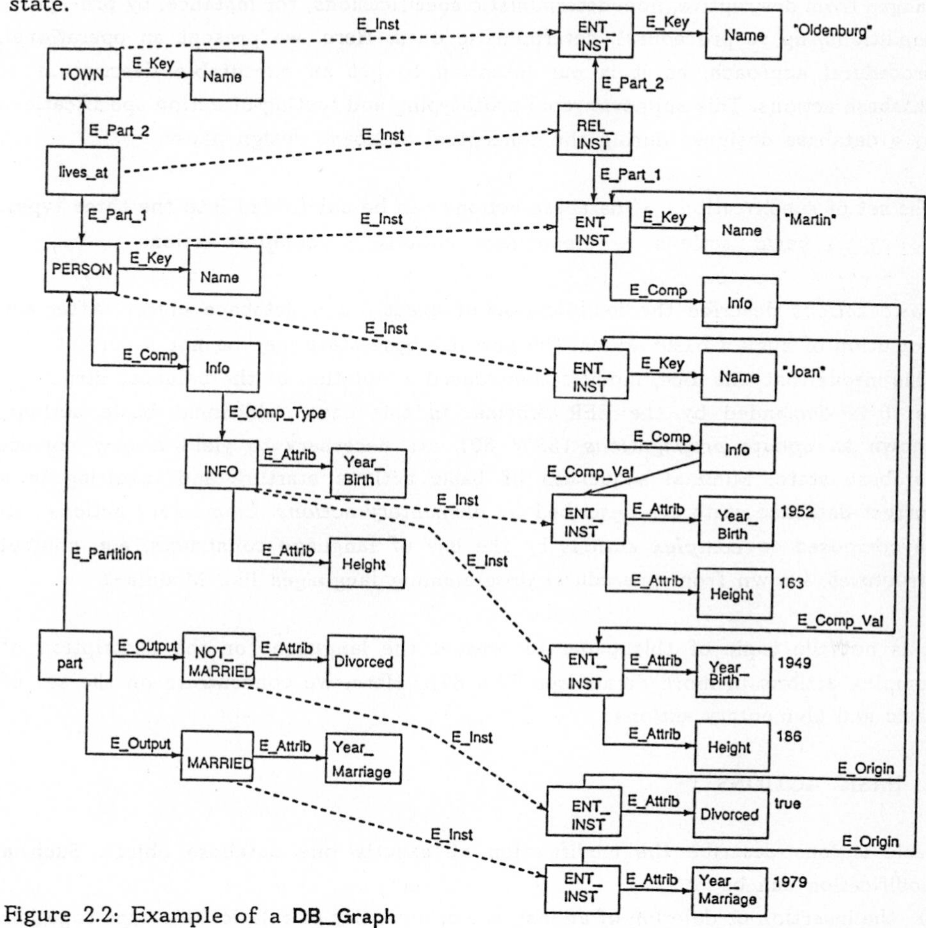

Figure 2.2: Example of a DB_Graph

3. ACTIONS ON AN EXTENDED ENTITY-RELATIONSHIP DATABASE

Up to now, we have described how the static structure of a database can be modelled. Using the concepts of our EER model, an EER schema fixes the possible structure of objects and their interrelations in a database. But, the specification of the static structure of a database is only a first step in modelling a database. A database is not a dead, unchangeable read-only memory of objects, but an alive, often changing storage, where objects can be inserted, deleted, or updated. Of course, all these modifications have to regard the structural restrictions specified by the EER schema.

In the literature, there exist several approaches to specify those infinite sets of allowed modifications of a current system state by a finite description. The scope ranges from descriptive, non-deterministic specifications, for instance, by pre-/post-conditions, up to procedural, deterministic ones. Here, we present an operational, procedural approach, as it is our intention to get an executable description of database actions. This supports rapid prototyping and testing of action specifications by a database designer during the conceptual database design phase.

The set of specifications of database actions can be subdivided into the three types:

basic actions - *elementary* actions - *complex* actions

Basic actions describe the modification of exactly one database object. After the execution of such a basic action, the new database state may be not a correct one. This means that this local modification caused a violation of the database structure, as it is demanded by the EER schema. In this case, additional basic actions, known as *update propagations* [SSW 80], are necessary to yield a new correct database state. Minimal sequences of basic actions starting and resulting in a correct database state are described by *elementary actions*. Elementary actions can be composed to *complex actions* by the use of language constructs, e.g. control structures, known from procedural programming languages like Modula-2.

It is not the topic of this paper to present the language for the description of complex actions in more detail (see [Wo 89]). Here, we concentrate on the set of basic and elementary actions.

3.1 BASIC ACTIONS

Basic actions describe the modification of exactly one database object. Such a modification can be

(i) the insertion or deletion of an instance of an entity or relationship type together with all attribute values,
(ii) the addition or removal of a component of an instance of an entity type,
(iii) the insertion or deletion of the membership of a database object in a certain type construction, or
(iv) the update of attribute values of existing database objects.

For our example, the signature of some of these basic actions have the following form:

(i) **basic_insert_PERSON (name : string) : PERSON**
basic_delete_PERSON (p : PERSON)
basic_insert_lives_at (p : PERSON; t : TOWN) : lives_at
(ii) **basic_add_comp_INFO (p : PERSON; year_of_birth : int; height : int) : INFO**
basic_remove_comp_INFO (p : PERSON)
(iii) **basic_insert_cons_PERSON_MARRIED(p : PERSON; year_of_marriage : int) : MARRIED**
basic_delete_cons_MARRIED (m : MARRIED)
(iv) **basic_update_INFO.Height (i : INFO; height : int)**

All insert actions are functions which yield as result a modified database state, and, additionally, a reference to the inserted instance of an entity or relationship type. All attributes of the entity or relationship type occur as formal parameters of the basic insert action.
All delete actions as well as the insertion of a relationship instance only require references to instances as actual parameters to denote the database objects which are relevant for the execution of this action.

All these basic actions are implicitly given for each entity or relationship type of the specified EER schema. This means that the syntax, i.e. the signature, and, as we will see, also the semantics of basic actions can automatically be derived from the EER schema.

We have shown in section 2 how database states can be represented by attributed graphs. As the execution of a basic database action modifies the current database state, this execution can be viewed as the application of an appropriate graph transformation rule.
For example, the semantics of the action **basic_insert_PERSON** can be described by the following graph transformation rule:

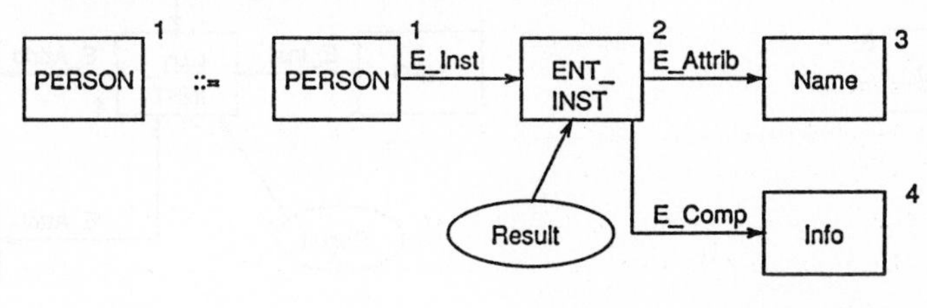

Figure 3.1

Each graph transformation rule consists of the following four components: two graphs, termed the *left-hand* and *right-hand* side of the rule, an *embedding transformation,* and a sequence of *attribute assignments.* These components control the application of a rule in several steps. The first step is to identify a subgraph in the current **DB_Graph** (cf. representation 2 in section 2), which is isomorphic to the left-hand side. The corresponding subgraph is removed in the second step of the application of a graph replacement rule, and replaced by a graph which is isomorphic to the right-hand side. In the third step, the newly inserted graph has to be connected appropriately to the **DB_Graph** by additional edges according to the given embedding transformation. Here, we only need the identical embedding. Therefore, it is not contained in Figure 3.1 (for further details see [ELS 87]). In the last step, the attribute values of the inserted nodes have to be set according to the specified attribute assignments.

Our approach to graph replacements bases on [Na 79], and on an extended version described in [ELS 87] and [En 86], where the reader can find a complete description.

Sometimes, it is useful to restrict the applicability of a graph replacement rule to a specific subgraph in the host graph. Therefore, we extend the approach of [ELS 87] by the introduction of *node-valued parameters* and *node-valued functions.* This means that the application of a graph replacement rule may yield as result a certain node of the host graph. This is expressed in the right-hand side by a "**Result**"-node (cf. Figure 3.1). Furthermore, node-valued parameters are allowed. For example, the rule for the addition of a component of type INFO has as parameter the node p of an instance of a person:

rule basic_add_comp_INFO (p : PERSON; year_of_birth : int; height : int) : INFO

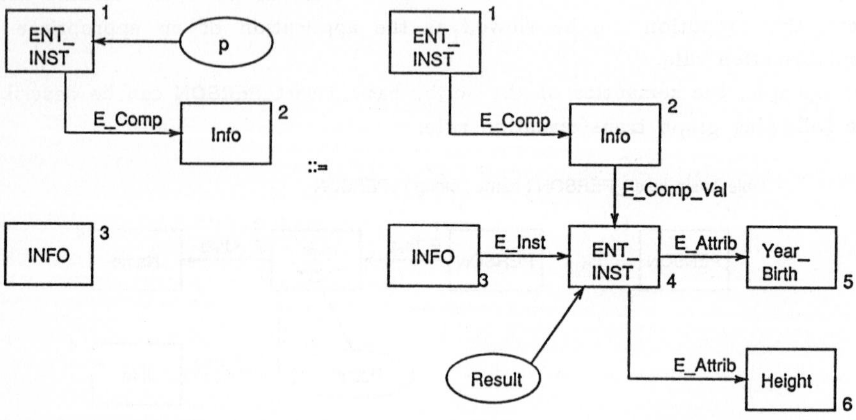

ENT_ INST 1
p
E_Comp
Info 2
INFO 3

::=

ENT_ INST 1
E_Comp
Info 2
E_Comp_Val
INFO 3
E_Inst
ENT_ INST 4
E_Attrib
Year_ Birth 5
Result
E_Attrib
Height 6

attr_value(5) := year_of_birth
attr_value(6) := height

Figure 3.2

As an example for a basic delete action, we give the rule for the removal of the component INFO from a person. The current person is denoted by the node-valued parameter p.

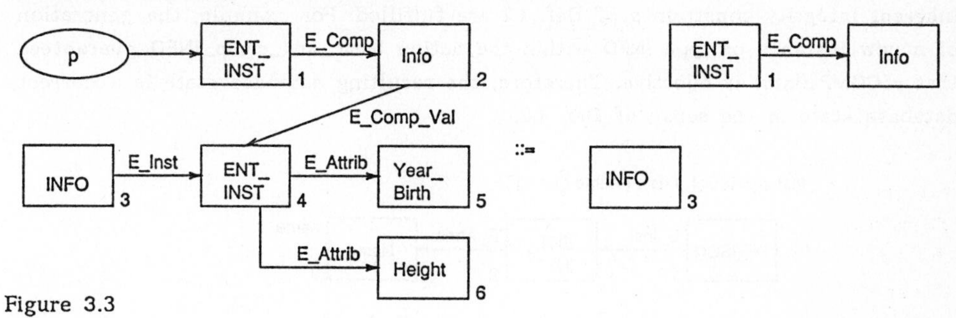

Figure 3.3

It is obvious that all these graph replacement rules for the description of basic actions on a database can automatically be derived from the specified EER schema and the corresponding graph representation. In this sense, a database designer, who has specified a database schema, has also implicitly specified all schema dependent basic actions. But, as we have already mentioned, the execution of a basic action may yield an incorrect database state. Therefore, basic actions have to be composed to elementary actions.

3.2 ELEMENTARY ACTIONS

Let us illustrate such an elementary action by the example of the insertion of a married person (cf. Fig. 3.4).

```
elem action insert_MARRIED ( name : string;
                             year_of_birth : int; height : int; year_of_marriage : int );
objects p : PERSON; i : INFO; m : MARRIED;
begin
  if not PERSON_exists ( name ) then
    p := basic_insert_PERSON ( name );
    i := basic_add_comp_INFO ( p, year_of_birth, height );
    m := basic_insert_cons_PERSON_MARRIED ( p, year_of_marriage );
  end
end
```

Figure 3.4: Elementary action for the insertion of a married person

Each person is uniquely identified by the key attribute name. So, at first, it is checked whether a person with this name already exists. The semantics of this test can be given by a subgraph test (cf. Fig. 3.5). Afterwards, a sequence of three basic actions has to be executed, to insert a new instance of type **PERSON**, to add as component an instance of type **INFO**, and to add a new instance of type **MARRIED** to express the partition. Because of the generation of these three new instances, the schema inherent integrity constraints of Def. 1.2 are fulfilled. For example, the generation of a new instance of type **INFO** within the action **basic_add_comp_INFO** guarantees that μ[**COMP**](**Info**) is bijective. Therefore, the resulting database state is a correct database state in the sense of Def. 1.2.

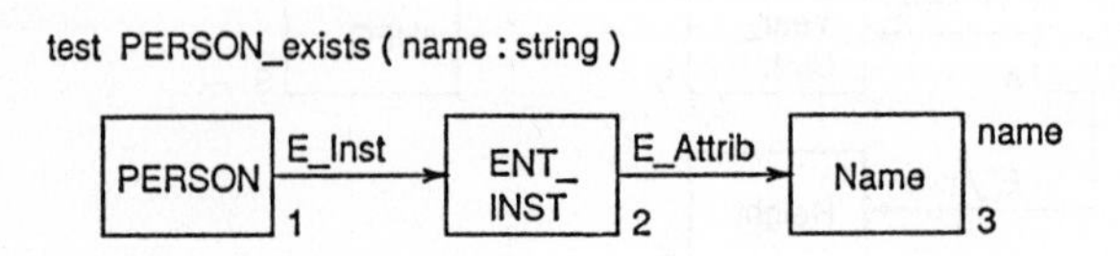

Figure 3.5: Example of a subgraph test

As each call of a basic action can be viewed as the application of a graph replacement rule and each boolean expression as the execution of a subgraph test, the whole elementary action describes a programmed graph replacement in the sense of [ELS 87], yielding an operational semantics for elementary actions.

In case of the deletion of a person, similar update propagation actions have to be executed to yield a new correct database state (cf. Figure. 3.6). Here, it has to be checked whether this person is a married or not-married person, and the existing one has to be deleted. In addition to basic modifying operations, there are basic read operations like **fetch_PERSON**, which yield as result a reference (or in the graph representation a node) to a database object. Furthermore, a person may participate in the relationship **lives_at**, where all occurrences have to be deleted, too.

```
elem action delete_PERSON ( name : string );
objects  p : PERSON;  m : MARRIED; n : NOT_MARRIED;
         la_set : set ( lives_at);
begin
  if PERSON_exists ( name ) then
    p := fetch_PERSON ( name );
    if MARRIED_exists ( p ) then
      m := fetch_PERSON_as_MARRIED ( name );
      basic_delete_cons_MARRIED ( m );
    else
      n := fetch_PERSON_as_NOT_MARRIED ( name );
      basic_delete_cons_NOT_MARRIED ( n );
    end;
```

```
        la_set := fetch_relship_lives_at_PERSON ( p );
        forall la in la_set do
          basic_delete_lives_at ( la )
        end;
        basic_remove_comp_INFO ( p );
        basic_delete_PERSON ( p )
      end
    end;
```

Figure 3.6: Elementary action for the deletion of a person

In both cases, only a subdiagram (or subgraph) of the current EER schema, termed *propagation subgraph*, is involved in the corresponding elementary action. For example, in the case of the insertion of an instance of an entity type **e**, instances of all entity types in the chain of partitions starting at a non-constructed entity type and ending at this entity type **e** have to be inserted. Furthermore, for each newly inserted instance all attribute values have to be set, and also all dependent components have to be inserted.

While the general proceeding for the insertion of a new instance is always the same, the shape of concrete elementary actions totally depend on the current database schema. Therefore, it is possible to build a generator for elementary actions, which automatically derives elementary actions from a given database schema. Because of our representation of database schemes by attributed graphs, this generator is a set of special graph algorithms. Each algorithm describes a traversal of the propagation subgraph of the **Schema_Graph**, which corresponds to the elementary action currently to be generated.

Figure 3.7 sketches the algorithm to generate an elementary insert action. This algorithm can be subdivided into four parts:
- In part 1, the chain of type constructions is computed, starting at the current entity type, indicated by **ent_name**, and ending at a non-constructed entity type.
- In part 2, the frame of an elementary insert action is generated (cf. Fig. 3.4), and the call of a basic insert action for an instance of the first entity type in the type construction chain is generated.
- In part 3, all components of this entity type are collected by an access (**AG_...**) to the attributed graph **Schema_Graph**, and appropriate basic actions to add components are generated. This generation may be recursive, if the component contains further components.
- In part 4, all entity types along the chain of type constructions are handled.

```
procedure  Gen_elem_insert_entity ( ent_name : string );
var var_name : string;
    ent_chain, components : list ( nodes );
    basic_ent, input_ent, output_ent : node;
begin
   /* part 1 */
   compute_partition_chain ( ent_name, ent_chain );
   /* part 2 */
   basic_ent := head ( ent_chain );
   Gen_frame_elem_insert_entity ( ent_name, basic_ent );
   Gen_basic_insert_entity ( basic_ent, var_name );
   /* part 3 */
   components := AG_get_target_node_list ( basic_ent, E_Comp );
   forall comp in components do
      Gen_add_component ( comp, var_name )
   end;
   /* part 4 */
   input_ent := basic_ent;
   while not empty ( ent_chain ) do
      output_ent := head ( ent_chain );
      Gen_handle_partition ( input_ent, output_ent, var_name );
      input_ent := output_ent
   end
end
```

Figure 3.7

A detailed description of the generators for elementary actions can be found in [Wo 89]. It is the topic of current research to prove that the generated elementary actions together with the graph replacement rules for basic actions describe transitions between correct database states, as they were defined in section 1.

4. CONCLUSIONS

Let us summarize the ideas of this paper:
We have shown that the description of the static structure of a database can be used to derive automatically corresponding modifying actions, termed elementary actions, which describe correct state transitions. The semantics of these elementary actions was given by programmed graph replacements. For this purpose, we have shown how a database state can be represented by an attributed graph, so that the semantics of basic actions, i.e. the constituents of elementary actions, can be given by the application of graph replacement rules.

These automatically derived elementary actions can then be used by a database designer to compose complex actions, by which greater parts of the database state are modified and which model specific actions of an application area.

The language for complex actions, the generator for elementary actions, and a corresponding interpreter are realized within CADDY ([Wo 89], [Sc 90]). CADDY is a database design environment, which supports a user in designing and testing a database on a conceptual level ([EHHLE 89]).

REFERENCES

[AbH 87] Abiteboul, S. / Hull, R.: *IFO - A Formal Semantic Database Model.* ACM Transactions on Database Systems 1987, Vol. 12, No. 4 (525 - 565)

[Ch 76] Chen, P.P.: *The Entity-Relationship Model - Towards a Unified View of Data.* ACM Transactions on Database Systems 1976, Vol. 1, No. 1 (9 - 36)

[EHHLE 89] Engels, G. / Hohenstein, U. / Hülsmann, K. / Löhr-Richter, P. / Ehrich, H.-D.: *CADDY: Computer-Aided Design of Non-Standard Databases.* In: N. Madhavji, H. Weber, W. Schäfer (eds.): Int. Conf. on System Development Environments & Factories. Berlin, May 1989. Pitman Publ., London 1990

[EK 80] Ehrig, H. / Kreowski, H.-J.: Applications of Graph Grammar Theory to Consistency, Synchronization, and Scheduling in Data Base Systems. In: Information Systems, Vol. 5, 1980, (225-238)

[ELS 87] Engels, G. / Lewerentz, C. / Schäfer, W.: Graph Grammar Engineering: A Software Specification Method. In: [ENRR 87], (186-201)

[En 86] Engels, G.: Graphen als zentrale Datenstrukturen in einer Software-Entwicklungsumgebung, Fortschrittber. VDI, Nr. 62, Düsseldorf, VDI-Verlag, 1986

[ENR 83] Ehrig, H. / Nagl, M. / Rozenberg, G. (eds.): Graph Grammars and Their Application to Computer Science, 2nd Intern. Workshop, LNCS 153, Berlin, Springer 1983

[ENRR 87] Ehrig, H. / Nagl, M. / Rozenberg, G. / Rosenfeld, A. (eds.): Graph Gram- mars and Their Application to Computer Science, 3rd Intern. Workshop, Warrenton (Virginia) 1986, LNCS 291, Berlin, Springer 1987

[ERA 80] P.P. Chen (ed.): *Proc. of the 1st Intern. Conference on Entity-Relationship Approach.* Los Angeles (California), 1980

[ERA 88] C. Batini (ed.): *Proc. of the 7th Int. Conference on Entity-Relationship Approach.* Rome (Italy) 1988

[ERA 89] F. Lochovski (ed.): *Proc. of the 8th Int. Conference on Entity-Relationship Approach.* Toronto (Canada) 1989

[EWH 85] Elmasri, R.A. / Weeldreyer, J. / Hevner, A.: *The Category Concept: An Extension to the Entity-Relationship Model.* Data & Knowledge Engineering 1985, Vol. 1 (75 - 116)

[FV 83] Furtado, A.L. / Veloso, P.: Specification of Data Bases Through Rewriting Rules. In: [ENR 83], (102 - 114)

[Gö 88] Göttler, H.: Graphgrammatiken in der Softwaretechnik. Informatik-Fachberichte 178, Berlin, Springer 1988

[HaM 81] Hammer, M. / McLeod, D.: *Database Description with SDM: A Semantic Database Model.* ACM Transactions on Database Systems 1981, Vol. 6, No. 3 (351 - 386)

[HK 87] Hull, R. / King, R.: *Semantic Database Modeling: Survey, Applications, and Research Issues.* ACM Computing Surv. 1987, Vol. 19, No. 3 (201 - 260)

[HNSE 87] Hohenstein, U. / Neugebauer, L. / Saake, G. / Ehrich, H.-D.: *Three-Level Specification Using an Extended Entity-Relationship Model.* In R.R. Wagner, R. Traunmüller, H.C. Mayr (eds.): Informationsbedarfsermittlung und -analyse für den Entwurf von Informationssystemen. Informatik-Fachberichte Band 143, Springer 1987 (58 - 88)

[HoGo 88] Hohenstein, U. / Gogolla, M.: *A Calculus for an Extended Entity-Relationship Model Incorporating Arbitrary Data Operations and Aggregate Functions.* In: [ERA 88] (129 -148)

[LyK 86] Lyngbaek, P. / Kent, W.: *A Data Modeling Methodology for the Design and Implementation of Information Systems.* In K.R. Dittrich, U. Dayal (ed.): Proc. of the Int. Workshop on Object-Oriented Database Systems, Pacific Grove (California) 1986 (6 - 17)

[MMR 86] Makowski, J.A. / Markowitz, V.M. / Rotics, N.: *Entity-Relationship Consistency for Relational Schemes.* In G. Ausiello, P. Atzeni (eds.): Proc. International Conference on Database Theory ICDT 1986, Springer LNCS 243 (306 - 322)

[MyW 80] Mylopoulos, J. / Wong, H.K.T.: *Some Features of the TAXIS Data Model.* In: Proc. 6th International Conference on Very Large Data Bases 1980, Montreal (Canada) (399 - 410)

[Na 79] Nagl, M. : Graph-Grammatiken: Theorie, Implementierung, Anwendungen. Braunschweig, Vieweg 1979

[Na 87] Nagl, M.: *A Software Development Environment Based on Graph Technology.* In: [ENRR 87], (458 - 478)

[PRYS 89] Parent, C. / Rolin, H. / Yètongon, K. / Spaccapietra, S.: *An ER Calculus for the Entity- Relationship Complex Model.* In: [ERA 89]

[Sc 90] Schmidt, R. : Entwurf und Implementierung eines Interpreters für eine Sprache zur Beschreibung schema-abhängiger Aktionen in einem erweiterten Entity-Relationship Modell, Diploma Thesis, TU Braunschweig, 1990

[SSW 80] Scheuermann, P. / Schiffner, G. / Weber, H. : *Abstraction Capabilities and Invariant Properties Modelling within the Entity-Relationship Approach.* In: [ERA 80], (121 - 140)

[TYF 86] Teorey, T.J. / Yang, D. / Fry, J.P.: *A Logical Design Methodology for Relational Databases Using the Extended Entity-Relationship Model.* ACM Computing Surveys 1986, Vol. 18, No. 2 (197 - 222)

[Wo 89] Wolff, M.: Eine Sprache zur Beschreibung schema-abhängiger Aktionen in einem erweiterten Entity-Relationship-Modell, Diploma Thesis, TU Braunschweig, 1989

PHYSICALLY-BASED GRAPHICAL INTERPRETATION OF MARKER CELLWORK L-SYSTEMS

F. David Fracchia and Przemyslaw Prusinkiewicz

Department of Computer Science
University of Regina
Regina, Saskatchewan, CANADA S4S 0A2

ABSTRACT: Map L-systems with dynamic interpretation have been successfully applied to the modeling of the development of two-dimensional cell layers [3, 4]. We extend this technique to three-dimensional cellular structures. The seminal notion of three-dimensional *cyclic edge-label-controlled OL-systems*, termed *cellworks*, was introduced by A. Lindenmayer [8]. We provide an alternative definition of cellworks using *markers*, and use it as a formal basis for a simulation program. Cell geometry is viewed as the result of mechanical cell interactions due to osmotic pressure and wall tension. Developmental sequences can be animated by considering periods of continuous expansion delimited by instantaneous cell divisions. As an example, the method is applied to visualize the development of a three-dimensional epidermal cell layer.

Keywords: computer graphics, mathematical modeling in biology, simulation, visualization of development, map L-system, cellwork L-system, dynamic model.

CONTENTS

0. INTRODUCTION

An important issue in plant morphology is the study of cell division patterns, that is, the spatial and temporal organization of cell divisions in tissues. In the past, the modeling of cellular structures focused mainly on the development of branching and nonbranching filaments, represented by string and bracketed L-systems [7], and two-dimensional planar and spherical cell layers whose topology was described by map L-systems [10]. Such methods are described in [1, 15]. This paper presents a method for simulating and visualizing the development of three-dimensional multicellular structures.

The practical motivation for this work is related to two applications. As a *research tool*, graphical simulations make it possible to study the impact of cell divisions on cell arrangement and global shape formation. As a *visualization tool*, simulations provide a method for presenting features that cannot be captured using time-lapse photography. For example, pseudocolor may be introduced to distinguish groups of cells descending from a specific ancestor or to indicate cell age. Inconspicuous structural elements, such as new division walls, can be emphasized.

The modeling method consists of two stages. First, the *topology* of the cell division patterns is expressed using the formalism of *cellwork L-systems.* At this stage, the neighborhood relations between cells are established, but the cell shapes remain unspecified. Next, cell *geometry* is modeled using a dynamic method that takes into account the osmotic pressure inside the cells and the tension of cell walls. The development can be animated by considering periods of continuous cell expansion, delimited by instantaneous cell divisions.

This paper is organized as follows. Section 1. focuses on the simulation of cellular development at the topological level. After a brief survey of previous three-dimensional models, the formalism of marker-based cellwork L-systems is proposed to describe cell neighborhood. Section 2. presents a dynamic model for the specification of cell geometry, given the topology. In section 3., the method is applied to model the development of epidermal cells. Section 4. discusses open problems.

1. THREE-DIMENSIONAL MODELS AND CELLWORK L-SYSTEMS

Various models have been proposed for the modeling of three-dimensional cellular structures. Rules whose main control elements were cell walls have been informally presented by Korn [6] and the Lücks [11]. Double wall stereomap generating systems were introduced by the Lücks [12] to model walls in three-dimensional space, but were somewhat difficult to interpret geometrically. Recently, the Lücks [13] presented the formalism of double-wall cellwork L-systems for modeling plant meristems.

We propose a method which extends the notion of two-dimensional single-wall marker map L-systems to three-dimensions, based on the structures operated on by the cyclic cellwork L-systems introduced by Lindenmayer [8]. An initial, more restricted version of our method was considered in [4].

1.1. Cellworks

In order to capture the structure of three-dimensional cellular tissues, Lindenmayer [8] proposed an extension of map L-systems called *cellwork* L-systems. The notion of a cellwork is characterized as follows.

- A cellwork is a finite set of *cells.* Each cell is surrounded by one or more *walls* (faces).
- Each wall is surrounded by a boundary consisting of a finite, circular sequence of *edges* which meet at *vertices.*
- Walls cannot intersect without forming an edge, although there can be walls without edges (in the case of cells shaped as spheres or tori).
- Every wall is part of the boundary of a cell, and the set of walls is connected.
- Each edge has one or two vertices associated with it. Edges cannot cross without forming a vertex and there are no vertices without an associated edge.
- Every edge is a part of the boundary of a wall, and the set of edges is connected.

1.2. mBPCOL-systems

The process of cell division can be expressed as cellwork rewriting. This notion is an extension of map rewriting. Several map-rewriting systems have been described in the past [9]. To capture the development of three-dimensional structures we extend two-dimensional mBPMOL-systems, proposed by Nakamura, Lindenmayer, and Aizawa [14] as a refinement of the basic concept of map L-systems introduced by Lindenmayer and Rozenberg [10], to the formalism of *marker Binary Propagating Cellwork OL-systems.* The name is derived as follows. A *cellwork OL-system* is a parallel rewriting system which operates on cellworks and modifies cells irrespective of the states of other neighboring cells (a *context-free* mechanism). The system is *binary* in that a cell can split into at most two daughter cells. It is *propagating* in the sense that edges cannot be erased, thus cells cannot fuse or die. The *markers* represent a technique for specifying the positions of inserted edges used to split the walls and divide cells.

An mBPCOL-system $\mathcal{G}$ is defined by a finite alphabet of *edge labels* Σ, a finite alphabet of *wall labels* Γ, a *starting cellwork* ω, and a finite set of *edge productions* P. The initial cellwork ω is specified as a list of walls and their bounding edges. Edges may be directed or neutral, indicated by the presence or absence of arrows above edge labels. Each production is of the form $A : \beta \rightarrow \alpha$, where the edge $A \in \Sigma$ is the *predecessor*; the string $\beta \in \{\Gamma^{+}, *\}$ is a list of *applicable walls* ($*$ denotes all walls); and the string α, composed of edge labels from Σ, wall labels from Γ, and symbols [and], is the *successor.* The sequence of symbols outside the square brackets describes the subdivision pattern of the successor. Pairs of matching brackets [and] delimit *markers* which specify possible attachment sites for new edges and walls. Arrows indicate the directions of the successor edges and markers with respect to the predecessor edge. For successor edges, the right arrow indicates a direction consistent with the predecessor edge, the left arrow indicates the opposite direction, and no arrow is neutral. In the case of markers, the right arrow indicates an outward orientation from the predecessor edge, the left arrow indicates an inward orientation, and no arrow is neutral. The list β contains all walls into which a marker should be inserted. In addition to the labels for edges and markers, a successor specifies the labels of walls which may be created as a result of production application.

For example, production $\vec{A}$: $14 \rightarrow \vec{D}\,\overleftarrow{C}_2[\vec{E}_5]_3\vec{B}\,F$ applies to the edge A if it belongs to one or more walls labeled 1 or 4 (Figure 1a). The predecessor edge is subdivided into four edges D, C, B and F. During a derivation step, marker E is introduced into all walls of type 1 or 4 which share edge A (Figure 1b), and can be connected with a matching marker inserted into the same wall by another production. As a result, the wall will split into two. The daughter wall created before the matched marker in the direction of the predecessor edge A will be labeled 2, and the wall formed after the marker will be labeled 3 (Figure 1c). Markers E can be connected only if both productions assign labels to the daughter walls in a consistent way. Otherwise, the markers are considered non-matching

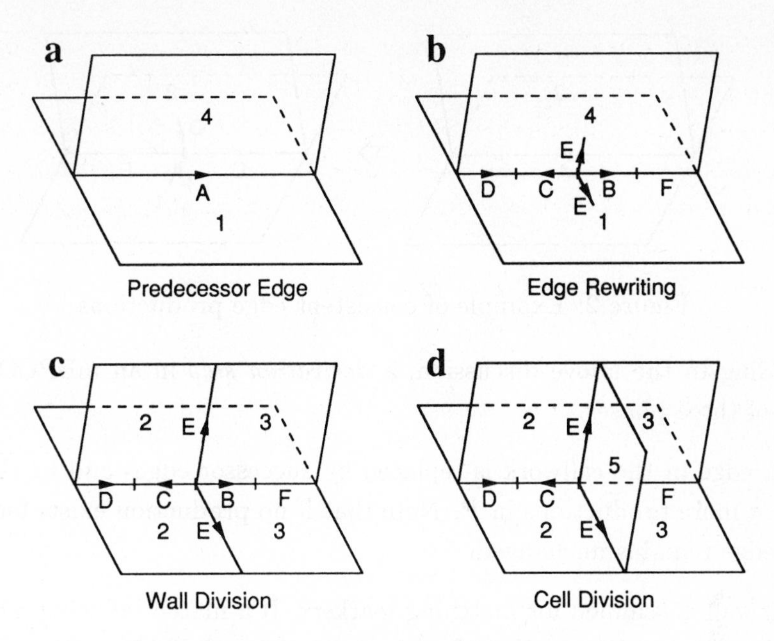

Figure 1: The phases of a derivation step.

and are discarded. If several walls bounding a cell split in such a way that the sequence of new edges forms a closed contour, a new wall bound by these edges may be created. In order for this to occur, all markers involved must specify the same label for the new wall, 5 in this example (Figure 1d).

The limitation of the scope of a production to specific walls may create a consistency problem while rewriting edges. For instance, assume that walls 1 and 2 share edge A, and the following productions are in P:

$$\begin{aligned} p_1 &: \vec{A}{:}\,1 \rightarrow \vec{C}\overleftarrow{E} \\ p_2 &: \vec{A}{:}\,2 \rightarrow A\vec{B} \end{aligned}$$

Productions p_1 and p_2 are inconsistent since they specify two different partitions of the same edge. We assume the mBPCOL-systems under consideration are free of such inconsistencies. This does not preclude the possibility of applying several productions simultaneously to the same edge. For example, a production pair,

$$\begin{aligned} p_1 &: \vec{A}{:}\,1 \rightarrow \vec{C}_2[\vec{F}_3]_4\overleftarrow{E} \\ p_2 &: \vec{A}{:}\,2 \rightarrow \vec{C}_5[\overleftarrow{D}_6]_7\overleftarrow{E}, \end{aligned}$$

consistently divides edge A into segments C and E, although the markers inserted into walls 1 and 2 are different (Figure 2).

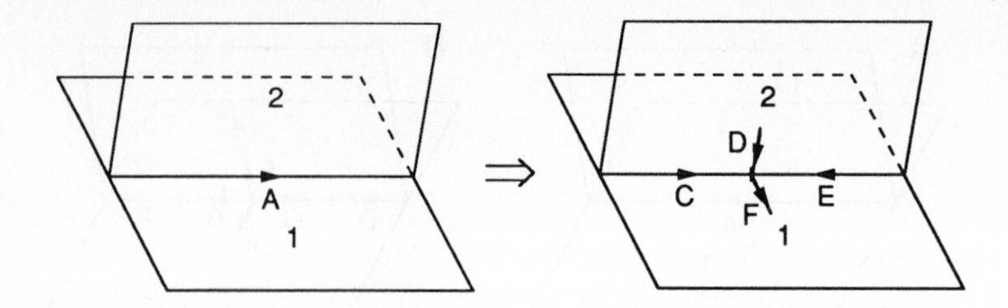

Figure 2: Example of consistent edge productions.

According to the above discussion, a *derivation step* in an mBPCOL-system consists of three phases.

1. Each edge in the cellwork is replaced by successor edges and markers using one or more productions in P. Note that if no production exists for an edge, the edge remains unchanged.

2. Each wall is scanned for matching markers. If a match inducing a consistent labeling of daughter walls is found, the wall is subdivided. The selection of matching markers is done by the system. Unused markers are discarded.

3. Each cell is scanned for a circular sequence of new division edges having the same wall label. If such a sequence is found, it is used to bound the new wall which will divide the cell into two daughter cells. If different possibilities exist, the edges are selected by the system.

A wall may be subdivided more than once as long as new division edges do not intersect and a consistent labeling of daughter walls is possible. In contrast, a cell may be divided only once in any derivation step.

For example, Figure 3 presents a three-dimensional cellwork L-system. In the first derivation step, production p_1 divides walls labeled 1, and production p_2 divides walls labeled 2. The inserted edges form a cycle that divides the cell with a new wall labeled 2. In the subsequent steps this process is repeated, generating a pattern of alternating division walls. Production p_3 introduces the necessary delay.

A more complex example is the construction of a Sierpiński tetrahedron, which is a three-dimensional extension of the Sierpiński gasket described in [16]. The cellwork L-system is given in Figure 4. It has been simplified by the addition of superscripts, for example A^i for $i = \{0, 1, 2\}$ replaces three edge labels (the total number of edge labels involved is 38). Also, productions without markers that

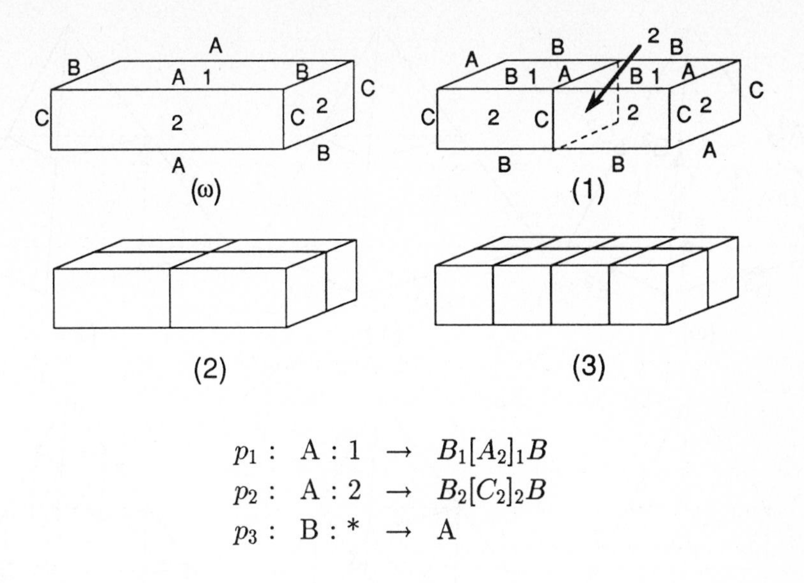

$$
\begin{array}{llll}
p_1: & A:1 & \rightarrow & B_1[A_2]_1B \\
p_2: & A:2 & \rightarrow & B_2[C_2]_2B \\
p_3: & B:* & \rightarrow & A
\end{array}
$$

Figure 3: Example of a cellwork L-system.

match those yielding markers for a particular edge are not shown. For example, production

$$p_{18}: \ \vec{d}^0 : 12 \rightarrow \vec{D}^1{}_1[\vec{B}^1{}_1]_{21}[\vec{E}^1{}_1]_2\vec{d}^1$$

yields markers for edge $\vec{d}^0$ contained in walls 1 and 2 and has matching production

$$m_{18}: \ \vec{d}^0 : 3 \rightarrow \vec{D}^1\vec{d}^1$$

which does not yield markers for walls labeled 3. Such matching productions are necessary to ensure the consistent replacement of edges. The productions in the cellwork L-system are applied as follows. Productions $p_1 - p_6$ are responsible for the first division (Figure 4(1)) which results in a new tetrahedron appearing at the top of the structure (given the orientation of the initial tetrahedron in (ω)). The next division (2) occurs at the left hand corner and is the result of productions p_{13}, p_{18} and p_{23}. The next two divisions ((3) and (4)) occur counterclockwise (viewed from the top) at the remaining corners. Division three results from the application of productions p_{15}, p_{19} and p_{21}, while the fourth division is determined by productions p_{16}, p_{22} and p_{25}. The remaining productions delay the modification of edge labels such that after four successive divisions, the initial tetrahedron is divided into four tetrahedrons having the same initial edge and wall labels, and a central octahedron which does not divide. The process is then repeated for each tetrahedron, as seen in derivation (8).

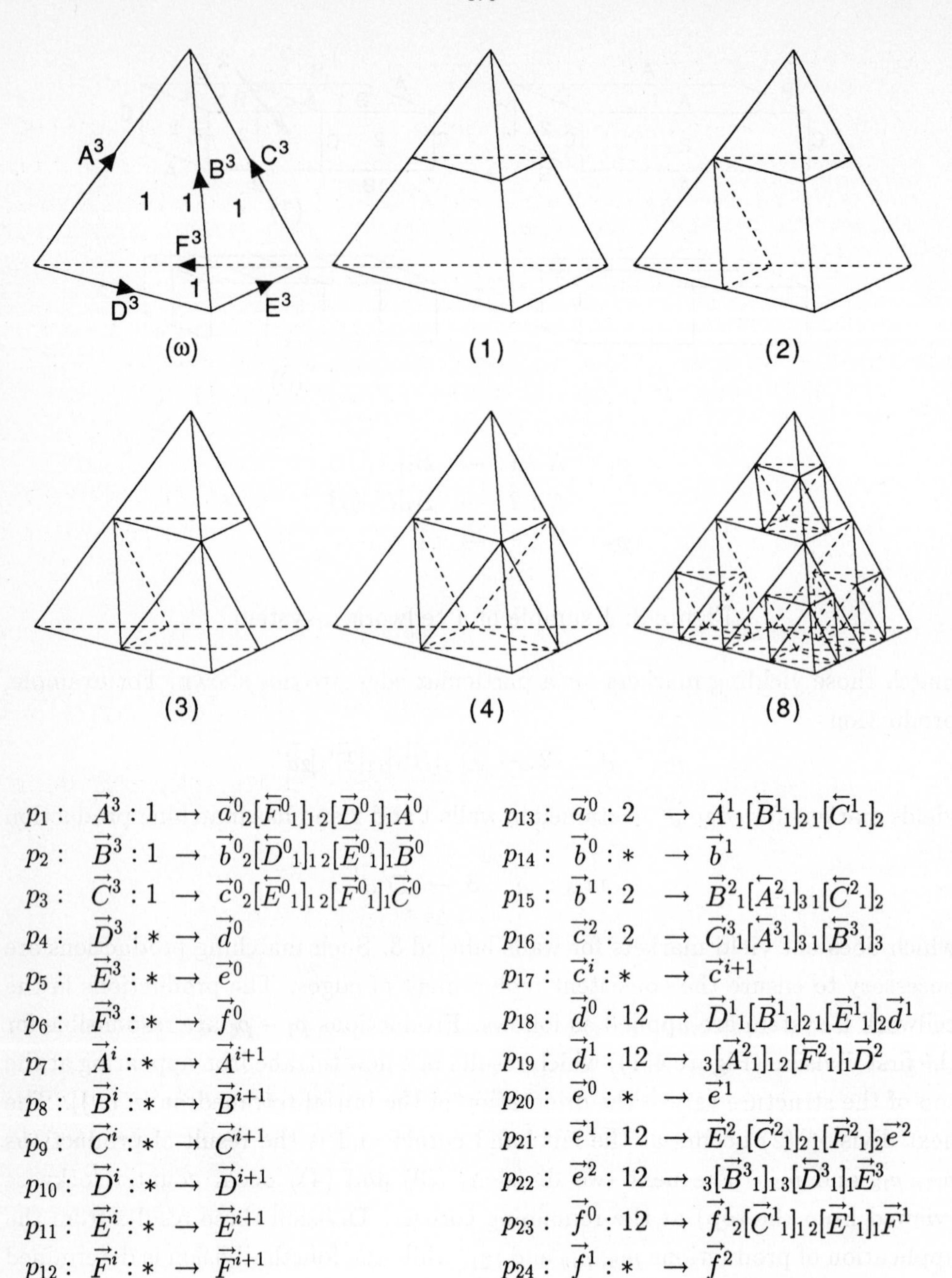

$p_1 : \overrightarrow{A}^3 : 1 \rightarrow \overrightarrow{a}^0{}_2[\overleftarrow{F}^0{}_1]_{1\,2}[\overrightarrow{D}^0{}_1]_1\overrightarrow{A}^0$

$p_2 : \overrightarrow{B}^3 : 1 \rightarrow \overrightarrow{b}^0{}_2[\overleftarrow{D}^0{}_1]_{1\,2}[\overrightarrow{E}^0{}_1]_1\overrightarrow{B}^0$

$p_3 : \overrightarrow{C}^3 : 1 \rightarrow \overrightarrow{c}^0{}_2[\overleftarrow{E}^0{}_1]_{1\,2}[\overrightarrow{F}^0{}_1]_1\overrightarrow{C}^0$

$p_4 : \overrightarrow{D}^3 : * \rightarrow \overrightarrow{d}^0$

$p_5 : \overrightarrow{E}^3 : * \rightarrow \overrightarrow{e}^0$

$p_6 : \overrightarrow{F}^3 : * \rightarrow \overrightarrow{f}^0$

$p_7 : \overrightarrow{A}^i : * \rightarrow \overrightarrow{A}^{i+1}$

$p_8 : \overrightarrow{B}^i : * \rightarrow \overrightarrow{B}^{i+1}$

$p_9 : \overrightarrow{C}^i : * \rightarrow \overrightarrow{C}^{i+1}$

$p_{10} : \overrightarrow{D}^i : * \rightarrow \overrightarrow{D}^{i+1}$

$p_{11} : \overrightarrow{E}^i : * \rightarrow \overrightarrow{E}^{i+1}$

$p_{12} : \overrightarrow{F}^i : * \rightarrow \overrightarrow{F}^{i+1}$

$p_{13} : \overrightarrow{a}^0 : 2 \rightarrow \overrightarrow{A}^1{}_1[\overleftarrow{B}^1{}_1]_{2\,1}[\overleftarrow{C}^1{}_1]_2$

$p_{14} : \overrightarrow{b}^0 : * \rightarrow \overrightarrow{b}^1$

$p_{15} : \overrightarrow{b}^1 : 2 \rightarrow \overrightarrow{B}^2{}_1[\overleftarrow{A}^2{}_1]_{3\,1}[\overleftarrow{C}^2{}_1]_2$

$p_{16} : \overrightarrow{c}^2 : 2 \rightarrow \overrightarrow{C}^3{}_1[\overleftarrow{A}^3{}_1]_{3\,1}[\overleftarrow{B}^3{}_1]_3$

$p_{17} : \overrightarrow{c}^i : * \rightarrow \overrightarrow{c}^{i+1}$

$p_{18} : \overrightarrow{d}^0 : 12 \rightarrow \overrightarrow{D}^1{}_1[\overrightarrow{B}^1{}_1]_{2\,1}[\overrightarrow{E}^1{}_1]_2\overrightarrow{d}^1$

$p_{19} : \overrightarrow{d}^1 : 12 \rightarrow {}_3[\overrightarrow{A}^2{}_1]_{1\,2}[\overleftarrow{F}^2{}_1]_1\overrightarrow{D}^2$

$p_{20} : \overrightarrow{e}^0 : * \rightarrow \overrightarrow{e}^1$

$p_{21} : \overrightarrow{e}^1 : 12 \rightarrow \overrightarrow{E}^2{}_1[\overrightarrow{C}^2{}_1]_{2\,1}[\overrightarrow{F}^2{}_1]_2\overrightarrow{e}^2$

$p_{22} : \overrightarrow{e}^2 : 12 \rightarrow {}_3[\overrightarrow{B}^3{}_1]_{1\,3}[\overleftarrow{D}^3{}_1]_1\overrightarrow{E}^3$

$p_{23} : \overrightarrow{f}^0 : 12 \rightarrow \overrightarrow{f}^1{}_2[\overrightarrow{C}^1{}_1]_{1\,2}[\overrightarrow{E}^1{}_1]_1\overrightarrow{F}^1$

$p_{24} : \overrightarrow{f}^1 : * \rightarrow \overrightarrow{f}^2$

$p_{25} : \overrightarrow{f}^2 : 12 \rightarrow \overrightarrow{F}^3{}_1[\overrightarrow{A}^3{}_1]_{3\,1}[\overrightarrow{D}^3{}_1]_3$

Figure 4: The Sierpiński tetrahedron.

2. DYNAMIC INTERPRETATION

Cellworks are topological objects without inherent geometric properties. In order to visualize them, some method for assigning geometric interpretation must be applied. Assuming the dynamic point of view, the shape of cells and thus the shape of the entire organism result from the action of forces. The unbalanced forces due to cell divisions cause the gradual modification of cell shapes until an equilibrium is reached. At this point, new cell divisions occur, and expansion resumes. The dynamic method is an extension of a similar approach used to model two-dimensional cell layers described by map L-systems [3, 4].

The dynamic interpretation method is based on the following assumptions:

- the structure is represented as a three-dimensional network of masses corresponding to cell vertices, connected by springs which correspond to cell edges,
- the springs are always straight and obey Hooke's law,
- for the purpose of force calculations, walls can be approximated by flat polygons,
- the cells exert pressure on their bounding walls; the pressure on a wall is directly proportional to the wall area and inversely proportional to the cell volume,
- the pressure on a wall is divided evenly between the wall vertices,
- the motion of masses is damped, and
- other forces are not considered.

The position of each vertex, and thus the shape of the structure, is computed as follows. As long as an equilibrium is not reached, unbalanced forces put masses in motion. The total force $\vec{F_T}$ acting on a vertex X is given by the formula:

$$\vec{F_T} = \sum_{e \in E} \vec{F_e} + \sum_{w \in W} \vec{F_w} + \vec{F_d},$$

where $\vec{F_e}$ are forces contributed by the set of edges E incident to X, $\vec{F_w}$ are forces contributed by the set of walls W incident to X, and $\vec{F_d} = -b\vec{v}$ is a damping force. The forces $\vec{F_e}$ act along the cell edges and represent wall *tension.* The magnitude is determined by Hooke's law, $F_e = -k(l - l_0)$, where k is the spring constant, l is the current spring length, and l_0 is the rest length. The forces $\vec{F_w}$ are due to the *pressure* exerted by the cells on their bounding walls. The total force of pressure exerted by a cell on a wall w has direction normal to w and is equal to $p \cdot A$, where p is the internal cell pressure and A is the wall area. The pressure p is assumed to be inversely proportional to the cell volume, $p \sim V^{-1}$, which corresponds to the equation describing osmotic pressure (with constant solute concentration and temperature). The area A of a wall is found by tesselating it into triangles and summing the areas of each triangle. The volume V of a cell is calculated by tesselating the cell into tetrahedra.

The force $\vec{F}_T$ acts on the mass at the cellwork vertex. Newton's second law of motion applies,

$$m\frac{d^2\vec{x}}{dt^2} = \vec{F}_T,$$

where $\vec{x}$ is the vertex position. If the entire structure has N vertices, we obtain a system of $2N$ differential equations,

$$m_i\frac{d\vec{v}_i}{dt} = \vec{F}_{T_i}(\vec{x}_1, \cdots, \vec{x}_N, \vec{v}_i), \qquad \frac{d\vec{x}_i}{dt} = \vec{v}_i,$$

where $i = 1, 2, \ldots, N$. The task is to find the sequence of positions $\vec{x}_1, \ldots, \vec{x}_N$ at given time intervals, assuming that the functions $\vec{F}_{T_i}$ and the initial values of all variables $\vec{x}_1^0, \ldots, \vec{x}_N^0$ and $\vec{v}_1^0, \ldots, \vec{v}_N^0$ are known. These initial values are determined as follows.

- Coordinates of the vertices of the starting cellwork are included in the input data for the simulation.
- Positions of existing vertices are preserved through a derivation step. New vertices partition the divided edges into segments of equal length. The initial velocities of all vertices are set to zero.

The system of differential equations with the initial values given above represents an *initial value problem.* It can be solved numerically using the *forward (explicit) Euler method* [2]. To this end, the differential equations are rewritten using finite increments $\Delta\vec{v}_i$, $\Delta\vec{x}_i$ and Δt,

$$\Delta\vec{v}_i^k = \frac{1}{m_i}\vec{F}_{T_i}\left(\vec{x}_1^k, \cdots, \vec{x}_N^k, \vec{v}_i^k\right)\Delta t, \qquad \Delta\vec{x}_i^k = \vec{v}_i^k\Delta t,$$

where the superscripts $k = 0, 1, 2, \ldots$ indicate the progress of time, $t = k\Delta t$. The position and velocity of a point i after time increment Δt are expressed as follows:

$$\begin{aligned}\vec{v}_i^{k+1} &= \vec{v}_i^k + \Delta\vec{v}_i^k \\ \vec{x}_i^{k+1} &= \vec{x}_i^k + \Delta\vec{v}_x^k\end{aligned}$$

The iterative computation of the velocities $\vec{v}_i^{k+1}$ and positions $\vec{x}_i^{k+1}$ is carried out for consecutive values of index k until all increments $\Delta\vec{v}_i^k$ and $\Delta\vec{x}_i^k$ fall below a threshold value. This indicates that the equilibrium state has been approximated to the desired accuracy. The next derivation step is then performed. A system of equations corresponding to the new cellwork topology is created, and the search for an equilibrium state resumes. In such a way, the developmental process is simulated as periods of continuous cell expansion, delimited by instantaneous cell divisions. Continuity of cell shapes during divisions is preserved by the rule which sets the initial positions of vertices. The dynamic method is illustrated by the example in the following section.

3. DEVELOPMENT OF EPIDERMAL CELLS

A division pattern that frequently occurs in epidermal cell structures is described by the cellwork L-system in Figure 5, based on a cyclic cellwork L-system developed by Lindenmayer [8]. Productions p_1, p_2, p_6 and p_7 are responsible for cell divisions, while the remaining productions change the states of edges for future divisions (delays). The resulting pattern consists of staggered divisions of sister cells such that all cells remain hexagonal and form a three-dimensional cell layer. The dynamic model for cellwork interpretation produces regular hexagonal cells without the specification of edge growth rates and exact division angles (as in [8]).

4. CONCLUSIONS

This paper presents a modeling method for three-dimensional cellular structures. Cell topology is captured by mBPCOL-systems, while the geometry results from a dynamic model that takes into account internal cell pressure and wall tension. The method is illustrated by a biological example.

The present formalism of cellwork L-systems imposes a restriction on how walls are allowed to subdivide. That is, a wall may subdivide more than once as long as the new division edges do not intersect. This will cause problems in the case where two neighbor cells divide in one derivation step along a shared wall such that the division edges of that wall cross each other. We are certain this case will arise many times while modeling three-dimensional biological structures. One solution to this problem is to introduce markers which themselves contain edge and marker labels into the cellwork L-system (hierarchical marker system). On

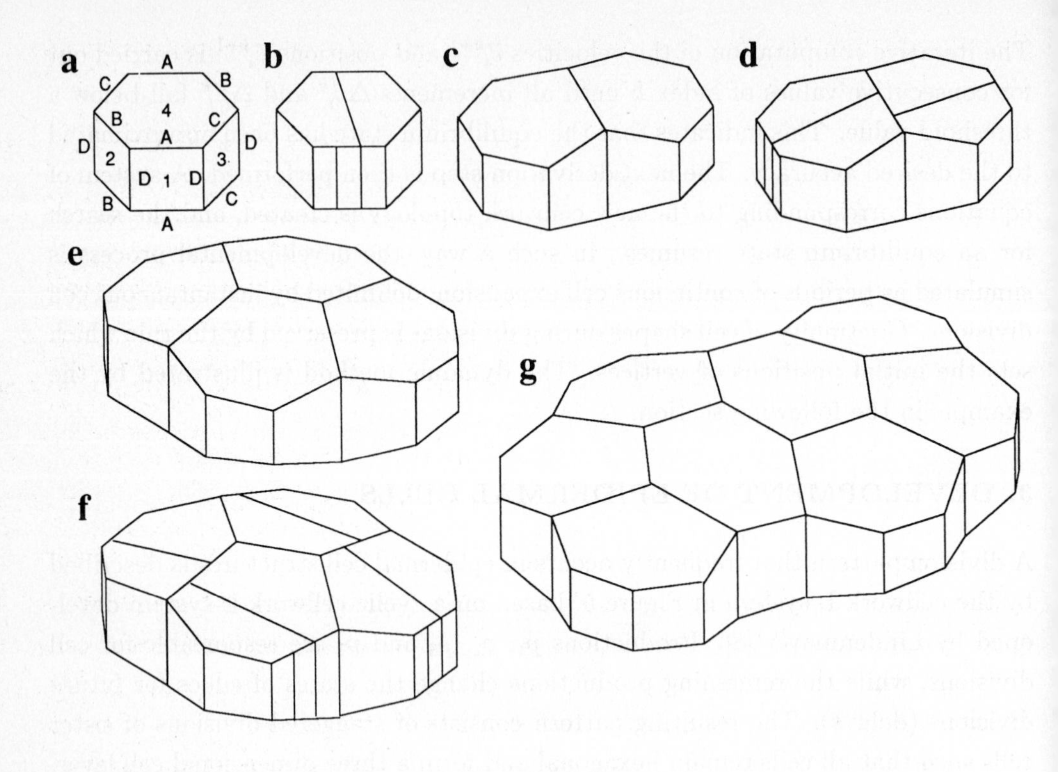

$$
\begin{array}{rlcl}
p_1: & A:123 & \rightarrow & C_3[E_1]_2B_2[D_1]_3C \\
p_2: & A:4 & \rightarrow & CB_4[F_1]_4C \\
p_3: & B:* & \rightarrow & A \\
p_4: & C:* & \rightarrow & B \\
p_5: & E:* & \rightarrow & D \\
p_6: & F:123 & \rightarrow & HGH \\
p_7: & F:4 & \rightarrow & H_4[F_1]_4G_4[F_1]_4H \\
p_8: & G:* & \rightarrow & F \\
p_9: & H:* & \rightarrow & G
\end{array}
$$

Figure 5: Developmental sequence of epidermal cells: (a) The starting cellwork; (b), (d) and (f) cellworks immediately after cell divisions; (c), (e) and (g) the corresponding cellworks at equilibrium.

the other hand intersections could be detected at the geometric level resulting in the construction of new vertices at intersection points.

Double-wall cellwork L-systems have been proposed by the Lücks [13] for the modeling of plant meristems. It may be expected that, as in the two-dimensional case, three-dimensional double-wall systems have the advantage of being more convenient than single-wall systems when describing cell development, however, single-wall systems are simpler to implement. The translation of double-wall systems to single-wall systems may also parallel the two-dimensional case.

The dynamic method for determining cell shapes involves many arbitrary assumptions, such as equal distribution of pressure between the wall vertices, and reduction of wall tension to forces acting along the wall edges. It is tempting to introduce more sophisticated assumptions concerning physical properties of cells and their components. At this time we are not aware of biological observations which would provide a solid basis for such refinements.

The lack of data presents an obstacle to the modeling of three-dimensional structures using mBPCOL-systems. For example, we attempted to model the development of a root of *Azolla pinnata* presented in [5] and frequently quoted in biological literature, but the available description was too general to be captured in the form of an mBPCOL-system. Specifically, the development of the segments of the root could not be determined. Only the development of the outer surface of segments was distinguishable. Assuming such data was available, there is also the problem of inferring the cellwork L-system.

ACKNOWLEDGEMENTS

We are deeply indebted to Professor Lindenmayer for inspiring discussions and comments on cellwork L-systems. The reported research has been supported by an operating grant, equipment grants and a scholarship from the Natural Sciences and Engineering Research Council of Canada. Facilities of the Department of Computer Science, University of Regina, were also essential. All support is gratefully acknowledged. Thanks also to the referees for their helpful comments and suggestions.

REFERENCES

[1] M. J. M. de Boer. *Analysis and computer generation of division patterns in cell layers using developmental algorithms.* PhD thesis, University of Utrecht, the Netherlands, 1989.

[2] L. Fox and D. F. Mayers. *Numerical solution of ordinary differential equations.* Chapman and Hall, London, 1987.

[3] F. D. Fracchia, P. Prusinkiewicz, and M. J. M. de Boer. Animation of the development of multicellular structures. In N. Magnenat-Thalmann and D. Thalmann, editors, *Computer Animation '90*, pages 3–18, Tokyo, 1990. Springer-Verlag.

[4] F. D. Fracchia, P. Prusinkiewicz, and M. J. M. de Boer. Visualization of the development of multicellular structures. In *Proceedings of Graphics Interface '90*, pages 267–277, 1990.

[5] B. E. S. Gunning. Microtubules and cytomorphogenesis in a developing organ: The root primordium of *Azolla pinnata.* In O. Kiermayer, editor, *Cytomorphogenesis in plants*, Cell Biology Monographs 8, pages 301–325. Springer-Verlag, Wien, 1981.

[6] R. W. Korn. Positional specificity within plant cells. *J. Theoretical Biology*, 95:543–568, 1982.

[7] A. Lindenmayer. Mathematical models for cellular interaction in development, Parts I and II. *Journal of Theoretical Biology*, 18:280–315, 1968.

[8] A. Lindenmayer. Models for plant tissue development with cell division orientation regulated by preprophase bands of microtubules. *Differentiation*, 26:1–10, 1984.

[9] A. Lindenmayer. An introduction to parallel map generating systems. In H. Ehrig, M. Nagl, A. Rosenfeld, and G. Rozenberg, editors, *Graph-grammars and their application to computer science*, pages 27–40. Springer-Verlag, 1987. Lecture Notes in Comp. Sci. 291.

[10] A. Lindenmayer and G. Rozenberg. Parallel generation of maps: developmental systems for cell layers. In V. Claus, H. Ehrig, and G. Rozenberg, editors, *Graph-grammars and their application to computer science and biology*, pages 301–316. Springer-Verlag, Berlin, 1979. Lecture Notes in Comp. Sci. 73.

[11] H. B. Lück and J. Lück. Vers une metrie des graphes evolutifs, representatifs d'ensembles cellulaires. In H. Le Guyader and T. Moulin, editors, *Actes du premier seminaire de l'Ecole de Biologie Théorique du CNRS*, pages 373–398. Ecole Nat. Sup. de Techn. Avanc., Paris, 1981.

[12] J. Lück and H. B. Lück. 3-dimensional plant bodies by double wall map and stereomap systems. In H. Ehrig, M. Nagl, and G. Rozenberg, editors, *Graph-grammars and their application to computer science*, pages 219–231. Springer-Verlag, 1983. Lecture Notes in Comp. Sci. 153.

[13] J. Lück and H. B. Lück. Double-wall cellwork systems for plant meristems. In this volume, 1990.

[14] A. Nakamura, A. Lindenmayer, and K. Aizawa. Some systems for map generation. In G. Rozenberg and A. Salomaa, editors, *The Book of L*, pages 323–332. Springer-Verlag, Berlin, 1986.

[15] P. Prusinkiewicz and J. S. Hanan. *Lindenmayer systems, fractals, and plants.* Springer-Verlag, New York, 1989. Lecture Notes in Biomathematics 79.

[16] W. Sierpiński. Sur une courbe dont tout point est un point de ramification. *Comptes Rendus hebdomadaires des séances de l'Académie des Sciences*, 160:302–305, 1915. Reprinted in W. Sierpiński, *Oeuvres choisies*, S. Hartman et al., editors, pages 99–106, PWN – Éditions Scientifiques de Pologne, Warsaw, 1975.

Dactl: An Experimental Graph Rewriting Language

J.R.W.Glauert, J.R.Kennaway and M.R.Sleep*
Declarative Systems Project, School of Information Systems,
University of East Anglia, Norwich NR4 7TJ, U.K.

Abstract: Dactl is an experimental language programming language based on fine grain graph transformations. It was developed in the context of a large parallel reduction machine project. The design of the language is outlined, and examples given of its use both as a compiler target language and as a programming language. Dactl has a formal semantics and stable implementations on a number of platforms.

Keywords: Dactl, graph rewriting, computational model, parallel architecture, reduction machine, compiler target language, declarative language, categorical semantics.

CONTENTS

1 Introduction.

Term (or tree) rewriting systems have proved useful both as specifications and — though less commonly — as practical systems for symbolic computation (see [HO82] for a practical system with a sound theoretical underpinning). Klop [Klo90] and Dershowitz and Jouannaud [Der89] provide comprehensive treatments of term rewriting theory, which is now reasonably well understood.

The idea of studying transformation systems based on graphs (as opposed to trees) dates back at least to [Ros72], and a significant body of theory has been developed, most notably by the

---

* This work was partially supported by ESPRIT project no. 2025 (European Declarative System) and basic research action no. 3074 (Semagraph). J.R. Kennaway was also partially supported by an SERC Advanced Fellowship, and by SERC grant no. GR/F 91582.

Berlin school of Ehrig and others: [Ehr89] gives an authoritative overview.

Practical uses of 'graph rewriting' date back at least to Wadsworth [Wad71], which develops a graph based representation of lambda terms and an associated implementation method for normal order evaluation of lambda calculus expressions. The relation between tree and graph rewriting has been studied in some detail [Sta80a, Sta80b, Bar87, Hof88, Far90]. The main result is that sharing implementations produce the correct semantics at least for orthogonal (also called regular) term rewrite systems.

New generation Logic languages of the committed choice variety (for example Concurrent Prolog [Sha86] and Parlog [Gre87]) may be viewed as specialised graph rewriting languages, as may actor models such as DyNe [Ken85]. More recently Lafont [Laf89] has proposed an interaction net model of computation which again may be viewed as specialised graph rewriting, whose constraints are inspired by Girard's work on Linear Logic.

In 1983 the authors undertook an ambitious project aimed at designing a common model of computation which would be general enough to support a range of more restricted computational models such as those required for functional, logic and actor-like languages. The primary aim of the project was to produce a common target language (CTL) for a range of symbolic processing languages, particularly functional languages and committed choice logic languages. The project chose *graph rewriting* as the basis for its work.

The main success of the project was the design and implementation of a general model of computation based on graph rewriting. The model is called DACTL (for Declarative Alvey Compiler Target Language). The main failure of the project was that it proved difficult within the timescale to develop the compiler technology necessary for Dactl to act as an *efficient* CTL: we seriously underestimated the work needed here. Nevertheless, it was possible to demonstrate working compilers for a surprisingly wide range of languages including HOPE, LISP, PARLOG, GHC, ML and CLEAN within the timescale [Ham88, Gla88a, Gla88b, Ken90a].

The CTL motivations of the Dactl project are now mainly historical. What remains is one of the few genuine graph rewriting language implementations in existence. There is a stable, reasonably engineered implementation of Dactl for the Sun with modular compilation facilities and a comprehensive Unix interface, and a more recent implementation for Macintosh computers which is in regular use. Our experience of the language design process, together with our experience in using Dactl in its present form suggest that others may find it useful as an experimental tool for exploring practical graph rewriting systems.

The paper is organised as follows. The remainder of this introductory section outlines the main features of Dactl, and briefly describes the history of the Dactl project. The body of the paper consists of a more detailed description of Dactl, and a variety of illustrations of its use. Finally, the relationship between the operational semantics of a Dactl rewrite and categorical semantics of graph rewriting is briefly discussed.

1.1 Main features of the Dactl language

a. Dactl graphs are *term graphs* in the sense of [Bar87]. That is, every node has a symbol (or label) together with zero or more directed out-arcs to other nodes. In fact, Dactl graphs are more general, as they may be cyclic. Thus Dactl nodes together with their symbols and out-arcs correspond to the labelled *hyperedges* used to model term graph rewriting in the Jungle evaluation model developed more recently by Hoffmann and Plump[Hof88].

b. A Dactl rewrite is *atomic*. This is expressed by requiring that every valid outcome of a Dactl computation must correspond to an outcome which could be reached by sequential execution. The great benefit of atomicity is that invariance of properties across individual rules also holds for all valid executions. The cost is that an implementation must ensure that co-existing conflicting rewrites are not executed concurrently. This may be done for example by locking critical nodes. For certain classes of rule systems, it is possible to show that no locking is needed to ensure the correctness of concurrent execution of rewrites[Ken88].

c. A Dactl rewrite may contain a multiple reassignment of out-arcs (called *redirections* in Dactl terminology). It is this feature which gives Dactl much of its expressive power, allowing non-declarative behaviours to be expressed.

d. There is no built-in evaluation strategy in Dactl. In particular, evaluation order is not directed by pattern-matching, in contrast to the usual situation in functional languages.

e. Evaluation order is instead determined by *control markings* on the nodes and the arcs of Dactl graphs. These allow a wide range of evaluation strategies and synchronization conditions to be expressed. The control markings are an integral part of Dactl: a graph which contains no control markings is not rewritable according to Dactl semantics, even if the graph contains redexes in the usual TRS sense. Techniques for generating appropriate markings automatically are reported by Kennaway [Ken90a], and Hammond and Papadopoulos [Ham88].

f. Dactl supports separate compilation, and a classification scheme for symbol usage which allows the writer of a Dactl module to constrain external use of exported symbols by appropriate symbol class declarations.

g. The implementation supports a comprehensive interface to Unix.

h. The implementation gathers statistics and execution traces corresponding to both sequential and parallel execution.

Whilst it is clear that the design could be improved, we believe that this is best delayed until there is significantly more experience with the present design. The current definitive reference document for Dactl is Final Specification of Dactl [Gla88c], obtainable from the authors.

1.2 Project History

In 1983 a number of ad-hoc meetings were held in the U.K. in an attempt to identify a common basis for the development of parallel machines suited to the needs of the 'new generation' languages, particularly those based on logic formalisms (the *logic* languages) and those based on the lambda calculus (the *functional* languages).

In April 1984 the U.K. Alvey directorate sponsored a meeting at the Royal Society of key workers from academia and industry to consider a proposal to develop a common model of parallel computation. The meeting identified strong polarization between those who believed that efficiency was paramount, and those who believed that the benefits of working towards a common model outweighed potential performance drawbacks. It was recognised that many specialist parallel architectures would evolve which required specialist interfaces The outcome of the meeting was a decision to proceed with a project whose aims were limited to identifying a common model for 'declarative' languages.

By May 1985, work had reached the stage where a preliminary proposal for such an interface could be given limited circulation. This was followed in September 1985 by the first release of a reference interpreter for the preliminary version of the interface, which was given the title Dactl (for Declarative Alvey Compiler Target Language).

During this period a consortium involving ICL, Imperial College, Manchester University and East Anglia was formed to develop this early work in the context of the Alvey Flagship[Wat87] project which focussed on declarative languages and parallelism. Alvey funded work on Dactl, based at UEA, began in May 1986. The technical aims of the work were primarily concerned with developing a precisely defined graph rewriting model of computation and exploring its properties, working closely with the Flagship team. The development of the model was to be expressed as a series of reports defining the model, together with a number of releases of reference interpreters.

Work on the more formal aspects was aided by collaboration with the Dutch Parallel Reduction Machine Project led by Prof.H.P.Barendregt. This collaboration led to a number of joint publications[Bar87,Bar89], including a paper specifying a common abstract model of graph rewriting called LEAN (which, apart from syntax, is essentially Dactl without the control and synchronization markings).

This formal work with the Dutch, together with modified requirements input from the Flagship team, led to a fundamental redesign of the computational model. A specification of the core model resulting from this work was released in March 1987 (the Core Dactl report) and accepted by Flagship shortly afterwards. A preliminary definition of a revised design of Dactl was completed in June 1987. This was augmented by a release of the design in December 1987 which included a very detailed UNIX interface.

The final design of the language was released early in January 1989, and a consistent supporting version of the reference interpreter followed shortly afterwards.

2 The Dactl Graph Rewriting Model of Computation.

We start by considering the canonical representation of graphs used by Dactl, and then describe the form of rewriting rules in the language. Later we discuss the details of rewriting and control of the rewriting process.

2.1 Dactl Graphs

A Dactl program manipulates *directed graphs*. Each *node* is labelled with a *symbol* which may be interpreted as a function, predicate, or constructor according to the requirements of the computation being implemented. From nodes will originate an *ordered* sequence of zero or more directed *arcs* leading to *successor nodes*. Graphs may be cyclic and need not be connected, but there is a distinguished node in the graph known as the *root*. When considering the final form of a graph, only nodes reachable from the root are (by definition) of interest. Hence unreachable nodes which cannot affect the final form may be removed from the graph along with their successor arcs.

The following examples give the textual and pictorial representation of two graphs.

Example 1: A DAG

Shorthand textual form: `Append[s:Cons[0 Nil] s]`

Equivalent longhand textual form: `r: Append[s s], s: Cons[z n], z: 0, n: Nil`

The longhand form is a tabular representation of the graph. A node is made up of a symbol and a list of the identifiers of successor nodes. Each node is given an identifier beginning with a lower-case letter. Integers and identifiers beginning with an upper-case letter represent symbols. The root of the graph is taken to be the first node specified. In the shorthand form we may combine the definition of a node with one of its occurrences and may omit unnecessary node identifiers.

Pictorial form:

```
        r:Append
           ||
           vv
         s:Cons
        /      \
    z:0        n:Nil
```

Note that a graph equivalent to a ground term may be described without using node identifiers.

Example 2: A Cyclic Structure

Shorthand textual form: c: Cons[1 c]

Equivalent longhand textual form: c: Cons[o c], o: 1

Pictorial form:

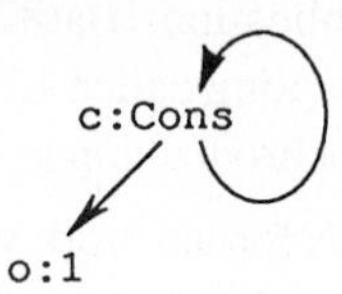

2.2 Dactl Rewriting Rules

The reduction relation for a Dactl system is described by a set of rewriting rules which describe *graph transformations*.

The left-hand side of a rule consists of a *pattern* which is a generalisation of a Dactl graph. Any Dactl graph as described above is a Dactl pattern. In addition, a pattern may contain special *pattern symbols*, which match a class of symbols, and *pattern operators*. The simplest special pattern symbol is ANY, which identifies a variable node. The pattern operators of Dactl are +, - and & and represent union, difference and intersection respectively.

Before rewriting can take place, it is necessary to establish a *match* between a subgraph of the program graph called a *redex*, and the pattern of a rule. Formally, this means identifying a structure preserving mapping between the nodes on the pattern and the graph undergoing rewriting.

The right-hand side of a rule includes the *contractum graph*, and a number of *redirections*. Rewriting involves building the contractum, a copy of the right hand side of the rule, and connecting it into the original graph according to the *redirections* specified as part of the rule. Very frequently only a single redirection of the root is intended, and the syntax of Dactl provides a special connective => between the left and right hand side of a Dactl rule for this purpose.

Besides manipulating the structure of the graph, Dactl rules also manipulate 'control markings' on the graph, which govern execution order. A discussion of control markings is delayed until section 2.4.

The following example rules model the appending of lists. In shorthand form, the rules are:

```
Append[Cons[h t] y]    =>    Cons[h Append[t y]]   |
Append[Nil y]    =>    y
```

It is a principle of Dactl design that the meaning of all shorthand is given by translation to longhand canonical form. For the above rules, this involves a tabular listing of the nodes in the pattern and the contractum, and the explicit inclusion of redirections, which take the syntactic form of conventional assignment statements.

```
r: Append[c y], c: Cons[h t], h: ANY, t: ANY, y: ANY
        ->    s: Cons[h b], b: Append[t y], r:=s    |
a: Append[n y], n: Nil, y: ANY    ->    a:=y
```

2.3 Graph rewriting in Dactl

A single Dactl rewrite takes place in three phases, namely *match*, *build*, and *redirect*.

The *match* phase identifies a graph homomorphism — that is, a structure-preserving mapping — between the pattern and the graph undergoing rewriting.

The *build* phase adds new nodes (as specified by the contractum) to the graph. Where the contractum contains occurrences of variable nodes these are replaced by their bindings

The *redirect* phase performs redirections which change the destination of some arcs of the original graph, allowing the 'gluing in' of new contractum nodes into the existing graph.

These phases are described in more detail below.

2.3.1. Matching

A match is a graph homomorphism from the pattern of a rule to the program graph. Structure is preserved by this mapping, except at variable nodes. There is a match between the pattern of the rule:

```
r: Append[c:Cons[h:ANY t:ANY] y:ANY]    =>    s:Cons[h b:Append[t y]]
```

and the following example graph:

```
Cons[0 r:Append[c:Cons[h:1 t:Nil] y:Nil]]
```

The subgraph of the program graph onto which the nodes of the pattern are mapped is known as the *redex*. The root of the pattern has a special significance. We say that a pattern *matches at a node* if the node in question is the root of a redex for the pattern. The matching process identifies bindings for the variable nodes in the pattern: in the example given, the binding is {y–>y, h–>h, t–>t} and takes this simple form because of the careful choice of node identifiers in the subject graph.

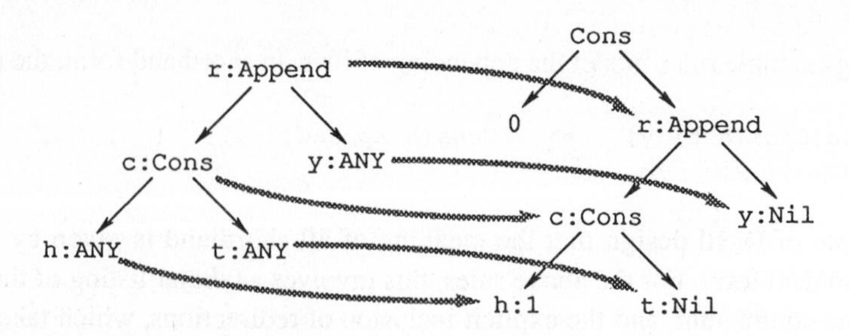

Note that the atomic rewriting principle of Dactl allows us to ignore the operation of other parallel rewriting activities in the graph in describing a rewrite.

2.3.2. Building

The second phase of rewriting builds a copy of the contractum of the rule matched. The contractum contains no pattern symbols, but may contain occurrences of identifiers from the pattern. During building, such occurrences become arcs to the corresponding nodes matched in the first phase. After building, the example graph has the form:

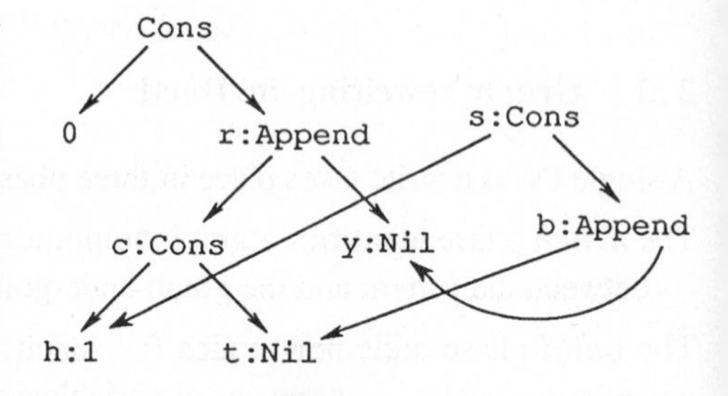


```
Cons[0 r:Append[c:Cons[h:1 t:Nil] y:Nil]],
s: Cons[h b:Append[t y]]
```

where the new graph is on the second line. The build phase is missing for 'selector' rules, such as the second rule of `Append`.

2.3.3. Redirection

The build phase allows the new portion of the graph to contain references to parts of the subject graph. The purpose of the redirection phase is to allow references in the old graph to be changed consistently to refer to parts of the new structure. Very general transformations are possible, as any or all of the nodes identified by pattern variables may be redirected within a single atomic rewrite.

Following a rewrite according to the first rule of the Append example, we expect to find references to the new `Cons` node, `s`, in place of references to the `Append` node. In other words, we wish all arcs referencing the root node, `r`, to be *redirected* to reference `s`. Hence, all occurrences of `r` as a successor of another node are replaced by `s`. The resulting graph, including all the 'garbage' not reachable from the root, is:

```
Cons[0 s:Cons[h:1 b:Append[t:Nil
y:Nil]]],
r: Append[c:Cons[h t] y]
```

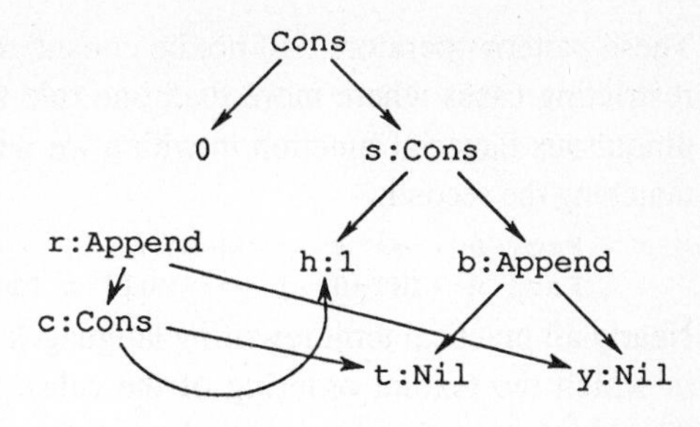

In this case the original Append node may be garbage collected. A common implementation technique is to *overwrite* the Append node with the contents of s, thus avoiding in practice the overhead of supporting genuine physical redirection of pointers.

Performing the redirections is the final phase of rewriting. Nodes from which arcs are redirected are always nodes from the original graph. The target of the redirection may be in the original graph, or in the contractum as in our example. The graph is left in a state where a redex for the second rule exists. As before, we give the rule:

```
a: Append[n:Nil y:ANY] => y
```

and the graph, showing the match:

```
Cons[0 Cons[1 a:Append[n:Nil y:Nil]]]
```

No graph is built in the redirection phase of rewriting. The right hand side of the rule indicates that the graph should be rewritten so that in place of the `Append` node at the root of the redex, we now see the node referenced by the second argument, `y`. This parallels term rewriting theory in which, following a corresponding rewrite, the original parent of `a` would find the subterm rooted at `y` as its direct descendant in place of `a`. The effect of replacing all references to `a` by `y` is the following graph:

```
Cons[0 Cons[1 y:Nil]], a:Append[n:Nil y]
```

No references to the `Append` node will now remain. It is common for implementations to re-use the node by changing it to an *indirection* to `y`.

To ensure that the overwriting and indirection techniques are always correct, we impose a restriction on legal rules. Whenever a rule contains two redirections a:=b, c:=d, the pattern of the rule must be such as to make it impossible for the source node of one redirection to match the same node as the target of the other.

2.3.4. Pattern Operators

The Dactl rules illustrated so far are left-linear: the patterns can be expressed in shorthand form with no repeated node identifiers. However, Dactl *does* allow repeated pattern identifiers in which case each occurrence must match the same program graph node. This interpretation arises naturally from the definition of matching in terms of a graph homomorphism.

In addition to the pattern symbol ANY, Dactl also provides three pattern operators. These are expressed in infix notation. If Π and Σ are patterns, then the following are also patterns:

$$(\Pi + \Sigma) \qquad (\Pi - \Sigma) \qquad (\Pi \& \Sigma)$$

A pattern can be regarded as defining a set, being the set of all Dactl graphs which will match the pattern with the root of the graph being the root of the redex. The first two pattern operators act like union and difference operators on such sets. Hence, the first form of pattern matches at a node if either Π or Σ matches at the node and the second requires that Π matches, but Σ does not. The & operator corresponds to set intersection.

These pattern operators will not be considered in detail in this paper, but they prove useful in restricting cases where more than one rule may apply. For illustration, we give rules for the ubiquitous factorial function in which we wish to exclude graphs matching the first rule from matching the second:

```
Fac[ 0 ] => 1                |
Fac[ n:(INT-0) ] => IMul[ n Fac[ ISub[n 1] ] ]
```

Nearly all practical term rewriting languages are designed as priority rewrite systems [BEG87], in which the textual ordering of the rules expresses rule priority. In Dactl it is possible to disambiguate such rule systems by careful use of pattern expressions, and Dactl is unusual in this respect.

However, priority rewrite semantics is both common and convenient, and so syntactic sugar is provided in Dactl to support it. Rules separated by a semi-colon are matched in order, whereas rules separated by | may be dealt with in any order.

```
Fac[ 0 ] => 1        ;
Fac[ n:INT ] => IMul[ n Fac[ ISub[n 1] ] ]
```

There is an implicit use of the pattern difference operator to exclude graphs matching the pattern of the first rule from matching the second rule. The obvious implementation, which has significant performance benefits, is to consider the rules in the given textual order.

The graph rewriting framework, based on pattern-matching rules, gives great expressive power to Dactl. *Ambiguity* required by languages allowing non-deterministic results is expressed by systems in which redexes overlap. When disjoint redexes exist in a graph, *concurrency* can be exploited by an implementation. Imperative and Object-oriented code which manipulates program *state* can be implemented using rules which redirect references to nodes not at the redex root. This technique can also be used to implement logic variables. The graphical basis enables *sharing* of subterms to be expressed and exploited.

2.4 Control of evaluation in Dactl

Dactl is concerned with control of evaluation as well as the properties of an abstract rewriting system. During the design of Dactl we observed that differences in evaluation strategy markedly affect both the 'look and feel' of a given language, and also the semantics. Even the more modern 'lazy' functional languages include some operational rules about the way in which pattern matching is handled in their semantics. Laville [Lav87], Kennaway [Ken90c] and others have examined this problem, but in terms of language design the expression of strategy and control remains a subtle problem. This is partly a human factors problem of course, and hence not amenable to theory in its present state.

Faced with these problems we decided to include fine grain control and synchronization markings as an integral part of Dactl. It was recognised at the time that contemporary technology for parallel computing was easiest to exploit using coarse grain parallelism: our work on the ZAPP architecture[McB87] did just that in the context of transputers. But technology advances, and fine grain parallelism may look much more realistic quite soon.

The intuitive basis of Dactl's control markings is that each agent executing a Dactl rule is a *locus of control* of some process. In the von Neumann model, there is exactly one process and exactly one locus of control. Each instruction appoints a unique successor. In a Dactl rule, zero or more successor 'control loci' may be appointed either by creating new nodes active during the build phase, or by activating nodes in the original graph during the activate phase. The single syntactic token * serves for both purposes in Dactl.

An active node can be thought of as a process, and processes need to communicate and synchronise. This leads to Dactl's notification markings on arcs, which specify reverse communication paths in the graph, and suspension markings on nodes to enable the expression of processes suspended awaiting a certain number of events.

A fundamental design question for Dactl was 'when should rewriting agents communicate?'. The decision taken was to adopt a single, very simple rule:

Dactl rewriting agents communicate when a match is attempted at an active node, and the match fails.

Intuitively, failure to match indicates that some sort of temporary normal form has been reached: it's not in general a normal form, because future Dactl rewrites may make it a redex. But failure to match at a node is a key event, worth signalling to all who have marked references to the node and this is the principle adopted in Dactl.

2.4.1. Dactl Markings

Dactl encodes the strategy in a pattern of *markings* on the nodes and arcs of the program graph. There are two forms of node marking and a single form of arc marking:

The most important *node* marking is the *activation* denoted by *. *Only activated nodes are considered as the starting points of rule matching.* Once a Dactl program graph contains no activations, execution is complete and the graph viewed from the root is the result of computation. It may, or may not, be in normal form with respect to the rewrite rules stripped of markings.

The marking # indicates a suspended node, and a node may have one or more such markings enabling it to await a specified number of notification events before becoming active again. The node concerned will not be considered for matching immediately, but will be reconsidered when the corresponding number of notifications is received.

The arc marking ^ is used in conjunction with # for such synchronization purposes. It indicates a notification path between the target of the arc and its source. When evaluation of the target is complete in the sense that rule matching fails (for example because it has been redirected to a node with a constructor symbol) the arc marking is removed along with a # marking on the source node, if present. When the last # is removed, it is replaced by *, thus making the node active. This supports a model of evaluation close to dataflow since operators wait until their operands are available before being rewritten.

The operator * may also be used on the right hand side before the identifier of a node matched by the pattern. This is taken as an instruction to activate the corresponding node if it is currently neither active nor suspended.

All the markings, including both uses of * are illustrated by the following version of the `Append` rules:

```
Append[Cons[h t] y] => #Cons[h ^*Append[t y]]    |
Append[Nil y] => *y
```

2.4.2. Dactl Rewriting with Markings

If matching succeeds at an active node, the * marking is removed, the contractum is built, required nodes are activated, and redirections are performed. Most rules redirect references to the original root node which may then be garbage collected.

The criterion for notification is *failure to match* an active node to any rule. The activation marker is removed from the node and any notifications required by direct ancestors are performed. Matching failure most commonly occurs with constructor nodes for which there are no rules. To return a result therefore, a rule will usually redirect its root to an activated constructor node. Considering the marked rules given above with a new graph of the form:

```
a:*Append[k:Cons[o:1 n:Nil] k]
```

We see that there is a redex for the first rule with the node `k` in the graph matched by more than one part of the pattern. It is perfectly consistent for a tree-structured pattern to match a graph with sharing. After rewriting, the structure is:

```
m:#Cons[o:1 ^b:*Append[n:Nil k:Cons[o n]]], a:Append[k k]
```

The original node `a` is now garbage and can be removed. There is now a redex for the second rule and the graph is rewritten to:

```
m:#Cons[o:1 ^k:*Cons[o n:Nil]], b:Append[n k]
```

The node `b` is now garbage. Evaluation is now complete, but notification of the ancestors of an `Append` node is delayed until the whole operation is complete. This is achieved by suspending the `Cons` node `m` waiting for evaluation of the rest of the list. `Cons` is a constructor so matching fails and the parent node, `m`, is notified and hence activated:

```
m:*Cons[o:1 k:Cons[o n:Nil]]
```

Again, the `Cons` node, `m`, will match nothing, so the final graph will be:

```
m:Cons[o:1 k:Cons[o n:Nil]]
```

The examples used display no concurrency during evaluation. Concurrency arises when the right-hand side of a rule contains several active nodes so that the rule nominates many successors to receive control. Also, several nodes may be suspended awaiting notification from the same node, and all will be activated once it is evaluated.

2.4.3. Module structure

Sets of Dactl rules may be grouped into modules. Symbols declared in a module may be private to that module, or exported. When a symbol is declared, it is given an access class. Available classes are:

GENERAL	permits unrestricted use
REWRITABLE	may occur as the principal symbol of a rule
OVERWRITABLE	may be redirected when matched as a non-principal node of a rule
CREATABLE	may not be redirected

If a symbol is exported, it is also given a PUBLIC access class. This controls the use of the symbols within importing modules. The public class of a symbol is always at least as restrictive as the declaration class.

3 Some Dactl Examples.

The following examples are complete working source codes which have been tested on the Dactl reference interpreter.

3.1 Sorting

```
MODULE SortModule;
IMPORTS Arithmetic; Logic; Lists;
SYMBOL REWRITABLE PUBLIC CREATABLE Sort;
SYMBOL REWRITABLE Insert;Compare;
RULE
      Sort[x:Nil]  =>  *x;
      Sort[Cons[h t]]  =>  #Insert[h ^ *Sort[t]];

      Insert[n x:Cons[h t]]  =>  #Compare[ ^*IGt[n h] n x];
      Insert[n Nil]  =>  #Cons[ ^ *n Nil];
      Insert[n x:(ANY-Nil-Cons[ANY ANY])]  => #Insert[n ^ *x];

      Compare[True n x:Cons[h t]]  =>  #Cons[h ^*Insert[n t]];
      Compare[False n x]  =>  *Cons[n x];
ENDMODULE SortModule;
```

The Sort symbol is declared in a manner typical of symbols intended to represent functions. It is REWRITABLE, to allow the rules for Sort above, but exported only as CREATABLE. This allows other modules to create Sort nodes, but not to add more rules for the Sort symbol. In addition, no rule in any module can use the Sort symbol in its left hand side other than at the root. The symbols Insert and Compare are not exported, thus cannot be mentioned in other modules. The imported modules define the other symbols used here. For example, Cons and Nil are declared in Lists as CREATABLE symbols, and exported as CREATABLE. Thus nodes bearing these symbols can never be redirected. This information is useful to the implementor of Dactl.

Notionally, the modules Arithmetic, Logic, and Lists are thought of as ordinary Dactl modules, containing Dactl rules defining various basic constructors and functions, but they are normally implemented directly by a Dactl implementation.

The following module illustrates the use of the sort module.

```
MODULE SortTest;
IMPORTS SortModule;
RULE
      INITIAL => *Sort[Cons[5 Cons[2 Cons[9 Cons[3 Cons[1 Nil]]]]]];
ENDMODULE SortTest;
```

3.2 A simple Head Normal Form reducer for Combinatory Logic

Combinatory logic is a simple term rewrite system which is both important in the foundations of mathematics [Sch24] and useful as the basis of graph rewriting implementations of functional languages [Tur79]. We show a description of combinatory logic in Dactl.

We use explicit binary application, and define patterns for HNF (Head Normal Form) and also for redexes. This allows us to write just 3 rules, two for the redex cases and one for the (ANY–Redex–Hnf) case.

```
MODULE SK1;
SYMBOL REWRITABLE PUBLIC CREATABLE Ap;
SYMBOL CREATABLE PUBLIC CREATABLE S; K;
PATTERN PUBLIC Hnf = (S+K+Ap[(S+K) ANY]+Ap[Ap[S ANY] ANY]);
PATTERN Redex = ( Ap[Ap[K ANY] ANY] + Ap[Ap[Ap[S ANY] ANY] ANY] );
RULE
      Ap[Ap[K x] y]   =>   *x;
      Ap[Ap[Ap[S f] g] x]   =>   *Ap[Ap[f x] Ap[g x]];
      (Ap[x y]&(ANY-Redex-Hnf))   =>   #Ap[ ^x y], *x;
ENDMODULE SK1;
```

If we activate a combinatory term in the presence of these rules, it will notify if and when it reaches a head normal form. For example,

```
*Ap[ Ap[ Ap[ S S ] K ] Ap[ Ap[ K S ] K ] ]
```

(representing the combinatory term which would conventionally be denoted by SSK(KSK)) will when evaluated by these rules notify when it reaches the form

```
Ap[ Ap[ S x ] Ap[ K x ] ], x:Ap[ Ap[ K S ] K ]
```

(representing the term S x (K x) where x=KSK). To obtain a normalising reducer we add the following rules:

```
Ap[K x]   ->   *x;
Ap[S x]   ->   *x;
Ap[Ap [S f] g]   ->   *f, *g;
```

The left-hand sides of these rules together match all terms in head normal form, and only such terms. They do not do any redirection, but merely activate their components. The effect is to obtain the normal form of any term which has one; however, the parents of such a computation will not be notified when it has completed. A notifying solution can be obtained at the expense of slightly complicating the rule set.

4 Translating other languages to Dactl.

The translation of strongly sequential term rewrite systems to Dactl (complete with correct control markings) is described in [Ken90a]. Translation schemes for both functional and logic languages to Dactl are described in [Ham88]. Here we give some brief examples intended to illustrate the techniques used.

4.1 Strict Evaluation

Here are the rules for the Append function from section 2.4.1.

```
Append[Cons[h t] y] => #Cons[h ^**Append[t y]]   |
Append[Nil y] => *y
```

If the first argument to Append has yet to be evaluated to a list, then these rules will fail to match. An extra rule of the following form would ensure evaluation:

```
Append[x y]  =>  #Append[^*x y]
```

Although the first argument will be evaluated, rewriting may complete without evaluating the second argument to a list. To force the function to be strict, we should add rules to coerce the result to a list. The full set would be:

```
RULE
        Append[Cons[h t] y]  =>  #Cons[h ^*Append[t y]] |
        Append[Nil y]  =>  *ForceList[y]        ;
        Append[x y]  =>  #Append[^*x y]         ;
RULE
        ForceList[n:Nil]  =>  *n        |
        ForceList[Cons[h t]]  =>  #Cons[h ^*ForceList[t]]       ;
        ForceList[a]  =>  #ForceList[^*a]   ;
```

4.2 Lazy Evaluation

It is common to use a more lazy form of evaluation to *head-normal form*. Roughly, this means that evaluation proceeds until the outermost node of the expression is a constructor. The rules would be:

```
RULE
        Append[Cons[h t] y]  =>  *Cons[h Append[t y]]     |
        Append[Nil y]  =>  *y   ;
```

The recursive application of Append is not reduced by the first rule. The result of the second rule is activated, however, and should reduce to a constructor. A default rule could be employed in case the first argument was not a Cons or Nil. Techniques have been developed at UEA for the translation of the functional language Clean which ensure that arguments have always been evaluated sufficiently so that such default rules are not needed[Ken90a].

4.3 Early Completion

To allow for stream parallelism an early completion scheme can be used. In this, we notify when a head-normal form has been produced, but continue to evaluate to normal form at the same time:

```
RULE
        Append[Cons[h t] y] =>  *Cons[h *Append[t y]]    |
        Append[Nil y]  =>  *ForceList[y]        ;
        Append[x y]  =>  #Append[^*x y]         ;
RULE
        ForceList[n:Nil]  =>  *n  |
        ForceList[Cons[h t]] =>  *Cons[h *ForceList[t]] ;
        ForceList[a]  =>  #ForceList[^*a]    ;
```

4.4 Examples from moded Logic programming

Translating flat concurrent logic languages to Dactl has been studied at UEA and by the Parlog group at Imperial College. We illustrate the flavour of these translations using an alternative technique for Append. The rule will attempt to instantiate the third argument with the result of appending the second argument to the first. The symbol `Var` indicates an uninstantiated variable. (Note that `Var` is not a special symbol in Dactl: we could have chosen other symbols for this example. It is the rule system, together with the graph rewriting semantics of Dactl, which make `Var` have some of the properties of a logic variable.) This may be replaced by a data value by redirection of arcs to a *non-root* node. This is the basis of many techniques in Dactl for handling logic and object-oriented programming.

The rule set used would be as follows:

```
RULE
        root:Append[Nil y:(ANY-root) v:Var]   =>   *Succ, v:=*y  |
        Append[Cons[h t] y v:Var]   =>   *Append[t y n:Var],
                                         v:=*Cons[h n]           |
        Append[x:Var y v:Var]   =>   #Append[^x y v]              ;
        Append[x y v]   =>   *Fail                                ;
```

In the first case the variable, `v`, is instantiated to the second argument. By firing it, we force a notification to be passed to any node suspended waiting for this instantiation. The second rule is similar, but we notify those awaiting `v` with a `Cons` node whose second argument is a variable which will be instantiated eventually by the evaluation of the new `Append`.

The third rule suspends on the variable, but does not activate it. Activation would be pointless since there are no rules for `Var` and notification would return immediately. However, notification will be provoked eventually by the rule which instantiates the variable. The final, failure case indicates that the first argument is some data value, but not a list, or that the third argument is not a variable. We will illustrate evaluation of the following graph:

```
Res[r], *Append[a Cons[2 Nil] r:Var], *Append[Nil Cons[1 Nil] a:Var]
```

Taking the first active node:

```
=>      Res[r], #Append[^a Cons[2 Nil] r:Var],
        *Append[Nil c:Cons[1 Nil] a:Var]
```

The term suspends until the variable is instantiated:

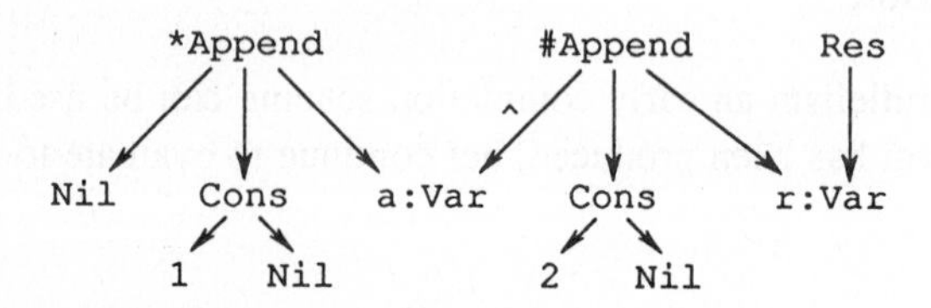


```
=>      Res[r], #Append[^c Cons[2 Nil] r:Var],
        *Succ, c:*Cons[1 Nil]
```

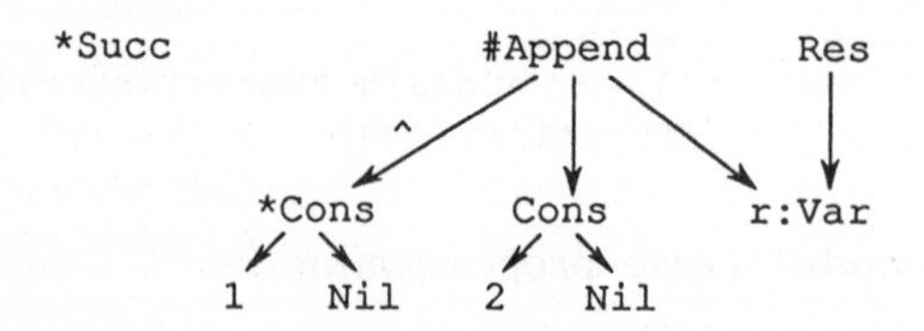


```
=>      Res[r],
        *Append[c:Cons[h:1 t:Nil] y:Cons[2 Nil] r:Var]
```

```
=>      Res[d], *Append[t:Nil y:Cons[2 Nil] n:Var],
        d:*Cons[h:1 n]
```

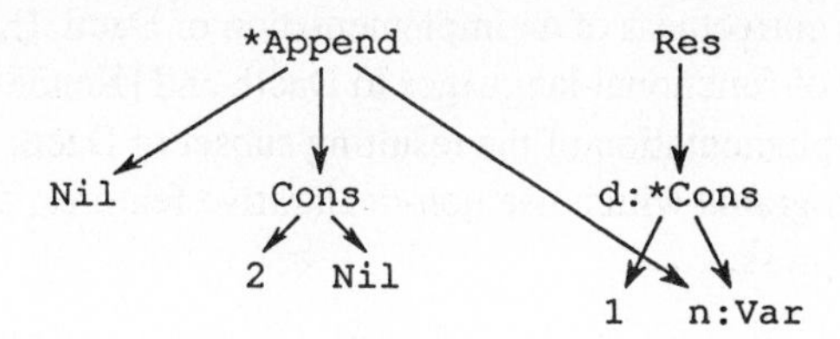


```
=>      Res[Cons[1 y]], *Succ, y:*Cons[2 Nil]
```

And Finally:

```
=>      Res[Cons[1 Cons[2 Nil]]]
```

5 Categorical Semantics.

The semantics of Dactl can be divided into two parts:

a. the semantics of an individual graph rewrite.

b. the semantics associated with Dactl's control markings.

In [Ken90b] one of us has presented a category-theoretic definition of graph rewriting in the category of jungles. It unifies the two previous categorical models of [Rao84,Ken87] and [Ehr79]. Our concern in that paper was to describe term graph rewriting, but in fact the definitions given there are general enough to describe Dactl rewrites as well, at least for part (a). In fact, the semantics thus obtained is the "overwriting" which we described in section @ as an optimisation, a pleasing agreement between the abstract mathematical semantics and a standard implementation technique.

We have not yet addressed the question of a more mathematical semantics for control markings. One approach is to consider them as function symbols in their own right. This has been done in [Eek86] for similar concepts in term rewriting systems. However, this might result merely in an intractable encoding of their role, a more direct treatment being preferable.

6 Conclusion.

We have described a practical language of graph rewriting, and given a wide range of examples of its use. These range from graph manipulation algorithms to translations from functional and logic languages. The semantics of an individual Dactl rewrite agrees with that obtained from the categorical construction of [Ken90b]. Both the design and implementations of Dactl are stable.

There are several directions for future research. The categorical semantics of [Ken90b] can describe Dactl's graph transformations, but does not currently deal with control markings. It would be desirable to so extend it, either by representing markings as function symbols on extra nodes, or — more directly — by using a different category of graphs in which each edge is labelled by a control mark, and each node by a function symbol and a control mark.

It is also essential to have proof techniques for Dactl programs, and initial work on this is in progress. There are two basic issues here: the correctness of a Dactl program, given the semantics of Dactl, and the correctness of an implementation of Dactl. [Ken90a] has proven the correctness of a translation of functional languages to Dactl, and [Ken88] has demonstrated the correctness of a parallel implementation of the resulting subset of Dactl. This work needs to be extended to cover Dactl programs which use non-declarative features, and initial work in this direction is currently in progress.

7 Acknowledgements.

Nic Holt, Mike Reeve and Ian Watson made major contributions to the design of Dactl. Kevin Hammond designed the Unix interface. The implementation work was done mainly by Geoff Somner, and more recently by Ian King. Much of the early work on Dactl was supported by SERC grant no. GR/D59502. Ian King contributed with helpful comments on early drafts of this paper.

8 References.

LNCS = Lecture Notes in Computer Science (pub. Springer-Verlag).

[Bar87] Barendregt, H.P., van Eekelen, M.C.J.D., Glauert, J.R.W., Kennaway, J.R., Plasmeijer, M.J., and Sleep, M.R., "Term graph rewriting", Proc. PARLE conference, LNCS, 259, 141–158, 1987.

[Bar89] Barendregt, H.P., van Eekelen, M.C.J.D., Glauert, J.R.W., Kennaway, J.R., Plasmeijer, M.J., and Sleep, M.R., "Lean: An Intermediate Language Based on Graph Rewriting" Parallel Computing, 9, 163-177, 1989.

[BEG87] Baeten J.C.M, Bergstra J.A. and Klop J.W., "Term Rewriting Systems with Priorities", Rewriting Techniques and Applications, Bordeaux, France, LNCS, 257, 83–94, 1987.

[Der89] Dershowitz, N., and Jouannaud, J.P., "Rewrite Systems", Chap. 15. in Handbook of Theoretical Computer Science, B, North-Holland, 1989.

[Eek86] van Eekelen, M.C.J.D. and Plasmeijer, M.J., "Specification of reduction strategies in term rewriting systems", in Proc. Workshop on Graph Reduction, LNCS, 279, 215-239, 1986.

[Ehr79] Ehrig H, "Tutorial introduction to the algebraic approach of graph grammars", in LNCS, 73, 83–94, 1979.

[Ehr89] Ehrig, H., and Löwe, M., (eds), "GRA GRA: Computing by Graph Transformation. ESPRIT Basic Research Working Group 3299". Report no. 89/14, Technische Universität Berlin, Fachbereich 20, Informatik, Franklinstraße 28/29, D-1000 Berlin 10, 1989.

[Far90] Farmer, W.M., Ramsdell, J.D., and Watro, R.J., "A correctness proof for combinator reduction with cycles", ACM TOPLAS, 12, 123-134, 1990.

[Gla88a] Glauert, J.R.W., Hammond, K., Kennaway, J.R., and Papadopoulos, G.A., "Using Dactl to Implement Declarative Languages", Proc. CONPAR 88, 1988.

[Gla88b] Glauert, J.R.W., and Papadopoulos, G.A., "A Parallel Implementation of GHC", Proc. International Conference on Fifth Generation Computer Systems 1988. ICOT, Tokyo, 1988.

[Gla88c] Glauert, J.R.W., Kennaway, J.R., Sleep, M.R., and Somner, G.W., "Final Specification of Dactl", Report SYS-C88-11, School of Information Systems, University of East Anglia, Norwich, U.K., 1988.

[Gre87] Gregory, S., "Parallel Logic Programming in PARLOG – The Language and its Implementation", Addison-Wesley, London, 1987.

[Ham88] Hammond, K., and Papadopoulos, G.A., "Parallel Implementations of Declarative Languages based on Graph Rewriting" UK IT 88 ("Alvey") Conference Publication, IEE, 1988.

[HO82] Hoffmann C. and O'Donnell M.J., "Programming with equations", ACM Transactions on Programming Languages and Systems, 83-112, 1982.

[Hof88] Hoffmann, B., and Plump, D., "Jungle Evaluation for Efficient Term Rewriting", Proc. Joint Workshop on Algebraic and Logic Programming, Mathematical Research, 49, 191-203, Akademie-Verlag, Berlin, 1988.

[Ken85] Kennaway, J.R. and Sleep, M.R. Syntax and informal semantics of DyNe. in The Analysis of Concurrent Systems, LNCS, 207, 1985.

[Ken87] Kennaway, J.R., "On 'On graph rewritings'", Theor. Comp. Sci., 52, 37–58, 1987

[Ken88] Kennaway, J.R., "The correctness of an implementation of functional Dactl by parallel rewriting", UK IT 88 ("Alvey") Conference Publication, IEE, 1988.

[Ken90a] Kennaway, J.R., "Implementing Term Rewrite Languages in Dactl", Theor. Comp. Sci., 72, 225-250, 1990.

[Ken90b] Kennaway, J.R., "Graph rewriting in a category of partial morphisms", these proceedings, 1990.

[Ken90c] Kennaway, J.R., "The specificity rule for lazy pattern-matching in ambiguous term rewrite systems", Third European Symposium on Programming, LNCS 432, pp 256–270, 1990.

[Klo90] Klop, J.W., "Term rewriting systems", Chap. 6. in Handbook of Logic in Computer Science, 1, (eds. Abramsky, S., Gabbay, D., and Maibaum, T.), Oxford University Press, 1990.

[Laf89] Lafont. Y, "Interaction Nets", LIENS report, Paris, 1989 (also Proc. ACM Symposium on Principles of Programming Languages, 1990).

[Lav87] Laville, A., "Lazy pattern matching in the ML language", Report 664, INRIA, 1987.

[McB87] McBurney, D.L., and Sleep, M.R., "Transputer-based experiments with the ZAPP architecture", Proc. PARLE conference, LNCS, 258, 242–259, 1987.

[Pap89] Papadopoulos, G.A., "Parallel Implementation of Concurrent Logic Languages Using Graph Rewriting Techniques", Ph.D. Thesis, University of East Anglia, UK, 1989.

[Rao84] Raoult, J.C., "On graph rewritings", Theor. Comp. Sci., 32, 1–24, 1984.

[Ros72] Rosenfeld A and Milgram D.L., "Web automata and web grammars", Machine Intelligence 7 (1972), 307-324, 1972.

[Sch24] Schönfinkel, M., "Über die Bausteine der mathematischen Logik", Math. Annalen, 92, 305-316, 1924.

[Sha86] Shapiro, E.Y., "Concurrent Prolog: A Progress Report", Fundamentals of Artificial Intelligence - An Advanced Course, LNCS, 232, 1986.

[Sta80a] Staples, J., "Computation on graph-like expressions", Theor. Comp. Sci., 10, 171-185, 1980.

[Sta80b] Staples, J., "Optimal evaluations of graph-like expressions", Theor. Comp. Sci., 10, 297-316, 1980.

[Tur79] Turner, D.A., "A new implementation technique for applicative languages", Software: Practice and Experience, 9, 31-49, 1979.

[Wad71] Wadsworth, C.P., Semantics and pragmatics of the lambda-calculus, Ph.D. thesis, University of Oxford, 1971.

[Wat87] Watson, I, Sargeant, J., Watson, P., and Woods, V., "Flagship computational models and machine architecture", ICL Technical Journal, 5, 555–594, 1987.

Use Graph Grammars to Design CAD-Systems !

Herbert Göttler, Joachim Günther, Georg Nieskens

Lehrstuhl für Programmiersprachen
Universität Erlangen-Nürnberg
Martensstr. 3
D-8520 Erlangen
F R Germany

Tel.: ++49 +9131 857624
e-mail: goettler@informatik.uni-erlangen.de

Abstract: Graph grammars, especially when enriched with attributes, can be used as a powerful software engineering technique. The main idea behind this approach is: A problem domain is modelled by a graph, the representation graph, whose nodes correspond to the objects of the domain and whose edges to the relations between the objects, respectively. Typical operations which normally change the structure of the representation graph, like introducing new objects at a certain state of the problem description, or modifying relations between objects, are expressed by graph productions. Quantitative informations are handled by the attributes attached to the nodes of the representation graph. So, the implementation aspects are reduced to a very general and flexible data structure, namely graphs.

Keywords: CR-classification: D.2 software engineering, F.4.2 grammars and other rewriting systems, I.3.5 computational geometry and object modelling, I.3.6 methodology and techniques (interaction techniques), J.6 computer-aided engineering; additional keywords: object oriented program design, attributed graph grammars.

CONTENTS

1 *Introduction/Motivation/Overview*

This paper proposes a software engineering method based on graph grammars. Hopefully, it will turn out that the advocated method is not just useful for the design of CAD-systems. This area of application has been selected as an example only, because CAD-systems are especially suited since good ones incorporate three components: visualization (Pictures, diagrams, etc. are used to put, say, objects into a visible form.), sophisticated user interfaces (The user gets guidance; the system helps him/her to avoid errors.), and semantic postprocessing (The creation of the pictures is not the end of the effort in itself, as it is the case in ordinary drawing programs where the result of a session is more or less an unstructured set of pixels. Something is supposed to be done with the objects of a created picture, for instance, the momentum of a solid body represented by the picture has to be calculated.) In this sense, developing a (good) CAD-program for a specific application is some kind of a 'worst case' of a software engineering problem. If a method is suited for aiding the implementation of a CAD-system in a satisfactory way it must be a good one and it is worth the while to spend some time on it.

Nikolaus Wirth, the designer of PASCAL, once coined the title: "Algorithms + Data structures = Programs". There seems to be much truth in this statement but questions remain: For example, is it wise to develop an ad-hoc data structure for a specific problem or is it better to use a general one to which the problem domain can be mapped. A hint for a good decision comes from the real-world. Here, the common experience tells us that a product with, say, the smallest number of moving parts is the most reliable. An analogous observation holds true in the field of software engineering: A simple and straight forward program design with a minimal number of esoteric features is usually the most solid one with the desired quality. So we should strive for non-esoteric data structures and non-esoteric algorithms within a framework which prevents 'free-style programming', but, of course, without reducing computational power. What we suggest is: use graphs for data structures and use graph productions for describing the algorithms for the structural change of the graphs. Understanding the data structure and its changes is usually the most difficult part when reading a program! Other algorithmical aspects than structural changes, the 'quantitative' ones, are still expressed in a conventional way. Using graphs as data structures is certainly non-esoteric. One can find graphs in abundance in computer science. But what about graph grammars in general and graph productions in particular? Dear reader, they will loose the scent of esotericism in the following especially since an easy-to-understand way to depict them will be introduced.

In order to improve program quality, the software engineering community advocates the "object-oriented approach" to software design. The basic strategy is: Look at the problem domain and try to classify the objects ("data types") you see. If possible, introduce a hierarchic structure for the classes. (A subclass inherits all the properties of its superclass and possesses additional information necessary to specify the properties of the objects.) Then specify the "methods" (procedures, functions) for changing the states of the objects. The objects communicate by exchanging "messages". This means that some kind of information is sent to an object which then executes a suitable method. To a certain extent object-oriented programming can be realized by conventional programming languages, too. But, of course, it is important to have also syntactical means to express the features. The authors are strong believers in the value of the object-oriented approach.

What's left for a brief overview on the method described in this paper is to say something on the integration of the ideas stemming from the graph grammar approach with the object-oriented programming paradigm. But this comes very natural: The objects of the universe of discourse are modelled by the (typed) nodes of a graph; relations are modelled by the (untyped) edges of a graph. Along with a type comes a specific set of attributes to describe the properties of the objects in the class. The values of the attributes are derived from formulas associated with them.

Now the setting should be clear: To derive an easy-to-maintain program we advocate to describe the data structures in terms of attributed graphs and their changes in terms of attributed graph productions in an object-oriented manner. This way, the information comes to precisely the place where a reader (other than the author of the program) expects it and free-style programming is avoided. Since an interpreter for the graph grammar specification method in terms of attributed graph grammars is available, we have proposed not just a non-executable specification method but nothing less than *graph programming!*

2 *An Example*

Before stepping into theory, we want to present an example: the implementation of a 'composer' for designing kitchens. The task was, not to create a sophisticated CAD program that would be able to design anything, but a system that enables the user to fetch pre-assembled objects out of a library and arrange them according to certain 'syntactic' rules. (One could be: Don't put the refrigerator beside the oven!) The system is integrated into a project for illumination engineering that produces photorealistic images of complex illuminated scenes and allows "walking" through these 3D models.

The composer was realized by means of graphs and graph grammars. The (attributed) graph serves as a database for storing objects and relations as well as graphic data (e.g. position and color of an object) and information for the lighting calculation algorithms. A graph grammar describes the possible operations, i.e. the syntactic rules that allow the design of "correct" scenes only.

Fig. 1 shows the problem domain, a kitchen. The objects of the domain are the kitchen room and the furnishing. The problem domain is modelled by a graph, a part of which can be found in fig. 2. The nodes of the graph represent the objects (furniture), labelled edges describe relations between the objects ":has-element", ":instance-of". (The colon indicates to the LISP interpreter not to evaluate the following symbol, because it is supposed to be a keyword.) The system is implemented on a graphic workstation using the PHIGS-compatible graphic subsystem GMR ("Graphics Metafile Resource"). ("PHIGS" is the acronym for "programmer's hierarchical interactive graphics system", see [ISO].) There are mainly two kinds of objects, "structures" and "instance elements". A structure is a sequence of elements, which may be simple elements, like polylines and polygons, or instance elements. An instance element causes the elements of a structure to be 'reexecuted' after being transformed by some 4x4-matrix (very much like a procedure call in a high-level programming language) causing, for example, translation or rotation. So any object, e.g. a kitchen sink, has to be assembled only once, and may then be instanced repeatedly at different positions, in varying size and orientation.

Besides the nodes which model graphic objects, the representation graph contains additional nodes for administrative purposes. A "library node" groups the furniture of one product line, a "viewport node" holds the viewing parameters, i.e. reference point, gaze direction or refresh function. These administrative nodes have no representation in "the real world", i.e. the objects to be seen on the computer screen. Fig. 3 just shows the part of the graph representing the scene in fig. 1.

Now the graph operations get attention. Such a graph operation is, as can be seen in fig. 4, a graph, too. The X-connector divides it into four parts. (Section 3 will discuss this in detail.) To explain what will happen, here is just a brief informal description. When applied to a graph, nodes isomorphic to the left side of the X will be deleted (along with their attached edges) and the right side will be generated. The nodes below determine the context, i.e. where to apply the graph operation, and the nodes above describe how to connect the generated nodes to the existing graph. In our example a node of type element is created that represents an instance of a given structure. The representation graph in fig. 5 is the result of applying the graph operation in fig. 4 to the graph in fig. 3.

Representing the information of a scene by a graph is only part of the whole story. Additional, quantititative, information is necessary. It is expressed by means of attributes which are attached to the nodes. Now, a graph operation not only describes the structural change of a graph, but also the modifications of the attributes of the nodes. Since LISP is the language of the graph grammar implementation, this attribution is written down in a LISP (i.e. prefix) notation.

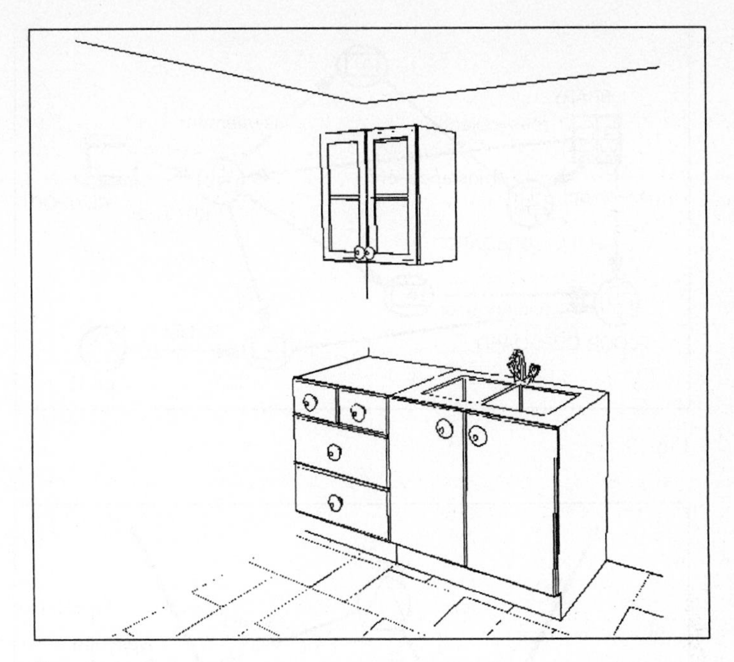

Fig. 1

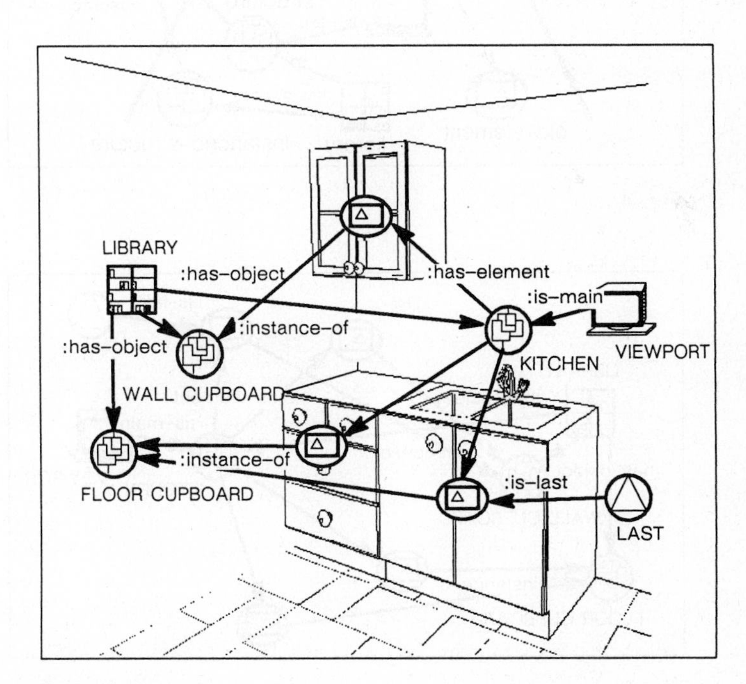
LIBRARY
:has-object
:has-element
:is-main
:instance-of
:has-object
WALL CUPBOARD
KITCHEN
VIEWPORT
:instance-of
FLOOR CUPBOARD
:is-last
LAST

Fig. 2

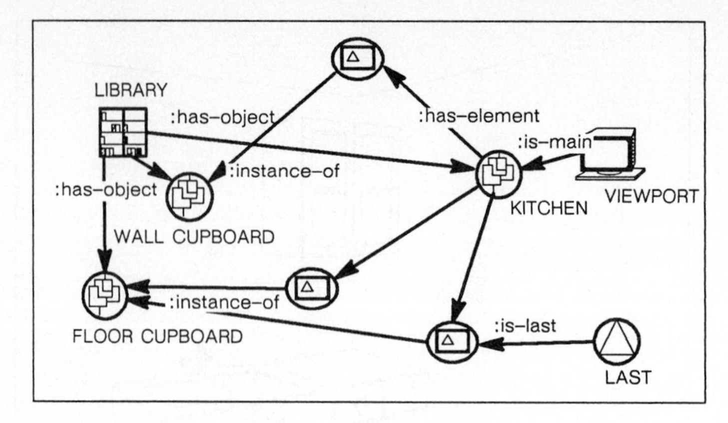

Fig. 3

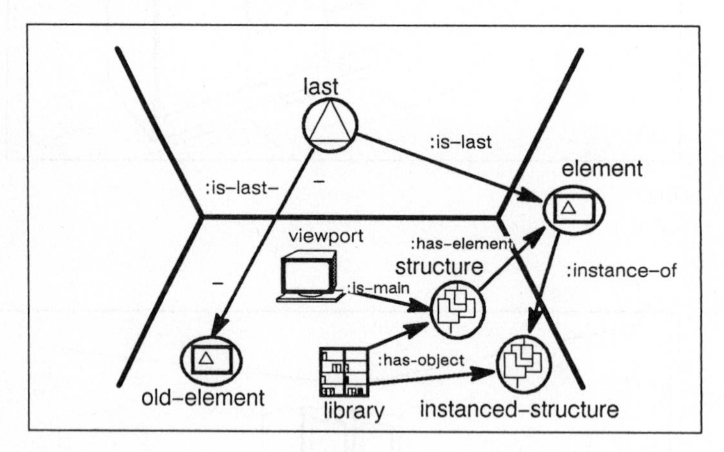

Fig. 4

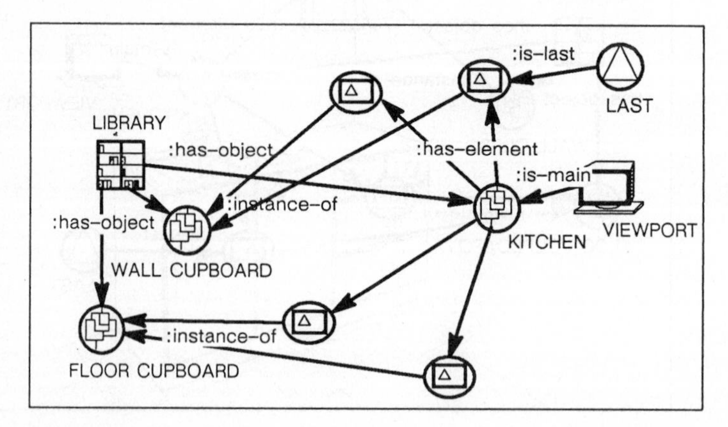

Fig. 5

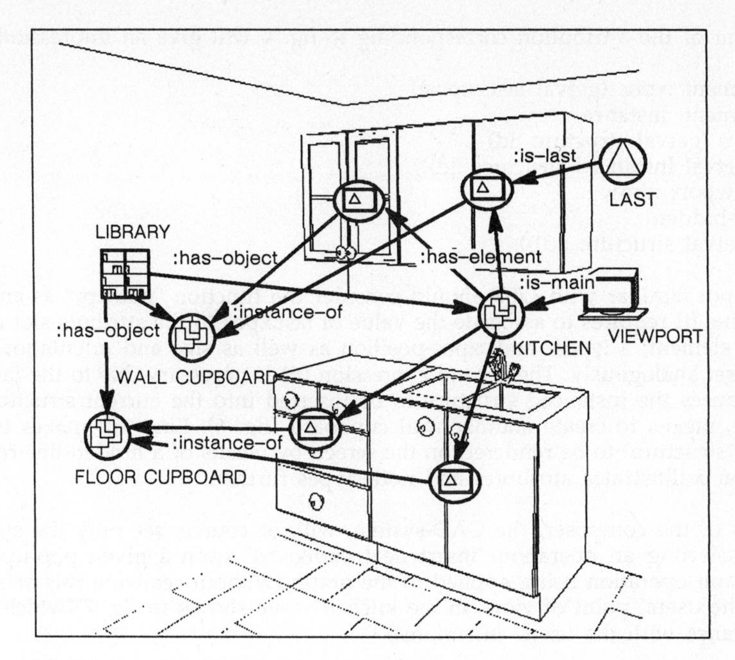
:is-last
LAST
LIBRARY
:has-object
:has-element
:is-main
:instance-of
:has-object
VIEWPORT
KITCHEN
WALL CUPBOARD
:instance-of
FLOOR CUPBOARD

Fig. 6

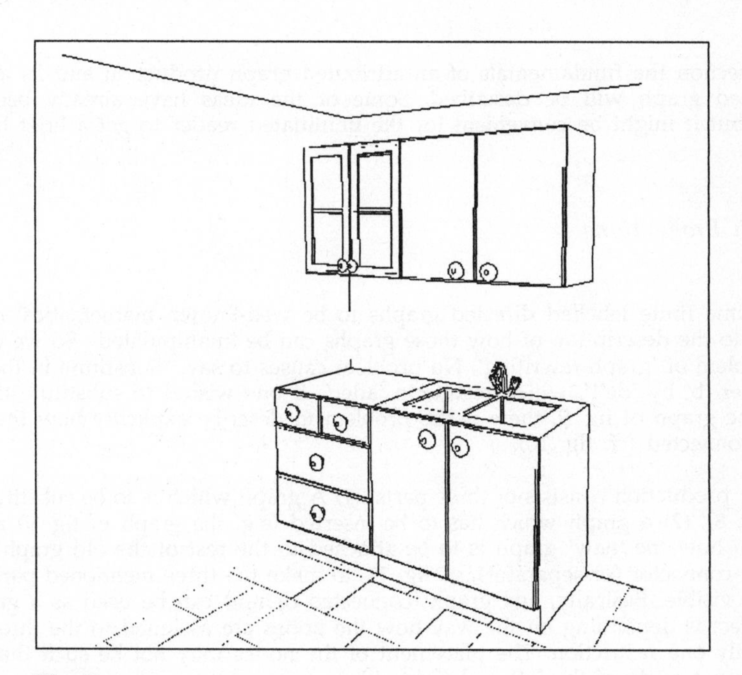

Fig. 7

A fragment of the attribution corresponding to fig. 4 will give an impression:

```
(putexpr element :xpos (getval last :xpos))                         (i)
(putexpr element :instance
    (instance (getval structure :id)
        (getval instanced-structure :id)))                          (ii)
(putexpr viewport :draw
    (render-hidden
        (getval structure :id)))                                    (iii)
```

Readers not familiar with LISP should consider the function "putexpr" as an assignment statement. Line (i) requires to associate the value of last.xpos to the attribute slot named :xpos of the node element; a :ypos- and :zpos-position as well as size and orientation of the new element are set analogously. The second expression (ii) is necessary due to the fact that GMR is used. It causes the instanced structure to be inserted into the current structure which, in our example, means to create another wall cupboard (fig. 6). Line (iii) makes the (abstract) GMR-object "structure" to be rendered on the screen by means of a hidden-line-removal algorithm. Section 3 illustrates attributes and node types further.

The users of the composer, the CAD-system, will, of course, see only the state in fig. 1, at first. By selecting an operation "insert next cupboard" from a given pop-up-menu, they cause the graph operation being applied to the graph (without realizing this at all!). This results, from the users' point of view, in the kitchen scene shown in fig. 7 (which is hoped to be in accordance with the users' intentions.)

3 *Theory behind*

In this section the fundamentals of an attributed graph production and its application to an attributed graph will be described. Some of the ideas have already been published elsewhere but it might be convenient for the uninitiated reader to get a brief introduction.

3.1 *Graph Productions*

We assume finite labelled directed graphs to be well-known mathematical concepts and confine us to the description of how those graphs can be 'manipulated'. So we get to the inherent problem of 'graph-rewriting'. No problem causes to say: "Substitute in the string 'abc' the character 'b' by 'de'!", which results in 'adec'. If one wishes to substitute the node 1 in fig. 8 by the graph of fig. 9, there is the problem to describe explicitly how the two graphs are to be connected (cf. fig. 10).

A graph production consists of three parts: (1) A graph which is to be substituted (e.g. node 1 in fig. 8), (2) a graph which has to be inserted (e.g. the graph of fig. 9) and (3) some prescription how the 'new' graph is to be attached to the rest of the old graph. We use the sign Y as a connector (or separator), cf. fig. 11, to make the three mentioned parts of a graph production visible. Basically, any graph (connected or not) can be used as a graph production. Its effect is depending on the way how the nodes are assigned to the three parts of Y. There is only one restriction: The placement of the nodes may not be such that there is an edge between a node of the left and right side.

We now apply the graph production r of fig. 11 to node 1 in fig. 8 as follows:

(1) If you find, in the given graph g, the left side of r, thus a node labelled with a, construct a new graph g′, resulting from g, by taking out this particular node (together with the adjacent edges) first.

(2) Insert into the partially constructed g′ the right side of r.

(3) Consider in r the edges cut by the right hand of Y independently from each other (In our example they are labelled with k, m and n, respectively.) Then r requires the following actions to take place:

(3k) Go back to g (fig. 8) and look for a node labelled with d which is connected with the node 1 to be taken out in the following manner: Start at 1 and go in the reverse direction of an (incoming) edge labelled with i to a node labelled with c. From this node advance, if possible, in the direction of a j-labelled edge to a node labelled with d. Thus we get via node 2 to node 4. Now draw an edge labelled with k from 4 to the f-labelled node in the partially constructed graph g′.

(3m) Go back to g and look for a node labelled with b which is connected with node 1 via a j-labelled edge starting in 1. This yields node 3. To this node we draw an edge labelled with m from the f-labelled node in the intermediate construction for g′.

(3n) Analogously to (3m) we get node 3. As fig. 11 says, we have to draw to node 3 from the e-labelled node 6 in the partially constructed g′ an edge labelled with n.

This completes the construction of g′. The result is shown in fig. 12 and is denoted by appl(r,g,1) (the application of r to g in 1).

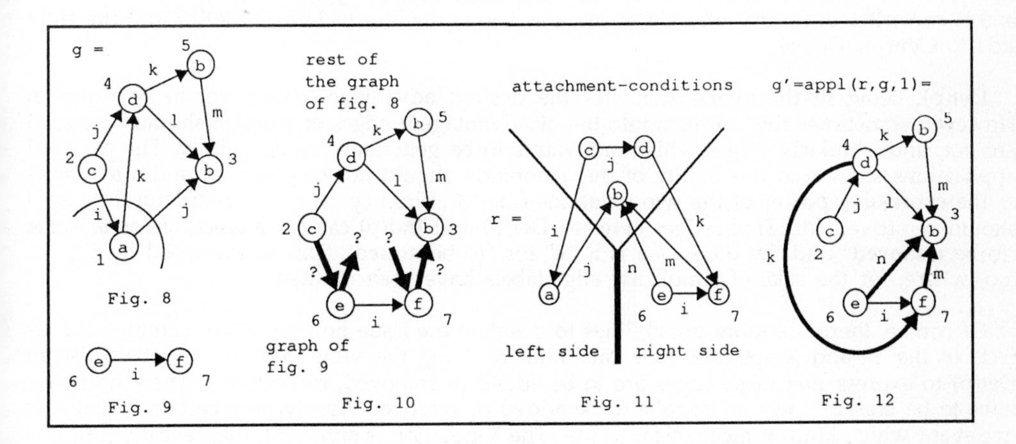

Fig. 8 Fig. 9 Fig. 10 Fig. 11 Fig. 12

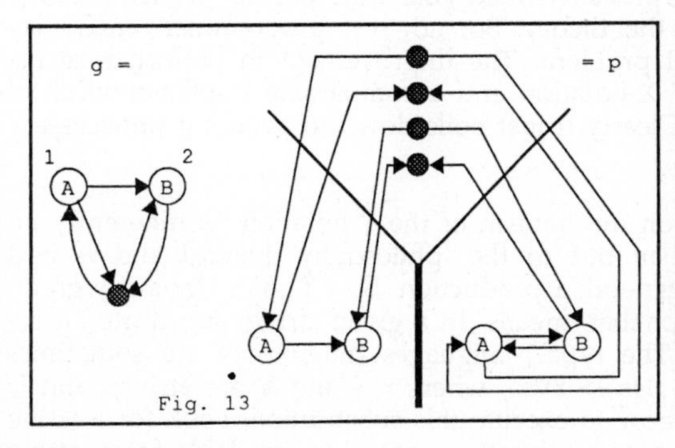

Fig. 13

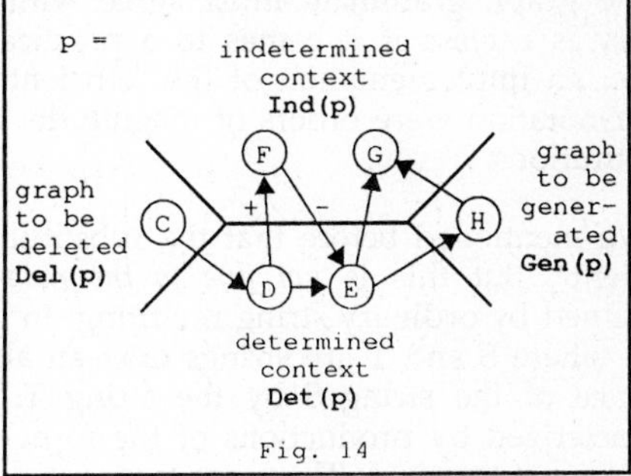

Fig. 14

One should notice that the substitution of a partial graph of g which is isomorphic to the left side of r will always take place if there is at least one. If there are more occurancies of the left side of r in g then one is chosen nondeterministically. But the connecting edges between the rest of g and the inserted right side of r are drawn iff the conditions exemplified in step (3), (3k), (3m) and (3n) hold, otherwise they are not drawn. If no isomorphic partial graph of the left side of r could be found in graph g, of course, nothing is changed in g. This is a trivial application.

Implementing the substitution mechanism exactly the way as it was described above is not wise. It is inherently inefficient. This can be demonstrated in fig. 13 where only the relevant labels are shown. Let's assume, only a second edge between node 1 and node 2 of graph g is to be constructed. The graph production p would accomplish this. However, according to the definitions above the nodes have to be taken out of the graph and then are resubstituted together with the additional edge. The upper part of p guarantees an identical embedding. In [GRA82] the problem was discussed how this unnecessary work can be avoided. The main idea is to identify in some way the parts which remain 'constant'.

Fig. 14, the X-notation of a graph production (called graph operation in the sequel), takes its rise from this consideration. The lower part of the X denotes the subgraph which would be in common to the left and to the right side in the Y-notation. When a X-shaped graph operation p is applied to a graph g, a partial graph is searched which is isomorphic to the graph consisting of Del(p)∪Det(p). Then in g the part isomorphic to Del(p) is deleted and Gen(p) is added. Similarly to the Y-notation the edges connecting the restgraph to the inserted one is constructed. So the left side of the Y is corresponding to Del(p)∪Det(p) and the right side to Gen(p)∪Det(p).

Det(p), being in the intersection, has the desired advantage having not to be restored. However, sometimes this can turn into the disadvantage if edges of g which should be erased are not, and, similarly, edges which one wants to be generated are not, either. The practical applications show that this ability of the Y-notation should not be given up and a synthesis of the expressive power of the two notations (the 'X-efficiency' and the 'Y-structural-power') should be strived for. Thus, edges between Det(p) and Ind(p) can get a label, one is a "-" for "to be removed" and the other one is a "+" for "to be added". This is expressed by fig. 14, too, where, for the sake of simplicity, edge-labels have been omitted.

Of course, there are more possibilities to describe the issue how to wisely combine the effects of the Y- and X-notation. One can think of using the same "+"- and "-"-labels also in Det(p) to express that these edges are to be added or removed, respectively. Then, no nodes have to be erased if just an edge is to be added or removed. Truely, to a certain extent it is irrelevant which kind of mechanism to use. The Y-notation is sufficient, theoretically. But one point has to be stressed! The intention of all the efforts was to prove the usefulness of the graph grammars approach for practical purposes. A main goal was that the implementation of the graph grammars must agree with the theory. So, no 'real programmer' could say theory is useless if it comes to a practical problem. The improvement in performance between an implementation of the "efficient" X-notation and a comparable implementation of the Y-notation were orders of magnitude. Clearly it just boils down to avoiding unnecessary substitutions.

We mentioned before that the substitution mechanism of the Y-notation "is inherently inefficient". But this is *not due to the notation* but to the 'philosophy' behind and is best explained by ordinary string rewriting: In general, a production S -> T of a Chomsky-grammar, where S and T are strings over an alphabet, means: In a given string, substitute an occurance of the string S by the string T. The type-1-languages (monotone) are sometimes characterized by productions of the form rUs -> rVs , where r, s and V are strings and U is some nonterminal. There are two views how to execute this substitution: Look for a string matching the left side rUs and wholly replace it by the right side, or, look for a string matching the left side but just replace the nonterminal U by V. The result is the same but the procedure is quite different. In this very sense, the Y- and the X-notation represent different views of context sensitve graph replacement.

3.2 *Type-Hierarchy and Attribution*

In a graph, the meaning of a node and the graphic object behind this node is not only modelled by the topological position of the node relative to its neighbours in the graph but also by a set of information tied to the node. This data (typically texts, identifiers, numbers, etc.) is put into (named) 'slots' associated to the given nodes. (The nodes may be regarded as 'records'.) The slots are called "attributes" of the nodes. The node type determines which slots are available.

An example of the "kitchen graph grammar" shows a typical node type definition:

```
(make-nodetype :furniture
    :attributes
    '( (:id 0)
       (:flags nil)
       (:xpos 0.0)
       (:ypos 0.0)
       (:width (calculate-width (getval :@ :id)))))
```

All nodes instantiated with the node type ":furniture" have five attributes, ":id", ":flags", ":xpos", ":ypos" and ":width", respectively. Nodes of type ":furniture" represent an element of the furniture in the kitchen example and the five attributes listed above are at least required to keep the necessary information about local and individual properties of the graphic object.

If there are more assumptions about the graphic object in question, one can define a subtype, which inherits not only the attributes of its ancestor, but also maintains more specific attribute information. Again, an example will be helpful to explain the situation:

```
(make-nodetype :wall-cupboard
    :attributes
    '( (:flags '(:drawers))
       (:depth 60.0))
    :parents '(:furniture))
```

Here, a wall-cupboard is to be modelled by the node type ":wall-cupboard". Nodes of that type inherit all attributes associated with node types listed after the item ":parents" in the definition. So ":wall-cupboard" inherits the attributes ":id", ":flags", ":xpos", ":ypos" and ":width". In addition, all nodes of type ":wall-cupboard" have an attribute called ":depth". The conflicting definition of another attribute named ":flags" is resolved by a simple priority-rule: The local definition of an attribute beats the value of the inherited attribute. In this case, the attribute value "nil" of the inherited ":flags" is overwritten by the local definition of the value '(:drawers) for the attribute ":flags".

Now we have described how a hierarchy of node types can be defined, and how every node type defines the set of attributes to be kept by a node. When building an attributed graph with an attributed graph grammar, all nodes generated by graph productions initialize their set of attributes with the attribute values given by the node type definition.

When generating new nodes, the graph operation defines the type of the new nodes. Then the attributes of types are copied to their new instances. In the example above, the node type ":furniture" not only defines constant values for attributes, but also an expression for its attribute ":width". Attribute expressions may use the values of other attributes of the same or different nodes to calculate their values. The reference to the value of another attribute is expressed by the function "getval". In terms of our example: ":width" is calculated by applying the function calculate-width to the value of the attribute ":id" of the same node (which is denoted by ":@").

Graph operations not only build up the topology of an attributed graph, but also define a new attribution for the nodes given to the operation. An example of an attributed graph operation named "insert" will be given below (refer to fig. 4, too):

```
(make-prod insert
    :below
    (viewport library instanced-structure structure old-element)
    :above (last)
    :right (element)
    :nodetypes
    (:viewporttype :librarytype :roomtype :structuretype :furniture :furniture :furniture)
    :edges ( (:is-last last old-element :-)
             (:is-last last element)
             (:has-object library structure)
             (:has-object library instanced-structure)
             (:has-element structure element)
             (:is-main viewport structure)
             (:instance-of element instanced-structure))
    :body ( (putexpr element :id
             (getval structure :id))
             (putexpr element :xpos
               (+ (getval old-element :width)
                  (getval old-element :xpos)))))
```

The new nodes are connected to a room node and a structure node. In addition, the attribute ":id" of the new node is identical to the ":id" of the structure node, and the new ":xpos" coordinate is the sum of the ":xpos"-coordinate of the old element and its width, such that the new element will be placed next to the previously inserted.

The attribute expressions are evaluated after the structural changes indicated by the graph operation are done. The order of the attribute instructions in a graph operation is not significant since the evaluation algorithm will choose an appropriate sequence.

More difficult attributions have been used in the kitchen example to calculate the position of furniture elements and to do layouting for the kitchen, taking into account several constraints provided by the application. (One could be: The total width of a contiguous row of furniture fits into the prescribed space of the kitchen.) However, to obtain unambiguous results when evaluating attributes, there must not be a cyclic dependency in the set of all attributes of a representation graph.

Some of the ideas presented here might remind some reader of approaches proposed by the artificial intelligence community for knowledge representation. In AI, frame based systems are used to model, say, part of the real world. Objects are represented by *frames* which are collections of *slots* holding the values pertaining to the objects. The attributed graph grammars are influenced by these ideas and incorporate them. However, we believe they are even superior to the frame philosophy. Their main advantage is that the edges of the representation graph can be used explicitly and not just the slot values of the objects if a change in the real world situation has to be reflected by the data structure.

3.3 *Application of an Attributed Graph Operation*

A graph operation describes how to modify a graph in a local environment. It gives no information about which particular nodes and edges of the actual graph will be affected. Given a graph operation p, it is clear that a graph g may contain none, exactly one, or several partial graphs isomorphic to $\mathrm{Del}(p)\cup\mathrm{Det}(p)$. When applying the graph operation to the graph g, the formal nodes of p have to be associated with actual nodes of g. This is controlled by 'graph programs'. There are several methods how to find these actual nodes within a program.

If one doesn't know anything about the current state of the graph, a function "find-subgraph" will provide a list of all partial graphs of g isomorphic to $\mathrm{Del}(p)\cup\mathrm{Det}(p)$. One of these partial graphs can be selected to apply the operation to. If certain properties of a node are known, the function "find-nodes" gives a list of all actual nodes matching these specifications. A call, say, (find-nodes :nodetype :element-t :value (:xpos (< :xpos 3.14))) will find all nodes whose type is :element-t and whose attribute :xpos has a value less than 3.14. Given one actual node n and an edge label l, one can determine all nodes connected to n via an l-labelled edge. By means of this operation "get-neighbours" a program can find its way through the representation graph and collect the particular actual nodes to be substituted for the formal nodes of the graph operation. Another method to access specific nodes is to assign them a globally known identifier which usually is defined at the time the graph programs are written. These so called "demon nodes" can then be found using the predefined identifiers.

The graph program, having determined actual nodes in the graph, will call the graph operation to be applied. The program supplies additional parameters required by the graph operation. Furthermore, if the graph operation will create new nodes (i.e. $\mathrm{Gen}(p)\neq\emptyset$), the program has to give identifiers to be associated with the new nodes. These identifiers may either be stated explicitly (in order to generate "demon nodes") or be created automatically. A graph program if free to use local variables and control structures to handle the application of the graph operations.

Given the representation graph g in fig. 3, the application of the graph operation p (fig. 4) works as follows. The nodes in fig. 3 named LIBRARY, VIEWPORT and LAST serve as demon nodes. A call to (get-neighbours 'last :is-last) returns the node corresponding to the last instance element in the GMR structure. Similarly the node representing the GMR main structure (i.e. the structure to be seen on the screen) can be found via the VIEWPORT demon and an edge labelled :is-main.

Since a graph operation is implemented as a LISP function, the application of a graph operation to a graph is a call to this function. The function name is simply the name of the graph operation, arguments to such a function are pairs of formal and actual node identifiers, and formal and actual parameters, respectively. First, the function (i.e. the graph operation) tests whether a given precondition holds. A precondition may assure, for example, that a cupboard will only be inserted if it is inside the room. In terms of LISP this might be some expression like (< (+ (getval last :xpos) (getval instanced-structure :width)) (getval room :x-size)). If the precondition fails, the graph operation will not be executed. If it holds, the correctness of node types, the existence of the actual nodes and the matching of actual and formal nodes and edges will be checked. If these tests hold, too, the 'real work' starts (as described in section 3.1).

"Geometric pathologies", like a solid body holds the same position as another one, cannot always be avoided by syntactical means, i.e. edge-relations of the underlying graph. But to guarantee the physical soundness of the construction, the precondition mechanism can be used.

The actual nodes corresponding to the nodes of Del(p) will be deleted. The fact is, they will not explicitly be removed or overwritten but just be marked as "deleted". This is necessary due to the fact that attributes of once deleted nodes have to stay alive since they may still be referenced by attributes of different nodes. The edges connecting to the nodes of Del(p) will be deleted next. Since edges are implemented not as entities but as lists attached to the nodes, the edges within Del(p) will automatically be gone when the connected nodes are deleted. Edges between Del(p) and Det(p) or Ind(p), respectively, will be removed by deleting a corresponding entry in the edge list of the nodes they connect.

Next, nodes corresponding to Gen(p) will be generated. The type of the nodes to be generated is usually determined by the node type of the corresponding formal node. However, a more specific node type may be delivered as an argument to the graph operation call. If the (formal) node type of a node to be created is, say, ":element-t" (as in the example in section 2), which stands for any GMR element, one may want to create a node of the more specific type ":cupboard-t" (which has ":element-t" in its parents' list, but may have additional attribute slots). Along with the nodes the incoming or outgoing edges are created. At this step also those edges between Ind(p) and Det(p) (or completely within Det(p), if defined so) that have a "+"- or "-"-label are generated or deleted, respectively.

After these structural changes are finished, the information attached to the nodes will be modified. The body of the graph operation is executed, i.e. the nodes' attributes are changed according to the specified rules. Note, attributes of newly created nodes inherit the values and expressions given in the node type definition. As mentioned before, only attributes of those nodes contained in the graph operation may be referenced or modified. The attribute instructions are, of course, free to use parameter values and common calculating functions. The attributes will be evaluated when the body of the graph operation has been processed (or even not until the graph program, which may contain calls to several graph operations, is finished).

3.4 *Evaluation of the Attributes*

For the implementation, it is necessary for each attribute of a node, to keep apart the expression for the evaluation of an attribute and the final result. When a graph operation is applied, first the structural changes take place, then the expressions of the attributes are changed according to the specified rules. Take the ":body"-definition of the graph operation "insert" in section 3.2 for an example. The function "putexpr" causes the expression "(getval structure :id)" to be copied into the expression slot of the attribute ":id" of the newly generated node "element". The value slots have not been changed yet.

After having processed all attribute instructions in a graph operation, every attribute still has an old value and some of the attributes have got new expressions. Since attributes with their old expression unchanged by the graph operation may depend on attributes with changed expressions, possibly every attribute in the graph may get a new value by re-evaluation, due to (indirect) dependencies.

The order of attribute evaluation is subject to the mutual dependencies of the attributes. Due to the fact that a graph and, thus, also the attribution is changed quite often during a session with an implemented CAD-system, only direct attribute dependencies are stored in the implementation. The attribute evaluation algorithm must find an order for replacing old attribute values by the ones obtained from evaluating attribute expressions. Since the attribute dependencies form a partial order, there is a simple algorithm to calculate correct values: Start at the minimal elements of the partial order and label all attributes that have already been evaluated with a "passed"-flag. Then compute only those expressions refering to attributes which have already been "passed". If an old value is replaced by the result of an expression, then again label the attribute in question as being "passed". Proceed this way until all the maximum elements are "passed", too.

4 *Related Work*

This described approach is not the only one, of course, which stresses the importance to model concepts by graphs. The IPSEN-project is a related one, cf. [NAG87]. However, there, the graph operations are not interpreted but just used for specification purposes. [SCHÜR90] is an approach to overcome this.

For the purpose of producing syntax directed diagram editors other approaches can be found in literature. There are table driven systems like the ones of [SOM87] or [TIC87]; a system which uses a grammar system (but not graph grammars), too, has been developed by [SZW89].

[SCHÜT87] is based on the work described in this paper. However, it is restricted to precedence graph grammars but allows attributed edges.

[GÖT82], [GÖT87], and [GÖT88] give a detailed description of the fundamentals of attributed graph grammars, their implementation, and their use in an integrated software engineering environment. Tools have been developed for dealing with the construction of graph operations, like an editor for the X-diagrams.

Modelling by graphs is also a common concept for CAD-systems, cf. [STR89]. However, the use of graph grammars for the manipulation of the graphical objects cannot be found there.

To overcome the disadvantages of LISP with regarding execution time and memory needs, a realization in the C programming language is forthcoming.

5 *Summary and Other Areas of Application*

The scenario for the base of a software engineering environment could look like introduced above. Using graph grammars yields better results: improved quality of programs and a standardized procedure for program development. The main advantage of graph grammars is the unified approach for the design of the data structures and the representation of the algorithms as graphs and graph operations, respectively.

The results proved graph grammars to be a software engineering method of its own, *graph grammar engineering*. It serves as a framework in which other software engineering principles can be incorporated. One, for example, is "stepwise refinement" at each phase of program development. When designing a CAD-system, the decision, say, on the exact placement of the graphical objects can be postponed for quite a long time. The placement could qualitatively be expressed first by relations of the type "neighbouring to". Then later in the course of developing the whole system this neighbouring-relation will be refined into formulas. Another principle, "object orientedness", is enforced, too. The nodes are instances of data types, the attributes possess their own evaluation procedures, and there is a inheritance mechanism.

Graph grammars have grown to such a state of maturity that they can also be used in a professional industrial environment. There are several joint projects between the University of Erlangen-Nürnberg and software companies, based on the method of attributed graph grammars, for the design of CAD-systems and 3D-modelling packages for interior design - see section 2 - and furnishing or lighting. In addition, editors for diagram techniques in the field of software engineering have been implemented, e.g. SDL, SADT. One of the requirements presented by the engineers was to provide an interactive system running on a workstation with mouse and high resolution graphics screen.

Acknowledgement: We would like to thank very much the reviewers for their helpful suggestions to clarify our ideas!

6 *Literature*

GÖTTLER, H.: "Attributed Graph Grammars for Graphics", Proc. "2nd Intern. Workshop on Graph Grammars and their Application to Computer Science", Osnabrück 1982 FRG, Lect. Notes in Computer Science Nr. 153, edited by H. EHRIG & M. NAGL & G. ROZENBERG, Springer Verlag, Heidelberg, 1982.

GÖTTLER, H.: "Graph Grammars and Diagram Editing", Proc. "3rd Intern. Workshop on Graph Grammars and their Application to Computer Science", Warrenton, VA. USA, Lect. Notes in Computer Science Nr. 291, edited by H. EHRIG & M. NAGL & G. ROZENBERG, Springer Verlag, Heidelberg, 1987.

GÖTTLER, H.: "Graphgrammatiken in der Softwaretechnik", Informatik Fachberichte, Nr. 178, Springer Verlag, Heidelberg, 1988.

GRABSKA, E.: "Pattern Synthesis by Means of Graph Theory", PhD-Thesis, Uniwersytet Jagiellonski, Krakow, 1982.

ISO: "PHIGS, Programmer's Hierarchical Interactive Graphics System", ISO DIS 9592, Dec. 1987.

MULLINS, S. & RINDERLE, J.: "Grammatical Approaches to Design", Proc. "!st International Workshop on Formal Methods in Engineering Design", edited by P. FITZHORN, Colorado State University, 1990.

NAGL, M.: "A Software Development Environment Based on Graph Technology", Proc. "3rd Intern. Workshop on Graph Grammars and their Application to Computer Science", Warrenton, VA. USA, Lect. Notes in Computer Science Nr. 291, edited by H. EHRIG & M. NAGL & G. ROZENBERG, Springer Verlag, Heidelberg, 1987.

SCHÜRR, A.: "PROGRESS: A VHL-Language Based on Graph Grammars", in this volume.

SCHÜTTE, A.: "Spezifikation und Generierung von Übersetzern für Graphsprachen durch attributierte Graphgrammatiken", EXpress Edition Verlag, Berlin, 1987.

SOMMERVILLE, I. & WELLAND, R. & BEER, S.: "Describing Software Design Methodologies", The Computer Journal, Vol. 30, No. 2, 1987.

SZWILLUS, G.: "Supporting Graphical Languages with Structure Editors", Proc. EUROGRAPHICS'89 (European Computer Graphics Conference and Exhibition), edited by W. HANSMANN & F.R.A. HOPGOOD & W. STRASSER, North-Holland, Amsterdam, 1989.

STRASSER, W.: "Theory and Practice of Geometric Modeling", Springer Verlag, Berlin, 1989.

TICHY, W.F. & NEWBERRY, F.J.: "Knowledge-Based Editors for Directed Graphs", Proc. ESEC'87 ("1st European Software Engineering Conference"), edited by NICHOLS, H.K. & SIMPSON, D., Lecture Notes in Computer Science Nr. 289, Springer Verlag, Heidelberg, 1987.

Collage Grammars *

Annegret Habel, Hans-Jörg Kreowski
Universität Bremen
Fachbereich Mathematik und Informatik
W-2800 Bremen 33

ABSTRACT: In this paper, we introduce and study the notion of collage grammars. A collage (in our sense) consists essentially of a set of parts being geometric objects and a set of hyperedges being subjects of further replacement. A set of collages represents a set of geometric patterns where each pattern is just the union of the parts of a collage. By overlay of the represented patterns, a set of collages yields a fractal pattern. Finally, collage grammars embody syntactic means for the generation of sets of collages in the usual way. As collages represent patterns that may be overlaid in addition, collage grammars provide also syntactic devices for the generation of sets of patterns as well as a certain type of fractal patterns.

Keywords: pattern generation, graph grammars, hyperedge-replacement grammars, fractal geometry, geometric patterns, fractal patterns, self-affine patterns

CONTENTS

1. Introduction

The generation and recognition of artificial pictures and patterns are challenging tasks in computer science and other applied areas. In the literature, one encounters quite a variety of syntactic approaches where classes of patterns are described by grammars (see, e.g., Feder [Fe 68+71], Shaw [Sh 69], Pfaltz and Rosenfeld [PR 69], Fu [Fu 74], Stiny and Gips [St 75, Gi 75], Gonzalez and Thomason [GT 78], Maurer, Rozenberg, and Welzl [MRW 82], Bunke [Bu 83], de Does and

* This work was partly supported by the ESPRIT Basic Research Working Group No. 3299: Computing by Graph Transformations.

Lindenmayer [DL 83], Prusinkiewicz, Sandness, and Hanan [Pr 86, PS 88, PH 89], Lauzzana and Pocock-Williams [LP 88], Caucal [Ca 89], and Dassow [Da 89]).

In this paper, a new theoretical framework for the study of syntactic pattern generation is outlined. The notion of collage grammars is introduced and investigated as a graph-grammatical device for generating classes of patterns. The key structures are the collages which serve two purposes. On the one hand, a collage consists of a set of parts which form a pattern if they are put together. Patterns of interest are usually infinite sets of points probably with irregular and bizarre shapes. Nevertheless, representing a pattern as the overlay of some parts provides a finite description of the pattern if the number of parts is finite and if each part has a finite description (for example as a triangle, square, circuit, etc.). On the other hand, a collage consists of a set of hyperedges that are labeled by nonterminal symbols and attached to some points. Altogether, a collage is a kind of hypergraph representing a pattern. This allows one to adapt and employ the traditional ways of graph rewriting for the generation of collage languages. Actually, we propose an approach to collage rewriting which employs techniques known from hyperedge-replacement grammars (see, e.g., [Ha 89]) and generalizes and extends our earlier work on figure grammars (see [HK 88]).

A collage grammar provides three ways of pattern generation:
(1) the pattern underlying a collage which is derived from a start collage,
(2) the overlay of all patterns obtained in (1),
(3) considering an infinite derivation, the overlay of all patterns underlying occurring collages.
While the first kind yields patterns of ordinary Euclidean shapes if the parts are of such shapes, the other two possibilities may lead to rather bizarre and fractal patterns.

The consideration in this paper is restricted to collage grammars with a context-free kind of rules whose left-hand sides consist of single nonterminal labels (see sections 3 and 6 with preliminaries in section 2). The set of rules of such a collage grammar can be interpreted as a system of formal equations. In sections 4 and 5, we show that this system of equations has solutions on various levels:
(1) the family of sets of finite derivations (with common starting handle),
(2) the family of sets of infinite derivations (with common starting handle),
(3) the family of sets of collages derived from a common starting handle,
(4) the family of sets of patterns underlying the objects in (3),
(5) the family of overlays of the objects in (4),
(6) the family of sets of overlays of the patterns underlying the decorated collages occurring in infinite derivations.
The first case (see theorem 4.5) corresponds to the Context-Freeness Lemma for hyperedge-replacement grammars, the third case (see theorem 5.2) to the Fixed-Point Theorem. The other cases (see remark 5.1 and theorem 5.2) seem to be less familiar in such a setting. In sections 7 and 8, we point out some relations between our approach and the area of fractal geometry (see, e.g., Mandelbrot [Ma 82] and Barnsley [Ba 88] for the latter). In particular, it turns out that so-called Sierpinski grammars — generalizing the notion of Koch systems in the sense of Prusinkiewicz and Sandness [PS 88] — generate self-affine patterns.

Due to the limitation of space we omit the proofs. They can be found in [HK 90]. We would like to mention that other contributions in [EKR 91] concern also the generation and rule-based description of patterns from various points of view (cf. the papers by Brandenburg and Chytil [Br 91, BC 91], de Boer [Bo 91], Fracchia and Prusinkiewicz [FP 91], as well as Lück and Lück [LL 91]).

2. Collages and Patterns

In this section, we introduce the basic notions concerning collages and patterns. We choose so-called collages as terminal structures of interest. To generate sets of collages, they are decorated by hyperedges in intermediate steps. Each hyperedge has a label and an ordered finite set of tentacles, each of which is attached to a point, and is a place holder for a collage or — recursively — for another decorated collage. Each decorated collage has a sequence of pin points. If a hyperedge is replaced by a decorated collage, the decorated collage is transformed in such a way that the images of the pin points of the latter match with the points attached to the hyperedge.

2.1 Assumption

We assume that the reader is familiar with the elementary notions of Euclidean geometry (see, e.g., Coxeter [Co 89]). $\mathbb{R}$ denotes the set of real numbers and $\mathbb{R}^n$ the Euclidean space of dimension n for some $n \geq 1$. $\mathbb{R}^n$ is equipped with the ordinary distance function $dist : \mathbb{R}^n \times \mathbb{R}^n \to \mathbb{R}$.

2.2 Definition (collages)

1. A *collage* (in $\mathbb{R}^n$) is a pair $(PART, pin)$ where $PART \subseteq \mathcal{P}(\mathbb{R}^n)$ [1] is a set of *parts*, each $part \in PART$ being a set of points in $\mathbb{R}^n$, and $pin \in (\mathbb{R}^n)^*$ [2] is a sequence of *pin points*. The class of all collages is denoted by $\mathcal{C}$.
2. Let N be a set of *labels*. A *(hyperedge-)decorated collage* (over N) is a construct $C = (PART, EDGE, att, lab, pin)$ where $(PART, pin)$ is a collage, called the *collage underlying* C and denoted by $collage(C)$, $EDGE$ is a set of *hyperedges*, $att : EDGE \to (\mathbb{R}^n)^*$ is a mapping, called the *attachment*, and $lab : EDGE \to N$ is a mapping, called the *labeling*. C is said to be *finite* if $PART$ and $EDGE$ are finite. The class of all decorated collages over N is denoted by $\mathcal{C}(N)$.
3. Two decorated collages $C, C' \in \mathcal{C}(N)$ are said to be *isomorphic*, denoted by $C \cong C'$, if the underlying collages of C and C' are equal and if there is a bijective mapping $f : EDGE_C \to EDGE_{C'}$ such that $att_C(e) = att_{C'}(f(e))$ and $lab_C(e) = lab_{C'}(f(e))$ for all $e \in EDGE_C$.

Remark: 1. The components $PART$, $EDGE$, att, lab, and pin of a decorated collage C are also denoted by $PART_C$, $EDGE_C$, att_C, lab_C, and pin_C, respectively.
2. Each decorated collage C induces a set of points

$$POINT_C = points(pin_C) \cup \bigcup_{e \in EDGE_C} points(att_C(e))$$

where, for $s \in (\mathbb{R}^n)^*$, $points(s)$ denotes the set of points occurring in s.
3. A collage can be seen as a decorated collage C without hyperedges, i.e., $EDGE_C = \emptyset$ and att_C as well as lab_C being the empty mappings. In this sense, $\mathcal{C} \subseteq \mathcal{C}(N)$. In the description of decorated collages without hyperedges, we will drop the components $EDGE_C$, att_C, and lab_C.

2.3 Definition (patterns)

1. A *pattern* (in $\mathbb{R}^n$) is a pair $(POINT, pin)$ where $POINT \subseteq \mathbb{R}^n$ is a set of points and $pin \in (\mathbb{R}^n)^*$ is a sequence of *pin points*. The class of all patterns is denoted by $\mathcal{Q}$.
2. A *(hyperedge-)decorated pattern* (over N) is a construct $Q = (POINT, EDGE, att, lab, pin)$ where $(POINT, pin)$ is a pattern, called the *pattern underlying* Q and denoted by $pattern(Q)$, $EDGE$ is a set of *hyperedges*, $att : EDGE \to (\mathbb{R}^n)^*$ is a mapping, called the *attachment*, and $lab : EDGE \to N$ is a mapping, called the *labeling*. Q is said to be *finite* if the sets $POINT$ and $EDGE$ are finite. The class of all decorated patterns over N is denoted by $\mathcal{Q}(N)$.

---

[1] For a set X, $\mathcal{P}(X)$ denotes the powerset of X.
[2] For a set X, X^* denotes the set of all sequences over X, including the empty sequence λ.

3. Let $C \in \mathcal{C}(N)$ be a decorated collage. Then the *decorated pattern* of C, the *pattern underlying* C, and the *points* of C are given by

$$C^p = (\bigcup_{part \in PART_C} part\ ,\ EDGE_C, att_C, lab_C, pin_C),$$

$$pattern(C) = (\bigcup_{part \in PART_C} part\ ,\ pin_C),$$

$$points(C) = (\bigcup_{part \in PART_C} part).$$

Remark: 1. The components *POINT*, *EDGE*, *att*, *lab*, and *pin* of a decorated pattern Q are also denoted by $POINT_Q$, $EDGE_Q$, att_Q, lab_Q, and pin_Q, respectively.
2. In 2.3.3, two transformations concerning decorated collages are considered. The first construction transforms a decorated collages into a decorated pattern by collecting the points occurring in parts of a decorated collage; the second — more rigorous — construction transforms a decorated collages into a pattern by collecting the points occurring in parts of a decorated collage and forgetting the hyperedges. In this way, a finite (decorated) collage may represent an infinite (decorated) pattern. Clearly, we have $pattern(C^p) = pattern(collage(C)) = pattern(C)$.

We are mainly interested in the generation of patterns, which are usually infinite sets of points. On the other hand, we suggest the use of grammatical methods, which concern rule-based rewriting of finite objects. To bridge this gap, we have introduced the notion of a finite (decorated) collage consisting of a finite set of parts (where we assume that each part has some finite description even if it is an infinite set of points). So a finite collage can be the object of rewriting and, at the same time, represent a potentially infinite pattern by the union of parts. In other respects, the difference between collages and patterns is small. Each collage yields the underlying pattern, and each pattern $(POINT, pin)$ induces a collage $(PART, pin)$ where the parts are the singleton subsets of $POINT$. This transformation allows one to consider a pattern as a collage.

2.4 Example

An example of a decorated collage (in $\mathbb{R}^2$) is given in Fig. 2.1(a). It has five (partly overlapping) parts of the shape of triangles and two S-labeled hyperedges with four tentacles each. Let e be one of the hyperedges, then the tentacle i (for $i = 1, 2, 3, 4$) is attached to the i-th attachment point $att(e)_i$, i.e. $att(e) = att(e)_1 \dots att(e)_4$. The points 1 to 4 form the sequence of pin points (that span a square in this case).

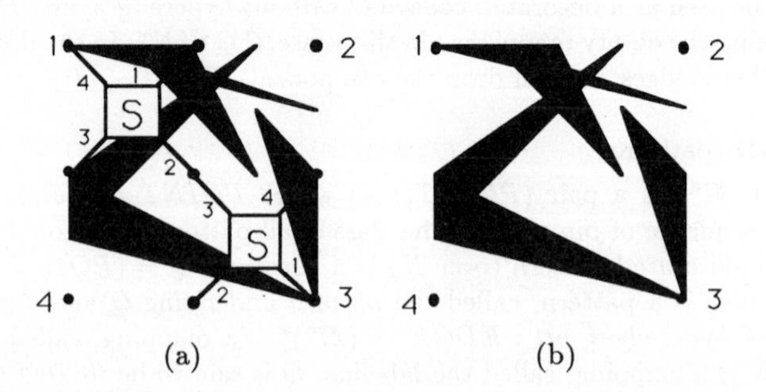

Fig. 2.1. A decorated collage and its underlying collage

In drawings, a set of parts is not distinguished from the overlay of the parts. Hence collages are not distinguished from patterns. For example, Fig. 2.1(b) shows the collage as well as the pattern underlying the decorated collage in Fig. 2.1(a).

Hyperedges in decorated collages serve as place holders for (decorated) collages. Hence the key construction is the replacement of some hyperedges in a decorated collage by (decorated) collages. While a hyperedge is attached to some points according to our conventions, a (decorated) collage has got some pin points. If there is a transformation which maps the pin points to the attached points of the hyperedge, the hyperedge may be replaced by the transformed (decorated) collage. The formal definition of hyperedge replacement makes use of three simpler constructions on decorated collages (resp. decorated patterns): hyperedge removal, transformation, and addition. Moreover, we give a construction of so-called handles and the notion of an empty collage which will be used later.

2.5 Construction

1. Let $C \in \mathcal{C}(N)$ and $B \subseteq EDGE_C$. Then the *removal* of B from C yields the decorated collage

$$C - B = (PART_C, EDGE_C - B\ ^{3)}, att, lab, pin_C)$$

with $att(e) = att_C(e)$ and $lab(e) = lab_C(e)$ for all $e \in EDGE_C - B$ [3]. In the special case $B = \{e\}$, we also write $C - e$ instead of $C - \{e\}$.
2. Let $C \in \mathcal{C}(N)$ and $t : \mathbb{R}^n \to \mathbb{R}^n$ be a mapping which will be referred to as a transformation. Then the *transformation* of C by t yields the decorated collage

$$t(C) = (t(PART_C)\ ^{4)}, EDGE_C, att, lab_C, pin)$$

with $att(e) = t(att_C(e))$ [4] for all $e \in EDGE_C$, and $pin = t(pin_C)$.
3. Let $C \in \mathcal{C}(N)$ and $Y \subseteq \mathcal{C}(N)$. Then the *addition* of Y to C yields the decorated collage

$$C + Y = (PART_C \cup \bigcup_{R \in Y} PART_R\ ,\ EDGE_C + \sum_{R \in Y} EDGE_R\ ^{3)}, att, lab, pin_C)$$

with $att(e) = att_C(e)$ and $lab(e) = lab_C(e)$ for all $e \in EDGE_C$ and $att(e) = att_R(e)$ and $lab(e) = lab_R(e)$ for all $e \in EDGE_R$, $R \in Y$.
4. Let $pin \in (\mathbb{R}^n)^*$ and L be a set of decorated collages with $pin_C = pin$ for all $C \in L$. Then the *overlay of* L is the decorated collage

$$overlay(L) = (\bigcup_{C \in L} PART_C\ ,\ \sum_{C \in L} EDGE_C, att, lab, pin)$$

with $att(e) = att_C(e)$ and $lab(e) = lab_C(e)$ for all $e \in EDGE_C$, $C \in L$.
5. Let $A \in N$ and $pin \in (\mathbb{R}^n)^*$. Then $(A, pin)^\bullet$ denotes the decorated collage $(\emptyset, \{e\}, att, lab, pin)$ with $lab(e) = A$ and $att(e) = pin$. It is called the *handle induced by* A and pin.
6. Let $pin \in (\mathbb{R}^n)^*$. Then $pin^\bullet$ denotes the collage $(\emptyset, pin)$ and is called the *empty collage* induced by pin.

Remark: 1. Removal removes some hyperedges without changing anything else. In particular, $pin_{C-B} = pin_C$.

---

[3] Given sets X, Y, Z with $X \subseteq Y$, $Y - X$ denotes the complement of X in Y, and $Y + Z$ denotes the disjoint union of Y and Z. Let X_i for $i \in I$ be sets, then $\sum_{i \in I} X_i$ denotes the disjoint union of the X_i.

[4] The mapping $t : \mathbb{R}^n \to \mathbb{R}^n$ can be extended to the following mappings:
- $t : \mathcal{P}(\mathbb{R}^n) \to \mathcal{P}(\mathbb{R}^n)$ by $t(part) = \{t(x) | x \in part\}$ for all $part \subseteq \mathbb{R}^n$,
- $t : \mathcal{P}(\mathcal{P}(\mathbb{R}^n)) \to \mathcal{P}(\mathcal{P}(\mathbb{R}^n))$ by $t(PART) = \{t(part) | part \in PART\}$ for all $PART \subseteq \mathcal{P}(\mathbb{R}^n)$,
- $t : (\mathbb{R}^n)^* \to (\mathbb{R}^n)^*$ by $t(x_1 \ldots x_n) = t(x_1) \ldots t(x_n)$ for all $x_i \in \mathbb{R}^n$, $i = 1, \ldots, n$.

2. The transformation transforms points and parts according to t. The set of hyperedges is not changed, their labels remain unchanged, but the hyperedges are attached to the transformed attached points. Note that $pin_{t(C)} = t(pin_C)$.
3. The addition of a set of decorated collages to a decorated collage C is asymmetric with respect to the choice of the pin points which are borrowed from C. Hence, $pin_{C+Y} = pin_C$.
4. The overlay of a non-empty set L of decorated collages can be obtained by an addition of a set of decorated collages to a decorated collage. The overlay of an infinite set L may yield an infinite decorated collage even if all elements of L are finite.
5. The constructions on decorated collages mentioned above (removal, transformation, addition, and overlay) also can be defined for decorated patterns. In the following, we will use the same notions for both types of constructions.

2.6 Definition (hyperedge replacement)

Let $TRANS$ be a set of transformations. Let $C \in \mathcal{C}(N)$, $B \subseteq EDGE_C$, and $(repl, trans)$ be a pair of mappings $repl : B \to \mathcal{C}(N)$, $trans : B \to TRANS$ with $att(e) = trans(e)(pin_{repl(e)})$ for all $e \in B$. Then the *replacement* of B in C through $(repl, trans)$ yields the decorated collage

$$REPL(C, repl, trans) = (C - B) + Y(B)$$

where $Y(B) = \{trans(e)(repl(e)) | e \in B\}$ denotes the set of decorated collages determined by $(repl, trans)$.

Remark: 1. Hyperedge replacement is a simple construction where some hyperedges are removed, the associated decorated collages are transformed in such a way that the images of the pin points match the points attached to the corresponding hyperedges, and the transformed decorated collages are added. Note that the pin points may restrict the choice of possible transformations.
2. The transformed decorated collages replacing hyperedges are fully embedded into the resulting decorated collage, but their pin points loose their status.
3. Hyperedge replacement is a construction which transforms a decorated collage into a decorated collage. Accordingly, hyperedge replacement can be defined for decorated patterns yielding a decorated pattern. In the following, we will use the same notion for both types of replacements.

3. Collage Grammars

In this section, we introduce collage grammars as collage-manipulating and collage-language generating devices. Based on hyperedge replacement as introduced in the preceding section, one can derive (decorated) collages from decorated collages by applying productions of a simple form. A production is given by a label $A \in N$ and a decorated collage $R \in \mathcal{C}(N)$. It may be applied to a hyperedge e with label A provided that there is a transformation from a given set $TRANS$ of admissible transformations which maps the pin points of R to the attached points of e. The result of the application is obtained by replacing the hyperedge by the transformed image of R. More generally, several productions may be applied in parallel. Besides the ordinary notion of a derivation and a generated language we introduce the notion of an infinite derivation which will be used as a representation of a fractal pattern.

3.1 General Assumption

Let $TRANS$ be a set of transformations $t : \mathbb{R}^n \to \mathbb{R}^n$.

Remark: Various sets of transformations may be considered, e.g., the set of *isometries*, the set of *central dilatations*, the set of *similarity transformations*, or the set of *affine transformations*.

3.2 Definitions (productions and derivations)

1. Let N be a set of labels. A *production* (over N) is a pair $p = (A, R)$ with $A \in N$ and $R \in \mathcal{C}(N)$. A is called the *left-hand side* of p and is denoted by $lhs(p)$. R is called the *right-hand side* and is denoted by $rhs(p)$.
2. Let $C \in \mathcal{C}(N)$, $B \subseteq EDGE_C$, and P be a set of productions (over N). Then a pair $(prod, trans)$ of mappings $prod : B \to P$ and $trans : B \to TRANS$ is called a *base on* B *in* C if $lab_C(e) = lhs(prod(e))$ and $att_C(e) = trans(e)(pin_{rhs(prod(e))})$ for all $e \in B$, and if $trans$ is the only mapping with the latter property.
3. Let $C, C' \in \mathcal{C}(N)$ and $(prod, trans)$ be a base on B in C. Then C *directly derives* C' *through* $(prod, trans)$ if $C' \cong REPL(C, repl, trans)$ with $repl(e) = rhs(prod(e))$ for all $e \in B$. A direct derivation is denoted by $C \Longrightarrow C'$ through $(prod, trans)$, $C \Longrightarrow_P C'$, or $C \Longrightarrow C'$.
4. A sequence of direct derivations of the form $C_0 \Longrightarrow_P C_1 \Longrightarrow_P \ldots \Longrightarrow_P C_k$ is called a *derivation* from C_0 to C_k and is denoted by $C_0 \Longrightarrow_P^* C_k$ or $C_0 \Longrightarrow^* C_k$.
5. A family $d = (C_i \Longrightarrow_P C_{i+1})_{i \geq 0}$ of direct derivations $C_i \Longrightarrow_P C_{i+1}$ for $i \geq 0$ is called an *infinite derivation* (over P).

Remark: 1. The uniqueness requirement for $trans$ (see 3.2.2) makes sure that there are no two ways to apply the same production to some hyperedge with different results.
2. Each derivation $C_0 \Longrightarrow C_1 \Longrightarrow \ldots \Longrightarrow C_k$ in the sense of 3.2.4 can be considered as an infinite derivation $(C_i \Longrightarrow C_{i+1})_{i \geq 0}$ by elongating it by dummy steps, this is direct derivations $C_i \Longrightarrow C_{i+1}$ for $i \geq k$ through the empty base. In particular, this means that $C_k \cong C_i$ for $i \geq k$.

Using the introduced concepts of productions and derivations, collage grammars and collage languages can be introduced in the usual way. Since collages represent patterns and the overlay of patterns yields a pattern, collage grammars specify also pattern languages and single (fractal) patterns.

3.3 Definitions (collage grammars and languages)

1. A *collage grammar* is a system $CG = (N, P, Z)$ where N is a finite set of *nonterminals*, P is a finite set of *productions* (over N) with finite right-hand sides, and $Z \in \mathcal{C}(N)$ is a finite decorated collage, called the *axiom*.
2. The *collage language generated by* CG consists of all collages which can be derived from Z by applying productions of P:

$$L^c(CG) = \{C \in \mathcal{C} \mid Z \overset{*}{\underset{P}{\Longrightarrow}} C\}.$$

3. The *pattern language generated by* CG is given by the patterns underlying the generated collages:

$$L^p(CG) = \{pattern(C) \mid C \in L^c(CG)\}.$$

4. The *fractal pattern generated by* CG is given by the overlay of all generated patterns:

$$L^{op}(CG) = overlay(L^p(CG)).$$

Remark: 1. Collage grammars as introduced above are based on the replacement of hyperedges. They are closely related to hyperedge-replacement grammars as studied, e.g., in [Ha 89]. Moreover, collage grammars generalize our figure grammars investigated in [HK 88].
2. The collage language generated by CG depends on the set of admissible transformations. To emphasize this aspect, we sometimes write $L^c_{TRANS}(CG)$ instead of $L^c(CG)$ to denote the collage language generated by CG with respect $TRANS$. Let us mention that $TRANS \subseteq TRANS'$ implies $L^c_{TRANS}(CG) \subseteq L^c_{TRANS'}(CG)$.
3. Let $CG = (N, P, Z)$ be a collage grammar. Then $pin_C = pin_Z$ for all $C \in L^c(CG)$. Therefore, the overlay of $L^c(CG)$ is well-defined. Accordingly, the overlay of $L^p(CG)$ is well-defined.

3.4 Definition

Let $CG = (N, P, Z)$ be a collage grammar and $L^{der}(CG) = \{(C_i \Longrightarrow_P C_{i+1})_{i \geq 0} | C_0 = Z\}$ be the set of infinite derivations of CG starting with Z. Then the *set of fractal patterns generated by* the infinite derivations of CG is given by the set of overlays of the patterns underlying the decorated collages occurring in infinite derivations:

$$L^{pder}(CG) = \{overlay(L^p(der)) | der \in L^{der}(CG)\}$$

with $L^p((C_i \Longrightarrow C_{i+1})_{i \geq 0}) = \{pattern(C_i) | i \geq 0\}$.

3.5 Example

Let $CG = (\{S\}, P, Z)$ be the collage grammar with the axiom and the productions as shown in Fig. 3.1.

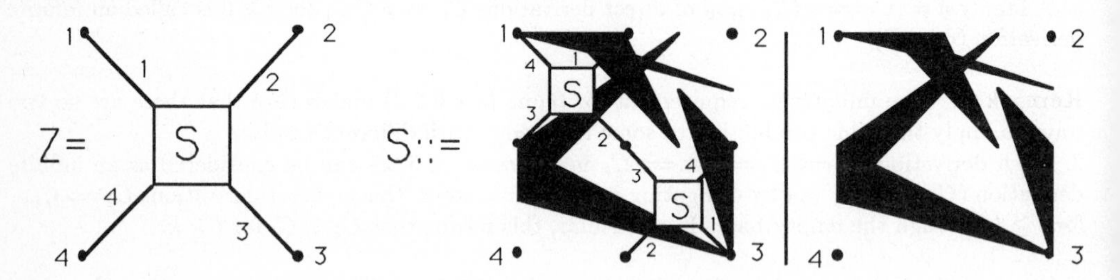

Fig. 3.1. Axiom and productions of CG

Let $SIM(2)$ (the set of all similarity transformations of $\mathbb{R}^2$) be the set of admissible transformations. Then the following derivation steps are possible: Starting from the axiom Z, the first production may be applied yielding a decorated collage with two S-labeled hyperedges isomorphic to the right-hand side. Using a combination of a contraction (with magnification factor $\frac{1}{2}$), a rotation (with angle of rotation 90° resp. 180°), and a translation one may replace these two hyperedges. In this manner one may generate collages (patterns) of the following form:

Fig. 3.2. Collages (patterns) generated by CG

4. Context-Freeness Lemmata

The replacements of different hyperedges of a decorated collage are independent of each other. This leads to two context-freeness lemmata characterizing the derivations of a collage grammar, that start in handles, in a recursive way. The first lemma corresponds to well-known results for various types of context-free grammars including hyperedge-replacement graph grammars. The second lemma is a corresponding result for infinite derivations. Both results lead to characterizations of languages, collages, and patterns generated by collage grammars that will be interpreted as fixed points in the next section.

4.1 General Assumption

Let $TRANS$ be a set of transformations $t : \mathbb{R}^n \to \mathbb{R}^n$ forming a group (with respect to functional composition as group operation, the identity transformation as unit element, and the inverse transformations as inverse elements).

Remark: The sets $AFF(n)$ and $SIM(n)$ of all affine resp. similarity transformations of $\mathbb{R}^n$ form groups.

For technical reasons, we show a normal form for collage grammars which uses the general assumption 4.1 and will simplify the reasoning later on.

4.2 Definition (proper collage grammars)

A collage grammar $CG = (N, P, Z)$ is said to be *proper* if

- CG is *normalized*, i.e., for all $C \in \mathcal{C}(N)$ occurring in CG (i.e., the right-hand sides as well as the axiom), all $e \in EDGE_C$, and all $p \in P$ with $lhs(p) = lab_C(e)$, there is a transformation $t \in TRANS$ such that $t(pin_{rhs(p)}) = att_C(e)$, and, whenever, two productions have the same left-hand side, then their right-hand sides have the same sequences of pin points,
- CG is *terminating*, i.e., for all $A \in N$, there is a production $(A, R) \in P$ and a derivation $R \overset{*}{\Longrightarrow}_P C$ with $C \in \mathcal{C}$.

Remark: 1. Let CG be proper. Then, for all $C \in \mathcal{C}(N)$ derivable from Z, all $e \in EDGE_C$, and all $p \in P$ with $lhs(p) = lab_C(e)$, p is *applicable* to e, i.e., there is a transformation $t \in TRANS$ such that $t(pin_{rhs(p)}) = att_C(e)$.
2. Each proper collage grammar $CG = (N, P, Z)$ defines a mapping $AX : N \to \mathcal{C}(N)$ with $AX(A) = (A, pin_R)^\bullet$ for some production $(A, R) \in P$. In the following this mapping is said to be the *axiom assignment induced* by CG.
3. Each proper collage grammar $CG = (N, P, Z)$ defines a family of proper collage grammars $CG(A) = (N, P, AX(A))$ and families $L^x(CG)(A) = L^x(CG(A))$ for $x \in \{c, p, op, pder\}$ $(A \in N)$.

4.3 Normal-Form Theorem

For each collage grammar CG, a proper collage grammar CG'' can be constructed such that $L^c(CG') = L^c(CG)$.

4.4 Lemma (restriction and embedding)

1. Let $d = (C_i \Longrightarrow C_{i+1})_{i \geq 0}$ be an infinite derivation with direct derivation steps $C_i \Longrightarrow C_{i+1}$ through $(prod_i : B_i \to P, trans_i : B_i \to TRANS)$. Moreover, let $C'_0 \subseteq C_0$ [5] . Then there is an infinite derivation $RESTRICT(d, C'_0) = (C'_i \Longrightarrow C'_{i+1})_{i \geq 0}$ with direct derivations $C'_i \Longrightarrow C'_{i+1}$ through $(prod'_i, trans'_i)$, called the *restriction* of d to C'_0, where, for $i \geq 0$, $C'_i \subseteq C_i$ and $prod'_i$ and $trans'_i$ are the restrictions of $prod_i$ and $trans_i$ to $B'_i = B_i \cap E_{C'_i}$.
2. Let $C \in \mathcal{C}(N)$ be a decorated collage, $B \subseteq EDGE_C$, $der : B \to DER$ be a mapping assigning an infinite derivation $der(e) = (C(e)_i \Longrightarrow C(e)_{i+1})_{i \geq 0}$ with direct derivations $C(e)_i \Longrightarrow C(e)_{i+1}$ through $(prod(e)_i, trans(e)_i)$ to each $e \in B$, and $trans : B \to TRANS$ be a mapping assigning a transformation $trans(e)$ to each $e \in B$ such that, for $e \in B$, $trans(e)(pin_{C(e)_0}) = att_C(e)$. Then there is an infinite derivation $EMBED(C, der, trans) = (C_i \Longrightarrow C_{i+1})_{i \geq 0}$ with direct derivations $C_i \Longrightarrow C_{i+1}$ through $(prod_i, trans_i)$, called the *embedding* of der into C, where, for $i \geq 0$, $C_i = REPL(C, repl_i, trans)$ with $repl_i(e) = C(e)_i$ for $e \in B$, $prod_i = \sum_{e \in B} prod(e)_i$, and $trans_i = \sum_{e \in B} trans(e)_i$ [6] .

[5] Let $C, C' \in \mathcal{C}(N)$. Then $C' \subseteq C$ denotes that C' is a *decorated subcollage* of C, i.e., $PART_{C'} \subseteq PART_C$, $EDGE_{C'} \subseteq EDGE_C$, and $att_{C'}$ and $lab_{C'}$ are the restrictions of att_C and lab_C to $EDGE_{C'}$, respectively. Note that there are no restrictions for the sequences of pin points.

[6] Let, for $i \in I$, $f_i : A_i \to B$ be a mapping. Then $\sum_{i \in I} f_i$ denotes the mapping from $\sum_{i \in I} A_i$ to B defined by $(\sum_{i \in I} f_i)(x) = f_j(x)$ for $x \in A_j$ and $j \in I$.

If a derivation starts in the handle $AX(A)$ for some $A \in N$, the first non-trivial step applies a production to the only hyperedge and derives the right-hand side R of some production (to get this, the uniqueness property in definition 3.2.2 is essential). The remaining derivation can be restricted to the subhandles of R. Due to the lemma above, it can be reconstructed as the embedding of all the restrictions into R. Moreover, each restriction can be transformed into a derivation starting in the handle $AX(B)$ for some $B \in N$ according to our assumptions on the transformations and proper grammars. This reasoning leads to a recursive version of the lemma above, which is called Context-Freeness Lemma (see 4.5). It also applies to finite derivations because they may be considered as the special case of infinite derivations where only a finite number of steps are rewriting a non-empty set of hyperedges.

4.5 Theorem (Context-Freeness Lemma I)

Let $CG = (N, P, Z)$ be a proper collage grammar and $AX : N \to \mathcal{C}(N)$ be the induced axiom assignment. Moreover, let $A \in N$ and $C \in \mathcal{C}$. Then there is a derivation $AX(A) \Longrightarrow^* C$ if and only if there is a production $(A, R) \in P$ and, for each $e \in EDGE_R$, there is a derivation $AX(lab_R(e)) \Longrightarrow^* C(e)$ and a transformation $trans(e)$ such that $C = REPL(R, repl, trans)$ with $repl(e) = C(e)$ for $e \in EDGE_R$.

Remark: Theorem 4.5 can be shown in analogy to the case of hyperedge-replacement grammars (see, e.g., [Ha 89]).

4.6 Corollary

Let $CG = (N, P, Z)$ be a proper collage grammar and $EQ : N \to \mathcal{C}(N)$ the mapping given by $EQ(A) = \{R' | ((A, R) \in P \text{ and } R' \cong R\}$ for $A \in N$. Then

1. $L^c(CG)(A) = \bigcup_{R \in EQ(A)} \left\{ REPL(R, repl, trans) \;\middle|\; \begin{array}{l} repl(e) \in L^c(CG)(lab_R(e)) \\ trans(e) \in TRANS \end{array} \right\}$,
2. $L^p(CG)(A) = \bigcup_{R \in EQ(A)} \left\{ REPL(R^p, repl, trans) \;\middle|\; \begin{array}{l} repl(e) \in L^p(CG)(lab_R(e)) \\ trans(e) \in TRANS \end{array} \right\}$,
3. $L^{op}(CG)(A) = overlay(\bigcup_{R \in EQ(A)} \left\{ REPL(R^p, repl, trans) \;\middle|\; \begin{array}{l} repl(e) = L^{op}(CG)(lab_R(e)) \\ trans(e) \in TRANS \end{array} \right\})$

where the properties for $repl(e)$ and $trans(e)$ are required for all $e \in EDGE_R$.

Now we are going to state a similar result for infinite derivations which leads to characterizations of the sets of infinite derivations (starting in the same axiom) and classes of languages represented by them. Each infinite derivation defines languages of collages resp. patterns underlying the decorated collages occurring in the derivation. Consequently, the set of all infinite derivations starting in the same axiom represents classes of collage and pattern languages.

4.7 Theorem (Context-Freeness Lemma II)

Let $CG = (N, P, Z)$ be a proper collage grammar and $AX : N \to \mathcal{C}(N)$ the induced axiom assignment. Moreover, let $A \in N$. Then there is an infinite derivation $d = (C_i \Longrightarrow C_{i+1})_{i \geq 0}$ with $C_0 = AX(A)$ if and only if there is a direct derivation $AX(A) \Longrightarrow C_1$ with $C_1 \cong R$ for some $(A, R) \in P$ and, for each $e \in EDGE_{C_1}$, there is an infinite derivation $der(e) = (C_i(e) \Longrightarrow C_{i+1}(e))_{i \geq 1}$ with $C_1(e) = AX(lab_{C_1}(e))$ and a transformation $trans(e)$ such that $EMBED(C_1, der, trans) = (C_i \Longrightarrow C_{i+1})_{i \geq 1}$.

Remark: The proof is essentially the same as the proof of theorem 4.5.

4.8 Corollary

Let $CG = (N, P, Z)$ be a proper collage grammar and $EQ : N \to \mathcal{P}(\mathcal{C}(N))$ the mapping given by $EQ(A) = \{R' | ((A, R) \in P \text{ and } R' \cong R\}$ for $A \in N$. Then

$$L^{p_{der}}(CG)(A) = \bigcup_{R \in EQ(A)} \left\{ REPL(R^p, repl, trans) \;\middle|\; \begin{array}{l} repl(e) \in L^{p_{der}}(CG)(lab_R(e)) \\ trans(e) \in TRANS \end{array} \right\}$$

where the properties for $repl(e)$ and $trans(e)$ are required for all $e \in EDGE_R$.

5. A Fixed-Point Theorem

The corollaries 4.6 and 4.8 present characterizations of languages, collages, and patterns associated to a collage grammar. The results can be strengthened if the rules of a collage grammar are considered as a set of formal equations. Then it turns out that some of the languages and patterns associated to the grammar are the least fixed points of the equations with respect to suitable domains.

5.1 Definition (equation systems, solutions, and least fixed points)

1. A mapping $EQ : N \to \mathcal{P}(\mathcal{C}(N))$ is said to be an *equation system* over N. For $A \in N$, the pair $(A, EQ(A))$ is called an *equation* over N and is written by

$$A = EQ(A).$$

The set N is said to be the set of *indeterminates* or *unknowns* of EQ.
2. A family $F : N \to X$ (with an extension $F : \mathcal{P}(\mathcal{C}(N)) \to X$ for some suitable set X) is said to be a *fixed point* (or *solution*) of EQ if, for all $A \in N$,

$$F(A) = F(EQ(A)).$$

3. A family $F : N \to X$ is a *least fixed point* of EQ if F is a fixed point of EQ, and, if F' is any other fixed point of EQ, then $F(A) \subseteq F'(A)$ for all $A \in N$ where $\subseteq$ is a suitable partial order on X.

Remark: 1. Each proper collage grammar $CG = (N, P, Z)$ induces an equation system EQ given by $EQ(A) = \{R' | (A, R) \in P \text{ and } R \cong R'\}$ for $A \in N$.
2. Each proper collage grammar $CG = (N, P, Z)$ induces a family $L^c(CG) : N \to X^c$ of sets of collages ($X^c = \mathcal{P}(\mathcal{C})$). Defining the extension of a family $L^c : N \to X^c$ of sets of collages by

$$L^c(Y) = \bigcup_{R \in Y} \left\{ REPL(R, repl, trans) \;\middle|\; \begin{array}{l} repl(e) \in L^c(lab_R(e)) \\ trans(e) \in TRANS \end{array} \right\},$$

$L^c(CG)$ turns out to be a fixed point of the equation system EQ induced by CG (cf. corollary 4.6).
3. CG induces a family $L^p(CG) : N \to X^p$ of sets of patterns ($X^p = \mathcal{P}(\mathcal{Q})$). Defining the extension of a family $L^p : N \to X^p$ of sets of patterns by

$$L^p(Y) = \bigcup_{R \in Y} \left\{ REPL(R^p, repl, trans) \;\middle|\; \begin{array}{l} repl(e) \in L^p(lab_R(e)) \\ trans(e) \in TRANS \end{array} \right\},$$

$L^p(CG)$ turns out to be a fixed point of the equation system EQ induced by CG (cf. corollary 4.6).
4. CG induces a family $L^{op}(CG) : N \to X^{op}$ of overlays of underlying patterns ($X^{op} = \mathcal{Q}$ with the inclusion of patterns as partial order). Defining the extension of a family $L^{op} : N \to X^{op}$ of patterns by

$$L^{op}(Y) = overlay(\bigcup_{R \in Y} \left\{ REPL(R^p, repl, trans) \;\middle|\; \begin{array}{l} repl(e) = L^{op}(lab_R(e)) \\ trans(e) \in TRANS \end{array} \right\}),$$

$L^{op}(CG)$ turns out to be a fixed point of the equation system EQ induced by CG (cf. corollary 4.6).
5. CG induces a family $L^{pder}(CG) : N \to X^{pder}$ of sets of overlays of the patterns underlying the decorated collages occurring in the infinite derivations ($X^{pder} = \mathcal{P}(\mathcal{Q})$). Defining the extension of a family $L^{pder} : N \to X^{pder}$ of sets of patterns by

$$L^{pder}(Y) = \bigcup_{R \in Y} \left\{ REPL(R^p, repl, trans) \;\middle|\; \begin{array}{l} repl(e) \in L^{pder}(lab_R(e)) \\ trans(e) \in TRANS \end{array} \right\},$$

$L^{pder}(CG)$ turns out to be a fixed point of the equation system EQ induced by CG (cf. corollary 4.8).

5.2 Fixed-Point Theorem

Let $CG = (N, P, Z)$ be a proper collage grammar, $EQ : N \to \mathcal{P}(\mathcal{C}(N))$ be the equation system induced by CG, and $L^x(CG) : N \to X^x$ for $x \in \{c, p, op\}$ be one of the families generated by CG. Then $L^x(CG)$ is the least fixed point of EQ.

Proof: By remark 5.1, the families $L^x(CG)$ with $x \in \{c, p, op\}$ are fixed points of EQ. Let now L^x be an arbitrary fixed point of EQ. Then, by induction, it can be shown that, for all $A \in N$, $W \in L^x(CG)(A)$ implies $W \in L^x(A)$. Hence, for $x \in \{c, p, op\}$, $L^x(CG)$ is the least fixed point of EQ. □

The construction of least fixed points provides an alternative method for generating collages (patterns) of the generated collage (pattern) language. This may be illustrated by the following example.

5.3 Example

Let $CG = (\{S\}, P, Z)$ be the collage grammar with the axiom and the productions as shown in Fig. 5.1.

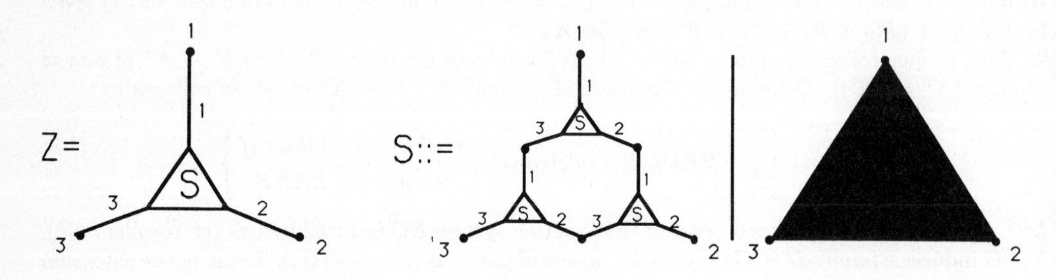

Fig. 5.1. Axiom and productions of CG

Moreover, let $SIM(2)$ (the set of all similarity transformations) be the set of admissible transformations. Then the following collages are derivable from $AX(S)$:

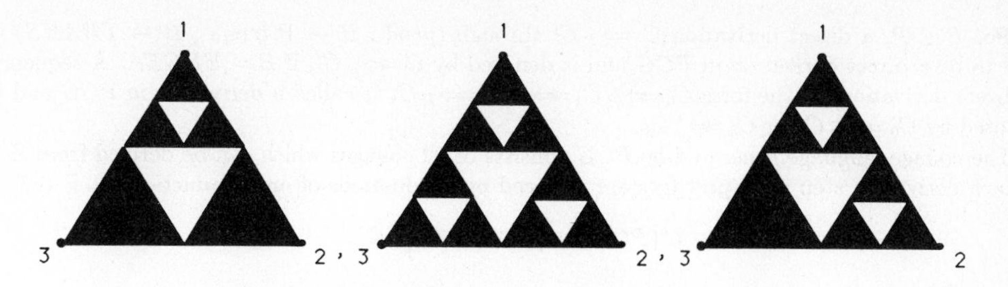

Fig. 5.2. Collages C_1, C_2, and C_3 derivable from $AX(S)$

We may use this for constructing other collages derivable from $AX(S)$. Obviously, there is a direct derivation $AX(S) \Longrightarrow R$ where R denotes the right-hand side of the first production. Moreover, there are derivations $AX(S) \Longrightarrow^* C_1$, $AX(S) \Longrightarrow^* C_2$, and $AX(S) \Longrightarrow^* C_3$ to the collages presented in Fig. 5.2. Embedding these derivations into the decorated collage R with the three hyperedges e_1, e_2, e_3, we obtain a derivation $R \Longrightarrow^* C$ where $C = REPL(R, repl, trans)$ with $repl(e_i) = C_i$ and $trans(e_i)$ is the contraction (with magnification factor $\frac{1}{2}$) combined with a suitable translation $(i = 1, 2, 3)$. Hence C can be derived from $AX(S)$ as well. It can be constructed directly from R, C_1, C_2, and C_3 and looks as follows:

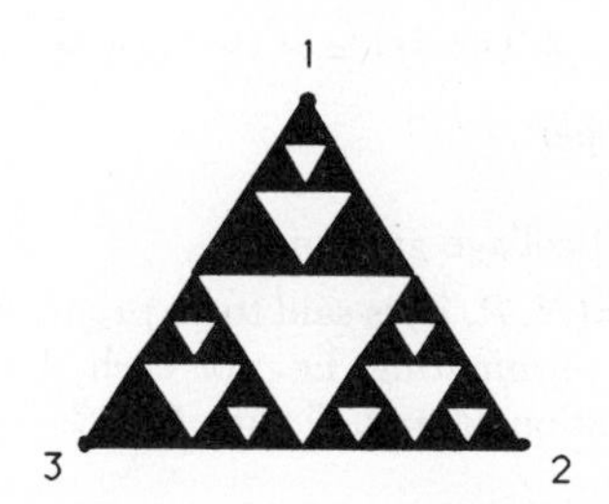

Fig. 5.3. Collage C derivable from $AX(S)$

6. Parallel Collage Grammars

The collage grammar $CG = (N, P, Z)$ considered in example 5.3 allows us to generate highly regular collages, called Sierpinski triangles. But in 5.3 it is shown, too, that there are several other collages derivable from the axiom. Although in each derivation step all hyperedges may be replaced, different productions may be used for different hyperedges. If one is interested in generating the Sierpinski triangles only, one has to use an ETOL-mode of rewriting, this means for this example:

(1) Replace all hyperedges in each step.
(2) Use only one of the productions in each step.

This is the reason for considering parallel collage grammars beside the usual ones. In parallel collage grammars, we have sets of production sets; in each derivation step we choose one set and replace all hyperedges by applying productions of this set. With respect to our example, the parallel collage grammar $PCG = (N, \{\{p_1\}, \{p_2\}\}, Z)$, where p_1 and p_2 denote the first and second production of CG, respectively, generates the set of Sierpinski triangles.

6.1 Definitions (parallel collage grammars)

1. A *parallel collage grammar* is a system $PCG = (N, \mathcal{P}, Z)$ where N is a finite set of *nonterminals*, $\mathcal{P}$ is a finite set of finite production sets over N and $Z \in \mathcal{C}(N)$ is the axiom.

2. For $P \in \mathcal{P}$, a direct derivation $C \Longrightarrow_P C'$ through $(prod : B \rightarrow P, trans : B \rightarrow TRANS)$ is said to be a *direct derivation* in PCG and is denoted by $C \mathrel{=\!\!\bullet\!\!\Rightarrow}_{\mathcal{P}} C'$, if $B = EDGE_C$. A sequence of direct derivations of the form $C_0 \mathrel{=\!\!\bullet\!\!\Rightarrow}_{\mathcal{P}} C_1 \mathrel{=\!\!\bullet\!\!\Rightarrow}_{\mathcal{P}} \ldots \mathrel{=\!\!\bullet\!\!\Rightarrow}_{\mathcal{P}} C_k$ is called a *derivation in* PCG and is denoted by $C_0 \mathrel{=\!\!\bullet\!\!\Rightarrow}^*_{\mathcal{P}} C_k$ or $C_0 \mathrel{=\!\!\bullet\!\!\Rightarrow}^* C_k$.
3. The *collage language generated by* PCG consists of all collages which can be derived from Z if in each derivation step all hyperedges are replaced by productions of one production set $P \in \mathcal{P}$:

$$L^c(PCG) = \{C \in \mathcal{C} \mid Z \mathrel{\underset{\mathcal{P}}{\overset{*}{=\!\!\bullet\!\!\Rightarrow}}} C\}.$$

Remark: 1. Parallel collage grammars employ the mode of rewriting of ETOL systems (see, e.g., Herman and Rozenberg [HR 75]).
2. Analogously to collage grammars, a parallel collage grammar not only specifies a collage language $L^c(PCG)$, but also a pattern language $L^p(PCG)$, a fractal pattern $L^{op}(PCG)$, and a set of fractal patterns $L^{pder}(PCG)$.
3. Parallel collage grammars and collage grammars are related in the following way:
(a) For each collage grammar $CG = (N, P, Z)$, the parallel collage grammar $PCG(CG) = (N, \{P\}, Z)$ generates the same language, i.e., $L^c(CG) = L^c(PCG(CG))$.
(b) For each parallel collage grammar $PCG = (N, \mathcal{P}, C)$, the *underlying* collage grammar $CG(PCG) = (N, \bigcup_{P \in \mathcal{P}} P, Z)$ generates a language including the language of PCG, i.e.,

$$L^c(PCG) \subseteq L^c(CG(PCG)).$$

In general, the inclusion is proper.

6.2 Definition (proper parallel collage grammars)

A parallel collage grammar $PCG = (N, \mathcal{P}, Z)$ is said to be *proper* if the underlying collage grammar $CG(PCG)$ is proper and PCG is *terminating*, i.e., for each $A \in N$ and each $P \in \mathcal{P}$, there is a production $(A, R) \in P$ and a derivation $R \mathrel{=\!\!\bullet\!\!\Rightarrow}^*_{\mathcal{P}} C$ with $C \in \mathcal{C}$.

Remark: 1. For each parallel collage grammar PCG, a parallel collage grammar PCG' with proper underlying collage grammar can be constructed such that $L^c(PCG') = L^c(PCG)$.
2. A parallel collage grammar $PCG = (N, \mathcal{P}, Z)$ with proper underlying collage grammar is proper provided that, for each $A \in N$ and each $P \in \mathcal{P}$, there is a production $(A, R) \in P$ and there is a set $P \in \mathcal{P}$ such that, for each $A \in N$, there is a production $(A, R) \in P$ with $R \in \mathcal{C}$.
3. Each proper parallel collage grammar $PCG = (N, \mathcal{P}, Z)$ defines a mapping $AX : N \rightarrow \mathcal{C}(N)$ with $AX(A) = (A, pin_R)^\bullet$ for some production $(A, R) \in \bigcup_{P \in \mathcal{P}} P$. This mapping is said to be the *axiom assignment induced* by PCG.
4. Each proper parallel collage grammar $PCG = (N, \mathcal{P}, Z)$ defines a family of proper parallel collage grammars $PCG(A) = (N, \mathcal{P}, AX(A))$ and families $L^x(PCG)(A) = L^x(PCG(A))$ for $x \in \{c, p, op, pder\}$ $(A \in N)$.

6.3 Lemma

Let PCG be a proper parallel collage grammar. Then PCG and $CG(PCG)$ generate the same fractal patterns, i.e.,

$$L^{op}(PCG) = L^{op}(CG(PCG)).$$

7. Generation of Self-Affine Patterns

Self-affinity and self-similarity are fundamental concepts of fractal geometry (see, e.g., Mandelbrot [Ma 82] and Barnsley [Ba 88]). For technical reasons, we extend these notions to patterns and sets of patterns.

7.1 Definition (self-affinity and self-similarity)

1. A set $POINT \subseteq \mathbb{R}^n$ of points is said to be *self-affine* if there are affine mappings $a_i : \mathbb{R}^n \to \mathbb{R}^n$ for $i = 1, \ldots, k$ with $k \geq 2$ such that

$$POINT = \bigcup_{i=1}^{k} a_i(POINT) \quad \text{and}$$

$$[a_i(POINT)] \cap [a_j(POINT)] \subseteq \delta[a_i(POINT)] \cap \delta[a_j(POINT)]$$

for all $i \neq j$. ($[POINT]$ denotes the convex hull of the set $POINT$ of points and $\delta[POINT]$ its boundary.)
2. A pattern $(POINT, pin)$ with $points(pin) \subseteq POINT \subseteq [points(pin)]$ is said to be *self-affine* if $POINT$ is self-affine with respect to the affine mappings a_i for $i = 1, \ldots, k$, and in addition,

$$points(pin) \subseteq \bigcup_{i=1}^{k} a_i(points(pin)).$$

3. A set L of patterns (with common sequence pin_L of pin points) is *self-affine* if the pattern $overlay(L)$ is self-affine.
4. A self-affine set of points, (pattern, set of patterns) is *self-similar* if each involved affine mapping is a similarity mapping.

In the following, we show that a certain type of parallel collage grammars, so called non-overlapping Sierpinski grammars, generate self-affine languages. Koch systems as introduced by Prusinkiewicz and Sandness [PS 88] are special cases of this type of grammars. According to lemma 6.3, a parallel collage grammar and its underlying collage grammar yield the same overlay patterns of the generated languages. This lemma and the Fixed-Point Theorem are the keys to the result on self-affinity. This means that the result could be formulated without reference to parallel collage grammars. Nevertheless, we introduced the parallel mode of rewriting, because it seems to be closer related to the usual considerations in fractal geometry than the sequential mode of rewriting.

7.2 Definition (Sierpinski grammars)

1. A parallel collage grammar $PCG = (N, \mathcal{P}, Z)$ is of *Sierpinski type* or a *Sierpinski grammar* if $N = \{S\}$ for some symbol S, $Z = (S, pin)^\bullet$ for some sequence $pin \in (\mathbb{R}^n)^*$, and $\mathcal{P} = \{\{(S, R)\}, \{(S, pin^\bullet)\}\}$, and $R \in \mathcal{C}(N)$ is a decorated collage subject to the following conditions
(1) $EDGE_R$ contains more than one hyperedge,
(2) $PART_R = \{\{point\} \mid point \in \bigcup_{e \in EDGE_R} points(att_R(e))\}$,
(3) $pin_R = pin$ and $points(pin) \subseteq \bigcup_{e \in EDGE_R} points(att_R(e)) \subseteq [points(pin)]$,
(4) there is uniquely one mapping $a : EDGE_R \to AFF(n)$ assigning an affine mapping $a(e)$ with $a(e)(pin_R) = att_R(e)$ to each hyperedge $e \in EDGE_R$.
2. A Sierpinski grammar $PCG = (N, \mathcal{P}, Z)$ is said to be *non-overlapping* if

$$[points(att_R(e))] \cap [points(att_R(e'))] \subseteq \delta[points(att_R(e))] \cap \delta[points(att_R(e'))]$$

for all $e, e' \in EDGE_R$ with $e \neq e'$.

Remark: Each Sierpinski grammar is proper.

7.3 Example

The parallel collage grammar $PCG = (\{S\}, \{\{p_1\}, \{p_2\}\}, Z)$ with the axiom Z and the productions p_1 and p_2 as shown in Fig. 7.1 is a non-overlapping Sierpinski grammar.

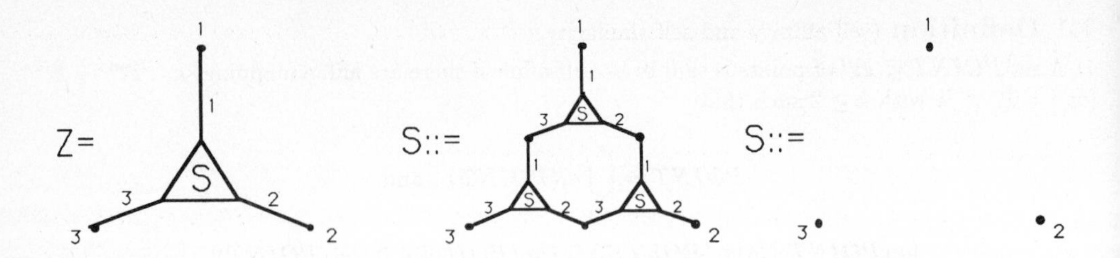

Fig. 7.1. Axiom and productions of a non-overlapping Sierpinski grammar

7.4 Theorem (non-overlapping Sierpinski grammars generate self-affine pattern languages)
Let PCG be a non-overlapping Sierpinski grammar and $AFF(n)$ be the set of affine transformations of $\mathbb{R}^n$. Then the pattern language $L^p(PCG)$ (with respect to $AFF(n)$) is self-affine.

To a certain extent, the converse result holds, too. If a pattern is self-affine, then we can construct a Sierpinski grammar generating a self-affine subpattern of the given one.

7.5 Construction (of a Sierpinski grammar)
Let $(POINT, pin)$ be a self-affine pattern with respect to the affine mappings a_i for $i = 1, \ldots, k$. Then the *induced Sierpinski grammar* $PCG(POINT, pin)$ is given by $PCG(POINT, pin) = (\{S\}, \{\{(S, R)\}, \{(S, pin^\bullet)\}\}, (S, pin)^\bullet)$ where R is the decorated collage

$$R = (\left\{ \{point\} \,\middle|\, point \in \bigcup_{i=1}^{k} a_i(points(pin)) \right\}, \emptyset, \{e_1, \ldots, e_k\}, att_R, lab_R, pin)$$

with $att_R(e_i) = a_i(pin)$ and $lab_R(e_i) = S$ for $i = 1, \ldots, k$.

Remark: The induced Sierpinski grammar $PCG(POINT, pin)$ is non-overlapping.

7.6 Theorem (generation of a self-affine subpattern)
Let $(POINT, pin)$ be a self-affine pattern and $PCG(POINT, pin)$ be the induced Sierpinski grammar. Then $L^p(PCG(POINT, pin))$ is self-affine and $L^{op}(PCG(POINT, pin)) \subseteq (POINT, pin)$.

8. Generation of Mutually Affine Patterns

In the previous section, we have shown that the Fixed-Point Theorem can be used to establish the self-affinity of certain generated patterns. The more general situation of the Fixed-Point Theorem motivates a generalization of the notion of self-affine sets to so-called mutually affine sets.

8.1 Definition (mutually affine sets)

1. The sets $POINT_1, \ldots, POINT_m \subseteq \mathbb{R}^n$ for some $m \geq 1$ are *mutually affine* if, for each $l \in \{1, \ldots, m\}$, there are a set $POINT_{0l}$ and affine mappings $a^l_{11}, \ldots, a^l_{k(1,l)1}, \ldots, a^l_{1m}, \ldots, a^l_{k(m,l)m}$ such that

$$POINT_l = POINT_{0l} \cup \bigcup_{j=1}^{m} (\bigcup_{i=1}^{k(j,l)} a^l_{ij}(POINT_j)).$$

2. The patterns $PATTERN_1, \ldots, PATTERN_m$ in $\mathbb{R}^n$ with $PATTERN_l = (POINT_l, pin_l)$ for $l = 1, \ldots, m$ are *mutually affine* if $POINT_1, \ldots, POINT_m$ are mutually affine.

3. For $l = 1, \ldots, m$, let L_i be a set of patterns (with common sequence pin_{L_i} of distinguished points). Then $L_1, \ldots, L_m$ are *mutually affine* if the patterns $overlay(L_1), \ldots, overlay(L_m)$ are mutually affine.

Remark: Every self-affine pattern is mutually affine.

8.2 Theorem (proper parallel collage grammars generate mutually affine pattern languages)
Let $PCG = (N, \mathcal{P}, Z)$ be a proper parallel collage grammar and $N = \{A_1, \ldots, A_m\}$. Then the pattern languages $L^p(PCG)(A_1), \ldots, L^p(PCG)(A_m)$ are mutually affine.

Remark: We get a corresponding result for proper collage grammars because of Lemma 6.3: Let $CG = (N, P, Z)$ be a collage grammar and $N = \{A_1, \ldots, A_m\}$. Then the pattern languages $L^p(CG)(A_1), \ldots, L^p(CG)(A_m)$ are mutually affine.

9. Conclusion

This paper presents the first steps towards a systematic theory of collage grammars as syntactic devices for the generation of (classes of) patterns. The further study should include the following topics:
(1) In [HK 88], we discussed so-called figure grammars that may be seen as special cases of collage grammars where all involved parts are lines whose end points belong to a distinguished point component. Obviously, arbitrary parts are more flexible than lines concerning the composition of patterns. But how can the difference be made more precise?
(2) The notion of a shape grammar (see, e.g., Stiny [St 75], Gips [Gi 75], and Lauzzana and Pocock-Williams [LP 88]) provides another well-known way to describe patterns syntactically. How are they related to collage grammars?
(3) Maurer, Rozenberg, and Welzl [MRW 82] (see also Dassow, Brandenburg, and Chytil [Da 89, BC 91]) introduce the concept of chain grammars that generate sets of strings which can be interpreted as line drawings. In a similar, more sophisticated way, Prusinkiewicz, Hanan, and Fracchia [Pr 86, PH 89, FP 91] and Lück and Lück [LL 91] use L-systems. How are these approaches related to collage grammars?
(4) Clearly, it would be interesting to learn how collage grammars are related to other pattern-generation approaches based on graph grammars (see, e.g., de Does and Lindenmayer [DL 83], Bunke [Bu 83], and de Boer [Bo 91]).
(5) We have suggested to define fractal patterns by infinite derivations as well as by generated languages where the first way seems to be more flexible and less redundant. But the relationship between fractal geometry and collage grammars studied in chapters 7 and 8 concerns the patterns described by generated languages. What about infinite derivations in this respect?
(6) We obtain fractal patterns from an infinite set or sequence of regularly shaped patterns by overlay. The relationship between collage grammars and fractal geometry discussed in section 7 is based on this rather clumsy kind of limit constructions. We hope that there are more sophisticated limits that lead to better results.
(7) How may fractal patterns or languages of patterns look that cannot be generated by context-free collage grammars? Which results besides the Fixed-Point Theorem reveal the structure of generated collage languages?
(8) Collage grammars specify (classes of) patterns. Is it possible in addition to deduce certain properties of the generated objects inspecting the grammars? For example, under which conditions is it decidable for collage grammars whether the generated patterns are space-filling or consist of infinitely many disconnected areas or have infinitely many holes?
(9) The considerations in the chapters 2 to 6 do not depend essentially on the assumed Euclidean space $\mathbb{R}^n$, but could be done for any set of transformations on an arbitrary, but fixed space. Are

there interesting examples for other spaces than the $\mathbb{R}^n$? In particular, it may be necessary to replace $\mathbb{R}^n$ by the set of pixels if it comes to the implementation of the techniques introduced in this paper.

We think that a grammatical approach to pattern generation provides a lot of fascinating questions. Hopefully, the readers will share our view.

Acknowledgment

We are very grateful to Hartmut Ehrig, Joost Engelfriet, Raymond Lauzzana, and Frieder Nake whose valuable suggestions and comments led to various improvements.

10. References

[Ba 88] M. Barnsley: Fractals Everywhere, Academic Press, Boston 1988

[Br 91] F. Brandenburg: Layout Graph Grammars, to appear in [EKR 91]

[BC 91] F. Brandenburg, M. Chytil: Syntax-Directed Translation of Picture Words, to appear in [EKR 91]

[Bo 91] M. de Boer: Modelling and Simulation of the Development of Cellular Layers with Map Lindenmayer Systems, to appear in [EKR 91]

[Bu 83] H. Bunke: Graph Grammars as a Generative Tool in Image Understanding, Lect. Not. Comp. Sci. 153, 8–19, 1983

[Ca 89] D. Caucal: Pattern Graphs, Techn. Report, Université de Rennes 1, 1989

[Co 89] H.S.M. Coxeter: Introduction to Geometry, Second Edition, Wiley Classics Library Edition, John Wiley & Sons, New York 1989

[Da 89] J. Dassow: Graph-Thoeretic Properties and Chain Code Picture Languages, J. Inf. Process. Cybern. EIK 25, 8/9, 423-433, 1989

[DL 83] M. de Does, A. Lindenmayer: Algorithms for the Generation and Drawing of Maps Representing Cell Clones, Lect. Not. Comp. Sci. 153, 39–58, 1983

[EKR 91] H. Ehrig, H.-J. Kreowski, G. Rozenberg (Eds.): Graph-Grammars and Their Application to Computer Science, to appear in Lect. Not. Comp. Sci.

[Fe 68] J. Feder: Languages of Encoded Line Patterns, Inform. Contr. 13, 230–244, 1968

[Fe 71] J. Feder: Plex Languages, Inform. Sci. 3, 225–241, 1971

[FP 91] D. Fracchia, P. Prusinkiewicz: Physically-Based Graphical Interpretation of Map L-Systems and Cellworks, to appear in [EKR 91]

[Fu 74] K.S. Fu: Syntactic Methods in Pattern Recognition, Academic Press, New York 1974

[Gi 75] J. Gips: Shape Grammars and Their Uses, Artifical Perception, Shape Generation and Computer Aesthetics, Birkhäuser Verlag, Basel/Stuttgart 1975

[GT 78] R.C. Gonzalez, M.G. Thomason: Syntactic Pattern Recognition, Addison-Wesley, Reading, Mass., 1978

[Ha 89] A. Habel: Hyperedge-Replacement: Grammars and Languages, Ph. D. Thesis, Bremen 1989, to appear in Lect. Not. Comp. Sci.

[HK 88] A. Habel, H.-J. Kreowski: Pretty Patterns Produced by Hyperedge Replacement, Lect. Not. Comp. Sci. 314, 32–45, 1988

[HK 90] A. Habel, H.-J. Kreowski: Collages and Patterns Generated by Hyperedge Replacement, Techn. Report No. 15/90, Univ. Bremen, Dept. of Computer Science, 1990

[HR 75] G.T. Herman, G. Rozenberg: Development Systems and Languages, North Holland/American Elsevier, New York 1975

[LP 88] R.G. Lauzzana, L. Pocock-Williams: A Rule System for Analysis in the Visual Arts, Leonardo, Vol. 21, No. 4, 445-452, 1988

[LL 91] J. Lück, H.B. Lück: 3D Cellular Networks with Double-Walls, Representative of the Development of Plant Tissues, to appear in [EKR 91]

[Ma 82] B.B. Mandelbrot: The Fractal Geometry of Nature, W.H. Freeman and Company, New York 1982

[MRW 82] H.A. Maurer, G. Rozenberg, E. Welzl: Using String Languages to Describe Picture Languages, Information and Control 54, 155–185, 1982

[PR 69] J.L. Pfaltz, A. Rosenfeld: Web Grammars, Proc. Int. Joint Conf. Art. Intelligence, 609–619, 1969

[Pr 86] P. Prusinkiewicz: Graphical Applications of L-Systems, Proc. of Graphics Interface '86 — Vision Interface '86, 247–253, 1986

[PH 89] P. Prusinkiewicz, J. Hanan: Lindenmayer Systems, Fractals, and Plants, Lecture Notes in Biomathematics 79, Springer, New York 1989

[PS 88] P. Prusinkiewicz, G Sandness: Koch Curves as Attractors and Repellers, IEEE Computer Graphics and Applications 8, No. 6, 26–40, 1988

[Sh 69] A.C. Shaw: A Formal Description Schema as a Basis for Picture Processing Systems, Inf. Contr. 14, 9–52, 1969

[St 75] G. Stiny: Pictorial and Formal Aspects of Shape and Shape Grammars, Birkhäuser Verlag, Basel/Stuttgart 1975

The four musicians: analogies and expert systems - a graphic approach

L. Hess[1], IME, Rio de Janiero, Brasil
B. Mayoh, Computer Science Department, Aarhus University, Aarhus, Denmark

In their paper "Graph rewriting with unification and composition" in the last GraGra conference (PEM) Parisi-Presicce, Ehrig and Montanari suggested that graph grammars might be useful in rule based expert systems. The idea is that graphs capture the relationships between facts, while graph productions capture rules for deriving new facts. In this paper we develop this idea using "graphics" (HM) instead of the usual arc and node labelled graphs. Graphics have the advantage of incorporating variables directly (pointed out to one of the authors by Ehrig) but they seem to have the apparent disadvantage that arcs are neither directed nor labelled.

Section 1 describes how graphics can capture the information in the labels on directed arcs, so familiar from the semantic nets, conceptual schemes and other knowledge representations in data bases and expert systems. Section 2 describes how graphic productions can capture "rule" information: in traditional IF-THEN rules, in Prolog rules with assert and retract, and in type reasoning in inheritance hierarchies. Section 3 shows how graphic productions can also capture reasoning by analogy, not just logical reasoning. Section 4 gives the general definition of graphic morphism in such a way that category of graphics is free from the pushout problems that plague both (PEM) and (HM). The final section collects the arguments for the controversial messages of this paper:

(1) for accurate modelling of knowledge one should formalise in categories that have all pushouts,
(2) one should consider more subtle notions of *occurrence* than the usual "morphism from the left side of a rule",
(3) one should consider notions of productions that can capture reasoning by analogy,
(4) graphics can represent knowledge accurately.

#1 Graphics & directed arc labels

A graphic is a graph where all vertices get elements of a Σ-algebra as extra labels. More precisely:

Definition 1 A **graphic** G over a Σ-algebra $\mathcal{A}$ and set L consists of 4 functions

[1] Conference participation financed by faberj and IBM Brasil.

s,t: Edges => Vertices
lab: Vertices => L
atr: Vertices => A

L A
lab atr
Vertices
s t
Edges

As an example of a graphic consider figure 1:

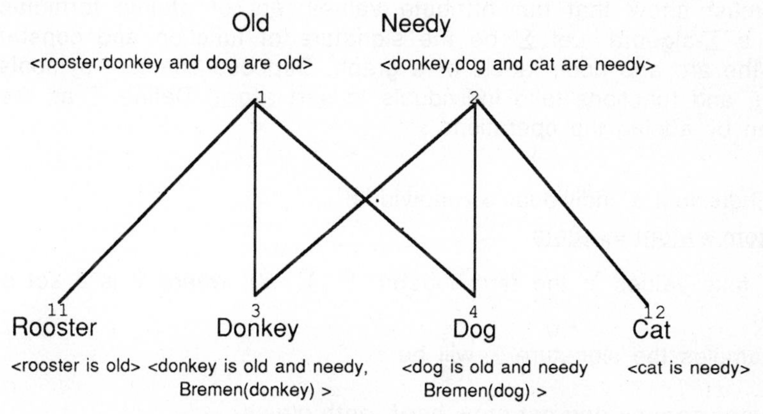

fig. 1 Graphic D1 over SIGMA

This represents a small database with a binary relation "is" and a unary relation "Bremen". In the style of (PEM) this database would be represented by the graph

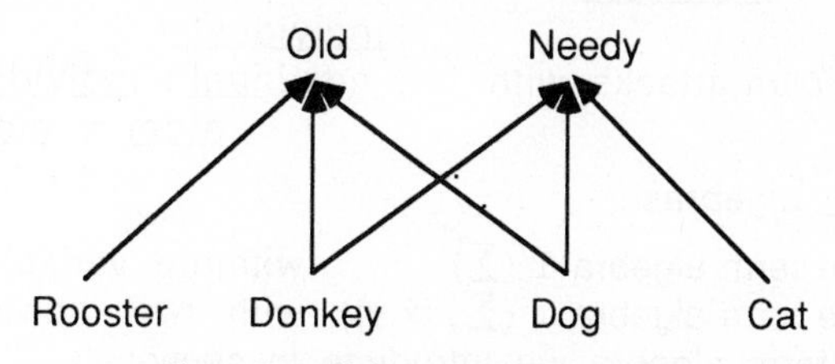

fig. 2 SC-graph for D1

where the arcs should be labelled "is" and there should be two loops labelled "Bremen".

This example illustrates our general method of converting label information on directed arcs to atomic formulas in attribute values:

— each arc label becomes a predicate symbol
— each node label becomes a constant symbol
— each arc becomes an atomic fact
— each atomic fact is attached to the nodes at the end of the corresponding arc as part of their attribute value.

For the sake of readability we use infix notation for binary predicates and obvious linguistic conventions,so

<is(rooster,old),is(donkey,old),is(dog,old)>

becomes < rooster,donkey and dog are old>. To make our later definition of graphic morphisms absolutely clear, we also number the vertices in our representations of graphics.

Remark As there is no reason why arc labels should be binary predicate symbols,our conversion method also works for directed hypergraphs. Some data base systems (Ge,Ul) use such hypergraphs for "representing conceptual knowledge".

We must show that our attribute values,sets of atomic formulas, are elements of a Σ-algebra. Let Σ' be the signature of function and constant symbols given by the arc and node labels of a graph. Suppose constant symbols have sort individual, and functions take individuals to sort atom. Define Σ as the extension of Σ' given by adding the operations

, : individual x individual => individual

• : atom x atom => atom.

Attributes take values in the term algebra $T(\Sigma, V)$ where V is a set of individual variables.

In our examples the signature Σ will be :

rooster,donkey,dog,cat,crow,hoof,teeth,claws,
judge,monster,assassin,witch,judgement,club,knife,nails,
old,needy,worried,creature,domestic,cottage : individual
, : individual x individual => individual
≈ : individual => individual
Bremen : individual => atom
is,uses ,Musician,attacks_with : individual x individual => atom
• : atom x atom => atom

We will use three Σ-algebras:

SIGMA	the term algebra $T(\Sigma)$	with no variables
SIGVAR	the term algebra $T(\Sigma, \{x,y\})$	with two individual variables
ROBBER	a term algebra we introduce in section 3.	

Graphics over these algebras will have the terms of sort individual as their label set L.

The kinds of graphs used in (PEM) for representing relationships between facts are the SC-graphs where one has preordered sets,CA and CN, and functions

s,t: Arcs => Nodes
arc_colour : Arcs => CA
node_colour: Nodes => CN

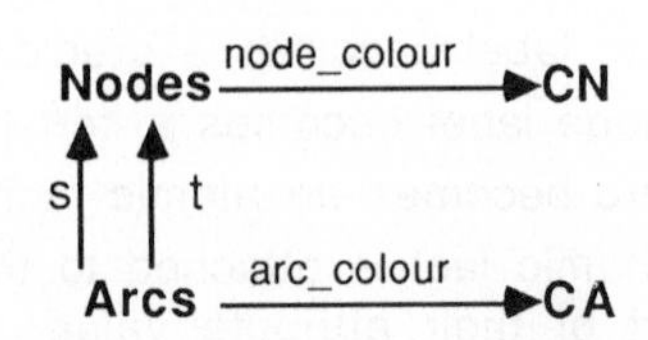

Any SC-graph can be converted into a graphic. One can take CN as L and define Σ as: an individual constant for each graph node, a function symbol Ca: Node*Node-> atom for each arc colour in CA, and a conjunction operator "$\cdot$". As the Σ -algebra A one can take the term algebra $T(\Sigma)$. For each node n in the SC-graph, the attribute value is the formula set:

Ca(m,n) for each arc from m to n with colour Ca

Ca'(n,m) for each arc from n to m with colour Ca'

and its label is its nodecolour. In the same way that we added "$\cdot$" earlier, we can convert the preorders in CA and CN into equations

Cn = Cn,Cn'	for Cn $\leq$ Cn'
Ca(m,n) =Ca(m,n).Ca'(m,n)	for Ca $\leq$ Ca'.

Remark We use "$\leq$" for both preorders (reflexive and transitive relations) and x~y for the "interchangeability" relation: $x \leq y$ and $y \leq x$. When the preorders are trivial ($x \leq y$ iff x=y), we get the usual labelled graphs. If CN has only one element and CA has the trivial preorder, then we have the much studied labelled transition systems. When the preorders are flat ($x \leq y$ iff x=y or y=top)we get the partially coloured graphs with top as a new colour for "unknown","absent" or "transparent". The sets CA and CN can be lattices, unified algebras(Mo), or Boolean algebras.

In her work on type inheritance hierarchies (Pa), Lin Padgham has introduced an interesting kind of graph in which node labels have a lattice structure. These graphs, which she uses to clarify the traditional Clyde, Tweety and Nixon examples, can also be converted to graphics. Her graphs capture the distinctions made in earlier type inheritance hierarchies; in particular they capture Etherington's distinction in (Et) between "strict is a", "default is a", "strict is not a", "default is not a", and "exception" arcs. In (Pa) nodes are labelled by "sets of characteristics" and there are several kinds of nodes. There are two kinds of arcs: directed arcs for $\leq$ in the label lattice (usually inclusion), and undirected arcs for "inconsistent labels". As an example consider

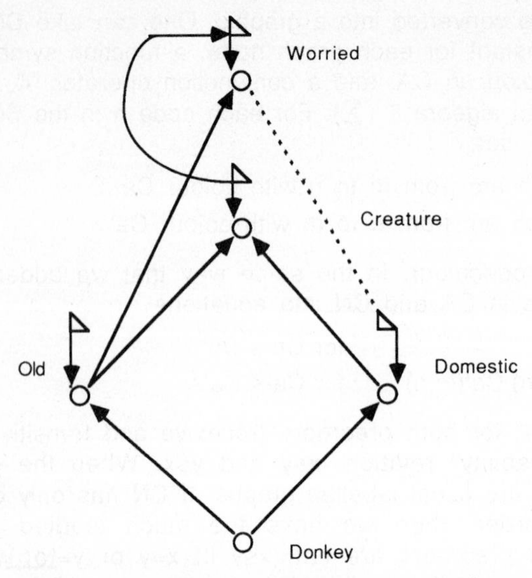

fig. 3 Type inheritance graph

where circles represent "core nodes" and triangles represent "default nodes". Here the undirected arc represents

'Typical domestic creatures never worry'

and the directed arcs represent such beliefs as

' Old creatures always worry'

' Creatures often worry'

There are two ways of converting type inheritance hierarchies into graphics. One can introduce new binary function symbols in Σ so our example becomes the graphic

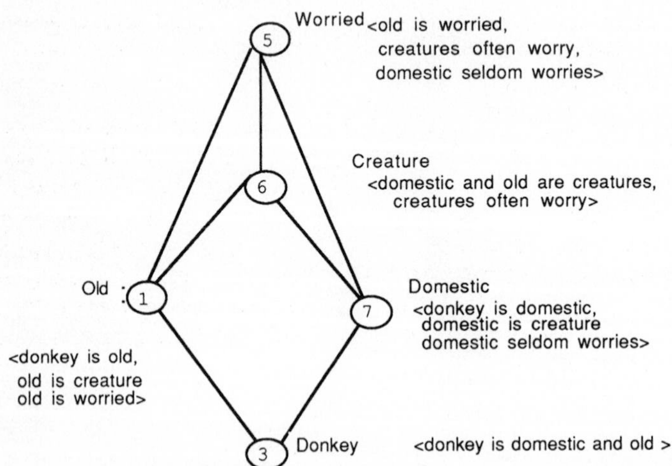

fig. 4 Graphic D2 over an extension of SIGMA

Alternatively one can introduce a unary function symbol '≈' so our example becomes the graphic

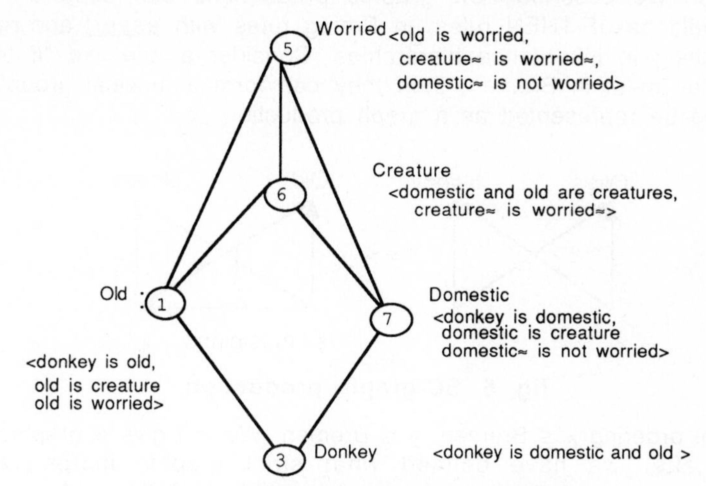

fig. 5 Graphic D3 over SIGMA

In the next section we show how graphic productions can capture type inheritance reasoning.

From the logical programming viewpoint what we have done in this section is very natural. A graphic can be built on any collection C of facts in a logical programming language such as PROLOG. For each constant or variable cv, there is a vertex in G with label cv. The attribute value of the vertex cv is the set of facts in C that mention cv. There is an edge between cv1 and cv2, whenever there is some fact in C that mentions both cv1 and cv2. For an example the reader can take any graphic in this section as G, and the union of the attribute values at its vertices as C. Note also that one can have vertices in a graphic for Prolog predicate symbols. The attribute value for such a vertex is the atomic formulas using the corresponding predicate.We shall often use this option,our decision that Bremen' should be a predicate and 'old' an individual was arbitrary.

Comment: an alternative approach (CM) to the conversion of logic programs to graph grammars should be mentioned. Given a collection C of atomic facts in a logical programming language such as PROLOG, one can introduce (1) vertices for constants and variables as above (2) labelled hyperedges for each atomic fact in C. In this approach the "clauses" in a logic program become context-free hypergraph productions. The methods in this section convert the hypergraph for C into the graphic for C, but the underlying categories for hypergraphs and graphics differ. The category for graphics (based on indexed categories) has pushouts, but not pushout complements.

#2 Productions and rules

In this section we describe how graphic productions can capture "rule" information: in traditional IF-THEN rules, in Prolog rules with assert and retract , and in type reasoning in inheritance hierarchies. Consider a rule like "if two old and needy creatures meet in Bremen, then they can form a musical group". In (PEM) this rule would be represented as a graph production

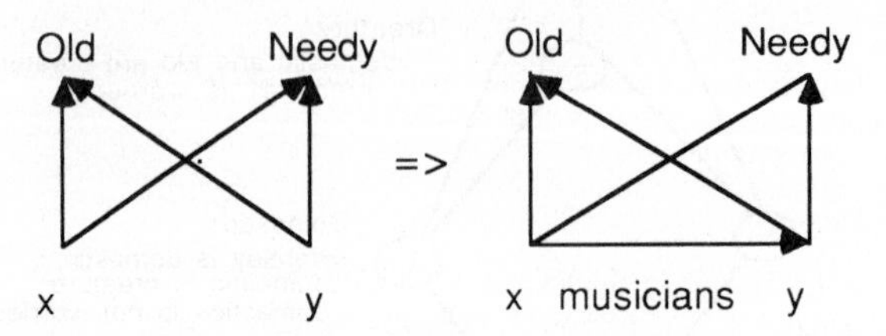

fig. 6 SC-graph production

and the partial ordering x ≤ Bremen, y ≤ Bremen. We will give a graphic production for this rule,after we have defined what strict graphic morphisms are. All morphisms of graphics have three parts: a morphism of the underlying graphs,a function on labels, and a Σ-algebra homomorphism.

Definition 2 A *strict morphism* from a graphic (s,t,lab,atr,L,A) to a graphic (s',t',lab',atr',L',A') consists of a graph morphism (ve,ed) and

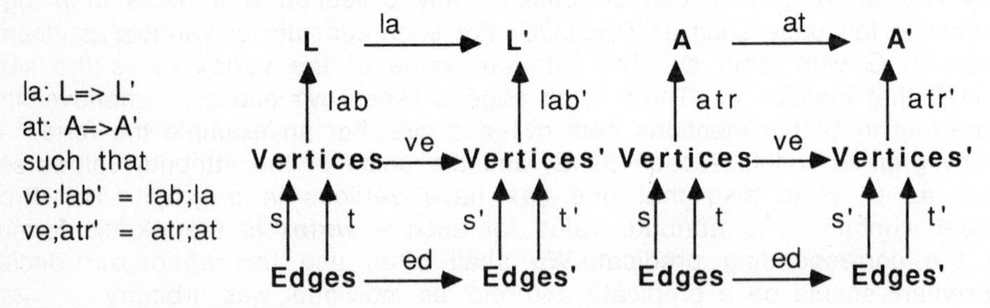

where at is a Σ-algebra homomorphism. A *matching* is a strict morphism that has an injective vertex function.

Strict graphic morphisms suffice for all our examples, but technical problems force us to define a more general notion of graphic morphism in the last section. Let us show that there is a strict morphism from the graphic

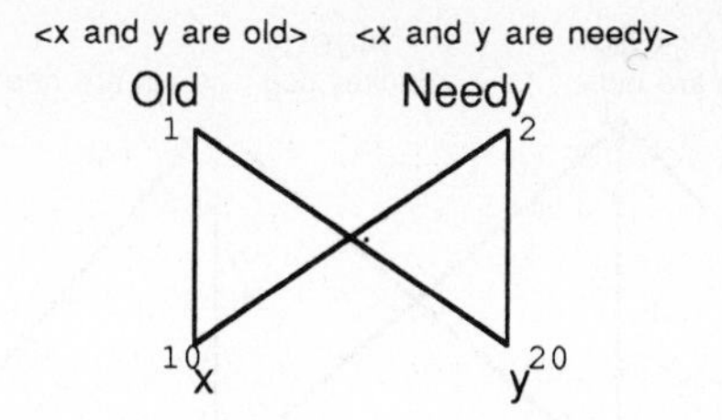

fig. 7 Graphic L1 over SIGVAR

to the graphic R1

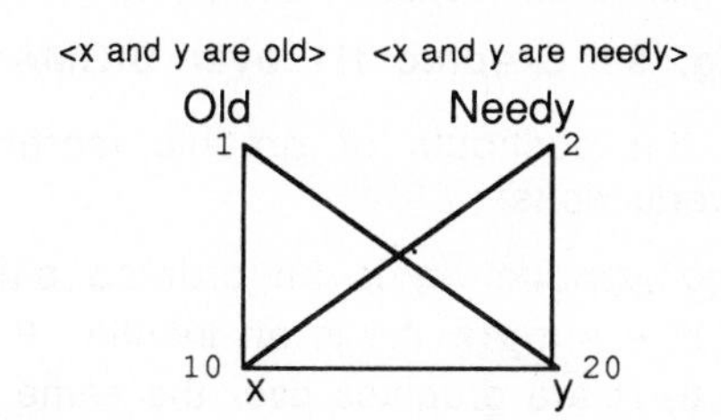

fig. 8 Graphic R1 over SIGVAR

The obvious embedding as labelled graphs and the definition:

at(S) = if Bremen(x) or Bremen(y) in S then S U{Musician(x,y)} else S

give the required graphic morphism r1 from L1 to R1. This graphic morphism is a "matching" because its vertex function is the identity function on {1,2,10,20}. Graphic morphisms can do many things: (1) add or identify vertices, (2) add or identify edges, (3) change vertex labels, (4) change attribute values. Our morphism r1: L1 => R1 illustrates only (2) and (4) and it is monotonic in that: it is an embedding of labelled graphs and at(S) always contains S.

There is an occurrence of L1 in our earlier graphic D1; if one substitutes 'donkey' for x and 'dog' for y, one gets a graphic morphism d: L1 => D1. The label part of the morphism is given by :

la(cv) = if cv is x then donkey else if cv is y then dog else cv

and this substitution gives the attribute part :

at(S) = the result of applying substitution la to S.

This graphic morphism is a "matching" because its vertex function

ve(1) = 1 ve(2)=2 ve(10) = 3 ve(20) = 4 is an injection.

The pushout of this graphic morphism d with the 'rule' morphism r1: L1 => R1 gives the graphic H1

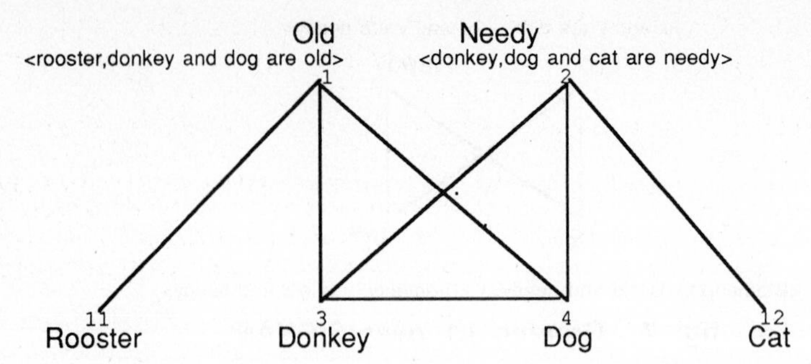

fig. 9 Graphic H1 over SIGMA

Later we will show that the pushouts of graphic morphisms always exist, but now we define graphic productions.

Definition 3 A **graphic production** is an ordered pair of graphic morphisms l: K => L and r: K => R. It is **simple** if l is an identity. It is **partial** if l is an embedding. It is **logical** if K, L, R are graphics over the same Σ-algebra.

The graphic production (l,r) can be applied when one has a graphic morphism d: K => D, and the pushout diagram

$$\begin{array}{ccccc} L & \overset{l}{\Leftarrow} & K & \overset{r}{\Rightarrow} & R \\ g\Downarrow & po & \Downarrow & po & \Downarrow h \\ G & \underset{g'}{\Leftarrow} & D & \underset{h'}{\Rightarrow} & H \end{array} \qquad \begin{array}{ccccc} L & \overset{l}{\Leftarrow} & K & \overset{r}{\Rightarrow} & R \\ & & \Downarrow & & \\ G & \underset{g'}{\Leftarrow} & D & \underset{h'}{\Rightarrow} & H \end{array}$$

fig.10 (normal) Application of a (logical) production

shows how the production transforms G to the graphic H. This double pushout definition of the result of applying a production is pleasantly general, but it leads to problems in practical implementations. If occurrence morphisms are matchings and/or productions are simple or partial, these problems are less severe (more on this in the section 4). Sometimes the application of graphic productions can give *surprises* , and it might be wise for an expert system to keep to normal applications.

Definition 4 The application of a logical graphic production (l,r) is **normal** if the label and attribute parts of the pushout morphisms

g: L=>G, h: R=>H are the same as those in d: K=>D.

Our example showed a normal application of the simple graphic production (id,r1: L1=>R1), but more general graphic productions are also useful in expert systems; to quote (PEM):

"This allows for a uniform treatment of 'forward chaining' systems, where changes in the working memory data produce a match for the left hand side of a rule which can then be applied, and 'backwards chaining' systems where rules are examined in search for a match with a fixed goal".

We should note the 'applicability' problem with graphic productions: to determine if a production (l:K=>L,r:K=>R) can be used to transform a graphic G, one needs the morphism d:K=>D - it is not enough to find a morphism g:L=>G. There may be zero, one or many morphisms d, whose pushout with l is g. Nevertheless for all simple productions and most productions, that arise in practice, there is a unique d for any g.

Our example of a graphic production is close both to the corresponding Prolog rule:

Musicians(x,y) :- Bremen(x),Old(x),Needy(x),Bremen(y),Old(y),Needy(y).

and to the running example of a rule in (PEM):

"if two women have the same mother, then they are sisters".

From the discussion at the end of the last section, it is clear that any 'pure' Prolog rule gives a simple graphic production. If the rule also uses assert, the corresponding graphic production is still simple because assert can only add new edges and extend attribute values. Only when a rule uses retract do we have to go beyond simple productions and let the graphic morphism l:K=>L remove edges and reduce attribute values. Note however that assert and retract have no influence on the unifier when a Prolog rule is applied. This corresponds to the fact that the substitution map is determined by the occurrence map d:K=>D when a graphic production is applied.

For another example of (PEM)'s "rules containing reasoning knowledge.represented by graph productions" let us look at type inheritance reasoning. Previous techniques for type reasoning (Et, Sa, To) are all captured in (Pa) where Padgham describes her version of type reasoning in a way that we can translate into graphic productions. One of these productions is

(INH) x __________ y => x ________________ y

< x is y> <x is y,P(y)> <x is y,P(x)> <x is y,P(y)>

Using this repeatedly on the graphic D3 in figure 5 gives

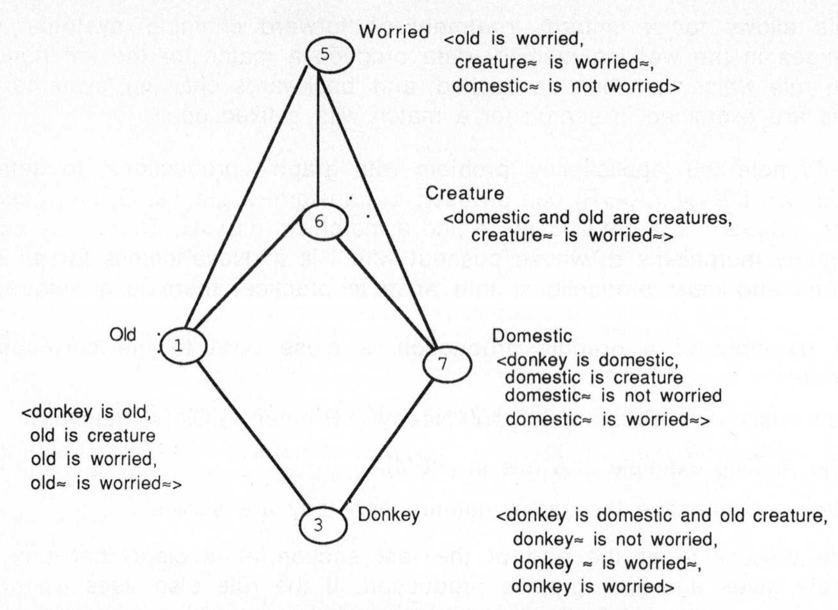
Worried
<old is worried,
creature≈ is worried≈,
domestic≈ is not worried>
5
Creature
<domestic and old are creatures,
creature≈ is worried≈>
6
Old
1
Domestic
<donkey is domestic,
domestic is creature
domestic≈ is not worried
domestic≈ is worried≈>
7
<donkey is old,
old is creature
old is worried,
old≈ is worried≈>
3
Donkey
<donkey is domestic and old creature,
donkey≈ is not worried,
donkey ≈ is worried≈,
donkey is worried>

fig. 11 Graphic H2 over SIGMA

Clearly we need productions like

(DEL) x__________________y => x __________ y
<x is y,x≈ is not y> <x.is y> <x is y> <x is y>

to remove 'default' information when it conflicts with 'core' information. Using this production we can delete the unwanted 'donkey≈ is not worried' from the attribute of 'Donkey' in the graphic H2. So far we have not seen graphic productions that create new vertices, so we give

y => x __________ y
<> <x is y> <x is y>

Using this production one can add old roosters, cats and dogs to the graphic H2. For any symbol in the signature $\sum$ we can introduce a production for introducing facts using the symbol into graphics.

#3 Analogies

There is a large literature on reasoning by analogy (Pr), and some of it is concerned with whether an analogy is a map of a situation G into a situation H or whether G and H must have a common structure or pattern D. This dispute is related to whether analogies should be modelled by simple graphic productions or whether we need productions that are not simple.

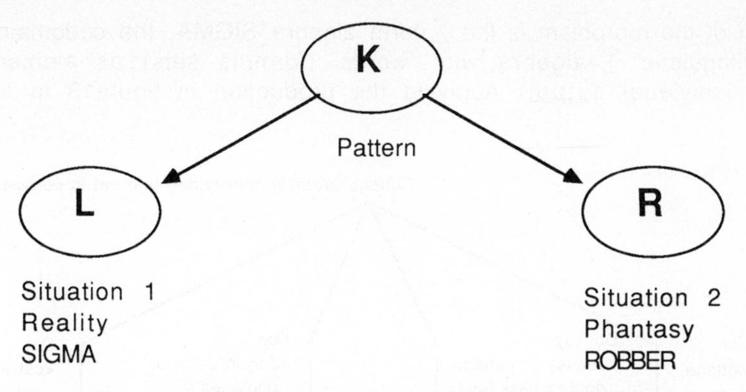

fig. 12 Analogy as a graphic production

Frequently situations can be described by graphics and a map from situation G into situation H can be described by a graphic morphism. Many natural analogies are pushouts of simple graphic productions.

Definition 5 A graphic production (l:K=>L, r:K=>R) is analogical if L and R are over different Σ-algebras.

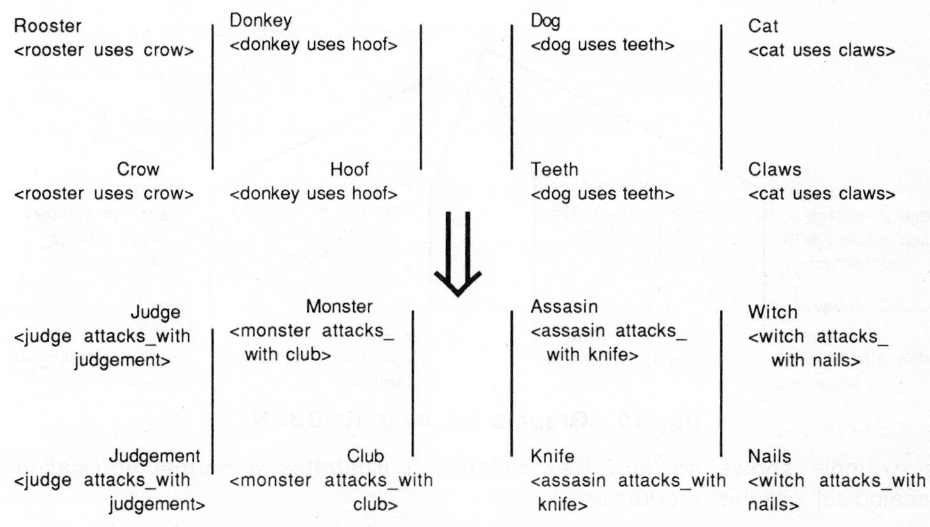

fig. 13 Analogical graphic production from L4 to R4

In this example the signature Σ is that given in section 1, the graph part of the morphism is trivial and the other parts are given by the Σ-algebra homomorphism at: SIGMA => ROBBER

rooster	->	judge	crow	->	judgement
donkey	->	monster	hoof	->	club
dog	->	assassin	teeth	->	knife
cat	->	witch	claws	->	nails
uses	->	attacks_with.			

The domain of the morphism is the Σ-term algebra SIGMA, the codomain ROBBER is also a 'linguistic' Σ-algebra with 'words' (formula sets) as elements of the carrier for 'individual' (atom) Applying the production in figure13 to the graphic D4

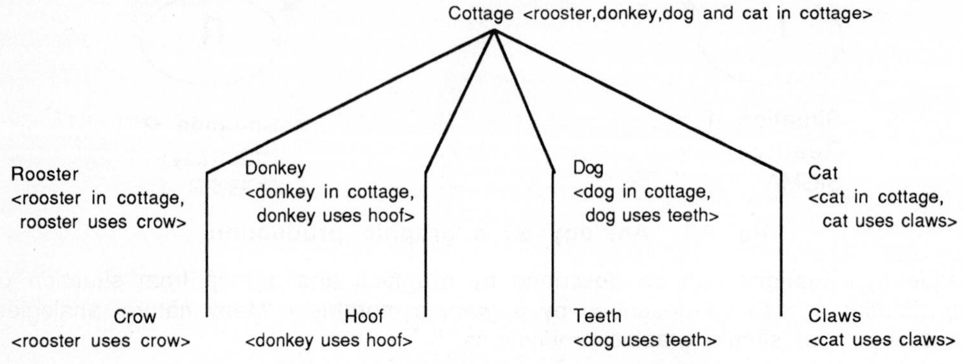

fig. 14 Graphic D4 over SIGMA

gives the graphic H4.

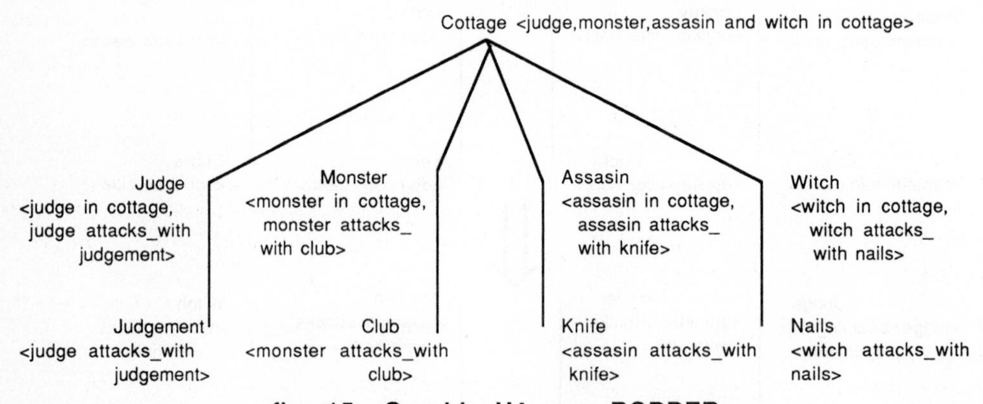

fig. 15 Graphic H4 over ROBBER

This example shows no *surprises* because it illustrates a normal application of an analogical graphic production.

Definition.6 The application of an analogical graphic production (l,r) is normal if the label and attribute parts of the pushout morphisms g': D=>G, h': D=>H are the same as those in l: K=>L, r: K=>R respectively.

A graphic can represent the internal problem relations by the Σ-algebra of the attributes and the edges of a graph. This enables graphic productions to relate problem models by generating analogies and identifying ambiguities in the resulting graphics. As an example one can have a relation "over" in a linguistic Σ-algebra for our Bremen musicians. Then one can give an analogical graphic production from this linguistic algebra to a vector space. If this space is "flat two space", the resulting image corresponds to the picture conventions of the middle ages; if this space is "perspective two space" (reduction of three space with projection implying volume), the resulting image corresponds to the conventions of the Renaissance. The spatial relation analogous to "over" is ambiguous - either "more distant than" or "on top of". This ambiguity can be eliminated by introducing another concept - "colour" say - in the Σ-algebra.

fig. 16 Bremen musician analogy
fragment of Middle Age painting(De)

Therefore we claim that graphics, because of their Σ-algebras of attributes can contribute to the delimitation of knowledge, the verification of indeterminate correspondences and the reduction of ambiguity in information (because analogous problems are not identical).

#4 Technicalities

In this section we define the precise notion of graphic morphism in such a way that pushouts of graphics always exist. The underlying idea is that graphic morphisms must be continuous in operations for 'gluing' labels and attributes.

We will assume that each label set L has a binary operation "," and each Σ-algebra A has a binary operation "·". We assume these binary operations are associative and commutative. We will write

, S for l1.l2.l3.....	when S = (l1,l2,l3,..)	is a subset of L
· S for a1,a2,a3.....	when S = (a1,a2,a3...)	is a subset of A
la: $\underline{L}$ => $\underline{L}$' for la: L => L'	such that la(l1,l2)	= la(l1) , la(l2)
at:$\underline{A}$ => $\underline{A}$' for at:A => A'	such that at(a1.a2)	= at(a1) · at(a2).

There is no loss of generality with our assumptions, because one can always replace L and each carrier domain of A by their power sets, so union is available for "·" and ",".

Definition.7 A *morphism* from a graphic (s,t,lab,atr,L,A) to a graphic (s',t',lab',atr',L',A') consists of a graph morphism (ve,ed) and la: L => L' such that ve;lab' = lab;la

at:A => A' such that ve;atr' = atr;at

where at is a Σ-algebra homomorphism, lab(v) is ,{lab(v')! ve(v') = ve(v)} and atr(v) is {atr(v')! ve(v') = ve(v)}.

Our earlier definition of strict graphic morphisms corresponds to the case when lab = lab and atr = atr, because ve(v) = ve(v') implies lab(v) = lab(v') and atr(v) = atr(v'). All of our examples have also had 'la' and 'at' generated from a total map on 'singletons', but partial maps also generate strict graphic morphisms.

Many graph grammar theoreticians want to restrict the morphisms in graph productions to monomorphisms - see the paper on "high level replacement systems" in these proceedings. If we accepted this restriction, the theory of graphics we are about to present would be somewhat simpler. Our definition of simple and partial graphic morphisms in #2 places a monic restriction on the left morphism of the production. Some graph grammar theoreticians want to restrict "occurrences" to partial morphisms. If both occurrences and productions are partial morphisms, then application of a production is given by a *single* pushout. This happens in a new category where pushouts always exist. Let us show that pushouts of graphic morphisms always exist, because the categories of unlabelled graphs, sets and Σ-algebras have pushouts. Suppose we have a graphic morphism r = <rve,red,rla,rat> from K to R and another graphic morphism k = <kve,ked,kla,kat> from K to D. One might expect trouble with the vertices in R+D that must be glued together. The pushout of the graph morphisms, <rve,red> and <kve,ked>, determines which vertices must be glued together. For each vertex v in K it also determines the label Klab(v) and the attribute Katr(v). Thus glued together vertices get the label la"(Klab(v)) and attribute at"(Katr(v)), where la" is the pushout of rla and kla, and at" is the pushout of rat and kat. This is the pushout of the graphic morphisms, r and k, because of the continuity requirements - ve;lab' = lab;la and ve;atr' = atr;at - in the definition of graphic morphisms. Thus pushouts of graphic morphisms always exist and the problems of (PEM) were caused by the fact that their 'g-substitutions' do not always have pushouts.

Conclusion

A formalism for knowledge representation should ensure that 'rule P is applicable in situation G' automatically implies that 'the result of applying P to G is well defined'. This is ensured by the widely accepted double pushout definition in figure 10, provided that the pushout of K->D and K->R always exists. For an accurate formalism for representing knowledge one should work in a category that has all pushouts, because checking for the existence of pushouts is unnatural. Our discussion of 'surprises' shows that the well defined results of applying rules may be rather strange, but insisting on normal applications (definitions 4 and 6) is an unnatural restriction on modellers imposed by lazy implementors. Another unnatural restriction on modellers is to insist that morphisms must be 'colour preserving'. Although the running example in this paper is simple and only chosen for its Bremen associations, most of the morphisms described (in particular the substitutions L1->D1 and R1->H1, and the analogies L4->D4 and R4->D4) do not preserve colours.

Yet another unnatural restriction on modellers is to insist on occurrences that have a well-defined 'pushout complement'. An occurrence of a rule should be more than a morphism L->G, it should also reveal the particular K->D that glues L into G. There is no reason whatsoever to insist on exactly one K->D for each L->G. This rather subtle notion of occurrence is similar to that in those papers in this proceedings that advocate the "single pushout" approach to graph grammars.

There is considerable evidence that most reasoning is by analogy, not by logic. Think of lawyers with their collections of cases or doctors with their vast experience of patients. Any formalism that cannot handle analogies and case based reasoning is seriously defective. In section 3 we have shown that analogies can be handled when knowledge is represented by graphics and graphic productions. In section 4 we showed that the category of graphics has all pushouts. In the rest of the paper we showed that graphics are sufficiently expressive and flexible that they can be taken seriously as a formalism for knowledge representation.

References

(BP) W. Bibel, B. Petkoff, *Artificial Intelligence methodology, systems, applications*, N. Holland 1985, ISBN0-4444-87743-6

(CM) A. Corradini, U. Montanari, *An algebraic representation of logic program computations*, Pisa TR-36/89, 1989.

(De) R. Delort, *Le Moyen Age*, Edita, Lausanne, 1972.

(ENRL) H. Ehrig, M. Nagl, G. Rozenberg, A. Rosenfeld, *Graph grammars and their application to computer science*. Springer LNCS 291.

(Et) D. W. Etherington, *Formalizing nonmonotonic reasoning systems*. Art. Int 31 (1987) 41-85.

(Ge) I. Georgescu, *The hypernets method for representing knowledge*, pp. 47-58, in (BP).

(HM) L. Hess, B. Mayoh, *Graphics and their grammars*, pp 232-249 in (ENRL).

(Mo) P. D. Mosses, *Unified algebras and institutions*. LICS89, Fourth Ann. Symp. Logic in Comp. Sci.

(Pa) L. Padgham, *A model and representation for type information and its use in reasoning with defaults*. Proc. AAAI 88 (1988) 409-414.

(PEM) F. Parisi-Presicce, H. Ehrig, U. Montanari, *Graph rewriting with unification and composition*, pp. 496-514 in (ENRL).

(Pr) A. Prieditis (ed.), *Analogica*. ISBN0-273-8780-0, Pitman 1983.

(Sa) E. Sandewall, *Non-monotonic inference rules for multiple inheritance with exceptions*. Proc. IEEE 74 (1986) 1345-1353.

(ST) D. Sannella, A. Tarlecki, *On observational equivalence and algebraic specification*. J. Comp. Sys. Sci 34 (1987) 150-178.

(To) D. S. Touretzky, *The mathematics of inheritance systems*. M.Kaufmann Pub. 1986.

(Ul) J. D. Ullman, *Principles of database systems*. ISBN0-7167-8069-0 Comp. Sci. Pr. 1982.

STRUCTURED TRANSFORMATIONS AND COMPUTATION GRAPHS FOR ACTOR GRAMMARS

D. Janssens
Department of Mathematics and Computer Science
Free University of Brussels, V.U.B., Pleinlaan 2, B–1050 Brussels, Belgium

G. Rozenberg
Department of Computer Science
Leiden University, Niels Bohrweg 1, PB 9512, 2300 RA Leiden, The Netherlands

ABSTRACT: Actor Grammars are a model of actor systems based on graph rewriting. Computation graphs model rewriting processes in actor grammars, and hence, computations in actor systems. The relationship between computation graphs and structured transformations, as introduced in [JR 89], is investigated. A structured transformation may be viewed as a description of the external effect of of a computation described by a computation graph.

Keywords: Actor systems, Graph grammars, Concurrency, Processes.

CONTENTS

0. INTRODUCTION

In [JR 87] and [JR 89] actor grammars have been introduced as a formal model for actor systems (see, e.g., [H 77] and [A 86]). In an actor grammar a configuration of an actor system is modeled by a configuration graph. Consequently the behaviour of the system is modeled by transformations of configuration graphs. To this aim the notion of a structured transformation is introduced in [JR 89], and it is demonstrated that the class of all structured transformations of a system can be constructed from primitive structured transformations, which model elementary steps of the system. Hence primitive structured transformations (called primitive event transformations in [JR 89]) suffice to specify all structured transformations of the system.

The aim of this paper is to introduce computation graphs as a representation of derivation processes in actor grammars, and to investigate the relationship between computation graphs and structured transformations. A computation graph gives a precise description of which primitive transformations are used, and which output nodes of one primitive transformation are used as input nodes of another primitive transformation. Hence computation graphs yield the same kind of description of "computation histories" for actor grammars as processes do for Petri Nets (see, e.g., [R 87]).

First we briefly describe the way actor grammars and structured transformations were introduced in [JR 89]. Then it is shown that, for each pair (g, C), where C is a computation graph and g is the initial graph of the rewriting process described by C, one can construct a structured transformation $\Gamma(g, C)$ that may be viewed as a description of the external effect of the corresponding rewriting process. For a given computation graph C, there exist in general configuration graphs g such that the derivation process described by C cannot be carried out starting from g (intuitively speaking, g may not have "enough edges"). We give a characterization of the class of pairs (g, C) that are valid in the sense that the derivation process described by C can indeed be carried out starting from g. Such pairs are called computations.

Observe that, in this way, one has two distinct ways of describing the behaviour of an actor grammar: on the one hand its class of structured transformations, and on the other hand the class of computations (g, C). It is shown that the two descriptions are related in a natural way: the class of structured transformations obtained by applying the construction Γ to the class of computations of an actor grammar X equals (up to isomorphism) the class of structured transformations associated with X (as defined in [JR 89]). Hence, when describing the dynamic behaviour of an actor system, one has the choice between the detailed description provided by computation graphs, or the description provided by structured transformations.

For formal proofs, a more complete investigation of the notions presented in this paper, and a demonstration of their use for comparing systems, we refer to [JLR 90].

1. PRELIMINARIES

In this section we recall some basic notions and terminology concerning graphs and sets. This allows us to set up a notation suitable for this paper.

Sets and Relations

Let A, B be sets and let $R \subseteq A \times B$. Then R^{-1} denotes the inverse of R, and, for each subset C of A,

$$R(C) = \{y \in B \mid \text{there exists an } x \in C \text{ such that } (x, y) \in R\}.$$

If $C = \{x\}$, then we write $R(x)$ instead of $R(\{x\})$. The domain and the range of R are denoted by $Dom(R)$ and $Ran(R)$, respectively. Hence $Dom(R) = R^{-1}(B)$ and $Ran(R) = R(A)$. If $R \subseteq A \times B$ and $R' \subseteq A' \times B'$, where $R \cap R' = \emptyset$, then $R \cup R'$ will sometimes be denoted as $R \oplus R'$.

Graphs

Let Σ and Δ be finite sets.

(1) A (Σ, Δ)-*labeled graph* is a system $g = (V, E, \phi)$ where V is a finite nonempty set (called the *set of nodes of* g), $E \subseteq V \times \Delta \times V$ (called the *set of edges* of g), and ϕ is a function from V into Σ (called the *node-labeling function of* g). For a (Σ, Δ)-labeled graph g, its set of nodes, its set of edges and its node-labeling function are denoted by $Nd(g)$, $Ed(g)$ and ϕ_g, respectively.

(2) Let g and h be (Σ, Δ)–labeled graphs.

(2.a) An *isomorphism from g onto h* is a bijection $\xi : Nd(g) \to Nd(h)$ such that $\phi_g = \phi_h \circ \xi$ and $Ed(h) = \{ (\xi(x), \delta, \xi(y)) \mid (x, \delta, y) \in Ed(g) \}$. The graphs g and h *are isomorphic* if there exists an isomorphism from g onto h. Note that an isomorphism as defined here preserves labels.

(2.b) h is a *subgraph of g* if $Nd(h) \subseteq Nd(g)$, $Ed(h) \subseteq Ed(g)$ and ϕ_h is the restriction of ϕ_g to $Nd(h)$. For a subset A of $Nd(g)$, the *subgraph of g induced by A* is the graph $(A, Ed(g) \cap (A \times \Delta \times A), \phi')$, where ϕ' is the restriction of ϕ_g to A.

(2.c) The graphs g and h are *disjoint* if $Nd(g) \cap Nd(h) = \emptyset$.

(2.d) Let g and h be disjoint. Then $g \oplus h$ is the (Σ, Δ)–labeled graph

$$(Nd(g) \cup Nd(h), Ed(g) \cup Ed(h), \phi_g \oplus \phi_h).$$

(3) Let g be a (Σ, Δ)–labeled graph.

(3.a) For each $v \in Nd(g)$, $({}^\bullet v)_g$ and $(v^\bullet)_g$ denote the sets defined by

$$({}^\bullet v)_g = \{ x \in Nd(g) \mid (x, \delta, v) \in Ed(g), \text{ for some } \delta \in \Delta \},$$

and

$$(v^\bullet)_g = \{ x \in Nd(g) \mid (v, \delta, x) \in Ed(g), \text{ for some } \delta \in \Delta \}.$$

When it is obvious which graph g is intended, then we often write ${}^\bullet v$ and $v^\bullet$ instead of $({}^\bullet v)_g$ and $(v^\bullet)_g$, respectively.

(3.b) For the purpose of this paper it is convenient to use the following notion of a path in a (Σ, Δ)–labeled graph. A *path in g* is a sequence $(v_0, \delta_0, v_1, \delta_1, \ldots, \delta_{n-1}, v_n)$, where $v_0, v_1, \ldots, v_n \in Nd(g)$, $\delta_0, \delta_1, \ldots, \delta_{n-1} \in \Delta$ and, for each $i \in \{0, \ldots, n-1\}$, $(v_i, \delta_i, v_{i+1}) \in Ed(g)$. $(v_0, \delta_0, \ldots, v_n)$ is a path *from* v_0 *to* v_n. Observe that we consider only directed paths, and we allow paths of length 0.

(3.c) Let $E \subseteq Nd(g) \times \Delta \times Nd(g)$. Then $Aug(E, g)$ denotes the (Σ, Δ)–labeled graph $(Nd(g), Ed(g) \cup E, \phi_g)$.

2. STRUCTURED TRANSFORMATIONS AND ACTOR GRAMMARS

In this section we briefly describe the way actor grammars were introduced in [JR 89]. We proceed as follows.

(1) First, we define the notions of an *actor vocabulary*, and a *configuration graph* (for a given vocabulary). An actor vocabulary simply formalizes the global supply of names to be used in a system. A configuration graph formalizes the notion of a "snapshot" of an actor system.

(2) Then we formalize the way that configuration graphs are transformed into each other. To do so, we introduce the notion of a *structured transformation*, we define operations on them enabling one to construct new structured transformations from given ones, and we finally introduce the notion of an actor grammar.

Before formally introducing actor vocabularies and configuration graphs , consider how a snapshot of an actor system can be described by a graph. The basic elements of the snapshot, i.e., the actors and messages present, can be represented by nodes. Local states of actors and values of messages can be represented by node labels. Acquaintance relationships between actors and messages are then represented by directed edges, and acquaintance names by edge labels. The relation "is the destination of", between actors and messages, can be represented by directed edges labeled by the "special symbol" $\not{C}$. The formal definitions of the notions of an actor vocabulary and a configuration graph capture these ideas.

Definition 2.1. An *actor vocabulary* is a 4–tuple $(\mathbf{A}, \mathbf{M}, AQ, MQ)$, where $\mathbf{A}$, $\mathbf{M}$, AQ and MQ are finite sets such that $\mathbf{A} \cap \mathbf{M} = \emptyset$, $AQ \cap MQ = \emptyset$ and $\not{C} \in MQ$. □

For an actor vocabulary $S = (\mathbf{A}, \mathbf{M}, AQ, MQ)$, the *set of node labels of* S, denoted by Σ_S, is the set $\mathbf{A} \cup \mathbf{M}$, and the *set of edge labels of* S, denoted by Δ_S, is the set $AQ \cup MQ$. An *S-graph* is a (Σ_S, Δ_S)–labeled graph. For an S–graph g, the *set of actor nodes of* g, denoted by $Act(g)$, is the set $\{ v \in Nd(g) \mid \phi_g(v) \in \mathbf{A}_S \}$ and the *set of message nodes of* g, denoted by $Msg(g)$, is the set $\{ v \in Nd(g) \mid \phi_g(v) \in \mathbf{M}_S \}$

Definition 2.2. Let $S = (\mathbf{A}, \mathbf{M}, AQ, MQ)$ be an actor vocabulary. An *S-configuration graph* is an S–graph g such that
(1) $Ed(g) \subseteq (Msg(g) \times MQ_S \times Act(g)) \cup (Act(g) \times AQ_S \times Act(g))$, and
(2) for each $v \in Nd(g)$ and each $\delta \in \Delta_S$, v has at most one δ–labeled outgoing edge. □

Example 2.1. Let S be the actor vocabulary $(\mathbf{A}, \mathbf{M}, AQ, MQ)$, where $\mathbf{A} = \{ a_1, a_2, a_3 \}$, $\mathbf{M} = \{ m_1, m_2, m_3 \}$, $AQ = \{ \alpha, \beta, \gamma \}$ and $MQ = \{ \sigma, \tau, \not{C} \}$. Then the graph of Figure 2.1 is an S–configuration graph .

Throughout the paper we use the following conventions for the pictorial representation of graphs.
(1) The identity of a node is denoted inside the corresponding box or circle.
(2) The label of a node is denoted outside the corresponding box or circle.

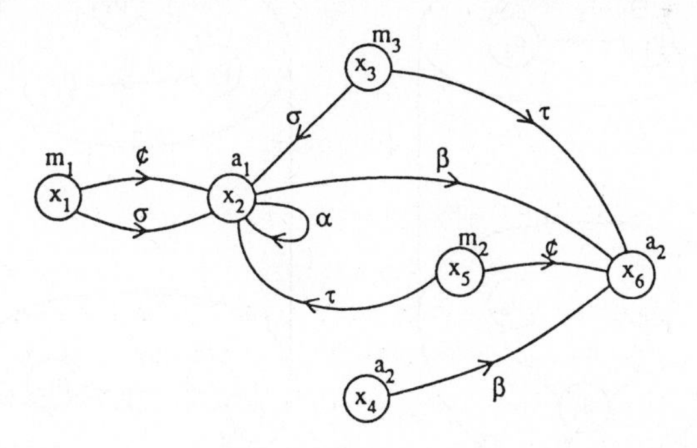

Figure 2.1.

The methodology for graph grammars used in [JR 89] is based on the notion of a structured transformation. A structured transformation is a description of a graph transformation. It describes not only the initial graph and the result graph of a transformation, but it also describes how the transformation will behave in a larger context: i.e., when its initial graph occurs as a subgraph of a larger graph. To this aim a structured transformation is equipped with two relations that specify how "potential" edges (belonging to the context) are transferred from the initial graph to the result graph. Using these relations, one defines operations that build larger structured transformations from smaller ones. Formally, a structured transformation is defined as follows.

Definition 2.3. Let Σ, Δ be finite sets. Let $d = (g, h, iden, Emb)$, where g and h are (Σ, Δ)–labeled graphs, *iden* is an injective partial function from $Nd(g)$ into $Nd(h)$, and $Emb \subseteq (Nd(g) \times \Delta) \times (Nd(h) \times \Delta)$. Then d is a (Σ, Δ)*–structured transformation* if and only if, for each $(x, \mu, y) \in Ed(g)$ and each $(v, \delta) \in Nd(h) \times \Delta$ such that $y \in Dom(iden)$ and $((x, \mu), (v, \delta)) \in Emb$, $(v, \delta, iden(y)) \in Ed(h)$. □

For a (Σ, Δ)–structured transformation $d = (g, h, iden, Emb)$, g is called the *initial graph of d*, h is called the *result graph of d*, *iden* is called the *identification function of d* and *Emb* is called the *embedding relation of d*. Also, g, h, *iden* and *Emb* are denoted by $in(d)$, $res(d)$, $iden_d$ and Emb_d, respectively. The class of all (Σ, Δ)–structured transformations is denoted by $Tr(\Sigma, \Delta)$. Structured transformations d and d' are *disjoint* if $in(d)$ and $in(d')$ are disjoint, and $res(d)$ and $res(d')$ are disjoint.

Example 2.2. Let S be the actor vocabulary from Example 2.1. Then π_1 and π_2, depicted in Figure 2.2, are (Σ_S, Δ_S)–structured transformations. The left half of the figure depicts π_1, the right half depicts π_2. The upper halfs depict $in(\pi_1)$ and $in(\pi_2)$, and the lower halfs depict $res(\pi_1)$ and $res(\pi_2)$. Moreover, $iden_{\pi_i}$ and Emb_{π_i} are defined as follows:

$iden_{\pi_1} = \{ (\rho_2, \rho_4) \}$, $Emb_{\pi_1} = \{ ((\rho_2, \alpha), (\rho_3, \not{c})), ((\rho_1, \tau), (\rho_5, \not{c})), ((\rho_2, \alpha), (\rho_4, \beta)) \}$,
$iden_{\pi_2} = \{ (\rho_8, \rho_9), (\rho_6, \rho_{10}) \}$, and $Emb_{\pi_2} = \{ ((\rho_6, \beta), (\rho_9, \gamma)), ((\rho_8, \gamma), (\rho_{10}, \beta)) \}$.

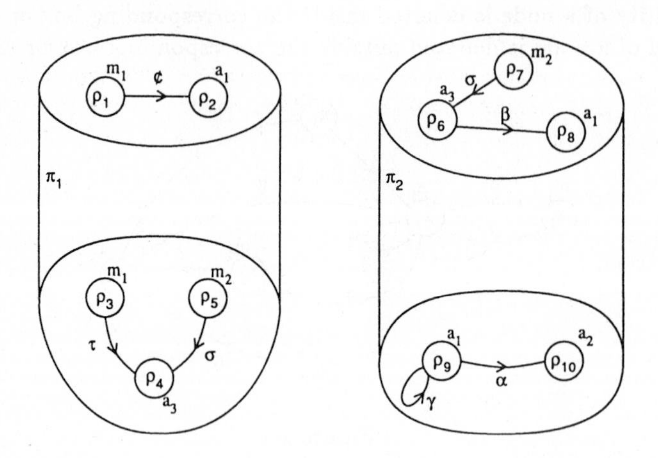

Figure 2.2.

Two structured transformations d, d' are *isomorphic* if they only differ in the choice of nodes for their initial and result graphs. It will be useful to represent isomorphisms between graphs by structured transformations. More precisely, for (Σ, Δ)–labeled graphs

g, h and an isomorphism ξ from g into h, ξ corresponds to the (Σ, Δ)–structured transformation (g, h, ξ, Emb), where

$$Emb = \{ ((x, \alpha), (\xi(x), \alpha)) \mid x \in Nd(g) \text{ and } \alpha \in \Delta \}.$$

One may view such a structured transformation as a description of rewriting process that is trivial in the sense that no rewriting takes place. The class of structured transformations obtained in this way from isomorphisms is denoted by $Iso(\Sigma, \Delta)$.

We now introduce three operations on structured transformations : concurrent composition, sequential composition and augmentation. The first two operations are rather straightforward, the third one requires a bit more explanation.

(1) Concurrent Composition

Let g_1, g_2, h_1 and h_2 be graphs such that g_1 and g_2 have disjoint sets of nodes, and h_1 and h_2 have disjoint sets of nodes. Let d_1, d_2 be structured transformations describing a transformation from g_1 into h_1, and from g_2 into h_2, respectively. Then the concurrent composition of d_1 and d_2, denoted by $d_1 \underline{cc}\, d_2$, is the structured transformation obtained by "putting together" d_1 and d_2 in such a way that they do not interfere with each other. The situation is depicted in Figure 2.3. The notion is formally defined as follows.

Definition 2.4. Let Σ, Δ be finite sets and let $d_1, d_2 \in Tr(\Sigma, \Delta)$ be disjoint. Then the *concurrent composition of* d_1 *and* d_2, denoted by $d_1 \underline{cc}\, d_2$, equals

$$(\, in(d_1) \oplus in(d_2)\, ,\, res(d_1) \oplus res(d_2)\, ,\, iden_{d_1} \oplus iden_{d_2}\, ,\, Emb_{d_1} \oplus Emb_{d_2}\,).$$

□

(2) Sequential Composition

Let g_1, g_2, g_3 be graphs and let d_1, d_2 be structured transformations describing transformations from g_1 into g_2 and from g_2 into g_3, respectively. The sequential composition of d_1 and d_2, denoted by $d_1 \underline{sq}\, d_2$, is the structured transformation obtained by "performing" first d_1 and then d_2. The situation is depicted in Figure 2.4.

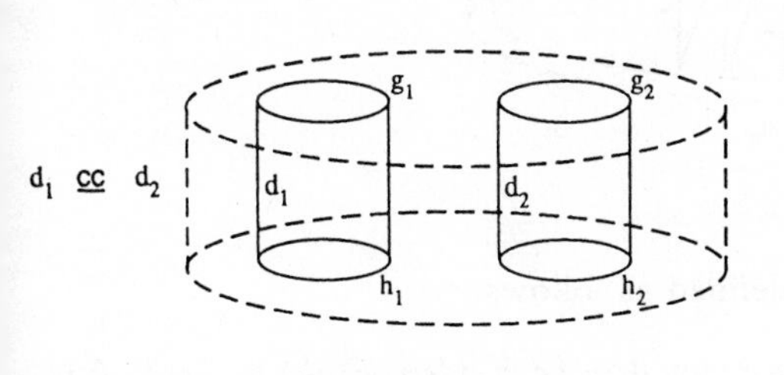

Figure 2.3.

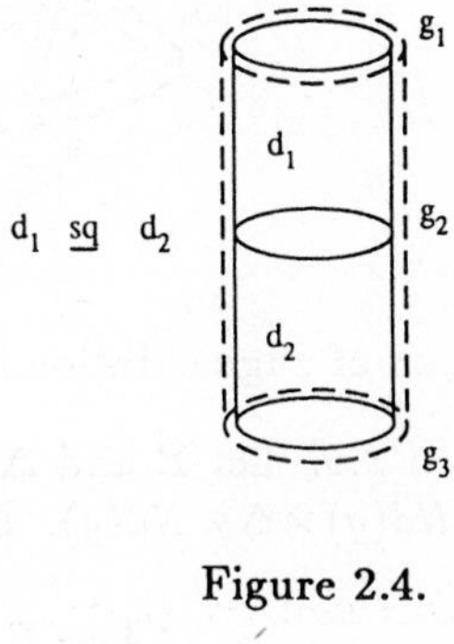

Figure 2.4.

The notion is formally defined as follows.

Definition 2.5. Let Σ, Δ be finite sets and let $d_1, d_2 \in Tr(\Sigma, \Delta)$ be such that $in(d_2) = res(d_1)$. The *sequential composition of* d_1 *and* d_2, denoted by $d_1 \; \underline{sq} \; d_2$, equals

$$(in(d_1) , res(d_2) , iden_{d_2} \circ iden_{d_1} , Emb_{d_2} \circ Emb_{d_1}).$$

□

(3) Augmentation

The operation of augmentation is somewhat different from the first two operations. To explain it, we first point out why $\underline{cc}$ and $\underline{sq}$ alone are not sufficient for our purposes. Assume that one wants to build a structured transformation describing the (simultaneous) rewriting of a number of subgraphs $g_1, \ldots, g_n$ of a graph g. Let g' denote the "remainder" of g (i.e., the part of g that is not involved in the rewriting). For each $i \in \{1 \ldots, n\}$, the rewriting of g_i (into a graph h_i, say) can be described by a structured transformation d_i, the initial graph of which is g_i and the result graph of which is h_i. The fact that g' is not changed can be described by a trivial structured transformation d'. It seems rather natural to use the concurrent composition of $d_1, \ldots, d_n$ and d' as a description of the intended transformation . However, it should be clear that this is not satisfactory: both the initial graph and the result graph of this concurrent composition consist of a number of subgraphs (one for each of $g_1, \ldots, g_n$ and g') which are not connected to each other. Since the desired structured transformation should have g as its initial graph, one needs an operation that "adds edges" to the initial graph of a structured transformation. The additional edges of the initial graph give rise to additional edges of the result graph: edges are "transferred" from the initial graph to the result graph. The situation is illustrated in Figure 2.5. This process of adding edges to the initial graph of a structured transformation, and adding the corresponding (transferred) edges to its result graph, is formalized by the operation of augmentation.

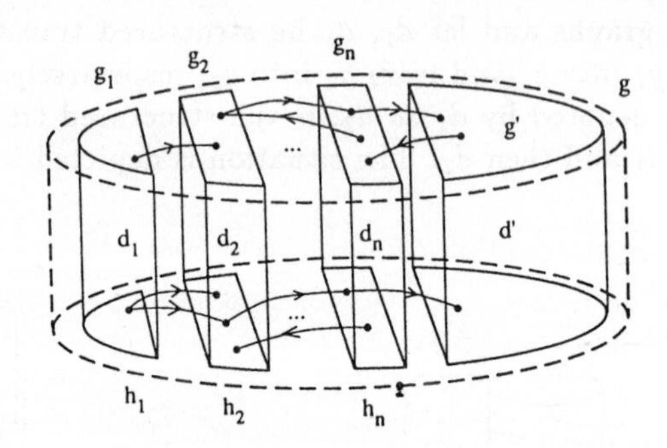

Figure 2.5.

The notion of augmentation is formally defined as follows.

Definition 2.6. Let Σ and Δ be finite sets, let $d = (g, h, iden, Emb) \in Tr(\Sigma, \Delta)$, and let $E \subseteq Nd(g) \times \Delta \times Nd(g)$. Then the E*-augmentation of* d, denoted by $Aug(E, d)$, equals

$$(Aug(E, g), Aug(E', h), iden, Emb),$$

where

$$E' = \{(v,\delta,w) \in Nd(h) \times \Delta \times Nd(h) \mid \text{there exists an } (x,\mu,y) \in E \text{ such that } ((x,\mu),(v,\delta)) \in Emb \text{ and } (y,w) \in iden\}.$$

□

Remark 2.1. One easily verifies that, when applied to (Σ, Δ)–structured transformations, the operations of concurrent composition, sequential composition and augmentation yield (Σ, Δ)–structured transformations. □

The class of (Σ, Δ)–structured transformations generated by (built from) a given set of "primitive" (Σ, Δ)–structured transformations is defined as follows.

Definition 2.7. Let Σ, Δ be finite sets and let P be a set of (Σ, Δ)–structured transformations . The *class of* (Σ, Δ, P)*–structured transformations*, denoted by $Tr(\Sigma, \Delta, P)$, is the smallest subclass X of $Tr(\Sigma, \Delta)$ such that

(1) $P \subseteq X$ and $Iso(\Sigma, \Delta) \subseteq X$, and
(2) X is closed under the operations $\underline{cc}$, $\underline{sq}$ and Aug. □

Actor grammars are obtained by restricting the structured transformations that are used as "primitive" transformations (i.e., as elements of P). As explained in [JR 89], structured transformations of this restricted form, called primitive event transformations, model the processing of a message by an actor in an actor system.

Definition 2.8. Let $S = (\mathbf{A}, \mathbf{M}, AQ, MQ)$ be an actor vocabulary and let $d \in Tr(\Sigma_S, \Delta_S)$. d is a *S–primitive event transformation* if $in(d)$ is a two–node graph such that the following conditions hold. Let $Nd(in(d)) = \{x, y\}$.

(1) $Ed(in(d)) = \{(x, \not{C}, y)\}$, $\phi_{in(d)}(x) \in \mathbf{M}$ and $\phi_{in(d)}(y) \in \mathbf{A}$,
(2) $res(d)$ is an S–configuration graph,
(3) $Dom(iden_d) = \{y\}$ and $iden_d(y) \in Act(res(d))$,
(4) Emb_d is injective, and, for each $((u,\mu), (v,\delta)) \in Emb_d$,
  (4.1) $\mu \neq \not{C}$,
  (4.2) $(u,\mu) \in (\{x\} \times MQ) \cup (\{y\} \times AQ)$,
  (4.3) $(v,\delta) \in (Msg(res(d))) \times MQ) \cup (Act(res(d))) \times AQ)$, and
  (4.4) v has no outgoing δ–labeled edge. □

Example 2.3. The S–structured transformation π_1 from Example 2.2 is a primitive S–event transformation; π_2 is not an S–primitive event transformation.

We finally define the notion of an actor grammar, and the class of structured transformations specified by it.

Definition 2.9. Let S be an actor vocabulary. An *S-actor grammar* is a pair $G = (P, Init)$ such that $Init$ is a set of S–configuration graphs and P is a finite set of S–primitive event transformations. □

Definition 2.10. Let S be an actor vocabulary and let $G = (P, Init)$ be an S–actor grammar. A *G-structured transformation* is a (Σ_S, Δ_S, P)–structured transformation d such that $in(d) \in Init$. □

3. COMPUTATION GRAPHS

In this section we introduce computation graphs as a representation of rewriting processes in actor grammars. A computation graph is a bipartite, directed, acyclic graph. The two kinds of nodes are distinguished by their node labels. One kind is labeled by the "primitive" structured transformations of the system – these nodes (called event–nodes) represent elementary actions (rewritings). The other kind is labeled by the "special" symbol $\underline{ob}$ – these nodes (called object–nodes) correspond to nodes of configuration graphs . The nodes of the initial and the result graphs of primitive structured transformations will be used as edge labels in computation graphs. In this way one relates incoming edges of an event–node to nodes of the initial graph of its label and outgoing edges of an event–node to nodes of the result graph of its label.

Throughout Sections 3 and 4, Σ and Δ are finite sets and P is a set of (Σ, Δ)–structured transformations. The elements of P will be used as "primitive" structured transformations (elementary actions). Hence the systems considered are more general than actor grammars, because their "primitive" structured transformations need not be primitive event transformations. The set

$$R = \bigcup_{\pi \in P} (Nd(in(\pi)) \cup Nd(res(\pi)))$$

will be used as set of edge labels. We assume for the sequel of the paper that, for each $\pi \in P$, $Nd(in(\pi)) \cap Nd(res(\pi)) = \emptyset$. For an element π of P, a *graph–version of* π is a $(P \cup \{\underline{ob}\}, R)$–labeled graph $\tilde{\pi}$ such that

$$Nd(\tilde{\pi}) = Nd(in(\pi)) \cup Nd(res(\pi)) \cup \{w\}, \text{ where } w \notin Nd(in(\pi)) \cup Nd(res(\pi)),$$
$$Ed(\tilde{\pi}) = \{(\rho, \rho, w) \mid \rho \in Nd(in(\pi))\} \cup \{(w, \rho, \rho) \mid \rho \in Nd(res(\pi))\},$$

and

$$\phi_{\tilde{\pi}} = \begin{cases} \underline{ob}, & \text{if } x \neq w, \\ \pi, & \text{if } x = w. \end{cases}$$

Example 3.1. Let π_1 be the structured transformation from Example 2.2 (see Fig. 2.2). Then the graph $\tilde{\pi}_1$, depicted in Figure 3.1, is a graph–version of π_1.

The notion of a computation graph is formally defined as follows.

Definition 3.1. A *(Σ, Δ, P)-computation graph* is a $(P \cup \{\underline{ob}\}, R)$–labeled graph C such that

(1) C is acyclic,
(2) for each $x \in Nd(C)$ such that $\phi_C(x) = \underline{ob}$, x has at most one incoming and at most one outgoing edge, and

(3) for each $v \in Nd(C)$ such that $\phi_C(v) \in P$, there exists a graph–version $\bar{\pi}$ of $\phi_C(v)$ such that $\bar{\pi}$ is isomorphic to the subgraph of C induced by ${}^\bullet v \cup \{v\} \cup v^\bullet$. □

Example 3.2. Let π_1 and π_2 be the structured transformations from Example 2.2. Then the graph of Figure 3.2 is a (Σ, Δ, P)–computation graph for $P = \{\pi_1, \pi_2\}$. Intuitively speaking, it represents a derivation process consisting of three rewritings: two rewritings corresponding to π_1 and one rewriting corresponding to π_2.

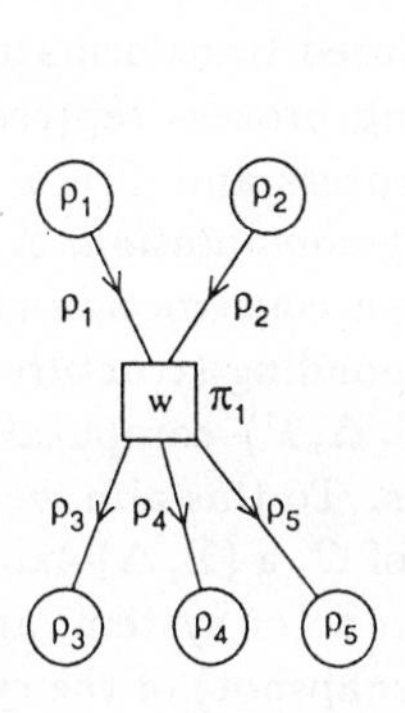

Figure 3.1

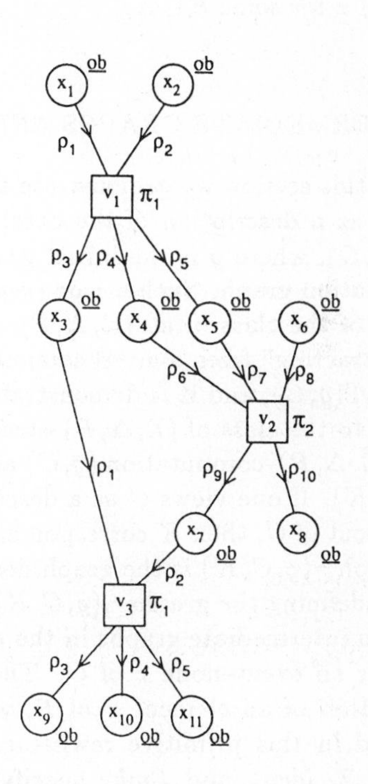

Figure 3.2

The following notions and notation will be useful in discussing computation graphs.

(1) The set of *event–nodes of* C, denoted by $Evn(C)$, is the set $\{v \in Nd(C) \mid \phi_C(v) \in P\}$, and the set of *object–nodes of* C, denoted by $Obn(C)$, is the set $\{v \in Nd(C) \mid \phi_C(v) = \underline{ob}\}$.

(2) $<_C$ denotes the partial order defined on $Nd(C)$ by the edges of C: for $x, y \in Nd(C)$, $x <_C y$ if and only if there exists a path from x to y in C. $Min(C)$ and $Max(C)$ denote the sets of minimal and maximal elements of $Nd(C)$ with respect to $<_C$, respectively. An *object–coset of* C is a set $K \subseteq Obn(C)$ such that, for each $x, y \in K$, neither $x <_C y$ nor $y <_C x$.

The way rewriting processes of actor grammars are described by computation graphs is similar to the way that the behaviour of elementary net systems (and condition/event systems) is described by occurrence nets in Petri Net theory (see, e.g., [R 87]). The cuts of an occurrence net may be viewed as descriptions of the intermediate situations (cases) that occur while the system is running. A similar interpretation can be given to the maximal object–cosets of, called *object–cuts*, of a computation graph. The notion of an object–cut is formally defined as follows.

Definition 3.2. Let C be a (Σ, Δ)–computation graph. An *object–cut of* C is an object–coset K of C such that, for each $x \in Nd(C) - K$, either $x <_C k$ for some $k \in K$, or $k <_C x$ for some $k \in K$. □

4. INTERMEDIATE GRAPHS AND COMPUTATIONS

In this section we demonstrate that a (Σ, Δ, P)–structured transformation may be viewed as a description of the external effect of a rewriting process represented by a pair (g, C), where g is the initial graph of the rewriting process and C is a (Σ, Δ, P)–computation graph. Such a pair (g, C) is called a (Σ, Δ, P)*–computation.* A characterization of the class of all (Σ, Δ, P)–computations is given, a construction is described for "extracting" from a given computation (g, C) the coresponding structured transformation $\Gamma(g, C)$, and it is demonstrated that the class of (Σ, Δ, P)–computations corresponds to the class of (Σ, Δ, P)–structured transformations. To this aim we construct, for a (Σ, Δ, P)–computation (g, C) and an object–coset K of C, a (Σ, Δ)–labeled graph $\gamma(g, C, K)$. If one views C as a description of a "run" of an actor system, and K is an object–cut of C, then K corresponds to a configuration (a snapshot) of the system, and the graph $\gamma(g, C, K)$ is the graph describing this configuration.

In defining the graphs $\gamma(g, C, K)$ one has to consider the way edges are transferred between intermediate graphs in the rewriting process described by (g, C). To this aim, consider an event–node x of C. The node x represents a primitive rewriting, i.e., an application of an element π of P, where π is the label $\phi_C(x)$ of x. Nodes of ${}^\bullet x$ are replaced in this primitive rewriting, and nodes of $x^\bullet$ are created. As explained in Section 2, $iden_\pi$ and Emb_π specify how incoming and outgoing edges, respectively, are transferred from ${}^\bullet x$ to $x^\bullet$. Hence edges are transferred via directed paths in C. Formally, one needs the following properties of directed paths in computation graphs.

Definition 4.1. Let C be a (Σ, Δ, P)–computation graph and let $p = (v_0, \rho_0, v_1, \rho_1, \dots, v_n)$ be a path in C such that $v_0, v_n \in Obn(C)$. Let, for each $i \in \{1, 3, 5, \dots, n-1\}$, π_i be the label of v_i.

(1) p is *iden–compatible* if and only if, for each event–node v_i on p, $\rho_i = iden_{\pi_i}(\rho_{i-1})$.
(2) For each $\alpha, \beta \in \Delta$, p is (α, β)*–compatible* if and only if there exist $\delta_0, \delta_2, \delta_4, \dots, \delta_n \in \Delta$ such that $\delta_0 = \alpha$, $\delta_n = \beta$ and, for each event–node v_i on p, $((\rho_{i-1}, \delta_{i-1}), (\rho_i, \delta_{i+1})) \in Emb_{\pi_i}$. □

For a (Σ, Δ, P)–computation graph C, for $x, y \in Obn(C)$ and for $\alpha, \beta \in \Delta$, we write $x \overset{id}{\underset{C}{\rightarrow}} y$ if there exists an *iden*–compatible path from x to y in C, and we write $x \overset{\alpha,\beta}{\underset{C}{\rightarrow}} y$ if there exists an (α, β)–compatible path from x to y in C.

Example 4.1. In the computation graph of Example 3.2 (Figure 3.2), the path $(x_6, \rho_8, v_2, \rho_9, x_7, \rho_2, v_3, \rho_4, x_{10})$ is *iden*–compatible and the path $(x_2, \rho_2, v_1, \rho_4, x_4, \rho_6, v_2, \rho_9, x_7)$ is (α, γ)–compatible.

We now define the graphs $\gamma(g, C, K)$. Intuitively speaking, $\gamma(g, C, K)$ has two kinds of edges:

(1) edges of g that are transferred from g to K via compatible paths of C, and
(2) edges that are created in a primitive rewriting corresponding to an event–node of C, and that, after their creation, are transferred to K via compatible paths of C.

Formally one has the following.

Definition 4.2. Let C be a (Σ, Δ, P)–computation graph, let g be a (Σ, Δ)–labeled graph such that $Nd(g) = Min(C)$, and let K be an object–coset of C. For an event-node v of C, let π_v denote its label. Then $\gamma(g, C, K)$ is the (Σ, Δ)–labeled graph k such that

(1) $Nd(k) = K$,

(2)
$$\begin{aligned} Ed(k) = \{ (x, \beta, y) \in K \times \Delta \times K \mid\ & \text{there exists an edge } (u, \alpha, w) \in Ed(g) \\ & \text{such that } u \overset{\alpha,\beta}{\underset{C}{\rightarrow}} x \text{ and } w \overset{id}{\underset{C}{\rightarrow}} y \} \\ \cup \{ (x, \beta, y) \in K \times \Delta \times K \mid\ & \text{there exists an event node } v \in Evn(C), \text{ a label} \\ & \alpha \in \Delta \text{ and edges } (v, \rho_1, u), (v, \rho_2, w) \in Ed(C) \\ & \text{such that } (\rho_1, \alpha, \rho_2) \in Ed(res(\pi_v)), \\ & u \overset{\alpha,\beta}{\underset{C}{\rightarrow}} x \text{ and } w \overset{id}{\underset{C}{\rightarrow}} y \}, \end{aligned}$$

(3)
$$\phi_k(x) = \begin{cases} \phi_g(x), & \text{if } x \in Min(C), \\ \phi_{in(\pi_v)}(\rho), & \text{if } (v, \rho, x) \in Ed(C), \text{ for some } v \in Evn(C). \end{cases}$$

□

Example 4.2. Let g be the graph depicted in Figure 4.1, let C be the computation graph of Example 3.2 (Fig. 3.2) and let $K = \{ x_3, x_4, x_5, x_6 \}$. Then $\gamma(g, C, K)$ is the graph depicted in Figure 4.2.

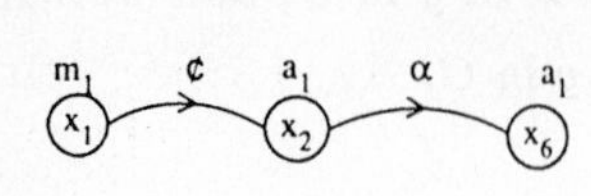

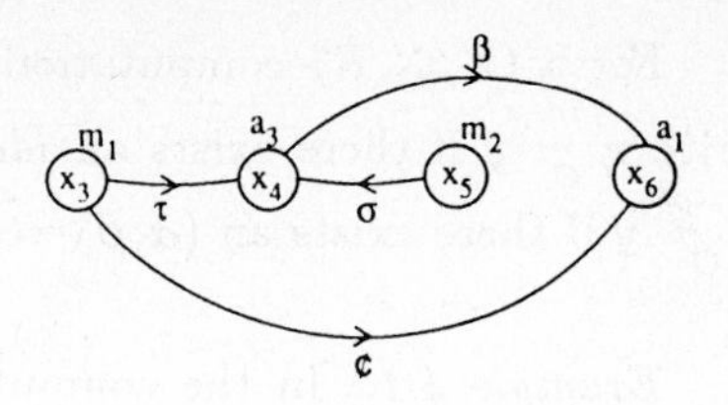

Figure 4.1. Figure 4.2.

We now introduce our second construction, Γ. It yields, for a given computation graph C and a given graph g such that $Min(C) = Nd(g)$, a structured transformation $\Gamma(g, C)$.

Definition 4.3. Let g be a (Σ, Δ)–labeled graph and let C be a (Σ, Δ, P)–computation graph such that $Min(C) = Nd(g)$. Then $\Gamma(g, C)$ is the 4–tuple $(in, res, iden, Emb)$, where

$$in = g,\ res = \gamma(g, C, Max(C)),$$
$$iden = \{ (x, y) \mid x \in Min(C),\ y \in Max(C) \text{ and } x \overset{id}{\underset{C}{\to}} y \}, \text{ and}$$
$$Emb = \{ ((x, \alpha), (y, \beta)) \mid x \in Min(C),\ y \in Max(C),\ \alpha, \beta \in \Delta \text{ and } x \overset{\alpha,\beta}{\underset{C}{\to}} y \}. \quad \square$$

Example 4.3. Let g be the graph from Figure 4.1 and let C be the computation graph of Example 3.2 (Fig. 3.2). Then $d = \Gamma(g, C)$ is the structured transformation depicted in Figure 4.3, where $iden_d = \{ (x_2, x_8), (x_6, x_{10}) \}$ and $Emb_d = \{ ((x_6, \gamma), (x_8, \beta)) \}$.

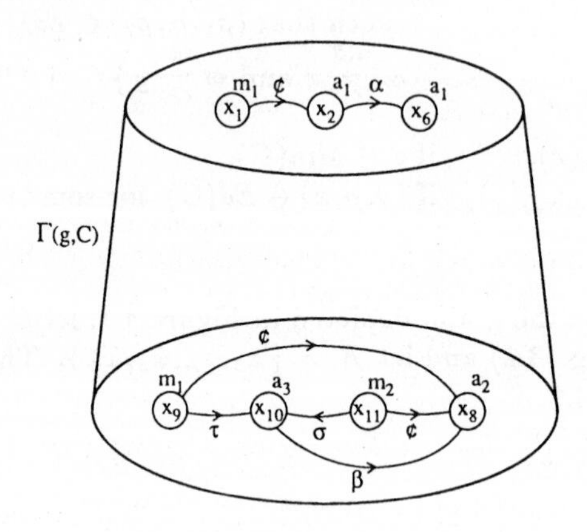

Figure 4.3.

Remark 4.1. It is easily verified that $\Gamma(g,C)$ is a (Σ,Δ)–structured transformation. □

The fact that, for an object–cut K of C, $\gamma(g,C,K)$ may be viewed as an intermediate graph of the rewriting process described by the pair (g,C) is illustrated by the following result. Intuitively speaking, one may divide C into two parts C_1 and C_2, according to K, where C_1 "precedes" K and C_2 "follows" K. Then the structured transformation corresponding to the pair (g,C) is the sequential composition of structured transformations corresponding to C_1 and C_2, respectively, where $\gamma(g,C,K)$ is used as the initial graph for C_2. Formally one has the following (for the proof we refer to [JLR 90]).

Theorem 4.1. Let g be a (Σ,Δ)–labeled graph, let C be a (Σ,Δ,P)–computation graph such that $Nd(g) = Min(C)$, and let K be an object–cut of C. Let

$$V_1 = \{\, x \in Nd(C) \mid x <_C k, \text{ for some } k \in K \,\} \cup K,$$

$$V_2 = \{\, x \in Nd(C) \mid k <_C x, \text{ for some } k \in K \,\} \cup K,$$

and let C_1, C_2 be the subgraphs of C induced by V_1 and V_2, respectively. Then

$$\Gamma(g,C) = \Gamma(g,C_1) \ \underline{sq}\ \Gamma(h,C_2),$$

where $h = \gamma(g,C_1,K)$. □

We now define the notion of a (Σ,Δ,P)–computation. Intuitively it can be understood as follows. An event-node v of a computation graph C represents a rewriting corresponding to the primitive structured transformation π which is the label of v; i.e., the initial graph of π is replaced by its result graph. Evidently, this can be done only if the nodes which are replaced are "connected" by a suitable graph structure. More precisely, this graph structure must correspond to the initial graph of π, or an augmentation of it. The nodes rewritten in the event corresponding to v are the nodes of ${}^\bullet v$; the graph structure connecting them is $\gamma(g,C,{}^\bullet v)$. Hence we have the following definition.

Definition 4.4. Let g be a (Σ,Δ)–labeled graph and let C be a (Σ,Δ,P)–computation graph. The pair (g,C) is a (Σ,Δ,P)–*computation* if and only if $Nd(g) = Min(C)$ and, for each $x \in Evn(C)$, the following holds. Let π denote the label of x. Then for each edge (ρ_1,α,ρ_2) of $in(\pi)$, the graph $\gamma(g,C,K)$ contains the edge (y_1,α,y_2), where $y_1,y_2 \in {}^\bullet x$ are such that $(y_1,\rho_1,x),(y_2,\rho_2,x) \in Ed(C)$. □

The class of all (Σ,Δ,P)–computations is denoted by $Comp(\Sigma,\Delta,P)$.

The following result shows that the class of (Σ,Δ,P)–computations corresponds to the class of (Σ,Δ,P)–structured transformations. For a formal proof we refer to [JLR 90].

Theorem 4.2.

(1) For each $(g, C) \in Comp(\Sigma, \Delta, P)$, $\Gamma(g, C) \in Tr(\Sigma, \Delta, P)$.

(2) For each $d \in Tr(\Sigma, \Delta, P)$, there exists a structured transformation $d' \in Tr(\Sigma, \Delta, P)$ and a (Σ, Δ, P)–computation $(g, C) \in Comp(\Sigma, \Delta, P)$ such that d and d' are isomorphic and $d' = \Gamma(g, C)$. □

Remark 4.2. For an actor grammar $G = (P, Init)$, a G–computation may be defined as a (Σ, Δ, P)–computation such that $g \in Init$). Obviously, the class of G–computations corresponds to the class of G–structured transformations. □

REFERENCES

[A 86] G. A. Agha, *Actors: A Model of Concurrent Computation in Distributed Systems*, M.I.T. Press, Cambridge, MA, 1986.

[H 77] C. Hewitt, Viewing Control Structures as Patterns of Passing Messages, *J. Artificial Intel.*, 8 (1977), 323-364.

[JLR 90] D. Janssens, M. Lens and G. Rozenberg, Computation Graphs for Actor Grammars, Technical Report, Dept. Mathematics and Computer Science, Vrije Universiteit Brussel, V.U.B. (1990).

[JR 87] D. Janssens and G. Rozenberg, Basic Notions of Actor Grammars: a Graph Grammar Model for Actor Computation, in *Graph Grammars and Their Application to Computer Science*, Lecture Notes in Computer Science, Vol. 291, Springer-Verlag, Berlin, 1987, 280-298.

[JR 89] D. Janssens and G. Rozenberg, Actor Grammars, *Math. Systems Theory*, 22 (1989), 75-107

[R 87] G. Rozenberg, Behaviour of Elementary Net Systems, in *Advances in Petri Nets 1986, Part I*, Lecture Notes in Computer Science, Vol. 254, Springer-Verlag, Berlin, 1987, 60-94.

Grammatical Inference Based on Hyperedge Replacement

Eric Jeltsch, Hans-Jörg Kreowski
Fachbereich Mathematik und Informatik
Universität Bremen
Postfach 33 04 40
D - 2800 Bremen 33

ABSTRACT: In this paper, a grammatical-inference algorithm is developed with finite sets of sample graphs as inputs and hyperedge-replacement grammars as outputs. In particular, the languages generated by inferred grammars contain the input samples. Essentially, the inference procedure iterates the application of an operation which decomposes hyperedge-replacement rules according to edge-disjoint coverings of the right-hand sides of the rules. The main result is a characterization of the inferred grammars as "samples-composing" meaning that each sample can be derived and each rule contributes to the generation of samples in a certain way.

Keywords: grammatical inference, hyperedge-replacement, graph grammars.

CONTENTS

1 Introduction

As a fundamental goal of pattern generation and recognition, one wants to find syntactic descriptions for certain types or sets of patterns in such a way that automatic generation or recognition of the patterns is provided (see, e.g., Fu [Fu 82] and Gonzalez and Thomason [GT 78]). The types or sets of patterns, that are of interest, may not be known completely, but only by some samples. This describes the basic situation of grammatical inference. Given a finite set S_+ of sample patterns, one is looking for a grammar G out of a predefined class $\mathcal{G}$ (or likewise for a syntactic device of some kind) such that the generated language $L(G)$ contains the samples, $S_+ \subseteq L(G)$. Moreover, there may be a finite set S_- of countersamples subject to the requirement that G does not generate any countersample, $L(G) \cap S_- = \emptyset$.

Fu and Booth [FB 75] solve this problem in the case that the patterns are encoded as strings and regular grammars are considered. A similar approach is studied in Barzdin [Ba 74] based on finite automata. In [CRA 76], Cook, Rosenfeld and Aronson introduce inference operations that infer context-free grammars from finite sets of strings. Our approach is inspired by the inference operations in those three publications (cf. also the related papers by Gold [Go 67] and Takada [Ta 88]). More complex patterns are considered by Jürgensen and Lindenmayer [JL 87] and Bartsch-Spörl [Ba 82]. The former authors develop inference procedures for finite sequences of branching structures that yield deterministic $0L$-systems. The latter author sketches an inference algorithm where a certain type of monotone (highly non-context-free) graph grammars is inferred from finite sets of directed labelled graphs.

In the present paper, a grammatical-inference algorithm is developed which constructs hyperedge-replacement grammars from sets of patterns that are represented by undirected unlabelled graphs. Hyperedge replacement (see, e.g., Bauderon and Courcelle [BC 87], Habel and Kreowski [HK 87a+b], Habel [Ha 89] and the introductory note by Drewes and Kreowski in this volume) provides a context-free mode of rewriting on graphs and hypergraphs with nice structural, combinatoric and algorithmic properties. Essentially, the inference procedure iterates the application of an operation which decomposes hyperedge-replacement productions according to edge-disjoint coverings of the right-hand sides of the productions. The decomposition of productions can be combined with renaming the nonterminal labels whereover they occur in a grammar. Moreover, a production may be removed from a grammar if its left-hand side derives its right-hand side by applying other productions of the grammar.

2 PRELIMINARIES

We assume that all samples to be considered are undirected unlabelled graphs. To generate sets of such graphs in a simple, but flexible way, graphs may be decorated by hyperedges in intermediate steps. The hyperedges serve as place holders subject to replacement by other decorated graphs. In other words, this paper is based on the framework of hyperedge replacement as sketched and discussed by Drewes and Kreowski in the tutorial section of this volume. The basic notions and notation can be found there. We only present a further example in this section.

2.1 Example

1. The *complete bipartite graph* $K_{m,n}$ for some $m, n \in \mathbb{N}$ consists of $m + n$ nodes such that each node out of m nodes is adjacent to each node out of the other n nodes. The following figure shows the $K_{3,i}$ for $i = 1, 2, 3, 4$.

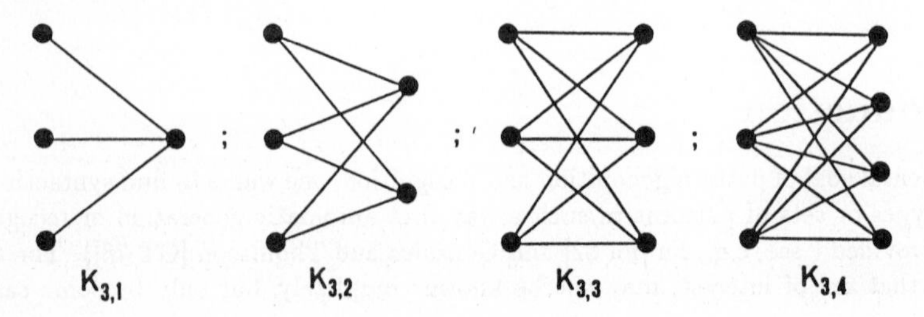

2. The set $\{K_{3,n} \mid n \geq 1\}$ can be generated by the hyperedge-replacement grammar $HRG = (\{D, K\}, \{p_1, p_2, p_3\}, Z)$ where the axiom Z and the three productions are given by the following figure:

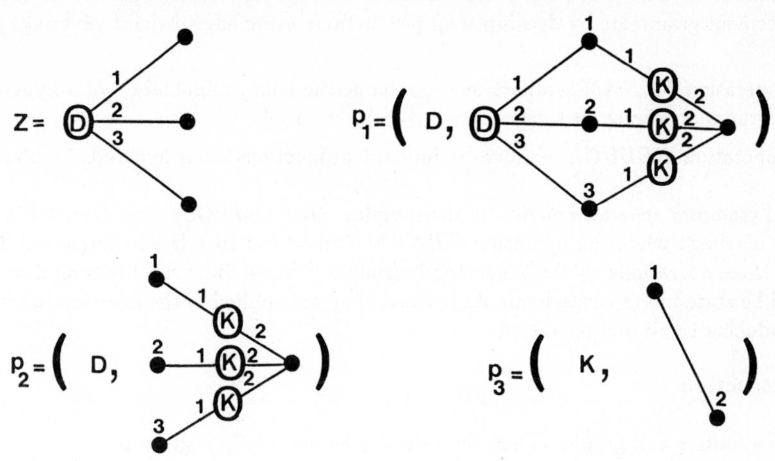

3. The axiom contains three nodes to which a D-edge (i.e. a hyperedge labelled by D) is incident. If p_1 is applied, the D-edge is replaced by another one incident to the same three nodes. The application of p_2 erases the D-edge such that neither p_1 nor p_2 can be applied afterwards. The application of p_3 does not affect the D-edge. Consequently, each sentential form contains at most one D-edge, which is incident to the three axiom nodes. Moreover, the application of p_1 or p_2 creates a new node that is adjacent to the three axiom nodes through K-edges. K-edges can be replaced by ordinary edges, eventually. Altogether, applying p_1 i-times, p_2 once and p_3 as long as possible generates $K_{3,i}$ for any $i \in I\!N$ (and nothing else can be generated). The following figure illustrates this situation where the label K is omitted.

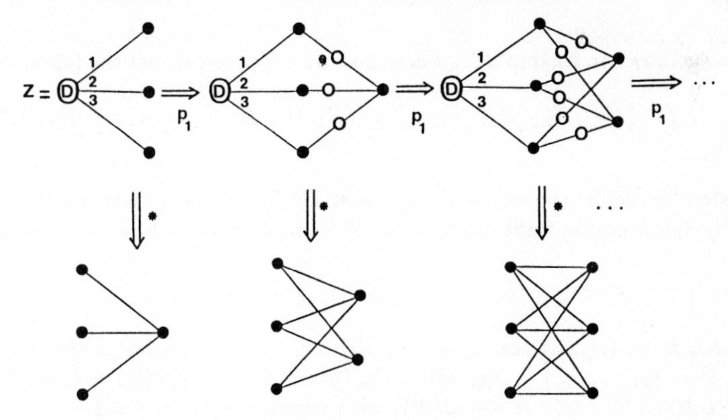

3 A GRAMMATICAL INFERENCE ALGORITHM

In this section, we develop a nondeterministic procedure that infers a hyperedge-replacement grammar from a finite set of graph samples in such a way that the language of the inferred grammars contains all the samples. The algorithm is based on four operations:

(1) The initial operation $INIT$ transforms a finite set of graphs into a hyperedge-replacement grammar taking the samples as right-hand sides of a single left-hand side.

(2) The operation $DECOMPOSE$ transforms hyperedge-replacement grammars into hyperedge-replacement grammars by decomposing productions using edge-disjoint coverings of right-hand sides.

(3) The operation $RENAME$ allows one to rename the nonterminal labels of a hyperedge-replacement grammar whereever they occur.

(4) The operation $REDUCE$ removes redundant productions from hyperedge-replacement grammars.

The initial grammar generates obviously the samples. $DECOMPOSE$ and $REDUCE$ preserve the generated language while the operation $RENAME$ may lead to a larger language. The effects of the latter three operations on the generated languages is based on their effects on direct derivations which will be stated in separate lemmata because they are applied in the next section, too. We start with introducing the initial operation.

3.1 Construction

Let S_+ be a finite set of graphs. Then the *initial grammar* of S_+ is given by

$$INIT(S_+) = (\{S\}, \{(S, M) \mid M \in S_+\}, (S, 0)^\bullet).$$

3.2 Observation

$$L(INIT(S_+)) = S_+.$$

Proof: Each derivation of $INIT(S_+)$ has the form $(S,0)^\bullet \stackrel{1}{\Longrightarrow} M$ for some $M \in S_+$. □

3.3 Example

If the four graphs drawn in Example 2.1.1 are chosen as samples, we get the following initial grammar:

$$INIT(\{K_{3.1}, K_{3.2}, K_{3.3}, K_{3.4}\}) = (\{S\}, \{(S, K_{3.i}) \mid i = 1,2,3,4\}, (S,0)^\bullet).$$

As the next step, we define the essential operation $DECOMPOSE$ which allows one to decompose productions by decomposing right-hand sides. The construction is based on the following lemma.

3.4 Lemma

Let $p = (A, R)$, $p_0 = (A, R_0)$ and $p_i = (A_i, R_i)$ for $i = 1, \ldots, n$ be productions with $type(R) = type(R_0)$, let $B = \{y_1, \ldots, y_n\} \subseteq Y_{R_0}$ with $lab_{R_0}(y_i) = A_i$ and $type(y_i) = type(R_i)$ for $i = 1, \ldots, n$, let $R = REPLACE(R_0, repl : B \longrightarrow \mathcal{G}(N))$ with $repl(y_i) = R_i$ for $i = 1, \ldots, n$. Let $G, H \in \mathcal{G}(N)$, and $y \in Y_G$ with $lab_G(y) = A$.

Then $G \underset{y,p}{\Longrightarrow} H$ if and only if
$G \underset{y,p_0}{\Longrightarrow} X_0 \underset{y_1,p_1}{\Longrightarrow} X_1 \underset{y_2,p_2}{\Longrightarrow} \ldots \underset{y_n,p_n}{\Longrightarrow} X_n = H$ for some $X_i \in \mathcal{G}(N)$ $(i = 0, \ldots, n)$.

Proof: The statement is a straightforward reformulation of basic facts known for hyperedge replacement.□

3.5 Construction

Let $HRG = (N, P, Z)$ be a hyperedge-replacement grammar and $(A, R) \in P$. Let N_{new} be a set of new labels not occurring in N, $\overline{N} = N \cup N_{new}$, $R_0 \in \mathcal{G}_{\overline{N}}$, $B = \{b \in Y_{R_0} \mid lab_{R_0}(b) \in N_{new}\}$, and let $repl : B \longrightarrow \mathcal{G}(N)$ be a mapping with $type(b) = type(repl(b))$ for all $b \in B$ such that $R = REPLACE(R_0, repl)$. Then we get the following grammar:

$$DECOMPOSE(HRG) = (\overline{N}, \overline{P}, Z)$$

with $\overline{P} = (P - \{(A, R)\}) \cup \{(A, R_0)\} \cup \{(lab_{R_0}(b), repl(b)) \mid b \in B\}$.

3.6 Theorem

$$L(HRG) = L(DECOMPOSE(HRG)).$$

Proof: Let $\overline{HRG} = DECOMPOSE(HRG)$ where $p = (A, R) \in P$ is decomposed into $p_0 = (A, R_0)$ and $p_i = (lab_{R_0}(y_i), repl(y_i))$ for all $i = 1, \ldots, n$ based on the mapping $repl : B \longrightarrow \mathcal{G}(N)$ with $B = \{y_1, \ldots, y_n\}$ and $R = REPLACE(R_0, repl)$. First, we are going to show $L(HRG) \subseteq L(\overline{HRG})$. Let $M \in L(HRG)$, then there is a derivation $s = (Z \overset{*}{\underset{P}{\Longrightarrow}} M)$. If $p = (A, R)$ is applied within s, then s can be decomposed into $Z \overset{*}{\underset{P}{\Longrightarrow}} G \underset{y,p}{\Longrightarrow} H \overset{*}{\underset{P}{\Longrightarrow}} M$ for some $y \in Y_G$. Using the lemma 3.4, we can replace the derivation step by a derivation of the form $G \underset{y,p_0}{\Longrightarrow} X_0 \underset{y_1,p_1}{\Longrightarrow} \ldots \underset{y_n,p_n}{\Longrightarrow} X_n = H$. By induction, we can replace all applications of p within s transforming s into a derivation $\overline{s} = (Z \overset{*}{\underset{\overline{P}}{\Longrightarrow}} M)$. In other words, $M \in L(\overline{HRG})$.
It remains to be show that $L(\overline{HRG}) \subseteq L(HRG)$. Let $M \in L(\overline{HRG})$, then there is a derivation $\overline{s} = (Z \overset{*}{\underset{\overline{P}}{\Longrightarrow}} M)$. If p_0 is not applied within $\overline{s}$, then also the p_i for $i = 1, \ldots, n$ are not applied because they need nonterminal hyperedges that are produced by p_0 only. In this case, $\overline{s}$ is already a derivation in HRG. Otherwise, there is a first ocurrence of p_0, and $\overline{s}$ can be decomposed into $Z \overset{*}{\underset{P}{\Longrightarrow}} G \underset{y,p_0}{\Longrightarrow} H \overset{*}{\underset{\overline{P}}{\Longrightarrow}} M$. Without loss of generality, we may assume that $B \subseteq Y_H$. For each $i = 1, \ldots, n$, y_i must be replaced later on by applying p_i because M is terminal. Because of the context-freeness, the derivation steps where these p_i are applied are independent of all its predecessors within the derivation $H \overset{*}{\underset{\overline{P}}{\Longrightarrow}} M$. In other words, we can assume that this derivation has the form

$$H = X_0 \underset{y_1,p_1}{\Longrightarrow} X_1 \underset{y_2,p_2}{\Longrightarrow} \ldots \underset{y_n,p_n}{\Longrightarrow} X_n = H \overset{*}{\underset{\overline{P}}{\Longrightarrow}} M.$$

Using Lemma 3.4, we can transform $\overline{s}$ into a derivation of the form

$$Z \overset{*}{\underset{P}{\Longrightarrow}} G \underset{y,p}{\Longrightarrow} H \overset{*}{\underset{\overline{P}}{\Longrightarrow}} M$$

with one ocurrence of p_0 less than in $\overline{s}$. By induction, $\overline{s}$ can be transformed into a derivation $Z \overset{*}{\underset{P}{\Longrightarrow}} M$ such that $M \in L(HRG)$. This completes the proof. □

3.7 Example

Consider the productions $r_i = (S, K_{3.i})$ for i=1,2,3,4 in Example 3.3. The production r_2 may be decomposed into the productions $r_{21} = (S, H_1)$ and $r_{22} = (D, H_2)$, where H_1 and H_2 are given in the following figure.

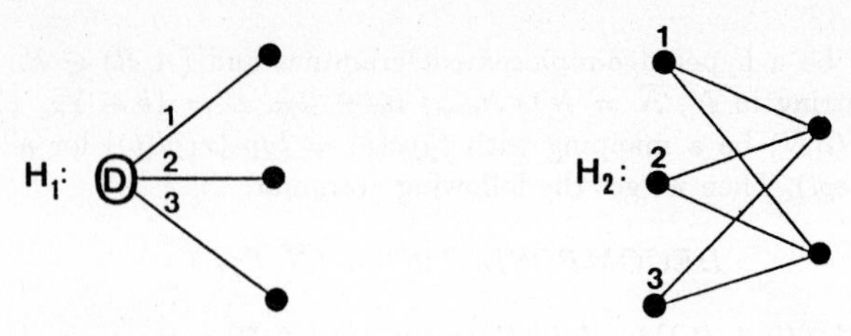

The rule r_{22} may be decomposed once more into the rules $r_{221} = (D, H_3)$ and $r_{222} = (D', H_4)$, where H_3 and H_4 are given in the following figure.

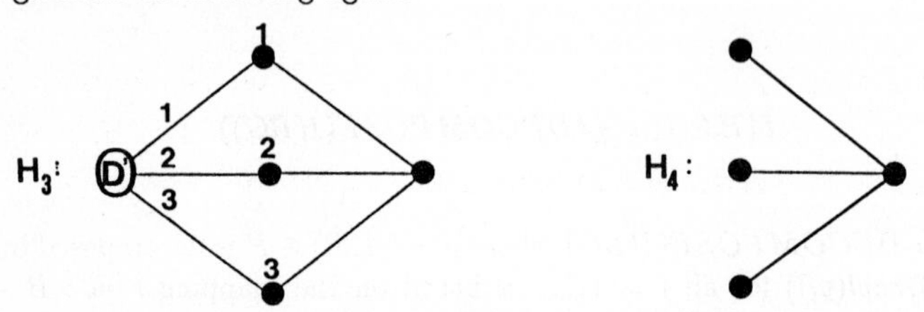

Therefore, the rules of the inferred grammar $\overline{HRG}$ are r_1, r_{21}, r_{221}, r_{222}, r_3, and r_4. In this grammar we have for example the following derivation:

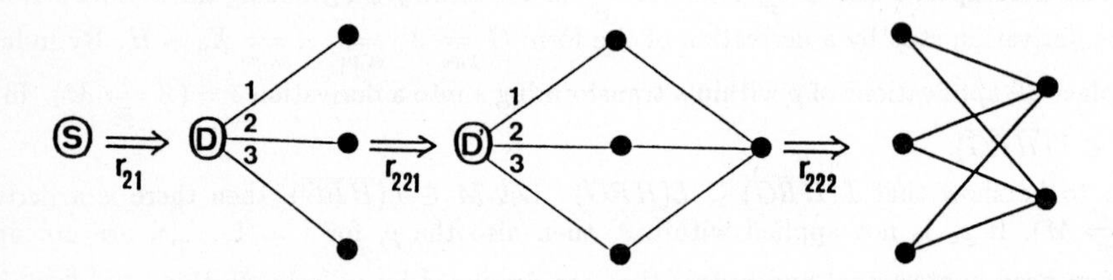

Now, we are going to define a renaming operation, which allows one to rename nonterminals within a hyperedge-replacement grammar. Note that this operation may identify nonterminals. While the *DECOMPOSE*-operation yields an equivalent grammar the *RENAME*-operation yields a grammar whose generated language may be properly included in the original language.

3.8 Construction

1. Let $\psi : N \longrightarrow N'$ be a mapping. Then each $H \in \mathcal{G}(N)$ can be transformed into $H' = (V_H, E_H, Y_H, att_H, lab', ext_H) \in \mathcal{G}_{N'}$ where $lab' : Y_H \longrightarrow N'$ is defined by $lab'(e) = \psi(lab_H(e))$ for all $e \in Y_H$. The only effect of this transformation is that each hyperedge of H with label A is relabeled by $\psi(A)$. H' may be denoted by $rename_\psi(H)$. Note that $rename_\psi(H) = H$ if $H \in \mathcal{G}$.
2. The renaming can be extended to grammars. Let $HRG = (N, P, Z)$ be a hyperedge replacement grammar and $\psi : N \longrightarrow N'$ a mapping for some set N' of labels. Then we get the following grammar
$$RENAME(HRG) = (N', P', Z')$$
with $P' = \{(\psi(A), rename_\psi(R)) \mid (A, R) \in P\}$ and $Z' = rename_\psi(Z)$.

3.9 Theorem

$$L(HRG) \subseteq L(RENAME(HRG)).$$

Proof: Let $M \in L(HRG)$. Then we have a derivation $Z = H_0 \Longrightarrow H_1 \Longrightarrow \ldots \Longrightarrow H_n = M$. Using Lemma 3.10 given below, this induces a derivation $Z' = H_0' \Longrightarrow H_1' \Longrightarrow \ldots \Longrightarrow H_n'$ with $H_i' = rename_\psi(H_i)$ for $i = 0, \ldots, n$ and some mapping $\psi : N \longrightarrow N'$. In particular, we have $H_n' = rename_\psi(H_n) = rename_\psi(M) = M$ because $M \in \mathcal{G}$. This implies $M \in L(RENAME(HRG))$. □

3.10 Lemma

Let $G, H \in \mathcal{G}(N)$, $(A, R) \in P$, $e \in Y_G$, and let $\psi : N \longrightarrow N'$ be a mapping.
Then $G \underset{e,(A,R)}{\Longrightarrow} H$ implies $rename_\psi(G) \underset{e,(\psi(A),rename_\psi(R))}{\Longrightarrow} rename_\psi(H)$.

Proof: Consider $G \underset{e,(A,R)}{\Longrightarrow} H$. Then $H = G \otimes_e R$. Let $G' = rename_\psi(G)$. By construction, we have $lab_{G'}(e) = \psi(lab_G(e)) = \psi(A)$ such that $(\psi(A), rename_\psi(R))$ is applicable to G', yielding H', i.e., $H' = G' \otimes_e rename_\psi(R)$. Obviously, we have $rename_\psi(H) = H'$. □

3.11 Example

We continue the discussion of Example 3.7. Consider the renaming $\psi : \{S, D, D'\} \longrightarrow \{S, D\}$ with $\psi(D) = \psi(D') = D$ and $\psi(S) = S$. Besides the productions r_1, r_{21}, r_3 and r_4, that are not changed by renaming, the grammar $RENAME(\overline{HRG})$ contains the rules $r_{223} = (D, rename_\psi(H_3))$ und $r_{224} = (\psi(D'), H_4) = (D, H_4)$ with

$rename_\psi(H_3) =$ 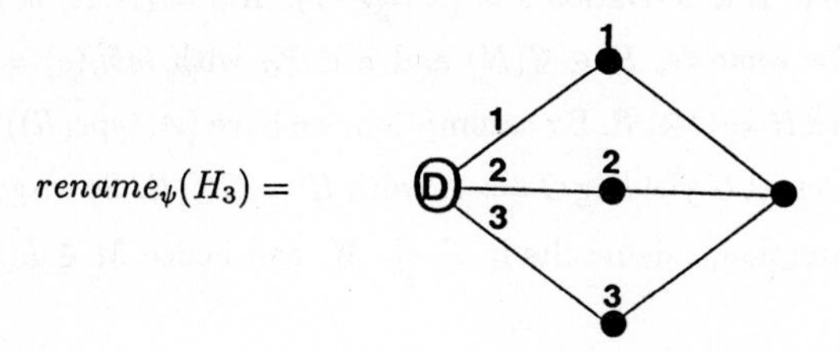

The new rule r_{223} can be applied to its own right-hand side as is demonstrated by the following derivation:

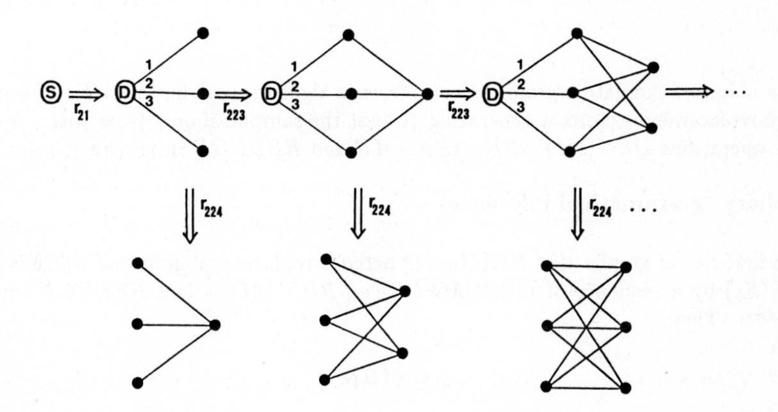

We see that $L(RENAME(\overline{HRG})) = \{K_{3.n} \mid n \geq 1\}$.

In particular, the right-hand side of the productions r_1, r_3, and r_4 can be derived from the axiom using other productions only. Thus we have used decomposition and renaming to derive a grammar that is equivalent to the grammar in Example 2.1.2.

As in the preceeding example, a production may be considered as redundant if its right-hand side can be derived from the decorated graph corresponding to its left-hand side by applying other productions. Removing redundant productions, reduces grammars while keeping the generated language.

3.12 Construction

Let $HRG = (N, P, Z)$ be a hyperedge-replacement grammar and let $(A, R) \in P$ be a production such that $(A, type(R))^\bullet \underset{P'}{\overset{*}{\Longrightarrow}} R$ for $P' = P - \{(A, R)\}$. Then we get the following grammar:

$$REDUCE(HRG) = (N, P', Z).$$

3.13 Theorem

$$L(HRG) = L(REDUCE(HRG)).$$

Proof: Let $HRG' = REDUCE(HRG)$. Clearly, $L(HRG') \subseteq L(HRG)$. We have to show $L(HRG) \subseteq L(HRG')$. Let $M \in L(HRG)$. Then there is a derivation $s = (Z \underset{P}{\overset{*}{\Longrightarrow}} M)$. If $p = (A, R)$ is applied, s has the form $Z \underset{P}{\overset{*}{\Longrightarrow}} G \underset{e,p}{\Longrightarrow} H \underset{P}{\overset{*}{\Longrightarrow}} M$ for some G, $H \in \mathcal{G}(N)$ and $e \in Y_G$ with $lab_G(e) = A$ and $type(e) = type(R)$. In particular, we have $H = G \otimes_e R$. By assumption, we have $(A, type(R))^\bullet \underset{P'}{\overset{*}{\Longrightarrow}} R$. which can be embedded into the larger context G yielding $G \underset{P'}{\overset{*}{\Longrightarrow}} H'$ with $H' = G \otimes_e R$. So we get a new derivation $Z \underset{P}{\overset{*}{\Longrightarrow}} G \underset{P'}{\overset{*}{\Longrightarrow}} H \underset{P}{\overset{*}{\Longrightarrow}} M$. By induction, this results in $Z \underset{P'}{\overset{*}{\Longrightarrow}} M$, and hence $M \in L(HRG')$. □

3.14 Example

According to the considerations in Example 3.11, $RENAME(\overline{HRG})$ can be reduced by the rules r_1, r_3 and r_4.

We are now able to formulate a grammatical-inference algorithm. A finite set S_+ of samples yields a hyperedge-replacement grammar generating at least the samples if one starts with $INIT(S_+)$ and iterates the operations $DECOMPOSE$, $RENAME$ and $REDUCE$ in arbitrary order.

3.15 Corollary (grammatical inference)

Let S_+ be a finite set of graphs. Let HRG be a hyperedge-replacement grammar which is constructed from $INIT(S_+)$ by a sequence of $DECOMPOSE-$, $RENAME-$ and $REDUCE-$operations in arbitrary order. Then:

$$S_+ \subseteq L(HRG).$$

Proof: The statement follows from 3.2, 3.5, 3.9 and 3.13. □

3.16 Example

Summarizing the considerations of the example, we get the following situation: Initializing the inference process by the samples $K_{3,1}$, $K_{3,2}$, $K_{3,3}$ and $K_{3,4}$ (see 2.1.1 and 3.2), two *DECOMPOSE*-steps (3.7), a *RENAME*-step (3.11) and three *REDUCE*-steps (3.14) yield a grammar which generates the complete bipartite graphs $K_{3,n}$ for $n \geq 1$ and is of similar simplicity as the one given in 2.1.2. In this particular case, the inference algorithm allows one to infer an interesting and reasonable generalization of the samples.

4 CHARACTERIZATION THEOREM

In this section, we characterize the hyperedge-replacement grammars that can be inferred using our inference algorithm. The characterization is based on three observations on the inferred grammars (cf. Lemma 4.3). (1) Each sample can be derived from the axiom of an inferred grammar. (2) Each production either being an initial one or one obtained by decompositions and renamings can be used for deriving one of the samples. (3) The axiom of an inferred grammar is $(S,0)^\bullet$ or some renaming of it because the axiom has this form initially and the *RENAME*–operation is the only one affecting the axiom. A grammar with the properties (1), (2) and (3) will be called samples-composing. We are going to show that a samples-composing hyperedge-replacement grammar can already be inferred (cf.Theorem 4.4). An easy consequence of the characterization is the decidability of the question for hyperedge-replacement grammars whether a given grammar is inferrable or not (cf. Corollary 4.5).

4.1 Definition

Let S_+ be a finite set of graphs. Let $HRG = (N, P, Z)$ be a hyperedge-replacement grammar with $Z = (A,0)^\bullet$ for some $A \in N$. Then HRG is called S_+-*composing* if there is a mapping $dec : S_+ \longrightarrow P^*$ with the following properties:
(1) For each $M \in S_+$, there is a derivation $Z \stackrel{*}{\Longrightarrow} M$ through $dec(M)$.
(2) For each $p \in P$, there is an $M_p \in S_+$ such that p occurs in $dec(M_p)$.

Remark: The first condition requires essentially $S_+ \subseteq L(HRG)$ where the mapping *dec* chooses one of the sequences of productions for each sample in such a way that the sample can be generated from the axiom by applying production in the choosen order. The second condition demands the existence of a choise which covers all productions. In other words, each production can be used in deriving one of the samples. In particular, the graph underlying a right-hand side of a production (i.e. the right-hand side without hyperedges and externe nodes) is a subgraph of a sample. All grammars in section 2 and 3 are samples-composing with respect to the four samples in 2.1.1.

4.2 Notation

Let S_+ be a finite set of graphs. Then $INFERENCE(S_+)$ may denote the set of grammars that can be inferred from S_+ using the algorithm in Corollary 3.15, i.e. grammars that are obtained from $INIT(S_+)$ by an arbitrary sequence of *DECOMPOSE*–, *RENAME*–, and *REDUCE*–operations.

4.3 Lemma (preservation of S_+-composability)

Let S_+ be a finite set of graphs and HRG be an S_+-composing hyperedge-replacement grammar. Then the grammars (1) $INIT(S_+)$, (2) $DECOMPOSE(HRG)$, (3) $RENAME(HRG)$ and (4) $REDUCE(HRG)$ are S_+-composing.

Proof: (1) The mapping $dec : S_+ \longrightarrow P^*$ given by $dec(M) = (S, M)$ satisfies obviously the requirements of Definition 4.1.

By assumption, we have $HRG = (N, P, (A,0)^\bullet)$ and $dec : S_+ \longrightarrow P^*$ with the properties stated in Definition 4.1.

(2) By construction $\overline{HRG} = DECOMPOSE(HRG) = (N \cup N_{new}, \overline{P}, (A,0)^\bullet)$ with

$$\overline{P} = (P - \{(X,R)\}) \cup \{(X,R_0)\} \cup \{(lab_{R_0}(b), repl(b)) \mid b \in B\}$$

and $repl : B \longrightarrow \mathcal{G}(N)$ with $B = \{b \in Y_{R_0} \mid lab_{R_0}(b) \in N_{new}\}$ and $R = REPLACE(R_0, repl)$. Consider now the mapping $\overline{dec} : S_+ \longrightarrow \overline{P}^*$ being defined by $\overline{dec}(M) = modify(dec(M))$ where $modify : P^* \longrightarrow \overline{P}^*$ is given by

$$modify(X,R) = (X,R_0)(lab_{R_0}(b_1), repl(b_1)) \dots (lab_{R_0}(b_k), repl(b_k))$$

with $B = \{b_1, \dots, b_k\}$, $modify(p) = p$ for $p \in P$ with $p \neq (X,R)$ and the usual extension to the free monoid. We know by definition that $(A,0)^\bullet \overset{*}{\Longrightarrow} M$ through $dec(M)$ for $M \in S_+$. If p_0 does not occurs in $dec(M)$, we have $\overline{dec}(M) = dec(M)$. If p_0 occurs in $dec(M)$, we can replace the corresponding derivation step $G \underset{(X,R)}{\Longrightarrow} H$ in $(A,0)^\bullet \overset{*}{\Longrightarrow} M$ by a derivation $G \overset{*}{\Longrightarrow} H$ through $modify(X,R)$ using Lemma 3.4. Altogether we get a derivation $(A,0)^\bullet \overset{*}{\Longrightarrow} M$ through $\overline{dec}(M)$. Moreover, we know that $p \in P$ occurs in some $dec(M_p)$. If $p \neq (X,R)$, this is true for $\overline{dec}(M_p)$, too. If $p = (X,R)$, $modify(X,R)$ is a substring of $\overline{dec}(M_p)$. Hence all the new productions occur in $\overline{dec}(M_p)$. In other words, $\overline{HRG}$ is S_+-composing.

(3) By construction, $HRG' = RENAME(HRG) = (N', P', (\psi(A), 0)^\bullet)$ for some renaming function $\psi : N \longrightarrow N'$ where $P' = \{rename(p) \mid p \in P\}$ with $rename(p) = (\psi(X), rename(R))$ for $p = (X,R) \in P$. Using the usual extension $rename : P^* \longrightarrow P'^*$, we define $dec' : S_+ \longrightarrow P'^*$ by $dec'(M) = rename(dec(M))$. By Lemma 3.10, dec' satisfies the requirements of Definition 4.1, and, therefore, HRG' is S_+-composing.

(4) By construction, $\overline{HRG} = REDUCE(HRG) = (N, P - \{p_0\}, (A,0)^\bullet)$ for some $p_0 = (X,R) \in P$ with $(X, type(R))^\bullet \overset{*}{\Longrightarrow} R$ through $w \in (P - \{p_0\})^*$. Then each derivation step $G \underset{p_0}{\Longrightarrow} H$ can be replaced by a derivation $G \overset{*}{\Longrightarrow} H$ through w. Hence it can be shown that $\overline{HRG}$ is S_+-composing as in Point 2.□

4.4 Theorem (Characterization)

Let S_+ be a finite set of graphs and HRG be a hyperedge-replacement grammar.

Then $HRG \in INFERENCE(S_+)$ if and only if HRG is S_+-composing.

Proof: We show first that HRG is S_+-composing if $HRG \in INFERENCE(S_+)$ by induction on the number of operations that are applied to infer HRG from S_+. Let this number be denoted by $\#HRG$. If $\#HRG = 1$, we have $HRG = INIT(S_+)$ which is S_+-composing due to Lemma 4.3(1). Assume now that the statement holds for $\#HRG = n$, and consider a HRG' with $\#HRG' = n+1$. Then we have three cases: $HRG' = DECOMPOSE(HRG)$ or $HRG' = RENAME(HRG)$ or $HRG' = REDUCE(HRG)$ for some $HRG \in INFERENCE(S_+)$ with $\#HRG = n$. By assumption, HRG is S_+-composing. Using Lemma 4.3(2)-(4), HRG' is S_+-composing, too.

For the other direction, let now $HRG = (N, P, (A,0)^\bullet)$ be S_+-composing due to a corresponding mapping $dec : S_+ \longrightarrow P^*$. We can assume $A = S$ because otherwise we may perform the renaming through $rename : N \longrightarrow N'$ with $N' = N \cup \{S\}$ defined by $rename(A) = S$ and $rename(X) = X$, for all $X \in N$ with $A \neq X$ if $S \notin N$ and $rename(A) = S$, $rename(S) = A$ and $rename(X) = X$ for $A \neq X \neq S$ otherwise (i.e. $N = N'$). We will show $HRG \in INFERENCE(S_+)$ by induction on n for $n = (\sum_{M \in S_+} length(dec(M))) - \#S_+$ where $\#S_+$ denotes the number of elements in S_+ and $length$ yields the length of a sequence. Note that $length(dec(M)) \geq 1$ for $M \in S_+$ because $M \in \mathcal{G}$, but $(S,0)^\bullet \notin \mathcal{G}$. If $n = 0$, $length(dec(M)) = 1$ for all $M \in S_+$. By Definition 4.1(1), the derivations $(S,0)^\bullet \stackrel{*}{\Longrightarrow} M$ through $dec(M)$ have length 1. Without loss of generality, we may assume that there can be no other rule such that $HRG = INIT(S_+) \in INFERENCE(S_+)$. We assume now that the statement holds for all $m < n$ for some $n \geq 1$. Consider an S_+-composing hyperedge-replacement grammar $HRG = (N, P, (S,0)^\bullet)$ with $(\sum_{M \in S_+} length(dec(M))) - \#S_+ = n$. Because $n \geq 1$, there is an $M_0 \in S_+$ with $length(dec(M_0)) > 1$. Let $(S,0)^\bullet = H_0 \underset{y_1,p_1}{\Longrightarrow} H_1 \underset{y_2,p_2}{\Longrightarrow} \ldots \underset{y_k,p_k}{\Longrightarrow} H_k = M_0$ with $k = length(dec(M_0))$ be the derivation due to Definition 4.1(1). There are productions $p_j = (A_j, R_j)$ with $R_j \notin \mathcal{G}$ (at least p_1 is of this kind). Let j_0 be the greatest index of this kind. Then $j_0 < k$ and, for $i = j_0+1, \ldots, k$, we have $p_i = (A_i, R_i)$ with $R_i \in \mathcal{G}$. Moreover, we can assume that each of these p_i replaces a hyperedge y_i of R_{j_0} (because otherwise it is independent of the j_0-step and can be shifted beyond the j_0-step).

First, we study the case that for $i = j_0+1, \ldots, k$, A_i does not occur in any rule other than p_{j_0} and p_i, that p_{j_0} does not occur in any $dec(M)$ for $M \neq M_0$ and that the A_i are pairwise distinct. Then we may consider the rule (A_{j_0}, R) with $R = REPLACE(R_{j_0}, repl)$ where $repl : Y_{R_{j_0}} \longrightarrow \mathcal{G}$ is given by $repl(y_i) = R_i$ for $i = j_0+1, \ldots, k$. We have $HRG = DECOMPOSE(\overline{HRG})$ with $\overline{HRG} = (\overline{N}, \overline{P}, (S,0)^\bullet)$, where $\overline{N} = (N - \{A_i \mid i = j_0+1, \ldots, k\})$, and $\overline{P} = (P - \{(A_i, R_i) \mid i = j_0, \ldots, k\} + \{(A_{j_0}, R)\})$. Define $\overline{dec} : S_+ \longrightarrow \overline{P}^*$ by $\overline{dec}(M_0) = p_1 \ldots p_{j_0-1}(A_{j_0}, R)$ and $\overline{dec}(M) = dec(M)$ otherwise. Obviously, $\overline{dec}$ satisfies the requirements of Definition 4.1 such that $\overline{HRG}$ is S_+-composing.

Moreover,

$$
\begin{aligned}
&(\sum_{M \in S_+} length(\overline{dec}(M))) - \#S_+ \\
&= (length(\overline{dec}(M_0)) + \sum_{M \neq M_0} length(dec(M))) - \#S_+ \\
&= (j_0 + \sum_{M \neq M_0} length(dec(M))) - \#S_+ \\
&< (k + \sum_{M \neq M_0} length(dec(M))) - \#S_+ \\
&= (\sum_{M \in S_+} length(dec(M))) - \#S_+ = n.
\end{aligned}
$$

So $\overline{HRG} \in INFERENCE(S_+)$ by the induction hypothesis, hence $HRG \in INFERENCE(S_+)$ as a decomposition of $\overline{HRG}$.

In the other case, i.e. A_i occurs in other rules or p_{j_0} occurs in some $dec(M)$ with $M \neq M_0$ or the A_i are not pairwise distinct, we can define an S_+-composing grammar HRG' with $HRG = RENAME(HRG')$ which may substitute HRG in the first case. So $HRG' \in INFERENCE(S_+)$ and, consequently, $HRG \in INFERENCE(S_+)$ as a renaming of HRG'. □

It is easy to see that the characterization theorem implies the decidability whether an arbitrary hyperedge-replacement grammar can be inferred or not.

4.5 Corollary

For all hyperedge-replacement grammars HRG and all finite sets S_+ of graphs, it is decidable whether $HRG \in INFERENCE(S_+)$ or not.

Proof: Due to Habel [Ha 89b], we can assume that HRG is in such a normal form that the length of derivations deriving some graph is bounded where the bound depends on the grammar and the size of the graph. In other words, we must inspect a given number of derivations only to check whether HRG is S_+-composing or not. Using Theorem 4.3, this yields the stated decidability.□

5 DISCUSSION

This paper presents our first attempt to combine the ideas and concepts of grammatical inference with the approach and theory of hyperedge replacement. In this way, graphs can be used as samples. The inferred grammars are context-free and generate at least all samples. As the basic inference operation, we employ a decomposition of productions (together with renaming of nonterminal labels). The class of inferred grammars is characterized independently of the inference operation establishing the main result of the paper.

In our opinion, these very first steps toward grammatical inference based on hyperedge replacement look promising. Nevertheless, further investigations are necessary before the usefulness of our approach becomes clear beyond doubts. Among the aspects that need intensive study are the following.

(1) The scheme behind our inference algorithm is as follows: Initial grammars are derived from the input samples, and the resulting grammars are transformed successively by grammar transformations. In general, each set of initializations and transformations yields an inference procedure and our algorithm is just one example. Considering different initializations and transformations, one may like to compare their power concerning inferred grammars and their generated languages. For instance, the class of inferred grammars does not change if the $REDUCE$-operation is forbidden in our case.

(2) The theory of hyperedge replacement provides further candidates for transformations. For instance, the filter theorems by Courcelle [Co 87+90], Lengauer and Wanke [LW 88] and Habel and Kreowski [HK 89] are of such a kind. They allow one to transform a given grammar into one which generates all members of the original language with a certain property if this property is inductive (in Courcelle's sense) or finite (in Lengauer's and Wanke's sense) or compatible (in Habel's and Kreowski's sense). It may be interesting to learn how filtering coexists with other inferring transformations.

(3) Because a right-hand side of a production may have many edge-disjoint coverings leading to many possible $DECOMPOSE$-steps, our inference algorithm is highly nondeterministic. Hence it is mandatory to look for means that dam the flood of inferred grammars. A first help may come from syntactic restrictions such as forbidding trivial decompositions or accepting the introduction of hyperedges of a certain type only. Unfortunately, syntactic means may not lead far enough.

(4) Again the theory of hyperedge replacement may help on a more semantic level. We may demand additional properties of the generated language such as: "All generated graphs are required to be planar, nearly all to be connected, generated graph should be 3-colorable, and only a finite number of generated graphs may be Hamiltonian". If the required properties are inductive, finite or compatible (cf. Point 2), the grammars that fulfil the requirements can be selected (see, e. g. Habel, Kreowski and Vogler [HKV 89]). Many graph properties are known to be inductive or finite or compatible, particularly, the question whether a graph is isomorphic to a predefined one or not. So we can forbid the generation of certain negative examples, a favorite kind of additional requirements in the area of grammatical inference.

(5) The idea of additional requirements as sketched in Point 4 reduces the number of grammars that are accepted as output, but not yet the number of constructed grammars. Even more interesting would be to find a class of properties that can be used to stop the inference process in certain branches at least. For example, if we require planarity of all generated graphs and infer a grammar which generates a nonplanar graph, it is meaningless to continue our inference process with this grammar because all successively inferred grammars generate the same or a larger language.

(6) Summarizing the reasoning of Point 3-5, one may look for suitable subclasses of the class of hyperedge-replacement grammars that allow one more efficient inference procedures. For instance, the class of linear grammars (the right-hand sides of their productions contain at most one hyperedge) seems to lead to drastic improvements.

(7) In the previous section, we have characterized the class of inferrable grammars. Is there a corresponding characterization of the generated languages ?

We should mention that first results concerning these seven points will appear in a forthcoming paper. Another open problem concerns the relationship of our approach to the languages identification in the limit in the sense of Gold [Go 67] and consistent identification in the sense of Jantke and Beick [JB 81].

Acknowledgement: We are very grateful to Lena Bonsiepen, Bruno Courcelle, Annegret Habel, Klaus-Peter Jantke and Brian Mayoh for many helpful suggestions and comments.

References

[Ba 74] J. M. Barzdin: Finite Automata: Synthesis and Behaviour, North-Holland, 1974.

[Ba 83] B. Bartsch-Spörl: Grammatical Inference of Graph-Grammars for Syntactic Pattern Recognition, Lect. Not. Comp. Sci. 153, 1-7, 1983.

[BC 87] M. Bauderon, B. Courcelle: Graph Expressions and Graph Rewriting, Math. Systems Theory 20, 83-127, 1987.

[Co 87] B. Courcelle: On Context-Free Sets of Graphs and Their Monadic Second-Order Theory, Lect. Not. Comp. Sci. 291, 133-146, 1987.

[Co 90] B. Courcelle: The Monadic Second-Order Logic of Graphs I Recognizable Sets of Finite Graphs, Information and Computation 85, 12-75, 1990.

[CRA 76] C. Cook, A. Rosenfeld, A. Aronson: Grammatical Inference by Hill Climbing, Informational Sciences 10, 59-80, 1976.

[Fu 82] K.S. Fu: Syntactic Pattern Recognition and Applications, Prentice-Hall, Englewood-Cliffs, N.J., 1982.

[FB 75] K.S. Fu, T.K. Booth: Grammatical Inference: Introduction and Survey Part I and II, IEEE-Trans. Syst. Man and Cyber. 5, 95-111 and 409-423, 1975.

[Go 67] E. M. Gold: Language Identification in the Limit, Information and Control 10, 447-474, 1967.

[GT 78] R.C. Gonzalez, M.G. Thomason: Syntactic Pattern Recognition, Addison-Wesley, Reading, Massachusetts, 1978.

[Ha 89] A. Habel: Hyperedge Replacement: Grammars and Languages, Ph. D. Thesis, Fachbereich Mathematik und Informatik, Universität Bremen, April 1989.

[HK 87a] A. Habel, H.-J. Kreowski: May We Introduce to You: Hyperedge Replacement, Lect. Not. Comp. Sci. 291, 15-26, 1987.

[HK 87b] A. Habel, H.-J. Kreowski: Some Structural Aspects of Hypergraph Languages Generated by Hyperedge Replacement, Lect. Not. Comp. Sci. 247, 207-219, 1987.

[HK 89] A. Habel, H.-J. Kreowski: Filtering Hyperedge-Replacement Languages Trough Compatible Properties, Lect. Not. Comp. Sci. 411, 107-120, 1989.

[HKV 89] A. Habel, H.-J. Kreowski, W. Vogler: Metatheorems for Decision Problems on Hyperedge Replacement Graph Languages, Acta Informatica 26, 657-677, 1989.

[JB 81] K. P. Jantke, H.-R.Beick: Combining Postulates of Naturalness in Inductive Inference, EIK 17, 465-484, 1981.

[JL 87] H. Jürgensen, A. Lindenmayer: Inference Algorithms for Developmental Systems with Cell Lineages, Bulletin of Mathematical Biology 49, Nr.1, 93-123, 1987.

[LW 88] T. Lengauer, E. Wanke: Efficient Analysis of Graph Properties on Context-Free Graph Languages, Lect. Not. Comp. Sci. 317, 379-393, 1988.

[Ta 88] Y. Takada: Grammatical Inference for Even Linear Languages Based on Control Sets, In: Proceedings of the ECAI 88, München, 375-377, 1988.

Specifying Concurrent Languages and Systems with Δ-GRAMMARS *

Simon M. Kaplan, Joseph P. Loyall, and Steven K. Goering
University of Illinois
1304 W. Springfield, Urbana, IL 61801 USA

ABSTRACT: This paper illustrates the use of graph grammars for specifying concurrent systems and languages. The model used in this paper, Δ-GRAMMARS, is rooted in existing graph grammar theory and provides a convenient framework in which to specify both static and dynamic concurrent systems. Our approach is illustrated by three examples.

KEYWORDS: graph grammars, Δ-GRAMMARS, concurrent languages, concurrent systems, visual, dynamic.

CONTENTS

0 Introduction

This paper illustrates the use of graph grammars for describing and specifying concurrent systems and languages. Graph grammars are well suited for such specifications because they have a strong theoretical base, support a visual intuition, and support specification of dynamic systems.

Most existing models for the specification of concurrent systems are either textual models, such as Actors [1], or static, visual models, such as Petri Nets [2] and CCS [21]. Concurrent systems have a two-dimensional structure, the flow of communication between processes and the flow of control within processes, that is difficult to express using the one-dimensional nature of text. Also many concurrent systems are dynamic, *i.e.* processes are created and destroyed and communication paths change during execution. Such dynamism is difficult to express using visual, but static, models.

The graph grammar model used in this paper, Δ-GRAMMARS, is based on Ehrig's double-pushout categorical construction for generalized graph-to-graph transformations [5] and is very similar to Göttler's X- and Y-grammar models [8, 9, 10]. The major differences are the addition of a *restriction* region to productions (effectively a visual guard on production application) and a theory of Kleene-like *-groups for Δ productions. Concurrent systems and languages are specified in Δ by representing the state of a program as a graph and representing state transitions by Δ-productions.

This paper is structured as follows: First, we present an overview of Δ-GRAMMARS and their relation to previous graph grammar research. Then we demonstrate the specification of concurrent systems and languages using Δ-GRAMMARS with three examples. The first is a specification for the *remote client-server* problem, in which connections among clients must be established in an arbitrary, dynamically changing network of clients and servers. This example illustrates how complex dynamic concurrent systems can be specified in Δ. The second example presents a Δ solution to the *dynamic dining philosophers* problem. The major focus of this example is on abstraction mechanisms in Δ so that a solution can be specified at multiple interconnected levels of abstraction. The third example shows how a concurrent language, Actors [1], can have its semantics defined using a Δ-GRAMMAR. The resulting semantics is lazy and highly concurrent.

*Supported in part by the National Science Foundation under grant CCR-8809479, by the Center for Supercomputing Research and Development at the University of Illinois and by AT&T through the Illinois Software Engineering Project. Steve Goering has been supported as an IBM fellow.

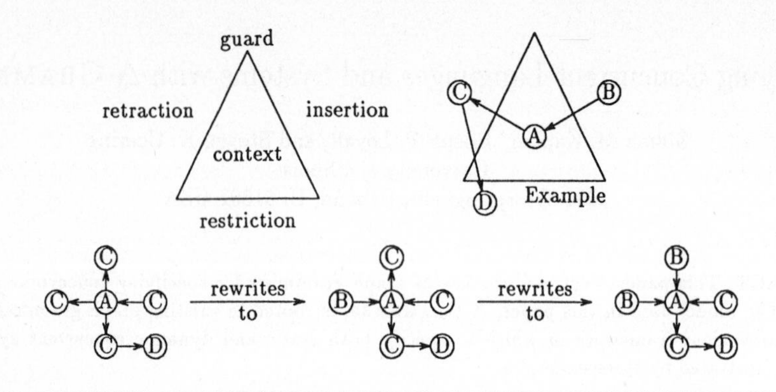

Figure 1: Form of Production

1 Δ-Grammars

In this section, we informally describe Δ-GRAMMARS. For the formal specifications and more detailed discussion, see [17, 7, 6, 18, 23]. We proceed in three parts. The first presents a brief informal overview; the second describes some of the features of Δ-GRAMMARS; and the third describes how Δ-GRAMMARS relate to existing graph grammar research.

The purpose of this section is not to give a formal overview of Δ-GRAMMARS. Formal definition of the basic concepts and their use in Δ-GRAMMARS can be found in the cited references.

1.1 Informal Overview

A Δ-GRAMMAR consists of an initial graph and a set of Δ-productions. Execution of a program specified in Δ consists of repeatedly choosing a Δ-production and applying it to the graph resulting from the previous transformation. Production application is said to *rewrite* the current graph at each step.

A Δ-production consists of five sections, represented by the triangular notation of Figure 1 (thus the name Δ-GRAMMARS).

- The *retraction*, the fragment of the graph removed during the rewrite, is written to the left of the triangle.
- The *insertion*, the fragment that is created and embedded into the graph during the rewrite, is written to the right of the triangle.
- The *context*, the fragment that is identified but not changed during the rewrite, is written in the center of the triangle. The context is common to both the left and right-hand sides of the production, *i.e.* the left-hand side of the production is the union of context and retraction, and the right-hand side is the union of context and insertion. The context is used to help identify the retraction and to indicate the part of the host graph to which the insertion is embedded.
- The *restriction*, the fragment of the graph that must not exist for the rewrite to occur, is written beneath the triangle. If the subgraph matching the context and retraction can be extended to match the restriction, then the production cannot be applied to that subgraph.
- The *guard*, a textually expressed condition that must be true for the rewrite to occur, is written above the apex of the triangle. A guard is a boolean expression over the variable labels which appear in the production.

While guards and restrictions serve a similar purpose — control over the applicability of the productions — they operate in orthogonal domains. The restriction is a restriction on the topology of the graph to which the production is being applied (it prohibits production application where certain 'shapes' can be found in the graph), while the guard is a restriction on the values of the labels on the nodes and edges of the graph.

A Δ-production p is applied to a graph g by the following steps:

1. A subgraph isomorphic to the left-hand side of p is identified in g.
2. If no isomorphic subgraph exists or if the isomorphic subgraph can be extended to match the left-hand side plus the restriction (if one exists), p cannot be applied to g.
3. The guard, if it exists, is evaluated. If it is false, p cannot be applied to g.
4. The elements of the subgraph isomorphic to the retraction are removed from g, leaving the *host* graph.
5. A graph isomorphic to the insertion, called the *daughter* graph, is instantiated.
6. The daughter graph is embedded into the host graph as shown by the edges between the context and insertion in p.

A sample Δ-production and its application to a graph is illustrated in Figure 1. The application of a Δ-production to a host graph is atomic with regard to the subgraph matching the context and retraction. Since it does not affect any part of the host graph not connected to the matched graph[1], any other part of the host graph may be modified at the same time. Because of this, many applications can occur simultaneously. Parallel production application is described formally in [6].

1.2 Features of Δ

Δ has many features that make it convenient to specify programs. Detailed discussion of Δ and these features is given in [7, 18].

- Labels that may be used in a graph are either names that are declared *a priori* or values of other known domains such as integers. Nodes and edges in graphs and Δ-productions may be either labeled or not labeled. During a rewrite, non-labeled elements of a Δ-production unify to non-labeled elements of a graph, and labeled elements unify to elements with the same labels.
- Variable labels may also be used, where each variable is restricted by a type system to only take on values from a subset of all legal labels. When the same variable is used in several places in a Δ-production, it must unify to the same value in each place for the rewrite to occur. Variable labels appear in Δ-productions as names preceded by a "?", *e.g.* "?x", "?y". Usually an intricate type system is unneeded, so that all variable labels may take on values of all constant labels. Where more distinction is needed, it is specified textually as a restriction on the range of values which can be bound to the variable label. There is a distinguished variable "?" that may be used multiple times in a Δ-production without requiring that those uses be unified to the same label during a rewrite. Using this feature reduces the number of variable names required and simplifies Δ specifications. Specific variable names are only used where a type restriction is needed or where unification of several graph elements is desired, *e.g.* where an element is retracted and an element of the same label is inserted with different connectivity. The variable is bound by its matching to the retracted element's label, and this binding is used to give the inserted element its label.
- Guards are boolean expressions over both type restrictions on variables and relations (predicates) on those variables.
- Subsets of elements in the contexts and retractions of Δ-productions can be grouped together into *folds.* During a rewrite multiple elements in a fold may map to the same graph element, as long as other restrictions (same label and same connectivity) are satisfied. Δ-productions are annotated by adding subscripts to labels such that elements with the same subscript belong to a common fold.
- Sets of elements in a Δ-production can be grouped together to form **-groups* (pronounced "star groups"). *-groups represent zero or more occurrences of the group of elements. A Δ-production p that contains a *-group is syntactic shorthand for an infinite sequence of Δ-productions with no *-groups, each with i occurrences of the group that is starred, $i = 0, 1, 2, \ldots$. Application of p to a graph g consists of applying the Δ-production from this sequence that matches a subgraph in g maximally. The semantic definition of *-groups is given in [6].
- Application of Δ-productions is *fair.* That is, if it is infinitely often possible to apply a Δ-production to a particular place in a graph, the Δ-production will be applied at that place an infinite number of times [6].

[1]Actually, just the part of the graph matched by the retraction.

1.3 Relation to Graph Grammar Theory

Δ-GRAMMARS are based on the double pushout categorical definition of graph grammars presented by Ehrig in [4, 5, 22]. Variable labels and unification of labels during application is based on the results in [23]. The restriction and guard regions are graphical and textual means of specifying the "application conditions" described by Ehrig and Habel in [3].

In the most general definition of Δ-GRAMMARS, there are several features that appear to violate conditions of the categorical definition of graph grammars and therefore break the double pushout construction. Specifically, removal of a node removes all its incident edges and nodes can be grouped into folds, each apparently violating the gluing condition described in [4]. However, these features are just sugaring on the basic Δ model that satisfies the double pushout. A Δ-production that contains folds is shorthand for several Δ-productions that do not contain any folds. Likewise, a Δ-production that removes a node and all its incident edges is equivalent to a Δ-production that matches all the incident edges with a *-group in the retraction. This is described in detail in [19].

Several concrete models of graph grammars have been developed. For example, NLC grammars [16, 24] have been developed as a clean, efficiently executable model. The major difficulties of executing graph grammars, graph isomorphism and specifying the embedding, are simplified in NLC grammars. Because of its single node replacement, establishing an isomorphism between the left-hand side of a production and a subgraph of the graph being rewritten is simply the matter of matching the single node in the left-hand side of the production to a single node in the graph and unifying their labels. The embedding in NLC grammars is usually specified as n-NCE embedding, *i.e.* within a *neighborhood* of distance n from the node being rewritten. NLC grammars are completely encompassed by Δ-GRAMMARS (productions in NLC grammars are just Δ-productions with a single node in the retraction), but there are many applications to which NLC grammars are not suited. Any rewrites requiring the cooperation and deletion of several distant nodes requires the cooperation of several NLC productions and are difficult to represent in NLC grammars.

Hyperedge replacement grammars [11, 12, 20] emphasize edges as the important elements in graphs, using nodes merely as sockets to glue edges together. Hyperedge replacement grammars, therefore, center rewriting around the replacement of a hyperedge in a graph. Graphs contain two kinds of nodes, terminal nodes that can never be rewritten, representing "socket" nodes in a hypergraph, and nonterminal nodes, representing hyperedges, that can be rewritten corresponding to the way hyperedges can be rewritten. Embedding in the immediate neighborhood of a nonterminal node (or hyperedge) is simplified by the fact that no two nonterminal nodes are adjacent, *i.e.* no hyperedge can be connected to another hyperedge without being "glued" together by a "socket" node. Hyperedge replacement grammars can be represented by Δ-GRAMMARS in a similar way that NLC grammars can, but there are many Δ-GRAMMARS that are difficult or impossible to represent using hyperedge replacement grammars.

Δ-GRAMMARS are most similar in nature to Y-grammars [8, 9] and X-grammars [10]. Productions in a Y-grammar are written around a Y, with the left-hand side to the left of the stem, the right-hand side to the right of the stem, and the embedding specification between the uprights. During application of a Y-production, the embedding specification functions as *optional context*, *i.e.* matching the embedding specification portion of the production is optional.

Y-grammars suffer from a lack of *required context*. That is, if a portion of the graph *must* exist for the rewrite to occur but it is left unchanged, that portion must occur in the left-hand side and again in the right-hand side of the production being applied, effectively being removed and then reinserted by the rewrite. For the embedding to be correct, all elements connected to this portion must also be included in both the left- and right-hand sides, and so on. X-grammars solves this problem. X-productions are written around an X, with left-hand side, right-hand side, and optional context as in Y-productions, but with required context written in the bottom part of the X.

Required context is provided in Δ-productions by the context region. Optional context is provided by *-groups that appear in the context region, since elements matching the elements in the *-groups might or might not exist and, if they exist, are unchanged by the transformation. Therefore Δ-GRAMMARS encompass Y- and X-grammars. Δ-productions also have application conditions in the form of guards and *forbidden context* (restrictions).

Other graph grammar models have been used to specify concurrent languages [14, 13] and distributed systems [25].

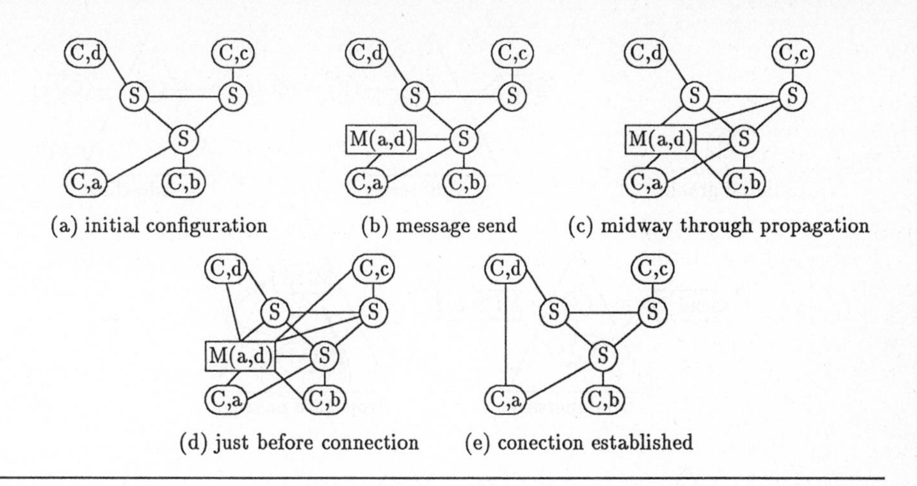

(a) initial configuration (b) message send (c) midway through propagation

(d) just before connection (e) conection established

Figure 2: Client-Server Example: Making the Connection

2 Clients and Servers

The purpose of this paper is to show how graph grammars, specifically the Δ model, can be used to specify solutions to a range of problems in software engineering. Of particular interest here are problems involving the specification of concurrent systems and languages. We now turn our attention to showing how three such problems — the client/server problem, the dynamic dining philosophers problem and a semantics for the concurrent language Actors — can be specified in Δ.

In this section we focus on the client/server problem.

Concurrent problems intuitively lend themselves to graphical representations, such as a representation of processes as nodes in a graph and communication links as edges between nodes. Graphical representations easily capture the two-dimensional relationship between the flow of data between processes and the flow of control within processes.

A Δ-specification of a concurrent problem solution is composed of a representation of the state of the solution program as a graph and a representation of state transitions as graph transformations in the form of Δ-productions.

Here is an informal specification of the client/server problem:

> Consider a message based system with an arbitrary number of nameservers and clients. The number of nameservers and clients may change dynamically. Nameservers may be connected together in any arbitrary topology and clients are attached to exactly one nameserver each. A nameserver may take requests from any of its clients and may make requests for services to one or more other nameservers. A client may make a request to its nameserver to be connected to another client in order to exchange data. The nameserver problem is to allow clients to connect to other clients in order to facilitate this exchange.

In more detail, a graphical representation of an example system configuration (state) is shown in Figure 2(a). Client (C, a) (*i.e.* the client with attribute a) requests connection to the client with attribute d. Such a client does exist in the system – (C, d), but neither (C, a) or its nameserver know this because (C, d) is attached to a remote nameserver. The connection is requested by sending a message to client (C, a)'s nameserver (Figure 2(b)), which then propagates the request through the system until it reaches (C, d) (Figures 2(c) and (d)). When this client is identified, the connection is established (Figure 2(e)).

A problem of this complexity requires many levels of interaction. There is the *network* level, where clients and servers interact by propagating messages and establishing connections; there is a *global* level where external events, such as human interaction or node failure, triggers the addition or deletion of clients and servers; and there is a *machine* level, where computation internal to clients and servers occurs, possibly

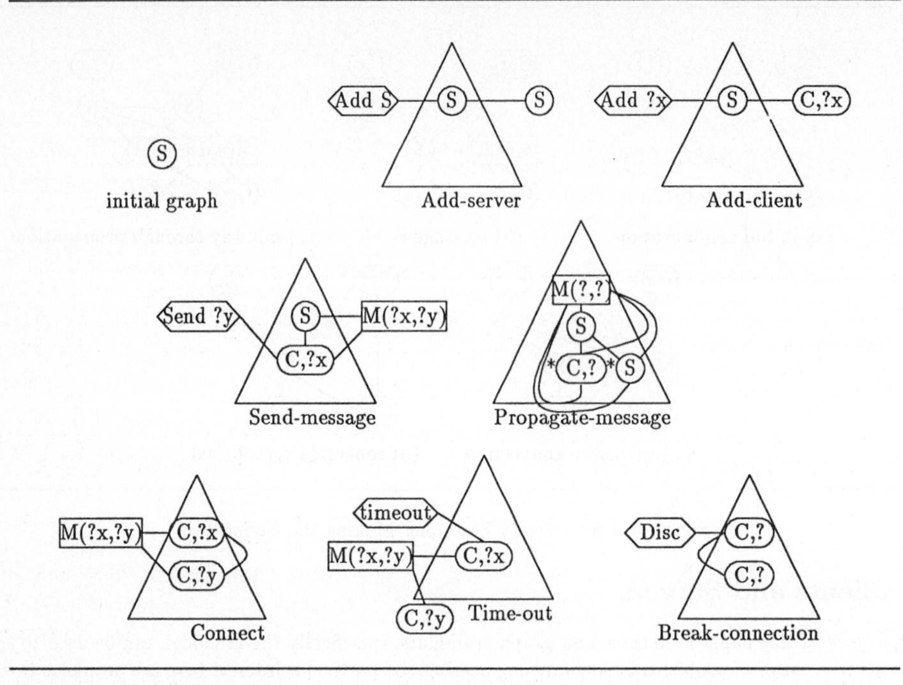

Figure 3: Client-Server Example: The Δ-specification

causing message sends, connection requests, and even timeouts and disconnections. We intend to specify only the network level here, ignoring the details of the other levels. In Section 3 we will explain how multiple levels and their interactions can be specified. Hexagonal nodes (triggers, explained in Section 3.1) will specify an interface point between levels.

The Δ-specification of the network level is shown in Figure 3. The initial configuration is just a single server with no clients. Δ-productions specify graph transformations that add clients to servers, add servers to servers, connect existing servers (so the server configuration can be more than just tree-like), and send connection requests from clients to servers. These productions are all straightforward.

The `Propagate-message` production propagates a message to all clients and servers connected to a particular client. Despite its seeming complexity, it is actually very simple:

- The client C with attribute a has sent a message indicating that it wants to connect to a client with attribute b. This message has reached server S.
- Server S has other servers and clients in its neighborhood.
- The effect of the production is to propagate the message to all clients and servers which have not already received it (the restriction ensures that clients and servers do not receive the message more than once). This is an example of the specification writer imposing a constraint on the implementation, and is necessary because an arbitrary topology of servers can include cycles.
- Notice the *-groups (one containing the `(C,?)`-node and its incident edges, the other containing an `S`-node and its incident edges) must match maximally, meaning that the message is propagated to *all* adjacent clients and servers that have not already received it.

The transformation that establishes a connection and halts all further message forwarding once a matching client has been found is represented by the production `Connect`. The unification in the graph matching phase of the rewriting ensures that the attribute of the message and that of the client match. The message is in the retraction; this indicates that the message vertex (and therefore all edges incident on it) are automatically removed as part of the connection establishment process. The actual connection is shown in the insertion part of the production.

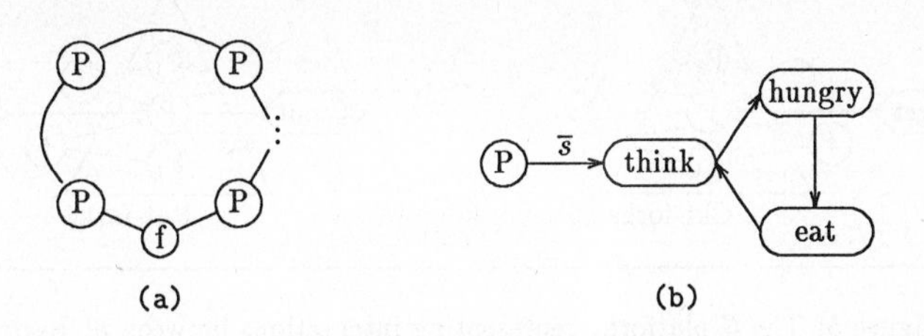

Figure 4: Two views of a dining philosophers system. The global view (a) expresses only the interaction between philosophers while the internal view (b) shows the state of each individual philosopher.

Application of the `Break-connection` production produces the obvious results. Application of this production is triggered by a different level, either an internal event in one of the clients or an external (system) event.

The final production, `Time-out`, implements the requirement that if, after some amount of time, the match has not succeeded, the client (or an external event) might abort the connection request. The restriction of `Time-out` that the connection must not be about to succeed is placed for efficiency and to demonstrate the utility of restriction regions.

Note that many clients can be requesting connections at the same time (*i.e.* there is no restriction on the number of active messages in the system). It may be useful to restrict a client from requesting more than one connection at a time; placing a restriction on `Send-message` would accomplish this.

3 Dynamic Dining Philosophers

An informal specification of the static dining philosophers problem is:

> Consider a number of philosophers sitting at a table, with a single fork between each adjacent pair. Each philosopher spends his time at the table thinking (meditating) for some finite time and eating for some finite time, infinitely alternating between the two states. When a philosopher wants to start eating after he has been thinking for some time, he must pick up both adjacent forks (he is eating spaghetti). Since there is contention for each fork by the two philosophers it is between, the problem is to specify the system in such a way that there is no deadlock (a state in which no philosopher can eat).

We will extend this to the dynamic problem in Section 3.3

The state of a solution program can be naturally expressed using a graph representing the status of the table at a given time. `P`-nodes represent philosophers seated at the table and `f`-nodes represent forks that are on the table and therefore available for use by an adjacent philosopher. There is at most one fork between any two philosophers. A sample state graph is illustrated in Figure 4(a).

Each philosopher is in one of three states; he is either thinking, hungry (waiting to eat), or eating. This is also naturally expressed in graphical notation, as shown in Figure 4(b). Each `P`-node has a finite state machine with those three states and an $\overline{s}$-edge pointing to the current state of the philosopher.

This view of the state provides a nice modular division of the solution of the dining philosopher problem. Dividing the functionality of the system in this way allows us to graphically specify the interaction between philosophers separate from the state of an individual philosopher. Doing so allows the system to be examined at different levels. At the global level, philosophers pick up their forks, hold them a while, and put them down. At the local level, a philosopher thinks for a while, gets hungry, and eats.

Remember that we divided the solution of the client-server problem into separate levels in a similar way. For Δ to be useful for specifying solutions to concurrent problems, we must be able to specify solutions in a systematic manner, dividing problems into subproblems, building Δ-specifications for each subproblem separately, and combining these specifications to provide a complete, modular solution to the main problem.

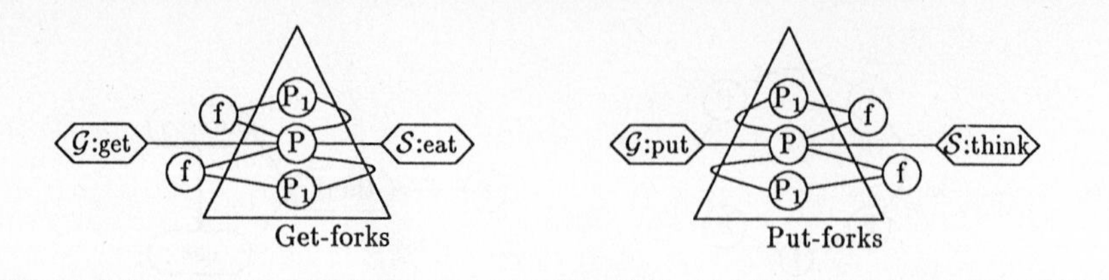

Figure 5: The $\mathcal{G}$ platform, representing interactions between philosophers.

Such a solution would be able to be modified and upgraded by adding new modules or by modifying specific modules without affecting other modules.

3.1 Platforms and Triggers

Although no one would argue that a program couldn't be written as one large unit, it is much easier to write and to understand if it is divided into modules, each a part of some logical partition of the problem. Likewise, specifying a program in Δ is easier if the main problem is divided into smaller, more manageable subproblems, and a set of Δ-productions is developed for each of these subproblems. These independent sets of Δ-productions are called platforms and the interface between them is provided by triggers. A trigger is a node with a special shape that provides the interface between platforms (we use hexagons to represent triggers). An *input trigger* is a trigger that appears in the precondition of every Δ-production in the platform. A platform is *called* by placement of the platform's input triggers into the graph[2]. *Output triggers* are triggers that appear in the insertions of Δ-productions in the platform, indicating the platform *calling* another platform.

Since the labels of nodes in the graph can be tuples of arbitrary structure, the labels of triggers in a platform can contain variables. Parameters are passed to a platform by unification of the label of the trigger in the graph with the labels of triggers in the platform.

The placement of a trigger in the graph does not guarantee that a useful transformation of the graph will occur at the location of the trigger. The platform is called by placement of the trigger but elements isomorphic to the rest of the elements in the preconditions of some of the platform's Δ-productions must exist (and elements matching the corresponding restrictions must not exist) in order for the Δ-productions to be applicable.

3.2 A Modular Solution to the Dining Philosophers Problem

The Δ-productions to specify the dining philosophers system are broken into two platforms accordingly. The $\mathcal{G}$ platform, shown in Figure 5, performs the global graph transformations, *i.e.* picking up and putting down the forks. The $\mathcal{S}$ platform, shown in Figure 6, performs the local transformations, *i.e.* each philosopher's state changes. The two communicate by placing parameterized triggers into the graph. The shape (hexagonal) indicates that the node is a trigger and the label contains a key, $\mathcal{G}$ or $\mathcal{S}$, indicating the platform being called and an actual parameter, *e.g.* `get` and `think`, indicating the action to be taken. The two platforms must communicate since some state changes cause a global action, *e.g.* transition to the hungry state requires that the philosopher try to pick up the forks, and some global actions cause a state change, *e.g.* picking up the forks allows the philosopher to eat.

The subscripts on the labels of P-nodes in `Get-forks` and `Put-forks` indicate that the nodes are in a single fold. That is, during application of `Get-forks` or `Put-forks` the P_1-nodes nodes might match a single P-node in the state graph, so the grammar works when there are only two philosophers at the table.

The `Stall` production models the need for arbitrary time delays. It can be applied any time a philosopher is eating or thinking but performs no transformation. Without this Δ-production, a philosopher would start eating and then immediately indicate that he is done eating and put down the forks. Likewise, thinking

[2]This is using the function analogy of functional languages. To use an actors system analogy, we would say, "A message is sent to a platform by placing the platform's input trigger into the graph."

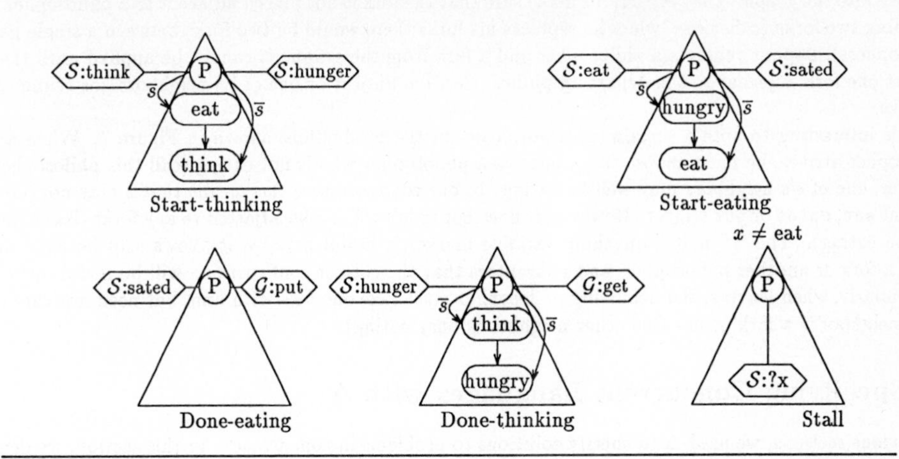

Figure 6: The $\mathcal{S}$ platform, representing state changes for an individual philosopher.

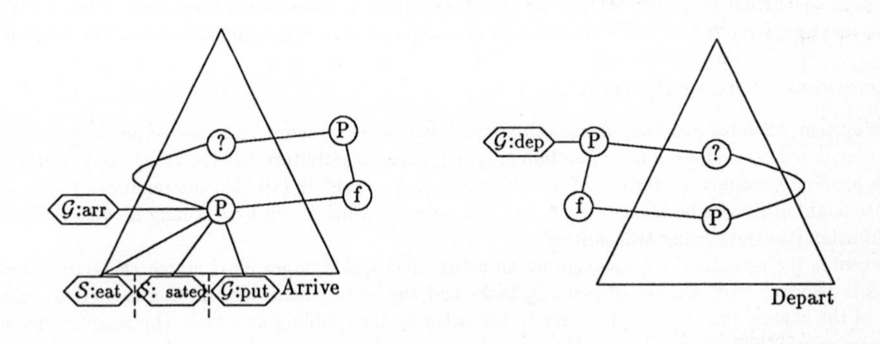

Figure 7: New productions to allow philosophers to enter and leave.

would take no time. The `Stall` production can be applied any finite number of times before another Δ-production in the $\mathcal{S}$ platform is applied, so that thinking and eating can take arbitrary time. The guarantee of fairness of production application in Δ [7] ensures that the `Stall` production can never be applied an infinite number of times. The guard ensures that a philosopher can start eating immediately when he has both forks, *i.e.* a philosopher is only hungry as long as he is trying to pick up his forks.

The specification is deadlock-free since picking up both forks is an atomic action, *i.e.* at any time a philosopher is eating or one of his neighbors is eating, and fairness ensures that an eating philosopher eventually relinquishes his forks.

3.3 Stepwise Refinement of the Specification

Suppose we now extend the problem, making the dining philosophers system dynamic, *i.e.* allowing new philosophers to enter and sit at the table and allowing existing philosophers to leave the table. Since this affects the global state of the table, but not the internal state of philosophers in the system, updating the specification only requires updating the $\mathcal{G}$ platform.

The two Δ-productions that represent philosophers arriving at and leaving the system are in Figure 7. These are in the $\mathcal{G}$ platform, since a $\mathcal{G}$ trigger is the only trigger that appears in the precondition of each Δ-production. Notice that, except for the triggers, the Δ-productions are mirror images of one another. This fits the intuitive notion that leaving should be the opposite of arriving. `Arrive` inserts a new philosopher and fork into the graph. The restrictions make sure that the fork is not placed adjacent to a philosopher that is holding two forks (otherwise, when he replaces his forks there would be two forks between a single pair of philosophers). Depart removes a philosopher and a fork from the graph. It cannot be applied until there is at least one fork adjacent to the departing philosopher, *i.e.* until it or one of its neighbors has relinquished its forks.

It is interesting to note a certain 'polymorphism' in the productions shown in Figure 7. When a new philosopher arrives, he must choose to sit next to a philosopher who is not eating (call this philosopher s); however, one of s's neighbors may well be eating. In our configuration, this means that s may not have an incident eat, sated or put trigger. However, it does not require that s be adjacent to any forks (its neighbors may be eating). The '?' match-anything variable marker is useful here, as it allows s to be adjacent to either a fork or another philosopher, and guarantees that the correct configuration will be maintained.

Similarly, when leaving, the departing philosopher must take one fork with him, but need not care if his other neighbor is a fork or not (his other neighbor is busy eating).

4 Specifying Concurrent Languages with Δ

In previous sections, we used Δ to specify solutions to problems in concurrency. In this section, we demonstrate how Δ is useful for specifying entire languages. Although Δ is powerful enough to specify any language, it is clear that it is more useful for specifying languages with a graphical intuition, such as concurrent languages and object-based languages. We illustrate this by presenting a specification for a most general object-based concurrent language, Actors [1]. First we briefly describe actor languages. Then we specify the actor language by giving a Δ-GRAMMAR and initial graph that define the behavior of the language.

4.1 Overview of Actor Systems

An actor system, an actor program in execution, consists of a set of *actors* and a set of pending *tasks*. Each actor is a mail *address* and a *behavior function* mapping tasks to activities that the actor can perform. Each task is a pending message. It consists of an *essence* (called a "tag" in [1]) that distinguishes it from other tasks that might otherwise be identical, a *target* (the actor to which it must eventually be delivered), and a *communication* (the data being transmitted).

An *event* is the reception of a message by an actor. The task's target must match the actor's address. The task is removed from the set of pending tasks and the actor's behavior function is used to calculate the rest of the change that the event causes to the actor system: adding new tasks (by sending messages), adding new actors (by creating new addresses and giving each of them an initial behavior), and specifying the next behavior of the current actor (or more correctly, a new actor with the same mail address).

An event is atomic. The actor and task involved provide all the data to calculate the event's effect on the actor system. Both are deleted (although the actor is always replaced by another actor with the same address, and this is usually described as being the next behavior of the same actor) and new actors and tasks to add to the system state are computed functionally. Though formally events are atomic, any realization has to implement them as sets of discrete activities. There is a lot of internal concurrency possible because the model does not specify a sequence on these activities. There is also external concurrency, events can be processed in parallel so long as no actor is used more than once at any given time.

4.2 An Actor Language

In this section, we present a most general actor language (hereafter referred to simply as *actors*) that we are specifying, based on [1]. Throughout the remainder of this discussion, we use the following abbreviations:

n	name of any kind
B	behavior name
L	list of acquaintance names
S	statement

The abstract syntax for actors is:

Prog	$\Rightarrow$	$B \mid Beh\ Prog$	[Behavior name and list of Beh]
Beh	$\Rightarrow$	$B\ L\ L\ S$	[Behavior definition]

L	$\Rightarrow$	ϵ | n L	[List of actual/formal param names]
B	$\Rightarrow$	n	
S	$\Rightarrow$	ϵ	[no op]
		| S S	[concurrent eval]
		| n B L	[create actor]
		| n n S S	[conditional]
		| n L	[send message]
		| B L	[becomes]

By the first production, an actor program is a single behavior name and a list of behavior definitions. Execution begins with a single actor called the receptionist with initial behavior as indicated by this behavior name. The receptionist is known, by definition, to the outside world so that the actor system can receive messages (*i.e.* input) from the outside. The initial receptionist behavior has no parameters so that the receptionist is instantiated with that behavior (without needing further data) to form the starting state for the actor program.

Each behavior definition is tagged with a name and has two lists of parameters and a statement. The first parameter list (elements called *creation acquaintances*) names the data received from an actor's creator when the actor is created. The second parameter list (elements called *message acquaintances*) names the data accompanying a message reception. The statement is the *body* of the behavior that is executed when an event occurs, *i.e.* a message is received.

Parameter passing is through a positional copying of data, where the length of the actual and formal parameter lists must match. All data are acquaintances with scope of one behavior instantiation. Besides creation and message acquaintances, there is the special acquaintance "`self`" that refers to the actor's own mail address, and there are *local acquaintances* bound by `create` statements to newly created actors.

To simplify presentation, we ignore error conditions and present a specification for only correct actor programs.

4.3 The Specification of Actors

The specification of actors consists of the *state graph* representing the state of the actor program, and the twelve productions (shown in Figures 8 and 9) that specify rules for transforming the state graph according to the actor execution model. Performing these transformations executes the actor program.

The initial state graph is a simple compilation of the abstract syntax tree of an actor program. Uses of behavior names (by `create` and `becomes` statements) are linked to the definitions of those behaviors and nodes representing the receptionist are created.

Each actor in the system is represented in the state graph by an `A`-node (its mail address), a set of $\overline{m}$-edges (its mailbag contents), and a set of $\mathcal{X}$-nodes (one for each message received and, hence, behavior instantiated). Since a `becomes` statement can be executed concurrently with other statements, a single actor can be executing several behaviors at once.

Each execution of a behavior by an actor is called an *execution context* and is represented by an $\mathcal{X}$-node and a set of outdirected edges. When a behavior is instantiated (by a `becomes` or `create` statement) there are three edges: a `self`-edge which is an acquaintance usable by the code that also gives the actor address from which a message must be received, a $\overline{b}$-edge recording the behavior the context must exhibit, and a $\overline{p}$-edge recording the actual parameters. $\overline{c}$-edges are used to control the copying of parameters[3].

Execution of an actor program is performed by transformations of the state graph as specified by Δ-productions. The transformation rules represented by these Δ-productions are applied to the state graph to control interpretation of the actor system. The execution is explained in two parts. First, the way that parameters are passed so that an actor can start executing a behavior and receive a message is explained, and second, the execution of the six kinds of statements is described.

Actor Behavior Startup The Δ-productions specifying the rules to startup a behavior are shown in Figure 8. We describe each of the transformations they represent individually.

[3] The receptionist starts with two outdirected edges the first time. There are no actual parameters in the system and no formal parameters in its initial behavior definition, so the step of copying data to instantiate it is omitted. Thus the edge recording actual parameters is not used.

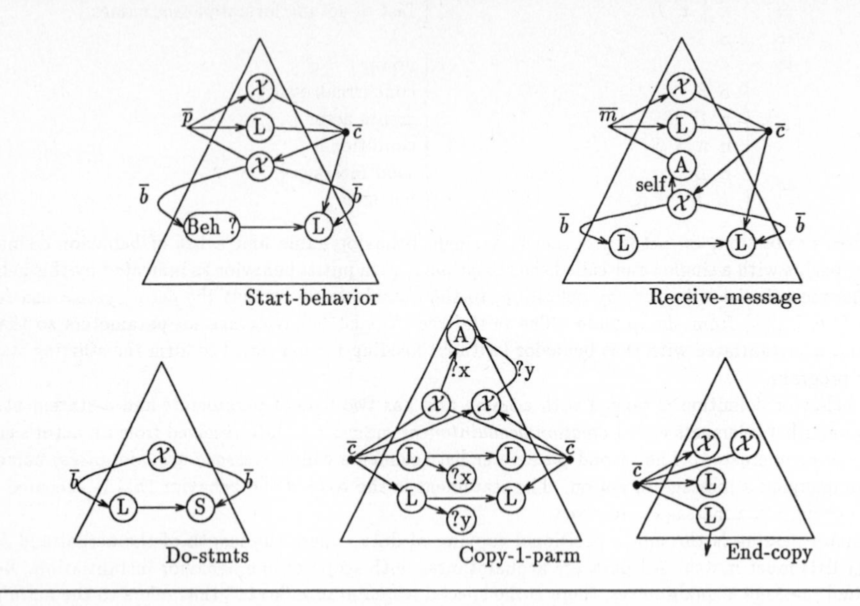

Figure 8: Starting a behavior and receiving a message.

Start-behavior — The $\overline{p}$-edge indicating instantiation of a behavior is removed and a $\overline{c}$-edge is inserted to control copying of parameters (of creation acquaintances). A $\overline{c}$-edge has two sources and two targets. Its sources are an L-node giving the list of actual parameter names yet to be copied and an execution context giving values for each of these names. Its targets are an L-node giving the list of formal parameter names yet to be copied and an execution context that will receive values for each of these names. The $\overline{b}$-edge is advanced. Parameter copying is done concurrently with continued execution controlled by the $\overline{b}$-edge.

Copy-1-parm — The $\overline{c}$-edge is advanced one name down each of the parameter lists and a new acquaintance edge is created in the receiving execution context. The new edge is labeled with the name in the formal parameter list and its target is the acquaintance of the actual parameter of the old execution context.

End-copy — Removes the $\overline{c}$-edge when all parameters have been copied.

Receive-message — Removes one $\overline{m}$-edge (representing a message in the actor's mailbag). A $\overline{c}$-edge is inserted to copy the actual parameters (in the form of acquaintances) in the message to the receiving behavior's (execution context's) formal parameters (message acquaintances). The $\overline{b}$-edge is advanced so that execution of the body of the behavior is done concurrently with parameter copying.

Do-stmts — Advances the $\overline{b}$-edge to begin execution of the body of the behavior.

Statement Execution Productions representing statement execution are given in Figure 9. We now explain the transformations they represent individually. Each $\overline{b}$-edge represents a separate thread of execution in the behavior.

Null — the $\overline{b}$-edge is retracted. This terminates a thread of execution, reducing the amount of concurrency.

Concurrent-eval — execution proceeds concurrently on both sub-statements, each controlled by a separate $\overline{b}$-edge.

Send — the $\overline{b}$-edge is removed and an $\overline{m}$-edge is placed in the graph, representing delivery of a message to the actor pointed to by the named acquaintance. Delivery of messages is immediate, but arrival-order nondeterminism is achieved because delivered messages are not queued. Reception (*i.e.* arrival) may be in any (fair) order among messages that have been delivered to an actor.

To apply the Send production an acquaintance edge of the correct name must exist pointing from the $\mathcal{X}$-node to the target actor's A-node. If this edge has not yet been created, application of this production

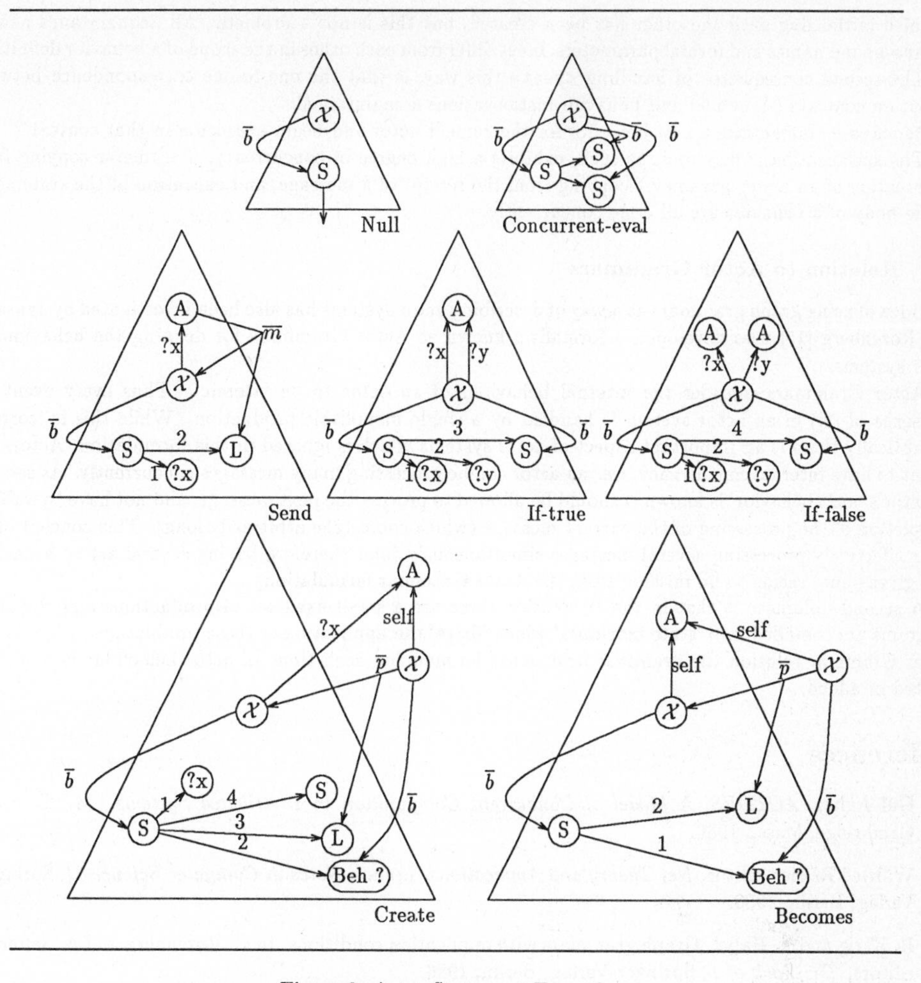

Figure 9: Actor Statement Execution.

is blocked until parameter copying (of both creation and message acquaintances) has proceeded far enough. A similar situation exists with the `If-true` and `If-false` productions.

`If-true` — tests for equality of addresses. The production is only applicable if both of the named acquaintances point to the same actor address. The $\overline{b}$-edge controlling execution then proceeds down the "then" branch.

`If-false` — the production is only applicable if the named acquaintances point to different actor addresses. The two `A`-nodes cannot match to the same graph node. If it applies, execution continues down the "else" branch.

`Create` — create a new actor address and an execution context to instantiate the first behavior for that actor. Also create a $\overline{p}$-edge to pass actual parameters to the new actor. Bind the new actor to a named acquaintance and continue execution of the current context concurrently with the new context.

Adding the new acquaintance to the current execution context rather than creating a new execution context (to extend by the acquaintance) has two results. First, the name is visible outside the scope of the `create` statement; the scope of the `create`-bound name is the entire behavior. Thus the code fragment

```
create e = behavior(f); create f = behavior(e)
```

makes sense. Two actors are created and each is passed the other as a parameter. Neither will be able to finish initializing until the other has been created, but this is not a problem. All acquaintance names, `create`-bound names and formal parameters, must differ from each other in the scope of a behavior definition.

The second consequence of handling `create` this way, is that the one-to-one correspondence between execution contexts ($\mathcal{X}$-nodes) and behavior instantiations is maintained.

`Becomes` — instantiate a new behavior for the current actor and begin execution in that context.

The specification is lazy and, as such, exhibits a high degree of concurrency. Parameter copying from the creation of an actor, parameter copying from the receipt of a message, and execution of the statements in the body of a behavior are all done concurrently.

4.4 Relation to Actor Grammars

The idea of using graph grammars as a way of describing actor systems has also been investigated by Janssens and Rozenberg [15] who developed a formalism known as Actor Grammars for defining the behaviour of actor systems.

Actor Grammars consider the internal behaviour of an actor to be 'atomic'. Thus every event (in the sense of [1]) in an actor system is handled by a single monolithic production. While this is 'correct' semantically there is an important aspect of actor systems which is ignored in this formulation. Actors are meant to have internal concurrency, *i.e.* an actor can be processing many messages concurrently. As soon as an actor's 'next behavior' is known it should be allowed to process the next message, and not have to wait for completion of the processing of the current message (which could take arbitrarily long). This concept of an actor effectively processing several messages simultaneously (and therefore having several active behaviors at a given time) seems to be missing from the Actor Grammar formulation.

A second difference is that in the Δ solution there are a small fixed set of productions and the actor programs are compiled into 'state machines' which 'drive' the application of these productions, while in the Actor Grammar solution the grammar itself must be modified each time an actor behaviour is modified, deleted or added.

References

[1] Gul Agha. *ACTORS: A Model of Concurrent Computation in Distributed Systems.* M.I.T. Press, Cambridge, Mass., 1986.

[2] Wilfried Brauer, editor. *Net Theory and Applications, Lecture Notes in Computer Science 84*, Springer-Verlag, Berlin, 1980.

[3] H. Ehrig and A. Habel. Graph grammars with application conditions. In G. Rozenberg and A. Salomaa, editors, *The Book of L*, Springer-Verlag, Berlin, 1986.

[4] Hartmut Ehrig. Introduction to the algebraic theory of graph grammars (a survey). In Volker Claus, Hartmut Ehrig, and Grzegorz Rozenberg, editors, *Graph Grammars and their Application to Computer Science and Biology, LNCS 73*, pages 1–69, Springer-Verlag, Heidelberg, 1979.

[5] Hartmut Ehrig. Tutorial introduction to the algebraic approach of graph-grammars. In Hartmut Ehrig, Manfred Nagl, Grzegorz Rozenberg, and Azriel Rosenfeld, editors, *Graph Grammars and their Application to Computer Science, LNCS 291*, pages 3–14, Springer-Verlag, 1987.

[6] Steven K. Goering. *A Graph-Grammar Approach to Concurrent Programming.* PhD thesis, University of Illinois at Urbana-Champaign, 1990. Tech. report UIUCDCS-R-90-1576.

[7] Steven K. Goering, Simon M. Kaplan, and Joseph P. Loyall. Theoretical properties of Δ. Book in preparation.

[8] H. Göttler. Semantical description by two-level graph-grammars for quasihierarchical graphs. In *Proc. 'Workshop WG 78 on Graphtheoretical Concepts in Comp. Science', Applied Computer Science, 13*, Hanser-Verlag, Munich, 1979.

[9] Herbert Göttler. Attributed graph grammars for graphics. In Hartmut Ehrig, Manfred Nagl, and Grzegorz Rozenberg, editors, *Graph Grammars and their Application to Computer Science, LNCS 153*, pages 130–142, Springer-Verlag, 1982.

[10] Herbert Göttler. Graph grammars and diagram editing. In Hartmut Ehrig, Manfred Nagl, Grzegorz Rozenberg, and Azriel Rosenfeld, editors, *Graph Grammars and their Application to Computer Science, LNCS 291*, pages 216–231, Springer-Verlag, 1987.

[11] Annegret Habel. *Hyperedge Replacement: Grammars and Languages.* PhD thesis, University of Bremen, 1989.

[12] Annegret Habel and Hans-Jorg Kreowski. May we introduce to you: hyperedge replacement. In Hartmut Ehrig, Manfred Nagl, Grzegorz Rozenberg, and Azriel Rosenfeld, editors, *Graph Grammars and their Application to Computer Science, LNCS 291*, pages 15–26, Springer-Verlag, 1987.

[13] Manfred Jackel. Ada-concurrency specified by graph grammars. In Hartmut Ehrig, Manfred Nagl, Grzegorz Rozenberg, and Azriel Rosenfeld, editors, *Graph Grammars and their Application to Computer Science, LNCS 291*, pages 262–279, Springer-Verlag, 1987.

[14] Manfred Jackel. *Specification of the Concurrent Constructs of Ada by Graph Grammars (in German).* PhD thesis, University of Osnabrück, 1986.

[15] D. Janssens and Grzegorz Rozenberg. Basic notions of actor grammars. In Hartmut Ehrig, Manfred Nagl, Grzegorz Rozenberg, and Azriel Rosenfeld, editors, *Graph Grammars and their Application to Computer Science, LNCS 291*, pages 280–298, Springer-Verlag, 1987.

[16] Dirk Janssens and Grzegorz Rozenberg. Graph grammars with node-label control and rewriting. In Hartmut Ehrig, Manfred Nagl, and Grzegorz Rozenberg, editors, *Graph Grammars and their Application to Computer Science, LNCS 153*, pages 186–205, Springer-Verlag, 1982.

[17] Simon M. Kaplan. Foundations of visual languages for object-based concurrent programming. In Gul Agha, Peter Wegner, and Akinori Yonezawa, editors, *Object-Based Concurrent Programming*, Addison-Wesley, to appear 1990.

[18] Simon M. Kaplan, Steven K. Goering, Joseph P. Loyall, and Roy H. Campbell. Δ *Working Papers.* Technical Report UIUCDCS-R-90-1597, University of Illinois Department of Computer Science, April 1990.

[19] Joseph P. Loyall. *Specification of Concurrent Programs Using Graph Grammars.* PhD thesis, University of Illinois, To appear, 1991.

[20] Michael Main and Grzegorz Rozenberg. Fundamentals of edge-label controlled graph grammars. In Hartmut Ehrig, Manfred Nagl, Grzegorz Rozenberg, and Azriel Rosenfeld, editors, *Graph Grammars and their Application to Computer Science, LNCS 291*, pages 411–426, Springer-Verlag, 1987.

[21] R. Milner. A calculus of communicating systems. In *Lecture Notes in Computer Science, LNCS 92*, Springer-Verlag, Berlin, 1980.

[22] M. Nagl. *Graph Grammars: Theory, Application, Implementation (in German).* Verlag, Wiesbaden: Vieweg, 1979.

[23] Francesco Parisi-Presicce, Hartmut Ehrig, and Ugo Montanari. Graph rewriting with unification and composition. In Hartmut Ehrig, Manfred Nagl, Grzegorz Rozenberg, and Azriel Rosenfeld, editors, *Graph Grammars and their Application to Computer Science, LNCS 291*, pages 496–514, Springer-Verlag, 1987.

[24] Grzegorz Rozenberg. An introduction to the NLC way of rewriting graphs. In Hartmut Ehrig, Manfred Nagl, Grzegorz Rozenberg, and Azriel Rosenfeld, editors, *Graph Grammars and their Application to Computer Science, LNCS 291*, pages 55–66, Springer-Verlag, 1987.

[25] H.J. Schneider. Describing distributed systems by categorical graph grammars. In *LNCS 411*, pages 121–135, Springer-Verlag, Heidelberg, 1989.

Graph rewriting in some categories of partial morphisms

Richard Kennaway*
School of Information Systems, University of East Anglia, Norwich NR4 7TJ, U.K.

Abstract: We present a definition of term graph rewriting as the taking of a pushout in a category of partial morphisms, adapting the rather ad hoc definitions we gave in [Ken87] so as to use a standard category-theoretic concept of partial morphism. This single-pushout construction is shown to coincide with the well-known double-pushout description of graph rewriting whenever the latter is defined. In general, the conditions for the single pushout to exist are weaker than those required for the double pushout. In some categories of graphs, no conditions at all are necessary.

Keywords: graph rewriting, partial morphism, hypergraph, term graph, jungle, category, double pushout, single pushout.

CONTENTS

1. Two definitions of graph rewriting.

There is a long tradition of graph rewriting (e.g. [EPS73, ER80, PEM86, and others]) in which the concept is defined in the following category-theoretic way: A rewrite rule is a pair of morphisms L←K→R in a category of graphs. L is called the left hand side of the rule, R is the right hand side, and K is the interface graph. An occurrence of this rule in a graph G is a morphism L→G, and reduction is performed by constructing the remaining components of the

* This work was partially supported by ESPRIT basic research action no. 3074 (Semagraph), SERC grant no. GR/F 91582, and an SERC Advanced Fellowship.

diagram of Figure 1, so as to make both squares pushouts. Conditions are imposed on the rule and the occurrence to ensure that this can be done uniquely, up to isomorphism.

In [Rao84, Ken87], a definition of graph rewriting is given using a more complicated category of graphs but a simpler definition of rewriting. A rule is a morphism L→R, an occurrence of this rule in a graph G is a morphism L→G (of a restricted class), and the rewrite is performed by taking the pushout shown in Figure 2.

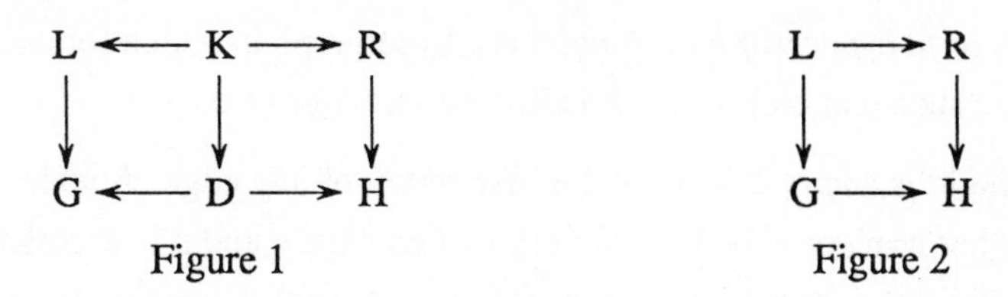

Figure 1 Figure 2

The pushout does not always exist; conditions are imposed on L→R and L→G to ensure that it does.

We show that both constructions can be seen as examples of a more general construction in categories of partial morphisms, in which both constructions are single pushouts.

In section 2 we introduce some categories of hypergraphs. In section 3 we describe the categorical concept of a partial morphism. In sections 4 and 5 we describe the relationship between the double- and single-pushout constructions in two categories of hypergraphs, proving that they coincide whenever the former is defined. We conclude by sketching some possible further developments of the theme.

2. Hypergraphs.

There are many different definitions of graphs in the literature. They may be directed or undirected, have edges or hyperedges, have labels of some sort on their nodes or edges, and so on. These technicalities for the most part make little difference to the theory. The particular kinds of graph we study here are just those most closely related to our desired applications.

2.1. DEFINITION. A *hypergraph* G over a set of function symbols Λ consists of a finite set of *nodes* N_G, a finite set of *hyperedges* E_G, a *labelling function* $L_G{:}E_G \to \Lambda$, and a *connection map* $c_G{:}E_G \to N_G^+$, where N_G^+ denotes the set of nonempty tuples of members of N_G. The members of $c_G e$ are the (first, second, etc.) *vertexes* of e. Each member of Λ is assumed to have an *arity* (a non-negative number) associated with it: the size of the tuple $c_G(e)$ is required to be one more than the arity of $L_G(e)$. Thus e.g. a hyperedge with one vertex is labelled by a zeroary symbol. The use of arities is merely a technical convenience.

A morphism $f{:}G \to H$ of hypergraphs consists of a function $N_f{:}N_G \to N_H$ and a function

E_f:$E_G \rightarrow E_H$ such that $c_H(E_f e) = N_f(c_G e)$ and $L_H(E_f e) = L_G e$. In the first condition, we have implicitly extended N_f to tuples in the obvious way.

$\mathfrak{H}(\Lambda)$ is the category of hypergraphs over Λ. Where the identity of Λ is not important, we may omit it. □

Also of interest to us is a subcategory of $\mathfrak{H}$, which — for reasons which will become clear — has properties quite different to $\mathfrak{H}$.

2.2. DEFINITION. A *term hypergraph* (or *jungle*) is a hypergraph in which for each node n there is at most one edge e such that $c(e)_1 = n$. e is called the *out-edge* of n.

A node of a jungle is *empty* if it is not the first vertex of any edge. A node n is *accessible from* a node n' if either n=n', or n' is the first vertex of an edge e and n is accessible from some other vertex of e. Note that accessibility is only defined for jungles, not for all hypergraphs.

$\mathfrak{J}(\Lambda)$ is the category of jungles over Λ. □

The importance of jungles is that they provide a natural representation of terms as graphs. Figure 3 should make the correspondence clear. The small italic numbers indicate the ordering of vertexes of each hyperedge. Note that empty nodes represent free variables, and that the identity of variables is expressed by identity of nodes of the jungle, without any need for a separate set of variable labels.

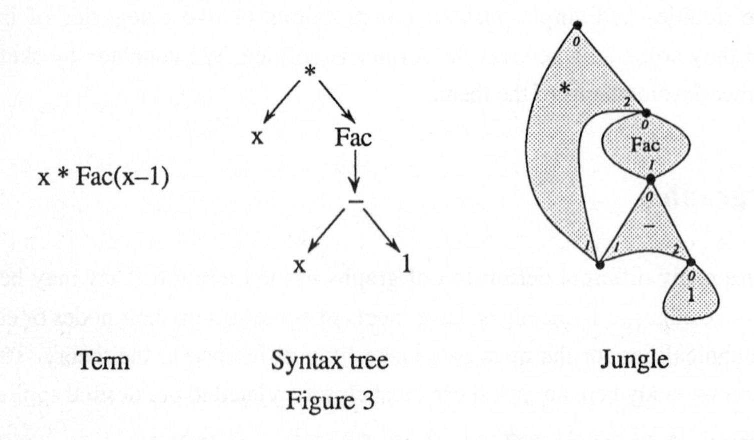

Figure 3

In [HP88], the double-pushout construction is applied to the category of jungles, to obtain a description of the usual graph-rewriting implementation of term rewriting. (We omit the merge phase of a jungle rewrite here, which automatically coalesces isomorphic subgraphs of the resulting jungle, and which [HP88] does not use the double pushout to describe.)

We shall define a single-pushout construction in categories of partial morphisms derived from $\mathfrak{H}$ and $\mathfrak{J}$, and show that they coincide with the standard double-pushout definitions.

3. Partial morphisms.

In this section we recall the concept of partial morphism in category theory, and study the existence and construction of pushouts in categories of partial morphisms. For reasons of space, we assume familiarity with some basic concepts of category theory and their fundamental properties: subobjects, limits, colimits, pushouts, and pullbacks.

Given a category **C**, we can (if **C** satisfies certain conditions) construct a category $\wp$(**C**) of "partial morphisms" of **C** [Rob88]. The objects of $\wp$(**C**) are those of **C**. An arrow of $\wp$(**C**) from A to B is an arrow of **C** from some subobject of A to B. Thus it is a pair $A \leftarrowtail K \rightarrow B$, where $K \rightarrowtail A$ is a monomorphism. More precisely, it is an equivalence class of such pairs: $A \leftarrowtail K_0 \rightarrow B$ and $A \leftarrowtail K_1 \rightarrow B$ represent the same arrow of $\wp$(**C**) if there is an isomorphism making Figure 4 commute. In particular, $K_0 \rightarrowtail A$ and $K_1 \rightarrowtail A$ must be (representatives of) the same subobject of A. We write $K \stackrel{\subseteq}{\rightarrow} A$ for a subobject of A, or a monomorphism of **C** considered as representing some subobject of A.

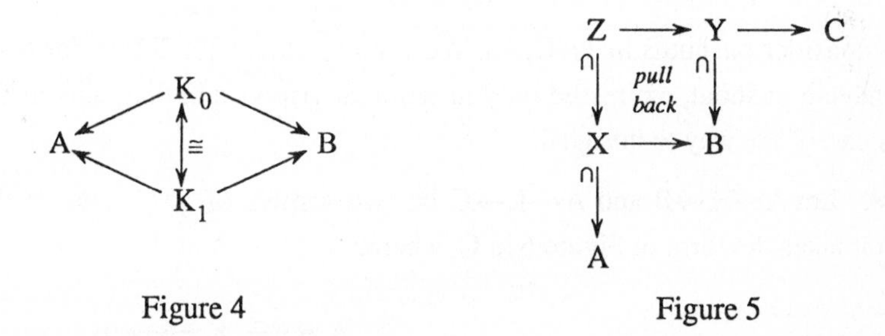

Figure 4 Figure 5

Morphisms $A \stackrel{\supseteq}{\leftarrow} X \rightarrow B$ and $B \stackrel{\supseteq}{\leftarrow} Y \rightarrow C$ are composed as in Figure 5. The square ZYXB is a pullback. **C** is assumed to have all such pullbacks. Standard properties of the pullback imply that $Z \rightarrow X$ is monic. A pullback of this form is called an *inverse image* (of $Y \stackrel{\subseteq}{\rightarrow} B$ along $X \rightarrow B$). Thus we assume that **C** has inverse images.

More generally, we may wish to restrict attention to some subclass $\mathcal{M}$ of the subobjects of **C**. $\wp$(**C**,$\mathcal{M}$) is the subcategory of $\wp$(**C**) with the same objects, but in which $A \stackrel{\supseteq}{\leftarrow} K \rightarrow B$ is a morphism iff $B \stackrel{\subseteq}{\rightarrow} A$ is in $\mathcal{M}$. It is necessary that $\mathcal{M}$ contain all isomorphisms and be closed under composition and inverse images. Such an $\mathcal{M}$ will be called *admissible*. Note that $\mathcal{M}$ is (the set of arrows of) a subcategory of **C**.

Given $f:X \stackrel{\subseteq}{\rightarrow} A$ in $\mathcal{M}$ and $g:X \rightarrow B$ in **C**, we denote the morphism $A \stackrel{\supseteq}{\leftarrow} X \rightarrow B$ in $\wp$(**C**,$\mathcal{M}$) by $(f,g):A \rightarrow B$, or if we wish to indicate X explicitly, by (f,X,g).

Some simple useful properties of $\wp$(**C**,$\mathcal{M}$) are collected in the next theorem.

3.1. THEOREM. (i) There is an embedding of **C** into $\wp$(**C**,$\mathcal{M}$), which maps $f:A \rightarrow B$ to $(id_A,f):A \stackrel{\equiv}{\leftarrow} A \rightarrow B$.

(ii) There is a forgetful functor from $\mathcal{M}^{op}$ (the opposite category of monomorphisms in $\mathcal{M}$) to $\wp(\mathbf{C},\mathcal{M})$, mapping $f{:}B\to A$ to $(f,id_A){:}A\leftarrow B \xrightarrow{=} B$. □

We call these embeddings the *covariant* and *contravariant* embeddings respectively.

(iii) An arrow $(f,g){:}A\to B$ of $\wp(\mathbf{C},\mathcal{M})$ is a monomorphism iff f is an isomorphism and g is a monomorphism.

(iv) $\wp(\wp(\mathbf{C},\mathcal{M}),\mathcal{M})$ is equivalent to $\wp(\mathbf{C},\mathcal{M})$ (where the second occurrence of $\mathcal{M}$ denotes the image of $\mathcal{M}$ in $\wp(\mathbf{C},\mathcal{M})$ by the inclusion of C in $\wp(\mathbf{C},\mathcal{M})$.

(v) The embedding of C in $\wp(\mathbf{C},\mathcal{M})$ preserves all pullbacks that exist in C.

PROOF. All routine. Note that (iv) is an easy consequence of (iii), and (v) is immediate from the fact that $(id,h)\cdot(f,g) = (f,h\cdot g)$. □

However, note that the embedding of (v) does not preserve products. Indeed, when C has products, the embedding can only preserve them if $\mathcal{M}$ consists only of isomorphisms [Rob88], and then $\wp(\mathbf{C},\mathcal{M})$ is equivalent to C.

We now consider pushouts in $\wp(\mathbf{C},\mathcal{M})$. We want to obtain conditions for two arrows of $\wp(\mathbf{C},\mathcal{M})$ to have a pushout, expressed only in terms of arrows and diagrams in C. The next theorem goes part of the way to this goal.

3.2. THEOREM. Let $A\xleftarrow{\supseteq}K\to B$ and $A\xleftarrow{\supseteq}L\to C$ be two arrows of $\wp(\mathbf{C},\mathcal{M})$. If they have a pushout, then it takes the form of Figure 6 in C, where:

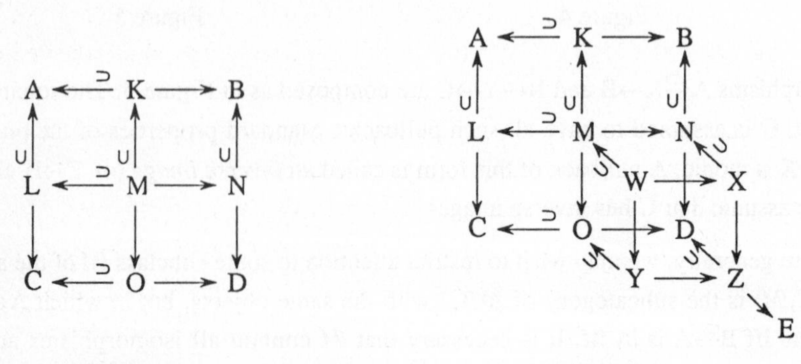

Figure 6

Figure 7

(i) the diagram commutes, and the top right and bottom left squares are pullbacks in C.

(ii) $M\xrightarrow{\subseteq}A$ (the diagonal of the top left square) is the largest possible subobject of A in $\mathcal{M}$ subject to (i).

(iii) Given (i) and (ii), $N\xrightarrow{\subseteq}B$ and $O\xrightarrow{\subseteq}C$ are the largest subobjects of B and C which pull back to $M\xrightarrow{\subseteq}K$ and $M\xrightarrow{\subseteq}L$ along $K\to B$ and $L\to C$ respectively.

(iv) MNOD is a pushout in $\wp(\mathbf{C},\mathcal{M})$.

Conversely, if a diagram satisfying (i)–(iv) exists, then it is a pushout in $\wp(\mathbf{C},\mathcal{M})$.

PROOF. (i) This is just the condition for the border of the diagram to be a commutative square in $\wp(\mathbf{C},\mathcal{M})$.

(ii) If there were a similar diagram with a larger or incomparable $M' \xrightarrow{\subseteq} A$, then that diagram, considered as a commutative square in $\wp(\mathbf{C},\mathcal{M})$, could not factorise through the one with M. The latter would therefore not be a pushout in $\wp(\mathbf{C},\mathcal{M})$.

(iii) Suppose (i) and (ii) are satisfied, and this condition were false. That is, one could find subobjects $N' \xrightarrow{\subseteq} B$ and $O' \xrightarrow{\subseteq} C$, pulling back to M and forming a commutative square in **C** with some arrows $N' \rightarrow D'$ and $O' \rightarrow D'$, but such that one or both of these two subobjects failed to be contained in $N \xrightarrow{\subseteq} B$ and $O \xrightarrow{\subseteq} C$ respectively. Then the resulting diagram, as a commutative square in $\wp(\mathbf{C},\mathcal{M})$, could not factorise through the given diagram.

(iv) Let M, N, and O be chosen satisfying (i)-(iii). Suppose MNOD is not a pushout in $\wp(\mathbf{C},\mathcal{M})$. Then there is a pair of arrows of $\wp(\mathbf{C},\mathcal{M})$ from N and O to some P which factorise not at all, or not uniquely, through $N \rightarrow D$ and $O \rightarrow D$. One then finds that the composition of these arrows with $B \xleftarrow{\supseteq} N \xrightarrow{=} N$ and $C \xleftarrow{\supseteq} O \xrightarrow{=} O$ provides a similar refutation of the pushout property for the outer border.

For the converse, suppose we have a diagram such as Figure 6 satisfying (i)–(iv). Let $B \xleftarrow{\supseteq} X \rightarrow E$ and $C \xleftarrow{\supseteq} Y \rightarrow E$ be two arrows of $\wp(\mathbf{C},\mathcal{M})$ forming a commutative square with $A \xleftarrow{\supseteq} K \rightarrow B$ and $A \xleftarrow{\supseteq} L \rightarrow C$. Then we can form a diagram of the same shape with M, N, O, and D replaced by W, X, Y, and Z, in which the top right and bottom left squares are pullbacks in **C**. By condition (ii), $W \rightarrow K$ and $W \rightarrow L$ must factorise through M. By condition (iii), $X \rightarrow B$ and $Y \rightarrow C$ must then factorise through N and O. Because certain of the arrows are monomorphisms, these factorisations are unique. The pushout property of MNOD in $\wp(\mathbf{C},\mathcal{M})$ then uniquely provides the remainder of Figure 7, in which all squares with just two opposite sides marked $\subset$ are pullbacks in **C**. Ths constitutes a unique factorisation of $B \xleftarrow{\supseteq} X \rightarrow E$ and $C \xleftarrow{\supseteq} Y \rightarrow E$ through $B \xleftarrow{\supseteq} N \rightarrow D$ and $C \xleftarrow{\supseteq} O \rightarrow D$. Therefore the border of the original diagram is a pushout in $\wp(\mathbf{C},\mathcal{M})$. □

The last condition of this theorem still refers to constructions in $\wp(\mathbf{C},\mathcal{M})$. To complete our characterisation, we seek conditions for a commutative square in **C** to be a pushout in $\wp(\mathbf{C},\mathcal{M})$. The next definition provides this.

3.3. DEFINITION. Let Figure 8 be a pushout square in **C**. This is a *hereditary* pushout (with respect to $\mathcal{M}$) if for every extension of this diagram as in Figure 9, where the monomorphisms are in $\mathcal{M}$ and EFAB and EAGC are pullbacks:

(i) there is a further extension, unique up to isomorphism, of the form of Figure 10, such that

$H \stackrel{\subseteq}{\to} D$ is in $\mathcal{M}$ and FBHD and GHCD are pullbacks, and

(ii) in this diagram, EFGH is a pushout.

If **C** has all pushouts, and those pushouts are hereditary, we say that **C** *has hereditary pushouts.* □

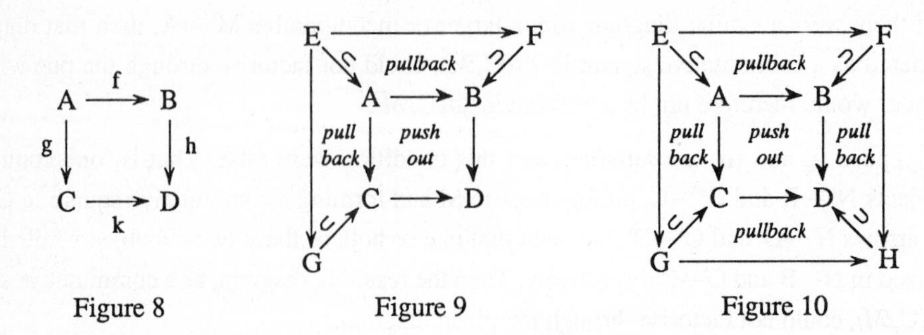

Figure 8 Figure 9 Figure 10

The property of having hereditary pushouts can be explained informally. In general, it is characteristic of categories whose objects have no internal "long-range" structure. The category of sets (with $\mathcal{M}$ the set of all monomorphisms) has hereditary pushouts, as does $\mathfrak{H}$. However, $\mathfrak{J}$ does not. Notice that in $\mathfrak{H}$, one can take any hypergraph, and glue together some of its nodes in any way, and obtain another object of $\mathfrak{H}$. In $\mathfrak{J}$ this is not so — if one glues together two nodes of a jungle, then to obtain a new jungle one must then glue together corresponding successors of those nodes, and their successors, and so on; and if one finds a mismatch of hyperedge labels, one cannot perform the gluing at all. A simple example of a non-hereditary pushout in $\mathfrak{J}$ is given by taking A = •, and B, C, and D to be the graph with one node, and one hyperedge of arity 0 labelled F. This uniquely determines the arrows of Figure 8, and this square is a pushout in $\mathfrak{J}$. However, A→B and A→C have no pushout in $\wp'(\mathfrak{J})$. For let X = •, Y = F, E = F, and consider the unique partial morphisms $B \stackrel{\supseteq}{\leftarrow} X \to E$ and $C \stackrel{\supseteq}{\leftarrow} Y \to E$. These form a commutative square in $\wp'(\mathfrak{J})$ with A→B and A→C. But there is no monomorphism $Z \stackrel{\subseteq}{\to} D$ whose pullbacks along B→D and C→D are $X \stackrel{\subseteq}{\to} B$ and $Y \stackrel{\subseteq}{\to} C$.

3.4. THEOREM. A commutative square in **C** is a pushout in $\wp(\mathbf{C},\mathcal{M})$ iff it is a hereditary pushout in **C**.

PROOF *If*: Let Figure 8 be a hereditary pushout in **C**. Let $\hat{p}:B \stackrel{\supseteq}{\leftarrow} F \to K$ and $\hat{q}:C \stackrel{\supseteq}{\leftarrow} G \to K$ be arrows of $\wp(\mathbf{C},\mathcal{M})$ such that $\hat{p}\cdot f = \hat{q}\cdot g$. This means that we have Figure 9, together with arrows F→K and G→K making a commutative square EFGK. We must show that there is a unique partial morphism $\hat{k}:D \stackrel{\supseteq}{\leftarrow} H \to K$ such that $\hat{k}\cdot(id_C,h) = \hat{p}$ and $\hat{k}\cdot(id_C,j) = \hat{q}$. Interpreted in **C**, this means that we require the extension of Figure 9 to Figure 10, such that F→K and G→K factor uniquely through F→H and G→H. But the extension to Figure 10 is given by hereditariness of Figure 8, and the factorisation bby the pushout property of EFGH in **C**.

Only if: Let Figure 8 in **C** be a pushout in $\wp(\mathbf{C},\mathcal{M})$. Suppose it were not a pushout in **C**. Then there would exist arrows $B \to K$ and $C \to K$ for which there was either no factorisation through an arrow $D \to K$ or more than one.

Any factorisation in **C** gives a factorisation in $\wp(\mathbf{C},\mathcal{M})$ by the covariant embedding. If there is more than one in **C**, there is more than one in $\wp(\mathbf{C},\mathcal{M})$.

Suppose there is no factorisation in **C**. Since ABCD is a pushout in $\wp(\mathbf{C},\mathcal{M})$, there must be a partial morphism $D \overset{\supseteq}{\leftarrow} X \to K$ through which $B \to K$ and $C \to K$ factorise in $\wp(\mathbf{C},\mathcal{M})$. But then $D \overset{\supseteq}{\leftarrow} X \overset{\subseteq}{\to} D$ and $D \overset{=}{\leftarrow} D \overset{=}{\to} D$ are distinct partial morphisms through which both $B \to D$ and $C \to D$ factor in $\wp(\mathbf{C},\mathcal{M})$, contradicting the pushout property in $\wp(\mathbf{C},\mathcal{M})$.

Therefore ABCD is a pushout in $\wp(\mathbf{C},\mathcal{M})$. For hereditariness, consider an extension of Figure 8 to Figure 9. Then the partial morphisms $C \overset{\supseteq}{\leftarrow} G \overset{\subseteq}{\to} C \to D$ and $B \overset{\supseteq}{\leftarrow} F \overset{\subseteq}{\to} B \to D$ mak a commuting square in $\wp(\mathbf{C},\mathcal{M})$ with $A \to B$ and $A \to C$. By the pushout property of ABCD in $\wp(\mathbf{C},\mathcal{M})$, there must be a unique morphism $D \overset{\supseteq}{\leftarrow} H \to D$ making a commutative diagram as in Figure 10 (plus the arrow $H \to D$), in which GHCD and FHBD are pullbacks, and $G \to H \to D$ and $F \to H \to D$ are the morphisms $G \overset{\subseteq}{\to} C \to K$ and $F \overset{\subseteq}{\to} B \to K$. It remains to show that $G \to H$ and $F \to H$ are the pushout in **C** of $E \to F$ and $E \to G$.

First, we show that $H \overset{\subseteq}{\to} D$ is the unique smallest subobject of D which pulls back to F and G. If it were not, there would be another subobject $H' \overset{\subseteq}{\to} D$ pulling back to F and G and not containing $H \overset{\subseteq}{\to} D$. But then there would be a second factorisation of $B \overset{\supseteq}{\leftarrow} F \to D$ and $C \overset{\supseteq}{\leftarrow} G \to D$ through the intersection of these subobjects, contradicting the pushout property.

Now let $F \to D'$ and $G \to D'$ be any two morphisms of **C** such that $E \to F \to D' = E \to G \to D'$. Then $B \overset{\supseteq}{\leftarrow} F \to D'$ and $C \overset{\supseteq}{\leftarrow} F \to D'$ make a commuting square in $\wp(\mathbf{C},\mathcal{M})$ with $A \to B$ and $A \to C$. Therefore there is a unique factorisation of $B \overset{\supseteq}{\leftarrow} F \to D'$ and $C \overset{\supseteq}{\leftarrow} F \to D'$ through D, which must consist of a factorisation in **C** of $F \to D'$ and $G \to D'$ through H. Conversely, any factorisation in **C** of $F \to D'$ and $G \to D'$ through H gives a factorisation in $\wp(\mathbf{C},\mathcal{M})$ of $B \overset{\supseteq}{\leftarrow} F \to D'$ and $C \overset{\supseteq}{\leftarrow} F \to D'$ through D, and different factorisations in **C** give different factorisations in $\wp(\mathbf{C},\mathcal{M})$. Therefore the factorisation in **C** is unique, and $F \to H$ and $G \to H$ are the pushout in **C** of $E \to F$ and $E \to G$. ☐

3.5. Corollary. If the set of subobjects of any object of **C**, ordered by inclusion, is a complete lattice, and **C** has hereditary pushouts, then $\wp(\mathbf{C},\mathcal{M})$ also has these properties.

Proof. The condition on subobjects ensures that the choices of M, N, and O specified in steps (ii) and (iii) of the anteprevious theorem can always be made. The h.p. condition does the same for step (iv). Therefore $\wp(\mathbf{C},\mathcal{M})$ has pushouts.

By theorem 3.1(iv), $\wp(\wp(\mathbf{C},\mathcal{M}),\mathcal{M})$ is equivalent to $\wp(\mathbf{C},\mathcal{M})$, therefore $\wp(\wp(\mathbf{C},\mathcal{M}),\mathcal{M})$ has pushouts, and they are the same, modulo this equivalence, as those of $\wp(\mathbf{C},\mathcal{M})$. Hence by the previous theorem the pushouts of $\wp(\mathbf{C},\mathcal{M})$ are hereditary.

Finally, a subobject of A in $\wp(C,\mathcal{M})$ is just a subobject of A in C, embedded covariantly. Hence the inclusion ordering of subobjects of A in $\wp(C,\mathcal{M})$ is isomorphic to that in C. □

4. Comparison of double- and single-pushout rewriting in $\wp(\mathfrak{H})$.

Recall the double-pushout construction of Figure 1. Sufficient conditions for the construction to be uniquely determined by the rule L←K→R and the occurrence L→G are these (see e.g. [ER80, PEM86]):

DANGLE: if e is a hyperedge of G not in $g(E_L)$, n is a node of L, and g(n) is a vertex of e, then n is in l(K). (In words:if an edge of G is outside the range of the occurrence, then none of its vertexes are deleted by the rewrite.)

IDENT: if n_1 and n_2 are two distinct nodes of L such that $g(n_1) = g(n_2)$, then n_1 and n_2 are both in $l(N_K)$. (In words: a rewrite may not both delete and not delete the same node, nor delete the same node twice.)

FAST: l:K→L is monic. (In words: a rewrite may not attempt to pull apart a node of G into multiple copies.)

It is often also required (for other reasons) that K→R be monic; but we do not need this hypothesis here.

That K→L is monic implies that L←K→R can be read as a partial morphism in $\wp(\mathfrak{H})$. L→G can be read as a total morphism in $\wp(\mathfrak{H})$. Taking the pushout of these partial morphisms defines the result of single-pushout rewriting. Let the pushout graph be H'.

4.1. THEOREM. If the conditions for double-pushout rewriting hold, then H'=H.

PROOF. We first state some properties of $\mathfrak{H}$, the proof of which is routine.

LEMMA. $\mathfrak{H}$ has all finite limits, all finite colimits, and hereditary pushouts. The subobjects of an object of $\mathfrak{H}$ form a complete lattice under the inclusion ordering. □

$$\begin{array}{ccccc} L & \xleftarrow{l} & K & \xrightarrow{r} & R \\ \| & & \uparrow\cup & & \uparrow\cup \\ L & \xleftarrow{l'} & K' & \xrightarrow{r'} & R' \\ g\downarrow & & d'\downarrow & & \downarrow h' \\ G & \longleftarrow & D' & \longrightarrow & H' \end{array}$$

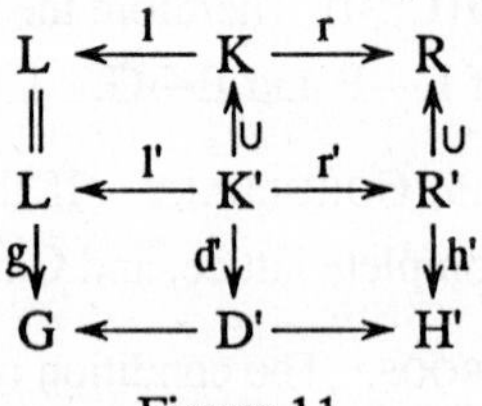

Figure 11

From corollary 3.5, we deduce that $\wp(\mathfrak{H})$ has pushouts. So H' always exists. The single pushout diagram, drawn in terms of morphisms of C, has the form of Figure 11. We first show that the double pushout conditions imply that the arrows K'⊆→K and R'⊆→R are in fact isomorphisms (and without loss of generality can be taken to be identities).

Because of the identity arrow L=→L, the construction of theorem 3.2 shows that we can take K'=K in step (ii) provided l:K⊆→L is a pullback along L→G of some subobject D'⊆→G. Then we take D'⊆→G to be the largest such subobject . It is easily verified that such a D'⊆→G exists iff the following condition holds:

IDENT′: if n_1 and n_2 are two distinct nodes of L such that $g(n_1) = g(n_2)$, then either n_1 and n_2 are both in $l(N_K)$, or neither is. (In words: a rewrite may not both delete and not delete the same node.)

But this condition is implied by IDENT. The top half of Figure 11 therefore consists of two copies of the rule, linked by three identities.

Since the squares KRDH and KRD'H' are both pushouts of $\mathfrak{H}$, to show that H'=H it is sufficient that the arrows K→D and D→G are equal to K→D' and D'→G respectively. That is, the theorem holds if the pushout and pullback complements of l along g coincide.

Using the informal notation g(L) to denote the set of nodes and edges in the range of g, and set-theoretic operations, the pushout complement is defined by $D = (G - g(L)) \cup g(l(K))$ (see e.g. [PEM86], prop. 3.13).

We defined the pullback complement to be the largest subobject of G pulling back along L→G to K. This is just the graph obtained by deleting from G every node and edge in the range of g but outside the range of g·l. That is, $D' = G - g(L - l(K))$. But IDENT implies that $g(L - l(K))$ and $g(l(K))$ are disjoint. Therefore $g(L - l(K)) = g(L) - g(l(K))$, and $D' = G - (g(L) - g(l(K))) = G - g(L) \cup g(l(K)) = D$. □

We thus have the following description of graph rewriting in $\mathfrak{H}$: a rewrite rule is a partial morphism, an occurrence is a total morphism, and a reduction is the pushout of a rule and an occurrence. Every redex can be reduced, without further conditions, and the result is determined up to isomorphism. When the conditions for double-pushout rewriting are satisfied, the single- and double-pushout constructions coincide.

The single-pushout construction may be seen as improving on the double-pushout, as it does not require any gluing conditions. Any occurrence of a rewrite rule can be reduced. However, note that the arrow R→H which results will not necessarily be a total morphism. One might reasonably desire that after a rewrite, the resulting graph contains somewhere a copy of the right hand side of the rule. The IDENT′ condition introduced above will ensure this.

IDENT′ is a weakened version of the IDENT gluing condition. DANGLE can be dispensed with entirely. DANGLE is violated when a rule deletes a node, and the graph contains a hyperedge connected to that node and not matched or deleted by the rule. In such a situation, the single-pushout construction will delete all such dangling hyperedges as well.

Essentially the same construction, but specialised to a particular category of graphs rather than formulated as an instance of the general construction, has independently been discovered by Löwe and Ehrig [LE90].

5. Comparison of double- and single-pushout rewriting of jungles.

Jungle rewriting [HP88] and the rewriting of [Rao84, Ken87] can be described by a categories of partial morphisms based on $\mathfrak{J}$, using a restricted class of monomorphisms. We note that the jungles of [HP88] have node labels as well as edge labels, but this difference is not significant. All of [HP88, Rao84, Ken87] for technical reasons restrict attention to cyclic graphs, but for the construction presented here this is unnecessary.

5.1. DEFINITION. $\wp'(\mathfrak{J})$ is $\wp(\mathfrak{J},\mathcal{M})$, where $\mathcal{M}$ is the set of monomorphisms of $\mathfrak{J}$ which are surjective on node-sets. □

$\wp'(\mathfrak{J})$ does not have such nice properties as $\wp(\mathfrak{H})$. Some which it does have a summariesed here; the proof is routine.

5.2. LEMMA. $\mathfrak{J}$ has all finite limits, and all finite consistent colimits (where a *consistent colimit* is a colimit of a digram on which there exists at least one cocone). The subobjects of an object of $\mathfrak{J}$ form a complete lattice under the inclusion ordering. □

However, $\mathfrak{J}$ does not have all pushouts, nor are all those which it has hereditary. However, the ones which we need are.

5.3. DEFINITION. A morphism f:A→B in $\mathfrak{J}$ is *strict* if it maps empty nodes of A to empty nodes of B. □

5.4. THEOREM. Let Figure 8 be a pushout square in $\mathfrak{J}$. It is hereditary if f or g is strict.

PROOF. Suppose the pushout is not hereditary. Then the diagram can be extended as in Figure 9 in such a way that it cannot be further extended to Figure 10. Because $\mathfrak{J}$ has consistent pushouts, E→F and E→G have a pushout H, and the pushout property lets us uniquely factorise F→B→D and G→C→D through H. Thus hereditariness can fail only because either H→D is not monic, or either of the squares FBHD or GCHD is not a pullback. We show that both alternatives lead to a failure of strictness of f and of g.

Suppose H→D is not monic. Then there are two edges or two nodes (for short: two components) of H which are mapped to the same component of D. Let the two components of H be η_1 and η_2, mapped to δ in D.

η_1 and η_2 must be in the union of the images of F→H and G→H; let ϕ_1 and ϕ_2 be components of F or G mapped to η_1 and η_2 respectively. ϕ_1 and ϕ_2 also belong to B or C through the inclusions of F and G. The mappings h and k map both ϕ_1 and ϕ_2 to δ. For this to be possible, there must be ancestor nodes v_1 and v_2 of ϕ_1 and ϕ_2 and a single node α of A which is mapped to both v_1 and v_2 by f and g. We see that one of ϕ_1 and ϕ_2 must be in B and one in C; then we have $v_1 = f(v)$, $v_2 = g(v)$. Now choose v_1 and v_2 so as to minimise the distance from v_i to ϕ_i (i=1,2); α will then be an empty node of A which is mapped to a nonempty node of both B and C. Thus neither f nor g is strict.

Suppose H$\to$D is monic, but one of the squares — without loss of generality let it be FBHD — is not a pullback. Then there is some component η of H which is not in the range of F$\hookrightarrow$B$\to$D, but is in the range of B$\to$D. Let β be mapped to η by B$\to$D. By a standard property of pushouts, η must be in the range of G$\to$H, hence there is a χ in C mapped to η by C$\to$D. For F and G to have a common pullback E, it is necessary that neither β nor χ be in the range of f or g. Yet β and χ are glued together by the pushout of f and g, therefore they must have ancestor nodes β_1 and χ_1 which are the images by f and g of a node α of A. But then as in the previous case, we obtain a counterexample to strictness of f or g. □

5.5. COROLLARY. Given arrows $f=(f',f''):A \overset{\supseteq}{\leftarrow} X \to B$ and $g=(g',g''):A \overset{\supseteq}{\leftarrow} Y \to C$ of $\wp'(\mathfrak{J})$, if g is total and f'' is strict, then f and g have a pushout. □

This corollary tells us that all term graph redexes can be reduced. A proof similar to that in the previous section shows that the result is the same as the double-pushout definition of [HP88].

Note that [HP88] adds an extra step to the reduction process: the result of the double-pushout is "collapsed" by gluing together all isomorphic subgraphs so as to minimise the size of the resulting graph. But this is done separately from the double-pushout, which is all that we are concerned with here, so we omit it. We conjecture that if one considers a category $\mathfrak{J}\mathfrak{J}$ of "strong jungles" (that is, the subcategory of $\mathfrak{J}$ containing only the "maximally collapsed" jungles), then rewriting in $\mathfrak{J}\mathfrak{J}$ will perform this collapsing automatically.

We briefly remark on the relation with [Ken87]. That paper presents a category $\mathbf{G}^{\mathbf{p}}$ of graphs and partial morphisms. The graphs of $\mathbf{G}^{\mathbf{p}}$ are just the hypergraphs of $\mathfrak{J}$ (although the formulation in [Ken87] is different). An arrow from a graph G to a graph H is a function f from the nodes of G to the nodes of H, together with a set S of nodes of G, such that for each nonempty node n in S, if e is the out-edge of n then f(e) has the same function symbol as e and f(c(e)) = c(f(e)). The idea is that S is the set of nodes at which f is known to be homomorphic, while outside S, f may behave arbitrarily.

$\mathbf{G}^{\mathbf{p}}$ has more arrows than $\wp'(\mathfrak{J})$. For example, in $\mathbf{G}^{\mathbf{p}}$ there are two arrows from the graph with one node and no hyperedges to itself (depending on whether S is empty or contains that one node), but in $\wp'(\mathfrak{J})$ there is only one. As a result of such apparently minor differences, the construction of pushouts is much more complicated in $\mathbf{G}^{\mathbf{p}}$ than in $\wp'(\mathfrak{J})$, and in certain cases involving cyclic graphs, a pushout exists in $\wp'(\mathfrak{J})$ when it does not in $\mathbf{G}^{\mathbf{p}}$. This is why [Ken87] in fact excluded cyclic graphs. We regard $\wp'(\mathfrak{J})$ as a better category to work in for the purpose of term graph rewriting.

6. Further developments.

6.1. Parallel composition of rewrites.

Suppose we have two redexes in a graph G, as in Figure 12 (where all the arrows are in $\wp(\mathfrak{H})$). We can define their *parallel composition* as follows. Take the pushout of $G \to H_1$ and $G \to H_2$, obtaining a graph H, which will be the result of the parallel composition. To see H as the result of a reduction of G, take the pushout U of the pullback S of the two occurrences, giving Figure 13. (Because $L_1 \to G$ and $L_2 \to G$ are total morphisms, it is easy to show that this is always possible.) To obtain Figure 14, factor the two pushout squares $L_iR_iGH_i$ through U, and complete a cube of pushouts.

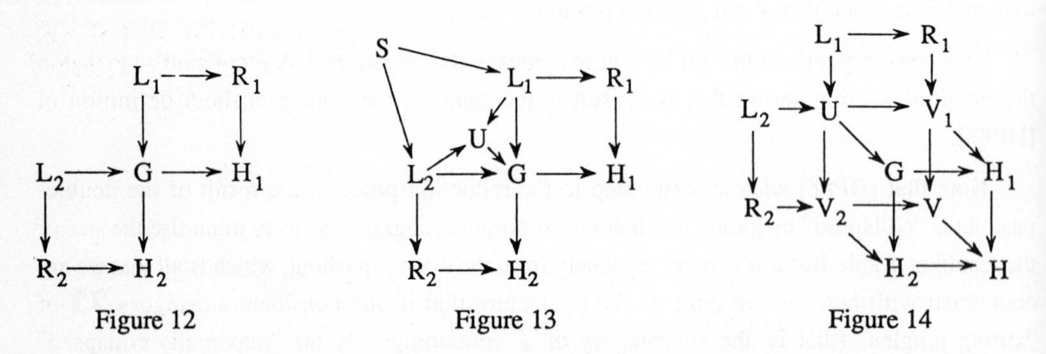

Figure 12 Figure 13 Figure 14

If the occurrences of L_1 and L_2 in G are disjoint (i.e. S is the empty graph), then $U \to V$ is the same as the sum of $L_1 \to R_1$ and $L_2 \to R_2$, and reduction by $U \to V$ is equivalent to reduction by the two original rules in either order.

When the occurrences are not disjoint, it may be that no sequence of single rewrites can take G to H. Thus adding parallel composition to a rewrite system can change its reduction relation. Whether the results of parallel composition of rewrites are useful is a subject for further study. Dactl is a graph rewrite language intended for parallel implementation (see [GHKPS88, GKS89], and in this volume [GKS90]). Its definition currently stipulates that the result of a computation must be identical to the result of some series of rewrites. This serialisability requirement was imposed to facilitate reasoning about Dactl programs. However, if the semantics of parallel rewrites is, as above, well-defined and mathematically simple, it may be unnecessary to insist on serialisability, but also allow parallel compositions. This may facilitate implementation, while retaining reasoning power.

6.2. Generalising term graph rewriting.

A term graph rewrite rule only changes one hyperedge of a jungle, either by deleting it and gluing its principal vertex to some other vertex, or by replacing it by a new hyperedge and possibly other new nodes and hyperedges. But in $\wp'(\mathfrak{J})$ it is just as easy to formulate rules which perform more complicated transformations, which are still inspired by the term graph rewriting paradigm. Dactl allows such rules. We illustrate this by a simple example, and its representation in $\wp'(\mathfrak{J})$.

Assignment can be performed by the following Dactl rule:

```
z:Assign[ x w:Cell[y] ]  ->  z := Done, w := Cell[x];
```

The rule is intended to store x into the cell w, replacing its previous contents y, and providing a sign to its caller that this has been done by rewriting the node z. This is represented in $\wp'(\mathfrak{J})$ by the partial morphism:

$$\text{z:Assign(x,w:Cell(y))} \quad \hookleftarrow \quad \text{z, x, w, y} \quad \rightarrow \quad \text{z:Done, w:Cell(x), y}$$

We have omitted consideration of Dactl's control markings. Apart from this, the rewrite in $\wp'(\mathfrak{J})$ has the same effect as its definition according to the Dactl specification [GKS89] (although that specification predates the work presented here).

7. Concluding remarks.

The traditional double-pushout definition of graph rewriting can be simplified to a single pushout in a category of partial graph morphisms. When hereditary pushouts exist in the base category of total morphisms and the subobjects of an object form a complete lattice (as is the case for the usual category of hypergraphs), all pushouts exist in the category of partial morphisms, without any need for the usual "gluing conditions".

The jungle category $\wp'(\mathfrak{J})$ can support rewriting more general than that which implements term rewriting: for example, rewrites which change several hyperedges at once. Such rewrites are the basis of the graph rewrite language Dactl. The description of graph rewriting presented here has suggested directions for the further development of Dactl. We expect this work to continue.

Finally, we note that although in this paper we have only looked at categories of graphs, our general definition of rewriting by pushouts of partial morphisms is capable of much wider application.

References

[EPS73] H. Ehrig, M. Pfender, and H.J. Schneider "Graph-grammars: an algebraic approach", Proc. IEEE Conf. on Automata and Switching Theory, 167–180, 1973.

[ER80] H. Ehrig and B.K. Rosen "Parallelism and concurrency of graph manipulations", *Theor. Comp. Sci.*, 11, 247–275, 1980.

[FRW90] W.M.Farmer, J.D.Ramsdell, and R.J.Watro, "A correctness proof for combinator reduction with cycles", ACM TOPLAS, **12**, n.1, 123-134, January 1990.

[GKS89] J.R.W. Glauert, J.R. Kennaway, and M.R. Sleep "Final specification of Dactl", Report SYS-C88-11, University of East Anglia, Norwich, U.K., 1989

[GKS90] J.R.W.Glauert, J.R.Kennaway and M.R.Sleep "Dactl: An Experimental Graph Rewriting Language", these proceedings, 1990.

[GHKPS88] J.R.W. Glauert, K. Hammond, J.R. Kennaway, G.A. Papadopoulos, and M.R. Sleep "Dactl: some introductory papers", Report SYS-C88-08, University of East Anglia, Norwich, U.K., 1988

[HP88] B. Hoffmann and D. Plump "Jungle evaluation for efficient term rewriting", Report 4/88, Fachbereich Mathematik und Informatik, Universität Bremen, Postfach 330 440, D-2800 Bremen 33, Germany, 1988. An earlier version appeared in Proc. Int. Workshop on Algebraic and Logic Programming, 1988. *Mathematical Research*, **49**. (Akademie-Verlag, Berlin, 1988).

[Ken87] J.R. Kennaway "On 'On graph rewritings'", Th. Comp. Sci. **52**, 37–58, 1987.

[KKSV9–] J.R. Kennaway, J.W. Klop, M.R. Sleep, and F.-J. de Vries "Transfinite reductions in orthogonal term rewrite systems" (in preparation, 199–).

[LE90] M. Löwe and H. Ehrig "Algebraic approach to graph transformation based on single pushout derivations" (unpublished, 1990).

[PEM86] F. Parisi-Presicce, H. Ehrig, and U. Montanari "Graph rewriting with unification and composition", Proc. 3rd Int. Workshop on Graph Grammars, LNCS 291, 496–514, Springer-Verlag, 1986.

[Rao84] J.C. Raoult "On graph rewritings", Th. Comp. Sci., **32**, 1–24, 1984.

[Rob88] E. Robinson and G. Rosolini "Categories of partial maps", Inf. & Comp., **79**, 95–130, 1988.

Application of Graph Grammars to Rule-based Systems[1]

MARTIN KORFF

Computer Science Department, Technical University of Berlin,
Franklinstr. 28/29, Sekr. FR 6-1; D-1000 Berlin 10

Abstract: Graph grammars can easily be considered as models for rule-based systems where solving state-space problems essentially requires searching. Since many AI problems are naturally graphical, graph grammars could narrow the usual gap between a problem and its formal specification.

In order to be able to solve such problems in practice one must reduce the effort of search. Here, for graph grammar specifications, the idea of precomputing its rules allows to prune the corresponding search-trees safely by explicitly pointing to those subtrees which are contained in others. Moreover, for some rules it becomes possible to use the information of a rule's former for to predict its later non-applicability, thus avoiding some redundant, expensive applicability tests. The example of solving a domino game based on breadth-first search demonstrates that indeed some remarkable reductions can be obtained.

Keywords: Graph grammar, rule-based system, rule independency, search-space reduction

1 Introduction

This paper was very much inspired by [MoPa 87] presented at the last GraGra-workshop in 1986. Their ideas were put into a formal applicational framework and continued; a discussion of their results finally lead to considerable improvements. This work is intended to encourage the use of Algebraic Graph Transformations (AGT) according to [Ehr 79] for specifying and solving search-space problems in the sense of [Ri 83] for example. For more details and a full discussion see [Ko 90].

Following, e.g., [Ri 83], a production or rule-based system consists of a problem *specification* and an additionally given *algorithmic control* component. The first is essentially based on the notion of a production rule (`precondition` → `action`). Unlike to procedures in conventional languages, it is potentially possible to undertake the `action` w.r.t. a global database whenever the rule's `precondition` is satisfied. The decision whether it will then indeed be applied depends on the second component, the inference algorithm.

Graph grammars augmented with a search algorithm fit this concept ideally. Following the maxim that *the closer the formalism to the problem's appearance, its structure, the greater the probability that the formalization expresses what we really intend*, graph grammars gain a promising new motivation for to *specify* problems where states and rules can naturally be modeled using graphs and graph production rules. This becomes especially important in areas which so far have hardly been formally investigated, in particular for many problems from AI.

A rule-based system specification of a problem basically consists of a number of rules and some initial state; solving such problems then requires to search for an additionally, but implicitly given final state. Very quickly the exhaustive or blind search runs into an exploding number of

[1]This work is partly supported by the projects "Computing by Graph Transformations (GraGra)" and "Kategorielle Methoden in Topologie und Informatik (KAMITI)"

generated states which i.g. cannot practically be managed. The usual answer to this problem in AI is to prune the search-tree, i.e. to select only some of all possible expansions or derivations.

Heuristics, as the most effective concept of pruning in AI, must be as specific to the problem as possible and can seldomly be treated formally. So, we will follow the complementary approach of precomputing given rules, a technique termed *knowledge-compilation* in AI.

Often for some sequence of rules we observe that by applying any other permutation of this sequence at most the same states can be derived. If this is true for some rules and arbitrary input-graphs we define the rules to be *semi-commutative* which then allows the safe pruning of all permutations of derivations without sacrifying completeness or correctness. Explicitly pointing on collapsed parts in an arbitrary search-tree can be considered as a search-space reduction.

The problem of finding an occurrence for some rule has been of considerable interest for term rewriting. As it is even more expensive and important for graphs, we formally define a second measure of a cost of search: for each rule and each state we count the calls on the procedure of finding the corresponding occurrences. A simple improvement can be obtained whenever we find those rules where the non-applicability of the second to the result of the first can be predicted, provided the second had been non-applicable before. Such rules will then be defined to be *monotonic.* Note, that skipping an applicability test preassumes that it has failed before.

The definitions of semi-commutative or monotonic rules cannot directly be used for a precomputation on rules, since they are universally quantified over graphs. Therefore, we look for effectively computable, or *syntactical,* criteria for the above properties. We start by asking how occurrences may relate and how these relations can effectively be determined. The answer will then be given as the (finite) set of all canonical relations of graph production rules.

It turns out that the classical notion of sequentially, in contrast to parallel, independent rules can be used as sufficient syntactical criteria. The investigation finally leads to the **main theorems** about *sc-* and *m-independency* being optimal syntactical criteria for semi-commutative and monotonic rules. The paper closes by giving a small example where all the above properties can be used to optimize the breadth-first search considerably.

It should be noted that the techniques presented can not only be used for prototyping but also for automatically hinting at inconsistencies of some specification, e.g. by reporting potential interactions of rules, or for giving support on formal proofs on specification inherent properties, which are most likely to be inductive proofs. Moreover, they are of interest for other applications of graph grammars.

2 The Model

Formal notions on graphs and graph manipulations on which the following investigations are based can be found in, e.g., [Ehr 79]. The well known algebraic approach is typically characterized by the double pushout (PO) diagram for the notion of derivation:

$$\begin{array}{ccccc} L_i & \xleftarrow{l_i} & K_i & \xrightarrow{r_i} & R_i \\ \downarrow g_i & (PO)_1 & \downarrow d_i & (PO)_2 & \downarrow h_i \\ G_i & \longleftarrow & D_i & \longrightarrow & H_i \end{array}$$

Some notational conventions and a brief survey on our notions is given below.

Generally, for graphs we assume some global coloring alphabet (C_N, C_A) on nodes and arcs, large enough to cover at least the contextly named graphs. A graph G consists of a set of nodes G_N and a set of arcs G_A supplied with the usual functions $s_G, t_G : A_G \to N_G$ for source and target mapping, node and arc coloring $m_N : N \to C_N$ resp. $m_A : A \to C_A$. Graph morphisms are defined as homomorphisms between graphs.

An *occurrence map* is defined to be a graph morphism which additionally satisfies the *gluing condition* (see [Ehr 79]), i.e.

$$\{x \in L \quad | \exists y \in L : \quad x \neq y \wedge g(x) = g(y)\} \subseteq lK \qquad \text{(Identification cond.) and}$$
$$\{n \in L_N \quad | \exists e \in (G \setminus gL)_A : \quad s(e) = g(n) \vee t(e) = g(n)\} \subseteq (lK)_N \qquad \text{(Dangling condition).}$$

Note that a rule p_i will only be applicable to a graph G_i iff the *left hand map* g_i turns out to be an occurrence map; $G \overset{(p_i,g_i)}{\Longrightarrow} H$ then denotes the corresponding derivation.

Let $arcs : G_N \rightarrow \mathcal{P}(G_A)$ be the function defined on the nodes of a given graph G yielding all its in- and outgoing arcs: $arcs(n) = \{a \in G_A | n = s_G(a) \vee n = t_G(a)\}$.

Whenever we are going to refer to some graph in a certain context of derivations, we will follow the convention of using the indexed uppercase letters, e.g. L_i for the left hand side of a rule p_i. *Graph morphisms* will always be denoted by lowercase letters. By default, an index (i) will be used to distinguish between rules within a given context; it does *not* refer to the i-th in a family of rules (P) and may be dropped if ambiguities are not possible. NG_i stands for the *N*on-*G*luing items of the left hand side of rule p_i, i.e. $L_i \setminus l_iK_i$, NG_i^{-1} for $R_i \setminus r_iK_i$.

The notion of a *'derivation bag'* stands for the leaves of a bunch of derivations based on a family of rules, abstracting from the tree structure of the corresponding derivations, i.e. the *derivation bag*[2] $P_J(\mathcal{G})$ w.r.t. a graph bag $\mathcal{G}$ and a family of rules $(p_j)_{j \in J}$ is:

$$P_J(\mathcal{G}) := \left\{ H | \exists G \in \mathcal{G}, j \in J, g_j : L_j \rightarrow G \text{ such that } G \overset{(p_j,g_j)}{\Longrightarrow} H \right\}$$

We allow ourself to notationally omit set-inclusions, thus being able to write $P_j(G)$ for $J = \{j\}$ and $\mathcal{G} = \{G\}$. Since the associativity axiom holds for derivation bags we also omit parentheses.

The framework of algebraic graph transformations (AGT) will now be used for modeling a certain kind of state-space problems. In *database systems* a goal is described procedurally by a given sequence of rules, a transaction. A *process* also executes in a single sequence. Contrastingly, we will consider only those models where a computational goal essentially requires searching: generally, one does neither know which nor how many states must be examined in order to find a solution. Moreover, we will look at only those problems, qualified by the attribute 'plain', where a solution consists of at most a single branch of the derivation tree. In any case, different computational outputs may be required, e.g. a complete solution branch (derivation), a goal state (graph) or merely a *yes/no*-answer. Note, that these problems differ from *reductionary* search-space problems too, which can only be solved by finding a whole subtree of the search tree (As an example for the latter take PROLOG-problems, where the goal forms the root and some PROLOG-facts form the leaves of a resulting tree).

Definition 2.1 A **(plain search-space AGT-)problem** is given by $PP = (Ax, P, F, C)$ with a graph Ax, called *initial database* or *axiom*, a family of graph production rules P, the *rule-base*, a unary predicate function on graphs F, the *filter*, and application conditions[3] C, called *constraints*.

---

[2] Multiset

[3] see [EhrHa 85]

Remark:

▷ The only application condition we will require here for all rules is the gluing condition which is a matter of course in the double pushout approach to graph transformation (see above). A consideration of other conditions aiming for similar results is delayed to the future. ⋆

Definition 2.2 A **solution** of an AGT-problem $PP = (Ax, P, F, C)$, is any Ax-reachable graph G, i.e. $Ax \stackrel{*P}{\Longrightarrow} G$, for which $F(G)$ is true. ⋆

Definition 2.3 Given a **class** of AGT-problems $PPC = (\mathcal{AX}, P, F, C)$, i.e. (Ax, P, F, C) is an AGT-problem for each $Ax \in \mathcal{AX}$, an algorithm S is called **search algorithm** (w.r.t. PPC), iff supplied with $Ax \in \mathcal{AX}$, the algorithm terminates with either a graph or the message FAIL. The result of S is called **correct** if it is indeed a solution of (Ax, P, F, C) or FAIL otherwise. ⋆

Definition 2.4 A **production-system** for AGT-problems, briefly an AGT-PS, is given by (PPC, S), i.e. a class of AGT-problems PPC and a search algorithm S. ⋆

According to the basic executional model of production systems, termed the *recognition-act-cycle* (cf. [ShiTsu 84]), derivations of a graph G based on production rule p are to be done in two steps: first one has to get the information of how to apply p to G, which can then be used to derive the new graph H, derived from G. In terms of AGT notions this means to preassume two primitive functions:

Recognize: graph × rule → occurrencelist: $(G, L \leftarrow K \rightarrow R) \mapsto gl$
with $gl = g_1, \ldots, g_m$ being the repetitionfree list of all occurrences $g_j : L \rightarrow G$; this freely allows to respect a certain strategic ordering of occurrences.

Derive: graph × occurrence → graph: $(G, g) \mapsto H$
with $G \stackrel{g}{\Longrightarrow} H$.[4]

Really hard problems arise for executing each of the two above operations, as well as for evaluating the filter F on graphs. The latter has been omitted here since, due to a lack of information about it, we assume its costs to be unreducable. All other operations are neglectably cheap. (see [Ko 90]). So, the following notion of cost covers both, the problem of a search-space-reduction (N_2 (pruning)), as well as efforts to determine an applicability of rules (N_1).

Definition 2.5 Given an AGT-PS = $((\mathcal{AX}, P, F, C), S)$, then on termination the **cost** of the search algorithm S based on the primitive functions Recognize and Derive is defined to be (N_1, N_2) with N_1 and N_2 being the number of calls for Recognize and Derive.

Cost (N_1, N_2) is said to be **not greater** than (N_1', N_2') iff $N_i \leq N_i'$ for $i = 1, 2$. ⋆

Since we do not want to assume a certain search strategy, i.e. some search algorithm, we aim for a notion of optimization which guarantees that every search algorithm can be improved in that its cost will not increase.

Definition 2.6 Given an AGT-PS = (PPC, S), and a search algorithm O w.r.t. PPC, then we call O **optimization** w.r.t. S iff O yields the *correct* solution w.r.t. to some AGT-problem of PPC whenever S does and the *cost of O is not greater* than that of S.

The optimization is called **tight** iff O yields always the *same* solution as S. ⋆

[4] A type 'occurrence' is assumed to carry the information about the corresponding rule. So 'rule' could be excluded from the functionality of derive.

3 Interaction of Rules

As it has already been outlined, our approach of optimizing search algorithms shall be based on properties of rules which must safely allow to avoid calls to the corresponding primitive functions, thus reducing the cost of the algorithm. Note that the properties below cannot directly be precomputed because they universally quantify over (the infinite set of) graphs.

Let $G \stackrel{p_1}{\Longrightarrow} H$ be a direct derivation. Furthermore, assume that one can `recognize` k different occurrences of rule p_2 in G. Then we can predict that at most k occurrences of p_2 in H can be found, provided p_1 is (increasingly) 'k'-monotonic w.r.t. p_2:

Definition 3.1 Rule p_1 is called (increasingly) 'k'-monotonic w.r.t. rule p_2 iff

$$|P_2(G)| = k \stackrel{\text{implies}}{\Longrightarrow} (\forall_{H_1 \in P_1(G)} \ |P_2(H_1)| \leq k). \qquad \star$$

Inverting the $\stackrel{\text{implies}}{\Longrightarrow}$ -arrow leads to a definition of monotonicity qualified as 'decreasingly', which could be used as a k-applicability prediction on backtracking.
As a special case for $k = 0$ we obtain monotonic rules:

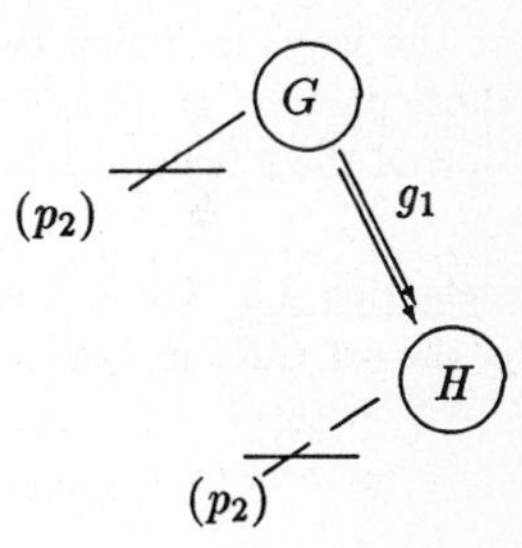

Definition 3.2 Rule p_1 is said to be **(increasingly) monotonic** w.r.t. rule p_2 iff

$$P_2(G) = \emptyset \stackrel{\text{implies}}{\Longrightarrow} P_2P_1(G) = \emptyset \qquad \star$$

This means that in case of a rule p_1 being monotonic w.r.t. a rule p_2 it holds for an arbitrary (direct) derivation $G \stackrel{g_1}{\Longrightarrow} H$ via p_1 that p_2 can only be applied to H if it could already be applied to G.

Fact 3.3 Given rules $\{p_1 \ldots p_m\}$, each of which is k-monotonic w.r.t. a rule p_q, then we have for every $H_{s_l} \in P_{s_l} \ldots P_{s_1}(G)$ with $1 \leq l$, $s_j \in \{1 \ldots m\}$, $j = 1 \ldots l$:

$$|P_q(G)| = k \stackrel{\text{implies}}{\Longrightarrow} k \geq |P_q(H_{s_l})| \qquad \star$$

Proof. By induction on l. □

Whereas monotonicity allows to eventually skip some `recognize`-calls, the second relation between rules, called (semi-)commutativity, hints to those situations where some `derive`-calls can safely be avoided.

Definition 3.4 Rule p_1 is said to be **semi-commutative** w.r.t. rule p_2 iff $P_2P_1(G) \subseteq P_1P_2(G)$ ⋆

Definition 3.5 Rule p_1 is said to be **commutative** w.r.t. rule p_2 iff $P_2P_1(G) = P_1P_2(G)$ ⋆

Fact 3.6 Given rules $p_1 \ldots p_m$ with p_i semi-commutative w.r.t. p_j, for $1 \leq i < j \leq m$, then we have for any permutation $s_1 \ldots s_m$ of $1 \ldots m$ $\qquad P_m \ldots P_1(G) \subseteq P_{s_m} \ldots P_{s_1}(G)$. ⋆

Proof. By induction on m. □

We easily obtain $P_m \ldots P_1(G) = P_{s_m} \ldots P_{s_1}(G)$ for any permutation $p_{s_1} \ldots p_{s_m}$ of rules $p_1 \ldots p_m$, provided they are pairwise commutative.

In order to let a search-algorithm take advantage of a *precomputation pass* having distinguished those rules which are monotonic or semi-commutative, appropiate, i.e. *effectively computable*, criteria must be found. Besides, in general, we cannot hope to find exact *characterizations* for the notions above, but *only sufficient* conditions, so we introduce the following concepts:

Definition 3.7 A **syntactical monotonicity ((semi)-commutativity) criterion** is an effectively computable binary predicate on rules telling whether these rules are monotonic ((semi)-commutative).

Given two syntactical monotonicity ((semi-)commutativity) criteria SC and SC_b, the latter is said to be **better** iff $SC \subset SC_b$.

A syntactical monotonicity ((semi-)commutativity) criterion is said to be **optimal** iff there is no better syntactical monotonicity ((semi-)commutativity) criterion. Optimality of a criterion assumes that the global alphabet is large enough.[5] ⋆

Asking how an interaction of rules can effectively be characterized, we will start by looking at the ways in which two rules may overlap in a derivation. To be more precise, we ask how those parts in graphs can be characterized, where the images of the right hand side of a first and that of the left hand side of a second rule, both taking respect to the gluing condition, intersect.

Definition 3.8 Given two graph morphisms $r : K_1 \to R$ and $l : K_2 \to L$ the gluing relation set is the set $\widetilde{GRS}(r,l)$ of relations

$$\widetilde{gr} = (\widetilde{gr}_A \subseteq R_A \times L_A, \widetilde{gr}_N \subseteq R_N \times L_N)$$

on nodes and arcs, such that each $\widetilde{gr} \in \widetilde{GRS}(r,l)$ also satisfies,
for $a, a' \in R, \quad b, b' \in L, \quad \overline{a} \in NG_1^{-1} = (R \setminus rK_1), \quad \overline{b} \in NG_2 = (L \setminus lK_2)$:

$$[Ax1] \quad a \ \widetilde{gr}_N \ b \quad \overset{\text{implies}}{\Longrightarrow} \quad m_N(a) = m_N(b)$$
$$[Ax2] \quad a \ \widetilde{gr}_A \ b \quad \overset{\text{implies}}{\Longrightarrow} \quad m_A(a) = m_A(b)$$
$$[Ax3] \quad a \ \widetilde{gr}_A \ b \quad \overset{\text{implies}}{\Longrightarrow} \quad (s(a) \ \widetilde{gr}_N \ s(b)) \quad \wedge \quad (t(a) \ \widetilde{gr}_N \ t(b))$$

$$[Ax4] \quad a \ \widetilde{gr}_N \ \overline{b} \quad \overset{\text{implies}}{\Longrightarrow} \quad \forall_{e_a \in arcs(a)} \exists_{e_b \in arcs(\overline{b})} \quad e_a \ \widetilde{gr}_A \ e_b$$
$$[Ax5] \quad \overline{a} \ \widetilde{gr}_N \ b \quad \overset{\text{implies}}{\Longrightarrow} \quad \forall_{e_b \in arcs(b)} \exists_{e_a \in arcs(\overline{a})} \quad e_a \ \widetilde{gr}_A \ e_b$$

$$[Ax6] \quad a \ \widetilde{gr} \ \overline{b} \wedge a \ \widetilde{gr} \ b' \quad \overset{\text{implies}}{\Longrightarrow} \quad \overline{b} = b'$$
$$[Ax7] \quad \overline{a} \ \widetilde{gr} \ b \wedge a' \ \widetilde{gr} \ b \quad \overset{\text{implies}}{\Longrightarrow} \quad \overline{a} = a'$$

Remark:

- ▹ Whenever we want to speak about a gluing relation set without explicitly referring to the corresponding maps (r,l) then we feel free to write just $\widetilde{GRS}$.
- ▹ A $\widetilde{gr} \in \widetilde{GRS}$ with $\widetilde{gr} \subseteq rK_1 \times lK_2$ is called an *interface* relation.

⋆

[5] The qualifier 'optimal' can be justified despite allowing an arbitrary enlarged alphabet for a comparison, because we can practically assume that the required (extra) color, which will be necessary for proving theorems 4.8 and 4.10, can already be found in the original global alphabet (Compare section 2).

Fact 3.9 Given two rules p_1 and p_2, the gluing relation set $\widetilde{GRS}(r_1, l_2)$ contains exactly the pullbacks of all $R_1 \xrightarrow{h_1} H_1 \xleftarrow{g_2} L_2$ with h_1 and g_2 being occurrence maps. ★

Remark:

▹ The fact that h_1 and g_2 are occurrence maps is equivalent to the existence of a derivation sequence $G \overset{p_1}{\Longrightarrow} H_1 \overset{p_2}{\Longrightarrow} H_2$, and hence the corresponding double gluing diagrams (POs).

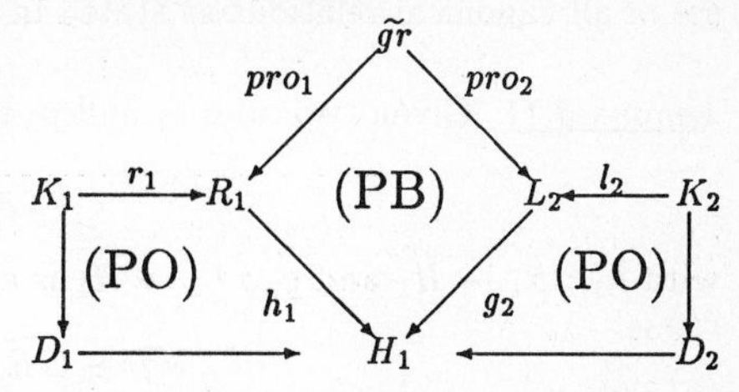

Proof. Axioms $[Ax1] - [Ax3]$ ensure that each $\widetilde{gr}$ is pullback (graph) of the right and left hand maps $R_1 \xrightarrow{h_1} H_1 \xleftarrow{g_2} L_2$, but not sufficient for any $\widetilde{gr}$ being the pullback of two *occurrences* h_1 and g_2; the latter requires the gluing i.e. the dangling and the identification condition to be satisfied w.r.t. h_1 and g_2 (see section 1), which is exactly ensured by the axioms $[Ax4, 5]$ as well as $[Ax6, 7]$ respectively. □

The following definition is based on [Ehr 79] where a relation R for a pair of production was defined as a pair of morphisms $R_1 \leftarrow R \rightarrow L_2$ such that there are pushout complements of $K_1 \rightarrow R_1 \leftarrow R$ and $R \rightarrow L_2 \leftarrow K_2$ (compare also [EhrHaRo 86]).

Definition 3.10 Let R be a relation for a pair of productions p_1, p_2. Let $G_1^+ \overset{p_1}{\Longrightarrow} H_1^+ \overset{p_2}{\Longrightarrow} H_2^+$ be the *minimal derivation sequence* with H_1^+ constructed as pushout of R (see figure below) and K_* the PB-object of $D_1^+ \longrightarrow H_1^+ \longleftarrow D_2^+$.

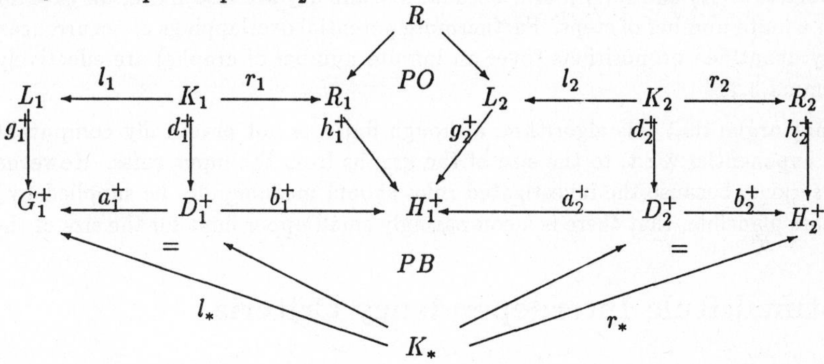

Then the production $p_* = (L_* \xleftarrow{l_*} K_* \xrightarrow{r_*} R_*)$ with $L_* = G_1^+$, $R_* = H_2^+$ and $l_* = K_* \longrightarrow D_1^+ \longrightarrow L_*$ as well as $r_* = K_* \longrightarrow D_2^+ \longrightarrow R_*$ is called ***R*-concurrent production** of p_1 and p_2, written $p_1 *_R p_2$. A direct derivation $G \overset{p_1 *_R p_2}{\Longrightarrow} X$ is called ***R*-concurrent derivation.** ★

Gluing relations $\widetilde{gr} \in \widetilde{GRS}(r_1, l_2)$ are closely related to R-relations: The pullback of two occurrence maps h_1 and g_2 as above, called *canonical relation*, was shown to be a special R-relation for a pair of productions p_1, p_2 in [Ehr 79]. So, fact 3.9 ensures that any derivation based on a pair of rules p_1, p_2 is $\widetilde{gr}$-related for some canonical relation $\widetilde{gr} \in \widetilde{GRS}(r_1, l_2)$. Moreover by the Concurrency-Theorem in [Ehr 79], every derivation sequence via a pair of rules bijectively corresponds to a direct derivation via some $\widetilde{gr}$-concurrent rule for the corresponding gluing relation.

Since we are particularly interested in *effectively* determining whether rules may overlap in a certain way, it is most important that rule-dependencies can be characterized by the (finite) set of all canonical relations as stated in the lemma below.

Lemma 3.11 Given two rules p_1 and p_2 and subgraphs $S_R \subseteq R_1$, $S_L \subseteq L_2$, then the proposition

$$\forall G \overset{p_1}{\Longrightarrow} H_1 \overset{p_2}{\Longrightarrow} H_2 \quad h_1 S_R \cap g_2 S_L = \emptyset$$

with $h_1 : R_1 \to H_1$ and $g_2 : L_2 \to H_1$ is equivalent to

$$\forall \widetilde{gr} \in \widetilde{GRS}(r_1, l_2) \quad \widetilde{gr} \subseteq S_R \times S_L$$

★

and can thus effectively be decided via the corresponding gluing relation set.

Proof.

'$\Longleftarrow$' — by contraposition. Assume there is an $\widetilde{gr} \in \widetilde{GRS}$ such that $\widetilde{gr} \not\subseteq S_R \times S_L$ then the corresponding minimal derivation sequence (see def. 3.10) with $R_1 \overset{h_1}{\longrightarrow} H_1^+ \overset{g_2}{\longleftarrow} L_2$ being PO of $\widetilde{gr}$ does *not* satisfy $h_1 S_R \cap g_2 S_L = \emptyset$.

'$\Longrightarrow$' — by contraposition. Assume there is a derivation sequence $G \overset{p_1}{\Longrightarrow} H_1 \overset{p_2}{\Longrightarrow} H_2$ such that $h_1 S_R \cap g_2 S_L \neq \emptyset$, then this immediately implies that the corresponding gluing (canonical) relation $\widetilde{gr} \in \widetilde{GRS}$ does *not* satisfy $\widetilde{gr} \subseteq S_R \times S_L$.

□

Obviously, this characterization of the gluing relation set is constructive. Since each of the relevant graphs R, L is finite and so are the sets $A = R_A \times L_A$ and $N = R_N \times L_N$. This implies finite powersets $\mathcal{P}(A)$ and $\mathcal{P}(N)$, and because K_1 and K_2 are also finite, all $\widetilde{gr}$-axioms can be checked in a finite number of steps. Furthermore potential overlappings of occurrences defined as universally quantified propositions (over an infinite number of graphs) are effectively decidable due to Lemma 3.11.

One may argue that this algorithm, although finite, is not practically computable, because its cost is exponential w.r.t. to the size of the graphs from the input rules. However, this does not seem serious, because the investigated rules should meaningfully be supplied by some user, so it appears plausible, that there is a comparingly small upper limit for the size of these graphs.

4 Optimal Rule-Interdependency Criteria

Can we use the 'classical' notions of *parallel* or *sequentially independency* on rules as suitable criteria for monotonicicty or semi-commutativity? If this is possible, are they just sufficient or also necessary, i.e. optimal? By Lemma 3.11, both can be determined effectively, hence they are syntactical.

Definition 4.1 A rule p_1 is said to be **parallel independent**, or short p-independent, of rule p_2 iff

$$\forall L_1 \overset{g_1}{\longrightarrow} G \overset{g_2}{\longleftarrow} L_2 : \quad g_1 L_1 \cap g_2 L_2 \subseteq g_1 l_1 K_1 \cap g_2 l_2 K_2,$$

where g_1, g_2 satisfy the gluing condition. ★

Guess: 4.2 **P-independency** is a syntactical k-monotonicity criterion. ★

Counterexample. Rules $p_1 = (A, \emptyset, B)$ and $p_2 = (B, \emptyset, \emptyset)$, which are clearly p-independent, are not monotonic due to a graph G consisting of an A-colored node only. □

Definition 4.3 A rule p_1 is said to be **sequentially independent**, or short s-independent, of rule p_2 iff $\forall R_1 \xrightarrow{h_1} H_1 \xleftarrow{g_2} L_2 : \quad h_1R_1 \cap g_2L_2 \subseteq h_1r_1K_1 \cap g_2l_2K_2,$ where h_1, g_2 satisfy the gluing condition. ★

Fact 4.4 **S-independency** is a syntactical k-monotonicity criterion. ★

Fact 4.5 **S-independency** is a syntactical semi-commutativity criterion. ★

Proof. (4.4 and 4.5) By the Parallelism-Theorem (cf. [Ehr 79]) and Lemma 3.11. □

As an example showing that s-independency is only a weak monotonicity criterion, consider the rules $p_1 = (A, \emptyset, A)$ and $p_2 = (A, \emptyset, \emptyset)$. As one can easy check p_1 is monotonic w.r.t. p_2, although p_1 is *not* s-independent of p_2. By replacing p_1 by $(A \longleftarrow B, \emptyset, A)$ one looses the above monotonicity. So we conclude firstly that items in the *left hand side* of the first rule, like the A, must be preserved w.r.t. the right and secondly that this 'replacement' of items of L_1 by some of R_1 must be *compatible* with the graph structure (no edge on A w.r.t. p_2). Exactly this compatibility requirement caused the difficulties in the approach in [MoPa 87] (cf. [Ko 90]).

Concluding it holds, roughly spoken, that two rules are independent if for all derivation sequences the overlap of the first and the second rule w.r.t. the intermediate graph will either be preserved or otherwise all relevant items being deleted must immediately be rebuild by the first rule. The latter may be considered as the problem of how to enlarge the interface part K_1 of the first rule such that an occurrence map of the second rule can be lengthened or traced across all previous derivations via the first rule, i.e. for each derivation $G_1 \overset{p_1}{\Longrightarrow} H_1 \overset{p_2}{\Longrightarrow} H_2$ an occurrence map $g_2' : L_2 \rightarrow G_1$ must be constructible from $g_2 : L_2 \rightarrow H_1$. This idea lead to a notion of *tracingly independent* rules in [Ko 90], which could be shown to improve s-independency as a syntactical k-monotonicity criterion. But still there are cases where a compatible replacement cannot be covered by t-independency. Fortunately, for a mere non-application prediction ($k = 0$) this direct approach can optimally be improved by canonically testing critical cases.

Lemma 4.6 (Monotonicity Lemma) Given rules p_1, p_2 and p_3, it holds w.r.t. the global coloring alphabet eventually enlarged by one element, that,

$$\left.\begin{array}{l} \forall \tilde{gr} \in \widetilde{GRS}(r_1, l_2) \text{ and } p_* = p_1 *_{\tilde{gr}} p_2 = L_* \xleftarrow{l_*} K_* \xrightarrow{r_*} R_* \\ \text{there is an occurrence map } g_3^+ : L_3 \rightarrow L_* \\ \text{such that} \quad g_3^+(NG_3) \subseteq g_*(NG_*) \qquad \}(UCC) \end{array}\right\}(SynR)$$

is equivalent to

$$\forall G_1 \quad (\exists G_1 \overset{p_1}{\Longrightarrow} H_1 \overset{p_2}{\Longrightarrow} H_2 \quad \overset{\text{implies}}{\Longrightarrow} \quad \exists G_1 \overset{p_3}{\Longrightarrow} H_3) \qquad \}(SemR)$$

Remarks:

- ▹ Recall, NG_i denotes the set $(L_i \setminus l_iK_i)$ w.r.t. some rule p_i. Analogously this holds for p_*.
- ▹ (SemR) is shorthand for *Semantical Relation*
  (SynR) is shorthand for *Syntactical Relation*
  (UCC) is shorthand for *UnCriticalness Condition*
- ▹ $(SemR)$, which can be read as "*if p_1 and p_2 can sequentially be applied to G_1 then p_3 must be applicable to G_1*", is equivalent to: $\quad P_3(G_1) = \emptyset \quad \overset{\text{implies}}{\Longrightarrow} \quad P_2P_1(G_1) = \emptyset$

★

Proof. For each derivation sequence $G_1 \stackrel{p_1}{\Longrightarrow} H_1 \stackrel{p_2}{\Longrightarrow} H_2$ there is some overlapping of occurrences h_1R_1 and g_2L_2 in H_1 and thus the derivation is R-related and, in the category $\mathit{CGRAPHS}$, also $\widetilde{gr}$-related for some canonical relation $\widetilde{gr} \in \widetilde{GRS}(r_1, l_2)$. Now, in particular, there is an occurrence map $g_* : L_* \to G_1$ of the corresponding canonical production $p_* = p_1 *_{\widetilde{gr}} p_2 = L_* \stackrel{l_*}{\longleftarrow} K_* \stackrel{r_*}{\longrightarrow} R_*$.

Let g_3^+ be an occurrence map from the left hand side of p_3 into that of the corresponding canonical production. Then there is a graph morphism $g_3 = g_* g_3^+ : L_3 \to G_1$. W.r.t. g_3^+ there are neither dangling nor identification points in NG_3, and due to $g_3^+(NG_3) \subseteq g_*(NG_*)$ this will also be valid for g_3. So, from the gluing condition being satisfied we conclude the existence of a derivation $G_1 \stackrel{p_3}{\Longrightarrow} H_3$.

Assuming that there is no such occurrence map g_3^+ for some $\widetilde{gr} \in \widetilde{GRS}(r_1, l_2)$, contradicts the existence of a derivation $L_* \stackrel{p_3}{\Longrightarrow} H_3$, although we have a ($\widetilde{gr}$-concurrent) derivation $L_* \stackrel{p_*}{\Longrightarrow} R_*$. If, otherwise, for some p_* we only have occurrence maps as $g_3^+ : L_3 \to L_*$ such that the uncriticalness condition (UCC) is violated there must be a node $i \in NG_3$ such that $g_3^+(i) = x \notin g_*(NG_*)$ (the existence of such an arc implies that of a node).
We now construct a graph G from L_* (see minimal derivation sequence in definition 3.10) by adding a single arc e and a single node y with $s(e) = x$, $t(e) = y$ colored such that $\forall_{k \in L_3} m_N(y) \neq m_N(k)$ which is always possible if we preassume a disjointly enlarged (node) coloring alphabet. As an occurrence map $g' : L_* \to G$ w.r.t. p_* we take the inclusion, which truly satisfies the gluing condition, since g' is injective and x which is the only dangling point is not in NG_*, i.e. a gluing item. Hence, we have a derivation from G' based on g', i.e. a derivation sequence $G' \stackrel{p_1}{\Longrightarrow} H_1' \stackrel{p_2}{\Longrightarrow} H_2'$, but none via p_3 based on $g_3 = g' \circ g_3^+$ because i is a nongluing dangling point. Moreover, any other potential occurrence map $g_3' : L_3 \to G$ must map some $l \in L_3$ onto y which is impossible because y is colored different to all l. □

Definition 4.7 Given rules p_1 and p_2, p_2 is said to be **monotonicly independent** of p_1, briefly m-independent(p_1, p_2), iff for each *non-interface* relation $\widetilde{gr} \in \widetilde{GRS}(r_1, l_2)$ there is an occurrence map $g_2^* : L_2 \to L_*$ with $p_* = p_2 *_{\widetilde{gr}} p_1 \equiv L_* \stackrel{l_*}{\longleftarrow} K_* \stackrel{r_*}{\longrightarrow} R_*$ such that the *Uncriticalness Condition*, i.e. $g_2^*(NG_2) \subseteq NG_*$, is satisfied. ⋆

THEOREM 4.8 M-independence is an optimal syntactical monotonicity criterion.[6] ⋆

Proof. This is an immediate consequence of the monotonicity lemma 4.6 by taking p_1, p_2 and p_2 again instead of p_1, p_2 and p_3; the restriction to relations which are not interface relations is possible since for these the corresponding derivations are s-independent in which cases monotonicity is guaranteed. So, p_1, p_2 m-independent iff $\forall G: \quad P_2(G) = \emptyset \stackrel{\text{implies}}{\Longrightarrow} P_2P_1(G) = \emptyset$
□

The same idea as in the Monotonicity-Theorem 4.8 above applies to the semi-commutativity:

[6] see definitions 3.2 and 3.7.

Definition 4.9 Given rules p_1 and p_2, p_2 is said to be **sc-independent** of p_1 iff[7]
$\forall p_* = p_1 *_{\widetilde{gr}} p_2 \equiv L_* \xleftarrow{l_*} K_* \xrightarrow{r_*} R_*$
$\exists p'_* = p_2 *_{\widetilde{gr}'} p_1 \equiv L'_* \xleftarrow{l'_*} K'_* \xrightarrow{r'_*} R'_*$
with $\widetilde{gr} \in \widetilde{GRS}(r_1, l_2)$ and $\widetilde{gr}' \in \widetilde{GRS}(r_2, l_1)$ respectively,
such that $L_* \overset{p'_*}{\Longrightarrow} R_*$
and the corresponding occurrence map $g'_* : L'_* \to L_*$ satisfies $g'_*(NG'_*) \subseteq (NG_*)$

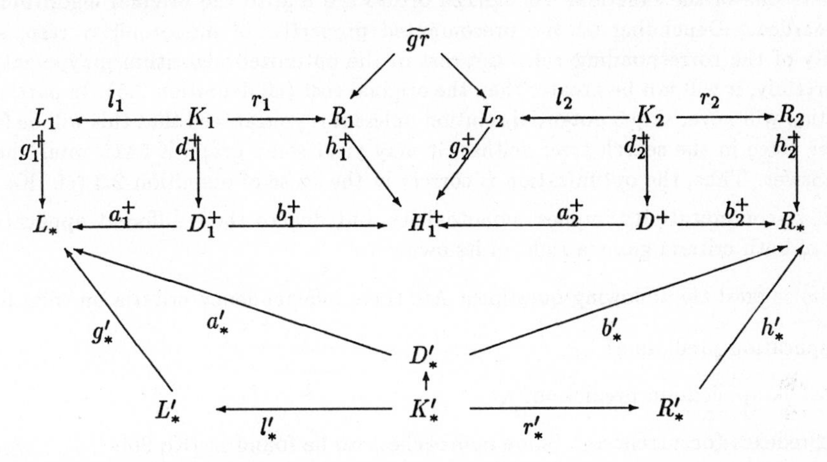

★

THEOREM 4.10 Sc-independence is an optimal syntactical semi-commutativity criterion.[8] ★

Proof. Whenever there is a derivation $G \overset{g_1}{\Longrightarrow} H_1 \overset{g_2}{\Longrightarrow} H_2$ with (p_1, p_2) being $\widetilde{gr}$-related and thus a derivation $G \overset{g_*}{\Longrightarrow} H_2$ via the $\widetilde{gr}$-concurrent production p_*, then there will be another gluing relation $\widetilde{gr}'$ for the pair (p_2, p_1) such that, there is a derivation $L_* \overset{p'_*}{\Longrightarrow} R_*$ with $g' : L'_* \to G$ satisfying $g'_*(NG'_*) \subseteq (NG_*)$; due to the monotonicity lemma 4.6, we then have an occurrence map $g_* \circ g'_* : L'_* \to G$ and by using PO-properties a derivation $G \overset{g'_*}{\Longrightarrow} H_2$ via the $\widetilde{gr}$-concurrent production p'_* thus there is a derivation $G \overset{p_2}{\Longrightarrow} H_1 \overset{p_1}{\Longrightarrow} H_2$. Obviously, all conditions are necessary which implies that there is no better criterion. □

5 Summary and Outlook

In theorem 4.8 we have shown that the definition of m-independency gives an effective method for predicting non-applicability of the corresponding rule provided that this rule could not have been applied before. In any of such situations by interpreting definition 3.2, a call to the `recognize`-function can safely be avoided. Fact 3.3 generalizes this to derivation sequences.

Similar, due to theorem 4.10 and by interpreting definition 3.4, the notion of sc-independency applies to cutting down the number of `derive`-calls for solving a problem by pointing to redundant derivation subtrees. Again, this can be generalized to derivation sequences due to fact 3.6.

[7] Analogously to NG'_* standing for $(L'_* \setminus l'_* K'_*)$, NG_*^{-1} abbreviates $(R_* \setminus r_* K_*)$.
[8] see definitions 3.4 and 3.7.

Moreover, preassuming that there is no information available about the graphs to which rules will be applied, these criteria are optimal in the sense of definition 2.6 due to theorems 4.8 and 4.10, i.e. due to being necessary and sufficient they can not be improved.

These results can be applied to rule-based systems based on graph grammars in the sense of definition 2.4: The basic idea is to slightly change the originally given search algorithm such that the resulting version turns out to be an *optimization* in the sense of definition 2.6.

Each call to one of the functions `recognize` or `derive` due to the original algorithm will simply be guarded. Depending on the precomputed properties of monotonicity resp. semi-commutativity of the corresponding rules the cost of the optimized algorithm may eventually decrease. Certainly, it will not be greater than the original cost (cf. definition 2.5). In particular, the optimization will never skip a potential solution unless it is guaranteed that this will be found at some other place in the search tree; neither, it may yield some graph if `FAIL` would be the appropiate answer. Thus, the optimization is correct in the sense of definition 2.3 (cf. [Ko 90]).

Note that sc-commutativity implies monotonicity, but due to their different applicational aspects each of both criteria gains a right of its own.

Our results suggest the following questions: Are there independency criteria on rules for

- Non-application prediction? $\surd$
- (At-most-)k-application prediction? $\surd$
- Tracing redexes (occurrences)? Some approaches can be found in [Ko 90].
- (Semi-)commutativity for rule-sequences? This essentially requires the associativity axiom for concurrent rules to be satisfied.
- Confluency? ...
- ...Knuth-Bendix completion? ...

Are there *optimal* criteria for these questions?

The presented criteria are optimal w.r.t. *all* graphs (see definition 3.2). So we should further ask: How are these criteria affected if one considers only those graphs, which are reachable from an axiom?

Any answer of the questions above may be a contribution to other applications of this Gra^2-formalism, e.g. Database-systems ([Kre 78]), Algebraic (ADT-)specifications ([Lö 89]) or Processes ([KreWi 83]).

In order to show how our results can be applied in practice, we will finally give a small example.

6 Domino — an Example

The game. A domino game is started with a number of single pieces, or stones, each carrying a number, e.g. 1, 2 or 3, on each of its two ends. The domino rule must ensure essentially that pieces will be connected only at those ends which carry equal numbers. Among the variety of possibilities, we define a goal to be any derivable configuration to which no more rules can be applied. Moreover, we allow circles, which brings us out of the class of problems which can directly be modeled using strings. Other examples, e.g. a solitaire called *fan*, were considered in [Ko 90].

A formalization. Graphs and correspondingly rules contain variables for node colors C_N. The idea is to denote rather a scheme of graphs, or rules, than single ones. Rules are given by an explicit left and right hand side; numbers at the border of items indicate gluing items.

So, a domino AGT-problem is given by $(\mathcal{AX}, P, F, \emptyset)$, where the termination predicate "GameOver", i.e. the **filter** F, will always be defined as matching any graph where none of the current rules can be applied to; additionally, a goal must consist of a single circle. An **axiom** $Ax \in \mathcal{AX}$ is given as a collection (set) of stones:[9]

•←(X)—(Y)→• (for $X, Y \in C_N$)

The connecting **rules** $P = (p_{(X,Y,Z)})_{X,Y,Z \in C_N}$ are given as:

•←(X)—(Y)→• •←(Y)—(Z)→• $\overset{p_1}{\Longrightarrow}$ •←(X)—(Y)→•←(Y)—(Z)→•
1 2 1 2

Clearly, stones do only need one unconnected end for make this rule applicable, and there is no interest in the tightening of the other end. Note, that $p(X,Y,Z)$ and $p(Z,Y,X)$ are isomorphic (in all components).

An independency examination. We observe, for any pair of (instantiated) rules $p_1 = p(X,Y,Z)$ and $p_2 = p(X',Y',Z')$, that

$$\neg \text{ s-indep}(p_1,p_2) \quad \textit{but} \quad \text{monotonic}(p_1,p_2) \quad \text{and} \quad \text{commutative}(p_1,p_2)$$

For rules p_1, p_2 such that $Y \in \{X', Z'\} \wedge Y' \in \{X, Z\}$ we get sequentially dependent derivations. In particular the non-gluing connection node ('•'), result of the 'linking process', and the rest of the doubly connected stone is common to both rule-occurrences.

Contrastingly, it can be determined that all rules are pairwise m-independent and hence, due to theorem 4.8 they are monotonic: whenever p_2 can be applied after p_1 could be to a graph G_1 then p_2 could also have been applied to G_1.

Moreover, they are pairwise sc-independent and due to theorem 4.10 also mutually semi-commutative and hence commutative: the order of application is irrelevant.

A domino problem solved searching breath-first. The standard breadth-first search algorithm as it can be found in, e.g., [ShiTsu 84] can be straightly adapted to the AGT case. A tight optimization in the sense of definition 2.6 based on facts 3.6 and 3.3 can be found as `Opt2search` in [Ko 90] using m- and sc-independency as syntactical criteria. On each state-graph occurrences for *all* rules will be recognized before a derivation will take place. Derivations will be executed in the order of rule-indices.[10]

As the coloring alphabet we assume $C_N = \{1,2,3\}$. The axiom shall consist of three stones [1:2], [2:3] and [3:1].

Then from the 15 rules $p(X,Y,Z)$, $X, Y, Z \in C_N$, only 3 can be applied to the axiom. According to figure 1 each of these derived states can then be expanded in two different ways

[9] nondirected arcs are shorthands for two opposingly directed arcs.

[10] When this paper was created, there were no computer supported graph rewriting systems at the TU Berlin. So, the following results are due to a hand-simulation.

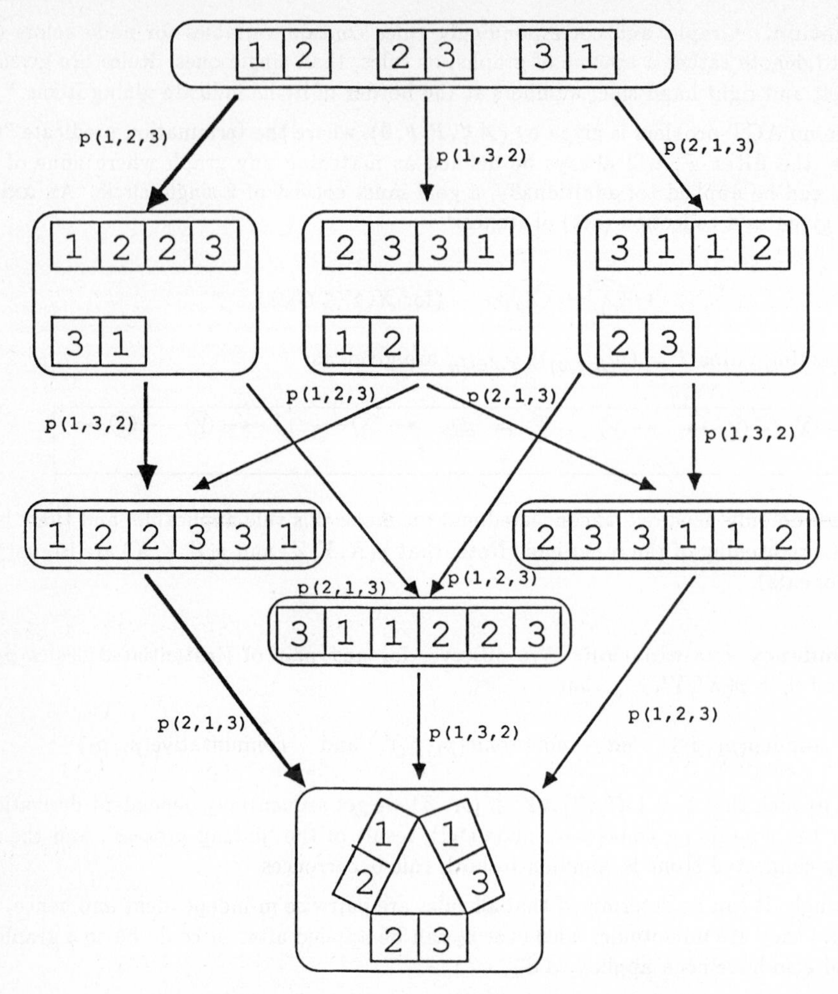

Arrows in bold show those derivations
which are actually made by the optimized search-algorithm

Figure 1: The (collapsed) search space of a domino problem

such that 6 states appear in the (uncollapsed) *search tree* on the corresponding level. The first expansion reaching the consecutive level will discover the goal state.

Summing up, this costs $15 \times 5 = 75$ calls on `Recognize`, 10 on `Derive` and 11 on `isGoal`. Using **monotonicity** and **sc-independency** of all rule combinations, only a single sequence will be generated, for which $15 + 3 + 2 = 20$, 3 and 4 calls are respectively needed, i.e. a **reduction by some** **(74,67,64)%** w.r.t. to the original costs can be achieved.

References

These are only the essential references for this paper, more can be found in [Ko 90].

[Ehr 79] H. Ehrig:
"Introduction to the Algebraic Theory of Graph Grammars" — LNCS 73, Springer Verlag, Berlin, pp. 1–69, 1979

[EhrHa 85] H. Ehrig, A. Habel:
"Graph Grammars with Application Conditions" — In "The Book of L", Springer Verlag, Berlin, pp. 87–100, 1985

[EhrHaRo 86] H. Ehrig, A. Habel, B.K. Rosen:
"Concurrent Transformations of Relational Structures" — Fundamenta Informatica, Vol IX (1), 1986

[Ko 90] M. Korff:
"Optimizations of Production Systems based on Algebraic Graph Transformations" — Technical Report 90/8, TU Berlin, 1990

[Kre 78] H.-J. Kreowski:
"Anwendungen der algebraischen Graphentheorie auf Konsistenz und Synchronisation in Datenbanksystemen" — Technical Report 78/15, TU Berlin, 1978

[KreWi 83] H.-J. Kreowski and A. Habel:
"Is Parallelism already Concurrency? Part II: Non-Sequential Processes in Graph Grammars" — LNCS 153, Springer Verlag, Berlin, pp. 360–380, 1987

[Lö 89] M. Löwe:
"Implementing Algebraic Specifications by Graph Transformation Systems" — Technical Report 89/26 of FB 20 at the TU Berlin, 1989

[MoPa 87] D. Moldovan and F. Parisi-Presicce:
"Parallelism Analysis in Rule-Based Systems Using Graph Grammars" — LNCS 291, Springer Verlag, Berlin, pp. 427–439, 1987

[Ja 74] P. Jackson, Jr.:
"Introduction to artificial intelligence" — Mason & Lipscomb Publishers, Inc., London, 1974

[Ni 74] N. Nilson:
"Principles of Artificial Intelligence" — Springer Verlag, Berlin, 1982

[Ri 83] E. Rich:
"Artificial Intelligence" — New York: McGraw-Hill, 1986

[ShiTsu 84] Y. Shirai, J. Tsujii:
"Artificial Intelligence: Concepts, Techniques and Applications" — John Wiley & Sons, 1984

Tree automata, tree decomposition and hyperedge replacement

Clemens Lautemann

Fachbereich Mathematik und Informatik
Johannes Gutenberg Universität
D–6500 Mainz

Abstract: Recent results concerning efficient solvability of graph problems on graphs with bounded tree–width and decidability of graph properties for hyperedge–replacement graph grammars are systematised by showing how they can be derived from recognisability of corresponding tree classes by finite tree automata, using only well–known techniques from tree-automata theory.

Keywords: graph grammar, tree automaton, graph algorithm, computational complexity, tree-width

Contents

0 Introduction

In recent years, for an ever growing number of computational problems on graphs which, in general, are infeasible, polynomial–time algorithms have been developed on an ever growing number of restricted graph classes. (For a survey of such results, see [Jo 85].) The problems include such celebrities as Hamiltonian cycles, vertex cover, k–colorability, and many others. Among the graph classes for which they are feasible we find important classes like series–parallel graphs, graphs of bounded band–width, outerplanar graphs, etc.
The techniques used to obtain these results are, in most cases, very similar. First it is shown that all graphs in the class under consideration can be decomposed in a certain way, and that such

a decomposition can be constructed in polynomial time. Then a polynomial (often even linear) algorithm is presented which solves the given problem on graphs that are represented by such a decomposition.
Obviously, when these similarities were discovered, the search for general formulations of this technique started, and for general conditions under which it can be applied. This research soon identified tree decomposition as a general description of the various decomposition techniques, and, consequently, the classes of graphs of bounded tree–width as a general description of the classes of graphs to which they can be applied, cf. [RS 86a]. Also, rather general classes of graph properties were described which can be decided in polynomial time on decomposition trees, hence on graphs with bounded tree–width, [Bo 88a, ALS 88]. Moreover, the same methods also yield efficient parallel solutions, so that, with certain restrictions, the same classes of problems belong to NC, for graphs of bounded tree–width, cf. [Bo 88b, La 88b].

In a parallel development, a similar generalisation process took place in connection with the decidability of graph properties for graph grammars. After it had been shown for a number of graph properties, using very similar techniques, that it can be decided whether a given graph grammar of a certain type generates some (or only) graphs with this property, attempts were made to find general formulations of graph properties for which these techniques work. This search for generality has been most successful for "hyperedge–replacement graph grammars", where "finite", "compatible", and "monadic–second order" properties were shown to be decidable, cf. [LW 88, HKV 87, Co 90]. Although these two developments have been largely independent, some authors did recognise the close connection between the respective methods and results. In particular, Courcelle discusses both, decidability for hyperedge–replacement grammars, and efficient decidability for graphs of bounded tree–width of monadic–second order properties, cf. [Co 88].

In this paper, both questions, feasibility for graphs of bounded tree–width and decidability for hyperedge–replacement grammars, are reduced to a single one: recognisability of a certain set of trees by a finite tree automaton.
In this way, many results like the ones cited above can be derived in a uniform way from well–known facts about tree automata. This approach reveals the close conceptual connections between the two types of problems, and should lead to a better understanding of the methods used for their solution. Furthermore, it leads to further generalisations, e.g., by considering tree automata with an infinite state set.

In detail, the paper is composed as follows. In Section 2, we summarise the definitions of hyperedge–replacement grammars, decomposition trees, etc., and some basic facts about these notions.
Section 3 contains a brief introduction to finite tree automata and we show how they can be used in order to solve problems on graphs of bounded tree–width or decide graph properties for hyperedge–replacement grammars. As the central result of the paper, we show that recognisability of the class of derivation trees which represent graphs with property Π implies decidability of Π for hyperedge–replacement grammars, and, similarly, recognisability of the class of decomposition trees which represent graphs with property Π implies efficient solvability of Π on graphs of bounded tree–width. In Section 4 these results are applied in order to derive some of the known general decidability and feasibility results mentioned above. In particular, we show that finite and compatible graph properties are both feasible for graphs with bounded tree–width and decidable for hyperedge–replacement grammars. Furthermore, we show the same for graph properties which can be expressed in monadic–second order logic. In all cases, our approach leads to simplified proofs, both conceptually and technically.

It must be pointed out that the idea of using tree automata in this context is not new. In particular, Courcelle's investigations of monadic–second order graph properties in [Co 90, Co 88] are based entirely on tree automata. However, in place of the purely algebraic treatment given there, I have chosen a more algorithmic approach, as seems appropriate given the algorithmic nature of the problems considered here.

Due to space restrictions, the presentation has been kept rather concise. Proofs had to be omitted in many places, in others only the main ideas are indicated. The interested reader is referred to the full paper [La 90b].

1 Preliminaries

1.1 Trees, graphs, and hypergraphs

All trees in this paper will be directed, rooted trees with node labels. Hence a tree T is given by a tuple $(N_T, r_T, A_T, \lambda_T)$, where

- N_T is a set of *nodes*,
- $r_T \in N_T$ is the *root* of T,
- $A_T \subseteq (N_T \times N_T) \setminus \{(n,n) \mid n \in N_T\}$ is a set of *arcs*, and
- $\lambda_T: N_T \to \Sigma$, for some set Σ. $\lambda_T(n)$ is the *label* of n.

Throughout this paper, we will make use of an infinite alphabet Ω_∞, together with a rank function $rk: \Omega_\infty \to \mathbb{N}$, such that for all $r \in \mathbb{N}$, $rk^{-1}(r)$ is infinite. $rk(B)$ is called the *rank* of B, the elements of Ω_∞ are called *nonterminals*.

Let $l \geq 0$. An *l-hypergraph* $H = (V_H, E_H, Y_H, \mu_H, ports_H, L_H)$ consists of

- a finite set of *vertices*, V_H,
- a set E_H of two–element subsets of V_H, called *edges*,
- a finite set, Y_H, the elements of which are called *hyperedges*,
- a labeling function, $\mu_H: Y_H \to \Omega_\infty$,
  $rk(\mu_H(y))$ is called the *rank* of y,
- a mapping $ports_H: Y_H \to V_H^*$, where, for every $y \in Y_H$, $ports_H(y)$ is a list of $rk(\mu_H(y))$ distinct vertices; these vertices are called the *ports* of y, and we will denote the set of all ports of y by $\{ports_H(y)\}$, and
- a list L_H of l distinct vertices of H, called the *outer vertices* of H. The set of all outer vertices of H will be denoted by $\{L_H\}$.

If $Y_H = \emptyset$, we will say that H is an l–graph and identify H and (V_H, E_H, L_H).
We will drop the prefix "l–", if l is irrelevant, or clear from the context.
Note that we speak of *nodes* and *arcs* of trees, but *vertices* and *edges* of (hyper)graphs.

We will denote the set of all (l-)graphs by Γ (resp. Γ_l), the set of all (l–)graphs with vertices from a set V will be denoted by $\Gamma(V)$ (resp. $\Gamma_l(V)$).

Two l–hypergraphs, H and H', are *isomorphic*, $H \cong H'$, if there are bijections $\pi_V: V_H \to V_{H'}$, and $\pi_Y: Y_H \to Y_{H'}$ such that $E_{H'} = \{\{\pi_V(v), \pi_V(w)\} \mid \{v,w\} \in E_H\}$, $\pi_V^*(L_H) = L_{H'}$,[1] and for all $y \in Y_H$

- $ports_{H'}(\pi_Y(y)) = \pi_V^*(ports_H(y))$, and
- $\mu_{H'}(\pi_Y(y)) = \mu_H(y)$.

The pair (π_V, π_Y) is called an *isomorphism.*

Let G, H be (hyper)graphs. We say that H is a *sub(hyper)graph* of G if $V_H \subseteq V_G$, $E_H \subseteq E_G$, $Y_H \subseteq Y_G$, and, for all $y \in Y_H$, $\mu_H(y) = \mu_G(y)$ and $ports_H(y) = ports_G(y)$.

[1] π_V^* is the canonical extension of π_V from V_H to V_H^*.

1.2 Hyperedge replacement graph grammars

We can combine a hypergraph H and an l–hypergraph G by removing some hyperedge y of rank l from H and replacing it by G, identifying $ports_H(y)$ with L_G. Intuitively, hyperedges can be considered "sockets", while outer vertices figure as "plugs"; the combination of H and G is then achieved by plugging G into H. More precisely, the new hypergraph is obtained as follows.
First of all, we will require that

- $V_G \cap V_H = \{L_G\}$,
- $ports_H(y) = L_G$, and
- $Y_G \cap Y_H = \emptyset$.

In that case (and only then), a hypergraph $H(y \leftarrow G)$ is defined as
$H(y \leftarrow G) := (V_H \cup V_G, E_H \cup E_G, (Y_H \backslash \{y\}) \cup Y_G, \mu, ports, L_H)$,
where $\mu(z) = \begin{cases} \mu_H(z) & \text{if } z \in Y_H \\ \mu_G(z) & \text{if } z \in Y_G \end{cases}$, and $ports(z) = \begin{cases} ports_H(z) & \text{if } z \in Y_H \\ ports_G(z) & \text{if } z \in Y_G. \end{cases}$

This operation of plugging one hypergraph into another is essential for the derivation process of hyperedge replacement. The following lemma states that it is both commutative and associative[2]. (Note that writing $H(y \leftarrow G)$ implies that the three consistency conditions listed above are satisfied, and that y is in Y_H and has rank $|L_G|$.)

1.2.1 Lemma

- Let $y_1, y_2 \in Y_H$. Then $H(y_1 \leftarrow G_1)(y_2 \leftarrow G_2) = H(y_2 \leftarrow G_2)(y_1 \leftarrow G_1)$.
- Let $y_1 \in Y_H, y_2 \in Y_{G_1}$. Then $H(y_1 \leftarrow G_1)(y_2 \leftarrow G_2) = H(y_1 \leftarrow G_1(y_2 \leftarrow G_2))$. □

Henceforth, instead of $H(y_1 \leftarrow G_1)(\ldots)(y_m \leftarrow G_m)$, we will simply write
$H(y_1 \leftarrow G_1, \ldots, y_m \leftarrow G_m)$.
Hyperedge–replacement has been used as a basis for graph rewriting for some time. As can be expected from Lemma 1.2.1, hyperedge–replacement graph grammars behave in many ways like context–free grammars.
Although all our results can also be shown for sets of directed, labeled hypergraphs , in order to avoid an unduely complicated formalism we will only deal with sets of simple graphs. Accordingly, hyperedge–replacement graph grammars as defined below generate simple graphs only. Apart from this restriction, our definition is essentially equivalent to the one given in, e.g., [Ha 89].

1.2.2 Definition (hyperedge–replacement grammar)
A hyperedge–replacement graph grammar, short HRG, $\mathcal{G} = (\Omega_\mathcal{G}, P_\mathcal{G}, S_\mathcal{G})$, is given by

- a finite set, $\Omega_\mathcal{G} \subseteq \Omega_\infty$, of nonterminals,
- a finite set, $P_\mathcal{G}$, of *rules* of the form $p = B \rightarrow H$, where
  - $lhs(p) := B \in \Omega_\mathcal{G}$ and
  - $rhs(p) := H$ is an $rk(B)$–hypergraph, and
- a *start symbol* $S \in \Omega_\mathcal{G}$, $rk(S) = 0$.

The *width* and the *order* of $\mathcal{G}$ are the maximal number of vertices and the maximal rank, respectively, of any nonterminal in $\Omega_\mathcal{G}$. □

[2]Courcelle speaks of confluence and associativity in [Co 87].

1.2.3 Definition (derivation tree)

Let $\mathcal{G} = (\Omega, P, S)$ be a hyperedge-replacement grammar. Let $Y_{\mathcal{G}} := \{\epsilon\} \mathring{\cup} \mathring{\bigcup}_{p \in P} Y_{rhs(p)}$.[3]
A tree $T = (N, r, A, \lambda)$ is a *$\mathcal{G}$-derivation tree* if

- $\lambda: N \to Y_{\mathcal{G}} \times P_{\mathcal{G}}$,
- for all $n \in N$, if $\lambda(n) = (y, B \to H)$ then there is an injection $h_n: Y_H \to N$ such that
  - $(n, n') \in A \iff n' \in h_n(Y_H)$,
  - if $n' = h_n(y')$ then $\lambda(n') = (y', B' \to H')$ and $\mu_H(y) = B'$, for some $B' \to H' \in P_{\mathcal{G}}$. □

1.2.4 Definition ($G(T)$, $\mathcal{L}(\mathcal{G})$)

Let $\mathcal{G} = (\Omega, P, S)$ be a hyperedge-replacement grammar, let $T = (N, r, A, \lambda)$ be a $\mathcal{G}$-derivation tree, and let $\lambda(r) = (y, B \to H)$. The *l-graph represented by T*, $G(T)$, is defined as follows:

- if $Y_H = \emptyset$ then $G(T) = H$,
- otherwise, let $Y_H = \{y_1, \ldots, y_s\}$, and let $h_r: Y_H \to N$ be as in 1.2.3. Then
  $G(T) = H(y_1 \leftarrow G(T(h_r(y_1))), \ldots, y_s \leftarrow G(T(h_r(y_s))))$.

We say that a graph G *can be derived from* $B \in \Omega_{\mathcal{G}}$, $B \to^+_{\mathcal{G}} G$, if there is a $\mathcal{G}$-derivation tree T such that $G(T) \cong G$ and $B = lhs(p)$, where $\lambda_T(r_T) = (\epsilon, p)$. If $B = S_{\mathcal{G}}$ then we say that T is a *basic* derivation tree for G. The set of all basic derivation trees is denoted by $\mathcal{T}_{\mathcal{G}}$.
The *graph language generated by* $\mathcal{G}$ is defined as $\mathcal{L}(\mathcal{G}) := \{G \in \Gamma \;/\; S_{\mathcal{G}} \to^+_{\mathcal{G}} G\}$.
A graph property[4] Π is *decidable* for a class $\mathcal{C}$ of hyperedge-replacement grammars if there is an algorithm, which, for every $\mathcal{G} \in \mathcal{C}$ decides whether $\mathcal{L}(\mathcal{G}) \cap \Pi = \emptyset$. □

Similarly to context-free string grammars, HRGs can be effectively normalised in a number of ways. In particular, we can assure that every nonterminal production creates a new vertex, or that the right-hand side of every nonterminal production contains exactly two hyperedges. Combining these two constructions, we arrive at a "binary reduced normal form" which we will use throughout.

1.2.5 Fact

There is an effective transformation which constructs, for every HRG $\mathcal{G}$, a HRG $\mathcal{G}'$ such that $\mathcal{L}(\mathcal{G}') = \mathcal{L}(\mathcal{G})$, every $\mathcal{G}'$-derivation tree is binary, and for every basic $\mathcal{G}'$-derivation tree T we have $|N_T| \leq 2|V_{G(T)}|$. □

Whereas making trees binary is mainly a matter of technical convenience, the size of derivation trees plays a crucial part in complexity considerations.
Examples and algebraic and structural results concerning hyperedge-replacement grammars can be found in, e.g., [Ha 89], [BC 87].

1.3 Decomposition trees

Many efficient algorithms on restricted graph classes, such as *series-parallel* graphs or *Halin* graphs, are based on some sort of decomposition technique. A rather general formulation of this technique is tree-decomposition, as introduced by Robertson and Seymour in [RS 86a]. Based on it the authors

---

[3] $\mathring{\cup}$ denotes the disjoint union.

[4] I.e. a predicate on graphs which is closed under isomorphism. We will identify a predicate with the set of those graphs for which it holds.

define the notion of tree–width and show that the disjoint paths problem has a polynomial–time solution on any set of graphs of bounded tree–width.[5]
Our definition below differs from those in the literature mainly in one aspect: rather than starting from a graph and defining its tree–decomposition, we will start with a labeled tree and define how to compose a graph from it. This leads to exactly the same notion of tree–width as in [RS 86a], but is more in line with our use of trees in this paper.

1.3.1 Definition (decomposition tree)

Let $k \in \mathbb{N}$. A *k–decomposition tree* is a tree $T = (N, r, A, \lambda)$, where $\lambda: N \rightarrow \bigcup_{l \leq k} \Gamma_l(\{1, \ldots, k\})$, and for every arc $(n, m) \in A_T$ we have $\{L_{\lambda(m)}\} \subseteq V_{\lambda(n)}$. The set of all k–decomposition trees is denoted by $\mathcal{T}_k$. □

A decomposition tree T represents an l–graph $G(T)$, which can be constructed from the labels $\lambda_T(n)$ by identifying, for every arc $(n, m) \in A_T$, outer vertices of $\lambda_T(m)$ with equally named vertices of $\lambda_T(n)$.

1.3.2 Definition ($G(T)$, tree–width)

1. Let $T=(N, r, A, \lambda)$ be a k–decomposition tree, and let $M_T = \{(n, i) \in N \times \{1, \ldots, k\} \,/\, i \in V_{\lambda(n)}\}$.
   Let $\sim_T$ be the equivalence relation on M_T induced by (i.e., $\sim_T$ is the transitive, symmetric and reflexive closure of) the relation
   $(n, i) \sim'_T (m, j) : \Longleftrightarrow \big(i = j$ and $(n, m) \in A$ and $j \in L_{\lambda(m)}\big)$.
   The l–graph *represented by* T is then defined to be the graph $G(T) = (V, E, L)$, where
   - $V = \{[(n, i)]_{\sim_T} \,/\, (n, i) \in M_T\}$,
   - $E = \{\{[(n, i)]_{\sim_T}, [(n, j)]_{\sim_T}\} \,/\, n \in N_T,\ \{i, j\} \in E_{\lambda(n)}\}$, and
   - $L = [(r, i_1)]_{\sim_T} \cdots [(r, i_l)]_{\sim_T}$, where $i_1 \cdots i_l = L_{\lambda(r)}$.

   T is called a *k–decomposition of* (any graph isomorphic to) $G(T)$.
2. The *tree–width* of a graph G is the smallest number k such that G has a $(k+1)$–decomposition. The set of all graphs of tree–width at most $k - 1$ is denoted by $\Gamma^{tw(k)}$. □

There is a straightforward recursive construction for $G(T)$ (or more precisely, for an isomorphic copy of it): starting from the leaves of T form, at every node n, the union of n's label, G, and of isomorphic copies of the graphs represented by the subtrees below n, which intersect only in those vertices that are represented in G.

1.3.3 Facts

1. Every graph on m vertices has tree-width at most $m - 1$.
2. The complete graph on m vertices has tree–width exactly $m - 1$; thus the classes of graphs of tree–width at most k, for $k \geq 0$, form a strict hierarchy.
3. If a graph G with m vertices has tree–width $\leq k-1$, then it has a k–decomposition T with at most m nodes.
4. For every k–decomposition tree T on m nodes there is a *binary* k–decomposition tree T' on at most $2m - 1$ nodes such that $G(T) = G(T')$.

---

[5]A survey of algorithmic techniques for graphs of bounded tree-width, as well as equivalent formulations of this concept are given in [Ar 85], a number of examples can be found in [Bo 86, Bo 88c]. Other related work includes [ALS 88, Bo 88a, La 88a].

5. If T is a k–decomposition tree for G, then the set of vertices represented at a node of T forms a separator of G. Thus G has separators of size at most k.
6. $\Gamma^{tw(k)}$ can be generated by a HRG $\mathcal{G}_k = (\Omega, P_k, B_0)$ as follows. Let $\Omega = \{B_0, \ldots, B_k\}$, and let $rk(B_i) = i$, for $i = 0, \ldots k$. Let $\mathcal{H}_i$ consist of all i–hypergraphs H such that
   - $V_H \subseteq \{1, \ldots, k\}$
   - $y \in Y_H \Longrightarrow \mu(y) \in \Omega$
   - for all $y_1, y_2 \in Y_H$, if $ports_H(y_1) = ports_H(y_2)$ then $y_1 = y_2$.

   P_k then consists of all rules $B_i \rightarrow H$, where $H \in \mathcal{H}_i$, $i = 0, \ldots, k$. □

In view of Remark 1.3.3.4 we will henceforth assume all decomposition trees to be binary.

As Remark 1.3.3.5 shows, a k–decomposition tree T represents $G(T)$ by way of separators of size $\leq k$. This can be exploited algorithmically, leading to linear–time algorithms on k–decomposition trees for a large number of, otherwise difficult, problems.
However, if a graph G is not given by a decomposition tree, in order to apply such algorithms, a tree–decomposition of G has to be constructed.

1.3.4 Fact ([RS 86b, ACP 87, La 88a])

For every $k \in \mathbb{N}$ there is a polynomial–time algorithm which decides, for every graph G, whether G has tree–width $\leq k$, and, if so, constructs a k–decomposition of G.[6] □

The grammar $\mathcal{G}_k$ of Fact 1.3.3.6 has an efficient parallel parsing algorithm. This leads to the following strengthening of Fact 1.3.4.

1.3.5 Fact ([La 90a])

$\Gamma^{tw(k)} \in$LOGCFL, for every $k \in \mathbb{N}$. □

Here, LOGCFL is the class of sets that are log–space reducible to context–free languages. This class is contained in P and plays an important part in parallel complexity theory. It has been characterised as the class of all those sets that can be recognised by a log–space bounded *alternating Turing machine* (ATM) with polynomial tree size, cf. [Ru 80]. This characterisation is particularly useful for our purposes, in particular, Fact 1.3.5 was proved with an ATM in [La 90a]. This paper also contains a brief review of ATMs and LOGCFL, for a general survey of parallel complexity, see [Co 81].

2 Finite tree automata

Tree automata were introduced by Thatcher and Wright and by Doner ([TW 65, Do 70]) as an extension of the concept of finite automata from strings to terms. Soon a rich structural theory developed, to a considerable extent by extension of finite (string–)automata theory. A good introduction to the field is the book of Gécseg and Steinby [GS 84], for an early informal survey, see [Th 73].

[6] The algorithms given in [ACP 87] and [La 88a]) can be made uniform in k, i.e., they can be formulated as algorithms with two inputs, k, and G. (Note that in this form, the problem is NP–complete, cf. [ACP 87].) Their running time is of order $n^{O(k)}$. The algorithm implicit in [RS 86a] is of order n^3 – only the (very big) constant depends on k – but is not uniform, i.e., only the existence of such an algorithm for every k is proven, but no construction is given to actually obtain one.

2.1 Definitions and basic facts

Both structures that we will apply tree automata to, derivation trees and decomposition trees, are unordered, and can without loss of generality be assumed to be binary. It will therefore be sufficient to consider symmetric tree automata on binary trees.
Before introducing tree automata, we will compile some further notation on trees.

2.1.1 Definition (Σ–tree)
Let Σ be a finite alphabet. A Σ–tree is a binary tree $T = (N, r, A, \lambda)$, where $\lambda: N \rightarrow \Sigma$. The set of all Σ–trees is denoted by $\mathcal{T}_\Sigma$. For every $\sigma \in \Sigma$, we also write σ for the Σ–tree $(\{r\}, r, \emptyset, (r \mapsto \sigma))$[7].
A *partial* Σ–tree $R = (N, r, A, \lambda, c)$ is a Σ–tree (N, r, A, λ) together with a distinguished leaf c. The set of all partial Σ–trees is denoted by $\mathcal{PT}_\Sigma$. □

Next, we define some operations on, and homomorphisms between trees.

2.1.2 Definition (operations on trees, congruence)
Let Σ and Ω be finite alphabets.

1. Let $S, T \in \mathcal{T}_\Sigma$, $\sigma \in \Sigma$. Then $\sigma(S, T)$ is the Σ–tree (N, r, A, λ), where
$N = N_S \mathring{\cup} N_T \mathring{\cup} \{r\}$,
$A = A_S \mathring{\cup} A_T \mathring{\cup} \{(r, r_S), (r, r_T)\}$,
$$\lambda(n) = \begin{cases} \sigma & \text{if } n = r \\ \lambda_S(n) & \text{if } n \in N_S \\ \lambda_T(n) & \text{if } n \in N_T. \end{cases}$$
2. If R is a partial Σ–tree and T is a (partial) Σ–tree then the (partial) Σ–tree $R \circ T$ is defined as (N, r, A, λ) (respectively, (N, r, A, λ, c)), where
$N = (N_R \backslash \{c_R\}) \mathring{\cup} N_T$,
$r = r_R$,
$A = (A_R \backslash \{(m, c_R) \ / \ m \in N_R\}) \mathring{\cup} A_T \mathring{\cup} \{(m, r_T) \ / \ (m, c_R) \in A_R\}$,
$$\lambda(n) = \begin{cases} \lambda_R(n) & \text{if } n \in N_R \\ \lambda_T(n) & \text{if } n \in N_T \end{cases}$$
$c = c_T$ (if T is partial).
3. For a partial Σ–tree R and Σ–trees S, T the Σ–tree $R(S, T)$ is defined as $R \circ (\lambda_R(c_R)(S, T))$.
4. An equivalence relation $\sim$ on $\mathcal{T}_\Sigma$ is called a *congruence* if, for all $\sigma \in \Sigma$, $T_1, T_2, S_1, S_2 \in \mathcal{T}_\Sigma$, we have $T_1 \sim S_1 \wedge T_2 \sim S_2 \Longrightarrow \sigma(T_1, T_2) \sim \sigma(S_1, S_2)$. □

2.1.3 Remarks

1. $\sigma(S, T)$ consists of (disjoint copies of) S and T, joined together by a root with label σ.
2. $R \circ T$ is obtained from R by replacing c_R by T.
3. $R(S, T)$ is obtained by connecting the distinguished node of R with the roots of S and T. In particular, if R is the partial tree on a single node with label σ then $R(S, T) = \sigma(S, T)$. □

2.1.4 Facts

1. Let R, S be partial Σ–trees, and let T be a (partial) Σ–tree. Then $(R \circ S) \circ T = R \circ (S \circ T)$.
2. If $\sim$ is a congruence and $T_1 \sim S_1$, $T_2 \sim S_2$ then $R(T_1, T_2) \sim R(S_1, S_2)$, and $R \circ T_1 \sim R \circ S_1$, for every partial Σ–tree R. □

We proceed now with the definition of finite tree automata.

[7] $(r \mapsto \sigma)$ denotes the mapping $\lambda: \{r\} \rightarrow \Sigma$ with $\lambda(r) = \sigma$.

2.1.5 Definition (finite tree automaton, FTA)

A *finite tree automaton (FTA)* F is given by a quadrupel $(\Sigma, Q_F, \delta_F, Q_F^0)$, where

- Σ is a finite set, called the *input alphabet*,
- Q_F is a finite set of *states*,
- $\delta_F: (\Sigma \cup (\Sigma \times Q_F \times Q_F)) \rightarrow Q_F$ is a mapping, called the *transition function*,
- $Q_F^0 \subseteq Q_F$ is the set of *accepting* states.

F is called *symmetric* if δ_F is symmetric, i.e., if for all $\sigma \in \Sigma$, and all $q_1, q_2 \in Q_F$, $\delta_F(\sigma, q_1, q_2) = \delta_F(\sigma, q_2, q_1)$.
A FTA with alphabet Σ is also called a Σ–FTA. □

Having been developed as a term–recognising device, general finite tree automata work on ordered trees. In our context, however, trees are always unordered, and we will therefore only consider symmetric tree automata, without specifically saying so.

Informally, a tree automaton F processes a Σ–tree $T = (N, r, A, \lambda)$ as follows: on every node n of T there is a processor, $P(n)$, which can read $\lambda(n)$, and decide if n is a leaf. If n is not a leaf $P(n)$ can read the states sent by processors on the children of n. As soon as $P(n)$ has all the information needed, i.e., $\lambda(n)$ (and the states q_1, q_2 sent by the processors on n's children, if n is not a leaf) it sends the state $\delta_F(\lambda(n))$ (resp. $\delta_F(\lambda(n), q_1, q_2)$) up to its parent node.
This form of tree automaton is often called "bottom–up", (or "frontier–to–root" in [GS 84]).

2.1.6 Definition (recognisability)

Let $T = (N, r, A, \lambda)$ be a Σ–tree, and let $F = (\Sigma, Q, \delta, Q^0)$ be a FTA. Then F *induces* a Q–labeling of T, i.e., a mapping $\gamma_T^F: N_T \rightarrow Q$, as follows:

If n is a leaf then $\gamma_T^F(n) := \delta(\lambda(n))$.

If n has children n_1, n_2 then $\gamma_T^F(n) := \delta(\lambda(n), \gamma_T^F(n_1), \gamma_T^F(n_2))$.

F *accepts* or *recognises* T if $\gamma_T^F(r) \in Q^0$.
The set of Σ–trees recognised by a FTA F is denoted $\mathcal{L}(F)$.
A set $\mathcal{T}$ of Σ–trees is said to be (FTA-)*recognisable*, if there is a FTA F such that $\mathcal{L}(F) = \mathcal{T}$. □

A straightforward application of this notion are recognisability of k–decomposition trees and of $\mathcal{G}$–derivation trees.

2.1.7 Facts

- For every $k \in \mathbb{N}$ there is a finite tree automaton F_k that recognises $\mathcal{T}_k$, the set of k–decomposition trees.
- For every hyperedge–replacement grammar $\mathcal{G}$ there is a finite tree automaton $F_{\mathcal{G}}$ that recognises $\mathcal{T}_{\mathcal{G}}$, the set of basic $\mathcal{G}$–derivation trees.
- The set of *all* $\mathcal{G}$–derivation trees is recognisable, for every hyperedge–replacement grammar $\mathcal{G}$.

□

The following facts about tree automata and recognisable sets hold for every finite alphabet Σ. Their proofs are either straightforward, or can be found in [GS 84].

2.1.8 Facts

1. A set $\mathcal{T}$ of Σ–trees is recognisable if and only if $\mathcal{T}$ is recognisable by a *nondeterministic* FTA (where δ is a relation rather than a function).
2. The set of all Σ–trees is recognisable.
3. There are algorithms which, for every pair F_1, F_2 of finite tree automata over Σ, construct a finite tree automaton which recognises $\mathcal{L}(F_1)\cup\mathcal{L}(F_2)$, $\mathcal{L}(F_1)\cap\mathcal{L}(F_2)$, and $\mathcal{L}(F_1)\backslash\mathcal{L}(F_2)$, respectively.
4. There is an algorithm which decides for every Σ–FTA F whether $\mathcal{L}(F) = \emptyset$ □

Recognisable sets have been characterised in a number of ways, the following of which we will use below.

2.1.9 Fact

A set $\mathcal{T}$ of Σ–trees is recognisable if and only if $\mathcal{T}$ is the union of equivalence classes of a congruence of finite index on $\mathcal{T}_\Sigma$.[8] □

2.1.10 Remark

We will also need the effective version of the "if"–part of 2.1.9, namely: there is an algorithm which, given a decision procedure for a congruence $\sim$ of finite index, a set $\mathcal{R}$ of representatives for $\sim$, and a subset of $\mathcal{R}$ which characterises $\mathcal{T}$, constructs a FTA that recognises $\mathcal{T}$. □

2.2 Algorithmic consequences

Finite tree automata, like finite string automata, are very efficient devices.

2.2.1 Proposition

1. For every fixed FTA F, $\mathcal{L}(F)$ can be decided in linear time.
2. For every fixed FTA F, $\mathcal{L}(F)\in$LOGCFL. □

Notation

Let Π be a graph property, i.e., a set of graphs which is closed under isomorphism. For every $k\in\mathbb{N}$, let $\mathcal{T}_k^\Pi$ denote the set

$$\mathcal{T}_k^\Pi = \{T\in\mathcal{T}_k \;/\; G(T)\in\Pi\}.$$

□

The ATM constructed in the proof of 1.3.5 constructs a k–DT T for its input graph in the sense that at every node n of an accepting computation tree it processes a node n' of T, and that the children of n' are processed at successor nodes of n. Thus for every FTA F we can combine this ATM with the one of 2.2.1.2 to an ATM which accepts a graph if and only if it has a k–derivation tree which is accepted by F.

[8]The index of an equivalence relation is the number of equivalence classes it induces.

2.2.2 Theorem

Let $\mathcal{T}_k^\Pi$ be FTA–recognisable. Then $\Pi\cap\Gamma^{tw(k)}\in$LOGCFL. □

Notation

Let $\mathcal{G}$ be a hyperedge–replacement graph grammar, and let Π be a graph property. By $\mathcal{T}_\mathcal{G}^\Pi$, we denote the set

$$\mathcal{T}_\mathcal{G}^\Pi = \{T\in\mathcal{T}_\mathcal{G} \;/\; G(T)\in\Pi\}\,.$$

□

From Fact 2.1.8.4 we can deduce that Π is decidable for a class $\mathcal{C}$ of hyperedge–replacement grammars if for every $\mathcal{G}\in\mathcal{C}$ a FTA for $\mathcal{T}_\mathcal{G}^\Pi$ can be found.

2.2.3 Theorem

Let $\mathcal{C}$ be a class of hyperedge–replacement graph grammars. If there is an algorithm which, for every $\mathcal{G}\in\mathcal{C}$ constructs a FTA which recognises $\mathcal{T}_\mathcal{G}^\Pi$ then Π is decidable for $\mathcal{C}$. □

In fact, decidability of Π for the class of *all* hyperedge–replacement grammars follows already from effective recognisability of $\mathcal{T}_k^\Pi$ for all k. Intuitively, this is not surprising, since derivation trees and decomposition trees are very similar and, indeed, from every derivation tree a decomposition tree for the same graph can be constructed. This transformation, however, cannot be done in one bottom–up traversal of the tree, as vertices in the labels of different nodes of a decomposition tree T may well represent the same vertex of $G(T)$, in which case they must have the same name. On the other hand, proceeding from the root down to the leaves, a $\mathcal{G}$–derivation tree T is easily transformed into a k–decomposition tree T' for the same graph, where $k = width(\mathcal{G})$. If $T = \sigma(T_1, T_2)$ and $\sigma{=}(\epsilon, S{\rightarrow}H)$ is the root label of T let $G\in\Gamma(\{1,\ldots,k\})$ be any graph isomorphic to (V_H, E_H), and let π be the corresponding isomorphism. Then T' has root label G, and the subtrees T_1' and T_2' can now be labeled. Let $(y_i, B_i{\rightarrow}H_i)$ be the root label of T_i. In order to ensure that the outer vertices of r_{T_i} get their proper names, the list $\pi^*(ports_H(y_i))$ is passed on to T_i. Then an isomorphic copy G_i of $(V_{H_i}, E_{H_i}, L_{H_i})$ is chosen in such a way that $L_{H_i} = \pi^*(ports_H(y_i))$. When all the leaves have been reached by this procedure, a k–decomposition tree will have been constructed. Thus, if $\mathcal{T}_k^\Pi$ is recognisable by a (nondeterministic) top–down (or "root–to–frontier", cf. [GS 84]) tree automaton, then so is $\mathcal{T}_\mathcal{G}^\Pi$.
Similarly, every k–decomposition tree can be transformed top–down into a $\mathcal{G}_k$–derivation tree for the same graph, where $\mathcal{G}_k$ is the grammar of 1.3.3.6 which generates $\Gamma^{tw(k)}$. Hence from a top–down tree automaton which recognises $\mathcal{T}_{\mathcal{G}_k}^\Pi$ we can construct a top–down FTA which recognises $\mathcal{T}_k^\Pi$.
Since nondeterministic top–down and bottom–up tree automata recognise the same sets, we have the following result.

2.2.4 Proposition

There is an algorithm which, for every $k\in\mathbb{N}$ constructs a FTA that recognises $\mathcal{T}_k^\Pi$ if and only if there is an algorithm which constructs, for every hyperedge–replacement grammar $\mathcal{G}$, a FTA $F_\mathcal{G}$ that recognises $\mathcal{T}_\mathcal{G}^\Pi$. □

2.2.4.1 Corollary

If there is an algorithm which, for every $k\in\mathbb{N}$ constructs a FTA that recognises $\mathcal{T}_k^\Pi$ then Π is decidable for the class of all hyperedge–replacement graph grammars. □

3 Applications

In this section, we will use Theorems 2.2.2 and 2.2.3 in order to show that some general conditions on graph problems imply efficient solvability on graphs of tree–width $\leq k$, or decidability for hyperedge–replacement grammars. Although our complexity results are slightly stronger than what was known before (cf. [Bo 88b]), our main interest is in the uniform way of treating what seemed rather different approaches, thus revealing the tight connections between them.[9]

3.1 Compatible graph properties

First, we consider so–called *compatible graph properties*, introduced and investigated by Habel, Kreowski and Vogler in [HKV 87].

3.1.1 Definition (compatibility)

1. Let $\mathcal{C}$ be a class of hyperedge–replacement grammars, $I = (I_k)_{k\in\mathbb{N}}$ be an infinite family of finite index sets I_k, PROP a decidable predicate[10] defined on pairs (G, i) with $G\in\Gamma_k$ and $i\in I_k$, and PROP′ a decidable predicate on triples $(H, assign, i)$ where H is a k–hypergraph, $assign{:}Y_H\to\bigcup_{l\in\mathbb{N}} I_l$ is a mapping such that for all $y\in Y_H$ $assign(y)\in I_{rk(\mu_H(y))}$, and $i\in I_k$.
   Then PROP is called $\mathcal{C}$, PROP′–*compatible* if, for every hyperedge–replacement grammar $\mathcal{G}\in\mathcal{C}$ and every $\mathcal{G}$–derivation tree $T = (y, B\to H)(T_1, T_2)$, the following holds:
   PROP$(G(T), i)$ is true if and only if there are $j_1\in I_{l_1}$, $j_2\in I_{l_2}$ such that
   PROP′$(H, (y_1 \mapsto j_1,\ y_2 \mapsto j_2), i)$, PROP$(G(T_1), j_1)$ and PROP$(G(T_2), j_2)$ are true,
   where $Y_H = \{y_1, y_2\}$, and $l_i = rk(\mu_H(y_i))$, for i=1, 2.
2. A graph property Π is called $\mathcal{C}$–*compatible* if predicates PROP and PROP′ and an index $i_0\in I_0$ exist such that PROP is $\mathcal{C}$, PROP′–compatible and $G\in\Pi \iff$ PROP(G, i_0). Π is called *compatible*, if it is $\mathcal{C}$–compatible, where $\mathcal{C}$ is the class of all hyperedge–replacement graph grammars. □

3.1.2 Fact

The class of $\mathcal{C}$–compatible predicates is closed under Boolean operations. □

3.1.3 Proposition

Let $\mathcal{C}$ be a class of hyperedge–replacement grammars. If the graph property Π is $\mathcal{C}$–compatible then there is an algorithm which, for every $\mathcal{G}\in\mathcal{C}$ constructs a FTA $F^{\Pi}_{\mathcal{G}}$ that recognises $\mathcal{T}^{\Pi}_{\mathcal{G}}$.

Proof (sketch)

We will only give the construction of $F^{\Pi}_{\mathcal{G}}$.

By assumption, there are $I = (I_k)_{k\geq 0}$, $i_0\in I_0$, PROP and PROP′ as in Definition 3.1.1, such that $G\in\Pi \iff$ PROP(G, i_0) and PROP is $\mathcal{C}$, PROP′–compatible.

Let $\mathcal{G} = (\Omega, P, S)$ be a grammar of width k in $\mathcal{C}$, and let $Y_{\mathcal{G}} := \{\epsilon\}\mathring{\cup}\mathring{\bigcup}_{p\in P} Y_{rhs(p)}$.

Define $F = (\Sigma, Q, \delta, Q_0)$, where

$$\Sigma = Y_{\mathcal{G}}\times P,$$

$$Q = \bigcup_{l\leq k}\mathcal{P}(I_l),$$

$$Q_0 = \{J_0\subseteq I_0\ /\ i_0\in J_0\},$$

[9] It was shown recently that the conditions under consideration are essentially equivalent, cf. [HKL 89].

[10] We assume that all considered predicates are *closed under isomorphisms*, i.e., if a predicate Φ holds for a (hyper)graph H and $H \cong H'$, then Φ holds for H', too.

$\delta(y, B\rightarrow H) = \{i\in I_l \ / \ \text{PROP}(H,i)\}$, where $l=|L_H|$,
$\delta((y, B\rightarrow H), J_1, J_2) = \{i\in I_l \ / \ (\exists j_1\in J_1, j_2\in J_2) : \text{PROP}'(H,(y_1 \mapsto j_1, y_2 \mapsto j_2), i)\}$, where $l=|L_H$
and $Y_H = \{y_1, y_2\}$. □

Combining 3.1.3 with Theorems 2.2.2 and 2.2.3, respectively, we obtain the following.

3.1.3.1 Corollary

If Π is $\mathcal{C}$–compatible then Π is decidable for $\mathcal{C}$. □

3.1.3.2 Corollary

If Π is compatible then $\Pi\cap\Gamma^{tw(k)}\in$LOGCFL, for every $k\in\mathbb{N}$.

3.2 Finite graph properties

Lengauer and Wanke introduced *finite* graph properties and proved their decidability for hyperedge–replacement grammars[11] in [LW 88].

3.2.1 Definition ($G \odot H$)

Let G, H be k–graphs with outer vertex lists L_G, L_H, respectively. Then $G \odot H$ denotes the graph $G'(y \leftarrow H)$, where G' is the hypergraph $(V_G, E_G, \{y\}, (y \mapsto B), (y \mapsto L_G), <>)$. □

3.2.2 Definition (finite graph property)

Let Π be a graph property.

1. Two k–graphs G, G' are *replaceable* with respect to Π, denoted by $G \sim_k^\Pi G'$, if, for every k–graph H,
$$\Pi(H \odot G) = \Pi(H \odot G').$$
2. Π is called *k–finite* if $\sim_k^\Pi$ has finite index.
3. Π is called *finite* if it is k–finite for every $k\in\mathbb{N}$. □

In a similar way to $\sim_k^\Pi$ we can define an equivalence relation on k–decomposition trees.

3.2.3 Definition ($T \approx_k^\Pi S$)

Let $T, S\in\mathcal{T}_k$. Then

$$T \approx_k^\Pi S \iff \{L_{\lambda_S(r_S)}\}=\{L_{\lambda_T(r_T)}\} \text{ and } [G(R\circ T)\in\Pi \iff G(R\circ S)\in\Pi \text{ for all those } R\in\mathcal{P}\mathcal{T}_k \text{ for which } R\circ T, R\circ S\in\mathcal{T}_k].$$
□

Thus, T and S are equivalent if the graphs represented by them are replaceable with respect to Π when only contexts represented by partial k–decomposition trees are considered.
Using Fact 2.1.4.1 we can deduce that $\approx_k^\Pi$ is a congruence on $\mathcal{T}_k$.

3.2.4 Lemma

If the graph property Π is finite then $\approx_k^\Pi$ has finite index. □

Thus Fact 2.1.9 implies that any union of $\approx_k^\Pi$–classes is recognisable. Since $\mathcal{T}_k^\Pi$ is such a union, we get the following corollary.

[11] Lengauer and Wanke call them *context–free cellular graph grammars.*

3.2.4.1 Corollary

If Π is finite then $\Pi\cap\Gamma^{tw(k)}\in$LOGCFL, for all $k\in\mathbb{N}$. □

Decidability of Π for HRGs is somewhat more involved, as it requires the algorithmic construction of an FTA for $\mathcal{T}_k^\Pi$ for every k. Like Lengauer and Wanke in [LW 88], we need a computable upper bound on the index of $\sim_k^\Pi$ for all k.

3.2.5 Theorem

Let Π be a decidable finite graph property, and assume that there is a computable function $i:\mathbb{N}\rightarrow\mathbb{N}$ such that, for all $k\in\mathbb{N}$, $i(k)$ is an upper bound on the index of $\sim_k^\Pi$.
Then there is an algorithm which constructs, for every $k\in\mathbb{N}$, a FTA F_k^Π which recognises $\mathcal{T}_k^\Pi$.
Proof (sketch)
The construction is in two steps. First, the function i is used in order to construct a set of representatives for the equivalence classes of $\approx_k^\Pi$, and then, based on these representatives, F_k^Π is constructed.
If two k–DTs, S,T, are not congruent (mod $\approx_k^\Pi$) then $\{L_{\lambda(r_T)}\}\neq\{L_{\lambda(r_S)}\}$ or the l–graphs $G(S)$ and $G(T)$ are not replaceable. Since there are no more than 2^k different possibilities for $\{L_{\lambda(r_T)}\}$, $m=2^k\cdot\sum_{l\leq k}i(l)$ is an upper bound on the index of $\approx_k^\Pi$.
The proof of the following two facts is omitted here.
Claim 1 Every equivalence class of $\approx_k^\Pi$ contains a tree of height at most m.
Claim 2 Let $T\not\approx_k^\Pi S$. Then there is a partial tree R of height at most $2m$ such that

$$\neg[G(R\circ T)\in\Pi \iff G(R\circ S)\in\Pi].$$

Thus, in order to find a set $\mathcal{R}$ of representatives for all equivalence classes of $\approx_k^\Pi$ it is sufficient to check, for every pair (S,T) of k–decomposition trees of height $\leq m$, whether there is a partial k–decomposition tree R of height $\leq 2m$ such that $\neg\,[G(R\circ T)\in\Pi \iff G(R\circ T)\in\Pi]$. Since Π is decidable, $\mathcal{R}$ can be constructed by an algorithm. Furthermore, Claim 2 yields a decision procedure for $\approx_k^\Pi$, and $\mathcal{T}_k^\Pi$ is represented by $\mathcal{R}^\Pi:=\{T\in\mathcal{R}\ /\ G(T)\in\Pi\}$.
Thus all requirements of Remark 2.1.10 are satisfied, and F_k^Π can be constructed. □

3.2.5.1 Corollary

Under the conditions of Theorem 3.2.5, Π is decidable for the class of all hyperedge–replacement grammars. □

3.3 Graph properties expressible in monadic–second order logic

Monadic–second order logic has been used by Courcelle [Co 90] and by Arnborg, Lagergren and Seese [ALS 88] in order to describe classes of graph properties which are decidable for hyperedge–replacement grammars (Courcelle) or solvable in polynomial time on graphs with bounded tree–width (Arnborg et al.). We will now derive versions of their results by means of the methods developed above. To this end we will employ the inductive proof technique of [Do 70, TW 65] that led to their decidability results for certain monadic–second order theories.

3.3.1 Definition

1. Let L_0 be the language consisting of
   - $\neg, \vee, \wedge, \exists, \forall, =, \in$
   - individual variables $x_1, \ldots$, set variables $X_1, \ldots$
   - unary predicates *vertex*, *edge*
   - a ternary predicate *connects*

   A formula over L_0 is called a monadic–second order (mso–) formula (over L_0).
2. Let f be a mso–formula with exactly the free individual variables $x_1, \ldots, x_l$ and exactly the free set variables $X_1, \ldots, X_s$, let G be a graph, and let $p_1, \ldots, p_l \in V_G \cup E_G$, $P_1, \ldots, P_s \subseteq V_G \cup E_G$ be given. Then $(G, p_1, \ldots, p_l, P_1, \ldots, P_s) \models f$, read "$(G, p_1, \ldots, p_l, P_1, \ldots, P_s)$ *satisfies* f", if $f\left[(x_i|p_i)_{i=1}^{l}, (X_i|P_i)_{i=1}^{s}\right]$[12] holds true when individual variables range over $V_G \cup E_G$, set variables range over $\mathcal{P}(V_G \cup E_G)$ and the constants are interpreted as suggested by their names, i.e., $=$ as equality, $\in$ as set membership, $vertex(x)$ as $x \in V_G$, $edge(x)$ as $x \in E_G$, and $connects(x, y, z)$ as $x = \{y, z\} \in E_G$.
3. *counting monadic second order* (cmso–) formulae are formed just as mso–formulae, with additional unary set predicates $card_q^r$, one for every pair $(r, q) \in \mathbb{N}^2$, $0 \leq r < q$. The intended semantics of $card_q^r(P)$ is "P has r (mod q) elements", thus $card_q^r(X)$ holds for (G, P) if and only if $|P| = r$ (mod q).
4. For a closed cmso–formula f over L_0, the *graph property defined by* f, $\Pi(f)$, is defined as $\Pi(f) := \{G \in \Gamma \;/\; G \models f\}$. □

3.3.2 Remark

Cmso-formulae can express graph properties mso–formulae can not, e.g., the property "G has an even number of vertices", cf. [Co 90]. □

3.3.3 Theorem

There is an algorithm which constructs, for every $k \in \mathbb{N}$ and every closed cmso–formula f over L_0, a FTA F_k^f which recognises $\mathcal{T}_k^{\Pi(f)}$.

Proof (sketch)
We will construct F_k^f by induction on the structure of f. In order to do so , we have to extend our notion of k–decomposition tree to trees representing structures of the form $(G, p_1, \ldots, p_l, P_1, \ldots, P_s)$. This can easily be done by extending the labels of a decomposition tree so as to contain additional information in the form of two matrices over $\{0, 1\}$; a $(k + \binom{k}{2}) \times l$–matrix Ψ, and a $(k + \binom{k}{2}) \times s$–matrix Φ.
Let $e_1, \ldots, e_{\binom{k}{2}}$ be a numbering of all possible edges between vertices from $\{1, \ldots, k\}$. Then an entry $\Psi[i, j] = 1$ means that

- vertex number i represents object p_j, if $i \leq k$
- edge number $i - k$ represents object p_j, if $i > k$.

Similarly, Φ contains information about $P_1, \ldots, P_s$: $\Phi[i, j] = 1$ means that

- vertex number i represents an object in P_j, if $i \leq k$
- edge number $i - k$ represents an object in P_j, if $i > k$.

Obviously, these extended k–decomposition trees are, again, recognisable.
Table 1 below gives a description of F_k^f, for every form of f. □

[12] This denotes the formula f after substitution of p_i for x_i $(i = 1, \ldots, l)$, and P_i for X_i $(i = 1, \ldots, s)$.

form of f	F_k^f
$vertex(x_m)$	test that whenever $\Psi[i,m]=1$ then $i \leq k$
$edge(x_m)$	test that whenever $\Psi[i,m]=1$ then $i > k$
$connects(x_j, x_l, x_m)$	test whether, on some node, $\Psi[c,j]=\Psi[h,l]=\Psi[i,m]=1$ for some $c>k$, $h,i \leq k$ such that $e_{c-k}=\{h,i\}$.
$card_q^r(X_m)$	add up (mod q) the number of i such that $\Psi[i,m]=1$ (multiple counting can be avoided)
$x_j = x_m$	test on each node that $\Psi[i,j]=\Psi[i,m]$, for all i
$x_j \in X_m$	test on each node that $\Psi[i,j] \leq \Phi[i,m]$, for all i.
$\neg g$	FTA for complement of $\mathcal{L}(F_k^g)$
$g \vee h$	FTA for $\mathcal{L}(F_k^g) \cup \mathcal{L}(F_k^h)$
$\exists x_j g$	FTA which simulates F_k^g, "guessing" column j of Ψ
$\exists X_j g$	FTA which simulates F_k^g, "guessing" column j of Φ

Table 1: Description of F_k^f, for every form of f

3.3.3.1 Corollary

If f is a closed cmso–formula then $\Pi(f) \cap \Gamma^{tw(k)} \in$ LOGCFL, for all $k \in \mathbb{N}$. □

3.3.3.2 Corollary

If f is a closed cmso–formula then $\Pi(f)$ is decidable for the class of all hyperedge–replacement grammars. □

In [ALS 88], a result similar to Corollary 3.3.3.1 is formulated for so–called *extended monadic–second order* (emso–)formulae. These are formulae with two parts, a mso–formula f and an arithmetical predicate π. Such a formula (f, π) holds for a structure $(G, P_1, \ldots, P_s)$, if f holds for $(G, P_1, \ldots, P_s)$, and, additionally, $(|P_1|, \ldots, |P_s|)$ satisfies π.[13]

The fact that emso–definable graph properties can be decided efficiently for graphs of bounded tree–width can be shown in a way similar to the proof of Theorem 3.3.3. Some care, however, is necessary, since the automata needed here are not finite; in addition to the capabilities of a FTA they have to have some means of performing (unbounded) arithmetic. Thus we cannot employ the structural results about FTA (such as the existence of a FTA for the complement of a recognisable set), however, since the syntax for π is restricted in certain ways, all necessary constructions are still possible.

4 Conclusion

We have presented a unified approach to efficient algorithms on graphs of bounded tree–width, and to decision algorithms for hyperedge–replacement graph grammars. This approach is based on the notion of a finite tree automaton, and uses some of the structural theory developed for FTA. It is shown that the key to both, efficient solvability on $\Gamma^{tw(k)}$, and decidability for HRG's of a graph

[13] This is a simplified, and informal description of emso–formulae and their semantics, the precise definition is given in [ALS 88].

property Π, is the recognisability of $\mathcal{T}_k^{\Pi}$, and we have used this method to give new proofs for a number of known results.
Due to space limitations, the exposition had to be sketchy in places, and many proofs had to be omitted. A more detailed presentation of the same material can be found in [La 90b]. There, we also expand on the idea of extending the tree automaton to infinite state sets, which was touched on in the discussion concluding Section 3.

References

[Ar 85] S.Arnborg, *Efficient algorithms for combinatorial problems on graphs with bounded decomposability – a survey.* BIT 25 (1985), pp. 2–23.

[ACP 87] S.Arnborg, D.Corneil, A.Proskurowski, *Complexity of finding embeddings in a k-tree.* SIAM J.Alg.Disc.Meth. 8 (1987), pp.277–284.

[ALS 88] S.Arnborg, J.Lagergren, D.Seese, *Which problems are easy for tree-decomposable graphs?* LNCS 317, pp. 38–51.

[BC 87] M.Bauderon, B.Courcelle, A *Graph Expressions and graph rewriting.* Math. Systems Theory 20 (1987), pp. 83–127.

[Bo 86] H.L.Bodlaender, *Classes of graphs with bounded tree-width.* Report RUU–CS–86–22, Rijksuniversiteit Utrecht, vakgroep informatica, 1986.

[Bo 88a] H.L.Bodlaender, *Dynamic programming on graphs with bounded tree-width.* LNCS 317, pp. 105–118.

[Bo 88b] H.Bodlaender, *NC-algorithms for graphs with small treewidth.* Report RUU–CS–88–4, Rijksuniversiteit Utrecht, vakgroep informatica, 1988.

[Bo 88c] H.L.Bodlaender, *Planar graphs with bounded tree-width.* Report RUU–CS–88–14, Rijksuniversiteit Utrecht, vakgroep informatica, 1988.

[Co 81] S.A.Cook, *Towards a complexity theory of synchronous parallel computation.* L'Enseignement mathématique 27 (1981), pp. 99–124.

[Co 87] B.Courcelle, *An axiomatic definition of context-free rewriting and its application to NLC graph grammars.* TCS 55 (1987), pp. 141–181.

[Co 88] B.Courcelle, *The monadic second-order logic of graphs, III: Tree-width, forbidden minors, and complexity issues.* Preprint, Université Bordeaux 1, 1988.

[Co 90] B.Courcelle, *The monadic second-order logic of graphs, I: Recognizable sets of finite graphs.* I&C 85 (1990), pp. 12–75.

[Do 70] J.E.Doner, *Tree acceptors and some of their applications.* JCSS 4 (1970), pp. 406–451.

[GS 84] F.Gécseg, M.Steinby, *Tree automata.* Akadémiai Kiadó, Budapest 1984.

[Ha 89] A.Habel, *Hyperedge replacement: Grammars and languages.* Dissertation, Bremen 1989.

[HK 87] A.Habel, H.-J.Kreowski, *Some structural aspects of hypergraph languages generated by hyperedge replacement.* LNCS 247 (1987), pp. 207–219.

[HKL 89] A.Habel, H.-J.Kreowski, C.Lautemann, *A comparison of compatible, finite, and inductive graph properties.* Report No. 7/89, Fachbereich Mathematik/Informatik, Universität Bremen, 1989, to appear in TCS.

[HKV 87] A.Habel, H.-J.Kreowski, W.Vogler, *Metatheorems for decision problems on hyperedge–replacement graph languages.* Acta Informatica 26 (1989), pp. 657–677.

[HKV 89] A.Habel, H.-J.Kreowski, W.Vogler, *Decidable boundednesss problems for sets of graphs generated by hyperedge replacement.* To appear in TCS.

[HU 79] J.E.Hopcroft, J.D.Ullman, *Introduction to automata theory, languages, and computation.* Addison-Wesley, 1979.

[Jo 85] D.S. Johnson, *The* NP*–completeness column: An ongoing guide.* J. Algorithms 6 (1985), pp. 434–451.

[La 88a] C.Lautemann, *Decomposition trees: structured graph representation and efficient algorithms.* LNCS 299 (1988), pp. 28–39.

[La 88b] C.Lautemann, *Efficient algorithms on context-free graph languages.* LNCS 317 (1988), pp. 362–378.

[La 90a] C.Lautemann, *The complexity of graph languages generated by hyperedge replacement.* Acta Informatica 27, 399–421 (1990)

[La 90b] C.Lautemann, *Tree decomposition of graphs and tree automata.* Informatik–Bericht 2/90, Johannes–Gutenberg Universität Mainz, 1990.

[LW 88] T.Lengauer, E.Wanke, *Efficient analysis of graph properties on context–free graph languages (Extended abstract).* LNCS 317, pp. 379–393. Revised version as Report 45 (1989), Fachbereich Mathematik–Informatik, Universität–Gesamthochschule Paderborn.

[LWV 84] J.Y.-T.Leung, J.Witthof, O.Vornberger, *On some variations of the bandwidth minimization problem.* SIAM J. Comp. 13 (1984), pp. 650–667.

[RS 86a] N.Robertson, P.D.Seymour, *Graph minors. II. Algorithmic aspects of tree-width.* J. Algorithms 7 (1986), pp. 309–322.

[RS 86b] N.Robertson, P.D.Seymour, *Graph minors. XIII. The disjoint paths problem.* Preprint 1986.

[Ru 80] W.L.Ruzzo, *Tree–size bounded alternation.* JCSS 20 (1980), pp. 218–235.

[Th 73] J.W.Thatcher, *Tree automata: an informal survey.* In: "Currents in the theory of computing" (ed. A.V.Aho), Prentice–Hall 1973, pp.143–172.

[TW 65] J.W.Thatcher, J.B.Wright, *Generalized finite automata theory with an application to a decision problem of second–order logic.* Math. Syst. Theory 2 (1968), pp. 57–81

Recognizing Rooted Context-Free Flowgraph Languages In Polynomial Time

Ulrike Lichtblau
Fachbereich Informatik, Universität Oldenburg
Postfach 2503, D-2900 Oldenburg, F.R.Germany

ABSTRACT: We introduce context-free flowgraph grammars, which allow to replace single vertices together with their outgoing edges, thereby using the basic embedding idea of the algebraic graph grammar approach. For a naturally defined subclass of these, called the rooted grammars, an algorithm is presented which solves the language recognizing problem in time polynomial in the number of vertices of the input graphs.

Keywords: flowgraph grammar, Earley's recognizing algorithm, decompilation.

CONTENTS

1. Introduction

Graph grammars are a useful tool in many areas of computer science and there is a variety of approaches motivated by potential applications, differences mainly lying in the embedding of the right hand side of a production rule into a host graph. Surveys of the field appeared in [ClER 79], [EhNR 83] and [ENRR 87].
An important and interesting question is the time complexity of context-free graph languages. Polynomial recognizing algorithms are known for more or less strongly restricted generating devices ([Slis 82], [Kaul 86], [MCre 87], [Schu 87], [Laut 89], [Vogl 90]).
Flowgraph grammars originate from the application area of decompilation, i.e. the translation from lower level into higher level programming languages. In this context flowgraphs turn out to be a suitable representation of the control flow of source programs (cf. [FaKZ 76], [Lich 85]). Since flowgraph grammars define the stepwise generation of flowgraphs, the problem of decompiling can easily be reduced to the problem of parsing flowgraph languages.

Here, we introduce context-free flowgraph grammars. These grammars generate graphs by replacing a single vertex together with its outgoing edges by a graph and some additional leaving edges. The embedding is via corresponding interfaces on the two sides of a production rule. This follows the embedding idea of the algebraic graph grammar approach ([EhPS 73]). For special context-free flowgraph grammars, called rooted, we present an algorithm which solves the language recognizing problem. Its time complexity is polynomial in the number of vertices of the testgraphs. As usual, in computing the complexity the grammar is fixed.

This paper states results from [Lich 90]. All proofs can be found there, as well as a discussion of non-context-free grammars.

2. Preliminaries

Let $\mathcal{A}_V$ and $\mathcal{A}_E$ be fixed countable sets of **vertex labels** and **edge labels**, resp.

g = (V,E,s,t,l,m) is a **graph** if V and E are finite sets (**vertices** and **edges**), $s,t : E \to V$ are mappings (**source** and **target**) and $l : V \to \mathcal{A}_V$, $m : E \to \mathcal{A}_E$ are partial mappings (**vertex labeling** and **edge labeling**).
A graph g is a **single vertex graph** if #V = 1 and E = {}.

If g is a graph we denote its components by V_g, E_g etc. We proceed analogously for other structures defined below.

Let g be a graph.
For each vertex v we call $d^-(v) := \#t_g^{-1}(v)$ the **in-degree** of v, $d^+(v) := \#s_g^{-1}(v)$ its **out-degree** and $succ_g(v) := t_g(s_g^{-1}(v))$ the set of **successors** of v.
A finite sequence of vertices $<v_1,v_2,...,v_n>$ is a **path** in g if there are edges $e_1, ..., e_{n-1}$ such that $s_g(e_i) = v_i$ and $t_g(e_i) = v_{i+1}$ for each i, $1 \le i < n$.
A vertex r is called a **root** of g if for each vertex v there exists a path <r,...,v> in g.
g is a **tree** if V_g = {} or g has a root r such that $d^-(r) = 0$ and $d^-(v) = 1$ for each vertex $v \neq r$.

Let g and g´ be graphs.
$h = (h_V,h_E)$ is a **graph homomorphism** between g and g´ (abbr. $h : g \to g'$) if $h_V : V_g \to V_{g'}$ and $h_E : E_g \to E_{g'}$ are mappings such that $s_{g'}(h_E(e)) = h_V(s_g(e))$, $t_{g'}(h_E(e)) = h_V(t_g(e))$ and $m_{g'}(h_E(e)) = m_g(e)$ for each $e \in E_g$.
A graph homomorphism $h : g \to g'$ is a **graph isomorphism** if h_V and h_E are bijective and $l_{g'}(h_V(v)) = l_g(v)$ for each $v \in V_g$.
g and g´ are **isomorphic** (abbr. $g \equiv g'$) if there is a graph isomorphism $h : g \to g'$.

Let g be a graph and b be a tree.
b is a **spanning tree** of g if $V_b = V_g$, $E_b \subset E_g$, $s_b = s_{g|E_b}$, $t_b = t_{g|E_b}$, $l_b = l_g$ and $m_b = m_g$.
b is a **depth-first-search spanning tree** of g with root r if b results from a depth-first-search on g starting at r.

Let b be a tree.
b is called **ordered** if for each vertex v there is an ordering function on $succ_b(v)$.
If b is ordered, $succ_b(k,v)$ denotes the **k-th successor** of a vertex v.

3. Rooted Context-Free Flowgraph Languages

We start by defining context-free flowgraph grammars. A production rule of such a grammar says how to replace a vertex and its outgoing edges by a graph and some leaving edges. This is due to the observation that, when interpreting a graph as a flowgraph in some way, the smallest units in this graph are its vertices together with their outgoing edges.
On application of a production rule the embedding of its right hand side is based on an appropriate definition of interfaces on both sides of the rule. Concerning those edges which are leaving a graph the interface is given by the set of target vertices. With respect to entering edges it is defined by distinguishing a single vertex (which on the left hand side is the one to be replaced). A graph that is enlarged by this kind of embedding information is called a graph pattern.

$Y = (g,I,O,o)$ is a **graph pattern** if g is a graph, $I \in V_g$, $O \subset V_g$, $o : \{1,...,\#O\} \to O$ a bijection, such that $I \notin O$ and $d^+(v) = 0$ for each $v \in O$.
I is the **input vertex**, (O,o) the ordered set of **output vertices**.
A graph pattern Y is a **vertex pattern** if $V_g = I \cup O$, $succ(I) = O$ and $d^+(I) = \#O$.

Vertex patterns appear on the left hand side of production rules, graph patterns - with corresponding interfaces - on the right hand side. Note that the former contain neither loops nor multiple edges.

$GG = (N,T,S,P)$ is a **context-free flowgraph grammar** if N and T are disjoint finite sets (**non-terminals** and **terminals**), S is a single vertex graph (**start graph**) and P is a finite set of **production rules** $p = (Y,Z)$, where Y is a vertex pattern and Z is a graph pattern such that $l(I_Y) \in N$, $l(V_{g_Z}) \subset N \cup T$ and $\#O_Y = \#O_Z$.

A context-free flowgraph grammar GG is **rooted** if I_Z is a root of g_Z for each $(Y,Z) \in P_{GG}$.

Rooted context-free flowgraph grammars are the ones for which we present a polynomial recognizing algorithm. They ensure the existence of suitable runs through the graphs in the production rules as well as through any derived graph.

As an example of a rooted context-free flowgraph grammar consider $\mathcal{GG} = (\mathcal{N},\mathcal{T},\mathcal{S},\mathcal{P})$ where $\mathcal{N} = \{P,S,S^*,C,E\}$, $\mathcal{T} = \{s,c,(,)\}$, $\mathcal{S}$ is labeled by P and $\mathcal{P} = \{p_1,...,p_{18}\}$ is shown in Figure 1. Input vertices are marked by arrow heads there, output vertices are drawn as smaller circles and their ordering is from left to right.

A production rule may be applied to a graph g if there is an embedding of its left hand side into g, i.e. a graph homomorphism which is an isomorphism as long as output vertices are not considered. These are allowed to be identified with one another as well as with the input vertex, and their labels are ignored.
The application of a rule consists of replacing the left hand side by the right hand side, where the output vertices are not involved. The embedding, then, is straightforward along corresponding interface vertices. All edges formerly entering the left hand side are inherited to the input vertex of the right hand side. Every edge leaving the right hand side for the k-th output vertex is connected to the image of the k-th output vertex of the left hand side in the graph under consideration.

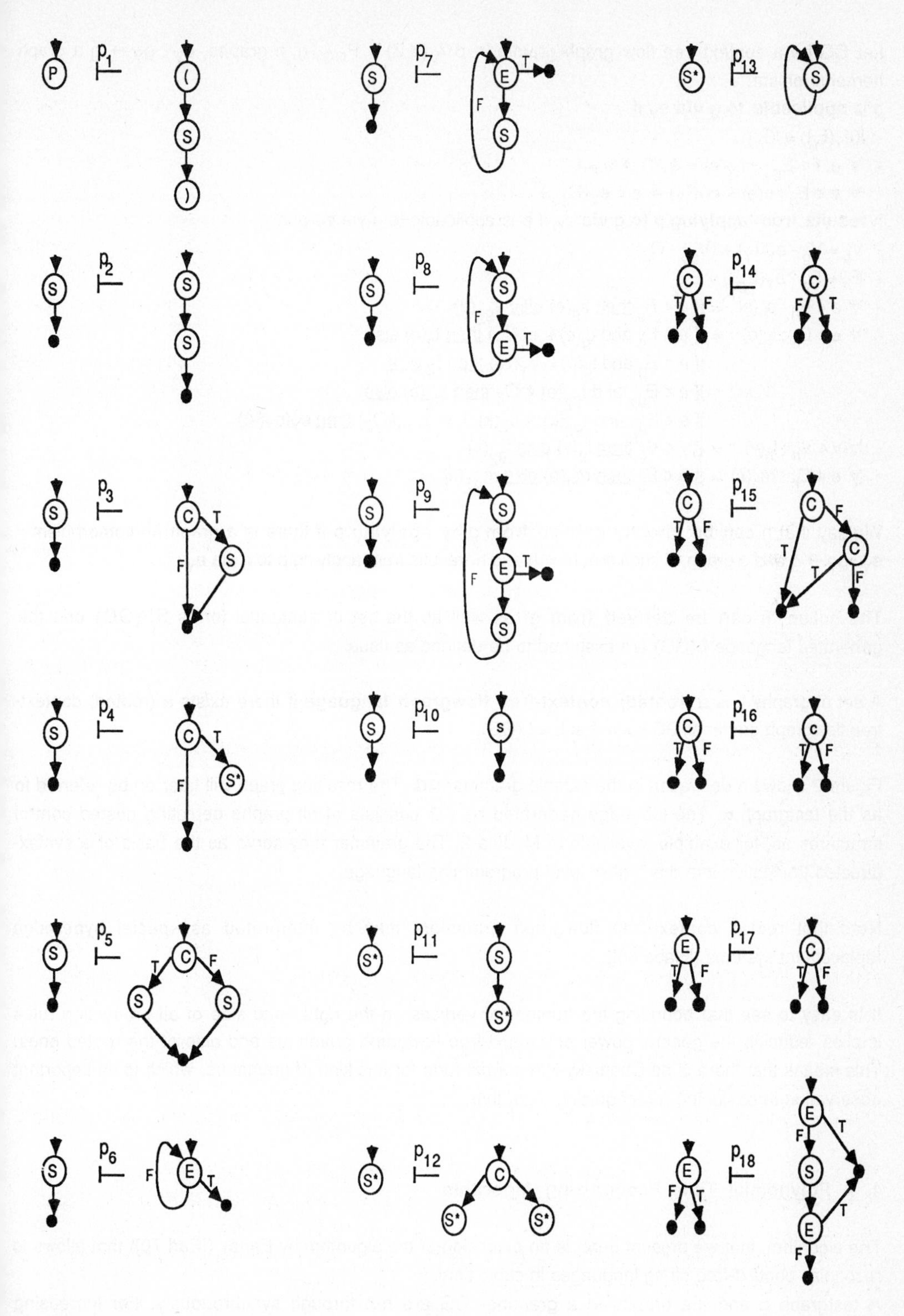

Figure 1

Let GG be a context-free flow graph grammar, $p = (Y,Z) \in P_{GG}$, g, h graphs, $e_Y : g_Y \to g$ a graph homomorphism.

p is **applicable to** g **via** e_Y if

- $l(e_Y(I_Y)) = l(I_Y)$
- $\forall\, e, f \in E_{g_Y} : e_Y(e) = e_Y(f) \Rightarrow e = f$
- $\forall\, e \in E_g : s(e) = e_Y(I_Y) \Rightarrow e \in e_Y(E_{g_Y})$.

h **results from applying** p **to** g **via** e_Y if p is applicable to g via e_Y and

- $V_h = V_g - e_Y(I_Y) + (V_{g_Z} - O_Z)$
- $E_h = E_g - e_Y(E_{g_Y}) + E_{g_Z}$
- $\forall\, e \in E_h : s_h(e) = \underline{if}\ e \in E_g\ \underline{then}\ s_g(e)\ \underline{else}\ s_{g_Z}(e)$
- $\forall\, e \in E_h : t_h(e) = \underline{if}\ e \in E_g \text{ and } t_g(e) \neq e_Y(I_Y)\ \underline{then}\ t_g(e)\ \underline{else}$
  $\underline{if}\ e \in E_g \text{ and } t_g(e) = e_Y(I_Y)\ \underline{then}\ I_Z\ \underline{else}$
  $\underline{if}\ e \in E_{g_Z} \text{ and } t_{g_Z}(e) \notin O_Z\ \underline{then}\ t_{g_Z}(e)\ \underline{else}$
  $\underline{if}\ e \in E_{g_Z} \text{ and } t_{g_Z}(e) = o_Z(k),\ k \in \{1,\dots,\#O_Z\}\ \underline{then}\ e_Y(o_Y(k))$
- $\forall\, v \in V_h : l_h(v) = \underline{if}\ v \in V_g\ \underline{then}\ l_g(v)\ \underline{else}\ l_{g_Z}(v)$
- $\forall\, e \in E_h : m_h(e) = \underline{if}\ e \in E_g\ \underline{then}\ m_g(e)\ \underline{else}\ m_{g_Z}(e)$.

We say that h **can be directly derived from** g **by applying** p if there is a graph homomorphism $e_Y : g_Y \to g$ and a graph h′ such that h′ ≡ h and h′ results from applying p to g via e_Y.

The notion "h **can be derived from** g" as well as the set of sentential forms **SF(GG)** and the generated language **L**(GG) are assumed to be defined as usual.

A set of graphs L is a **(rooted) context-free flowgraph language** if there exists a (rooted) context-free flowgraph grammar GG such that L = L(GG).

Figure 2 shows a derivation in the sample grammar $\mathfrak{GG}$. The resulting graph will later on be referred to as the testgraph $\mathfrak{g}$. The language generated by $\mathfrak{GG}$ consists of all graphs depicting nested control structures as, for example, available in Modula-2. The grammar may serve as the basis of a syntax-directed translation into this higher level programming language.

Note that rooted context-free flowgraph grammars may be interpreted as special hyperedge replacement systems ([Habe 89]).

It is easy to see that bounding the number of vertices on the right hand side of all production rules implies reducing the generic power of context-free flowgraph grammars and also of the rooted ones. This means that there is no Chomsky-like normal form for this kind of grammars, which is an important observation in designing a recognizing algorithm.

4. A Polynomial Time Recognizing Algorithm

The algorithm, that we present here, is an extension of the algorithm by Earley ([Earl 70]) that allows to recognize context-free string languages in cubic time.

A testgraph g and the graphs in a grammar GG are run through synchronously. For increasing subgraphs of g we construct the information from which vertices of which production rules they can be derived. All information of this kind is gathered in the parse table. We end up with the answer whether or not the whole of g can be derived from the start graph of GG.

During the realization of this idea two major problems arise. The first one is the question of how to appropriately sequentialize the testgraph and the production rules of the grammar. We choose the following solution: Ordered spanning trees are imposed on all graphs and then used as guides in running through. In fact, for technical reasons the testgraph is enriched by a depth-first-search spanning tree.

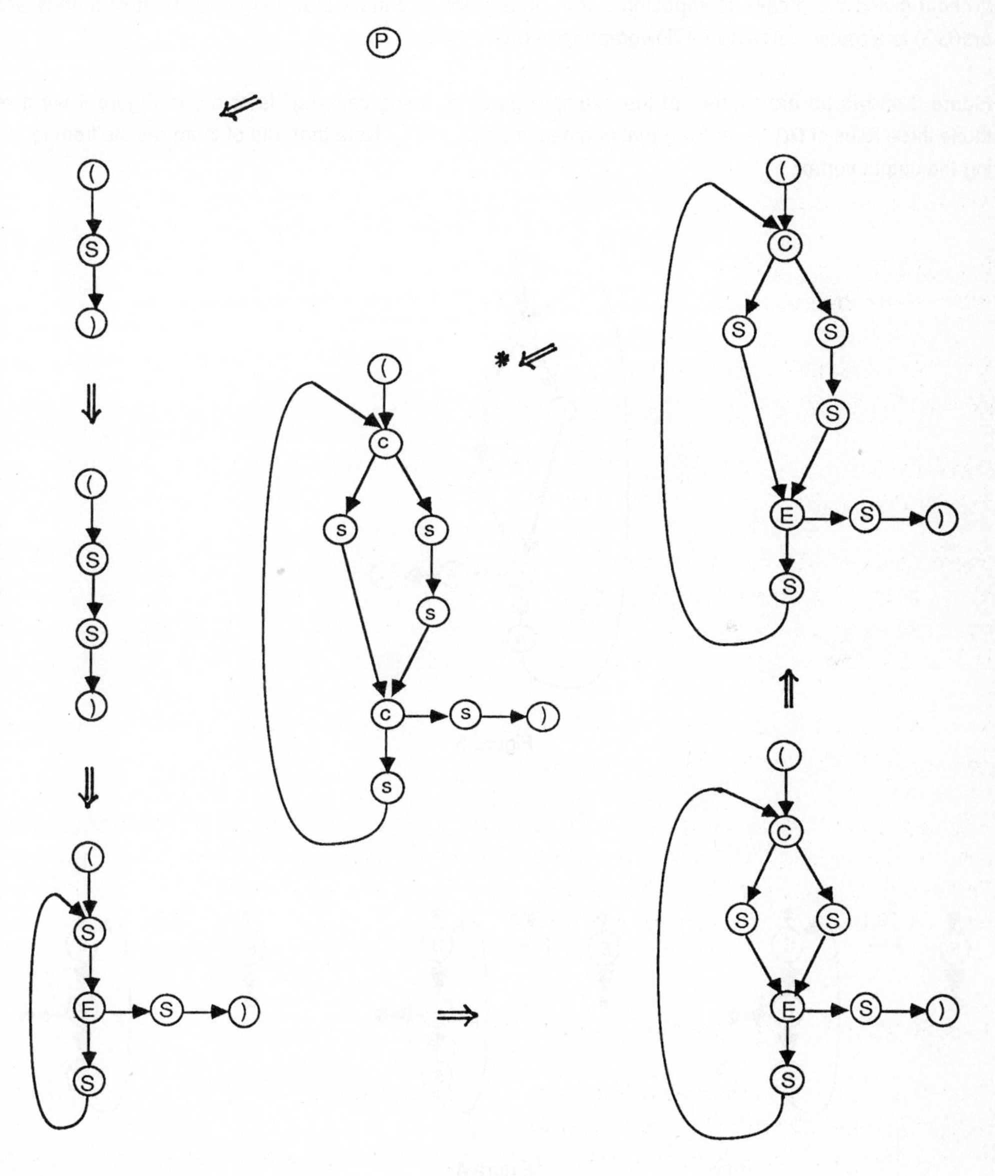

Figure 2

Let g be a graph and r a root of g.
Define **dfs(g,r)** = {g′| g′ results from g by imposing a depth-first-search spanning tree with root r on g}. Such a tree is denoted by b(g′).

Let GG be a rooted context-free flowgraph grammar.
Define **ord(GG)** = $(N_{GG}, T_{GG}, S_{GG}, P)$ where P = {(ord(Y),ord(Z))| (Y,Z) $\in P_{GG}$, ord(X) results from X $\in$ {Y,Z} by imposing an ordered spanning tree with root I_X on g_X and ord(Z) is compatible with ord(Y) with respect to the tree order of output vertices}.

Without giving the details of imposing a tree on a graph, we state that dfs(g,r) is a set of graphs and ord(GG) is a rooted context-free flowgraph grammar.

Figure 3 shows an enrichment of the example graph $\mathfrak{g}$, being called $\mathfrak{g}'$ later on. In Figure 4 we give those three rules of $\mathfrak{GG}'$ = ord($\mathfrak{GG}$) that originate from $p_9 \in P_{\mathfrak{GG}}$. Note that one of them results from ignoring the output vertex.

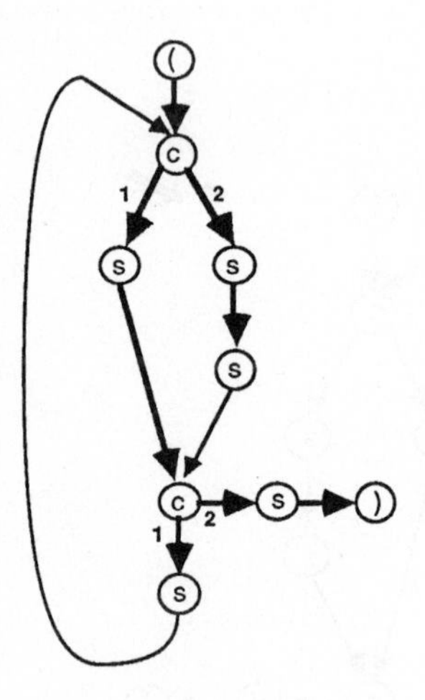

Figure 3

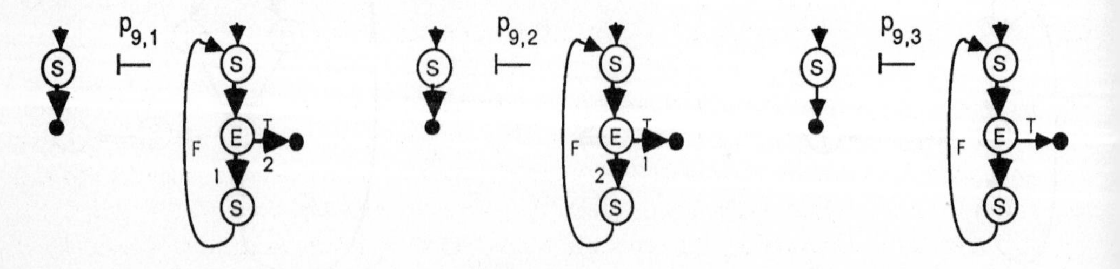

Figure 4

There is a polynomial time reduction algorithm of the problem of deciding $g \in L(GG)$ to the problem of deciding $g' \in L(ord(GG))$ for an arbitrary $g' \in dfs(g,r)$ for all roots r of g. As a first step consider

Lemma 1:
Let GG be a rooted context-free flowgraph grammar.
$L(GG) = \{g \mid \exists \text{ root } r \text{ of } g : \exists g' \in dfs(g,r) : g' \in L(ord(GG))\}$.

Lemma 1 ensures that one positive answer $g' \in L(ord(GG))$ for any root r of g and any $g' \in dfs(g,r)$ can be read as the answer $g \in L(GG)$. A negative answer $g \notin L(GG)$, however, can only be found by checking all enrichments of g. Lemma 2 shows that in this case, too, it is possible to restrict attention to one such graph for every root of g.

Lemma 2:
Let GG be a rooted context-free flowgraph grammar and g be a graph.
$\forall \text{ root } r \text{ of } g : (\exists g' \in dfs(g,r) : g' \in L(ord(GG))) \Leftrightarrow (\forall g' \in dfs(g,r) : g' \in L(ord(GG)))$.

These observations give rise to the following declaration of the recognizing algorithm's shell.

Let GG be a rooted context-free flowgraph grammar.
Algorithm **DECISION**(GG) is defined by
Input: A graph g.
Output: "$g \in L(GG)$" or "$g \notin L(GG)$".
Method: GG' := ord(GG);
for all roots r of g do
g' := arbitrary element of dfs(g,r); (*)
TABLE_CONSTRUCTION(M,GG',g')
od;
decide on the basis of all tables M.

Note that statement (*) can be realized by a depth-first-search on g starting at r.

It is an open question whether a similar reduction of the recognizing problem does exist for non-rooted context-free flowgraph grammars.

In the context of constructing the parse table of an enriched graph g' with respect to an enriched grammar GG' the second problem appears. A walk through the depth-first-search spanning tree does in general not obey the derivation structure of g'. This is very different from the string case and does obviously not depend on the method of sequentializing the original graphs.

We introduce a new state of production rules that are contained in the parse table. It is called the state of being left temporarily and accompanies the states of being under work and of being left finally, which are known from the string case. A rule is left temporarily when an output vertex is reached. This is a moment in which the synchronous walk through the testgraph leaves a derivation plateau that is revisited later on. On leaving a production rule p temporarily, the information gathered so far is propagated to those rules in which the next vertex corresponds to the left hand side of p. This is very similar to the propagation of the information of a rule that is left finally. However, a production rule which was left temporarily has to be taken under work again. This is at a point of time at which an appearance of its left hand side in another rule is revisited. Then, the corresponding entrance in the parse table must be identified, and to be able to do this we have to keep information that by far exceeds that needed in the string case.

The identifying information is called the inscription of a production rule. It consists of three components: the point in running through, a colouring of the vertices (which is the history of the run) and an assignment of numbers to the vertices (referring to vertices of the testgraph). Rules together with their inscriptions form the entries of the parse table.

Instead of giving formal definitions we discuss the example of an entry in the parse table of g' with respect to GG', that is depicted in Figure 5. First note that the inscription of the left hand side summarizes that of the right hand side. The union of all sets of assigned numbers determines the indices of the entry. αs point at the current vertices, vs at the next vertices. The dark colouring of the vertex labeled "S" together with the assignment of number 5 bears the information that it is proven that this vertex can create the subgraph of g′ that is induced by the 5th vertex. The light colouring of the vertex labeled "S" together with the assignment of number 2 and 3 says that it is supposed that this vertex can create a subgraph of g′ that contains at least the 2nd and the 3rd vertex. (In fact, this vertex does create the subgraph induced by the vertices 2, 3, 8 and 9. The last two of these have not been visited when computing the current entry.) The dark colouring of an output vertex together with an assignment of vertices indicates a possible embedding into a sentential form of GG′.

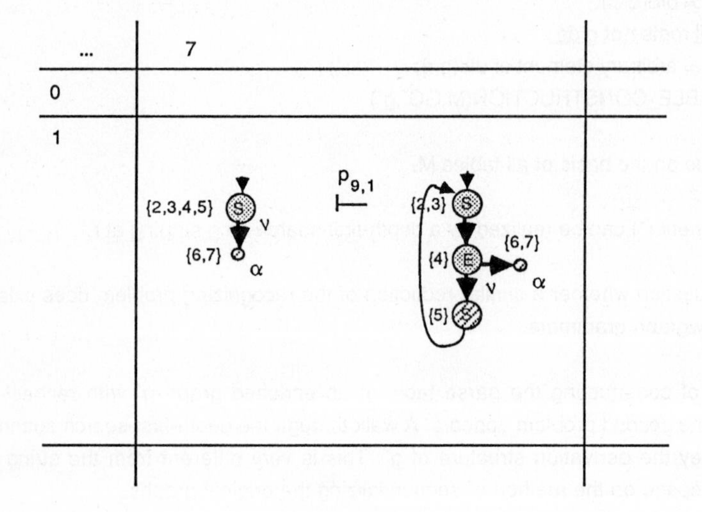

Figure 5

Given a parse table it is easy to decide whether or not a testgraph can be derived.

<u>Lemma 3:</u>

Let GG′ be an enriched rooted context-free flowgraph grammar, g′ an enriched graph, $n = \#V_{g'}$ and M the parse table of g′ with respect to GG′.

$g' \in L(GG') \Leftrightarrow \exists (Y,Z) \in M_{0n} : Y \equiv S_{GG}$ and all vertices in (Y,Z) are coloured dark.

Now we present a closer look at the construction of parse tables.

Let GG´ be an enriched rooted context-free flowgraph grammar, g´ an enriched graph and M a table.
Procedure **TABLE_CONSTRUCTION**(M,GG´,g) is defined by

```
Method:  TABLE_INIT(M,GG´);
         TABLE_CON(M,GG´,g´,root of b(g´)).
Where:   TABLE_INIT(M,GG´) is straightforward and
         TABLE_CON(M,GG´,g´,v), v a vertex, is:
         j := number of v in g´;
         v_scanner(j);
         for k := 1 to d+_b(g´)(v) do
            e_scanner(j);
            deactivator(j);
            predictor(j);
            TABLE_CON(M,GG´,g´,succ_b(g´)(k,v));
            reactivator(j,v);
            revisor(j,v)
         od;
         reverter(j,v);
         completer(j,v).
```

There, the nested procedures fulfill the following tasks.
v_sanner(j) simulates the first walk over the j-th vertex of the testgraph in the production rules and leaves these rules under work in the j-th column of the table.
e_scanner(j) does the same for the next tree edge to be visited in the testgraph.
revisor(j,v) simulates later walks over vertex v, where j again is the number of the column currently being built up and thus indicates the number of vertices visited so far.
reverter(j,v) simulates the last walk over vertex v, thereby testing all edges which do not belong to the testgraph´s spanning tree. (Note that this is a depth-first-search tree.)
deactivator(j) propagates the information gathered in a preliminarily left rule into other rules as discussed above.
completer(j,v) does the same for production rules that are left finally.
predictor(j) prepares the next step forward in the testgraph by taking into account all rules which might be visited next.
reactivator(j,v) does the same with respect to the next step backward, thus reviving rules which had been left preliminarily before.

We end with the main result.

<u>Theorem:</u>
Let GG be a rooted context-free flowgraph grammar.
Algorithm DECIDE(GG) is correct and its time complexity is polynomial in the number of vertices of the testgraphs.

A detailed proof, which is based on a further refinement of the algorithm, is given in [Lich 90].

References

[ClER79] Claus, V., Ehrig, H., Rozenberg, G. (eds.): Graph-Grammars and Their Application to Computer Science and Biology, LNCS 73, 1979.

[Earl 70] Earley,J.: An Efficient Context-Free Parsing Algorithm, Comm. ACM, Vol. 13, No. 2, 1970, 94 - 102.

[EhNR 83] Ehrig, H., Nagl, M., Rozenberg, G. (eds.): Graph Grammars and Their Application to Computer Science, LNCS153, 1983.

[EhPS 73] Ehrig, H., Pfender, M., Schneider, H.J.: Graph Grammars: An Algebraic Approach, Proc. 14th Annual IEEE Symposium on Switching and Automata Theory, 1973, 167 - 180.

[ENRR 87] Ehrig, H., Nagl, M., Rozenberg, G., Rosenfeld, A. (eds.): Graph-Grammars and Their Application to Computer Science, LNCS 291, 1987.

[FaKZ 76] Farrow, R., Kennedy, K., Zucconi, L.: Graph Grammars and Global Program Data Flow Analysis, Proc. 17th Annual IEEE Symposium on Foundations of Computer Science, 1976, 42 - 56.

[Habe 89] Habel, A.: Hyperedge Replacement: Grammars and Languages. Ph. D. Thesis, University of Bremen, 1989.

[Kaul 86] Kaul, M.: Syntaxanalyse von Graphen bei Präzedenz-Graph-Grammatiken. Ph. D. Thesis, University of Passau, 1986.

[Laut 89] Lautemann, C.: The complexity of graph languages generated by hyperedge replacement. Report No. 4/89, Fachbereich Mathematik und Informatik, University of Bremen, 1989.

[Lich 85] Lichtblau,U.: Decompilation of Control Structures by Means of Graph Transformations LNCS 185, 1985, 284 - 297.

[Lich 90] Lichtblau,U.: Flußgraphgrammatiken. Ph. D. Thesis, University of Oldenburg, 1990.

[MCre 87] McCreary, C.L.: An Algorithm for Parsing a Graph Grammar. Ph. D. Thesis, University of Colorado, 1987.

[Schu 87] Schuster, R.: Graphgrammatiken und Grapheinbettungen: Algorithmen und Komplexität. Ph. D. Thesis, University of Passau, 1987.

[Slis 82] Slisenko, A.O.: Context-Free Graph Grammars as a Tool for Describing Polynomial-time Subclasses of Hard Problems, Information processing Letters 14, 1982, 52 - 56.

[Vogl 90] Vogler, W.: Recognizing Edge Replacement Graph Languages in Cubic Time. Report TUM- I9017, Technical University of Munich, 1990.

Computing with Graph Relabelling Systems with Priorities

(Extended Abstract)

Igor Litovsky and Yves Métivier
Laboratoire Bordelais de Recherche en Informatique
U.R.A. 1304 du C.N.R.S.
E.N.S.E.R.B.
351, Cours de la libération
F-33405 Talence - FRANCE

ABSTRACT : in this paper, the computational power of the noetherian Graph Relabelling systems with Priorities (PGRS for short) is studied. The PGRS's are considered as recognizers for sets of graphs and for sets of 1-sourced graphs. We show that the PGRS's are strictly more powerful for the 1-sourced graphs than for the graphs. Furthermore every set of 1-sourced graphs definable in First Order Logic is recognizable by some PGRS.

Keywords : distributed algorithms, graphs, priority, recognizability, relabelling rules.

CONTENTS

0. Introduction

The Graph Rewriting Systems with Priorities (PGRS for short) have been introduced in [BLMS 89], where some illustrating examples show how the PGRS's allow to describe and to prove some classical algorithms on graphs and on distributed systems. Let us recall the features of this model. A PGRS is a finite set of rewriting rules which work on labelled graphs. These rewriting rules do not change the underlying graph, but change only the labels of vertices or the labels of edges. That is the reason why, in order to avoid any confusion, the Graph Rewriting Systems with Priorities are called Graph Relabelling Systems with Priorities in this paper. There is a partial order (called priority) on the set of rules : in the case when two occurrences of rewriting rules are overlapping, the rule (if any) with minor priority may not be applied ; whereas in the case of non-overlapping occurrences, any rule may be applied freely. That is this priority is purely local (which is a natural feature for expressing distributed computations), unlike many other models where the priority is global see for example the Markov algorithms [Ma 61] or the ordered grammars [Fr 68]. Hence the priority allows to locally control the order in which rules are applied, thus it allows to have invariants during the computing process, which is very useful to make proofs.

In this paper we study the computational power of the noetherian PGRS's. First we are interested in the effect of the priority : are the PGRS's strictly more powerful than the GRS's (i.e. the PGRS's without priority) ? We prove that, on the one hand the priority is necessary in the general case (i.e. the PGRS's are strictly more powerful than the GRS's), and that on the other hand it is useless whenever, working with graphs of bounded-degree, the degree of the vertices is known (i.e. in this case, the priority does not add expressive power).

Next we introduce the notion of *safe PGRS with respect to a given property P*. Since PGRS's encode algorithms on graphs, in many cases the requirement of confluence is not needed. For example, if we want to construct a spanning tree for a given graph, it is not forbidden that two computations produce two different spanning trees : the requirement is to obtain always some spanning tree, but not necessarily the same. The confluence is not for the graph obtained but for the property of being a spanning tree. In this way, we consider the PGRS's as recognizers for sets of graphs and we say that a PGRS is a *safe-recognizer*, when, for a given graph, any computation decides whether the graph is recognized or not. Then the class of sets of graphs safely recognizable by some PGRS is a boolean algebra. Since the class of graphs recognized does not seem to be very large, we consider graphs that have exactly one vertex distinguished (they are called 1-graphs as in [Co 90]). First we show that one cannot obtain a 1-graph from a graph and we prove that there is a gap between the recognizing power for graphs and the recognizing power for 1-graphs : each set of 1-graphs definable in First Order Logic is safely recognized by a PGRS, this is not true for graphs. Moreover some sets of 1-graphs safely recognizable by PGRS's are not definable in First Order Logic.

Due to the lack of place, the proofs have been omitted and may be found in [LM 90] which also contains the following complements of this paper. When one deals with computations on graphs, it is useful to have an ordering relation on the vertices (see [Im 87] and [Co 89]). So 1-graphs on which linear order is safely constructable by a PGRS are considered, then in this case, it is proved that every set of 1-graphs definable in Monadic Second Order Logic is safely recognizable by a PGRS and that the converse does not hold (in the general case of 1-graphs, no answer is given about the Monadic Second Order Logic). Next we are interested in the possibility to encode the orientation of an oriented graph in the underlying nonoriented graph, a problem investigated in [Co 89]. We only prove that one can simply encode the orientation when the graphs are of bounded degree or more generally for the graphs having a k-coloring ; in a same way the orientation may be encoded in linearly orderable graphs. Finally we are interested in particular graphs : trees and words. First for the trees, there is no difference between trees and rooted trees for the computation power of the PGRS's. Concerning the bounded-degree trees, the PGRS's are strictly more powerful than the usual tree automata (cf [GeSt 84]). Concerning the words, we show that the PGRS's recognize exactly the class of context-sensitive languages. This is not surprising by workspace considerations.

The paper is organized as follows. Part 1 contains definitions and notations. In part 2, we discuss the priority. Part 3 introduces the notion of safety. Part 4 concerns the PGRS's as recognizers and Part 5 deals with the 1-graphs.

1. Definitions and notation

a. Basic definitions on graphs

In this paper a (simple) *nonoriented graph* $G(V_G,E_G)$ is defined as a finite set V_G of vertices and a set E_G of edges (an *edge* is a subset of cardinality 2 or 1 of V_G). An *oriented graph* $G(V_G,A_G)$ is defined as a finite set V_G of vertices and a set A_G of arcs (an *arc* is an element of the cartesian product $V_G x V_G$). As much as possible we shall treat nonoriented graphs and oriented graphs simultaneously, so *graph* means nonoriented graph or oriented graph. Graphs are often denoted by G for short.

Let e = {v,v'} be an edge of a graph G, e and v are said to be *incident*. The *degree* of a vertex v in G is the number of incident edges to v. Let k be an integer, a graph G is a *k-bounded degree graph* if every vertex of G has a degree less than or equal to k.

A *path* in an oriented graph $G(V_G,A_G)$ is a sequence of vertices $v_0,...,v_k$, where each (v_i,v_{i+1}) is in A_G, k is the *length* of the path. A path is *simple* if the sequence is injective. In a nonoriented graph, paths are called *chains*. The *distance* between two vertices v, v' in a graph G is the length of a shortest path (or chain) between v and v'. A *cycle* is a simple path such that the edge $\{v_k,v_0\}$ is in E_G. A graph G is *connected* if there is a path between any two vertices. A nonoriented graph is a *tree* if it is connected and acyclic.

In the sequel E_G denotes a set of edges or a set of arcs. A *labelled graph* $G(V_G,E_G,\lambda)$ (or (G,λ) for short) is a graph $G(V_G,E_G)$ where $\lambda = (\lambda_V,\lambda_E)$ is a pair of mappings from V_G (resp. E_G) into a finite set of label C_V (resp. C_E). The mapping λ is a *labelling function* and $G(V_G,E_G)$ is the *underlying graph*. Let L be a vertex label (resp. edge label), any vertex (resp. edge) labelled by L is called a *L-vertex* (resp. *L-edge*) and $|L|_G$ denotes the number of L-vertices (resp. L-edges) in G. Let us note that all graphs may be considered as labelled graphs. The set of all labelled graphs is denoted by $\mathcal{G}$.

A *1-sourced-graph* (or simply a 1-graph) is a graph G which has a distinguished vertex called the source of G (cf [Co 90]). We consider in the following that the source is characterized by a special label.

A labelled graph $G(V_G,E_G,\lambda)$ is *isomorphic* to a labelled graph $G'(V_{G'},E_{G'},\lambda')$ if there is a pair $\varphi = (\varphi_V, \varphi_E)$ of one-to-one mapping $\varphi_V : V_G \longrightarrow V_{G'}$, $\varphi_E : E_G \longrightarrow E_{G'}$ such that $\varphi_E((s,t)) = (\varphi_V(s), \varphi_V(t))$ for every (s,t) in E_G and $\lambda_{V'} \circ \varphi_V = \lambda_V$ and $\lambda_{E'} \circ \varphi_E = \lambda_E$.

A labelled graph $G(V_G,E_G,\lambda)$ is a labelled *subgraph* of $G'(V_{G'},E_{G'},\lambda')$ if V_G is included in $V_{G'}$, and E_G is contained in $E_{G'}$, and λ is the restriction of λ' to V_G and to E_G. If a subgraph G of G' is isomorphic to a labelled graph G", G is called an *occurrence* of G" in G'. Two occurrences O and O' are *overlapping* in G, if the set of vertices $V_O \cap V_{O'}$ is not empty.

A subgraph G of a graph G' is a *connected* subgraph if there is a chain in G' between any pair of vertices of G. Let L be any label, a subgraph G is *L-connected* if G is a connected subgraph and all the vertices of G have a L-label.

b. Graph Relabelling Systems with Priority

A *graph relabelling rule* r, is a pair $((G_r,l),(G_r,l'))$ of two labelled connected graphs having the same underlying graph G_r, we write $(G_r,l) \xrightarrow[r]{} (G_r,l')$. If (G_l,l_l) is an occurrence of (G_r,l) in any graph (G,λ), (G,λ) may be rewritten in (G,λ') by r, which will be denoted by $(G,\lambda) \xrightarrow[r]{} (G,\lambda')$, where λ' is equal to λ except on G_l where λ' is equal to l'.

A *graph relabelling system with priority* (PGRS for short) is a finite set R of graph relabelling rules equiped with a partial order <, called *priority*, which is used in the following way : one may apply in (G,λ) a rule r on an occurrence O_r of (G_r,l) if O_r is not overlapping with an occurrence of a more prior rule of R. One writes $(G,\lambda) \xrightarrow[R]{} (G,\lambda')$, then we are interested in the reflexive-transitive closure of $\xrightarrow[R]{}$ denoted by $\xrightarrow[R]{*}$. In the following we consider only the sequential behaviour of the PGRS's, i.e. at each relabelling step, among all the occurrences which may be rewritten (w.r.t. the priority), one of them is choosen and the corresponding relabelling rule is applied.

A graph (G,λ) is said to be *irreducible* with respect to a PGRS R if (G,λ) does not contain an occurrence of any rule of R. Then each PGRS R induces a partial function $Irred_R$, from $\mathcal{G}$ (the set of all graphs) to $\mathbf{P}(\mathcal{G})$ (the power-set of $\mathcal{G}$) defined by :

$$\forall\ (G,\lambda) \in \mathcal{G},\ Irred_R(G,\lambda) = \{\ (G,\lambda') \ /\ (G,\lambda) \xrightarrow[R]{*} (G,\lambda') \text{ and } (G,\lambda') \text{ is irreducible w.r.t. R}\}$$

A PGRS R works only on graphs in some subset of $\mathfrak{G}$, the *domain* D(R) of the partial function $Irred_R$.

In the following, all the PGRS's will be assumed to be noetherian, hence the set $Irred_R(G)$ is not empty for every graph G in D(R). To prove that a given PGRS is noetherian, the usual way is to produce a noetherian order which is compatible with the given PGRS (see [BLMS 89]).

c. Logic for expressing graph properties

We recall here a few definitions about sets of graphs definable by logic formulas which may be found in [Co 90]. Any labelled graph may be defined as a logical structure $(V_G,E_G,(lab_{a,G})_{a \in C_V},(edg_{b,G})_{b \in C_E})$ where V_G is the set of vertices, E_G is the set of edges, C_V is a set of vertex labels and C_E is a set of edge labels, moreover the meaning of the predicates is the following :

$lab_{a,G}(v)$ is true iff the vertex v in G has label a,
$edg_{b,G}(e,v,v')$ is true iff e is the edge (v,v') in G and has label b.

To define sets of graphs one considers formulas built by using individual variables (vertex variables or edge variables), set variables (sets of vertices or sets of edges) and binary relation variables (subsets of $V_G x V_G$ or $E_G x E_G$ or $V_G x E_G$ or $V_G x E_G$).

Atomic formulas are the following :

1) x = x' where x, x' are two vertices or two edges,
2) $lab_a(v)$ where v is a vertex,
3) $edg_b(e,v,v')$ where e is an edge and v, v' are two vertices,
4) $x \in X$ where X is a set of vertices or a set of edges,
5) $(x,y) \in R$ where R is a binary relation included in a cartesian product XxY of two sets of vertices or edges.

A *First Order formula* is a formula formed with the above atomic formulas numbered from 1) to 3) together with the boolean connectives **OR, AND, NOT** and the individual quantifications $\forall x$, $\exists x$ (where x is a vertex or an edge).

A *Monadic Second Order formula* is a formula formed with the above atomic formulas numbered from 1) to 4) together with the boolean connectives **OR, AND, NOT**, the individual quantifications $\forall x$, $\exists x$ (where x is a vertex or an edge) and the set quantifications $\forall X$, $\exists X$ (where X is a set of vertices or a set of edges).

A *Second Order formula* is a formula formed with the above atomic formulas numbered from 1) to 5) together with the boolean connectives **OR, AND, NOT**, the individual quantifications $\forall x$, $\exists x$ (where x is a vertex or an edge), the set quantifications $\forall X$, $\exists X$ (where

X is a set of vertices or a set of edges) and the binary relation quantifications $\forall R$, $\exists R$ (where R is a binary relation).

Let X be F or MS or S, a set $\mathcal{S}$ of graphs is *definable* in XOL if there is a closed XOL-formula F (i.e. formula without free-variable) such that G belongs to $\mathcal{S}$ iff G satisfies F.

2. PGRS's versus GRS's

Here we compare the power of PGRS's with the power of GRS's (where the GRS's are PGRS's with the empty ordering). By the following example, the natural feeling that the PGRS's are strictly more powerful than the GRS's is confirmed.

Example 2.1. The following PGRS does not rewrite the edges and satisfies the following property **(P)** : *for each graph* G *and for each graph* (G,λ') *in Irred*(G), $\lambda'(V_G) = \{ L \}$ *iff G is a tree.*
No GRS which relabels only vertices can satisfy **(P)**.

First we give a PGRS which realizes **(P)**. This PGRS works by a leaf removing simulation. and Initially all vertices are labelled N.
The rules are given with a decreasing priority, and I means Interior, N means Neutral, L means Leaf, X and Y denote any label of {N,I}.

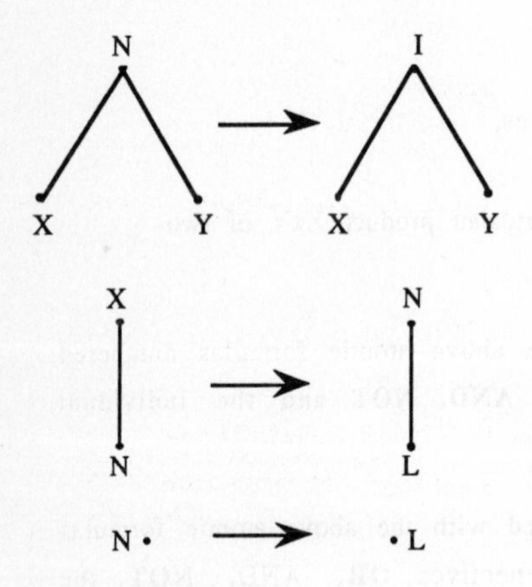

The termination argument is the strict decreasing with respect to the lexicographic ordering of the tuple of integers (|N|,|I|).
The invariants are :
. each L-connected subgraph is a tree
. each N-vertex has at most one neighbour in a L-connected component
. each I-vertex has at most one neighbour in a L-connected component

. each I-vertex has at least two N- or I-neighbours.
Furthermore the PGRS always halts after at most 2n-3 relabelling steps (where n denotes the cardinality of V_G).

On the other hand, it is not possible to satisfy (**P**) with a GRS which does not rewrite the edges. Indeed such a GRS $\mathfrak{R}$ would label in L a chain of length n Hence , whenever n is large enough, $\mathfrak{R}$ would also label in L a cycle of length n (which is not a tree) since $\mathfrak{R}$ does not relabel the edges. The cycle of L-vertices obtained is irreducible, this is a contradiction. ■

However we are going to see that for the bounded-degree graphs, whenever the degree of each vertex is computed, each PGRS may be simulated by a GRS. First we state that one can compute, by a GRS, the degrees of each vertex, when it is less than a given integer.

Proposition 2.1. *For each integer k, the following GRS R_k computes the degree of each vertex of degree at most k, and gives ∞ otherwise for all graphs.*

Proof. It is easy to see that the following GRS gives the result, when initially all the labels are 0 :

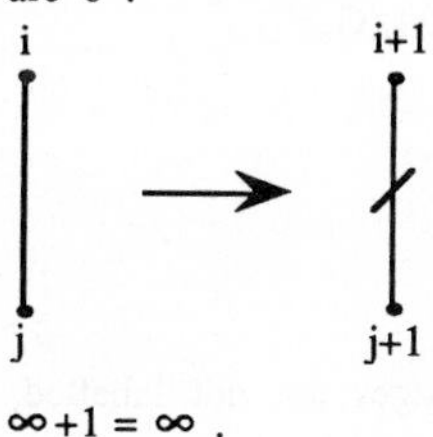

for each integer i and j in the set $\{0,1,...,k\} \cup \{\infty\}$, with $k+1 = \infty$ and $\infty+1 = \infty$.

■

Remark : one could compute the degree of each vertex with a GRS having infinitely many labels and rules.

Now we consider any PGRS R, working only on k-bounded-degree graphs.

Proposition 2.2. *With the assumption that one starts with k-bounded-degree graphs G such that for each vertex x of G, the degree of x in G is a component of the label of x, then each PGRS may be simulated by a GRS.*

Hence we have a decomposition of any PGRS :

Proposition 2.3. *Let R be a PGRS working on k-bounded-degree graphs. The PGRS's R and $R' \cup R_k$ give the same irreducible graphs where R' and R_k are two GRS's and the single priority is $r' < r$ for each r in R_k and each r' in R'.*

The previous priority means that PGRS R' works only on vertices with computed degree.

3. Safe PGRS's

The PGRS's are tools to encode and to prove algorithms on graphs or on distributed systems. From any input graph (G,λ), a PGRS gives an output graph (G,λ') such that (G,λ) and (G,λ') are linked by some relation $P((G,\lambda),(G,\lambda'))$ (for example λ' defines a spanning tree of G, or λ' is an independant set of G). Very often, for a given graph (G,λ), many graphs (G,λ') satisfy $P((G,\lambda),(G,\lambda'))$ (for example G may have many spanning trees, or may have many independant sets), however we seek only to obtain one of these. In this way the confluence of the PGRS R is not needed, but we want only that all irreducible graphs (G,λ') obtained from (G,λ) satisfy P. This leads to the notion of *safe PGRS with respect to a property P.*

Definition 3.1. Let R be a PGRS and P be a relation on $\mathcal{G}$. The system R is a *safe PGRS with respect to P* if

$\forall\ (G,\lambda) \in D(R),\ \exists\ (G,\lambda') \in \mathcal{G}\ /\ P((G,\lambda),(G,\lambda')) \Rightarrow \forall\ (G,\lambda') \in \mathrm{Irred}(G,\lambda),\ P((G,\lambda),(G,\lambda'))$

That is, each irreducible graph obtained from (G,λ) satisfies P.

Example 3.2. The spanning trees [BLMS 89].
We gives a PGRS **R** which works on labelled graphs (G,λ) such that the edges are not labelled, exactly one vertex v_R has a label R (as Root) and the other vertices have a label N.
Furthermore **R** is safe with respect to $P((G,\lambda),(G,\lambda'))$, defined by :
$P((G,\lambda),(G,\lambda'))$ iff $\lambda'(V_G) = \{R,F\}$, $\lambda'^{-1}(R) = \{v_R\}$, $\lambda'(E_G) = \{\_,t\}$ and the t-edges make a spanning tree of the underlying graph G.
The rules are given with a decreasing priority and A means Activ, W means Waiting, N means Neutral, F means Finished.

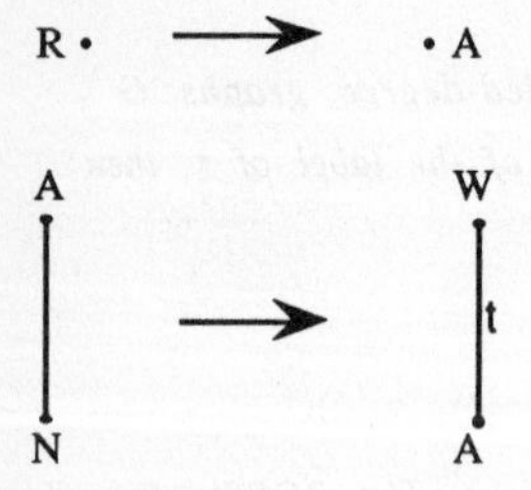

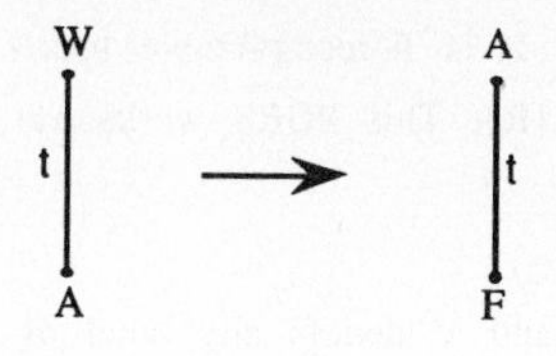

A • ⟶ • R

It is proved in [BLMS 89] that this PGRS is safe w.r.t. $P((G,\lambda),(G,\lambda'))$.

■

This notion of safe computation with respect to some property is used in Section 4 to define a safe recognizing mode (Definition 4.7), which is a natural extension of the deterministic recognizing modes for words or trees.

4. PGRS's as recognizers

Here we are interested in the ways of recognizing sets of graphs or sets of labelled graphs by using PGRS's. As every rewriting system, a PGRS can be considered as a recognizer by specifying a set of recognizing graphs. We give two definitions of recognizability. The first one (Definition 4.1) may be viewed as nondeterministic, in the sense that a graph is recognized when the answer "the graph is recognized" is given for some computation. The second one (Definition 4.7), called safe according to the previous section, is deterministic in the sense that any computation decides whether the graph is recognized or not.

Definition 4.1. Let R be a PGRS and T be a set of graphs. A graph (G,λ) is *recognized by* (R,T) if $(G,\lambda) \xrightarrow[R]{*} (G,\lambda')$ where (G,λ') is irreducible and belongs to T.

Then as usual the set of graphs recognized by (R,T) is defined by :

$L(R,T) = \{ (G,\lambda) \in D(R) \ / \ Irred((G,\lambda) \cap T \neq \emptyset \}$.

Let us note that T should have a "simple" structure otherwise every set of graphs may be recognized. We use the two criteria following :

the B(üchi)-recognizing criterion, where $T = \{ (G,\lambda') \ / \ \lambda'(G) \cap L \neq \emptyset \}$ for some set of labels L,

the M(üller)-recognizing criterion, where $T = \{ (G,\lambda') \ / \ \lambda'(G) \in F \}$ for some family of sets of labels F.

Example 4.2. From Example 2.1, the set of trees is M-recognizable. It is B-recognizable too. The following PGRS B-recognizes the trees with the set of labels L = {R}. This PGRS works by a leaf removing simulation as in example 2.1.
The rules are given with a decreasing priority.
(I means Interior, N means Neutral, L means Leaf, R means Root, X and Y denote any label of {N,I})
Initially all vertices are N.

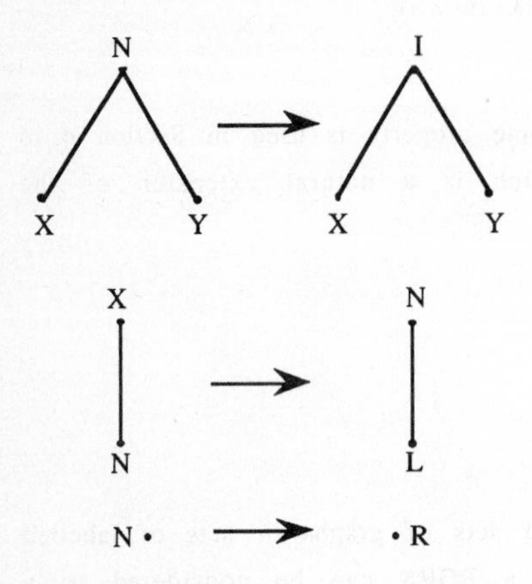

With the same arguments as in Example 2.1, one proves that this PGRS B-recognizes the set of trees. ■

Lemma 4.3. *The class of sets of graphs B-recognizable by the noetherian PGRS's is closed under union.*

Of course the B-recognizing mode is a particular case of the M-recognizing mode, moreover the converse holds.

Proposition 4.4. *A set of graphs is B-recognizable by a noetherian PGRS iff it is M-recognizable by a noetherian PGRS.*

Now we return to the need of the priority, and we only consider sets of k-bounded-degree graphs, then we obtain the following result which is stronger than Proposition 2.3.

Proposition 4.5. *Let L be a language of k-bounded-degree graphs,*
L is recognizable by a PGRS iff L is recognizable by a GRS.

The closure properties of the class of languages recognized by noetherian PGRS's are stated below.

Proposition 4.6. *The class of graphs languages recognizable by the noetherian PGRS's is*
- *closed under union*
- *closed under intersection*
- *not closed under complement.*

The non-closure under complement leads to another definition of recognizing mode, which has two advantages. First the recognized class makes a boolean algebra and second it is more efficient since any irreducible graph computed from a graph G decides whether G is recognized or not.

Definition 4.7. Let (R,T) be a noetherian PGRS-recognizer.
The system (R,T) is a *safe-recognizer* if R is safe with respect to the cartesian product L(R,T) x T, i. e. : $\forall$ $(G,\lambda) \in$ D(R), Irred$(G,\lambda) \subset$ T or Irred(G) $\cap$ T = $\varnothing$.

Hence we have immediatly :

Proposition 4.8. *The class of languages safely M-recognized by the noetherian PGRS's is a boolean algebra.*

This implies :

Corollary 4.9. *There is a language recognized by a PGRS which is not safely M-recognized by a PGRS.*

Remark : all languages safely M-recognized by a PGRS are recognized by a PGRS.

If we consider the class of sets of graphs recognizable (or safely recognizable) by a PGRS compared with the class of sets definable in First Order Logic, Monadic Second Order Logic or Second Order Logic, we note that they are not comparable, as shown below.

Properties of graphs	recognized by a PGRS	safely M-recognized by a PGRS
. **FOL**		
. exactly one a-label	no	no
. simple	yes	yes
. k-regular	yes	yes
. **MSOL**		
. connected	no	no
. 2-colorable	yes	no
. tree	yes	yes
. **SOL**		
. even number of vertices	yes	no
. odd number of vertices	no	no

That is the reason why we now consider graphs having one distinguished vertex, which yields more positive results.

5. Graphs versus 1-graphs

As in [Co 90] we use sourced-graphs, that is graphs which have distinguished vertices. Here we consider only graphs with one source, which is a starting point for the rewriting process. A first question is : starting from graphs, is it possible to obtain 1-graphs with a PGRS ? We prove that, starting from trees, it is possible with a PGRS to distinguish exactly one source that one can call the root. However, in the general case, starting from a graph it is not possible with a PGRS to distinguish exactly one vertex, that is starting from a graph one cannot obtain a 1-graph. Furthermore we are going to see that the power of PGRS's on graphs is weaker than on 1-graphs.

***Proposition** 5.1. There exists a PGRS which safely marks one root to any tree. No PGRS can safely mark one source to any graph.*

An important difference between graphs and 1-graphs is the following.

***Proposition** 5.2. There exists a PGRS which safely constructs an oriented rooted spanning tree from any 1-graph. No PGRS can safely construct a spanning tree from a graph.*

Furthermore the PGRS used to construct an oriented rooted spanning tree, has a nice property of local termination detection property : one root label (which may only appear in the source of 1-graphs) appears only at the last relabelling step. Hence this label can be the starting point of computation by another PGRS. Then the constructed rooted spanning tree allows to make successive traversals of the initial 1-graph. This feature is often used in the sequel to compute with PGRS's in 1-graphs.

So in a 1-graph, the source may control the computing process and centralize the results, hence it is not surprising that :

Proposition 5.3. *Let L be a set of 1-graphs,*
L is safely B-recognizable iff L is safely M-recognizable.

Like the class of sets of graphs safely recognizable by the PGRS's (Proposition 4.6), the class of sets of 1-graphs safely recognizable by the PGRS's is a boolean algebra. So we are interested in FOL-definable sets of graphs. First-Order formulas express local properties of graphs, as shown in [Ga 82], hence one may expect to recognize by PGRS's (which work locally) the FOL-definable sets of graphs. Unfortunately we have stated that the graphs which have exactly one vertex with label "a" (say) are not recognizable by a PGRS, hence all the FOL-definable sets of graphs are not PGRS-recognizable, however starting from 1-graphs the result holds :

Proposition 5.4. *Every FOL-definable set of 1-graphs is safely recognizable by a PGRS. Furthermore the number of relabelling steps is polynomial in the size of the graphs.*

To prove this result, one uses the fact that, starting from a given 1-graph G, a rooted spanning tree of G can be constructed, and then every subsets of V_G, which cardinality is a given integer, can be enumerated by some PGRS.

Moreover the converse of Proposition 5.4 does not hold, as shown by the following table, which suggests to investigate the Second Order Logic. This study is made for the linearly orderable 1-graphs in [LM 90], otherwise the question remains open.

Properties of graphs	graphs	1-sourced-graphs
. **FOL**		
. exactly one a-vertex	no	yes
. simple	yes	yes
. k-regular	yes	yes
. **MSOL**		
. connected	no	yes
. 2-colorable	no	yes
. k-colorable (k>2)	no	????
. hamiltonian	no	yes
. tree	yes	yes
. **SOL**		
. even number of vertices	no	yes
. as many a's as b's	no	yes

References

[Bi 89] M. Billaud, *Some Backtracking Graphs Algorithms expressed by Graph Rewriting Systems with Priorities*, Rapport Interne n° 8989, LaBRI, Univ. Bordeaux I.

[Bi 90] M. Billaud, *Un interpréteur pour les Systèmes de Réécriture de Graphes avec Priorités*, Rapport Interne n° 9040, LaBRI, Univ. Bordeaux I.

[BLMS 89] M. Billaud, P. Lafon, Y. Métivier and E. Sopena, *Graph Rewriting Systems with Priorities : Definitions and Applications*, Rapport Interne n° 8909, LaBRI, Univ. Bordeaux I.

[BLMS 89] M. Billaud, P. Lafon, Y. Métivier and E. Sopena, *Graph Rewriting Systems with Priorities*, in Graph-Theoretic Concepts in Computer Science, 15th Workshop on Graphs'89, LNCS n° 411, pp. 94-106.

[Co 88] B. Courcelle, *Some applications of logic of universal algebra, and of category theory to the theory of graph transformations*, Bulletin of E.A.T.C.S. n° 36 (1988), pp. 161-218.

[Co 89a] B. Courcelle, *The monadic second-order logic of graphs V : on closing the gap between definability and recognizability*, Rapport Interne n° 8991, LaBRI, Univ. Bordeaux I.

[Co 89b] B. Courcelle, *The monadic second-order logic of graphs VI : on several representations of graphs by relational stuctures*, Rapport Interne n° 8999, LaBRI, Univ. Bordeaux I.

[Co 90] B. Courcelle, *The monadic second-order logic of graphs I. Recognizable sets of finite graphs*, Information and Computation vol. 85 n° 1 (1990), pp. 12-75.

[Do 70] J. Doner, *Tree Acceptors and Some of Their Applications*, Journal of Computer and System Sciences 4 (1970), pp. 406-451.

[Fr 68] J. Fris, *Grammars with partial ordering of the rules*, Information Control 12 (1968), pp. 415-425.

[Ga 82] H. Gaifman, *On local and non-local properties*, Proc. of the Herbrand Symposium, Logic Colloquim'81, J. Stern ed., North-Holland Pub. Co. (1982), pp. 105-135.

[GeSt 84] F. Gécseg and M. Steinby, *Tree automata*, Akadémiai Kiado-Budapest (1984).

[HoUl 67] J.E. Hopcroft and J.D. Ullman, *Some Results on Tape-Bounded Turing Machines*, Journal of the Association for Computing Machinery, Vol. 16, n° 1 (1967), pp. 168-177.

[HoUl 79] J.E. Hopcroft and J.D. Ullman, *Introduction to Automata Theory, Languages, and Computation*, Addison-Wesley (1979).

[Im 87] N. Immerman, *Languages that capture complexity classes*, SIAM J. Comput., vol. n° 4 (1987), pp. 760-778.

[LM 90] I. Litovsky and Y. Métivier, *Computing with Graphs Rewriting Systems with Priorities*, Rapport Interne du LaBRI, Univ. Bordeaux I.

[Ma 61] A.A. Markov, *Theory of Algorithms*, Jerusalem : Israel Program for Scientific Translations (1961).

[Sa 73] A. Salomaa, *Formal Languages*, Academic Press (1979).

[Tu 84] W. Tutte, *Graph Theory*, Addison Wesley (1984).

[Ya 71] A. Yasuhara, *Recursive function theory & logic*, Academic Press (1971).

DOUBLE-WALL CELLWORK SYSTEMS FOR PLANT MERISTEMS

Jacqueline Lück and Hermann B. Lück
Laboratoire de Botanique analytique et Structuralisme végétal
Faculté des Sciences et Techniques de St-Jérôme,
13397 Marseille cedex 13, France

ABSTRACT: The development of the cellular pattern of plant meristems is simulated under rules specifying a constant positioning of division walls with respect to the previously introduced wall. The network, or cellwork, of the cell walls is, in analogy to real walls, represented by double-wall labeling.

The polyhedral boundary of a cell, when opened on the last division wall, becomes representable by a wall-map which procures the topological relationship of the walls viewed from within the cell. Cell divisions are defined as bipartitions of wall-maps. In addition, the edges introduced for the insertion of a new division wall induce by contact wall divisions in neighbour cells, increasing the number of their walls.

The double-wall cellwork system with wall labels consists of an alphabet of wall maps, subsets of map division rules representing cell divisions, and subsets of map transformation rules. The axiom is given by at least one wall-map. A cellwork derivation is obtained by rewriting in parallel all wall-maps describing the cells in a cellwork.

3D-systems are proposed as a tool to investigate the construction and behaviour of apices of ferns and mosses. Different developmental pathways may be traced back to differently oriented division walls. The simplest theoretical meristem, which functions by the helicoidal division of a tetrahedral apical cell, gives rise to a shoot composed of 3 rows of cells. If the derivative cell divides once again, it results 6 cell files.

Keywords: Cellwork-systems, L-systems, Plant development

CONTENTS

1. Introduction

Cell walls play an important role in plant growth and morphogenesis. Whereas growth is performed by cell wall extension and new wall formation, by both wall subdivision and wall addition, morphogenesis is closely related to the disposition of cells in their filiations. The problem of whether there exists a deterministic law for the positional laydown of new walls is still not solved. In any case, the position a new wall will be inserted is determined by microtubule pre-prophase bands very early in the cell cycle [Gunning *et al.*, 1978]. But their regular, repetitive positioning in cell filiations is controversial: a relationship between the position of new cell walls in mother and daughter cells has not been proved to exist. Nevertheless, in several cases and with the help of 2D map systems [cf. bibliography in Lindenmayer & Prusinkiewicz, 1988], it could be shown that the cellular pattern of apical meristems can be simulated under the assumptions of some rules specifying a precise and constant positioning of divisions walls from cell generation to cell generation. We suppose that this possible basic deterministic behaviour in the cell organization is present but (1) soon overwritten by physiological influences such as dorsiventrality, anisotropy in growth hormone distribution, etc., and (2) masked also by the fact that in the 2D case, most of the division orientations which occur in the third dimension of meristems cannot be taken into account.

Only few 3D developmental systems, able to simulate the development of plant meristems, have been conceived till now [Lindenmayer, 1984]. A recent annotated bibliography has been given by Lindenmayer & Prusinkiewicz [1988].

2. Networks or cellworks with double-walls

The network of plant cell walls, which has been called "cellwork" [Liu & Fu, 1981], can be described by different kinds of labelings, for example vertex labeling, edge labeling, face labeling or cell labeling. We adopt here a double face labeling. Why double-walls? A first motivation to adopt double labeling is that real cell walls are double; each cell constructs its own wall by apposition of cellulose microfibrils on the plasmalemma side.

The limit between two contiguous cells, the middle lamella, remains sometimes visible:

THE PLANT CELL WALL

IN A 2D OPTICAL CUT: THREE CELLS MEET AT A CORNER

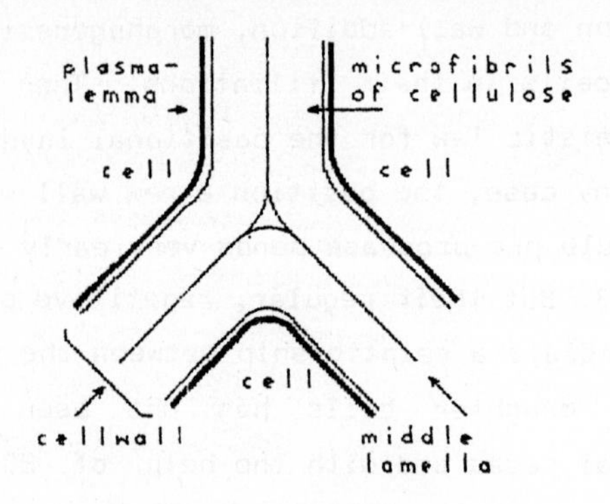

THE LABELING OF WALLS

IN 2D MAP REPRESENTATIONS: THREE CELLS MEET AT A CORNER

simple-edge labeling double-edge labeling

IN 3D CELLWORKS: THREE CELLS MEET AT AN EDGE

simple-wall labeling double-wall labeling

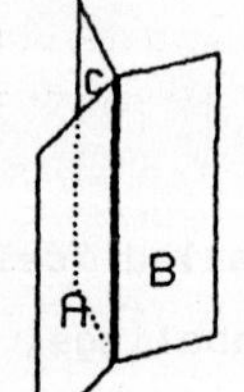

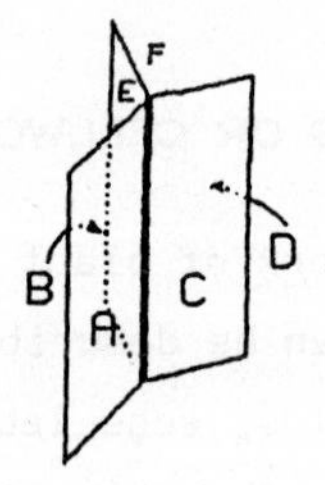

Secondly, the presently proposed 3D double-wall cellwork systems are constructed as an extension to 2D double-wall map systems [Lück & Lück, 1981 to 1989] which have proved to be useful for the construction of all possible developmental systems in a given framework. Classes of systems which account for typical botanical cellular behaviour could be defined [Lück & Lück, 1981, 1982]. It has been shown that the translation of double- to simple-wall labeled systems can be performed easily [De Boer & Lindenmayer, 1987; Lindenmayer, 1987]. The inconvenience of double labeling, especially for the derivations of maps or cellworks with geometrical interpretations [de Boer, 1990; Fracchia, *et al.*, 1990; Fracchia & Prusinkiewicz, 1990], can so be overcome.

3. Behaviour of meristems

Most of the apices of ferns and mosses are constructed by the helicoidal divisions of a tetrahedral apical cell. Its derivatives are called merophytes. The helix is counter-clockwise in the main axis of *Psilotum* [Hagemann, 1980] and reverses its direction from one branching order to the next one. This behaviour has been simulated by a map system [Lück & Lück, 1985]. But in *Dicranopteris nitida* [Hagemann, 1980] and *Angiopteris* [Imaichi, 1986] simulated by a double-wall system [Lück, *et al.*, 1988], the direction of the helix is the same in all branch orders. This behaviour is essentially related to the presence of *S* or *Z* tetrads [Lück, *et al.*, 1988]. Tetrahedral cells exist also at the tip of roots like those of *Azolla* [Gunning, 1981]. In ferns and mosses, the origin of leaves and branches is closely related to the apical helicoidal segmentation [Hébant-Mauri, 1977]. Different species of liverworts can be differentiated either according to the time delay necessary for the formation of new tetrahedral cells which in merophytes incept branches or according to physiological influences in the dorsiventral shoots leading to merophytes of different development [Crandall Stotler, 1972].

These differences may be traced back to one or two differently oriented division walls. Such investigation can only be done with the help of formal systems since, in the 3D case, the definition of the topological relationship between cells becomes very cumbersome quickly.

4. Construction of a cellwork system

4.1. Wall-map of a cell

A cell is supposed to have its own polyhedral boundary which, viewed from inside, has labeled faces, also called walls. If the cell boundary is opened, it can be represented by a map viewed from within the cell. For this purpose we cut off the last formed wall, called the division wall and look into the cell like looking into a box. The wall-map is bounded by the cut, and has an additionel region representing the division wall. As an example we take a tetrahedral apical cell (Fig.1a) with the division wall *D*. When opened, the wall-map gives the topological relationship of the walls with their interior labels (Fig.1b). An environment *e* surrounds the map border.

FIG. 1: 3D-VIEW WALL-MAP 1

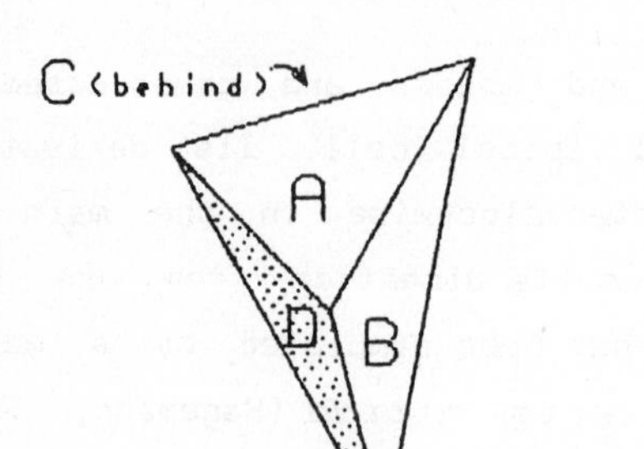

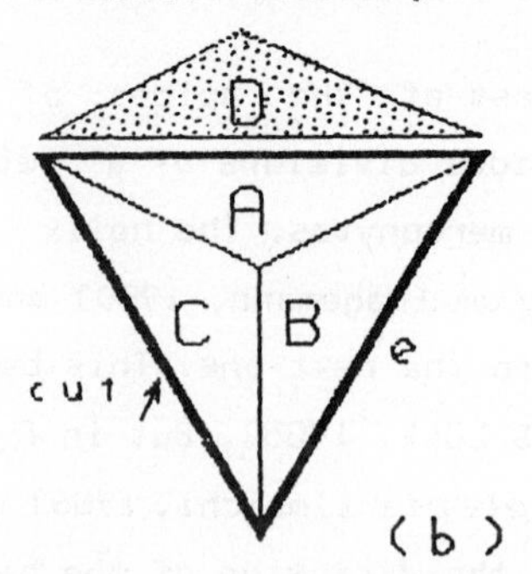

Such a wall-map is described by enumerating its walls; in the first step, each wall is labeled by a capital letter, followed, in parenthesis, by the clockwise sequence of neighbour walls. This sequence begins, if present, by the environmental label *e*. The list of walls becomes

A(eBC)
B(eCA)
C(eAB)
D, the division wall.

Then, the label *e* is replaced by *D*, the label of the division wall which is underlined; the neighbour wall sequence of *D* is given by the walls touching the map border, read counter-clockwise; it follows that the boundary of the apical tetrahedral cell *c* is described by a set of walls and a subscript indicating the number of faces:

$$c_4 = \begin{bmatrix} A(\underline{D}BC) \\ B(\underline{D}CA) \\ C(\underline{D}AB) \\ \underline{D}(ACB) \end{bmatrix}$$

4.2. CELL DIVISION

A tetrahedral apical cell divides into a new tetrahedral cell c_4 and a subsidiary cell c_5 called a merophyte. The cell division is denoted by

$$c_4 \longrightarrow c_4\, c_5 \;.$$

The division wall dw separating these two daughter cells is labeled D_1 on the side of the tetrahedral cell and D_2 on the side of the merophyte:

$$dw = {}^{D_1}/_{D_2}.$$

A cell division is represented by

(1) a bipartition of a wall-map of the dividing cell, and

(2) its repercussion in terms of wall divisions on neighbour cells.

4.2.1. Bipartition of a wall-map

There are several possibilities to partition a wall-map under the restriction that a region can at most split once and that the cut is, in the 3D boundary, a closed sequence of wall insertion edges. We retain the bipartition of Fig.2, which conforms to the segmentation of the apical tetrahedral cell. It can be described by a set of wall productions in which we define, at first, the wall labels:

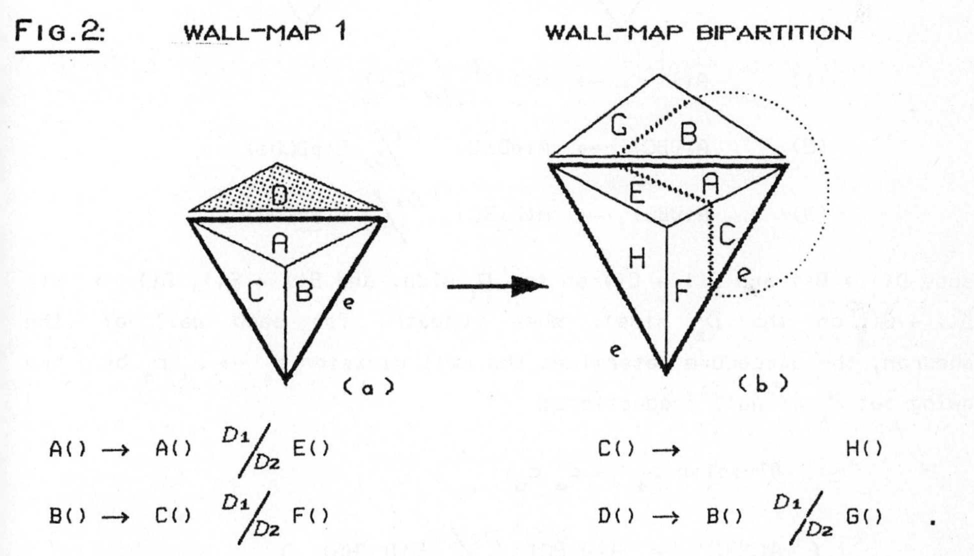

$$A() \longrightarrow A()\ {}^{D_1}/_{D_2}\ E() \qquad C() \longrightarrow H()$$

$$B() \longrightarrow C()\ {}^{D_1}/_{D_2}\ F() \qquad D() \longrightarrow B()\ {}^{D_1}/_{D_2}\ G() \;.$$

The slash (often called 'marker') indicates that the walls that preceed and follow it are separated by a <u>division-wall insertion edge</u>. The subscripts of the slash furnish the labels of the two faces of the division wall. All walls preceeding the slashes belong to the daughter cell with the division wall label at the left of the slash, and all walls following the slashes belong to the second daughter with the label at the right.

After specification of the labels as above, the <u>sequential wall</u>

<u>environment</u> of each new wall has to be specified. This sequence, in clockwise direction, is constructed in two further steps: it begins with e if the wall touches the map split, followed by the old environment, i.e. that of the mother cell. Then, the label e is replaced by the division wall label as indicated by the slash and underlined. The subsequent labels in the neighbour sequence are rewritten according to the wall productions. Explicitly, this procedure in three steps, i.e. specification (1) of the wall labels, (2) of their sequential mother wall environment, and (3) of their new wall environment, gives for the wall A :

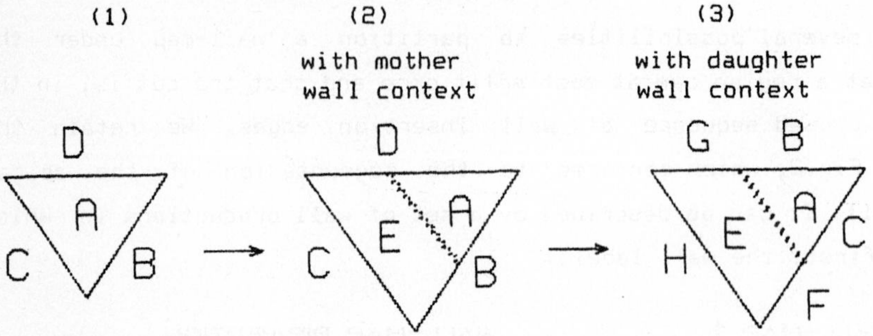

(1) (2) with mother wall context (3) with daughter wall context

$$
\begin{array}{llll}
(1) & A() \longrightarrow & A() \; {}^{D_1}\!/\!{}_{D_2} & E() \\
(2) & A(DBC) \longrightarrow & A(eD_1B) \; {}^{D_1}\!/\!{}_{D_2} & E(eBCD_1) \\
(3) & A(DBC) \longrightarrow & A(\underline{D_1}BC) \; {}^{D_1}\!/\!{}_{D_2} & E(\underline{D_2}FHG)
\end{array}
$$

(because D() → B() and B() → C() on the D_1 side, and B() → F(), C() → H() and D() → G() on the D_2 side). When repeated for each wall of the tetrahedron, the procedure determines the cell division $c_4 \longrightarrow c_4\, c_5$ by the following set P_1 of wall productions:

Division $c_4 \longrightarrow c_4\, c_5$:

$$
P_1 = \left[\begin{array}{lll}
A(\underline{D}BC) \longrightarrow & A(\underline{D}_1 BC) \; {}^{D_1}\!/\!{}_{D_2} & E(\underline{D}_2 FHG) \\
B(\underline{D}CA) \longrightarrow & C(\underline{D}_1 AB) \; {}^{D_1}\!/\!{}_{D_2} & F(\underline{D}_2 GHE) \\
C(\underline{D}AB) \longrightarrow & & H(GEF) \\
\underline{D}(ACB) \longrightarrow & B(\underline{D}_1 CA) \; {}^{D_1}\!/\!{}_{D_2} & G(\underline{D}_2 EHF)
\end{array} \right] .
$$

These productions allow the new wall-maps of the daughter cells c_4 and c_5 to be drawn. The wall-map 1 of c_4 is identical to that of the foregoing tetrahedral cell if the label D is replaced by D_1; it is given by the walls of the productions which preceed the slashes. The wall-map 2 of cell c_5, given by the walls which follow the slashes, becomes:

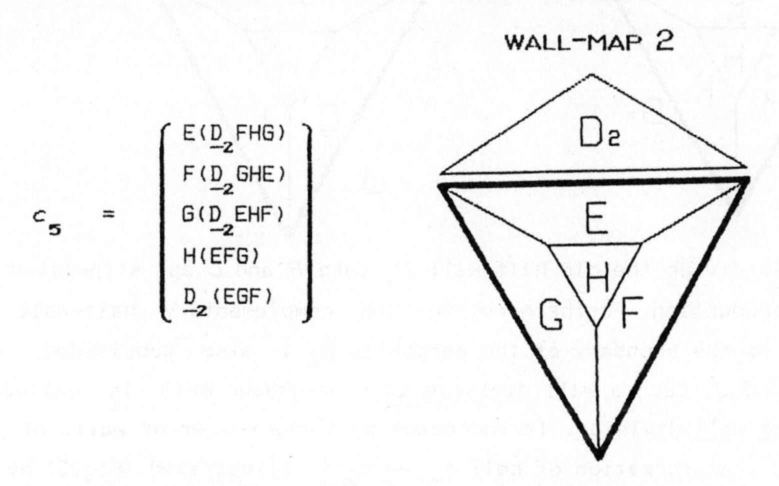

4.2.2. Contact induced wall divisions

It is assumed that the tetrahedral cell c_4 divides a second time according to the previous subset P of wall productions. The new division

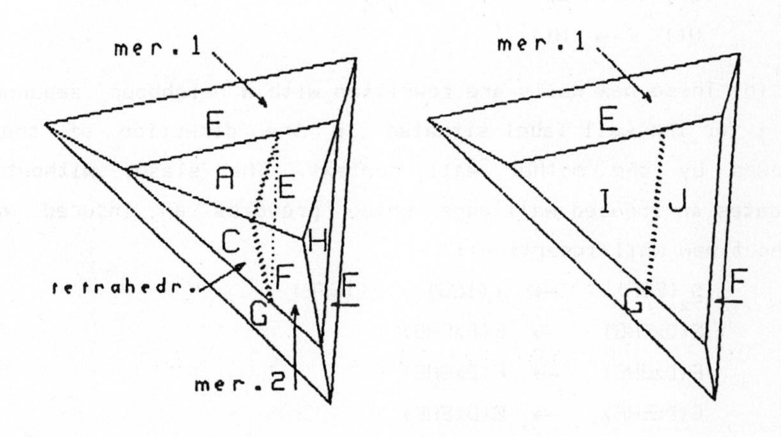

FIG.3: CONTACT INDUCED WALL DIVISION

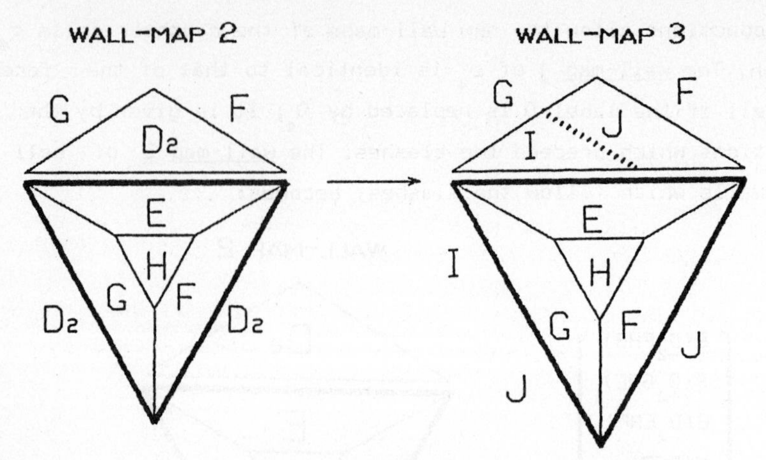

wall D_1/D_2 will divide the old half-wall D_1 into B and G as stipulated by the 4th wall production. Furthermore, the old complementary half-wall D_2 which belongs to the boundary of the merophyte c_5 is also subdivided, say into walls I and J. Such a wall division in a neighbour cell is called a <u>contact</u> <u>induced</u> <u>wall</u> <u>division</u>. It increases by 1 the number of walls of the merophyte. The transformation of cell $c_5 \rightarrow c_6$ is illustrated (Fig.3) by the transformation of <u>wall-map</u> <u>2</u> into <u>wall-map</u> <u>3</u>.

In order to obtain the subset P_2 of wall productions, which performs this map transformation, we follow again the three step procedure indicated for the map division productions, but in a slightly different manner:

(a) The new wall labels are introduced:

$\underline{D_2}() \rightarrow \underline{I}() \,/\, \underline{J}()$
E() $\rightarrow$ E()
F() $\rightarrow$ F()
G() $\rightarrow$ G()
H() $\rightarrow$ H() .

(b) These new walls are rewritten with a neighbour sequence beginning with i for the wall label situated in the direction of the new edge, followed by the mother wall context. The slash, without subscripts, indicates an induced wall edge which provokes an induced wall division (without new wall insertion):

$\underline{D_2}$(EGF) $\rightarrow$ $\underline{I}$(iEG) / $\underline{J}$(iGFE)
E(D2FHG) $\rightarrow$ E(D2FHG)
F(D2GHE) $\rightarrow$ F(D2GHE)
G(D2EHF) $\rightarrow$ G(D2EHF)
H(EFG) $\rightarrow$ H(EFG)

(c) The wall productions are rewritten by replacing

- in induced wall division productions:
  - the label i by the complementary underlined wall label indicated on the opposite side of the slash,
  - the subsequent labels in the neighbour sequence according to the wall productions of P_2,
- in simple-wall productions:
  - the division wall label D_2 by IJ or JI, if the mother wall is touched by an induced edge insertion; such walls are here E and G (they are given by the intersection of the mother environmental wall sequences of the daughter walls without the i: $\{E,G\} \cap \{G,F,E\} = \{E,G\}$),
  - the division wall by one of the labels I and J if the wall is not touched by an edge insertion; such a wall is here F (these walls belong to $\{E,G\} \cup \{G,F,E\} \setminus \{E,G\} \cap \{G,F,E\} = \{F\}$ and F belongs to $\mathcal{J}$).

It results in the following set of productions:

$$P_2 = \left[\begin{array}{lll} \underline{D}_2(EGF) & \rightarrow & \underline{I}(JEG) \;/\; \underline{J}(IGFE) \\ E(D_2FHG) & \rightarrow & E(IJFHG) \\ F(D_2GHE) & \rightarrow & F(JGHE) \\ G(D_2EHF) & \rightarrow & G(JIEHF) \\ H(EFG) & \rightarrow & H(EFG) \end{array}\right].$$

The subset of wall productions leading to the cell c is thus determined.

FIG.4: CONTACT INDUCED WALL DIVISION

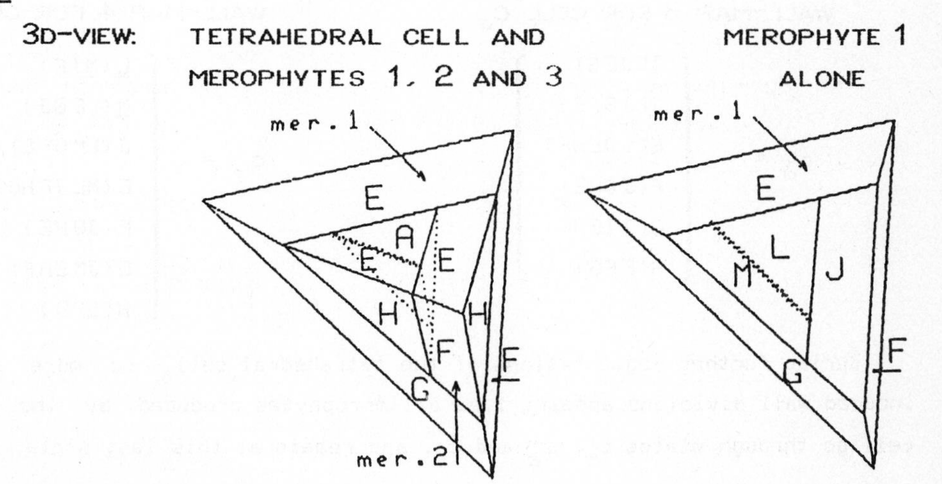

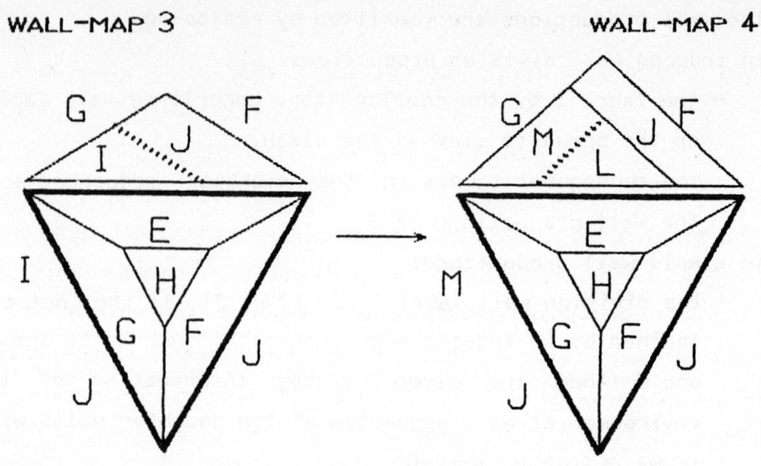

During the formation of the third merophyte, while the division of the tetrahedral cell follows again the rules of P_1, the second merophyte undergoes the transformation stipulated by the rules of P_2, and the first formed merophyte will become a cell c_7 with 7 faces by a further induced wall division (Fig.4). This is specified by the following set P_3 of wall productions, representative for the transformation of wall-map 3 into wall-map 4:

$$P_3 = \begin{bmatrix} \underline{I}(JEG) & \longrightarrow & \underline{L}(MJE) \;/\; \underline{M}(LEGJ) \\ \underline{J}(IGFE) & \longrightarrow & J(LMGFE) \\ E(IJEHF) & \longrightarrow & E(MLJFHG) \\ F(JGHE) & \longrightarrow & F(JGHE) \\ G(JIEHF) & \longrightarrow & G(JMEHF) \\ H(EFG) & \longrightarrow & H(EFG) \end{bmatrix}$$

WALL-MAP 3 FOR CELL c_6 | WALL-MAP 4 FOR CELL c_7

$$c_6 = \begin{bmatrix} \underline{I}(JEG) \\ \underline{J}(IGFE) \\ E(IJEHF) \\ F(JGHE) \\ G(JIEHF) \\ H(EFG) \end{bmatrix} \qquad c_7 = \begin{bmatrix} \underline{L}(\underline{MJ}E) \\ \underline{M}(LEGJ) \\ J(LMGFE) \\ E(MLJFHG) \\ F(JGHE) \\ G(JMEHF) \\ H(EFG) \end{bmatrix}$$

During further segmentations of the tetrahedral cell, no more contact induced wall divisions appear, i.e. all merophytes produced by the apical cell go through states c_5, c_6 and c_7, and remain at this last state.

4.3. Double-wall cellwork system

A double-wall cellwork system is defined by an alphabet of wall-maps, a set of map production rules and an axiom which is a cellwork consisting at least of one cell described by a wall-map. The wall-maps of the alphabet are specified by sets of wall labels, i.e. capitals followed, in parentheses, by the clockwise sequence of labels of immediate neighbour walls. Each wall-map describes the boundary of a 3-dimensional cell. The set of production rules contains subsets specifying wall-map transitions for contact induced wall divisions, and subsets specifying wall-map bipartitions representing cell divisions. In a cellwork, all cells are supposed to be convex.

The cellwork derivation is obtained by rewriting in parallel all wall-maps contained in a cellwork. New walls are inserted according to the circular sequence of edges specified by the cut of the bipartitioned maps. The exact topological relationship of cells is obtained by deriving in parallel the two complementary faces of a division wall. This furnishes the derived alphabet over double-walls. The cellwork system generates the topological cell organization of a 3-dimensionnal cellular tissue.

In the example given above, the alphabet contains 4 wall-maps (for cells c_4, c_5, c_6 and c_7); the map production rules consist in 1 subset (P_1) of map bipartition rules and 2 subsets (P_2 and P_3) of map transitions rules. The starting cell is a tetrahedron described by the wall-map 1. The cell derivation shows that the tetrahedron devides into a new tetrahedral cell and a sister cell called a merophyte. The merophyte goes through 2 map transitions in which wall divisions occur. From the state of a polyhedron with 5 faces, these cells will have 6 and 7 faces, successively.

The same example has been given for an edge labeled cellwork system [Lindenmayer, 1984]. In the double-wall cellwork system with markers proposed here, the derivation in parallel of the two faces of a division wall gives the key for its translation into a single-wall cellwork system with either face labels and markers, or edge labels and markers.

4.4. Geometrical interpretation of a cellwork system

The exact geometrical relationship between cells cannot be generated by the system as it does not contain geometrical algorithms. Nevertheless, wall-maps allow to sketch polyhedral shapes of the corresponding cells. The division wall derivation indicates the way to assemble them accurately. This is sufficient to construct, for each derivation step, a cellular body for biological representations like that of Fig.5., which is based on the previous example.

Fig.5 represents the simplest theoretical shoot generated by a tetrahedral cell. The shoot axis is build up by space-filling 7-sided cells arranged in 3 cell files (Fig.5a). Each new added cell is located higher

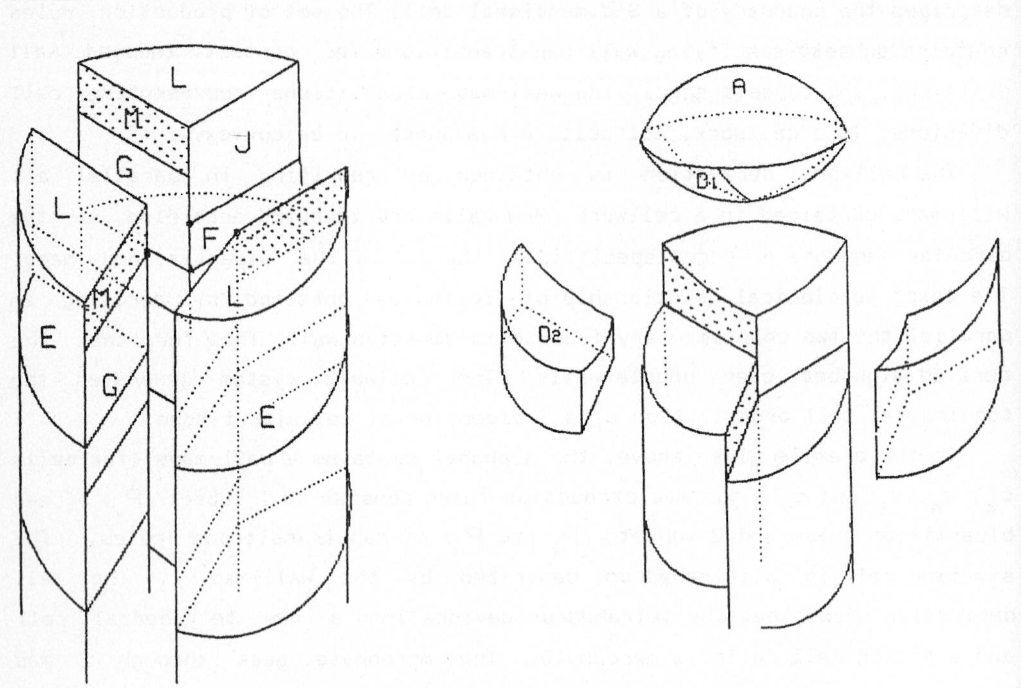

than its preceeding one, just as steps of a helicoidal stair (stippled walls *M* indicate steps). The shoot apex is composed of an apical cell and two derivatives in the transitory states c_5 and c_6 (Fig.5b). The divisions of the apical cell are parallel to its 3 embedded faces. Each division rotates by 120° with respect to the previous one. The produced merophytes do not divide further, but the number of their walls increases from 5 to 7.

5. TOWARDS THE MERISTEM OF THE FERN CERATOPTERIS

The very simple apical organization of the fern *Ceratopteris* (Fig.6 from Kny [1875 in Hébant-Mauri, 1977]) is approached if we allow the apical derivative also to divide. In this case, the apical tetrahedral cell is in fact 6-sided due to contact-induced wall divisions (highlighted by the wall labels with dashes). The corresponding cellwork system has an alphabet of 6 wall-maps over six cell types: an apical cell a_6 (the index giving the number of wall faces), its sister cell b_7, two transitory cells c_5 and d_8, and two cells in their final state, c_6 and d_{10}. The subsets of production rules can be deduced from following 2 map bipartitions and 4 map transformations, the axiom being a cell c_6 (as productions are not explicited, superscripts of wall labels replace the slashes):

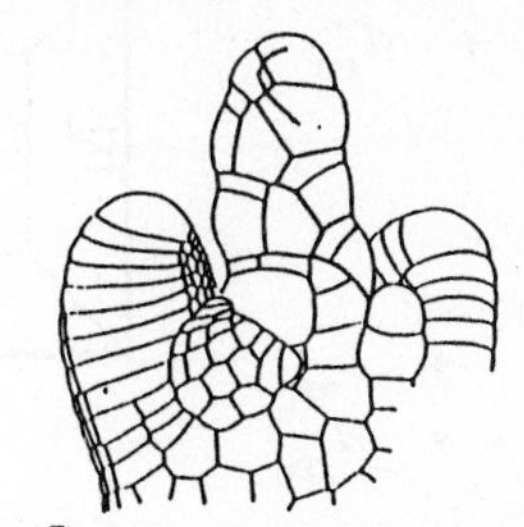

FIG.6: Apex of *Ceratopteris* after Kny [1875, in Hébant - Mauri, 1977]. Successive merophytes and two leaf primordia.

CELL PRODUCTIONS — CORRESPONDING MAP BIPARTITIONS OR MAP TRANSFORMATIONS

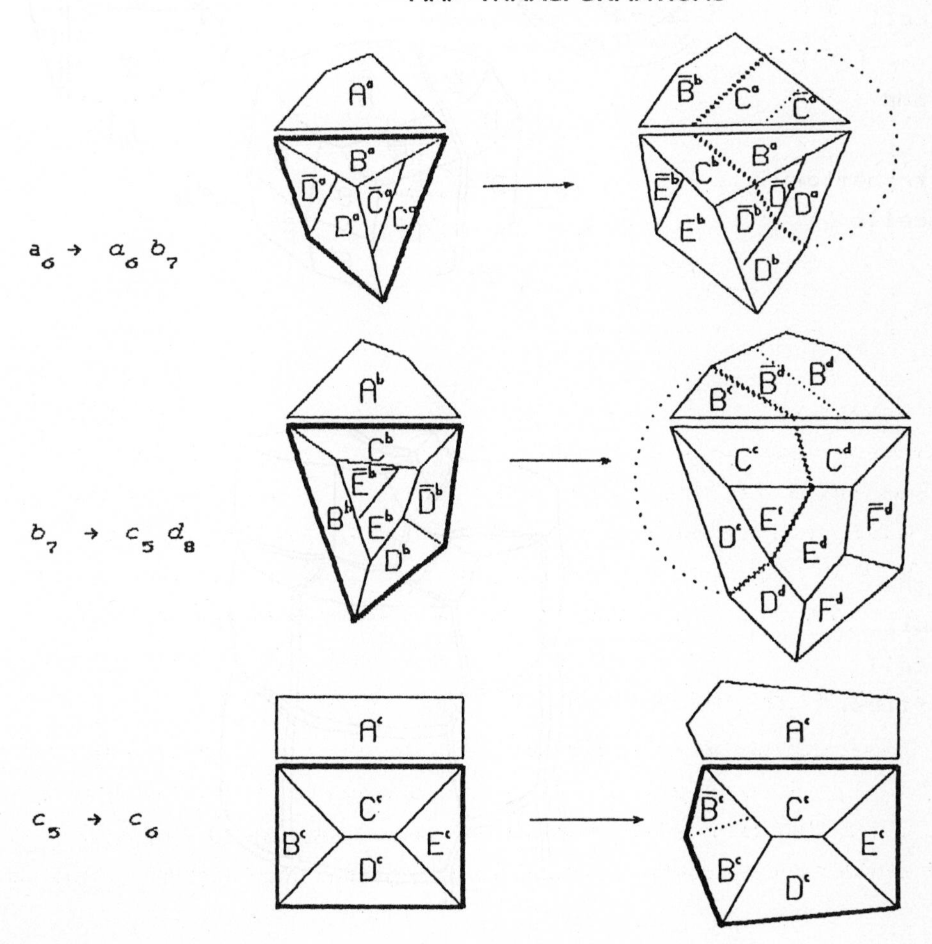

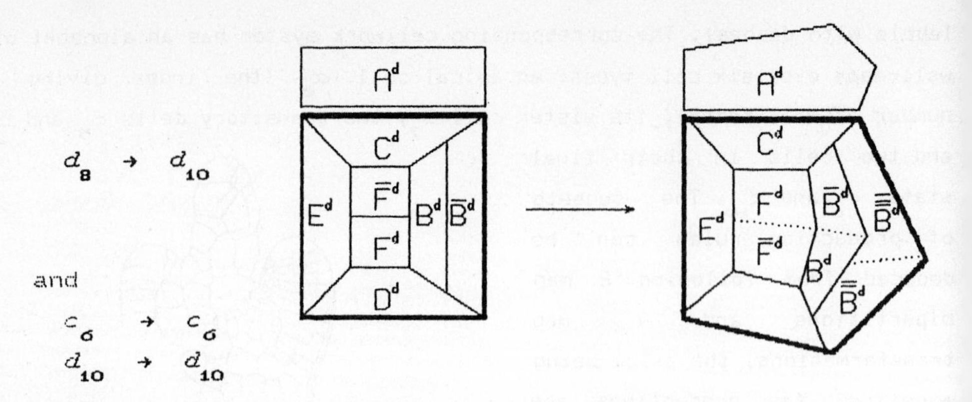

FIG.7: FIRST STEPS TOWARDS THE APEX OF CERATOPTERIS

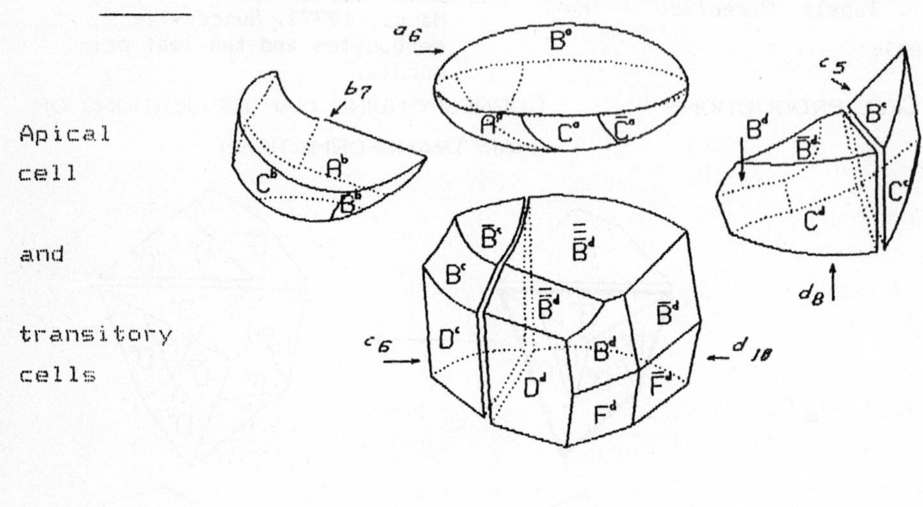

the
six
cell
files.

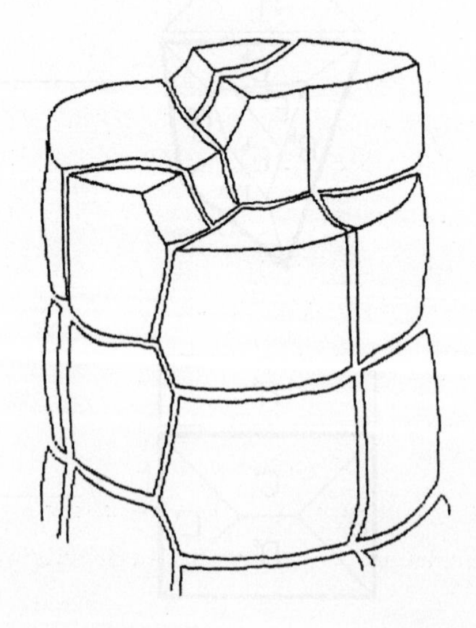

The geometrical interpretation leads to a shoot composed of an apex of 4 cells and an axis of 6 cell files, a packing with 6- and 10-sided cells (Fig.7). The assemblage of the cells is based on 22 cell contacts defined by the division wall derivation (10 contacts are represented in Fig.8). Fig.8a shows the double division wall A^{a}/A^{b} of the apical cell (stippled). In the next cell generation this wall generates 3 cell contacts, $C^{a}/\bar{B}^{c}$, $\bar{C}^{a}/B^{c}$, and B^{b}/B^{d} (Fig.8b). Those of the third cell generation, as shown by the Fig.8c, are $\bar{D}^{a}_{x+1}/\bar{B}^{c}_{x}$, $D^{a}_{x+1}/\bar{\bar{B}}^{d}_{x}$, $\bar{D}^{b}_{x+1}/B^{c}_{x}$, $D^{b}_{x+1}/\bar{\bar{B}}^{d}_{x}$, $D^{c}_{x+1}/\bar{B}^{d}_{x}$, and D^{d}_{x+1}/B^{d}_{x} (the two complementary faces of a double-wall belong now to cells of different generations as indicated by the subscript x). At the next step, these 6 double-walls become $\bar{E}^{b}_{x+2}/\bar{B}^{c}_{x}$, $E^{b}_{x+2}/\bar{\bar{B}}^{d}_{x}$, $\bar{F}^{d}_{x+2}/B^{c}_{x}$, $F^{d}_{x+2}/\bar{\bar{B}}^{d}_{x}$, $D^{c}_{x+1}/\bar{B}^{d}_{x}$, and D^{d}_{x+1}/B^{d}_{x}, respectively (faces of 4 from these 6 double-walls belong to cells with an age difference of 2 generations). The first 4 of these double-walls change in the following step to $E^{c}_{x+3}/\bar{B}^{c}_{x}$, $E^{d}_{x+3}/\bar{\bar{B}}^{d}_{x}$, $\bar{F}^{d}_{x+2}/B^{c}_{x}$, and $F^{d}_{x+2}/\bar{\bar{B}}^{d}_{x}$, respectively. Finally, the first 2 of these 4 walls change to $E^{c}_{x+3}/\bar{B}^{c}_{x}$ and $E^{d}_{x+3}/\bar{\bar{B}}^{d}_{x}$. In the cell files, the cell contacts are finally given by the 6 invariant double-walls which separate cells, differing in age by three generations.

FIG.8: WALL CONTACTS

(a)

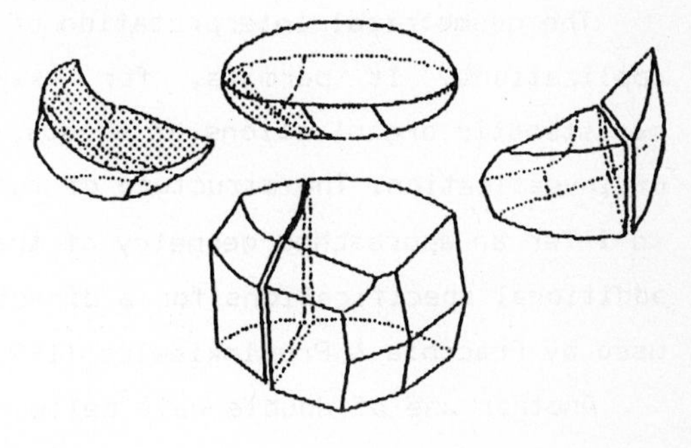

(b)

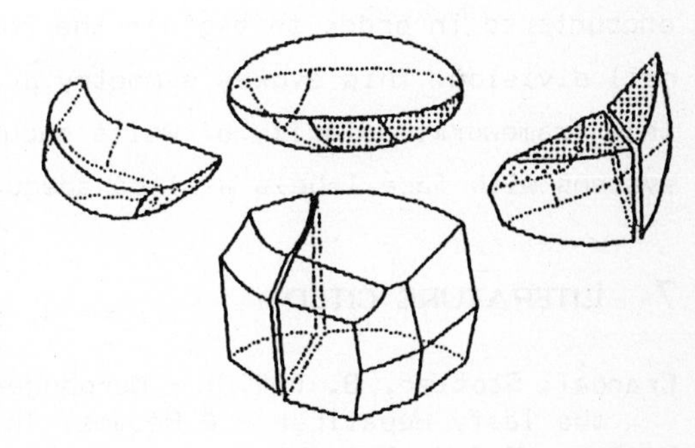

(c)

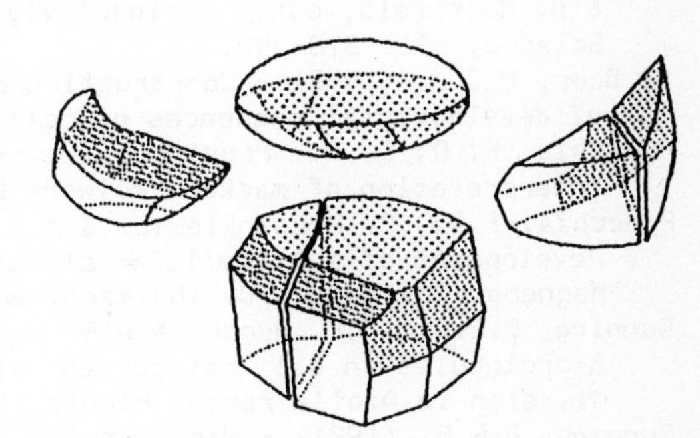

6. CONCLUSIONS

The double-wall cellwork systems presented in this paper are useful for system construction based on given cellular behaviour. They furnish a good insight into individual cellular form, cellular neighbourhood relationships and cellular assemblage. This is not the case by systems with single-wall labels. In return, their formalism is cumbersome. Once the double-wall derivations established, the formalism can be rendered concise by a translation of the double-wall labeling into a simple wall or edge labeling.

The geometrical interpretation of cellworks is necessary for biological applications. It permits, for example, the comparison of different meristematic organizations of shoots, both on the theoretical and real level of investigation. The structure of the system gives sufficient information to infer an approached geometry of the generated plant body. Nevertheless, additional specifications for a direct generation of shape like in systems used by Fracchia & Prusinkiewicz (1990) would be of help.

Another use of double-wall cellwork systems concerns the supramolecular structure of cell walls. They can simulate the rhythmic assemblage of cellulose microfibrils in the cell wall and help to solve problems encountered in order to explain the conservation of this assemblage during cell division. This evokes symmetry problems in relation to cell form. In this framework, polarity of walls should be considered. Coloured cellwork systems with face labels will be adequate.

7. Literature Cited

Crandall Stotler, B. (1972) - Morphogenetic patterns of branch formation in the leafy Hepaticae - A Résumé. The Bryologist 75: 381-403.

De Boer, M.J.M. & A. Lindenmayer (1987) - Map 0L-systems with edge label control: Comparison of marker and cyclic systems. In 'Graph Grammars and their Application to Computer Science', H. Ehrig, M. Nagl, G. Rozenberg & A. Rosenfeld, eds; Springer-Vlg. Berlin, Lect. Notes in Computer Science, 291: 378-392.

De Boer, M.J.M. (1990) - Construction on map 0L-systems for the generation of developmental sequences of cell patterns. (In this volume).

Fracchia, F. D. & P. Prusinkiewicz (1990) - Physically-based graphical interpretation of marker cellwork L-systems. (In this volume).

Fracchia, F.D., P. Prusinkiewicz & M.J.M. de Boer (1990) - Animation of the development of multicellular structures. In 'Computer Animation '90', N. Magnenat-Thalmann & D. Thalmann, eds; Springer-Vlg Tokyo : 3-18.

Gunning, B.E.S., J.E. Hugues & A.R. Hardham (1978) - Pre-prophase bands of microtubules in all categories of formative and proliferative cell division in Azolla roots. Planta 143: 145-160.

Gunning, B.E.S. (1981) - Microtubules and cytomorphogenesis in a developing organ: the root primordium of Azolla pinnata. In 'Cytomorphogenesis in Plants', O. Kiermayer, ed.; Springer-Vlg. Wien, Cell biology monographs 8: 301-325.

Hagemann, W. (1980) - Uber den Verzweigungsvorgang bei Psilotum und Selaginella mit Anmerkungen zum Begriff der Dichotomie. Plant Systematics und Evolution 133: 181-197.

Hébant-Mauri, R. (1977) - Segmentation apicale et initiation foliaire chez Ceratopteris thalictroides (Fougère leptosporangiée). Can. J. Bot. 55: 1820-1828.

Iamichi, R. (1986) - Surface-viewed shoot apex of Angiopteris lygodiifolia Ros. (Marattiaceae). Bot. Mag. Tokyo 99:309-317.

Lindenmayer, A. (1984) - Models for plant tissue development with cell division orientation regulated by preprophase bands of microtubules. Differentiation 26: 1-10.

Lindenmayer, A. (1987) - An introduction to parallel map generating systems. In 'Graph Grammars and their Application to Computer Science', H. Ehrig, M. Nagl, G. Rozenberg & A. Rosenfeld, eds; Springer-Vlg. Berlin, Lect. Notes in Computer Science 291: 27-40.

Lindenmayer, A. & P. Prusinkiewicz (1988) - Annotated bibliographie. In 'Artificial life, SFI studies in the Sciences of Complexity', C. Langton, ed.; Addison-Wesley Publ. Co.

Liu, H.L. & K.S. FU (1981) - Cellwork topology, its network duals, and some applications - three dimensional Karnaugh map and its virtual planar representation. Information Science 24: 93-109.

Lück, J. & H.B. Lück (1981) -Proposition d'une typologie de l'organisation cellulaire des tissus végétaux. Actes 1er Sém. de l'Ecole de Biol. théor., CNRS; H. Le Guyader et Th. Moulin, eds; ENSTA Paris : 335-371.

Lück, J. & H.B. Lück (1982) - Sur la structure de l'organisation tissulaire et son incidence sur la morphogenèse. Actes 2ème Sém. de Bio. théor. , H. Le Guyader ed.; Univ. Rouen : 385-397.

Lück, J. & H.B. Lück (1985) - Comparative plant morphogenesis founded on map and stereomap generating systems. In 'Dynamical Systems and Cellular Automata', J. Demongeot, E. Goles & M. Tchuente, eds; Academic Press, London: 111-121.

Lück, H.B. & J. Lück (1986) - Unconventional leaves (an application of map OL systems to biology). In 'The Book of L', G. Rozenberg & A. Salomaa, eds; Springer-Vlg. Berlin: 275-289.

Lück, H.B. & J. Lück (1987a) - From map systems to plant morphogenesis. In 'Mathematical Topics in Population Biology, Morphogenesis and Neurosciences', E. Teramoto & M. Yamaguti, eds; Springer-Vlg. Berlin, Lect. Notes in Biomathematics 71: 199-208.

Lück, H.B. & J. Lück (1987b) - Modélisation du fonctionnement d'un méristème par des L-systèmes et des systèmes de graphes et de cartes à réécriture parallèle. In 'Développement des végétaux, aspects théoriques et synthétiques', H. Le Guyader, ed.; Masson, Paris, Collection Biologie théorique 2: 375-395.

Lück, J. & H.B. Lück (1987c) - From OL and IL map systems to indeterminate and determinate growth in plant morphogenesis. In 'Graph Grammars and their Application to Computer Science', H. Ehrig, M. Nagl, G. Rozenberg & A. Rosenfeld, eds; Springer-Vlg. Berlin, Lect. Notes in Computer Science 291: 393-410.

Lück, J. & H.B. Lück (1987d) - Vers une nouvelle théorie des méristèmes apicaux des végétaux. In 'Biologie théorique, Solignac 1985', H.B. Lück, ed.; Ed. CNRS, Paris: 197-213.

Lück, J. & H.B. Lück (1989) - Modélisation de la croissance des tissus végétaux: l'arrêt des divisions cellulaires. In 'Biologie théorique, Solignac 1987', Y. Bouligand, ed.; Ed. du CNRS, Paris 123-140.

Lück, J., A. Lindenmayer & H.B. Lück (1988) - Models of cell tetrads and clones in meristematic cell layers. Bot. Gaz. 149: 127-141.

PROGRAMMED DERIVATIONS OF RELATIONAL STRUCTURES*

Andrea Maggiolo-Schettini
Dipartimento di Informatica, Università di Pisa
Corso Italia 40, 56100 Pisa, Italy

Józef Winkowski
Instytut Podstaw Informatyki PAN
00-901 Warszawa, PKiN, Skr.p. 22, Poland

ABSTRACT: Derivations of relational structures by applying productions are considered. The corresponding concepts are taken from the existing algebraic (category theory) approach to transformations of relational structures. An idea of programs for defining sets of derivations is introduced. A number of programming constructs is introduced to specify applications of productions in a controlled way. These constructs are given together with two mathematical semantics which allow a rigorous reasoning about programs.

Keywords: Relational structure, Production, Derivation, Quasiderivation, Program, Derivation behaviour, Configuration lattice, Trace, Trace behaviour.

CONTENTS

0. INTRODUCTION

Rewriting systems for deriving relational structures have been presented in a number of works (cf. [R] and [EKMRW], for example). In some cases these systems appeared more natural than rewriting systems of other types (like string or graph rewriting systems) to represent and describe manipulations of complex objects (cf. [MW2] and [MNT], for example). In this paper we are interested in definability of such manipulations.

In general, rewriting systems, even with highly context dependent productions, do not allow to control applications of productions in order to derive only objects with wanted properties. For example, there are no means of enforcing that some productions apply only after all the possible applications of some other productions. In this situation one may need a programmed way of applying productions.

In this paper we introduce programs of deriving relational structures by applying

* Research supported in part by Esprit Basic Research Working Group No. 3299.

productions. These programs are similar to those in [MW2] but with a richer variety of constructs and more adequate semantics. They are built by combining productions with the aid of programming constructs as sequential and parallel compositions, choice, and fixed-point operators. Programs may contain variables for which some concrete objects may be substituted.

A program represents a function from valuations of its variables to sets of derivations (one semantics) or to behaviours in the form of sets of traces (another semantics). Such a function can be obtained in a compositional way by combining the functions represented by program components.

The semantics in terms of sets of derivations is easier to understand and to deal with but it is based on a forced sequentialization of derivations which not necessarily require an order of application of productions. The semantics in terms of sets of traces is more complicated but it reflects precisely the relevant order of application of productions. In particular, it informs which applications may be concurrent.

1. DERIVATIONS OF RELATIONAL STRUCTURES

In this section we recall the concepts of a relational structure, a production, and a derivation, and we define some relations and operations on derivations.

We consider relational structures as in [EKMRW]. A relational structure is a set consisting of atoms and atomic formulas built by means of atoms and predicates. Predicates are given by a signature.

1.1. Definition. By a signature we mean Δ=(predicates$_\Delta$, arity$_\Delta$), where predicates$_\Delta$ is a set (of predicates) and arity$_\Delta$: predicates$_\Delta \rightarrow \omega$ (= {0,1, ...}) is a mapping (an arity function which assigns an arity to each predicate). ♦

1.2. Definition. By an atomic formula (abbr. : formula) of signature Δ we mean a sequence $\phi = (p, x_1, ..., x_n)$, where $p \in$ predicates$_\Delta$, $n =$ arity$_\Delta$(p), and $x_1, ..., x_n$ are arbitrary (not necessarily different) objects. We write such a formula as $p(x_1, ..., x_n)$ and call $x_1, ..., x_n$ atoms of ϕ. By atoms(ϕ) we denote the set of atoms of ϕ. Given a substitution h of some objects for the atoms of ϕ (and possibly for some other atoms), we write $h(x_1), ..., h(x_n)$ as $hx_1, ..., hx_n$, respectively, and by $h\phi$ we denote the formula $p(hx_1, ..., hx_n)$. Given a set S of formulas and possibly some other objects, and a substitution h of some objects for the objects occurring in S and in formulas of S, by hS we denote the set $\{hs : s \in S\}$. ♦

1.3. Definition. Given a signature Δ and a set X which does not contain any formula of signature Δ, by form(Δ,X) we denote the set of formulas of signature Δ with atoms in X, and by a relational structure (abbr. : r-structure) over Δ and X we mean any subset $S \subseteq X \cup$ form(Δ,X) such that atoms(ϕ)$\subseteq$S for each formula $\phi \in$S. Atoms of the formulas of S are called atoms of S. By atoms(S) and formulas(S) we denote the set of atoms and the set of formulas of S, respectively. By rstr(Δ,X) we denote the set of r-structures over Δ and X. ♦

1.4. Example. Trees with labelled edges can be represented as r-structures with nodes represented by atoms and labelled edges represented by formulas, each formula constructed from atoms representing the nodes of an edge and a predicate representing the corresponding label. In fig.1.1 we show a tree represented by an r-structure S={x,y,z,h(x,y),a(x,z)}. ♦

We consider productions which in the algebraic approach of [EKMRW] have the form p = (L⊇K⊆R) with the left-hand side L, the right-hand side R, and the gluing part K being r-structures. We assume K = L∩R and define a production by giving only its left- and right-hand sides.

1.5. Definition. A production for rstr(Δ,X) is π=(left(π),right(π)), where left (π) and right(π) are finite r-structures from rstr(Δ,X), called the left- and the right-hand side of π, respectively, such that left(π)∪right(π)≠∅. By prd(Δ,X) we denote the set of productions for rstr(Δ,X). ♦

1.6. Example. The pair π=({x,y,g(x,y)},{x,y,z,h(x,y),a(x,z)}) is a production with left(π)={x,y,g(x,y)} and right(π)={x,y,z,h(x,y),a(x,z)}. As is customary, we write π as

x,y,g(x,y) → x,y,z,h(x,y),a(x,z).

If viewed as a pair of trees with labelled edges as in 1.4 such a production can be also interpreted as a rule for edge replacement (see fig. 1.2). ♦

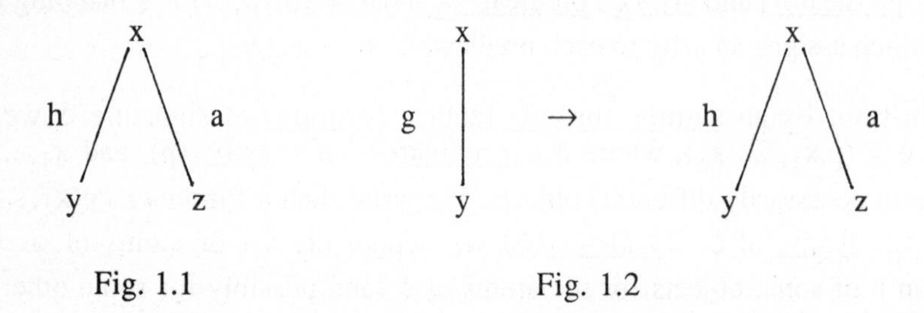

Fig. 1.1   Fig. 1.2

We consider derivations of r-structures by applying productions. These derivations consist of steps, or direct derivations, which in the algebraic approach are said to be injective and binatural (cf. [EKMRW], [MW1]). This reduces applications of productions to finding an occurrence of the left-hand side and replacing it by a suitable occurrence of the right-hand side. Such a pair of occurrences can be given by what we call an instance of a production.

1.7. Definition. An instance of a production π∈prd(Δ,X) is I=(prod(I),sub(I)), where prod(I)=π and sub(I) is a one-to-one substitution of some atoms from X for the atoms of the left- and right-hand sides of π. ♦

1.8. Definition. We say that an instance I of a production from prd(Δ,X) applies to an r-structure S$\in$ rstr(Δ,X) with an r-structure S'$\in$ rstr(Δ,X) as a result iff the following conditions are satisfied:

(1) sub(I)(left(prod(I))) $\subseteq$ S,

(2) D = (S-sub(I)(left(prod(I)))) $\cup$ sub(I)(left(prod(I)) $\cap$ right(prod(I))) is an r-structure,

(3) S'=D $\cup$ sub(I)(right(prod(I))). ♦

1.9. Definition. A direct derivation of S'$\in$ rstr(Δ,X) from S$\in$ rstr(Δ,X) via a production π from $\Pi \subseteq$ prd(Δ,X) is a triple $S \Rightarrow_I S'$, where I is an instance of π which applies to S with S' as the result. Any sequence $S_0 \Rightarrow_{I_1} S_1 \dots \Rightarrow_{I_m} S_m$ of direct derivations with m>0 is said to be a derivation of S_m from S_0 using productions from Π. ♦

1.10. Example. Consider the productions π = (x,y,g(x,y) $\rightarrow$ x,y,z,h(x,y),a(x,z)), π' = (x,y,f(x,y) $\rightarrow$ x,y,g(x,y)), π'' = (x,y,e(x,y) $\rightarrow$ x,y,f(x,y)), and the substitutions σ = [2/x,3/y,5/z] , σ' = [2/x,3/y], σ'' = [3/x,4/y]. The pairs $I_1 = (\pi',\sigma')$, $I_2 = (\pi'',\sigma'')$, $I_3 = (\pi,\sigma)$ are instances of π'',π',π, respectively. The instance I_1 applies to the structure S_0 = {1,2,3,4,b(1,2),f(2,3), e(3,4)} with S_1 = {1,2,3,4,b(1,2),g(2,3),e(3,4)} as a result, I_2 applies to S_1 with S_2 = {1,2,3,4,b(1,2),g(2,3),f(3,4)} as a result, and I_3 applies to S_2 with S_3 = {1,2,3,4,5,b(1,2),h(2,3),a(2,5),f(3,4)} as a result. In this manner we obtain a derivation $S_0 \Rightarrow_{I_1} S_1 \Rightarrow_{I_2} S_2 \Rightarrow_{I_3} S_3$. ♦

As we want to deal with programs which define sets of derivations, we need suitable operations on such sets. A standard way of introducing such operations would be to define them in terms of relations and operations on derivations. As for derivations there are no such natural relations and operations, we consider sequences of instances of productions of derivations, rather than derivations themselves, and use relations and operations on such sequences.

1.11. Definition. A sequence d = $(I_1, I_2, \dots, I_m)$ of instances of productions from Π is said to be a quasiderivation iff there exist r-structures $S_0, S_1, \dots, S_m \in$ rstr(Δ,X) such that $S_0 \Rightarrow_{I_1} S_1 \dots \Rightarrow_{I_m} S_m$ is a derivation. In this case we say that d applies to S_0 with S_m as a result. For convenience we consider also the empty quasiderivation $\varnothing$ meaning that it applies to each S$\in$ rstr(Δ,X) with S as a result. The set of quasiderivations is denoted by qder(Δ,X,Π). ♦

1.12. Definition. Given a quasiderivation d = $(I_1, I_2, \dots, I_m) \in$ qder(Δ,X,Π), we define dt(d), the data of d, as the set of $\phi \in X \cup$ form(Δ,X) such there exists a derivation $S_0 \Rightarrow_{I_1} S_1 \dots \Rightarrow_{I_m} S_m$ and i$\in$ {0, ..., m-1} such that $\phi \in (S_0 \cap \dots \cap S_i))$ $\cap$sub(I_{i+1})(left(prod(I_{i+1}))), and we define res(d), the result of d, as the set of $\phi \in X \cup$ form(Δ,X) such that there exists a derivation $S_0 \Rightarrow_{I_1} S_1 \dots \Rightarrow_{I_m} S_m$ and i$\in$ {0, ..., m-1} such that $\phi \in (S_{i+1} \cap ... \cap S_m) \cap$ sub(I_i)(right(prod(I_i))). For the empty quasiderivation $\varnothing$ we assume dt($\varnothing$)=res($\varnothing$)= $\varnothing$. ♦

Intuitively, a quasiderivation is the information about a derivation which is independent of a concrete context. The data (resp.: result) corresponds to the relevant part of the initial (resp.: resulting) r-structure. For example, for the derivation in 1.10 we have the quasiderivation d=(I_1,I_2,I_3) with dt(d)={2,3,4,f(2,3),e(3,4)} and res(d)= {2,3,4,5,h(2,3),a(2,5),f(3,4)}.

1.13. Proposition. A quasiderivation d = (I_1, I_2, ..., I_m) $\in$ qder(Δ,X,Π) applies to a structure S_0 with S_m as a result iff dt(d)$\subseteq S_0$, res(d)$\subseteq S_m$, and S_0-dt(d)=S_m-res(d).
Proof. Follows directly from the definition. ♦

1.14. Definition. Given a quasiderivation d = (J_1, ..., J_n) $\in$ qder(Δ,X,Π) we say that d is an interleaving of quasiderivations b = (H_1, ..., H_k) $\in$ qder(Δ,X,Π) and c = (I_1, ..., I_m) $\in$ qder(Δ,X,Π) iff there exists a decomposition of the sequence 1, ...,n into two subsequences f(1)< ...<f(k) and g(1)< ...<g(m), called the way of decomposing d into b and c, such that b = ($J_{f(1)}$, ..., $J_{f(k)}$) and c = ($J_{g(1)}$, ..., $J_{g(m)}$). If f(k)<g(1) then d becomes the usual concatenation of b and c. ♦

1.15. Example. The sequence d=(I_1,I_2,I_3) of instances of productions from 1.10 is a quasiderivation. Moreover the sequences b=(I_1,I_2) and c=(I_3) are quasiderivations (actually, there exist derivations $Q_0 \Rightarrow_{I_1} Q_1 \Rightarrow_{I_3} Q_2$ with Q_0={2,3,4,f(2,3),e(3,4),f(3,4)}, Q_1={2,3,4,g(2,3),e(3,4),f(3,4)}, Q_2={2,3,4,5,h(2,3),a(2,5),e(3,4),f(3,4)}, and $R_0 \Rightarrow_{I_1} R_1$ with R_0={2,3,4,g(2,3),e(3,4)}, R_1={2,3,4,g(2,3),f(3,4)}). Hence, d is an interleaving of b and c. ♦

1.16. Proposition. Given a one-to-one mapping r:X $\rightarrow$ X and a quasiderivation d=(I_1, ..., I_m) $\in$ qder(Δ,X,Π) the sequence rd = (rI_1, ..., rI_m) where rI_i denotes the instance of the production prod(I_i) with sub r(I_i), defined as rsub(I_i), is a quasiderivation in qder(Δ,X,Π).
Proof. If $S_0 \Rightarrow_{I_1} S_1 ... \Rightarrow_{I_m} S_m$ is a derivation, then also $rS_0 \Rightarrow_{I_1} rS_1 ... \Rightarrow_{I_m} rS_m$ is a derivation. ♦

The following property of members of a set of quasiderivations to be terminal allows us to define a concatenation of sets of quasiderivations.

1.17. Definition Given a nonempty set D$\subseteq$qder(Δ,X,Π), a quasiderivation d$\in$D is said to be terminal in D iff res(d)$\cap$dt(e)=$\varnothing$ for each quasiderivation e$\in$qder(Δ,X,Π) such that the concatenation of d and e is a member of D. ♦

The following definition allows us to combine sets of quasiderivations.

1.18. Definition Given nonempty sets $D_0, D_1, \ldots, D \subseteq \underline{qder}(\Delta,X,\Pi)$ and a one-to-one mapping $r:X \to X$, we define the following sets of quasiderivations:
(1) $D_0 \cup D_1$,
(2) $D_0 \| D_1 = \{d \in \underline{qder}(\Delta,X,\Pi) : d$ is an interleaving of some $d_0 \in D_0$ and $d_1 \in D_1\}$,
(3) $D_0 \cdot D_1 = \{d \in \underline{qder}(\Delta,X,\Pi) : d$ is a concatenation of a terminal $d_0 \in D_0$ and some $d_1 \in D_1\}$,
(4) $rD = \{d' \in \underline{qder}(\Delta,X,\Pi) : d' = rd$ for some $d \in D\}$. ♦

Note that the operation $(D_0, D_1) \mapsto D_0 \cdot D_1$ is continuous (w.r.t. the c.p.o. of set inclusion) only w.r.t. the second argument.

2. PROGRAMS

In some situations we would like to derive structures from structures by applying productions in a programmed way. With a typical situation of this type we have to do when using rewriting rules as operations of data base management systems. This case has been discussed in datails in [MW2]. Here we illustrate the idea of programmed derivations by showing that with such derivations one can simulate parallel rewriting systems. Consider as an example a step of a parallel rewriting system, the PDOL system "Callithamnion Roseum" (cf.[M]) with productions: $a \to bc, b \to b, c \to bd, d \to ed, e \to f, f \to g, g \to h(a), h \to h$. Interpreting strings with substrings in brackets as trees and representing such trees by r-structures as in 1.4 we may write these productions as:

$x,y,a(x,y) \to x,y,z,b(x,z),c(z,y)$ $\quad$ $x,y,c(x,y) \to x,y,z,b(x,z),d(z,y)$
$x,y,d(x,y) \to x,y,z,e(x,z),d(z,y)$ $\quad$ $x,y,g(x,y) \to x,y,z,h(x,y),a(x,z)$
$x,y,b(x,y) \to x,y,b(x,y)$ $\quad$ $x,y,e(x,y) \to x,y,f(x,y)$
$x,y,f(x,y) \to x,y,g(x,y)$ $\quad$ $x,y,h(x,y) \to x,y,h(x,y)$

A step of parallel rewriting with these productions can be realized in the framework of the sequential rewriting of section 1 by applying first, as long as possible, the following modified productions:

$\pi_1(x,y,z) = (x,y,a(x,y) \to x,y,z,\tilde{b}(x,z),\tilde{c}(z,y))$, $\pi_2(x,y,z) = (x,y,c(x,y) \to x,y,z,\tilde{b}(x,z), \tilde{d}(z,y))$,

$\pi_3(x,y,z) = (x,y,d(x,y) \to x,y,z,\tilde{e}(x,z),\tilde{d}(z,y))$, $\pi_4(x,y,z) = (x,y,g(x,y) \to x,y,z,\tilde{h}(x,y),\tilde{a}(x,z))$,

$\pi_5(x,y) = (x,y,b(x,y) \to x,y,\tilde{b}(x,y))$, $\quad$ $\pi_6(x,y) = (x,y,e(x,y) \to x,y,\tilde{f}(x,y))$,

$\pi_7(x,y) = (x,y,f(x,y) \to x,y,\tilde{g}(x,y))$, $\quad$ $\pi_8(x,y) = (x,y,h(x,y) \to x,y,\tilde{h}(x,y))$,

and by applying next, as long as possible, the following productions:

$\pi_9(x,y) = (x,y,\tilde{a}(x,y) \to x,y,a(x,y))$, $\quad$ $\pi_{10}(x,y) = (x,y,\tilde{b}(x,y) \to x,y,b(x,y))$,

$\pi_{11}(x,y) = (x,y,\tilde{c}(x,y) \to x,y,c(x,y))$, $\quad$ $\pi_{12}(x,y) = (x,y,\tilde{d}(x,y) \to x,y,d(x,y))$,

$\pi_{13}(x,y) = (x,y,\tilde{e}(x,y) \to x,y,e(x,y))$, $\quad$ $\pi_{14}(x,y) = (x,y,\tilde{f}(x,y) \to x,y,f(x,y))$,
$\pi_{15}(x,y) = (x,y,\tilde{g}(x,y) \to x,y,g(x,y))$, $\quad$ $\pi_{16}(x,y) = (x,y,\tilde{h}(x,y) \to x,y,h(x,y))$.

For each particular instance I of a production $\pi(x,y,\ldots)$ with a substitution $\underline{sub}(I)=[a/x, b/y,\ldots]$ we have a one-element set $[[\pi(x,y,\ldots)]](a,b,\ldots)$ of quasiderivations, namely the set consisting of the quasiderivation (I). The set of quasiderivations corresponding to all the possible instances of $\pi(x,y,...)$ can be defined as

$$[[\bigvee_{x,y,...}\pi(x,y,...)]] = \bigcup ([[\pi(x,y,...)]](a,b,...) \text{ for all } [a/x,b/y,...]).$$

The set of quasiderivations corresponding to all the possible instances of productions $\pi_1(x,y, ...)$ - $\pi_8(x,y, ...)$ can be defined as

$$[[\bigvee_{x,y,z}\pi_1(x,y,z)\vee ... \vee\bigvee_{x,y}\pi_8(x,y)]] = [[\bigvee_{x,y,z}\pi_1(x,y,z)]] \cup ... \cup[[\bigvee_{x,y}\pi_8(x,y)]].$$

The set of all possible interleavings of quasiderivations corresponding to instances of productions $\pi_1(x,y,z)$ - $\pi_8(x,y)$ can be defined as

$[[\mathrm{fix}_\xi(\xi \,\&(\bigvee_{x,y,z}\pi_1(x,y,z)\vee ... \vee\bigvee_{x,y}\pi_8(x,y)))]]$, where $\mathrm{fix}_\xi(\xi\&(\bigvee_{x,y,z}\pi_1(x,y,z)\vee ... \vee\bigvee_{x,y}\pi_8(x,y)))$ denotes the least set x of quasiderivations such that

$\xi = \xi \,\|\, [[\bigvee_{x,y,z}\pi_1(x,y,z)\vee ... \vee\bigvee_{x,y}\pi_8(x,y))]]$.

Similarly, the set of all possible interleavings of quasiderivations corresponding to instances of productions $\pi_9(x,y)$ - $\pi_{16}(x,y)$ can be defined as

$$[[\mathrm{fix}_\xi(\xi \,\&(\bigvee_{x,y}\pi_9(x,y)\vee ... \vee\bigvee_{x,y}\pi_{16}(x,y)))]].$$

Finally, the set of all possible quasiderivations which can be obtained by concatenating terminal quasiderivations corresponding to instances of productions $\pi_1(x,y,z)$ - $\pi_8(x,y)$ with quasiderivations corresponding to instances $\pi_9(x,y)$ - $\pi_{16}(x,y)$ can be defined as

$$[[\mathrm{fix}_\xi(\xi\&(\bigvee_{x,y,z}\pi_1(x,y,z)\vee...\vee\bigvee_{x,y}\pi_8(x,y)));\mathrm{fix}_\xi(\xi\&(\bigvee_{x,y}\pi_9(x,y)\vee...\vee\bigvee_{x,y}\pi_{16}(x,y)))]]$$

$$=[[\mathrm{fix}_\xi(\xi\&(\bigvee_{x,y,z}\pi_1(x,y,z)\vee...\vee\bigvee_{x,y}\pi_8(x,y)))]]\cdot[\,[[\mathrm{fix}_\xi(\xi\&(\bigvee_{x,y}\pi_9(x,y)\vee...\vee\bigvee_{x,y}\pi_{16}(x,y)))]].$$

The set of quasiderivations thus defined represents exactly the step of parallel rewriting as required, in the sense that each terminal quasiderivation in this set applies to the r-structure representing a string with a result as for one step in the original parallel rewriting system.

The set of quasiderivations which we have been defining in the form $[[e(x,y,...)]](v)$ can be viewed as meanings of the corresponding expressions $e(x,y,...)$ with variables x,y,... for the corresponding valuation v of variables. The expressions themselves can be viewed as programs for deriving r-structures by applying productions. Such programs and their semantic meanings can be defined formally as follows.

Let Δ,X,Π be as in section 1, and let X be a disjoint union of two sets, a set V of object variables, and a set C of constants. Let W, where $W\cap X=\emptyset$ be a set of behaviour variables.

2.1. Definition. A relational program (abbr.: r-program, or program) over Δ,V,C,W,Π is an expression which can be defined inductively as follows:

(1) each production $\pi \in \Pi \subseteq \underline{prd}(\Delta,X,)$ with atoms x,y,... from V is a program with objects variables x,y, ... ,
(2) each behaviour variable $\xi \in W$ is a program with the behaviour variable ξ,
(3) given a program $\pi_1(x_1,y_1, ...)$ with (object or behaviour) variables $x_1,y_1,...$ and a program $\pi_2(x_2,y_2, ...)$ with (object or behaviour) variables $x_2,y_2,...$,
$\pi_1(x_1,y_1,...) \vee \pi_2(x_2,y_2,...)$, $\pi_1(x_1,y_1,...) \,\&\, \pi_2(x_2,y_2,...)$,
$\pi_1(x_1,y_1,...) \,;\, \pi_2(x_2,y_2,...)$ are programs with variables $x_1,y_1,...,x_2,y_2,...$,
(4) given a one-to-one mapping $r : X \to X$ with $r(C) \subseteq C$ and a program $\pi(x,y,...)$ with object variables x,y,..., $(r\pi)(u,v,...) = \pi(r(x),r(y),...)$, where $\{u,v,...\} = V \cap r(\{x,y,...\})$, is a program with object variables u,v,...,
(5) given a program $\pi(x,y,...)$ with object variables x,y,..., $\bigvee_x \pi(x,y,...)$ is a program with object variables y,...; similarly, for $\bigvee_{x,y} \pi(x,y,...)$, etc.,
(6) given a program $\pi(\xi,\eta,...)$ with behaviour variables $\xi,\eta,...$, such that ξ does not occur in the first component of a subexpression of $\pi(\xi,\eta,...)$ of the form $e_1; e_2$, $fix_\xi \pi(\xi,\eta,...)$ is a program with behaviour variables $\eta,...$; similarly for $fix_{\xi,\eta} \pi(\xi,\eta,...)$, etc..

The set of programs over Δ,V,C,W,Π is denoted by $\underline{rprog}(\Delta,V,C,W,\Pi)$. ♦

2.2. Definition. The <u>meaning</u> of a program $\pi(x,y,...) \in \underline{rprog}(\Delta,V,C,W,\Pi)$ with (object or behaviour) variables x,y,... is a function $[[\pi(x,y,...)]]$ from $\underline{val}_{x,y,...}$, the set of possible valuations of x,y,..., to $\underline{powerset}(\underline{qder}(\Delta,X,\Pi))$, the set of sets of quasiderivations, where:
(1) $[[\pi(x,y,...)]](v) = \{(I)\}$ with $\underline{prod}(I) = \pi(x,y,...)$, and $\underline{sub}(I) = v$ for each production $\pi(x,y,...)$ with object variables x,y,..., and each $v \in \underline{val}_{x,y,...}$,
(2) $[[\xi]](v) = v(\xi)$ for each behaviour variable ξ and its valuation v,
(3) $[[\pi_1(x_1,y_1,...) \vee \pi_2(x_2,y_2,...)]](v) = [[\pi_1(x_1,y_1,...)]](v) \cup [[\pi_2(x_2,y_2,...)]](v)$,
$[[\pi_1(x_1,y_1,...) \,\&\, \pi_2(x_2,y_2,...)]](v) = [[\pi_1(x_1,y_1,...)]](v) \,\|\, [[\pi_2(x_2,y_2,...)]](v)$,
$[[\pi_1(x_1,y_1,...) \,;\, \pi_2(x_2,y_2,...)]](v) = [[\pi_1(x_1,y_1,...)]](v) \cdot [[\pi_2(x_2,y_2,...)]](v)$,
for all programs $\pi_1(x_1,y_1,...)$, $\pi_2(x_2,y_2,...)$, and valuations v of $x_1,y_1,...,x_2,y_2,...$,
(4) $[[r\pi(u,v,...)]](t) = [[\pi(x,y,...)]](r\,t)$, where r t denotes the superposition of r and t, for each one-to-one mapping $r : X \to X$ with $r(C) \subseteq C$ and each program $\pi(x,y, ...)$ with object variables x,y, ...with $V \cap r(\{x,y,...\}) = \{u,v,...\}$, and each valuation t of variables of $r\pi(u,v,...)$,
(5) $[[\bigvee_x \pi(x,y,...)]](v) = \bigcup([[\pi(x,y,...)]](v_c) : c \in C)$ where v_c is the function $v \cup \{(x,c)\}$, for each program $\pi(x,y,...)$ with object variables x,y,..., is a program with object variables x,y,..., and each valuation v of variables of $\bigvee_x \pi(x,y,...)$; similarly for $\bigvee_{x,y} \pi(x,y,...)$, etc.,

(6) $[[fix_{\xi}\pi(\xi,\eta,...)]](v)$ is the least set S such that $S = [[\pi(\xi,\eta,...)]](v_S)$, where $v_S = v \cup \{(\xi,S)\}$, for each program $\pi(\xi,\eta,...)$ with behaviour variables $\xi,\eta,...$ as in (6) of 2.1 and each valuation v of variables of $fix_{\xi}\pi(\xi,\eta,....)$; similarly for $fix_{\xi,\eta}\pi(\xi,\eta,...)$, etc.. ♦

3. BEHAVIOURS OF PROGRAMS

Programs of deriving relational structures have been introduced as definitions of sets of quasiderivations. Now we want to look at programs in a manner which could be a guideline for their possibly concurrent implementation. To this end we give the programs an additional semantics. In this semantics the meaning of a program is a behaviour which can be represented by a set of traces, where traces are isomorphism classes of structures similar to finite conflict-free labelled event structures in the sense of [W]. The formal definitions are as follows.

3.1. Definition. Given a set U, by a labelled set over U we mean a graph of an U-valued function, that is a set E of pairs (x,u) such that $u \in U$ and the relations $(x,u) \in E$ and $(x,v) \in E$ imply u=v. For each member e=(x,u) of E we write u as label(e).♦

3.2. Definition. Given a set U, by a configuration lattice (abbr.: c-lattice) over U we mean a nonempty finite set P of labelled sets over U that is closed w.r.t. intersections and unions of its subsets. Labelled sets $p \in P$ are called configurations of P. Members of such sets are called nodes of P. Given a configuration $p \in P$, each configuration $p' \in P$ such that $p' \subseteq p$ is called a subconfiguration of p. Given two nodes e,f of P, we say that e precedes f (resp.: e coincides with f) iff, for all configurations $p \in P$, the condition $f \in p$ implies (resp.: is equivalent to) the condition $e \in p$. ♦

Configuration lattices may be interpreted as activities with nodes standing for events and configurations standing for states of progress, where an event is a particular execution (occurrence) of an elementary action. More precisely, a node e=(x,u) stands for an execution (occurrence) denoted by x of an elementary action denoted by label(e)=u. In the case of activities of deriving relational structures we consider elementary actions of accepting atoms and formulas from an external world (written as accept ϕ) elementary actions of memorizing atoms and formulas which are produced (written as save ϕ), elementary actions of offering the memorized atoms and formulas to the external world (written as offer ϕ), and elementary actions of passing intermediate results from one step to another (written as transfer ϕ). The latter ones are considered as combinations of the respective complementary actions offer ϕ and accept ϕ in a sense like in CCS.

3.3. Example. The instance I_1 of the production $\pi_1(x,y,z)=(x,y,a(x,y) \rightarrow x,y,z, \tilde{b}(x,z),\tilde{c}(z,y))$ (cf. section 2) with $sub(I_1)=[1/x,2/y,3/z]$ can be regarded as an activity which consists of an indivisible step of accepting 1,2,a(1,2) and producing 1,2,3, $\tilde{b}$(1,3),$\tilde{c}$(3,2), followed by independent events of offering 1,2,3,$\tilde{b}$(1,3),$\tilde{c}$(3,2). We can represent such an activity by the c-lattice which consists of the labelled set defined by p(1) = accept 1, p(2) = accept 2, p(3) = accept a(1,2), p(4) = save 1, p(5) = save 2, p(6) = save 3, p(7) = save $\tilde{b}$(1,3), p(8) = save $\tilde{c}$(3,2), p(9) = offer 1, p(10) = offer 2, p(11) = offer 3, p(12) = offer $\tilde{b}$(1,3), p(13) = offer $\tilde{c}$(3,2), of the restrictions of p to the subsets containing {1,2,3,4,5,6,7,8}, and of ∅ (the empty set). Similarly for the instance I_2 of the production $\pi_{10}(x,y) = (x,y,\tilde{b}(x,y) \rightarrow x,y,b(x,y))$ with sub(I_2)=[1/x,3/y]. The respective c-lattices are shown in fig. 3.1, where each event is represented by an occurrence of the respective label in the figure, and where we omit the configurations resulting from the assumptions in 3.2. ♦

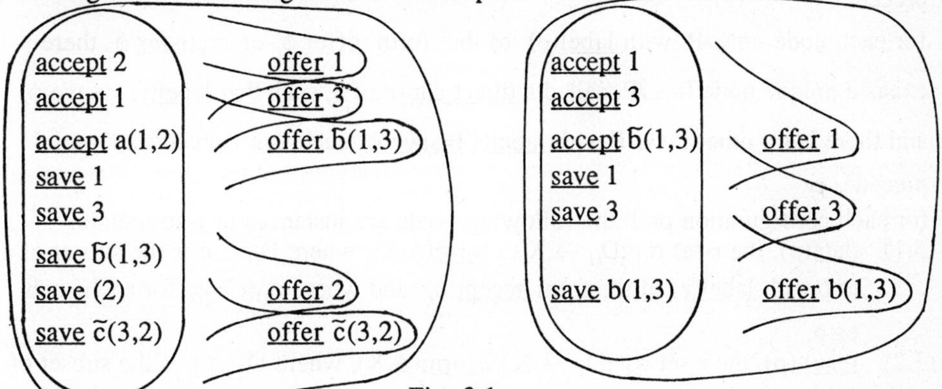

Fig. 3.1

3.4. Example. The quasiderivation (I_1,I_2), where I_1 and I_2 are as in 3.3, can be regarded as an activity consisting of the activity corresponding to I_1 followed by the activity corresponding to I_2, the latter with atoms 1,3 and formula $\tilde{b}$(1,3) obtained from the activity corresponding to I_1 (actions transfer 1, transfer 3 and transfer $\tilde{b}$(1,3)). We can represent such an activity by a c-lattice as shown in fig.3.2. ♦

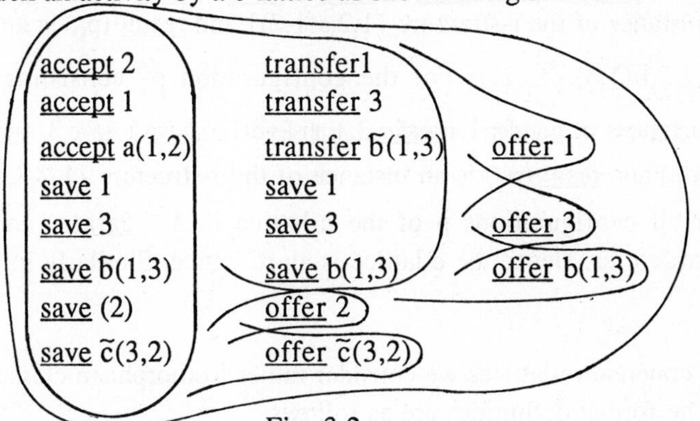

Fig. 3.2

In general, activities corresponding to quasiderivations can be represented by special c-lattices called relational c-lattices. Such c-lattices are defined as follows.

Let Δ be a signature and X a set of atoms as in 1.3.

3.5. Definition. Given a relational structure $S \in$ rstr(Δ,X), by an instance of S we mean a labelled set $\sigma : D \rightarrow X \cup$ form(Δ,X) such that σ is injective and $\sigma(D)=S$. ♦

3.6. Definition. By a relational c-lattice (abbr.: rc-lattice) we mean a c-lattice P over the set of labels of the form offer ϕ or accept ϕ or transfer ϕ or save ϕ, where $\phi \in X \cup$ form(Δ,X), such that:

(1) for each node $e \in \bigcup P$ with label(e) of the form offer ϕ_e there is no $f \in \bigcup P$ such that e precedes f,

(2) for each node $e \in \bigcup P$ with label(e) of the form offer ϕ_e or transfer ϕ_e there exists a unique node $f \in \bigcup P$, called a direct cause of e, such that label(f) = save ϕ_e and there is no other $f' \in \bigcup P$ with label(f')=save ϕ_e such that f precedes f' and f' precedes e,

(3) for each configuration $p \in P$ the following r-sets are instances of r-structures:

  (3.1) data(p), the r-set $\sigma_0 : D_0 \rightarrow X \cup$ form(Δ,X), where $D_0 \subseteq p$ is the subset of all $e \in p$ with label(e) of the form accept ϕ_e, and where $\sigma_0(e)=\phi_e$ for each such $e \in p$,

  (3.2) result(p), the r-set $\sigma_1 : D_1 \rightarrow X \cup$ form(Δ,X), where $D_1 \subseteq p$ is the subset of all $e \in p$ with label(e) of the form save ϕ_e and are not direct causes of any $f \in \bigcup P$ which has label(f)=transfer ϕ_e, and where $\sigma_1(e) = \phi_e$ for each such $e \in p$.

The universe of rc-lattices is denoted by rcl(Δ,X). ♦

3.7. Example. For the configuration p_0 consisting of the coincident occurrences of accept 1, accept 2, accept a(1,2), save 1, save 2, save 3, save $\breve{b}$(1,3), save $\tilde{c}$(3,2) in 3.4 data(p_0) is an instance of the r-structure {1,2,a(1,2)} and result (p_0) is an instance of the r-structure {1,2,3,$\breve{b}$(1,3),$\tilde{c}$(3,2)}. For the configuration p_1 consisting of p_0 and the coincident occurrences of transfer 1, transfer 3, transfer$\breve{b}$(1,3), save 1, save 3, save b(1,3) we have data(p_1)=data(p_0) and result(p_1) is an instance of the r-structure {1,2,3,$\tilde{c}$(3,2), b(1,3)}. In general, for all configurations p of the c-lattice in 3.4 data(p) and result(p) are instances of r-structures. Hence the c-lattice is an rc-lattice. Similarly for the c-lattice in 3.3. ♦

Instead of concrete rc-lattices we consider rather isomorphism classes of rc-lattices, called traces. The formal definitions are as follows.

3.8. Definition. Given two rc-lattices $P,Q \in \underline{rcl}(\Delta,X)$, we say that P and Q are isomorphic iff there exists a bijection $b: \cup P \rightarrow \cup Q$ (an isomorphism from P to Q) such that, for all $e,f \in \cup P$, $p \in P$, and $q \in Q$, the relation $f=b(e)$ implies $\underline{label}(f)=\underline{label}(e)$, $b(p) \in Q$, and $b^{-1}(q) \in P$. ♦

3.9. Definition. By a relational configuration trace (abbr.: rc-trace, or simply trace) we mean an isomorphism class of rc-lattices from $\underline{rcl}(\Delta,X)$. The trace which is the isomorphism class of an rc-lattice P is written as [P]. By $\varnothing$ we denote the trace $[\{\varnothing\}]$. By $\underline{rct}(\Delta,X)$ we denote the set of rc-traces.♦

For traces we have the following relations and operations which help us to define and construct behaviours of relational programs.

3.10. Definition. Given three traces t_0, t_1, $t \in \underline{rct}(\Delta,X)$, we say that t *consists* of t_0 and t_1 iff there exist rc-lattices P_0, P_1, P such that $t_0=[P_0]$, $t_1=[P_1]$, $t=[P]$ and a one-to-one correspondence $\alpha \subseteq (\cup P_0) \times (\cup P_1)$, called an association of P_0 with P_1, such that:

(1) for all $(e,f) \in \alpha$, $\underline{label}(e)$ and $\underline{label}(f)$ are complementary in the sense that there exists a unique $\phi_{ef} \in X \cup \underline{form}(\Delta,X)$ such that either $\underline{label}(e) = \underline{offer}\,\phi_{ef}$ and $\underline{label}(f) = \underline{accept}\,\phi_{ef}$ or $\underline{label}(e) = \underline{accept}\,\phi_{ef}$ and $\underline{label}(f) = \underline{offer}\,\phi_{ef}$,

(2) $p \in P$ iff $p=\{(((0,e),(1,f)),\underline{transfer}\,\phi_{ef}):(e,f) \in \alpha\} \cup \{((0,e), \underline{label}\,(e)) : e \in p_0-\alpha^{-1}(p_1)\} \cup \{((1,f),\underline{label}\,(f)) : f \in p_1 -\alpha(p_0)\}$ for some $p_0 \in P_0$ and $p_1 \in P_1$ such that $\alpha(p_0) \subseteq p_1$ and $\alpha^{-1}(p_1) \subseteq p_0$,

(3) $\cup P$ does not contain any cycle $e_0,e_1,\ldots,e_{n+1}=e_0$ such that $\underline{pr}_0(e_i)$ precedes $\underline{pr}_0(e_{i+1})$ or $\underline{pr}_1(e_i)$ precedes $\underline{pr}_1(e_{i+1})$ for all $i \in \{0,\ldots,n\}$ and $\underline{pr}_0(e_j)$ does not coincide with $\underline{pr}_0(e_{j+1})$ or $\underline{pr}_1(e_j)$ does not coincide with $\underline{pr}_1(e_{j+1})$ for some $j \in \{0,\ldots,n\}$, where $\underline{pr}_0(d)=e$ for some $d \in \cup P$ of the form $((0,e), \underline{label}(e))$, $\underline{pr}_0(d)=e$ and $\underline{pr}_1(d)=f$ for some $d \in \cup P$ of the form $((0,e),(1,f), \underline{transfer}\,\phi_{ef})$, $\underline{pr}_1(d)=f$ for some $d \in \cup P$ of the form $((1,f),\underline{label}(f))$.

If α contains only such (e,f) for which $\underline{label}(e)=\underline{offer}\,\phi_{ef}$ and $\underline{label}(f)=\underline{accept}\,\phi_{ef}$ and if it contains a pair (e,f) for each $e \in \cup P_0$ such that $\underline{label}(e)=\underline{offer}\,\phi_e$ and there exists $f \in \cup P_1$ with $\underline{label}(f)=\underline{accept}\,\phi_e$, then we say that t is a concatenation of t_0 and t_1. ♦

Intuitively, each configuration $p \in P$ is a result of combining some $p_0 \in P_0$ and $p_1 \in P_1$ such that $(e,f) \in p_0 \times p_1$ whenever $(e,f) \in \alpha$ and $e \in p_0$ or $f \in p_1$ and it consists of the following three disjoint sets: a set of events corresponding to $(e,f) \in \alpha \cap (p_0 \times p_1)$, a set of events corresponding to $e \in p_0$ such that $(e,f) \notin \alpha$ for any f, and a set of events corresponding to $f \in p_1$ such that $(e,f) \notin \alpha$ for any e. The condition (3) says that events which are not coincident or independent cannot become coincident. The concatenation corresponds to the case in which each $(e,f) \in \alpha$ represents a transfer of information from p_0 to p_1.

3.11. Example. The trace corresponding to the rc-lattice in 3.4 consists (and is a concatenation) of the traces corresponding to the rc-lattices in 3.3 for the production instances I_1 and I_2 (see fig.3.1 and 3.3). ♦

3.12. Proposition. Given any one-to-one mapping $r : X \rightarrow X$ and any trace $t \in \underline{rct}(\Delta,X)$, we have a unique trace $rt \in \underline{rct}(\Delta,X)$, called the result of relabelling t according to r, where rt=[rP] for some P such that t=[P] and rP standing for the set of $q=\{(e,r(\underline{label}(e))) : e \in p\}$ with $p \in P$ and where $r(\underline{offer}\ \phi)=\underline{offer}\ r\phi$, $r(\underline{accept}\ \phi)= \underline{accept}\ r\phi$, $r(\underline{transfer}\ \phi) = \underline{transfer}\ r\phi$, $r(\underline{save}\ \phi) = \underline{save}\ r\phi$.
Proof. Straightforward. ♦

Behaviours of relational programs are represented by sets of traces, called trace behaviours. The formal definition is as follows.

3.13. Definition. By a <u>relational trace behaviour</u> (abbr. : <u>rt-behaviour</u>, or simply <u>behaviour</u>) we mean any nonempty set of rc-traces. By $\underline{rtbeh}(\Delta,X)$ we denote the set of rt-behaviours. ♦

For example the set of prefixes of traces corresponding to rc-lattices in 3.3 and 3.4 are trace behaviours.

The following property of traces of a behaviour is important for defining a concatenation of trace behaviours.

3.14. Definition. Given a behaviour $B \in \underline{rtbeh}(\Delta,X)$, a trace $t \in B$ is said to be <u>terminal</u> in B iff for each trace $s \in B$ such that t is a prefix of s in the sense that s is a concatenation of t and some $t' \in \underline{rct}(\Delta,X)$ there exist rc-lattices P, Q such that t=[P], s=[Q], P is a prefix of Q in the sense that $P \subseteq Q$ and $q \in P$ for all $q \in Q$ with $q \subseteq \cup P$, and such that $\underline{result}(\cup P) \subseteq \underline{result}(\cup Q)$.♦

The following proposition allows us to combine trace behaviours.

3.15. Proposition. Given behaviours $B_0, B_1, ..., B \in \underline{rtbeh}(\Delta,X)$ and a one-to-one mapping $r : X \rightarrow X$, we define the following behaviours in $\underline{rtbeh}(\Delta,X)$:
(1) $B_0 \cup B_1$,
(2) $B_0 \ ||| \ B_1 = \{t \in \underline{rct}(\Delta,X) :$ t consists of some $t_0 \in B_0$ and $t_1 \in B_1\}$,
(3) $B_0 \circ B_1 = \{t \in \underline{rct}(\Delta,X) :$ t is a concatenation of a terminal $t_0 \in B_0$ and some $t_1 \in B_1\}$,
(4) $rB = \{t' \in \underline{rct}(\Delta,X) : t'=rt$ for some $t \in B\}$. ♦

Note that the operation $(B_0, B_1) \mapsto B_0 \circ B_1$ is continuous (w.r.t. the c.p.o. of set inclusion) only w.r.t. the second argument.

Now we have tools for defining trace behaviours of relational programs.

Let Δ, X, V, C, W, Π as in section 2. We start with defining a behaviour for an instance of a production following the idea in 3.3.

3.16. Proposition. Let I be an instance of a production $\pi \in \Pi$ with $\underline{sub}(I) : X \to C$ and let p be the function defined as

$$p(x) = \begin{cases} \underline{accept}\ \underline{sub}(I)y & \text{for } x=(0,y) \text{ with } y \in \underline{left}(\pi) \\ \underline{save}\ \underline{sub}(I)y & \text{for } x=(1,y) \text{ with } y \in \underline{right}(\pi) \\ \underline{offer}\ \underline{sub}(I)y & \text{for } x=(2,y) \text{ with } y \in \underline{right}(\pi) \end{cases}$$

for all $x \in D(\pi)=\{0\}\times\underline{left}(\pi) \cup \{1\}\times\underline{right}(\pi) \cup \{2\}\times\underline{right}(\pi)$. Then the set of restrictions of p to the subsets of $D(\pi)$ containing $\{0\}\times\underline{left}(\pi) \cup \{1\}\times\underline{right}(\pi)$, and of the empty set, constitute an rc-lattice L(I). The trace corresponding to this rc-lattice and its prefixes constitute a trace behaviour $\underline{beh}(I) \in \underline{rtbeh}(\Delta, X)$, called the behaviour of I.

Proof. Immediate from the construction. ♦

The behaviours of arbitrary relational programs can be defined as follows.

3.17. Definition. The behaviour of a program $\pi(x,y,\ldots) \in \underline{rprog}(\Delta,V,C,W,\Pi)$ with (object or behaviour) variables x,y,... is a function $[[[\pi(x,y,\ldots)]]]$ from $\underline{val}_{x,y,\ldots}$, the set of possible valuations of x,y,..., to $\underline{rtbeh}(\Delta,X)$, the universe of rt-behaviours over Δ,X, where:

(1) $[[[\pi(x,y,\ldots)]]](v) = \underline{beh}(I)$ with $\underline{prod}(I) = \pi(x,y,\ldots)$, and $\underline{sub}(I) = v$ for each production $\pi(x,y,\ldots)$ with object variables x,y,..., and each $v \in \underline{val}_{x,y,\ldots}$,

(2) $[[\xi]](v) = v(\xi)$ for each behaviour variable ξ and its valuation v (in $\underline{rtbeh}(\Delta,X)$),

(3) $[[[\pi_1(x_1,y_1,\ldots) \vee \pi_2(x_2,y_2,\ldots)]]](v) = [[[\pi_1(x_1,y_1,\ldots)]]](v) \cup [[[\pi_2(x_2,y_2,\ldots)]]](v)$,
$[[[\pi_1(x_1,y_1,\ldots) \& \pi_2(x_2,y_2,\ldots)]]](v) = [[[\pi_1(x_1,y_1,\ldots)]]](v) ||| [[[\pi_2(x_2,y_2,\ldots)]]](v)$,
$[[[\pi_1(x_1,y_1,\ldots)\ ;\ \pi_2(x_2,y_2,\ldots)]]](v) = [[[\pi_1(x_1,y_1,\ldots)]]](v) \circ [[[\pi_2(x_2,y_2,\ldots)]]](v)$,
for all programs $\pi_1(x_1,y_1,\ldots)$, $\pi_2(x_2,y_2,\ldots)$, and valuations v of $x_1,y_1,\ldots, x_2,y_2,\ldots$,

(4) $[[[r\pi(u,v,\ldots)]]](t) = [[[\pi(x,y,\ldots)]]](r\,t)$, where r t denotes the superposition of r and t, for each one-to-one mapping $r : X \to X$ with $r(C) \subseteq C$ and each program $\pi(x,y,\ldots)$ with object variables x,y, ...with $V \cap r(\{x,y,\ldots\})=\{u,v,\ldots\}$, and each valuation t of variables of $r\pi(u,v,\ldots)$,

(5) $[[[\bigvee_x \pi(x,y,\ldots)]]](v)= \bigcup([[[\pi(x,y,\ldots)]]](v_c):c \in C)$ where $v_c = v \cup \{(x,c)\}$, for each program $\pi(x,y,\ldots)$ with object variables x,y,..., is a program with object variables x,y,... and each valuation v of variables of $\bigvee_x \pi(x,y,\ldots)$; similarly, for $\bigvee_{x,y} \pi(x,y,\ldots)$, etc.,

(6) $[\![\![\mathrm{fix}_\xi\pi(\xi,\eta,...)]\!]\!](v)$ is the least rt-behaviour $B\in$ rtbeh(Δ,X) such that

$$B = [\![\![\pi(\xi,\eta,...)]\!]\!](v_P),$$

where $v_P=v \cup \{(\xi,B)\}$, for each program $\pi(\xi,\eta,...)$ with behaviour variables $\xi,\eta,...$ as in (6) of 2.1 and each valuation v of variables of $\mathrm{fix}_\xi\pi(\xi,\eta,...)$; similarly for $\mathrm{fix}_{\xi,\eta}\pi(\xi,\eta,...)$, etc.. ♦

4. COMPATIBILITY OF SEMANTICS.

In this section we show how the trace behaviours of relational programs (definable behaviours) represent the corresponding sets of derivations.

Let Δ,X,V,C,W,Π be as in section 2.

4.1. Definition. A definable behaviour is a behaviour which can be represented as $[\![\![\pi(x,y,...)]\!]\!](v)$ for a relational program $\pi(x,y, ...)$ with (object or behaviour) variables x,y,... and a valuation v of variables x,y,... . ♦

The relationship between definable behaviours and sets of quasiderivations is as follows.

4.2 Proposition. There exists a correspondence Γ between definable behaviours and sets of quasiderivations such that :

(1) $\Gamma(\mathrm{beh}(I)) = \{(I)\}$ for each instance I of a production from Π,

(2) $\Gamma(B_0 \cup B_1 \cup ...) = \Gamma(B_0) \cup \Gamma(B_1) \cup ...$,

$\Gamma(B_0 \mid\mid\mid B_1) = \Gamma(B_0) \mid\mid \Gamma(B_1)$,

$\Gamma(B_0 \circ B_1) = \Gamma(B_0) \cdot \Gamma(B_1)$,

for all definable behaviours B_0, B_1, ...,

(3) $\Gamma(rB)$=r$\Gamma(B)$ for each one-to-one mapping $r:X\to X$ and each definable behaviour B

This correspondence is given (as the additive extension) by a correspondence Γ_0 between maximal chains of configurations of rc-lattices representing traces of definable behaviours and quasiderivations. The correspondence Γ_0 is monotonic in the sense that $\Gamma_0(c)$ is a prefix of $\Gamma_0(c')$ whenever c is a maximal chain of $\downarrow p=\{q\in P$ such that $q\sqsubseteq p\}$ for some $p\in P$ such that c' is a maximal chain of P.

Proof outline. We start with the simple observation that each definable behaviour can be obtained from the behaviour corresponding to instances of productions with the aid of union and parallel and sequential compositions. Consequently, each trace of a definable behaviour can be decomposed with the aid of the relation "to consist of" into a finite number of irreducible traces corresponding to instances of productions. From the definitions it follows that such a decomposition is unique. This allows us to construct Γ_0 by a simple induction on the number of irreducible components of a trace.

For the only maximal chain c of the rc-lattice corresponding to an instance I of a production we define $\Gamma_0(c)$ as the quasiderivation (I). Suppose that Γ_0 is defined for the maximal chains of rc-lattices representing traces with at most m irreducible components.

Suppose that P is an rc-lattice such that t=[P] consists of m+1 irreducible components. Then we have rc-lattices P_0, P_1, and an association α as in 3.10, where t_0=[P_0] is a trace with m irreducible components and P_1=beh(I) for an instance I of a production. Moreover, due to the fact that P is a trace of a definable behaviour, we can choose P_0 and P_1 such that $\cup P_0 = \mathrm{pr}_0(p)$ for some $p \in P$ which belongs to every maximal chain of P. For each maximal chain c of P the projections $\mathrm{pr}_0(q)$ with $q \in c$ constitute a maximal chain c_0 of P_0. As t_0=[P_0] has m irreducible components, we have a quasiderivation $\Gamma_0(c_0)=(I_1, ..., I_m)$. On the other hand, from the properties of the decomposition of P into P_0 and P_1 we obtain that no $\phi \in$ res((I)) belongs to res((I_1, ..., I_m)) (otherwise result ($\cup$P) would not be an r-structure). Hence (I_1, ..., I_m,I) is a quasiderivation and we may define $\Gamma_0(c)=(I_1, ...,I_m,I)$. Eventually Γ_0 can be defined for all maximal chains of rc-lattices representing traces of definable behaviours. The monotonicity of Γ_0 follows easily from the construction. Given a definable behaviour B the corresponding set $\Gamma(B)$ of quasiderivations is defined as the set of all $\Gamma_0(c)$ with c being maximal chains of rc-lattices representing traces of B. Properties (1) and (3), and $\Gamma(B_0 \cup B_1 \cup ...) = \Gamma(B_0) \cup \Gamma(B_1) \cup ...$ follow directly from the definition. For $\Gamma(B_0 ||| B_1) \subseteq \Gamma(B_0) \,||\, \Gamma(B_1)$ suppose that c is a maximal chain of an rc-lattice P such that $[P] \in B_0 ||| B_1$. We can decompose [P] into irreducible components of the form [beh(J_1)], ..., [beh(J_n)] such that [P] consists of [$\downarrow p_n$] and [beh(J_n)] for some $p_n \in c$, [$\downarrow p_n$] consists of [$\downarrow p_{n-1}$] and [beh(J_{n-1})] for some $p_{n-1} \in c$, etc.. On the other hand, we can decompose [P] into $[Q] \in B_0$ and $[R] \in B_1$. Then we can decompose [Q] into irreducible components of the form [beh(H_1)], ..., [beh(H_k)] such that [Q] consists of [$\downarrow q_k$] and [beh(H_k)] for some configuration q_k of a maximal chain c_Q of Q, [$\downarrow q_k$] consists of [$\downarrow q_{k-1}$] and[beh(H_{k-1})] for some $q_{k-1} \in c_Q$, etc., and we can decompose [R] into irreducible components [beh(I_1)], ..., [beh(I_m)] such that [R] consists of [$\downarrow r_m$] and [beh(I_m)] for some configuration r_m of a maximal chain c_R of R, [$\downarrow r_m$] consists of [$\downarrow r_{m-1}$] and [beh(I_{m-1})] for some $r_{m-1} \in c_R$, etc.. As the irreducible components of [P] are determined uniquely, we have $H_1=J_{f(1)},...,H_k=J_{f(k)}$, $I_1=J_{g(1)},...,I_m=J_{g(m)}$ for a decomposition of the sequence 1, ..., n into f(1), ..., f(k) and g(1), ..., g(m). Assuming for c_Q and c_R the respective projections of c we obtain f(1)< ...< f(k) and g(1)< ...< g(m). Hence $\Gamma_0(c)=(J_1, ..., J_n)$ is an interleaving of $\Gamma_0(c_Q)$= (H_1, ..., H_k) and $\Gamma_0(c_R)=(I_1, ..., I_m)$, as required. For $\Gamma(B_0) \,||\, \Gamma(B_1) \subseteq \Gamma(B_0 ||| B_1)$ suppose that Q, R, c_Q, c_R, q_0,..., q_k, r_0,...,r_m, H_1,...,H_k, I_1, ...,I_m are as in previous part of the proof, and that e=(J_1, ..., J_n) is an interleaving of $\Gamma_0(c_Q)=(H_1, ..., H_k)$ and $\Gamma_0(c_R)=(I_1, ..., I_m)$ with decomposition of 1, ..., n into f(1)< ...<f(k) and g(1)< ... < g(m). We have to find the corresponding $[P] \in B_0 ||| B_1$ and c. For e we have a derivation $T_0 \Rightarrow_{J_1} T_1 ... \Rightarrow_{J_n} T_n$. Consider $j \in \{1, ..., n-1\}$ such that $T_{j-1} \Rightarrow_{J_{f(h)}} T_j \Rightarrow_{J_{g(i)}} T_{j+1}$ with j=f(h) and j+1=g(i) and $\phi \in \mathrm{res}((H_h)) \cap \mathrm{dt}((I_i))$. As e is a quasiderivation, we have

$\phi \in$ res$(\Gamma_0(c_Q)) \cap$ dt$(\Gamma_0(c_R))$. Hence, from the construction of Γ_0 it follows that there exists $e' \in q_{h+1}$ which is a direct cause of some $e_Q \in \bigcup Q$ with label(e_Q)=offer ϕ and $f_R \in r_{i+1}$ with label(f_R)=accept ϕ. Similarly, for $j \in \{1, ..., n-1\}$ such that $T_{j-1} \Rightarrow_{J_{g(i)}} T_j \Rightarrow_{J_{f(h)}} T_{j+1}$. Let α be the set of pairs (e_Q, f_R) thus obtained and let P denote the set as in (2) of 3.10. As the relations of precedence in Q and R is consistent with that in the quasiderivation e, P and α satisfy (1)-(3) of 3.10. On the other hand, for each $p \in P$, data(p) is an instance of an r-structure contained in T_0 and result(p) is an instance of an r-structure contained in some T_j, [P] is a trace which consists of [Q] and [R]. Moreover, one can construct a maximal chain c of P, as required. The relation $\Gamma(B_0 \circ B_1) = \Gamma(B_0) \cdot \Gamma(B_1)$ can be proved in a similar manner taking into account the fact that the terminality of quasiderivations corresponds to that of traces.♦

As a direct consequence of 4.1 we obtain the following result.

4.3 Theorem. For each relational program $\pi(x,y,...)$ with (object or behaviour) variables x,y,... and each valuation v of x,y,... we have

$$\Gamma([[[\pi(x,y,...)]]](v)) = [[\pi(x,y,...)]](v'),$$

where v' is obtained from v by replacing each $v(\xi)$ by $\Gamma(v(\xi))$ for each behaviour variable ξ. The correspondence Γ is a homomorphism w.r.t. the considered operations. ♦

References

[EKMRW] H.Ehrig, H.J.Kreowski, A. Maggiolo-Schettini, B.K. Rosen, J. Winkowski, Transformations of Structures: An Algebraic Approach, Math. Systems Theory 14 (1981) 305-334.

[MW1] A. Maggiolo-Schettini, J. Winkowski, Processes of Transforming Structures, J. Comput. System Sci. 24(1982) 245-282.

[MW2] A. Maggiolo-Schettini, J. Winkowski, Towards a Programming Language for Manipulating Relational Data Bases, in: D.Bjorner(ed.), Formal Description of Programming Concepts II, North-Holland, Amsterdam, 1983, 265-280.

[MNT] A. Maggiolo-Schettini, M.Napoli, G.Tortora, Web Structures: A Tool for Representing and Manipulating Programs, IEEE Trans. on Soft. Eng.14 (1988) 1621-1639.

[M] B.Mayoh, A Uniform Model for the Growth of Biological Organisms: Cooperating Sequential Processes, in: G.Rozenberg, A.Salomaa (eds.): The Book of L, Springer, Berlin, 1986, 291-301.

[R] V.Railich, Dynamics of Discrete Systems and Pattern Reproduction, J. Comput. System Sci. 11 (1975) 186-202.

[W] G.Winskel, An Event Structure Semantics for CCS and Related Languages, Springer LNCS 140, 1982, 561-576.

A Specification Environment for Graph Grammars *)

M. Nagl, A. Schürr
Lehrstuhl f. Informatik III
Aachen University of Technology
Ahornstr. 55, D-5100 Aachen, Germany

Abstract: Modelling environments (e.g. software development environments) offer tools which build up and maintain complex internal data structures. Therefore, before implementing such tools, it is advisable for the tool developer to formally specify the structure and the operations of these internal data structures. Graph Grammars as an operational specification method have been successfully used for this purpose for many years.

This paper describes an integrated set of tools for building up, maintaining, analyzing and executing such an internal specification described in the language PROGRESS, a graph grammar based VHL-Language. It sketches the characteristics of the PROGRESS environment, its software architecture, the specification of the environment in the language itself, and the implementation which, for the biggest part, have been generated.

The environment will be used for the following two main purposes: (i) for (syntax) checking PROGRESS specifications (at specification time) before implementing an efficiently working system which is equivalent to the specification, and (ii) for rapid prototyping purposes, i.e. for directly executing a specification.

The following paper has to be read in connection with /Sc 90a/ of this volume. Whereas /Sc 90a/ introduces the specification language PROGRESS and its methodological application to a small example the following paper concentrates on the environment's user characteristics and the realization of the environment.

Keywords: Operational specification of abstract data types by graph grammars; modelling environments, especially software development environments; static (type) checking; rapid prototyping; data modelling; integrated and incremental tools.

Contents

---

*) This project was/is partially supported by Stiftung Volkswagenwerk and DFG

0. Introduction and Motivation

Formal rewriting systems on edge and node labelled attributed graphs, in short *graph grammars*, have been used as an *operational specification method* for serious practical problems for about ten years now in our group. The internal behavior of many tools, mostly from the area of software development but also for text or library systems, have been specified using this method before starting to implement these tools. This means that specifications for abstract data types were produced, for which there are corresponding components in the software architecture of the final environment. In the context of software development environments mostly syntax–directed and incremental editors and analyzers have been specified, namely for the areas requirements engineering, programming in the large, programming in the small, and project management. The tools, developed according to these specifications, comprise an environment, called IPSEN, which is rather big in size for a university project.

Working with large graph grammar specifications the necessity came up to work out a specification methodology. This methodology is called *graph grammar engineering* /ELS 87/. The main ideas of this methodology are, (a) to carefully decompose a graph structure into its constituents, and (b) to specify the operations on these constituents and to combine them in order to get the specification of the abstract data type. Getting experience with this methodology a further result was that (c) this method can be applied to internal data structures of rather different application areas. Therefore, a rather uniform data modelling approach has been developed for and within the IPSEN project /Na 85/: A graph is always regarded to have a dominant tree part enriched by various context sensitive relations which can be derived from the underlying tree skeleton.

The applied *graph grammar* calculus and its corresponding *language* underwent different *stages* of *development*. Starting with /Na 79/, programmed graph rewriting was introduced /ELS 87/ by using deterministic control structures of imperative programming languages. Derived binary relations were introduced as special form of application conditions for graph productions and subgraph tests /Le 88b/. The language in its current form, called PROGRESS /Sc 89/, /Sc 90a,b/ adds two further main features: (i) There exists a separate declarative description of a graph's components, called graph schema (comprising definitions of directed edges and attributed nodes of different types as well as attribute evaluation rules), and (ii) programmed graph operations, called transactions, may be composed by using (non–deterministic) programming constructs and they may be parameterized with typed attribute–valued and type–valued parameters. The first extension was a major prerequisite for the implementation of tools checking context sensitive constraints, whereas the second one enables us to use a more declarative style of writing highly parameterized transactions. Therefore, both extensions have been made in order to facilitate the production of specifications which are easier to understand, to prove–read, and to maintain.

PROGRESS is a data modelling VHL–Language *supporting* different *programming/ specification paradigms*: It is data flow oriented by introducing directed attribute equations in the schema part. It is object–oriented by allowing us to describe the structure of nodes of different types in the schema part by a multiple inheritance class hierarchy. Furthermore, it is trivially rule based by using graph rewriting rules as basic graph changing operations. Finally, it is imperative as allowing transactions to be composed of (nondeterministic) control structures. To say it in another way, the major advantage of PROGRESS compared with its predecessors is that it allows to express a big portion of the specification in a declarative and nondeterministic way, therefore reducing the operational part to its very kernel.

In the past many large specifications have been written using PROGRESS or any of its predecessors. These specifications have been validated manually before building environments according to these specifications. Specification errors led to serious problems in the realization process. Now, using the PROGRESS language the schema part of the specification introduces redundancy allowing the formulation of context sensitive rules (static type checking rules). This improves the probability of a specification to be correct. Therefore, using the *PROGRESS environment* described in this paper, *specification errors* due to context free and, especially, due to context sensitive syntax of the specification language can be *detected immediately*.

As the specification is operational it can be executed before starting to realize an equivalent and efficient version of the specified component (the interpreter is one of the next tools of the environment to be developed). A PROGRESS specification together with the interpreter in the framework of the global architecture of a modelling environment (see section 3) is therefore a *rapid prototype* of the specified (tool/abstract data type) component.

Let us sketch the *history* of the specification environment *project* being described in this paper: Implementing graph grammars started some time ago /Na 79/. The first graph grammar tools, however, have more been a toy than a serious undertaking to handle practical applications by graph grammars. For some years following, we used graph grammar specifications only in a paper and pencil mode. In 1984 we decided to build tools for the specification process, mainly because they can be realized in very much the same way as the other environments we have built.

The *paper* is *organized* as follows: In the next two sections we motivate the tools of the PROGRESS environment and, thereby, introduce their characteristics. In section four we sketch the overall architecture of the environment which is the same as that of the IPSEN environment for which we mainly use PROGRESS specifications. The next section sketches that the central part of the PROGRESS environment can be specified by the language itself. In section 5 we give some figures about the realization of the PROGRESS environment. In the last section, finally, we summarize, compare with similar approaches, and list some future goals.

1. Building up, and Maintaining a Specification

The *PROGRESS environment* is an integrated set of tools devoted to build up, to maintain, and to execute a PROGRESS specification. It consists of a syntax-oriented (textual) editor, an analyzer, an instrumentor, and an interpreter, the latter two being under development. This and the next section motivates the wherefore of these tools, explains their use, and characterizes their behavior.

Fig. 1 shows a *cutout* of a *PROGRESS specification*. We see the editor window on the screen with a textual PROGRESS specification on the left side and possible commands on the right. We give here a very brief introduction to what is to be seen in this window, only to enable the reader to understand the following explanation (for details see /Sc 90a/). We see at (1) a production, at (2) a part of the schema, and at (3) a test of a specification. The schema part determines a node class hierarchy InnerNode is a ASTNode is a Node. The node class Node has an attribute Id. The edge type ToSon1 determines that edges of this type (label) have to start in nodes which belong to the class InnerNode and to end in nodes of the class ASTNode. The production (1) has a single node 1 in the left hand side and four nodes 1' to 4' in the right where node 1' is an identical copy of the node 1 with respect to its attribute values and its embedding in a host graph (determined by the increment 1'=1). From node 1' to 2', 3', and 4', respectively there are edges with type ToSon1, ToSon2 and ToSon3.

Building up and maintaining a PROGRESS specification is done by using a *structure-oriented editor*. This editor is command-driven. For example, the selection of a command to insert an edge type results in an immediate reaction on the screen by generating a template edge type increment (see (2)) with placeholders for the yet undetermined context free syntax parts (i) edge label, (ii) node class for the source node and (iii) node class for the target node. The concrete syntax of this edge type increment, namely the word symbol EDGE_TYPE, and the delimitors: ":", "->", and ";" are generated and, therefore, need not to be put in by the user. The placeholders can later on be expanded by the user such that the edge type declaration of (2) results. Editing is also possible in a free textual mode. For that, the user selects an increment, as small or big as he wants (even the whole specification), and changes it using a usual text editor. The increment is then parsed thereby detecting all its (context free PROGRESS syntax) errors.

As already told, the schema part allows many context sensitive checks. These checks are done by the *analyzer* which again works immediately and which is integrated to the editor. This means, that the user does not explicitly invoke the analyzer. Instead, the analyzer immediately tries to detect new errors if the user changes a spec or deletes errors if the user has corrected a spec. The analyzer has two interesting features: It works incrementally, i.e. it (i) internally analyses only those portions of the spec document which might be involved by the change and it (ii) highlights all erroneous parts (with respect to PROGRESS' context sensitive syntax/static semantics). But, it does not force the user to correct an error immediately after detection as it was done by earlier IPSEN tools /En 86/, /Sc 86/.

In Fig. 1 there are three *context sensitive errors*. The edge from 1' to 2' in the right hand side of the production at (4) has a wrong label as 1' and 2' are specified to be of node class ASTNode. On the other hand, the edge type ToSon1 at (2) is specified to start from a node class InnerNode. Furthermore, the application condition in this production at (5) determines, that the attribute designator No has to be equal to the attribute CurrIncr. However, node 1 being of the class ASTNode has no attribute No as to be seen from the node class definitions below (2). Finally, the applied occurrence of "Newincr" has no defining occurrence as it is wrongly spelled. It should be "NewIncr". As the increment 1' is the current one, the error message shown in Fig. 1 belongs to this increment. From the first commands in the menu we can see that (context-sensitive) errors and warnings (like NewIncr has no applied occurrence) can be hidden if the user wants them to be.

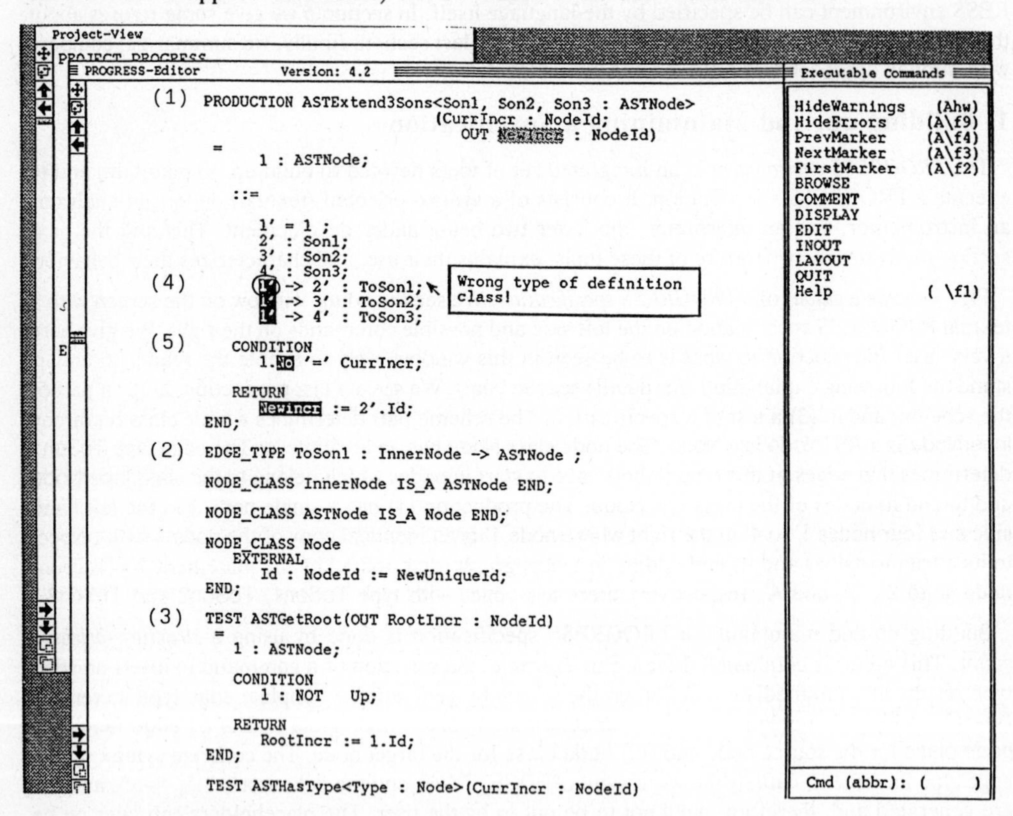

Fig. 1: The user interface of the editor/analyzer

A PROGRESS specification built up and maintained by the user using the editor/analyzer is continuously checked for context sensitive errors. As already told, context free errors are not possible in the command-driven mode or are detected during parsing. Therefore, the *user* has *full knowledge about* all *syntactical errors* of his specification. This holds even true for an incomplete spec, where many specification increments are not yet filled in. This is a big advantage compared to the situation where we used graph grammar specs in a paper and pencil mode. Specification errors are determined and, therefore, corrected earlier, thereby reducing their costs. Furthermore, starting to realize a tool for which a complete graph grammar spec was elaborated there is some guarantee that the specification is a reasonable starting point.

2. Using a Specification

Having a PROGRESS spec of the internal behavior of a tool, i.e. a specification of the corresponding abstract data type, the question now is *how to get* an *implementation* of this specification and of the tool. The first question is answered in this section, the second in the following. This question is important for all IPSEN and PROGRESS tools, and it can be answered equally.

There are *three ways* of a reasonable *translation* (cf. Fig. 2). All of them use the layer of a graph storage (cf. /BL 85/, /LS 88/), a non-standard data base system which is able to store graphs of arbitrary size and structure. The interface of this basic architectural component is on the usual level of handling graphs: insertion/deletion of nodes and edges with attributes, elementary graph queries etc.

Now, the first way of translation to be sketched here (cf. (1) in Fig. 2.b), is to translate a graph grammar specification *mechanically* but *manually* into an efficient program on top of a graph storage. This was done in earlier versions of IPSEN. The second solution (2) is to *automize* this process. This solution has been used to develop the current IPSEN prototype and also the PROGRESS environment. The translation, however, is not fully automatic: Based on basic components, other components are generated leaving some rest for handcoding.

The solution interesting within the context of this paper is given as (3) in Fig. 2.a. As the graph grammar specification is operational it can be *directly executed*, if there is an *interpreter* for PROGRESS. Such an interpreter is under development. A PROGRESS specification together with the interpreter is therefore an immediately available but less efficient implementation of the abstract data type, i.e. it is a rapid prototype.

The abstract data type's rapid prototype can be put into a fixed architectural framework of an interactive environment like the one sketched within the next section. This yields a *prototype* of the *interactive environment* the internal behavior of which was specified by a PROGRESS specification.

Another feature an integrated PROGRESS environment should have, which is also not available at this moment, is an *instrumentation tool*. PROGRESS specifications could be instrumented for two different purposes.

The first one is to introduce *global runtime conditions* which have to hold during execution. This is interesting inasmuch as it offers an opportunity to define a different nonoperational specification of the behavior of the abstract data type. Such a specification can be a positive one, i.e. that a certain graph situation has always to hold true, or a negative one, i.e. that a certain graph situation is not allowed to occur. It should be possible to map those conditions into a PROGRESS specification by a tool such that the above interpreter could watch their validity. Similar situations hold true for local runtime conditions which are nothing else than conditional breakpoints for an execution.

The other kind of instrumentation is to *monitor* the runtime *behavior* of a specification. For this, counters can be introduced counting the number of executions of certain constructs (transactions, control structures, productions, tests). The measurement of the storage consumption, on the other hand, is a service to be offered by the underlying graph storage.

Our final goal is to develop an *integrated environment* with all the aforementioned tools as it was done in IPSEN for the area "programming in the small" in /En 86/, /Sc 86/. Then, a possibly incomplete specification can be analyzed and instrumented, it can be executed, and, if running through a nonspecified part, the editor will be invoked such that the user can put in the missing increment allowing the execution to proceed. Finally, interpretation can be carried out in different granularity (from complex transactions down to rules/tests).

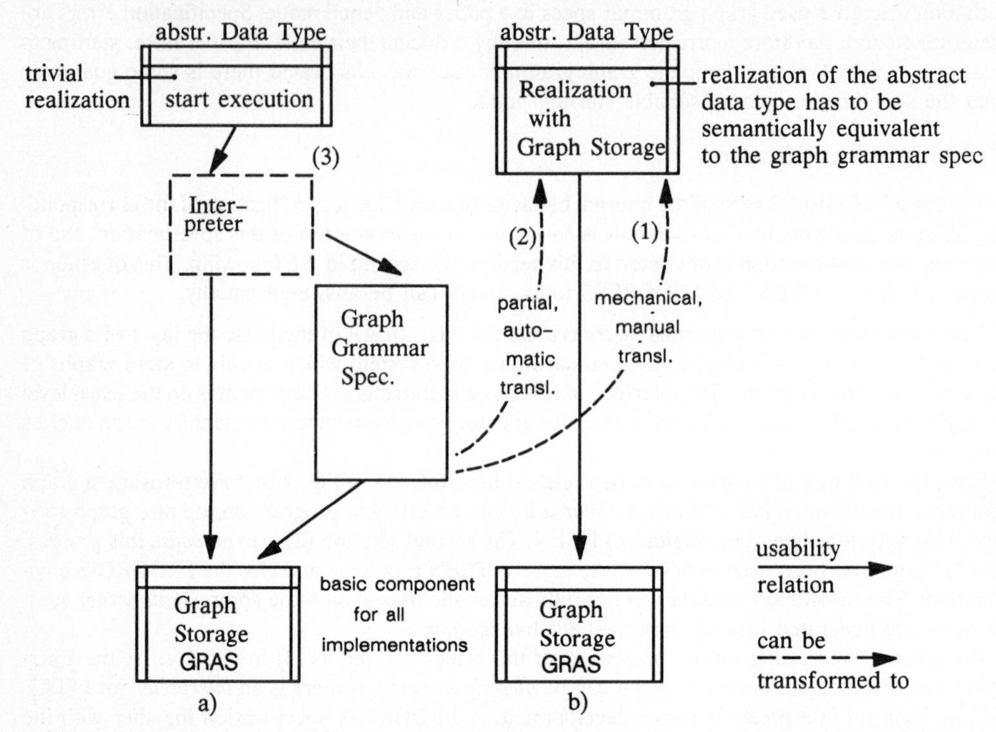

Fig. 2: Different ways of translating a PROGRESS specification

Now at the end of the two sections describing the user behavior of such an integrated PROGRESS environment let us *summarize* the *characteristics* of all *tools* (editor, analyzer, instrumentor, interpreter). All tools work immediate or interactive, syntax-oriented, in most cases command-driven, and in all cases incremental. This is the first logical bootstrap step in the PROGRESS project as all these characteristics could be found in former IPSEN tools.

3. The Environment's Software Architecture

This section sketches the *overall architecture* of the PROGRESS environment and, doing so, the second logical bootstrap step. In building different environments we detected that all these environments could be implemented using the same implementation technique: The structure of these environments (the architecture) is always nearly the same. Therefore, our previous experiences made within the IPSEN project had a strong influence on the development of the PROGRESS environment. (Conversely, the IPSEN project got profit by the PROGRESS undertaking as the machinery to generate components of the environment was developed within the PROGRESS environment and then adopted by the IPSEN project.)

The architecture of PROGRESS, currently consisting of about 150 modules, is roughly described by Fig. 3. This figure contains no basic components of the generation machinery which can be found inside the subsystems for the abstract data types for problem-oriented graphs (as PROGRESS Spec Graph Document here). The architecture consists of about *five layers* to be shortly explained (the architecture is restricted here only to one tool, namely the textual PROGRESS editor).

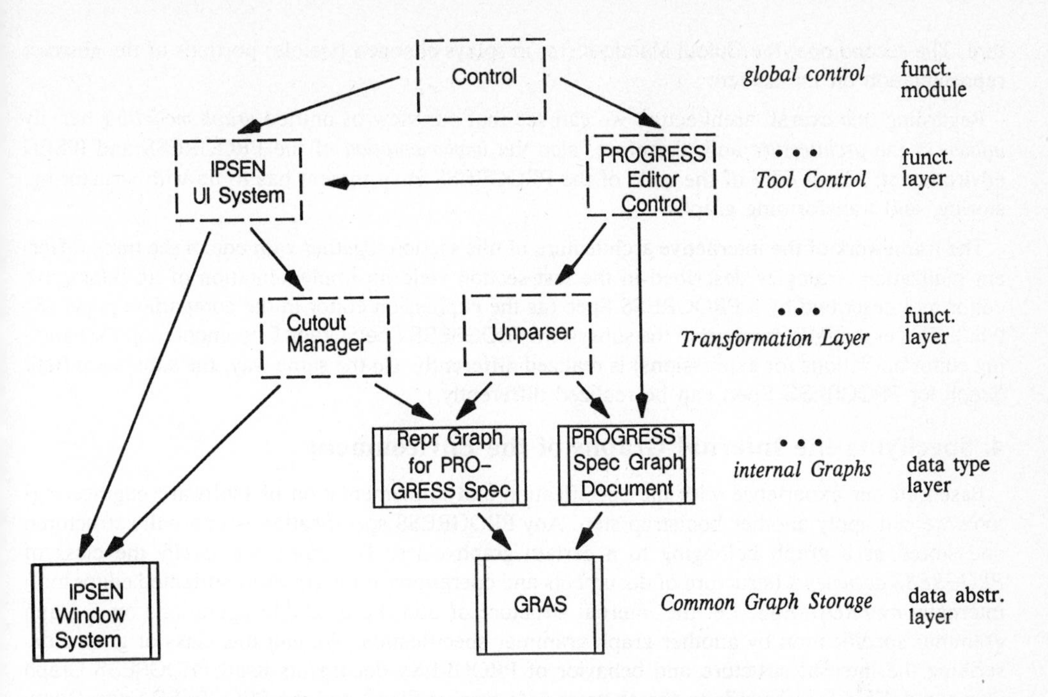

Fig. 3: Overall architecture of interactive environments following the graph technology approach

The first two layers are responsible for *control*, i.e. the global control of the dialogue which enables us to switch freely between different tools (Control) and the local dialogue within one tool (PROGRESS Editor Control). Both modules are built on top of a *user interface management* subsystem (IPSEN UI System) which controls the layout of windows, activates the updating of window contents and autonomously processes rather complex standard interaction with the environments's user like menu-based/ keystroke-based command selection, text editing etc. This subsystem has been implemented by using a small set of well-defined (IPSEN-specific) *window management resources* (IPSEN Window System) which are easy to implement on top of available standard window management systems.

The most interesting parts are the *data type subsystems* for *problem-oriented data structures*. In our case it is the subsystem PROGRESS Spec Graph Document containing the logical structure of a specification. From this logical graph a representation graph Repr Graph for PROGRESS Spec is built up which contains a specification's abstract representation (the representation is abstract with respect to all device-dependencies and rendering details as e.g. available fonts etc.).
To keep a separate *representation structure* apart from a specification's underlying logical structure is necessary due to the fact that a future version of the environment will support the simultaneous use of different representations (with different levels of detail, different layout, or even different concrete syntaxes of one (logical) specification). Both subsystems for graph classes are built on top of the graph storage GRAS.

Therefore, we need two (incrementally working) *transformations* to present a PROGRESS specification on the screen. The first one, the Unparser adapts a specification's abstract representation whenever the user is interactively changing the specification, i.e. its underlying logical graph structure. The second one, the Cutout Manager (re–)displays changed (visible) portions of the abstract representation on the screen.

Regarding this overall architecture we can see that our view of unified *graph modelling* heavily *influences* the *architecture* and, therefore, also the *implementation* of the PROGRESS and IPSEN environment: About 60% of the code of the PROGRESS environment has to do with structuring, storing, and transforming graphs!

The framework of the interactive architecture of this section together with one of the three different realization strategies described in the last section yield an implementation of an interactive editor tool described by a PROGRESS Spec (as the expression editor in the companion paper /Sc 90a/). This essentially means that the subsystem PROGRESS Spec Graph Document (e.g. for handling editor operations for expressions) is realized differently. (In the same way, the subsystem Repr Graph for PROGRESS Spec can be realized differently.)

4. Specifying the Internal Graphs of the Environment

Based on our experience with the specification and implementation of (software engineering) tools we can apply another bootstrap step: Any PROGRESS specification is internally structured and stored as a graph belonging to a certain graph class. Therefore, we *specify* the class of *PROGRESS documents* (structure of documents and operations of the structure–oriented editor tool) internally *by PROGRESS*, i.e. the internal structure of and the available operations on a graph grammar specification by another graph grammar specification. We call this class of graphs describing the internal structure and behavior of PROGRESS documents itself PROGRESS Graph Document. This corresponds to the abstract data type of Fig. 2 and the PROGRESS Spec Graph Document of Fig. 3. The editor operations on PROGRESS documents are like insert or delete a transaction, production, test, control structure etc. and are operationally specified by PROGRESS graph modifying transactions.

For this graph grammar specification we can apply the usual methodology which was called graph grammar engineering above. Expressed in PROGRESS, we first describe the *schema part* of *PROGRESS graph documents* by describing the structure and similarities of node types by a class hierarchy. Attribute equations determine, how attributes in the neighbourhood of a node are changed which belongs to a certain type. Edge types specify which directed labelled edges between which node types may occur.

In the *transaction* part, the *tree part* is determined first. There we have to specify all those operations which are used to manipulate the PROGRESS specification graph's underlying abstract syntax tree skeleton. Due to the fact that all these operations are very similar to each other, (standard situations are "replace a node by another one with 1, 2, 3, ... sons") their specification is based on a small set of transactions and productions which are parameterized with node type parameters (corresponding to the nonterminal classes of the PROGRESS language's syntax).

The *context sensitive parts* are added by context sensitive productions which typically have the following form: The left hand side contains two nodes between which a context sensitive edge is to be inserted. The nodes in the left hand side are determined by an application condition which, usually, is a derived binary relation of the tree part and which, therefore, can be expressed by a path expression. The similarities in handling the context free situations and the standard cases of context sensitive situations for rather different languages allows us to use two basic components for the implementation of (i) abstract syntax tree manipulation, (ii) binding of identifiers, and (iii) type checking within syntax–directed tools.

At the end of this section we sketch the *state* of *specification* of PROGRESS tools: The editor/analyzer has been specified by PROGRESS. The above mentioned instrumentation tools can be specified too. There is no difficulty in doing so. The executer can also be specified (as it was done for a part of Ada in /Ja 86/) which, again, has not been done up to now.

5. Implementing Means Generating

The *implementation* of the *PROGRESS environment* is in the following *state*: The editor with a textual user interface is finished. The same holds true for the analyzer. The interpreter is under development, and the implementation of the instrumentor has not yet been started (but can be traced back to the implementation of the other tools as it was done in IPSEN).

The PROGRESS editor/ analyzer is built on top of a *layer* of *basic components* consisting, for example, of a component for handling the standard tree rewriting situations (AST Graphs), of a component for handling the standard context sensitive situations like incrementally maintaining relations between defining and applied occurrences of identifiers or bookkeeping informations about all parts of a document which must be checked for their context sensitive correctness (Context Sensitive Graphs). This components can be found within any logical graph subsystem (as PROGRESS Spec Graph Document of Fig. 3) and, therefore, do not appear in Fig. 3.

Following the lines of our graph grammar engineering methodology, this basic layer can be used for the *implementation* of a *large class* of *language specific editors* (besides the PROGRESS editor/ analyzer in the IPSEN project various editors and analyzers have already been built on top of this layer /Ja 90/, /We 90/). It consists of about 124.000 lines of Modula-2 code from which 24.000 lines are used to implement GRAS, the underlying graph/object management system.

The *implementation* of the language specific parts of the *PROGRESS editor/ analyzer* consists of only 13.000 lines of Modula-2 code from which 10.000 lines are generated from an EBNF. This EBNF can be regarded as the tree part of the PROGRESS Graph Document specification. Only 3.000 lines of code of the editor/ analyzer including all type checking routines and some derived binary relationships necessary for binding applied to compatible defining occurrences of identifiers are handcoded.

The *PROGRESS interpreter* is *under development*. An incremental attribute evaluator, an important part of the interpreter's implementation, has been finished consisting of about 4.000 lines of Modula-2 code.

6. Summary, Comparison, and Future Work

Let us *summarize* the *essentials* of this *paper*:

We introduced an integrated environment for building up, maintaining, and executing graph grammar specifications of abstract data types. Using this environment PROGRESS specifications can be validated at specification time by getting a number of analyses and (in the future) by being able to execute this specification directly. The specification can be put in and maintained by a comfortable syntax-directed editor. It furthermore can be instrumented for execution. This PROGRESS environment will be used in the future for the specification of tools of interactive environments in different application areas.

For the specification and implementation of this environment a number of bootstrap steps have been applied. The characteristics of tools and the global architecture of the PROGRESS environment are the same as that of other interactive environments. The internal document classes of the PROGRESS environment can be specified using PROGRESS. An implementation technology was used which can also be applied for the development of other environments following the graph grammar engineering approach, e.g. in the IPSEN context. This implementation technology consists of a broad basic layer on top of which the specific graph structures of interactive tools are implemented. Thereby, the greatest part of the implementation is application independent. Large parts even of the language specific layer are not hand-coded but automatically produced by using the PROGRESS/ IPSEN editor generation machinery.

The *comparison* with *other approaches* is mainly determined by the features of the language, here PROGRESS, to be supported by the environment. The essential features of PROGRESS are the schema part allowing a number of context sensitive checks and the parameterization of transactions by node types (being instances of a class type inheritance hierarchy). Another graph grammar specification environment is described in /Gö 88/. The approach described there is different in many aspects (current research seems to go into a similar direction as in our project): Firstly, the supported language is typeless and has no schema definition capabilities. Thus, specifications produced within the environment are less declarative and less "reliable" as those produced within the PROGRESS environment. Secondly, the implementation language is Lisp which makes an implementation much easier (but less efficient). Therefore, the tools are usable for – and they were only aimed at – building tools for rapid prototyping. Thirdly, the implementation mechanisms sketched in this paper for getting efficient tools were out of the scope of that approach. Other references to be made here are modelling environments in the data base area (e.g. /Ce 83/, /LM 86/). The difference to those approaches mainly is that the data modelling specification language PROGRESS offers powerful rule based features for describing complex transactions.

The *future work* of PROGRESS and its environment is described as follows:
First, the interpreter has to be finished and the instrumentor tool has to be implemented. For the interpreter we are now developing an intermediate code for a graph machine (one basic operation is the attribute evaluation mentioned above). Interpretation then, similar to /En 86/, /Sc 86/, /ES 87/ for programming in the small, means (i) incremental translation of the internal PROGRESS graph into pieces of graph machine code and (ii) interpretation of the generated graph machine code. Another topic we are just starting to deal with is the development of a graphical user interface for the PROGRESS environment. Furthermore, we are improving the implementation machinery sketched in this paper, by trying to generate even more of the code of interactive tools, thereby reducing the handcoded parts. Finally, because PROGRESS is still in progress, the PROGRESS tools have to be adapted to new features of the language. One essential of this improvement of the language is to modularize PROGRESS specifications. This is a prerequisite to be able to specify integrated tools, not restricted to one internal class of documents.

Acknowledgements

The authors are indebted to C. Lewerentz who started again to work on tools for graph grammar specifications. They would like to thank R. Herbrecht, R. Spielmann and A. Zündorf, who contributed to the PROGRESS project. Furthermore, proofreading of A. Zündorf and the careful typing of A. Fleck and M. Hirsch are thankfully acknowledged. Finally, the remarks of two unknown reviewers led to considerable improvements of this paper.

7. References

/BL 85/ Th. Brandes/C. Lewerentz: GRAS: A Non-standard Data Base System within a Software Development, Proc. of the GTE Workshop on Software Engineering Environments for Programming in the Large, Harwichport, 113-121, (June 1985).

/Ce 83/ S. Ceri (Ed.): Methodology and Tools for Database Design, North-Holland (1983).

/EGNS 83/ G. Engels/R. Gall/M. Nagl/W. Schäfer: Software Specification using Graph Grammars, Computing 31, 317-346 (1983).

/ELS 87/ G. Engels/C. Lewerentz/W. Schäfer: Graph Grammar Engineering – A Software Specification Method, in Ehrig et al. (Eds.): Proc. 3rd Int. Workshop on Graph Grammars and Their Application to Computer Science, LNCS 153, 186-201, Berlin: Springer-Verlag (1987).

/En 86/ G. Engels: Graphs as Central Data Structures in a Software Development Environment (in German), Doctoral Thesis, Düsseldorf: VDI-Verlag (1986).

/ENS 86/ G. Engels/M. Nagl/W. Schäfer: On the Structure of Structures Editors for Different Applications, Proc. 2nd ACM Software Eng. Symp. on Pract. Software Development Environments, SIGPLAN Notices 22, 1, 190-198 (1986).

/ES 87/ G. Engels/A. Schürr: A Hybrid Interpreter in a Software Development Environment, Proc. 1st European Software Engineering Conf., LNCS 289, 87-96, Berlin: Springer-Verlag (1980).

/ES 89/ G. Engels/W. Schäfer: Program Development Environments - Concepts and Realization (in German), Stuttgart: Teubner Verlag (1989).

/Gö 88/ H. Göttler: Graph Grammars in Software Engineering (in German), IFB 178, Berlin: Springer-Verlag (1988).

/He 89/ R. Herbrecht: A Graph Grammar Editor (in German), Diploma-Thesis; RWTH Aachen (1989).

/HT 86/ S. Horwitz/T. Teitelbaum: Generating Editing Environments Based on Relations and Attributes, ACM TOPLAS 8, 4, 577-608 (1986).

/Ja 86/ M. Jackel: Formal Specification of the Concurrent Constructs of Ada by Graph Grammars, Doctoral Dissertation, Univ. of Osnabrueck (1986).

/Ja 91/ Th. Janning: Requirements Engineering and a Mapping to Programming in the Large, Dissertation forthcoming.

/Le 88a/ C. Lewerentz: Extended Programming in the Large in a Software Development Environment, Proc. 3rd ACM SIGPLAN/SIGSOFT Symp. on Practical Software Engineering Environments, Software Engineering Notes 13, 5, 173-182 (1988).

/Le 88b/ C. Lewerentz: Interactive Design of Large Program Systems (in German), Doctoral Thesis, IFB 194, Berlin: Springer-Verlag (1988).

/LM 86/ P.C. Lockemann/H.C. Mayr: Information System Design: Techniques and Software Support, in H.-J. Kugler (Ed.): Information Processing 86, 617-634, Elsevier Science Publ. (1986).

/LS 88/ C. Lewerentz/A. Schürr: GRAS, a Management System for Graph-like Documents, in C. Beeri et al. (Eds.): Prod. 3rd Int. Conf. on Data and Knowledge Bases, 19-31, Los Altos: Morgan Kaufmann Publishers (1988).

/Na 79/ M. Nagl: Graph Grammars: Theory, Applications, and Implementations (in German), Braunschweig: Vieweg-Verlag (1979).

/Na 85/ M. Nagl: Graph Technology Applied to a Software Project, in Rozenberg, Salomaa (Eds.): The Book of L, 303-322, Berlin: Springer-Verlag (1985).

/Re 84/ T. Reps: Generating Language-Based Environments, PH.D. Thesis, Cambridge, Mass.: MIT Press (1984).

/Sc 86/ W. Schäfer: An Integrated Software Development Environment: Concepts, Design and Implementation (in German) Doctoral Thesis, Düsseldorf; VDI-Verlag (1986).

/Sc 89/ A. Schürr: Introduction to PROGRESS, an Attribute Graph Grammar Based Specification Language, LNCS 411, 151-165 (1989).

/Sc 90a/ A. Schürr: PROGRESS: A VHL-Language Based on Graph Grammars, this volume.

/Sc 90b/ A. Schürr: Programming by Graph Rewriting Systems - Theoretical Foundations and the Corresponding Language (in German), Doctoral Dissertation, RWTH Aachen (1990).

/Sp 89/ R.Spielmann: Development of a Basic Layer for Graph Grammar Interpreters (in German), Diploma Thesis, RWTH Aachen (1989).

/We 90/ B. Westfechtel: Revision Control in an Integrated Software Development Environment (in German), Doctoral Dissertation, RWTH Aachen (1990).

/Zü 89/ A. Zündorf: Control Structures for the Specification Language PROGRESS (in German), Diploma Thesis, RWTH (1989).

THE THEORY OF GRAPHOIDS: A SURVEY

Azaria Paz*
Department of Computer Science
Technion - Israel Institute of Technology
Haifa 32000, Israel

A B S T R A C T: A survey type presentation, introducing the theory of Graphoids and their representation in graphs. Graphoids are ternary relations over a finite domain governed by a finite set of axioms. They are intended as models for the representation of irrelevance relations of the form $I(X,Z,Y)$ where (X,Z,Y) in I has the following interpretation: given that the *values* of the variables in Z are known, the *values* of the variables in Y can add no further information about the *values* of the variables in X.

Keywords: Axioms of Graph and Axioms of Graphoids, Completeness, Graphs, Graph Grammar, Decision Properties, Irrelevance Relation, Knowledge Representation, Probabilistic Distribution, Representation in Graphs, Soundness.

CONTENTS

1. Introduction

Graphoids are ternary relations over a finite domain that satisfy a finite set of axioms enabling symbolic derivations. Graphoids are intended as models for representing irrelevance relations consisting of statements of the form: given that the *values* of the variables in Z are known, the *values* of the variables in Y can add no further information about the *values* of the variables in X. Graphoids can be represented in graphs under a semantic to be defined in the text. Using a generative definition, one can define a Graphoid as a finite set of graphs which are derived from a *small* set of graphs (a core) by a

* This work was supported by The Fundation for the Promotion of Research at the Technion.

finite set of derivation rules. The rules induce modifications (in the previously generated graphs) and derivations of new graphs. Hence the connection to graph grammars. Graphoids may have applications to knowledge representation and may also provide a means for discovering independence relations between random variables by symbolic manipulation methods (rather than tedious calculations). They have many interesting properties and there are also some open problems which will be presented in the text.

2. Basic Definitions and Notations

Probabilistic distributions (denoted PDs) are defined in this paper over a finite set U of random variables. X,Y,Z, etc. denote disjoint subsets of U and x,y,x, etc. denote individual random variables. Every random variable is defined over a finite domain.

Let a,b,c be vectors of possible values of the sets of random variables X,Y,Z, correspondingly. We shall use the notation (the concatenation of X and Y stands for the union of the two sets):

$$P(XY) = P(X \mid Y)P(Y)$$

for the statement

$$(\forall a)(\forall b)P(X{=}a, Y{=}b) = P(X{=}a \mid Y{=}b)P(Y{=}b)$$

which is an identity over PDs.

Under the above notation the equality

$$P(XY \mid Z) = P(X \mid Z)P(Y \mid Z) \tag{1}$$

means that the set of variables X is independent on the set of variables Y given the set of variables Z (i.e., if the values of the variables in Z are known then the values of the variables in X may get are independent of the values of the variables in Y may get), for the given PD. The equality

$$P(XY) = P(X)P(Y) \tag{2}$$

means that the variables in X are (marginally) independent on the variables in Y for the given PD. The equality (1) represents a ternary relation over the set of variables U, to be denoted by I. We shall say that the sets X,Z,Y satisfy the relation I (notation: $I(X,Z,Y)$) if they satisfy the equality (1). Notice that every PD satisfies the following identity:

$$(\forall a) \sum_{Y=b} P(X=a, Y=b) = P(X=a) \tag{3}$$

or, under the above notation

$$\Sigma_Y P(XY) = P(X) . \tag{4}$$

3. Axioms for (marginal) PDs

Given any *PD*, it is easy to show that it satisfies certain properties which we will refer to as axioms. The following is an example. We use the simplified notation (X,Y) for the statement that the set of variables X is independent on the set of variables Y (and therefore satisfy (2) for the given *PD*). One can prove that

$$(X,Y) \wedge (XY,Z) \Rightarrow (X,YZ) \tag{5}$$

Proof: Using (2) we can express the left hand side in the form

(i) $P(XY) = P(X)P(Y)$

and

(ii) $P(XYZ) = P(XY)P(Z)$.

Applying (4) on (ii) we get (summation over X)

(iii) $P(YZ) = P(Y)P(Z)$

Substituting (i) into (ii) we get

(iv) $P(XYZ) = P(X)P(Y)P(Z)$

Substituting (iii) into the last formula we get

(v) $P(XYZ) = P(X)P(YZ)$

which by (2) and our notation is equivalent to the right hand side of (5).

Q.E.D.

4. The Graphoid Axioms

One can show, by methods similar to the methods used in the previous section, that any *PD* satisfies the 4 first axioms in the list of axioms below. If the *PD* has only strictly positive values (for any instantiation of the variables) then it satisfies the fifth axiom too. The 5 axioms listed are independent and any ternary relation satisfying them will be called a Graphoid (for reasons to be explained in the sequel). Thus the relations induced by strictly positive PDs are graphoids. The converse is not true. Recall that the notation $I(X,Z,Y)$ stands for

$$P(XY \mid Z) = P(X \mid Z) P(Y \mid Z) \stackrel{\Delta}{=} I(X,Z,Y) . \tag{6}$$

The Axioms

1. $I(X,Z,Y) \Rightarrow I(Y,Z,X)$: Symmetry ;

2. $I(X,Z,YW) \Rightarrow I(X,Z,Y) \wedge I(X,Z,W)$: Decomposition ;

3. $I(X,Z,YW) \Rightarrow I(X,ZY,W) \wedge I(X,ZW,Y)$: Weak Union ;

4. $I(X,Z,Y) \wedge I(X,ZY,W) \Rightarrow I(X,Z,YW)$: Contraction;

5. $I(X,ZY,W) \wedge I(X,ZW,Y) \Rightarrow I(X,Z,YW)$: Intersection.

Any relation which satisfies the first 4 axioms will be called a semigraphoid, while a relation satisfying axioms 1, 2, 3 and 5 will be called a pseudographoid [8 - Chapter 3].

5. Representation in Graphs

Given a graph $G = (V,E)$, let X,Y,Z be disjoint subsets of V. Define $I_G(X,Z,Y)$ to be the statement: Z is a cutset between X and Y in G. One can easily show that I_G is a graphoid thus enabling a representation of graphoids in graphs, under the above semantics, with vertices mapped into variables and independence represented by cutset separation.

To verify that I_G is a graphoid one must show that all the graphoid axioms hold in graphs, under the specified semantics. E.g. intersection means, for graphs, that: If in G, $Z \cup Y$ a cutset between X and W and $Z \cup W$ is a cutset between X and Y then Z is a cutset between X and $Y \cup W$. It is easy to see that every graph satisfies this property.

The representation of graphoids in graphs has certain advantages. There are exponentially many graphs over n vertices, thus allowing for a multitude of graphoids to be represented; the number of triplets (X,Z,Y) which can be represented in one graph may be exponential (corresponding to the number of cutsets existing that graph) and graphs are easy to store and manipulate in a data base.

On the other hand graphs satisfy two additional axioms which are not implied by the graphoid axioms. It follows that there are graphoids which cannot be represented in graphs. As a matter of fact, the relations I_G induced by graphs can be characterized in full. This is shown in the next section.

6. Graph-Isomorphism

A ternary relation I is graph-isomorphic, if there is a graph G, such that (X,Z,Y) is in I iff Z is a cutset between X and Y in G (we assume that the variables in the domain of I are 1-1 mapped onto the vertices of G).

The following characterization is known [7]:

Theorem: A relation I is graph-isomorphic iff it satisfies the following 5 axioms:

$I(X,Z,Y) \Rightarrow I(Y,Z,X)$: Symmetry ;

$I(X,Z,YW) \Rightarrow I(X,Z,Y)$: Decomposition;

$I(X,Z,Y) \Rightarrow (\forall W) I(X,ZW,Y)$: Strong Union;

$I(X,ZY,W) \wedge I(X,ZW,Y) \Rightarrow I(X,Z,Y)$: Intersection;

$I(X,Z,Y) \Rightarrow (\forall w)(I(X,Z,w) \vee I(w,Z,Y))$: Weak Transitivity.

Notice that all the graphoid axioms are either included or can be derived from the above axioms but not the other way around. The reader is encouraged to provide an interpretation of the above axioms in terms of cutset separation.

As an example, consider the following relation over $V = \{1,2,3,4\}$.

$$I = \{(1,4,3),(3,4,1),(2,\{1,3\},4),(4,\{1,3\},2)\}$$

It is easy to see that this relation is a graphoid: it is symmetric and all the other graphoid axioms are satisfied vacuously, since the l.h.s. configuration of the last 4 graphoid axioms is nonexistent in the set I. On the other hand, this relation cannot be represented in a graph since otherwise, by strong union, $(1,4,3)$ would imply that $(1,\{4,2\},3)$ is included in the relation which it is not.

Let B be a set of triplets belonging to a relation and let A be a set of axioms. The closure of B under A (notation $Cl_A(B)$) is defined to be the set of all triplets which can be derived from B via the axioms in A.

It has been shown that a graph G representing the closure under the graph-isomorph axioms of a polynomial set B of triplets can be constructed in polynomial time (in the number of vertices) [3].

It follows that it can be decided in polynomial time whether a given triplet t is in the closure, under the graph-isomorph axioms, of a given polynomial set of triplets.

7. Soundness and Completeness

As with regard to other logical models we shall refer to triplets (X,Z,Y) in a relation as statements which hold in the model if the triplet belongs to the relation induced by the model. Σ will denote a set of statements and σ a single statement. Let $\boldsymbol{P}$ denote a family of models and P an element in $\boldsymbol{P}$. The following notations will be used.

Given Σ and $\sigma \in \Sigma$ as above. If for every $P \in \boldsymbol{P}$, Σ holds in P implies that σ holds in P, then

$$\Sigma_P \models \sigma .$$

If σ can be derived from Σ via a set of axioms A, then

$$\Sigma_A \vdash \sigma .$$

Based on the above notation we can now introduce the definition below:

Definition of Soundness: A given set of axioms A is sound for a family of models $\boldsymbol{P}$ if for all Σ and σ

$$\Sigma \underset{A}{\vdash} \sigma \Rightarrow \Sigma_P \models \sigma .$$

Definition of Completeness: A given set of axioms A is complete for a family of models $\boldsymbol{P}$ if for all Σ and σ

$$\Sigma_P \models \sigma \Rightarrow \Sigma \underset{A}{\vdash} \sigma .$$

It has been shown that the graphoid axioms are sound for positive probabilistic distributions but are not complete [8 - Chapter 3] and [10]. It is reasonable to believe that any axioms, not implied by the graphoid set of axioms, but sound for PDs is quite complex and would occur only in PDs of some singular type.

8. Axioms for Marginal PDs are Sound and Complete

Marginal independencies are independencies of the form: "The set of variables X is independent on the set of variables Y". The interpretation for PDs is $P(XY) = P(X)P(Y)$.

The following set of 3 axioms has been shown sound and complete for marginal PDs (we shall use the notation $I(X,Y)$ for the induced relation) [2].

$I(X,Y) \Rightarrow I(Y,X)$: Symmetry,

$I(X,YZ) \Rightarrow I(X,Y)$: Decomposition,

$I(X,Y) \wedge I(XY,Z) \Rightarrow I(X,YZ)$: Mixing.

9. I-Mapness

Let R_1 and R_2 be two relations induced by 2 models M_1 and M_2. M_1 is an I-map of M_2 if every triplet in R_1 is a triplet in R_2. It has been shown in [7] that every graphoid M has a unique graph G such that:

(1) G is an I-map of M, and

(2) If G_1 is another I-map of M then the set of edges of G is a subset of the set of edges of G_1.

It has also been shown that every relation induced by a graph has a PD-model [3]. Thus graph-induced relations are a proper subset of the set of relations induced by PDs.

10. Decision Properties

Given a relation R (induced by some model) which is closed under a set of axioms A, we shall say that R is polynomially decidable if there is a polynomial algorithm which for any polynomial size Σ and σ, decides in polynomial time, whether $\Sigma_A \vdash \sigma$.

The relations induced by marginal PDs have been shown to be polynomially decidable [2].

Relations which are closed under the graph-isomorph axioms are polynomially decidable [3].

It is an open problem whether the graphoid relations are polynomially decidable (it is not known even if this decidability problem is in NP).

The related decidability problem described below has been shown to be NP-complete [5].

The CON-NOTCON Problem: Given a polynomial size set Σ and σ. Is there a graph G such that all the statements in Σ (statements of type 'NOT CON(NECTED)') are represented in G (and, possibly, some additional statements) but σ (a 'CON(NECTED)' type statement) is not represented in G

Another more complex but polynomially decidable problem will be shown in the next sections.

11. Representation in Families of Graphs

It has been shown in Section 6 that not every graphoid relation can be represented in a graph. One can also represent relations in families of graphs under the following semantics.

Let $\underline{G} = \{G_i\}$ be a family of graphs, where $G_i = (E_i, V_i)$ and $V_i \subseteq V$.

Let $I = \{(X_j, Z_j, Y_j)\}$ be a ternary relation where $X_j, Z_j, Y_j \subseteq V$. I is represented by $\underline{G}$ if every triplet in I is represented in some graph G_i and every triplet represented in some G_i is in I.

It is easy to see that every ternary relation can be represented in a family of graphs. This follows from the fact that every singly triplet (X,Z,Y) can be represented in the graph whose vertices are $X \cup Z \cup Y$ such that X, Z, Y are cliques, all the vertices in X are connected to all the vertices in Z and all the vertices in Y are connected to all the vertices in Z.

Representation in families of graphs is not very interesting unless one can show additional properties rendering decidability or other results.

12. The $\textcircled{\wedge}$ Operation on Graphs

Given two graphs $G_1 = (E_1, V_1)$ and $G_2 = (E_2, V_2)$ such that $V_1 \cap V_2 \neq \emptyset$. Construct the graph $G_3 = (E_3, V_3)$ where $V_3 = V_1 \cap V_2$ and $(a,b) \in V_3 \times V_3$ is <u>not</u> in E_3 iff either $V_3 - a - b$ is a cutset between a and b in G_1 or $V_3 - a - b$ is a cutset between a and b in G_2. Denote $G_3 = G_1 \textcircled{\wedge} G_2$.

Example

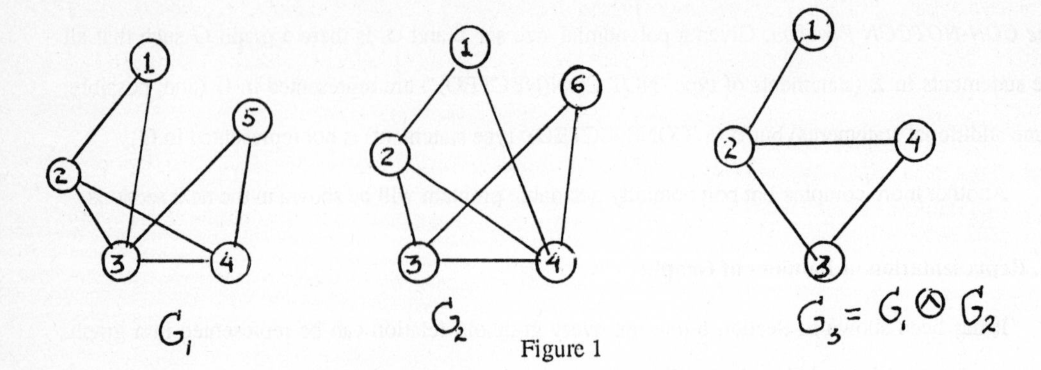

Figure 1

13. The Pseudographoid Graph-Language

A pseudographoid is a ternary relation which is closed under the subset of the graphoid axioms including Symmetry, Decomposition, Weak Union and Intersection (i.e., a pseudographoid satisfying contraction is a graphoid).

As every relation induced by a single graph is a graphoid, it is also a pseudographoid.

It has been shown in [6] that the pseudographoid closure of the relations represented by two graphs G_1 and G_2 can be represented in the set of graphs $\{G_1, G_2, G_1 \circledwedge G_2\}$. This induces the following algorithm for constructing a graph language representing the pseudographoid closure of a set of triplets $B = (t_1, ..., t_k)$.

(1) For every $t_i \in B$ construct G_i, the graph representing t_i (see Section 11) and add it to the graph language L (initiated to the empty set).

(2) Repeat until $L' \subseteq L$:
Construct $L' = \{G_r \circledwedge G_s : G_r, G_s \in L\}$ and set $L := L \cup L'$.

It has been shown [6] that the resulting graph language is a family of graphs representing the pseudographoid closure of the set B.

If the set of triplets is polynomial (in the number of variables) then one can show that the pseudographoid relation induced by it is polynomially decidable. The decision algorithm is based on the underlying graph language even though the size of the language may be exponential (in the number of variables involved). Thus, pseudographoids provide an example of a nontrivial but still decidable model. The problems of ascertaining whether graphoids or semigraphoids (satisfying the graphoid axioms except intersection) are polynomially decidable are not resolved yet.

The pseudographoid language provides also an example of production rules (ⓐ) creating a new graph out of 2 already created graphs (in contradiction to the regular production rules in graph grammars, where new graphs are created as a modification of previously created graphs).

14. Representation by DAGs

Another way of representing graphoids is by Directed Acyclic Graphs (DAGs). Again the vertices represent the variables in U. The definition of the representation is more complex and it takes into consideration the possibility of directing the arcs. There are three ways that a pair of arrows may meet at a vertex:

1: tail to tail, $X \leftarrow Z \rightarrow Y$

2: head to tail, $X \rightarrow Z \rightarrow Y$

3: head to head, $X \rightarrow Z \leftarrow Y$

Definitions:

(1) Tow arrows meeting head to tail or tail to tail at node α are said to be *blocked* by a set of S of vertices if α is in S.

(2) Two arrows meeting head to head at node α are *blocked* by S if neither α nor any of its descendents is in S.

(3) As undirected path P in a DAG G is said to be *d-separated* by a subset S of the vertices if at least one pair of successive arrows along P is blocked by S.

(4) Let X, Y and S be three disjoint sets of vertices in a DAG G. S is said to *d-separate* X from Y if all paths between X and Y are *d-separated* by S.

For example, in the graph in Figure 2 below, the triplet (2,1,3) is represented, as the set *{1} d−separates* between the vertices 2 and 3. On the other hand, the triplets (2,4,3) and (2,{1,5},3) are not represented.

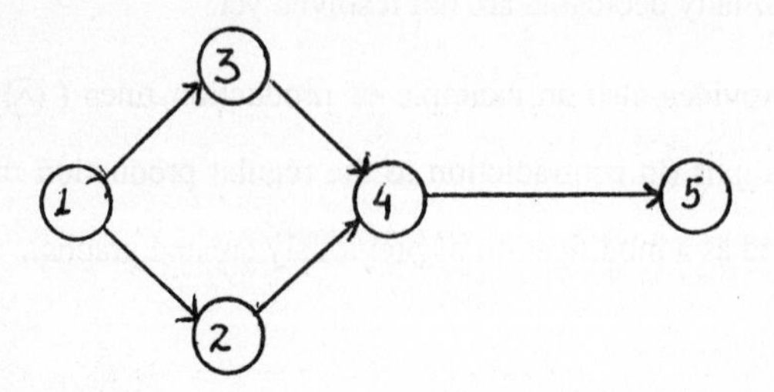

Figure 2

Families of DAGs and representation in such families can be defined in a way similar to the corresponding definitions for UGs.

The following properties of DAG representation have been shown:

(1) Every DAG induces a relation which is a graphoid but not every graphoid can be represented in a DAG [9].

(2) The family of relations induced by DAGs cannot be characterized by a finite set of axioms (contrary to the corresponding result for UGs) [1].

(3) For every n there exists a DAG with n vertices representing a relation which requires exponentially many UGs for its representation [5].

(4) For every n there exists a UG with n vertices representing a relation which requires exponentially many DAGs for its representation [5].

(5) Every DAG has a PD model [1] and every PD has a minimal DAG I-map [8, Chap. 3].

15. Annotated Graphs

A more powerful model for graphoid representation has been suggested recently, the model of annotated graphs [4]. That model has been shown to be stronger (with regard to its representation range) than both UGs and DAGs. The details of the model are too technical to be presented in a survey type paper and the interested reader is referred to [4] for its full description.

BIBLIOGRAPHY

[1] Geiger, D., *Towards the Formalization of Informal Dependencies,* (M.Sc. Thesis), UCLA Cognitive Systems Lab., Technical Report No. R-102.

[2] Geiger, D., Paz, A., and Pearl, J., "Axioms and algorithms for inferences involving probabilistic independence". To appear in *Information and Computation,* 1990.

[3] Geiger, D., *Graphoids: A Qualitative Framework for Probabilistic Inferences,* (Ph.D. Thesis), Cognitive Systems Lab., Technical Report No. R-142.

[4] Geva, R.Y. and Paz, A., *Representation of Irrelevance Relations by Annotated Graphs,* Technical Report #603, Computer Science Department, Technion-IIT, January 1990.

[5] Or, S., *Comparison of Graph-Representation Methods for Graphoids and the Membership Problem.* (M.Sc. Thesis), Technion-IIT, 1990.

[6] Paz, A., *A Full Characterization of Pseudographoids in Terms of Families of Undirected Graphs,* UCLA Cognitive Systems Laboratory, Technical Report 870055 (R-95), 1987.

[7] Pearl, J., and Paz, A., *GRAPHOIDS: A Graph-Based Logic for Reasoning about Relevance Relations, "Advances in Artificial Intelligence-II,"* edited by B. Du Boulay et al., Amsterdam: North-Holland.

[8] Pearl, J., *Probabilistic Reasoning in Intelligence Systems: Networks of Plausible Inference.* Chapter 3, Morgan-Kaufmann, San Mateo, CA. (1988).

[9] Pearl, J., and Verma, T.S., "The logic of representing dependencies by directed graphs". Proc. 6th Natl. Conf. on *AI (AAAI-87),* Seattle, 374-379, 1987.

[10] Studeny, M., "Attempts at Axiomatic Description of Conditional Independence," *Workshop on Uncertainty in Expert Systems,* Alsovice, Chechoslovakia, June 20-23, 1988.

Graph-Reducible Term Rewriting Systems *

Detlef Plump
Fachbereich Mathematik und Informatik
Universität Bremen, Postfach 33 04 40
2800 Bremen 33, Germany
det@informatik.uni-bremen.de

Abstract: Term rewriting is commonly implemented by graph reduction in order to improve efficiency. In general, however, graph reduction is not complete: a term may be not normalizable through graph derivations although a normal form exists. Term rewriting systems which permit a complete implementation by graph reduction are called graph-reducible. We show that the following property is sufficient for graph-reducibility: every term having a normal form can be normalized by parallel term rewrite steps in which a rule is applied to *all* occurrences of some subterm. As a consequence, a broad class of term rewriting systems which includes all terminating and all orthogonal systems can be shown to be graph-reducible.

Keywords: term rewriting, graph reduction, completeness of graph reduction

Contents

*Work supported by ESPRIT Basic Research Working Group #3264, *COMPASS*.

1 Introduction

Term rewriting serves as a model of computation in several areas of computing science such as algebraic specification, functional programming, and theorem proving. In practice, implementations of term rewriting usually represent terms by graphs in order to exploit the sharing of common subterms: repeated evaluations of the same subterm are avoided. However, sharing common subterms may lead to incompleteness. As an example, consider the following term rewriting system:

$$\begin{aligned} \mathtt{f(x)} &\to \mathtt{g(x,x)} \\ \mathtt{a} &\to \mathtt{b} \\ \mathtt{g(a,b)} &\to \mathtt{c} \\ \mathtt{g(b,b)} &\to \mathtt{f(a)} \end{aligned}$$

The term `f(a)` can be normalized by the rewrite sequence

$$\mathtt{f(a)} \to \mathtt{g(a,a)} \to \mathtt{g(a,b)} \to \mathtt{c}.$$

But no normal form can be reached when common subterms are shared:

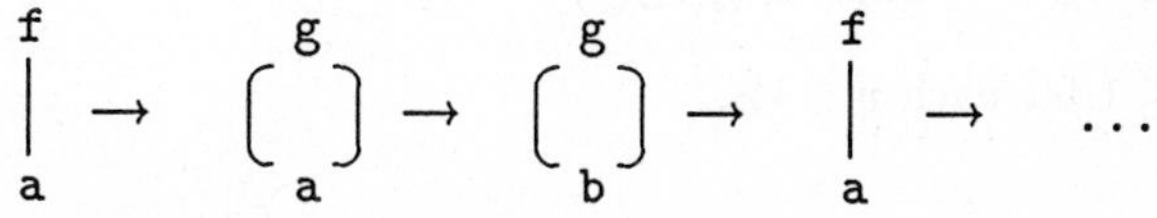

f | a → g ( ) a → g ( ) b → f | a → ...

This kind of incompleteness could be remedied by introducing graph rules which "unfold" shared subterms. But then the efficiency gained by sharing would get lost, and with it the very advantage of graph reduction. In this paper, we study the class of term rewriting systems that permit completeness despite sharing. The kind of completeness we are aiming at is normalizability: if a graph represents a term that has a normal form, then this graph must be reducible to some normal form. Following Barendregt et al. [BEGKPS 87], we call term rewriting systems with this property *graph-reducible*.

The problem with the term rewriting system shown above is that sharing excludes certain rewrite steps (e.g. $\mathtt{g(a,a)} \to \mathtt{g(a,b)}$). In contrast, we show in section 3 that parallel term rewrite steps of the following kind can always be implemented by graph reduction: a term is rewritten by applying a rule simultaneously to *all* occurrences of some subterm. Such a parallel step can be modelled on the graph level by maintaining a full sharing of common subterms.

Working with parallel term rewrite steps enables us to establish various sufficient conditions for graph-reducibility. In particular, it is straightforward to show that all terminating and all orthogonal term rewriting systems are graph-reducible. Moreover, normalizing term rewriting strategies that are defined in terms of parallel rewrite steps can effectively be translated into normalizing graph reduction strategies. Our classification of graph-reducible systems is complemented by examples which delimit subclasses from each other.

2 Jungle Evaluation

We review the jungle evaluation approach ([HKP 88], [HP 88]) as far as it is necessary for the following sections. We assume that the reader is familiar with basic notions of term rewriting (see for example [DJ 90], [Klo 90]).

Let $SIG = (S, F)$ be a signature, that is, S is a set of sorts and F is a set of function symbols such that each $f \in F$ has a string of argument sorts $s_1 \ldots s_n \in S^*$ and a result sort $s \in S$. A function symbol f is written $f : s_1 \ldots s_n \to s$ when the argument and result sorts matter.

A *hypergraph* $G = (V_G, E_G, s_G, t_G, l_G, m_G)$ over SIG consists of a finite set V_G of nodes, a finite set E_G of hyperedges (or edges for short), two mappings $s_G : E_G \to V_G^*$ and $t_G : E_G \to V_G^*$, assigning a string of source nodes and a string of target nodes to each hyperedge, and two mappings $l_G : V_G \to S$ and $m_G : E_G \to F$, labeling nodes with sorts and hyperedges with function symbols.

A hypergraph G over SIG is a *jungle* if

- the labeling of G is compatible with SIG, that is, for each $e \in E_G$, $m_G(e) = f : s_1 \ldots s_n \to s$ implies $l_G^*(s_G(e)) = s$ and $l_G^*(t_G(e)) = s_1 \ldots s_n$, [1]
- $outdegree_G(v) \leq 1$ for each $v \in V_G$,
- G is acyclic.

2.1 Example
Assume that SIG contains a sort **nat** and function symbols

$$\begin{aligned} &\mathtt{0} :\to \mathbf{nat}, \\ &\mathtt{succ}, \mathtt{fib} : \mathbf{nat} \to \mathbf{nat}, \\ &+ : \mathbf{nat}\ \mathbf{nat} \to \mathbf{nat}. \end{aligned}$$

Then the following hypergraph is a jungle over SIG.

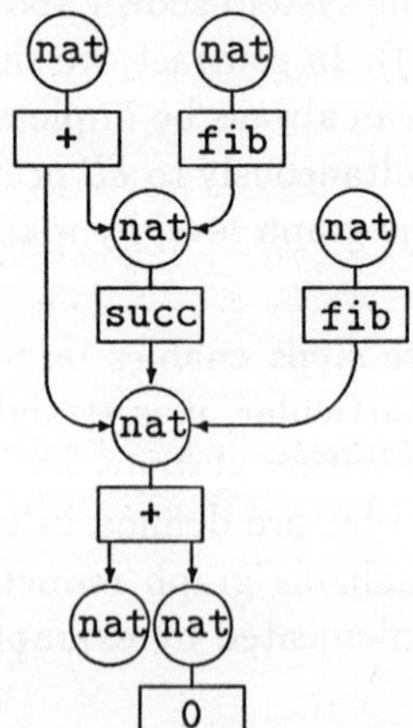

[1] Given a mapping $g : A \to B$, the extension $g^* : A^* \to B^*$ is defined by $g^*(\lambda) = \lambda$ and $g^*(aw) = g(a)g^*(w)$ for all $a \in A$, $w \in A^*$. Here λ denotes the empty string.

Nodes are drawn as circles and hyperedges as boxes, both with their labels written inside. A line without arrow-head connects a hyperedge with its unique source node, while arrows point to the target nodes (if there are any). The arrows are arranged from left to right in the order given by the target mapping. □

When *SIG*-terms shall be represented by jungles, nodes without outgoing hyperedge serve as variables. The set of all such nodes in a jungle G is denoted by VAR_G. (Note that each variable in VAR_G occurs only once in G, since variables are nodes rather than labels.) The term represented by a node v is then defined by

$$term_G(v) = \begin{cases} v & \text{if } v \in VAR_G, \\ m_G(e)term_G^*(t_G(e)) & \text{otherwise, where } e \text{ is the unique} \\ & \text{hyperedge with } s_G(e) = v. \end{cases}$$

For instance, if v is the left one of the two topmost nodes in the above example, then $term_G(v) = (\mathtt{x} + \mathtt{0}) + \mathtt{succ}(\mathtt{x} + \mathtt{0})$ (using infix notation) where $\mathtt{x}$ is the unique variable node in G.

For defining reduction rules and reduction steps we need structure preserving mappings between jungles. A *jungle morphism* $g : G \to H$ consists of two mappings $g_V : V_G \to V_H$ and $g_E : E_G \to E_H$ which preserve sources, targets, and labels, that is, $s_H \circ g_E = g_V \circ s_G$, $t_H \circ g_E = g_V^* \circ t_G$, $l_H \circ g_V = l_G$, and $m_H \circ g_E = m_G$.

Now we are ready to translate term rewrite rules into *evaluation rules.* For motivations and further details, however, we refer to [HP 88].

Assumption

For the rest of this paper we assume that $\mathcal{R}$ is an arbitrary term rewriting system, that is, each rule $l \to r$ in $\mathcal{R}$ consists of two terms l and r of equal sort such that l is not a variable and all variables in r occur already in l. □

Given a rule $l \to r$ from $\mathcal{R}$, the corresponding evaluation rule $(L \hookleftarrow K \xrightarrow{b} R)$ consists of three jungles L,K,R and jungle morphisms $K \hookrightarrow L$ and $b : K \to R$ such that:

- L is a variable-collapsed tree[2] with $term_L(root_L) = l$.
- K is the subjungle of L obtained by removing the edge outgoing from $root_L$.
- R is constructed from K as follows: if r is a variable, then $root_L$ is identified with the node r; otherwise, R is the disjoint union of K and a variable-collapsed tree R' with $term_{R'}(root_{R'}) = r$ where $root_{R'}$ is identified with $root_L$ and each $x \in VAR_{R'}$ is identified with its counterpart in VAR_K.
- $K \hookrightarrow L$ and $b : K \to R$ are inclusions with the possible exception that b identifies $root_L$ with some variable.

2.2 Example

The following picture shows the evaluation rule for the rewrite rule $\mathtt{succ(x)} + \mathtt{y} \to \mathtt{succ(x+y)}$ (where • depicts a $\mathtt{nat}$-labeled node):

[2] That is, there is a unique node $root_L$ with $indegree_L(root_L) = 0$ and only variables may have an indegree greater than 1.

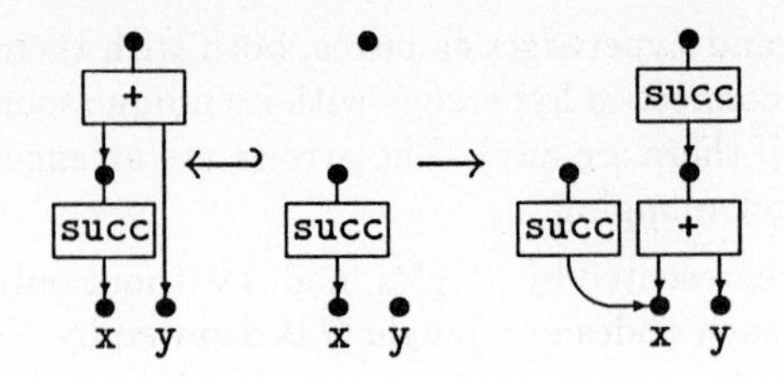

□

We assume from now on that every jungle G has a distinguished node $root_G$ which represents the term to be evaluated. Then the "relevant" part of G consists of all nodes and edges that are reachable from $root_G$. This subjungle is denoted by $G^\bullet$. The nodes and edges in $G - G^\bullet$ are said to be *garbage*. In the following we write $term(G)$ for $term_G(root_G)$.

The application of an evaluation rule $(L \hookleftarrow K \rightarrow R)$ to a jungle G requires to find an occurrence of L in $G^\bullet$ and works then analogously to the construction of K and R described above:

- Find a jungle morphism $g : L \rightarrow G^\bullet$.
- Remove the edge outgoing from $g(root_L)$, yielding a jungle D (with $root_D = root_G$).
- The resulting jungle H is obtained from D like R is constructed from K: if r is a variable, then $g(root_L)$ is identified with $g(r)$; otherwise, H is the disjoint union of D and the variable-collapsed tree R' where $root_{R'}$ is identified with $g(root_L)$ and each $x \in VAR_{R'}$ is identified with $g(x)$. Finally, $root_D$ becomes $root_H$.

We write $G \Rightarrow H$ if H is the result of applying an evaluation rule to G and call this an *evaluation step*. We also use the notation $G \Rightarrow_v H$ if v is the image of $root_L$ in G, that is, if $v = g(root_L)$. For the resulting jungle H its structure and labeling matters rather than the names of (non-variable) nodes. So every jungle H' essentially isomorphic[3] to H is allowed as a result of this evaluation step as well. We require, however, that variable nodes are not renamed.

For stating the effect of an evaluation step $G \Rightarrow H$ on $term(G)$ we need some more notions. Recall that the *positions* in a term t are strings of natural numbers denoting nodes in the tree representation of t and that for each position π in t, $t[\pi]$ is the subterm at π. π is called a *redex* if there is some rule $l \rightarrow r$ and some substitution σ such that $t[\pi] = \sigma(l)$.

Given a nonempty set $\Delta = \{\pi_1, \ldots, \pi_n\}$ of redexes in a term t such that $t[\pi_i] = t[\pi_j]$ for $1 \leq i, j \leq n$, we denote by $t \mathrel{+\!\!\!\!\!\rightarrow}_\Delta u$ the parallel rewrite step that reduces all the π_i simultaneously by the same rule.

For every jungle G there is a mapping $node_G$ which takes positions in $term(G)$ to nodes in $G^\bullet$ such that λ is mapped to $root_G$ and each position πi to the i-th target node

[3]Two jungles G and H are said to be ***essentially isomorphic*** if there is a jungle morphism $g : G^\bullet \rightarrow H^\bullet$ with g_V and g_E bijective. Note that we only compare the non-garbage parts of G and H.

of the unique edge outgoing from $node_G(\pi)$. For each position π, $node_G(\pi)$ represents $term(G)[\pi]$.

2.3 Evaluation Theorem ([HP 88])
Let $G \Rightarrow_v H$ be an evaluation step. Then

$$term(G) \underset{\Delta}{\twoheadrightarrow\!\!\!\!\!\!\!+} term(H)$$

with $\Delta = \{\pi \mid node_G(\pi) = v\}$. □

The Evaluation Theorem expresses correctness: evaluation steps perform term rewriting. But there is a problem with evaluation rules for non-left-linear rewrite rules: since the left-hand side L of such a rule is not a tree, there may be no jungle morphism from L to a jungle G although the underlying term rewrite rule is applicable to $term(G)$. This happens if a shared variable in L corresponds to different occurrences of a subterm in G. To overcome this problem we introduce *folding rules* which allow to compress a jungle such that equal subterms are represented by the same node.

For each function symbol in F there is a folding rule $(L \hookleftarrow K \xrightarrow{b} R)$:

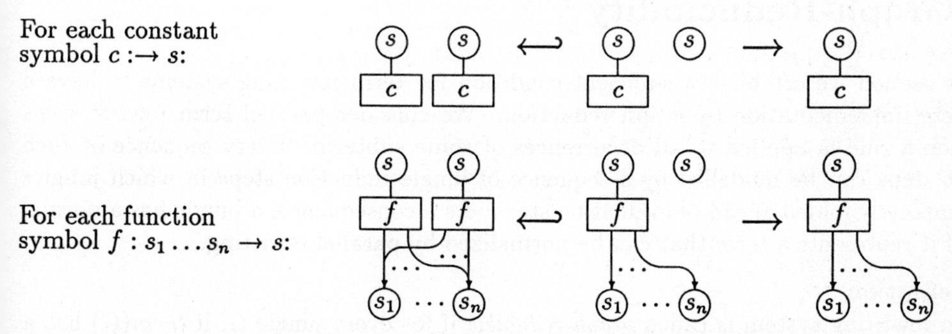

The application of a folding rule to a jungle G is called a *folding step* and works as follows:

- Find a jungle morphism $g : L \to G^\bullet$ with g_E injective.
- Remove $g(e)$, where e is the unique edge in $L - K$.
- Identify $g(v)$ and $g(v')$, where v and v' are the two source nodes in L.

If H is the resulting jungle, then this folding step is denoted by $G \rightsquigarrow H$. Again we allow every jungle essentially isomorphic to H as a result of this folding step (but require that variables are not renamed). A sequence $G \rightsquigarrow^* \tilde{G}$ of folding steps is called a *folding*.

A jungle G is said to be *fully collapsed* if no folding rule is applicable to G. Equivalently, G is fully collapsed if for all nodes v_1, v_2 in $G^\bullet$, $term_G(v_1) = term_G(v_2)$ implies $v_1 = v_2$. By applying folding rules as long as possible, every jungle can be transformed into a fully collapsed jungle.

2.4 Theorem ([HP 88])
For every jungle G there is a folding $G \rightsquigarrow^* \tilde{G}$ such that $\tilde{G}$ is fully collapsed. □

Note that folding rules are constructed in such a way that folding preserves terms: $G \rightsquigarrow^* \tilde{G}$ implies $term(G) = term(\tilde{G})$.

In the following we use evaluation and folding rules together and abbreviate $\Rightarrow \cup \rightsquigarrow$ by $\Rrightarrow$. A step $G \Rrightarrow H$ is called a *reduction step* and a sequence $G \Rrightarrow^* H$ of reduction steps is said to be a *reduction*.

A jungle H is a *normal form* (with respect to $\Rrightarrow$) if there is no reduction step $H \Rrightarrow J$. In this case H is a normal form *of* every jungle G with $G \Rrightarrow^* H$. A jungle G *has* a normal form if $G \Rrightarrow^* H$ for some normal form H. These notions are analogously defined for terms (by replacing $\Rrightarrow$ by $\rightarrow$).

2.5 Normal Form Theorem ([HP 88])
Let $G \Rrightarrow^* H$ be a reduction such that H is a normal form. Then $term(H)$ is a normal form of $term(G)$. □

3 Graph-Reducibility

In this section we establish a sufficient condition for term rewriting systems to have a complete implementation by graph reduction. We consider parallel term rewrite steps in which a rule is applied to all occurrences of some subterm. Every sequence of such rewrite steps can be modelled by a sequence of jungle reduction steps in which jungles are completely folded ahead of evaluation steps. As a consequence, a jungle has a normal form if it represents a term that can be normalized by parallel rewriting.

3.1 Definition
A term rewriting system is called *graph-reducible* if for every jungle G, if $term(G)$ has a normal form, then G has a normal form. [4] □

3.2 Theorem
The following is equivalent:

(i) $\mathcal{R}$ is graph-reducible.

(ii) For every jungle G, if $term(G)$ has a normal form, then there is a reduction $G \Rrightarrow^* H$ such that $term(H)$ is a normal form of $term(G)$.

Proof. Let G be a jungle such that $term(G)$ has a normal form.
(i) $\Rightarrow$ *(ii)*. If $\mathcal{R}$ is graph-reducible, then there is a reduction $G \Rrightarrow^* H$ such that H is a normal form. By Theorem 2.5, $term(H)$ is a normal form of $term(G)$.
(ii) $\Rightarrow$ *(i)*. By (ii) there is a reduction $G \Rrightarrow^* H$ such that $term(H)$ is a normal form of $term(G)$. A subsequent folding $H \rightsquigarrow^* \widetilde{H}$ yields a fully collapsed jungle $\widetilde{H}$ which is a

[4] The notion of graph-reducibility was introduced by Barendregt et al. [BEGKPS 87]. Their definition requires that *all* normal forms of a term can be computed by graph reduction. However, the results in [BEGKPS 87] that systems with "hypernormalizing" and "sib-normalizing" strategies are graph-reducible hold only for the weaker definition used here. See also section 4.

normal form by the Evaluation Theorem 2.3. (For if there were any jungle J with $\widetilde{H} \Rightarrow J$, then $term(H) = term(\widetilde{H}) \twoheadrightarrow\!\!\!\!\!+ \; term(J)$, so $term(H)$ would not be a normal form.) Thus $\mathcal{R}$ is graph-reducible. □

In order to give sufficient conditions for graph-reducibility, we consider the parallel application of a term rewrite rule to all occurrences of some subterm.

3.3 Definition

Given a term t and some redex π in t, $t \twoheadrightarrow\!\!\|_\pi u$ denotes a parallel rewrite step in which all redexes that represent the same subterm as π are reduced by the same rule. That is, $t \twoheadrightarrow\!\!\|_\pi u$ if and only if $t \twoheadrightarrow\!\!\!\!\!+_\Delta u$ with $\Delta = \{\pi' \mid t[\pi'] = t[\pi]\}$. □

By using folding rules it is possible to model arbitrary $\twoheadrightarrow\!\!\|$-steps by jungle reduction steps. This is achieved by completely folding a jungle before applying an evaluation rule.

3.4 Theorem

For every parallel term rewrite step $t \twoheadrightarrow\!\!\| u$ and every jungle G with $term(G) = t$ there is a reduction

$$G \overset{*}{\rightsquigarrow} \widetilde{G} \Rightarrow H$$

such that $term(H) = u$.

Proof. Let $l \to r$ be the rewrite rule and σ the substitution associated with $t \twoheadrightarrow\!\!\| u$. By Theorem 2.4 there is a folding $G \rightsquigarrow^* \widetilde{G}$ such that $\widetilde{G}$ is fully collapsed. Since $term(\widetilde{G}) = t$, there is a unique node v in $\widetilde{G}^\bullet$ with $term_{\widetilde{G}}(v) = \sigma(l)$. Let $(L \hookleftarrow K \to R)$ be the evaluation rule for $l \to r$. As shown in [HP 88], there is a jungle morphism $g : L \to \widetilde{G}^\bullet$ with $g(root_L) = v$, because $\widetilde{G}$ is fully collapsed. So there is an evaluation step $\widetilde{G} \Rightarrow_v H$. By the Evaluation Theorem 2.3 we have $t = term(\widetilde{G}) \twoheadrightarrow\!\!\!\!\!+_\Delta term(H)$ with $\Delta = \{\pi \mid node_{\widetilde{G}}(\pi) = v\}$. $node_{\widetilde{G}}$ takes each π with $t[\pi] = \sigma(l)$ to a node representing $\sigma(l)$. Since v is the only node in $\widetilde{G}^\bullet$ representing $\sigma(l)$, we conclude that $\Delta = \{\pi \mid t[\pi] = \sigma(l)\}$. Thus $t \twoheadrightarrow\!\!\| term(H)$ via $l \to r$ and σ, which implies $term(H) = u$. □

3.5 Corollary

Let $t \twoheadrightarrow\!\!\|^* u$ for some terms t, u. Then for every jungle G with $term(G) = t$ there is a reduction $G \Rrightarrow^* H$ such that $term(H) = u$. □

The above corollary suggests to consider the following kind of term rewriting systems.

3.6 Definition

A term rewriting system is called *parallelly normalizing* if for every term t having a normal form, there is a parallel rewrite sequence $t \twoheadrightarrow\!\!\|^* u$ to some normal form u. [5] □

Corollary 3.5 together with Theorem 3.2 yields the following result.

[5] Note that "parallelly normalizing" neither implies nor follows from the property that every term has a normal form. A term rewriting system with the latter property is called "normalizing" [DJ 90] or "weakly normalizing" [Klo 90].

3.7 Corollary
Parallelly normalizing term rewriting systems are graph-reducible.

Proof. Let $\mathcal{R}$ be parallelly normalizing and G be a jungle such that $term(G)$ has a normal form. Then $term(G) \twoheadrightarrow\!\!\|^* u$ for some normal form u. Hence, by Corollary 3.5, there is a reduction $G \Rrightarrow^* H$ with $term(H) = u$. With Theorem 3.2 it follows that $\mathcal{R}$ is graph-reducible. □

3.8 Example
Let $\mathcal{R}$ consist of the following three rules for the addition of natural numbers:

$$\begin{aligned} \mathtt{0} + \mathtt{x} &\to \mathtt{x} \\ \mathtt{succ(x)} + \mathtt{y} &\to \mathtt{succ(x+y)} \\ \mathtt{x} + \mathtt{y} &\to \mathtt{y} + \mathtt{x} \end{aligned}$$

Here a term t has a normal form if and only if there is at most one variable occurrence in t. The following procedure computes this normal form via innermost $\twoheadrightarrow\!\!\|$-steps (a redex π in t is said to be *innermost* if no proper subterm of $t[\pi]$ contains a redex):

while *there is some innermost redex* π *in* t **do**
begin
let $t[\pi] = u_1 + u_2$;
if u_1 *is a variable* **then** $t \twoheadrightarrow\!\!\|_\pi t'$ *by rule 3*
else $t \twoheadrightarrow\!\!\|_\pi t'$ *by rule 1 or 2;*
$t := t'$
end

The procedure terminates whenever the input term has a normal form, so $\mathcal{R}$ is parallelly normalizing and thus graph-reducible. □

The following example shows that the converse of Corollary 3.7 does not hold, that is, there are graph-reducible term rewriting systems which are not parallelly normalizing.

3.9 Example
Suppose that $\mathcal{R}$ is given as follows:

$$\begin{aligned} \mathtt{f(x,y)} &\to \mathtt{f(a,x)} \\ \mathtt{a} &\to \mathtt{b} \\ \mathtt{f(b,a)} &\to \mathtt{c} \end{aligned}$$

To see that $\mathcal{R}$ is graph-reducible, let t_1,t_2 be any terms and G be any jungle with $term(G) = \mathtt{f}(t_1, t_2)$. Then

$$G = [\mathtt{f}(t_1, t_2)] \Rightarrow [\mathtt{f}(\mathtt{a}, t_1)] \Rightarrow [\mathtt{f(a,a)}] \Rightarrow [\mathtt{f(b,a)}] \Rightarrow [\mathtt{c}]$$

where, for every term t, $[t]$ denotes some jungle with $term([t]) = t$. Note that the two occurrences of a in $\mathtt{f(a,a)}$ are represented by different nodes in $[\mathtt{f(a,a)}]$.

Although $\mathcal{R}$ is graph-reducible, it is not parallelly normalizing. f(b, b), for instance, cannot be normalized by $\twoheadrightarrow\!\!\|$-steps:

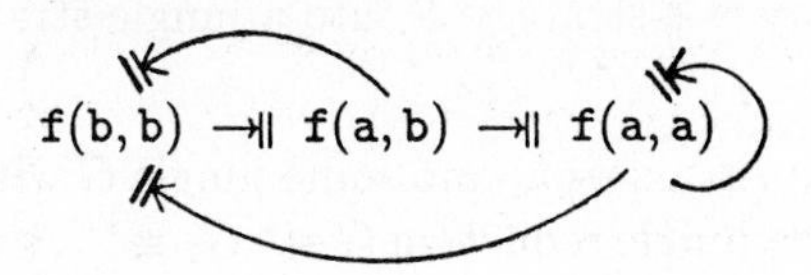

□

It should be noted that the above example does no longer work if we require that jungles are completely folded before an evaluation rule is applied. Indeed, if we consider graph-reducibility for this special kind of reduction, then a term rewriting system is graph-reducible if and only if it is parallelly normalizing.

4 Reduction Strategies

In evaluating terms or jungles we are interested to perform only steps that eventually lead to a normal form. So we need some control mechanism which selects for every term resp. jungle those steps that have to be performed next. This leads us to the investigation of reduction strategies.

Similar to Barendregt et al. [BEGKPS 87] we use an abstract definition of a strategy which can be applied to both term and graph rewriting.

4.1 Definition

1. Let $(B, \rightarrow)$ be an *abstract reduction system*, that is, B is a set and $\rightarrow$ is a binary relation on B. A *strategy* for $(B, \rightarrow)$ is a function $\mathcal{S}$ that maps each b in B to a set $\mathcal{S}(b)$ satisfying the following two conditions:

- Each element in $\mathcal{S}(b)$ is a reduction sequence of the form $b \rightarrow b_1 \rightarrow b_2 \rightarrow \ldots \rightarrow b_n$ with $n \geq 1$.
- $\mathcal{S}(b)$ is empty only if b is a normal form.

We write $b \rightarrow_{\mathcal{S}} b'$ if there is some sequence $b \rightarrow b_1 \rightarrow \ldots \rightarrow b_n = b'$ in $\mathcal{S}(b)$. $\mathcal{S}$ is called a *one-step strategy* if, for each b in B, all sequences in $\mathcal{S}(b)$ are of length 1. $\mathcal{S}$ is said to be *normalizing* if for each b in B having a normal form, there is no infinite sequence $b \rightarrow_{\mathcal{S}} b_1 \rightarrow_{\mathcal{S}} b_2 \rightarrow_{\mathcal{S}} \ldots$

2. A term rewriting strategy $\mathcal{S}$ is a $\twoheadrightarrow\!\!\|$*-strategy* if, for every term t, $\mathcal{S}(t)$ contains only sequences of the form $t \twoheadrightarrow\!\!\| t_1 \twoheadrightarrow\!\!\| t_2 \twoheadrightarrow\!\!\| \ldots \twoheadrightarrow\!\!\| t_n$. □

It is straightforward to verify the following characterization of parallelly normalizing systems.

4.2 Fact

A term rewriting system is parallelly normalizing if and only if it has a normalizing $\twoheadrightarrow\!\!\|$-strategy. □

By Corollary 3.7 it is in principle possible to compute term normal forms via jungle reductions if $\mathcal{R}$ is parallelly normalizing. To make this reality, however, we have to find a (computable) normalizing jungle reduction strategy. Fortunately, Theorem 3.4 allows us to effectively transform every $\twoheadrightarrow\!\!\|$-strategy $\mathcal{S}$ into a jungle strategy $\mathcal{S}_J$.

4.3 Definition

Given some sequence $t \twoheadrightarrow\!\!\| t_1 \twoheadrightarrow\!\!\| \dots \twoheadrightarrow\!\!\| t_n$ and some jungle G with $term(G) = t$, we denote by $red_G[t \twoheadrightarrow\!\!\| t_1 \twoheadrightarrow\!\!\| \dots \twoheadrightarrow\!\!\| t_n]$ a jungle reduction $G \Rrightarrow^* G_1 \Rrightarrow^* \dots \Rrightarrow^* G_n$ with $term(G_i) = t_i$, for $i = 1, \dots, n$. Such a reduction can be effectively constructed by Theorem 3.4.
Let now G be an arbitrary jungle. $\mathcal{S}_J(G)$ is defined as follows:

- $\mathcal{S}_J(G) = \emptyset$ if G is a normal form.
- $\mathcal{S}_J(G) = \{G \rightsquigarrow^* \tilde{G}\}$ if $term(G)$ is a normal form, but G is not fully collapsed. Here $G \rightsquigarrow^* \tilde{G}$ is some folding such that $\tilde{G}$ is fully collapsed.
- $\mathcal{S}_J(G) = \{red_G[term(G) \twoheadrightarrow\!\!\| t_1 \twoheadrightarrow\!\!\| \dots \twoheadrightarrow\!\!\| t_n] \mid term(G) \twoheadrightarrow\!\!\| t_1 \twoheadrightarrow\!\!\| \dots \twoheadrightarrow\!\!\| t_n \in \mathcal{S}(term(G)$ if neither of the first two cases applies.

4.4 Theorem

If $\mathcal{S}$ is a normalizing $\twoheadrightarrow\!\!\|$-strategy, then $\mathcal{S}_J$ is normalizing.

Proof. Let G be a jungle having a normal form. Suppose there were an infinite sequence $G \Rrightarrow_{\mathcal{S}_J} G_1 \Rrightarrow_{\mathcal{S}_J} G_2 \Rrightarrow_{\mathcal{S}_J} \dots$ Then $term(G) \to_{\mathcal{S}} term(G_1) \to_{\mathcal{S}} term(G_2) \to_{\mathcal{S}} \dots$ by the above construction. But $term(G)$ has a normal form by Theorem 2.5, so $\mathcal{S}$ would not be normalizing. □

If some term rewriting strategy $\mathcal{S}$ is known to be normalizing, one may try to "parallelize" $\mathcal{S}$ in order to obtain a normalizing $\twoheadrightarrow\!\!\|$-strategy. There is a natural choice for such a transformation if $\mathcal{S}$ is a one-step strategy.

4.5 Definition

Let $t \to u$ be a term rewrite step via some rule $l \to r$ and some redex π. Then $t \twoheadrightarrow\!\!\|_\pi \hat{u}$ via $l \to r$ is called the *parallelization* of $t \to u$. For every one-step strategy $\mathcal{S}$, the $\twoheadrightarrow\!\!\|$-strategy $\mathcal{S}_\|$ is defined by

$$\mathcal{S}_\|(t) = \{t \twoheadrightarrow\!\!\| \hat{u} \mid t \twoheadrightarrow\!\!\| \hat{u} \text{ is the parallelization of some step } t \to u \text{ in } \mathcal{S}(t)\}.$$

□

Examples for one-step strategies that are normalizing after parallelization are the so-called sib-normalizing strategies considered by Barendregt et al. [BEGKPS 87]. A one-step strategy $\mathcal{S}$ is said to be *sib-normalizing* if the extended strategy $\tilde{\mathcal{S}}$ defined by

$$\tilde{\mathcal{S}}(t) = \{t \underset{\mathcal{S}}{\to} t' \overset{*}{\to} t'' \mid t' \overset{*}{\to} t'' \text{ reduces siblings of the redex reduced by } t \underset{\mathcal{S}}{\to} t'\}$$

is normalizing. (A *sibling* of a redex π in a term t is a redex π' with $t[\pi] = t[\pi']$.) In [BEGKPS 87] it is shown that every left-linear term rewriting system that has a sib-normalizing strategy is graph-reducible.

4.6 Theorem
Every term rewriting system that has a sib-normalizing strategy is parallelly normalizing. The converse does not hold.

Proof. It is evident that $t \rightarrow_{\mathcal{S}_{\|}} t'$ implies $t \rightarrow_{\tilde{\mathcal{S}}} t'$, for every one-step strategy $\mathcal{S}$ and for all terms t, t'. Hence if $\mathcal{S}$ is sib-normalizing, then $\mathcal{S}_{\|}$ is a normalizing $\twoheadrightarrow_{\|}$-strategy. To show that the converse is not true, let $\mathcal{R}$ consist of the two rules $\mathtt{a} \rightarrow \mathtt{b}$ and $\mathtt{a} \rightarrow \mathtt{f(a,a)}$ [6]. Then the $\twoheadrightarrow_{\|}$-strategy which applies the first rule simultaneously to all occurrences of a in a term is normalizing. But for every one-step strategy $\mathcal{S}$ and for every term t that contains at least two occurrences of a, there is a step $t \rightarrow_{\tilde{\mathcal{S}}} t'$ such that t' again contains two occurrences of a. This is because after a step $t \rightarrow_{\mathcal{S}} \bar{t}$ there is an occurrences of a in $\bar{t}$ to which the second rule can be applied. Hence $\tilde{\mathcal{S}}$ is not normalizing. □

An interesting difference between $\tilde{\mathcal{S}}$ and $\mathcal{S}_{\|}$ is that (by definition) for $\tilde{\mathcal{S}}$ to be normalizing it is necessary that $\mathcal{S}$ is normalizing, while $\mathcal{S}_{\|}$ may be normalizing even if $\mathcal{S}$ is not. As an example for the latter, consider two terminating term rewriting systems $\mathcal{R}_0$ and $\mathcal{R}_1$ with disjoint function symbols. Then $\mathcal{R}_0 \cup \mathcal{R}_1$ may be non-terminating. In this case the strategy $\mathcal{S} : t \mapsto \{t \rightarrow t' \mid t' \text{ is an arbitrary term}\}$ is not normalizing (and hence not sib-normalizing) while $\mathcal{S}_{\|}$ is normalizing (cf. section 5.1).

5 Two Classes of Parallelly Normalizing Systems

In this section we show that both terminating and orthogonal term rewriting systems are parallelly normalizing. (In fact, the termination property can be relaxed to "parallel termination"). This gives us practical methods to prove graph-reducibility: many methods for proving termination are known (see [Der 87]), and orthogonality is a syntactic property which is simple to check.

5.1 Parallelly Terminating Systems

A term rewriting system is *terminating* if there is no infinite rewrite sequence $t_1 \rightarrow t_2 \rightarrow t_3 \rightarrow \ldots$ Clearly, in this case there is neither an infinite sequence of $\twoheadrightarrow_{\|}$-steps, so every terminating system is parallelly normalizing. This argument suggests the following weaker condition.

5.1 Definition
A term rewriting system is called *parallelly terminating* if there is no infinite sequence $t_1 \twoheadrightarrow_{\|} t_2 \twoheadrightarrow_{\|} t_3 \twoheadrightarrow_{\|} \ldots$ □

Hence for parallelly terminating systems every $\twoheadrightarrow_{\|}$-strategy is normalizing. The following example (from Toyama [Toy 87]) demonstrates that parallelly terminating systems are not necessarily terminating:

[6] This example is due to Michael Löwe.

$$\begin{aligned} \mathcal{R}_0: \; & \texttt{f(0,1,x)} \rightarrow \texttt{f(x,x,x)} \\ \mathcal{R}_1: \; & \texttt{or(x,y)} \rightarrow \texttt{x} \\ & \texttt{or(x,y)} \rightarrow \texttt{y} \end{aligned}$$

Although $\mathcal{R}_0$ and $\mathcal{R}_1$ are terminating, $\mathcal{R}_0 \cup \mathcal{R}_1$ is non-terminating since the term `f(or(0,1),or(0,1),or(0,1))` reduces in three steps to itself. In [Plu 91] it is shown that jungle reduction is terminating for $\mathcal{R}_0 \cup \mathcal{R}_1$, that is, there is no infinite sequence of $\Rrightarrow$-steps. With Theorem 3.4 it follows that $\mathcal{R}_0 \cup \mathcal{R}_1$ is parallelly terminating. Actually, the main result in [Plu 91] implies that a combined system $\mathcal{R}_0 \cup \mathcal{R}_1$ is parallelly terminating whenever $\mathcal{R}_0$ and $\mathcal{R}_1$ are terminating and when the left-hand sides of $\mathcal{R}_i$ have no common function symbols with the right-hand sides of $\mathcal{R}_{1-i}$ $(i = 0, 1)$.

5.2 Orthogonal Systems

Orthogonal (or *regular*) term rewriting systems are characterized by two properties: their rules are *left-linear* and *nonoverlapping*. That is, every variable occurs at most once in every left-hand side, and whenever there are two rules $l \rightarrow r$ and $l' \rightarrow r'$ such that l overlaps l' in some subterm u, then $l = u$ and $l \rightarrow r = l' \rightarrow r'$. (A term t *overlaps* a term t' in a subterm u of t, if u is not a variable and there are substitutions σ and σ' such that $\sigma(u) = \sigma'(t')$.)

An example for an orthogonal term rewriting system is *Combinatory Logic*:

$$\begin{aligned} & ((\texttt{S} \cdot \texttt{x}) \cdot \texttt{y}) \cdot \texttt{z} \rightarrow (\texttt{x} \cdot \texttt{z}) \cdot (\texttt{y} \cdot \texttt{z}) \\ & (\texttt{K} \cdot \texttt{x}) \cdot \texttt{y} \rightarrow \texttt{x} \\ & \texttt{I} \cdot \texttt{x} \rightarrow \texttt{x} \end{aligned}$$

The (nullary) function symbols `S`, `K`, `I` denote combinators while the binary function symbol $\cdot$ denotes function application. (See [Klo 90] for the importance of Combinatory Logic.)

It is well-known that the *parallel outermost* strategy is normalizing for orthogonal systems. This strategy reduces all the outermost redexes in a term in parallel. By a slight modification of the parallel outermost strategy we obtain a normalizing $\twoheadrightarrow\!\!\|$-strategy. Let t be any term not in normal form, and let $\Pi = \{\pi_1, \ldots, \pi_n\}$ be a set of outermost redexes in t such that

- $t[\pi_i] \neq t[\pi_j]$ for $i \neq j$, and
- for each outermost redex π in t there is some $\pi_i \in \Pi$ with $t[\pi] = t[\pi_i]$.

Then $\mathcal{S}(t) = \{t \twoheadrightarrow\!\!\|_{\pi_1} t_1 \twoheadrightarrow\!\!\|_{\pi_2} \ldots \twoheadrightarrow\!\!\|_{\pi_n} t_n\}$ defines a $\twoheadrightarrow\!\!\|$-strategy. (Note that $\mathcal{S}$ is well-defined by orthogonality: π_{i+1} is a redex in t_i for $i = 1, \ldots, n-1$.) $\mathcal{S}$ is normalizing by O'Donnell's result that "eventually outermost" rewrite sequences are normalizing (see [O'D 77]).

6 Conclusion

We can summarize our classification of graph-reducible term rewriting systems as follows:

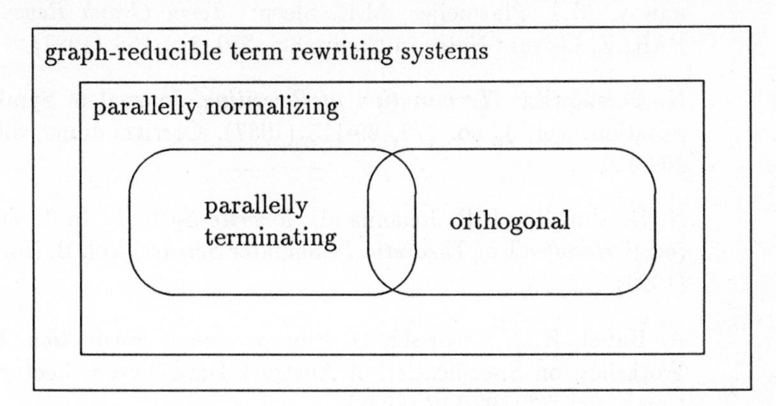

All the inclusions in this diagram are proper. (Example 3.8 shows a parallelly normalizing system which is neither parallelly terminating nor orthogonal.) Moreover, the example shown in the introduction demonstrates that confluence[7] is not sufficient for graph-reducibility.

Finally, we mention some topics for further research:

- Not all graph-reducible systems are parallelly normalizing. Is there a satisfactory *characterization* of graph-reducibility in terms of term rewriting notions?
- Are there sufficient *syntactic* conditions for graph-reducibility which go beyond orthogonality? If so, do these conditions give rise to efficient (or at least computable) normalizing strategies?
- Our definition of graph-reducibility can be strengthened by requiring that for every jungle G *all* normal forms of $term(G)$ can be computed by jungle reductions. This *strong graph-reducibility* clearly holds for graph-reducible systems in which every term has at most one normal form, so in particular for orthogonal systems. But what can be said for other classes of term rewriting systems? Terminating systems, for instance, are not strongly graph-reducible in general (a counterexample can be found in the long version of [HP 88]).

Acknowledgements

I wish to thank Berthold Hoffmann and Michael Löwe for comments on a preliminary version of this paper.

[7] A term rewriting system is called *confluent* if for all terms t, t_1, t_2 with $t_1 \; {}^*\!\!\leftarrow t \rightarrow^* t_2$ there is a term t_3 such that $t_1 \rightarrow^* t_3 \; {}^*\!\!\leftarrow t_2$.

References

[BEGKPS 87] H.P. Barendregt, M.C.J.D. van Eekelen, J.R.W. Glauert, J.R. Kennaway, M.J. Plasmeijer, M.R. Sleep: *Term Graph Rewriting.* Proc. PARLE, Lecture Notes in Comp. Sci. 259, 141-158 (1987)

[Der 87] N. Dershowitz: *Termination of Rewriting.* Journal of Symbolic Computation, vol. 3, no. 1/2, 69-115 (1987). Corrigendum: vol. 4, no. 3, 409-410

[DJ 90] N. Dershowitz, J.-P. Jouannaud: *Rewrite Systems.* In J. van Leeuwen (ed.): *Handbook of Theoretical Computer Science.* Vol. B, North-Holland (1990)

[HKP 88] A. Habel, H.-J. Kreowski, D. Plump: *Jungle Evaluation.* Proc. Fifth Workshop on Specification of Abstract Data Types, Lecture Notes in Comp. Sci. 332, 92-112 (1988)

[HP 88] B. Hoffmann, D. Plump: *Jungle Evaluation for Efficient Term Rewriting.* Proc. Algebraic and Logic Programming, Akademie-Verlag, Berlin, 191-203 (1988). Also in Lecture Notes in Comp. Sci. 343 (1989). Long version to appear in RAIRO Theoretical Informatics and Applications

[Klo 90] J.W. Klop: *Term Rewriting Systems — From Church-Rosser to Knuth-Bendix and Beyond.* Proc. ICALP '90, Lecture Notes in Comp. Sci. 443, 350-369 (1990)

[O'D 77] M.J. O'Donnell: *Computing in Systems Described by Equations.* Lecture Notes in Comp. Sci. 58 (1977)

[Plu 91] D. Plump: *Implementing Term Rewriting by Graph Reduction: Termination of Combined Systems.* Proc. 2nd Int. Workshop on Conditional and Typed Rewriting Systems, Montreal 1990, to appear in Lecture Notes in Comp. Sci.

[Toy 87] Y. Toyama: *Counterexamples to Termination for the Direct Sum of Term Rewriting Systems.* Information Process. Lett. 25, 141-143 (1987)

A NOTE ON GRAPH DECIMATION

Azriel Rosenfeld
Center for Automation Research
University of Maryland
College Park, Maryland 20742-3411, USA

ABSTRACT: To avoid paradoxes in parallel graph rewriting, it is desirable to forbid overlapping instances of a subgraph to be rewritten simultaneously. The selection of maximal sets of nonoverlapping instances corresponds to the selection of maximal independent sets of nodes in a derived graph. This note briefly examines the process of repeatedly selecting such sets of nodes. It shows that doing so "decimates" the graph in the sense that the number of nodes shrinks exponentially; but that unfortunately, the degree of the graph may grow exponentially.

Keywords: Parallel graph rewriting, maximal independent sets, graph decimation.

When a parallel rewriting rule $L \rightarrow R$ is applied to a graph, paradoxes may arise if overlapping instances of L can be rewritten simultaneously. It is preferable to apply the rule only to a maximal set of disjoint instances of L. This corresponds to a maximal independent set in the contracted graph whose nodes represent the instances of L, and in which two nodes are neighbors iff their instances overlap.

This note briefly examines the process of repeatedly selecting maximal independent sets of nodes in a graph. It shows that this process "decimates" the graph in the sense that the number of nodes shrinks exponentially; but that unfortunately, the degree of the graph may grow exponentially.

Definition: G' is a *decimation* of G if

a) The nodes of G' are a maximal independent set of the nodes of G (In other words: no two nodes of G' are neighbors in G; any node of G has distance at most 1 (in G) from a node of G'.)

b) Two nodes of G' are neighbors (in G') iff their distance in G is at most 3.

Proposition 1: If G is connected, so is G'.

Proof: Let P, Q be any two nodes of G'. If they have distance ≤ 3 in G, they are neighbors in G'. Let their distance in G be $m > 3$ and suppose all pairs of nodes of G' that have distance $< m$ in G are connected in G'. If a path of length m from P to Q in G has an intermediate node R in G', then by induction hypothesis P, R and R, Q are connected in G' and we are done. Otherwise, let the path be $P = P_0, P_1, P_2, \ldots, P_m = Q$, and let R be a neighbor of P_2 that is a node of G'. Then P and R have distance ≤ 3 in G, hence are neighbors in G'; and $R, P_2, \ldots, P_m = Q$ is a path of length $m-1$, so R, Q have distance $< m$ in G and so are connected in G', proving that P, Q are connected in G'. //

In fact, if nodes P, Q of G' are joined by a path of length m in G, they are joined by a path of length at most m in G'. Proof: If nodes $P_i, P_{i+1}, \ldots, P_j$ $(0 < i < j < n)$ on the path are not in G', each of them has a neighbor Q_{i+k} $(0 \leq k \leq j-i)$ that is in G', and since consecutive Q's have distance ≤ 3 in G, they must be neighbors in G' (or must be the same); similarly, Q_i is a neighbor of P_{i-1} and Q_j is a neighbor of P_{j+1}.

Proposition 2: Any G is a decimation of some G^*.

Proof: Construct G^* by inserting an "intermediate" node into every arc of G, and then construct $(G^*)'$ by deleting these extra nodes. Readily, this is a legal decimation and gives us G again. //

If G is connected (which we assume from now on), we cannot have $G' = G$ unless G consists of a single node. Thus if we iterate the process of decimation (notation: $G', G'', \ldots, G^{(k)}, \ldots$), we eventually obtain a graph consisting of a single node. (Evidently, if G has a nonempty set of nodes, so has G'.) In fact, we now show that if G has n nodes, this reduction to a single node requires only $O(\log n)$ iterations of decimation.

Proposition 3: If G has n nodes, G'' has fewer than $n/2$ nodes.

Proof: Let G' have $n - m$ nodes. None of these nodes were neighbors in G, and since G was connected, each of them was a neighbor of at least one of the m

deleted nodes. In G', the neighbors of any deleted node are a clique (since they have a common neighbor in G and so have distance 2 in G), and in G'', at most one node of any clique of G' can survive (since two neighbors in G' cannot both be in G''); hence G'' has at most m nodes. If $m \geq n/2$, G' has at most $n/2$ nodes, so G'' has fewer than $n/2$; and if $m < n/2$, G'' has fewer than $n/2$ nodes, since it has at most m. //

Corollary 4: If G has n nodes, $G^{2\log m}$ consists of a single node. //

The (maximum) degree of G' can be much higher than that of G. For example, let G be a rooted tree of out-degree d, and let P be a (sufficiently) interior node of G that belongs to G'. The children of P cannot be nodes of G', but every grandchild of P either is a node of G' or has a child that is a node of G', and all these nodes are neighbors of P in G'; hence P has degree at least d^2 in G'.

In fact, we can construct a rooted tree of maximum out-degree d that can be decimated repeatedly to yield rooted trees of maximum out-degrees d^2, d^4, $d^8, \ldots$; eventually we get a star of degree d^{2^h} (where the height of the original tree was 3^h), and this decimates to a single node. The number of nodes in the initial rooted tree at the levels below the root are successive powers of d, some of which are repeated. The number of repetitions of d^i depends on the highest power of 2 that divides i; if that power is 2^j, the number of repetitions is $(3^j+1)/2$. [Thus if i is odd (so $j=0$), d^i is repeated only once; if i is singly even ($j=1$), d^i is repeated twice; while for $j=2, 3, 4, \ldots$, the numbers of repetitions are 5, 14, 41, $\cdots$]. For example, the tree of height 27 (i.e. 27 levels below the root) has the following values of i:

$$0,1,2,2,3,4,4,4,4,4,5,6,6,7,8,8,8,8,8,8,8,8,8,8,8,8,8,8$$

Note that when two successive levels have different degrees, the nodes on the first level have out-degrees d; while if two successive levels have the same degree, the nodes on the first level have out-degrees 1. We decimate the tree by deleting all levels that are not multiples of 3 (where the root is level 0); evidently this is a legal decimation, and the result is still a rooted tree having 3^{h-1} levels. When we do this to the example given above, the levels in the resulting tree of height 9 have d^i nodes for $i=0,2,4,4,6,8,8,8,8,8$; thus the nodes at each level have out-degrees either d^2 or 1. Repeating the decimation yields a tree of height 3 whose levels have d^i nodes for $i=0,4,8,8$ (out-degrees d^4 or 1); and a final repetition yields a star (a tree of height 1) whose levels have d^i nodes for $i=0,8$ (out-degree d^8). In this example, by "stretching" the tree (so that most of its levels have out-degree 1), we have insured that nodes at the same level never become neighbors after the decimation [in fact, for every surviving level, the level above it (before the decimation) had the same number of nodes, so that no two nodes on the surviving level had distance ≤ 3], thus guaranteeing that the result of the decimation is still a tree.

We have seen that the decimation process shrinks a graph rapidly while preserving its connectedness, but does not preserve boundedness of its degree. The author and his colleagues have also considered alternatives to and restrictions on the decimation process that do preserve degree; work on the subject is in progress.

Acknowledgement

The author thanks Yung Kong, Walter Kropatsch, Peter Meer, Annick Montanvert, David Mount, and Angela Wu for their comments on and contributions to this note.

PROGRESS:

A VHL-Language Based on Graph Grammars

Andy Schürr

Lehrstuhl für Informatik III
Aachen University of Technology
Ahornstraße 55, D–5100 Aachen, Germany

Abstract: The Very High Level language **PROGRESS** presented within this paper is the first **statically typed** language which is based on the concepts of **PRO**grammed Graph **RE**-writing **S**y**S**tems. This language supports different programming paradigms by offering procedural and declarative programming constructs for the definition of integrity constraints, functional attribute dependencies, derived binary relationships, atomic graph rewrite rules, and complex graph transformations.
Both the language and its underlying formalism are based on experiences of about ten years with a **model–oriented** approach to the specification of document classes and document processing tools (of the **I**ntegrated **P**rogramming **S**upport **EN**viroment **IPSEN**). This approach, called **graph grammar engineering**, is characterized by using **attributed graphs** to model object structures. Programmed graph rewriting systems are used to specify operations in terms of their effect on these graph models.
This paper informally introduces PROGRESS' underlying graph grammar formalism and demonstrates its systematic use by specifying parts of a syntax–directed editor for a simple expression language. The construction of a **PROGRESS–specific environment** with an integrated set of tools is the topic of another paper within this volume (cf. /NS 90/).

Keywords: Model–oriented specification language, graph rewriting systems, attribute grammars, programmed graph grammars, statically typed language

Contents

---

*) This research was/is partially supported by Stiftung Volkswagenwerk and DFG.
A preliminary version of this paper appeared as /Sc 89/.

0. Introduction

Modern software systems for application areas like office automation and software engineering are usually highly interactive and deal with complex structured objects. The systematic development of these systems requires precise and readable descriptions of their desired behaviour. Therefore, many specification languages and methods have been introduced to produce formal descriptions of various aspects of a software system, such as the design of object structures, the effect of operations on objects, or the synchronization of concurrently executed tasks. Many of these languages use **special classes of graphs** as their underlying data models. Conceptual graphs /So 84/, (semantic) data base models /HK 87/, petri nets /GJ 82/, or attributed trees /Re 84/ are well-known examples of this kind.

Within the research project **IPSEN** a **graph grammar based specification method** has been used to model the internal structure of software documents and to produce executable specifications of corresponding document processing tools, as e.g. syntax-directed editors, static analyzers, or incremental compilers /Na 85/. The development of such a specification, which is termed '**programmed graph rewriting system**' consists of two closely related subtasks. The first one is to design a graph model for the corresponding complex object structure. The second one is to program object (graph) analyzing and modifying operations by composing sequences of subgraph tests and graph rewrite rules.

Based on experiences of about ten years with this IPSEN specific approach to the **formal specification and systematic development** of software, we were able to adapt the original formalism (introduced in /Na 79/) to the requirements of this application area (cf. /Le 88/). Furthermore, a method, called **graph grammar engineering**, has been developed for the construction of large rewriting systems in a systematic engineering-like manner (cf. /Na 85, EL 87/).

Parallel to the continuous evolution of the graph grammar formalism and the graph grammar engineering method, the design of a **graph grammar specification language** is in progress. The outcome of this design process is a language which tries to combine the advantages of

- **data definition languages** with respect to the definition of integrity constraints,
- **polymorphic/object-oriented programming languages** with respect to the concept of subtyping/multiple inheritance and the use of typed type-parameters,
- **attribute tree grammars** with respect to the definition of functional attribute dependencies,
- and **programmed graph rewriting systems** with respect to the definition of nondeterministic graph transformations

within one strongly-typed language. A first version of this language, named **PROGRESS** (for **PRO**grammed **G**raph **RE**writing **S**y**S**tems), has been fixed a few months ago, and a prototype of a **programming environment** for this version of the language is under development (cf. /NS 90/).

The main purpose of this paper is to survey the language PROGRESS. Therefore, the next section is dedicated to an informal introduction of our **graph grammar formalism** (more precisely a "programmed graph rewriting system" calculus). This section also introduces the running example which is used throughout the whole paper for demonstration purposes. Section 3 then presents a characteristical subset of the **language PROGRESS.** Section 4 is a brief survey of **related work** and section 5 contains a few remarks about **application areas** for PROGRESS and about the current stage of the PROGRESS **environment's** implementation. The last section, contains the conclusion and discusses **future development plans** for this graph grammar based specification language.

1. Attributed Graphs and Programmed Graph Rewriting Systems

Writing about PROGRESS one has to explain and discuss topics at least at two different levels. The first one is that of the language's underlying **graph grammar formalism** which builds a fundament for the semantic definition of the language. The second one is that of **language design issues** and comprises tasks like defining the abstract syntax, the name binding rules, and the concrete representations for the new language. To avoid a confusion of these two levels within the paper, we decided to dedicate this section completely to the informal introduction of our graph grammar formalism and to defer the presentation of the language itself to the following section.

The first subsection introduces the formalism's underlying data model – **attributed, node and edge labeled, directed graphs** – and demonstrates the mapping of complex object structures onto this kind of graphs. The second subsection sketches the specification of the functional behaviour of object (graph) processing tools by means of **graph rewrite rules** and **control diagrams.** Within this section and for the remainder of the paper we use a subset of the well-known applicative programming language 'Exp' as a running example (variants of this example may be found in /RT 84, JF 84/).

```
EXP ::= NIL_EXP | BINARY_EXP | CONST_EXP | DEF_EXP | NAME_EXP
        (* 'NIL_EXP' etc. are subphyla of the phylum 'EXP'. *)

NIL_EXP ::= NilExp ()            (* Placeholder for unexpanded subexpressions. *)

BINARY_EXP ::= Prod ( EXP, EXP ) | Quot ( EXP , EXP ) |  . . .

CONST_EXP ::= DecConst ( String )

DEF_EXP ::= Let ( NAME, EXP, EXP )(* Binds the first expression to applied occurrences
                                     of the name within the second expression. *)

NAME_EXP ::= NameUse ( String )  (* Applied  occurrence of an identifier. *)

NAME ::= NameDef ( String )      (* Defining occurrence of an identifier. *)
```

Figure 1 : Abstract syntax of the language 'Exp'.

1.1 Data Modeling with Attributed Graphs

This subsection introduces the class of **attributed, node and edge labeled, directed graphs** (in the sequel just called graphs). Based on the afore–mentioned 'Exp' language we will discuss how to represent the sentences of a certain programming language by the instances of a certain subclass of graphs.

Therefore, we start our explanations with the description of the language 'Exp'. This language characterizes all legal input sequences for a very primitive desk calculator. The well–known 'let' construct is used to name and reuse intermediate computation results. The usual scoping rules for block structured programming languages direct the binding of applied occurrences of names to their corresponding 'let'–definitions. Figure 1 defines the abstract syntax of the language in a manner of writing called the **operator / phylum** notation (cf. /No 87/). The somewhat artificial subphyla 'NIL_EXP', 'BINARY_EXP' etc. have been introduced to group operators with same properties and to emphasize the similarities between the abstract syntax description of a language and the PROGRESS declarations for a corresponding graph scheme (cf. section 2.).

Figure 2 presents the **text and graph representation** of a typical sentence of the language 'Exp'. This sentence still contains one unexpanded subexpression with an undefined value represented by '(<EXP>)'. As a consequence, the values of the two (sub–) expressions starting with 'let x = ...' are undefined, too. The computation of the subexpression 'let y = 2 in x*y' yields the value '16' due to the fact that 'x' is bound to '8' and 'y' to '2'.

let x = 8 in ⌊let x = ⌊let y = 2 in x * y in⌉(<EXP>) / 4⌉

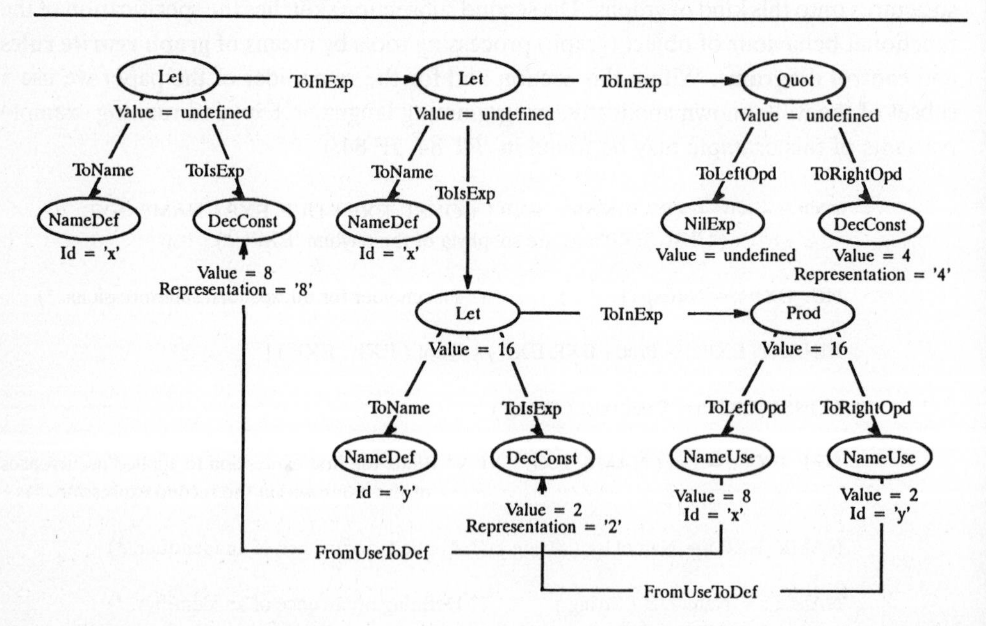

Figure 2 : Text and graph representation of an 'Exp' sentence.

Starting from the definition of the abstract syntax in figure 1, the **systematic development** of the graph representation has been directed by the following guide–lines (cf. also /EL 87/):

- Map the context free syntactic structure of an 'Exp' sentence onto a tree–like graph structure which is equivalent to the sentence's abstract syntax tree. This structure builds the skeleton of the graph representation and it contains a separate **labeled node** for the root of any subexpression of the sentence. Distinguish different types of subexpressions by labeling their root nodes with the operators 'NilExp', 'Sum' etc. of the abstract syntax definition.

- Represent the abstract syntax tree's context free relationships, only implicitly defined within the abstract syntax, by **labeled edges** with arbitrarily chosen labels ('ToLeftOpd', ...) and introduce an additional type of edges for any kind of context sensitive relationships. In our example we have to introduce edges labeled 'FromUseToDef' to bind all applied occurrences of names to their value defining expressions.

- Use **(external) node attributes** to hold instances of phyla whose values are atomic from the current point of view and which encode properties inherent to a single node. This holds true for all phyla which do not appear on the left–hand side of any rule of the abstract syntax (phyla whose instances represent lexical units). Within our example we have to attach a 'String' attribute to all nodes labeled with one of the operators 'DecConst', 'NameDef', and 'NameUse'. To emphasize a distinction between strings representing identifiers and strings representing numbers, we introduce an attribute called 'Representation' for nodes labeled 'DecConst' and an attribute called 'Id' for nodes labeled either 'NameDef' or 'NameUse'.

- Use **(derived) node attributes** to encode node properties usually concerning aspects of dynamic semantics. Such an attribute is called 'derived' – instead of 'external' – if and only if its value is defined by a directed equation. Within this equation other attributes of the same node or of adjacent nodes may be referenced. Thus, we are able to establish functional attribute dependencies like: The 'Value' attribute of a 'Prod' node must be equal to the product of the 'Value' attributes of its two operand nodes which are the sinks of the two outgoing 'ToLeftOpd' and 'ToRightOpd' edges.

1.2 'Programming' with Graph Rewrite Rules

In the previous subsection we explained PROGRESS' underlying data model and discussed the matter of systematically deriving graph representations for the sentences of the language 'Exp'. This subsection deals with the subject of **specifying complex operations** of tools like a syntax–directed editor, transforming one sentence of the language 'Exp' into another one of the same language. Thinking in terms of our data model these operations are nothing else but class preserving graph transformations.

let x = 8 in let x = let y = 2 in x * y in x / 4

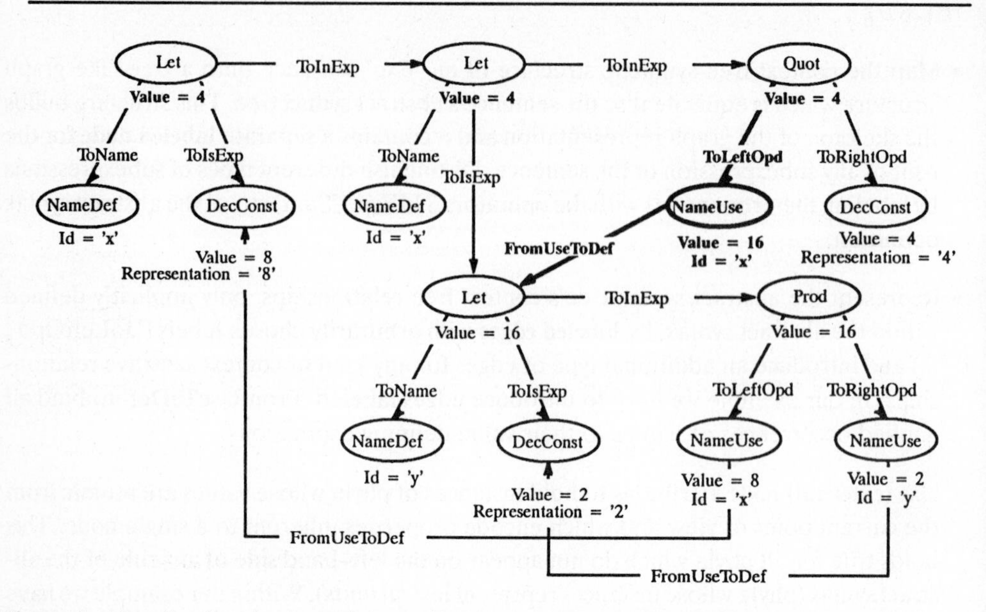

Figure 3 : Result of a rewrite rule's application to the graph of figure 2.

So–called (parametrized) **subgraph tests** and **graph rewrite rules** are the basic building blocks for the definition of complex graph transformations. The former are boolean functions which test for the occurrence of a certain subgraph (pattern) within a host graph, and the latter are graph transformations which search for a certain subgraph within a host graph and replace this subgraph by another one. **Control flow diagrams** (and subdiagrams), similar to those defined in /Na 79/ are used to compose complex graph transformations out of these tests and graph rewrite rules.

For a brief description of **graph rewrite rules** let us consider just one typical example. This is the rule "Replace an unexpanded subexpression by an applied occurrence of 'x' ". Figure 3 displays the effect of the application of such a rule onto the example of figure 2. The execution of this complex rule (specified in section 2.2, figure 10) may be divided into the following five steps:

- **Subgraph test**: Select any subgraph within the host graph complying with the aforementioned requests, i.e. in this case any node labeled 'NilExp' within our tree–like graph structure. If there is none, return with failure.
- **Subgraph replacement**: Erase the selected subgraph including all incoming and outgoing edges, i.e. the node labeled 'NilExp' and one edge labeled 'ToLeftOpd', and insert the nodes and edges of the new subgraph, i.e. the node labeled 'NameUse'.
- **Embedding transformation:** Connect the new subgraph with the remainder of the host graph by incoming and outgoing edges, i.e. by two new edges labeled 'ToLeftOpd' and

'FromUseToDef'. So-called path expressions are used to determine the sometimes far-away located sources and targets of embedding edges within the host graph.

- **Attribute transfer**: Assign values to the external attributes of the nodes of the new sub-graph, i.e. assign the value 'x' to the attribute 'Id' of the new node.
- **Attribute revaluation**: Compute the new values of all derived attributes within the new graph, i.e. the values of all 'Value' attributes, or at least of those derived attributes to which we have to assign new values.

For the presentation of a typical **control flow diagram** let us assume that we use several graph rewrite rules (instead of one complex rewrite rule) in order to transform the graph of figure 2 into the graph of figure 3. E.g., we have one parametrized graph rewrite rule ('CreateNameExp') to replace the placeholder node by an applied occurrence of 'x', and that we have another graph rewrite rule ('BindNameExp') which tries to bind the new applied occurrence to a visible definition of 'x'. In the case of failure of this second graph rewrite step a third rewrite rule will be activated in order to mark the applied occurrence of 'x' as to be erroneous. Figure 4 contains the corresponding control flow diagram (cf. figure 10).

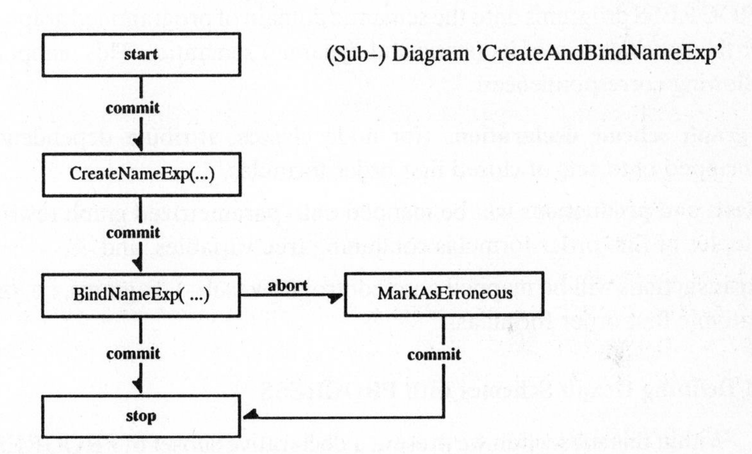

Figure 4 : Example of a control flow diagram (corr. to the transaction of fig. 10)

A **formal definition** of attributed graphs, graph classes, and (class preserving) programmed graph rewriting systems, relying on the fundamentals of a first order predicate logic, is beyond the scope of this paper but will be presented in /Sc 90/. There we use first order formulas to "encode"

- intensionally defined **graph properties** common to a certain class of graphs,
- directed, attributed, node– and edge labeled **graphs** themselves as instances of graph classes,
- left– and right–hand sides as well as additional pre– and postconditions of **graph rewrite rules**,
- and **control flow diagrams** as instances of a certain class of graphs.

And, based on (graph) morphisms defining "property preserving" correspondences betweeen sets of first order formulas, we are able to understand the process of graph rewriting as the construction of commutating pairs of diagrams (similar to those in /EH 86/).

2. The Specification Language PROGRESS

After this informal introduction to the basic principles of programmed graph rewriting systems, it's now time to focus our interest onto the specification language PROGRESS itself. For a first survey it should suffice to present only a typical subset of the language and to explain this subset by means of examples instead of formal definitions. The first subsection deals with the **declarative programming constructs** of PROGRESS which are used to specify graph schemes, whereas the second subsection deals with the definition of **atomic** graph queries and transformations, called **tests** and **productions**, and with the definiton of **complex** graph transformations, called **transactions.**

For a formal **definition** of the complete language's **syntax and semantics** the reader is referred to /Sc 90/. There we use an extended BNF for the specification of the language's (context free) syntax and we use a mapping from the domain of all syntactically correct PROGRESS programs onto the semantic domain of programmed graph rewriting systems for the specification of its (static and dynamic) semantics. This mapping establishes the following correspondences:

- **graph scheme declarations** (for node classes, attribute dependencies etc.) will be mapped onto sets of closed first order formulas,
- **tests** and **productions** will be mapped onto parametrized graph rewrite rules, i.e. onto tuples of first order formulas containing free variables, and
- **transactions** will be mapped onto control flow (sub-) diagrams, i.e. onto sets of closed atomic first order formulas.

2.1 Defining Graph Schemes with PROGRESS

Within this subsection we present a declarative subset of PROGRESS-constructs for the definition of so-called "graph schemes". Such a **graph scheme** is a set of graph properties common for a certain class of graphs. Thus, PROGRESS allows – in contrast to all other graph grammar based programming/specification languages, we know – to extract structural integrity constraints and functional attribute dependencies from graph rewrite rules and to denote them separately.

```
attribute_type Integer, String;

attribute_function undefined : Integer;
                   empty : String;
                   decToInt : ( String ) -> Integer;
                   mult : ( Integer, Integer ) -> Integer;
                   div : ( Integer, Integer ) -> Integer;
          ...
```

Figure 5: Declaration of attribute domains and functions.

```
node_class EXP
  derived Value : Integer;
end;

node_class NAME
  external Id : String;
end;

node_class DEF_EXP is_a EXP end;

edge_type ToName : DEF_EXP -> NAME;

edge_type ToIsExp : DEF_EXP -> EXP;

edge_type ToInExp : DEF_EXP -> EXP;

node_class NAME_EXP is_a NAME, EXP end;

edge_type FromUseToDef : NAME_EXP -> EXP;

node_class BINARY_EXP is_a EXP end;

edge_type ToLeftOpd, ToRightOpd : BINARY_EXP -> EXP;

node_class CONST_EXP is_a EXP
  external Representation : String;
end;

node_class NIL_EXP is_a EXP end;

node_class MARKER end;

edge_type ToMarkedExp : MARKER -> EXP;
```

Figure 6: Declaration of node classes, edge types, and attributes.

This was a necessary precondition for the development of a “powerful” **set of tools** which are able

- to prevent the **violation of integrity constraints** (by means of an incrementally working type checker) and
- to preserve all **functional attribute dependencies** (by means of an incrementally working attribute evaluator).

The scheme definition for a new class of graphs starts with the declaration of a set of attribute domains and a family of n–ary functions on these attribute domains (see figure 6). We shall skip the definition (implementation) of **attribute domains and functions** and assume a Modula–2 like host language for this purpose.

Based on these (incomplete) attribute declarations, type declarations of three different categories characterize our abstract data type’s graph representation. **Node and edge type declarations** are used to introduce type labels for nodes and edges, whereas **node**

class declarations play about the same role as phyla, their counterparts within the operator/phylum based description of the language 'Exp'.

Primarily, **node classes** are used to denote coercions of node types which have common properties (following the lines of IDL /NNG 90/). Thereby, they introduce the concepts of classification and specialization into our language, and they eliminate the needs for duplicating declarations by supporting the concept of **multiple inheritance** along the edges of a class hierarchy.

Additionally, node classes play the role of second order types. Being considered as types of node types, they support the controlled use of **formal (node) type parameters** within generic subgraph tests and graph rewrite rules (cf. figure 9 and 10). The class 'EXP' e.g. is the type of all those node types, whose nodes possess the derived 'Integer' attribute 'Value' (see figure 6 and 7). This holds true for all node types with the exception of 'NameDef' and 'Cursor'. The class 'BINARY_EXP', a subclass of the class 'EXP', is the type of all those node types, whose nodes may be sources of edges typed 'ToLeftOpd' and 'ToRightOpd'.

In the presence of the smallest class, the empty set, and the largest class, comprising all node types, this family of sets (the **class hierarchy**) has to form a **lattice** with respect to the ordering of sets by inclusion (corresponding to the 'is_a' relationship). This request enforces a disciplined use of the concept of multiple inheritance. Furthermore, it was a precondition for the development of a system of type compatibility constraints for PROGRESS which is simple to understand and to implement.

```
node_type NameDef: NAME end;

node_type NameUse : NAME_EXP
  Value := [ that -FromUseToDef->.Value | undefined ];
end;

node_type Let : DEF_EXP
  Value := that -ToInExp->.Value;
end;

node_type Prod : BINARY_EXP
  Value := mult( that -ToLeftOpd->.Value, that -ToRightOpd->.Value);
end;

node_type DecConst : CONST_EXP
  Value := decToInt( .Representation);
end;

node_type NilExp : NIL_EXP
  Value := undefined;
end;

node_type Cursor : MARKER end;
```

Figure 7: Declaration of node types, and attribute dependencies.

The main purpose of a **node type declaration** is to define the behaviour of the nodes of this type, i.e. the set of all directed attribute equations, locally used for the (re–) computation of the node's derived attribute values. Let us start our explanations of the definition of **functional attribute dependencies** with the declaration of the node type 'DecConst' (in figure 7). Its 'Value' attribute is equal to the result of the function 'decToInt' applied to its own 'Representation' attribute. To compute the 'Value' of a 'Prod' node, we have to multiply the corresponding attributes of its left and right operand, i.e. the sinks of the outgoing edges typed 'ToLeftOpd' and 'ToRightOpd', respectively. In this case we assume that any 'Prod' node is the source of exactly one edge of both types and we strictly prohibit the existence of more than one outgoing edge of both types (indicated by the key word '**that**').

Due to the fact that we cannot always guarantee the existence of at least one edge, we had to introduce the possibility to define **sequences of alternative attribute expressions.** The general rule for the evaluation of such a sequence is as follows: try to evaluate the first alternative. If this fails due to the absence of an (optional) edge, then try to evaluate the next alternative. We request that at least the evaluation of the last alternative succeeds. Therefore, the 'Value' of a 'NameUse' node is either the 'Value' of its defining expression or, in the absence of a corresponding definition, 'undefined'.

2.2 Defining Graph Queries and Transformations with PROGRESS

Being familiar with PROGRESS' declarative style of programming, we are now prepared to deal with the operational constructs of the specification language. Some of these constructs form a partly textual, partly graphic **query sublanguage** for the definition of graph traversals which is very similar to data base query languages like /EW 83/. Elements of this sublanguage, in the sequel called path expressions, are mainly used to denote rather complex context conditions within graph rewrite rules. Formal definitions of previous versions of this sublanguage are published in /EL 87, Le 88/.

Path expressions may either be considered to be derived binary relationships or to be (node–) set-valued functions. The path expression '–ToLeftOpd–>' e.g., used in figure 7 for the formulation of attribute dependencies, maps a set of nodes onto another node set; the elements of this second node set are sinks of 'ToLeftOpd' edges starting from nodes within the first node set. The expression '<–ToLeftOpd–' simply exchanges the roles of sinks and sources.

Figure 8 contains the **textual definition** of three parameterized functional abstractions of path expressions, using conditional repetition (in the form of '{ <condition> : <path–expression> }'), composition ('&'), and union ('**or**'), and one example of a **graphical definition.** These path expressions specify the binding relationships between applied occurrences of identifiers ('NAME_EXP' nodes) and their corresponding value defining expressions ('EXP' nodes). The abstraction 'nextValidDef' e.g., applied to a set of nodes, computes the set of all those nodes that may be reached from elements of the first node set by: (1) following 'ToLeftOpd', 'ToRightOpd', and 'ToIsExp' edges from sinks to sources (cf. declaration of 'father'), as long as there is no incoming 'ToInExp' edge at any node on this path, (2) and finally following a 'ToInExp' edge from sink to source.

```
path father : EXP -> EXP =
     <-ToLeftOpd- or <-ToRightOpd- or <-ToIsExp- or <-ToInExp-
end;

path nextValidDef : EXP -> DEF_EXP =
   (* Computes the root of the next valid surrounding definition. *)
     { not with <-ToInExp- : father }
     & <-ToInExp-
end;

path binding ( Id : String ) : EXP -> EXP =
   (* Computes the root of the next visible surrounding definition of the name 'Id' *)
     nextValidDef
   & { not with definition(Id) : nextValidDef }
     & -ToIsExp->
end;

path definition ( IdPar : String ) : DEF_EXP -> NAME =        1 => 2 in

   [1: DEF_EXP] --ToName--> [2: NAME]     (* Applied to a set of 'DEF_EXP' nodes
                                            it computes the set of all those
                                            appertaining 'NAME' nodes whose
   condition  2.Id = IdPar ;                'Id' attributes are equal to 'IdPar'. *)
end ;
```

Figure 8: Definition of graph traversals, using textual and graphical path expressions.

Skipping any further explanations concerning path expressions, we come to the specification (of a small subset) of the operations provided by the abstract data type 'ExpGraphs'. It is worth–while to notice that node type parameters keep this part of the specification independent from the existence of any particular node type with the exception of the node type 'Cursor'. A unique node of this type is used to mark that subtree within a graph, called **current subtree**, that should be affected by the application of a graph rewrite rule. Thus, the operational part of our specification becomes almost independant from the 'Exp'–language's abstract syntax, and the introduction of a new (binary) operator like 'div' or 'mod' only requires one additional node type declaration within our specification.

Let us start with the explanation of the operation 'Initialize' of figure 9. This operation creates the root of a tree–like subgraph if and only if the graph of interest contains no nodes of a type belonging to the class 'EXP'. We use the **test** 'ExpressionInGraph' to check this condition. The subsequent application of an instantiation of the graph rewrite rule (**production**) 'CreateExpressionRoot' never fails due to the fact that this rule has an empty matching condition. Being instantiated with any node type of the class 'NIL_EXP' its execution simply adds one node of this type, another node of the type 'Cursor', and an edge of the type 'ToMarkedExp' to a graph.

The production 'MoveCursorUp' is an example of a rewrite rule which **identifies (all) nodes** of its lefthand side with (all) nodes of its righthand side. An application of this rewrite rule neither deletes or modifies any node nor removes any not explicitly mentioned edge from the host graph. A successful application of this rule only redirects one edge of the type 'ToMarkedExp' so that its new sink is the father of its former sink, a condition that is expressed by the path operator 'father' within the rule's lefthand side.

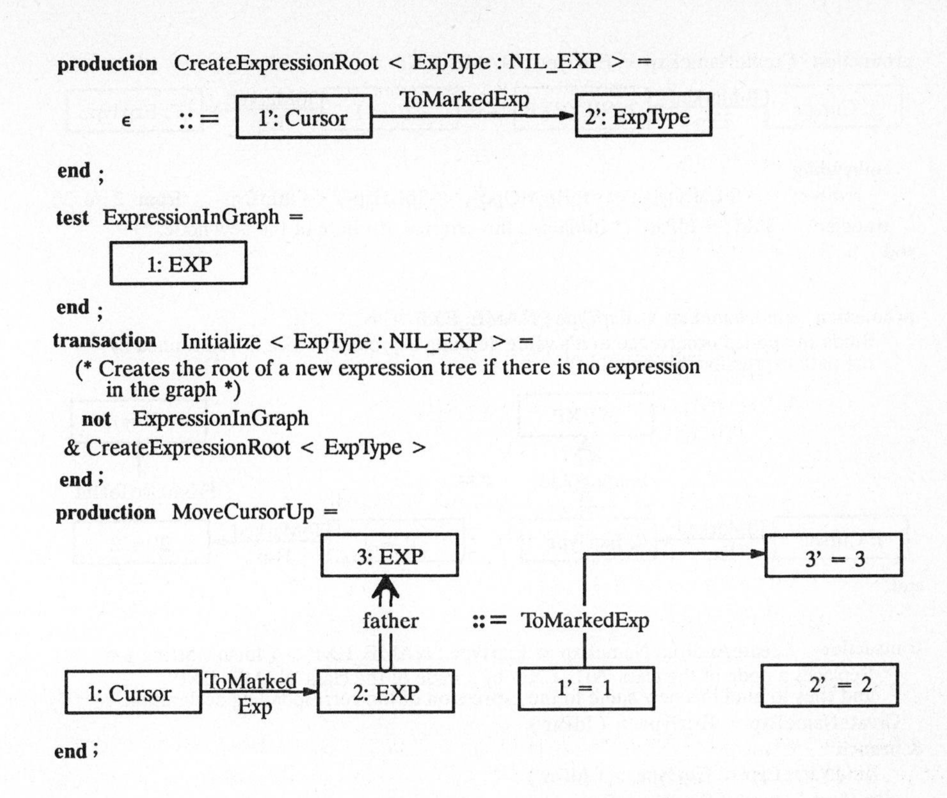

Figure 9: Declaration of simple productions, tests, and transactions.

(Complex) graph transformations (like 'Initialize' of figure 9 or 'CreateAndBindNameExp' of figure 10) are termed **transactions** in order to emphasize their **"atomic"** character. Similar to a single graph rewrite rule transactions support a nondeterministic way of programming and they either **commit** with a consistency preserving graph transformation or **abort** without any modifications of the graph they were applied to (cf. /Zü 89/). A transaction has to abort if one of its transactions or graph rewrite rules aborts. Abortion (failure) of a single graph rewrite rule occurs if the rule's lefthand side doesn't match with any subgraph of the graph it's applied to.

Thus the application of the transaction 'CreateAndBindNameExp' (of figure 10) either fails or consists of two graph rewrite steps. The first one causes the replacement of the current 'NIL_EXP' node by a new node of the class 'NAME_EXP'. The second one either binds the new applied occurrence to its value defining expression ('**branch** BindNameExp ...') or – in the case of failure of 'BindNameExp' – marks the new applied occurrence as to be erroneous. The execution of the second graph rewrite rule requires the evaluation of the non–trivial path operator 'binding(2.Id)' which determines the target of the 'FromUseToDef' edge, whereas the execution of the first graph rewrite rule contains an embedding transformation and an attribute transfer, the latter being the assignment of the new name's string representation to the corresponding node's 'Id' attribute.

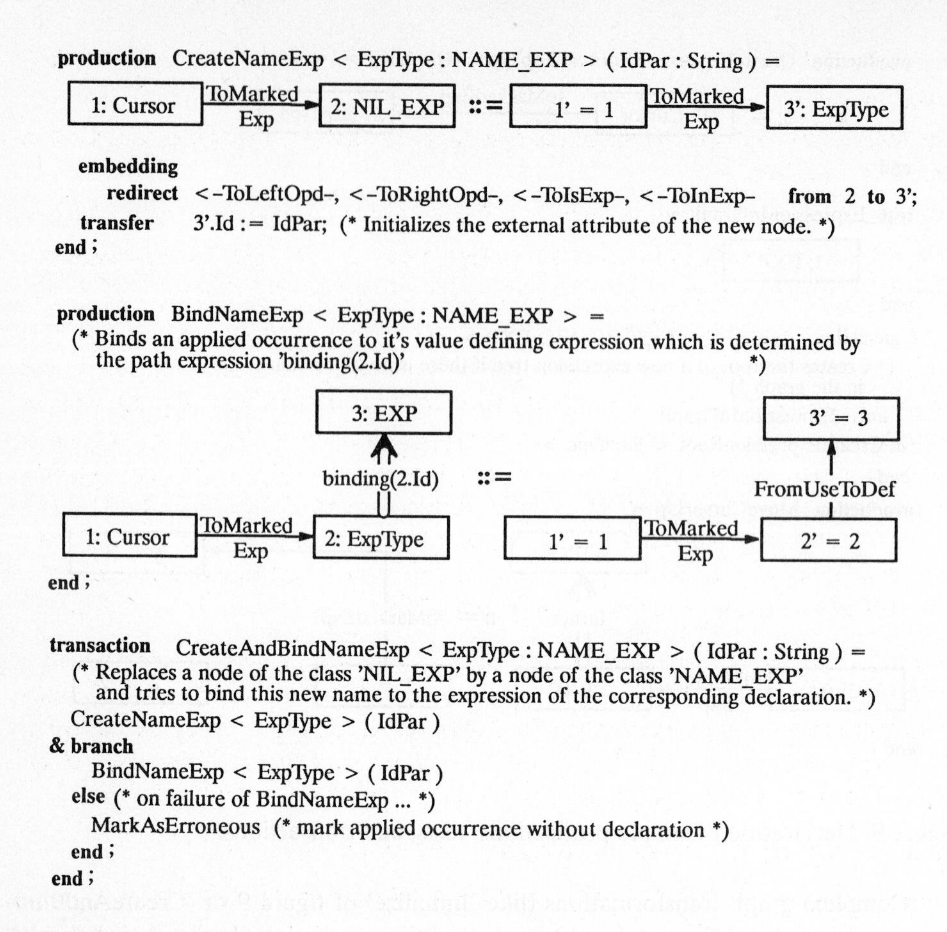

production CreateNameExp < ExpType : NAME_EXP > (IdPar : String) =
1: Cursor
ToMarked Exp
2: NIL_EXP
::=
1' = 1
ToMarked Exp
3': ExpType
embedding
redirect <-ToLeftOpd-, <-ToRightOpd-, <-ToIsExp-, <-ToInExp- from 2 to 3';
transfer 3'.Id := IdPar; (* Initializes the external attribute of the new node. *)
end ;
production BindNameExp < ExpType : NAME_EXP > =
(* Binds an applied occurrence to it's value defining expression which is determined by
the path expression 'binding(2.Id)' *)
3: EXP
binding(2.Id)
::=
3' = 3
FromUseToDef
1: Cursor
ToMarked Exp
2: ExpType
1' = 1
ToMarked Exp
2' = 2
end ;
transaction CreateAndBindNameExp < ExpType : NAME_EXP > (IdPar : String) =
(* Replaces a node of the class 'NIL_EXP' by a node of the class 'NAME_EXP'
and tries to bind this new name to the expression of the corresponding declaration. *)
CreateNameExp < ExpType > (IdPar)
& branch
BindNameExp < ExpType > (IdPar)
else (* on failure of BindNameExp ... *)
MarkAsErroneous (* mark applied occurrence without declaration *)
end ;
end ;

Figure 10: Specification of the graph transformation of section 1.

Both rewrite steps even may **trigger the revaluation of many derived attributes**. The application, explained in subsection 1.2 (transforming the graph of figure 2 to that of figure 3), for instance initiated the (re–) evaluation of four 'Value' attributes.

The above mentioned **embedding transformation** within the graph rewrite rule 'CreateNameExp' consists of one '**redirect**' clause. The purpose of this clause is to redirect any incoming edge belonging to the expression's abstract syntax tree skeleton from the former 'NIL_EXP' node to the new 'NAME_EXP' node. In addition to '**redirect**' clauses, '**remove**' and '**copy**' clauses may be used to remove edges from identically replaced nodes or to duplicate edges and attach them to the replacing subgraph's nodes.

3. Application areas and implementation efforts

Within the afore–mentioned IPSEN project the language PROGRESS and its predecessors have been used for rather different purposes like the specification/implementation of

- the functional behaviour of **syntax–directed editors** (e.g. a module–interconnection–language editor in /Le 88/),
- the internal structure and the interface operations of **abstract data types** (e.g. version graphs in /We 90/),
- optimizing **source–to–source transformations** (cf. /Zü 89/), and
- an incrementally working **source–to–target transformations** (from a module–interconnection–language document onto a Modula–2–document in /We 90/).

Due to the lack of PROGRESS–specific tools we had to produce all these specifications in a "paper and pencil" mode and we had to manually translate them into equivalent Modula–programs. Being concerned with the development of integrated programming support environments for more than a couple of years, it was obvious for us that we should try to develop another new **language–specific environment** for PROGRESS (cf. /NS 90/). This environment will be built on top of the X–Window based IPSEN–user–interface and the non–standard data/graph management system GRAS (cf. /LS 88/), and it will at least comprise a syntax–directed editor and a "hybrid" interpreter for PROGRESS (similar to the IPSEN environment's Modula–2 interpreter /ES 87/).

Up to now we have implemented two main components of the PROGRESS environment (cf. /NS 90/): The first one is a preliminary version of a **syntax–directed PROGRESS editor** (with a text–oriented interface) including a multiple–entry parser and an incrementally working type–checker (cf. /He 89, Zü 89/). The second one is a basic layer of the PROGRESS–interpreter, a graph (scheme) manager supporting (cf. /Sp 89/):

- dynamic checking of class–specific **integrity constraints**,
- **dynamic binding** of attribute evaluation rules to attribute designators,
- incremental **attribute (re–) evaluation** based on the so–called "naive" two phase algorithm (propagate 'affected' flags and then revaluate attribute values on demand) presented in /Re 84/, and
- arbitrary **undo/redo** of all graph modifications (a necessary precondition for the implementation of a "backtracking" graph–interpreter).

Up to now the PROGRESS environment's **implementation** on SUN workstations consists of about 140.000 lines of Modula–2 code which includes about 120.000 lines of language independant code (like the GRAS system) and about 10.000 lines of code produced by the IPSEN–editor–generator (cf. /Kö 90/).

4. Related Work

The work presented within this paper is an attempt to design a strongly typed language which is based on the concepts of **programmed graph rewriting systems** (as e.g. /Gö 88/) and supports a declarative style of programming for the description of object structures (like IDL /NNG 90/, or the languages surveyed in /HK 87/). Considering its expressiveness and its derivation history PROGRESS' underlying formalism belongs to the **set–theoretic branch** of the algorithmic graph grammar approach (cf. /Na 87/). But there are also some relationships between our formalism and the so-called **algebraic graph grammar approach** due to the fact that the definition of "graph rewriting" follows the lines of the constructive variant of the algebraic graph grammar approach. And last but not least we have to mention the work of Courcelle (cf. /Co 88/) who uses a **monadic second order logic** for the description of graph properties (but not for the description of graph rewrite rules).

The development of the language's **type concept** has been influenced by polymorphic programming languages like HOPE /BM 81/ which combine the flexibility of typeless languages with the reliability of strong and statically typed languages. By relying on the concept of a stratified type system with an infinite hierarchy of type universes (cf. /CZ 84/) we were able to incorporate **typed type parameters** into our language and to avoid the theoretical pitfalls of reflexive type systems with the 'Type is the type of all types including itself' assumption (cf. /MR 86/). The concept of **specialization** enables us to build complex class hierarchies. The idea to restrict this concept to the construction of **class lattices** has already been used within term or tree rewriting systems (cf. /CD 85/) and logic programming languages (cf. /AK 84/).

In order to facilitate a comparison of PROGRESS with those languages based on the more popular formalism of attribute (tree) grammars, we have used the well-known expression tree example for demonstration purposes. Therefore, it's necessary to emphasize that PROGRESS is **not restricted to** the specification of abstract data types with **tree-like representations** but even more adequate in the case of graph-like representations without any dominant tree-like substructure. Thus, it is almost impossible within the framework of attribute (tree) grammars to specify the restriction that a class hierarchy has to form a lattice, but it is a straightforward task to write an appropriate lattice test with PROGRESS (/He 89/).

Comparing the PROGRESS approach to model tree-like structures with that kind of modeling inherent to other attribute (tree) grammar approaches, we discover at least three principle differences:

- Derivation trees corresponding to subsequent applications of rewrite rules (productions) are neither used to represent object structures themselves nor to represent additional informations about these structures (like /Ka 85, Sc 87/). As a consequence, directed attribute equations are used to describe functional **attribute dependencies between adjacent nodes of a given graph,** and not to describe attribute flows along the edges of a derivation tree. Therefore, we believe (in contrast to /KG 89/) that the specification of attribute dependencies and the specification of graph rewrite rules should be kept separate from each other.

- A common disadvantage of our approach and the proposal in /KG 89/ is that **we are not able to identify directions like 'up' and 'down'** within arbitrary graphs. Therefore, we cannot adopt the classification of what we call 'derived' attributes into 'inherited' and 'synthesized' ones (cf. /Re 84/). Thus, the development of non-naive (incremental) attribute evaluation algorithms is much more difficult than for the case of attribute (tree) grammars. For basic work on this problem see /AC 87, Hu 87/.

- A third difference is also a direct consequence of the fact that we are not restricted to the world of trees: Our data model refrains the specificator from the somewhat artificially different treatment of relationships representing either the **context free or the context sensitive syntax** (usually called static semantics) of a sentence of the language 'Exp'. Using labeled directed edges for the representation of both kinds of relationships, we are not forced to put context sensitive informations into complex structured attributes (being a severe handicap for any incremental attribute evaluation algorithm), like /RT 84/, or to escape to another formalism, like /HT 86/.

5. Summary and Future Work

This paper contains an informal presentation of the new developped specification language **PROGRESS** (for writing **PRO**grammed **G**raph **RE**writing **S**ystem **S**pecifications), its underlying formalism, and a model oriented approach to the specification of abstract data types. This approach, called **graph grammar engineering**, is a model oriented one due to the fact that we use

- **attributed, node and edge labeled, directed graphs** to model complex object structures,
- and **programmed graph rewriting systems** to specify operations in terms of their effect on these graph models.

Having fixed the presented version of PROGRESS' **programming–in–the–small part** a few months ago, we are now starting to evaluate first experiences with the language and to develop the language's **programming–in–the–large part**, supporting the decomposition of large specifications into separate, reusable, and encapsulated subspecifications. Therefore, the presented version of PROGRESS may be classified as the programming–in–the–small kernel of a (very high level programming) language offering **declarative and procedural** elements for the definition of

- static and dynamic integrity constraints (based on a multiple–inheritance hierarchy of node classes),
- functional attribute dependencies (using directed equations),
- derived binary relationships (by means of so–called path expressions),
- atomic graph transformations (in the form of graph rewrite rules with complex preconditions),
- and complex graph transformations (using deterministic and nondeterministic control structures within "atomic" transactions).

A **formal definition** of this kernel is near–by completion (cf. /Sc 90/).

Last but not least, we have started to design and implement a **programming environment for PROGRESS**, based on our previous experiences with the construction of the **I**ntegrated **P**rogramming **S**upport **EN**vironment **IPSEN**. This environment will comprise at least a syntax–directed editor and an interpreter. Thus, in days to come we might be able to specify the functional behaviour of PROGRESS' environment with PROGRESS itself and to analyze and execute the environment's specification by means of its own implementation.

Acknowledgments

I am indebted to M. Nagl, C. Lewerentz and A. Zündorf for many stimulating discussions about PROGRESS, to B. Westfechtel for willingly producing large specifications with PROGRESS, and especially to U. Cordts, P. Heimann, P. Hormanns, R. Herbrecht, M. Lischewski, R. Spielmann, C. Weigmann, and A. Zündorf for designing and implementing first parts of the PROGRESS programming environment.

References

/AC 87/ Alpern B., Carle A., Rosen B., Sweeney P., Zadeck K.: *Incremental Evaluation of Attributed Graphs*, T. Report CS-87-29; Providence, Rhode Island: Brown University

/AK 84/ Ait-Kaci H.: *A Lattice-Theoretic Approach to Computation Based on a Calculus of Partially-Ordered Type Structures*, Ph.D. Thesis; Philadelphia: University of Pennsylvania

/BM 81/ Burstall R.M., MacQueen D.B., Sannella D.T.: *HOPE – An Experimental Applicative Language*, Technical Report CSR-62-80; Edinburgh University

/CD 85/ Cunningham R.J., Dick A.J.J.: Rewrite Systems on a Lattice of Types, in Acta Informatica 22, Berlin: Springer Verlag, 149-169

/Co 88/ Courcelle B., *The Monadic Second Order Theory of Graphs: Definable Sets of Finite Graphs*, in Leeuwen (ed.): *Proc. Int. Workshop on Graph-Theoretic Concepts in Computer Science (WG '88)*, LNCS 344; Berlin: Springer Verlag,

/CZ 84/ Constable R., Zlatin D.: *The Type Theory of PL/CV3*, in ACM TOPLAS, vol. 6, no. 1, 94-117

/EH 86/ Ehrig H., Habel A.: *Graph Grammars with Application Condition*, in Rozenberg, Salomaa (eds.): *The Book of L*, Berlin: Springer Verlag, 87-100

/EL 87/ Engels G., Lewerentz C., Schäfer W.: *Graph Grammar Engineering – A Software Specification Method*, in: Proc. 3rd Int. Workshop on *Graph Grammars and Their Application to Computer Science*, LNCS 153; Berlin: Springer Verlag, 186-201

/ES 87/ Engels G., Schürr A.: *A Hybrid Interpreter in a Software Development Environment*, in: Proc. TAPSOFT '85, LNCS 186; Berlin: Springer-Verlag, 179-193

/EW 83/ Elmasri R., Wiederhold G.: *GORDAS: A Formal High-Level Query Language for the Entity-Relationship Model*, in Chen (ed.): Entity-Relationship Approach to Information Modeling and Analysis, Amsterdam: Elsevier Science Publishers B.V. (North-Holland), 49-72

/GJ 82/ Genrich H.J., Janssens D., Rozenberg G., Thiagarajan P.S.: *Petri nets and their relation to graph grammars*, in Ehrig et al. (ed.): Proc. 2nd Int. Workshop on *Graph Grammars and Their Application to Computer Science*, LNCS 153; Berlin: Springer Verlag, 115-142

/Gö 88/ Göttler H.: *Graphgrammatiken in der Softwaretechnik*, IFB 178; Berlin: Springer-Verlag

/He 89/ Herbrecht R.: *Ein erweiterter Graphgrammatik-Editor*, Diploma Thesis; Aachen: University of Technology

/HT 86/ Horwitz S., Teitelbaum T.: *Generating Editing Environments Based on Relations and Attributes*, in Proc. ACM TOPLAS, vol. 8, no. 4, 577-608

/Hu 87/ Hudson S.E.: *Incremental Attribute Evaluation: An Algorithm for Lazy Evaluation in Graphs*, Technical Report TR 87-20; Tucson: University of Arizona

/HK 87/ Hull R., King R.: *Semantic Database Modeling: Survey, Applications, and Research Issues*, in: ACM Computing Surveys, vol. 19, No. 3, 201-260

/JF 84/ Johnson G.F., Fischer C.N.: *A Metalanguage and System for Nonlocal Incremental Attribute Evaluation in Language-Based Editors*, in: Proc. ACM Symp. POPL '84

/Ka 85/ Kaul M.: *Präzedenz-Graph-Grammatiken*, PH.D. Thesis; University of Passau

/KG 89/ Kaplan S.M., Goering St.K.: *Priority Controlled Incremental Attribute Evaluation in Attributed Graph Grammars*, in: Diaz, Orejas (eds.): Proc. TAPSOFT '89, vol. 1, LNCS 351, Berlin: Springer Verlag, pp.306-320

/Kö 90/ Köther R.: *Der EBNF-Editor/Generator*, Diploma Thesis; Aachen: University of Technology

/Le 88/ Lewerentz C.: *Interaktives Entwerfen großer Programmsysteme*, PH.D. Thesis, IFB 194; Berlin: Springer-Verlag;

/LS 88/ Lewerentz C., Schürr A.: *GRAS, a Management System for Graph-like Documents*, in Beeri et al. (eds.): Proc. 3rd Int. Conf. on Data and Knowledge Bases; Los Altos, California: Morgan Kaufmann Publishers Inc., 19-31

/MR 86/ Meyer A.M., Reinhold M.B.: *'Type' is not a type*, Proc. 13th ACM Symp. POPL '86, 287-295

/Na 79/ Nagl M.: *Graph-Grammatiken: Theorie, Implementierung, Anwendungen*; Braunschweig: Vieweg-Verlag

/Na 85/ Nagl M.: *Graph Technology Applied to a Software Project*, in: Rozenberg, Salomaa (eds): *The Book of L*; Berlin: Springer-Verlag, 303-322

/Na 87/ Nagl M.: *Set Theoretic Approaches to Graph Grammars*, in: Proc. 3rd Int. Workshop on Graph Grammars, LNCS 291; Berlin: Springer Verlag, 41-54

/NNG 90/ Nestor J.R., Newcomer J.M., Giannini P., Stone D.L.: *IDL: The Language and Its Implementation*; Engelwood Cliffs: Prentice Hall

/No 87/ Normark K.: *Transformations and Abstract Presentations in Language Development Environment*, Technical Report DAIMI PB-222; Aarhus University

/NS 90/ Nagl M., Schürr A.: *A Specification Environment for Graph Grammars*, this vol.

/Re 84/ Reps T.: *Generating Language-Based Environments*, PH.D. Thesis; Cambridge, Mass.: MIT Press

/RT 84/ Reps T., Teilbaum T.: *The Synthesizer Generator*, in: Proc. ACM SIGSOFT/SIGPLAN Symp. on Practical Software Development Environments, 42-48

/Sc 87/ Schütte A.: *Spezifikation und Generierung von Übersetzern für Graph-Sprachen durch attributierte Graph-Grammatiken*, PH.D. Thesis; Berlin: EXpress-Edition

/Sc 89/ Schürr A.: *Introduction to PROGRESS, an Attribute Graph Grammar Based Specification Language*, in: Nagl (ed.): Graph-Theoretic Concepts in Computer Science, WG '89, LNCS 411, Berlin: Springer Verlag, 151-165

/Sc 90/ Schürr A.: *Programmieren mit Graphersetzungssystemen: Theoretische Grundlagen und programmiersprachliche Einkleidung*, PH.D. Thesis (in preparation); Aachen: University of Technology

/So 84/: Sowa J.F.: *Conceptual Structures: Information Processing in Minds and Machines*; Reading, Mass.: Addison-Wesley

/Sp 89/ Spielmann R.: *Entwicklung einer Basisschicht für Gaphgrammatikinterpreter*, Diploma Thesis; Aachen: University of Technology

/We 90/ Westfechtel B.: *Revisionskontrolle in einer integrierten Softwareentwicklungsumgebung*, PH.D. Thesis (in preparation); Aachen: University of Technology

/Zü 89/ Zündorf A.: *Kontrollstrukturen für die Spezifikationssprache PROGRESS*, Diploma Thesis, Aachen: University of Technology

Movement of Objects in Configuration Spaces modelled by Graph Grammars

GABRIELE TAENTZER, HOLGER SCHWEEN

Computer Science Department, Technical University of Berlin,
Franklinstr. 28/29, Sekr. FR 6-1, D-1000 Berlin 10

Abstract

Configuration spaces are considered where arbitrary objects are placed in a two-dimensional discretized space and can move via translations and rotations within this space. If there is a motion of an object or even if several objects move in parallel the recognition of collisions between all objects is essential. To tackle these problems such configuration spaces are modelled by graphs and motions of objects in the configuration space by graph transformations. For modelling basic motions two-level graph transformations are used to create first object productions from a finite set of elementary productions which are then applied to configurations. These graph transformations model collision-free motions if the consistency of the configuration space is preserved.

Keywords: configuration, two-level graph transformation, parallel graph transformation

1 Introduction

Configuration spaces which consist of a two-dimensional discretized space and some arbitrary objects are taken into consideration. These objects are arbitrarily placed in the space which is considered to be consistent if the objects don't overlap each other. The space changes its state when an object moves. The configuration space should remain consistent after this motion. If more than one object move in parallel the configuration space should also be consistent after that.

The configuration space with arbitrary objects is modelled by a Cartesian grid structured graph. For modelling basic translations and rotations of objects two level graph transformations are used to create first object productions from a finite set of elementary productions that describe the motion. Afterwards these object produtions can be applied to configurations. To model more complex motions than the basic ones the object productions can be sequentially combined to new productions modelling trajectories. This is done by concurrent productions as presented in [EhRo80]. One approach to modelling the collision-free execution of motions is the preservation of consistency of the configuration space during the execution.

For modelling two-dimensional configuration spaces by graphs first a Cartesian grid is laid in the space and then this discretized space is mapped to a Cartesian graph. Each grid square is modelled by a node. If there is a part of an object in the square then the node is black otherwise it is white. Horizontally or vertically neighbouring squares are mapped to nodes which are connected by "h"-colored (horizontal) or "v"-colored (vertical) arcs respectively.

For modelling the possibilities of rotations an arbitrary fine polar grid is laid in the space such that the rotation point is covered by the center of the polar grid. Now each polar section is modelled by a node. If there is a part of an object in the section then the node is black otherwise it is white. If there are two polar sections which are neighboured radially or are in the same polar circle then there is an "r"-colored (radial) or "p"-colored (polar) arc resp. between the corresponding nodes.

Now we have various grids but no connection between them. So we say that the Cartesian grid is the base grid to which all polar grids refer. To model the position of a polar grid with respect to the Cartesian grid connection arcs are introduced. There are "c"-colored (connection) arcs between a polar node and those Cartesian nodes which model that area which covers just that polar section which is modelled by the polar node. All Cartesian nodes which are targets of connection arcs with a black source node are black, too. The other Cartesian nodes retain their colors.

1.1 Example

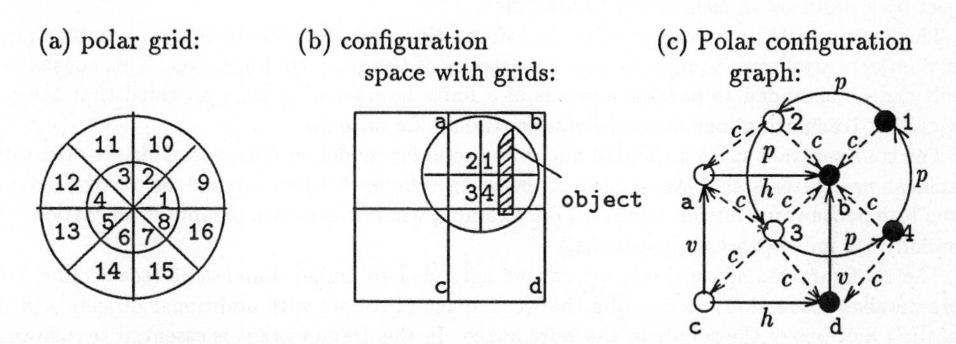

Each part of the polar grid which is supplied with a number is called a polar section. Polar sections 1 and 9 for example are neighboured radially and polar sections 1 and 2 are neighbours in the same polar circle.

We use graphs and graph transformations which are based on the algebraic graph grammar approach ([Eh79]) including pushout-stars which are used for parallel gluing constructions. However, the basic category GRAPH in [Eh79] is replaced by C-GRAPH that consists of C-graphs (configuration graphs) which are used to model configurations and C-graph morphisms that are defined between C-graphs. These graph morphisms do not preserve the colors in general: A mapping of a white node to a black node is permitted, but black nodes can only be mapped to black nodes. It can be noticed that there is a preorder on the node labels (i.e. w < b), so the category C-GRAPH is a special case of the category GRA_{SC} in [PEM87] or POGRAPHS in [EHLP91] which consists of graphs with preordered labels and label compatible graph morphisms.

The configuration space is not modelled by only one graph but by a graph star and its parallel gluing that yields the configuration graph. Such a graph star consists of a Cartesian grid structured graph which represents the configuration space without any object, a set of object graphs modelling arbitrary objects and some gluing graphs used to glue the object graphs and the grid graph together. [1]

The advantage of such a graph star in opposition to a single graph is first a formal possibility to describe the position of an object in the configuration space by a position morphism between the object graph and the grid graph. Then in a single graph there is the information missing that indicates which area belongs to which object if the information isn't coded into labels. It

[1] To describe the embedding of graphs in another graph similar techniques are used in [BEHL87] to model distributed graph grammars.

only models the information about occupied areas. Furthermore there is a categorical possibility to define the consistency of a configuration being satisfied if the objects don't overlap each other.

Basic motions like translations about one step and rotations about an elementary angle dependent on the fineness of the grids used are modelled by configuration transformations that preserve the consistency of a configuration. The transformation consists of two levels: First an object production for arbitrary objects is generated from a finite set of elementary productions for modelling basic motions like translations for one step or rotations about elementary angles. For describing these motions a finite set of elementary productions is only needed to generate an object production independent of the shape of this object. This generation is done by using parallel graph transformations ([Eh79]) applied to a part of the configuration graph. On the second level such an object production is applied to a configuration. The advantage of such two-level transformations is the complete description of an object production for any arbitrary object by a finite set of elementary productions.

The main results are: Configuration transformations modelling basic or more complex motions of objects are shown to preserve the consistency of the given configuration. This consistency result can be extended to parallel motions of a finite number of objects provided that the corresponding transformations are parallel independent for all pairs.

The configuration space presented might be useful for modelling bitmaps of object-orientated graphical user interfaces where collisions between different objects should be recognized and taken into account for further actions. (For example window-systems, computer animation, etc. especially for multi-processing systems.)

The configuration space presented can be extended to higher dimensions (see section 7 for more detailed discussion) to describe the work space of robots with additional objects and the (parallel) motions of the robots in the work space. In this framework it is essential to guarantee collision-free motions especially in cases where more than one robot can move in a work space.

By extending the concept of configuration spaces to allow underlying structures other than Cartesian and polar grids it is also possible to model several other situations, such as motions of trains in a railway system ([Wi81]) or motions of objects on assembly lines,etc. We intend to extend the theory to cover these kinds of applications.

2 Basic notions of graphs and graph transformations

This section introduces the category C–GRAPH which constitutes the framework for the following investigations.

2.1 Definition (C–graph) *A C-graph $G = (G_A, G_N, s_G, t_G, m_{GA}, m_{GN})$ consists of the sets of arcs and nodes, G_A and G_N, the mappings $s_G, t_G : G_A \longrightarrow G_N$, called source resp. target map, and the coloring maps for arcs and nodes, $m_{GA} : G_A \longrightarrow C_A$ and $m_{GN} : G_N \longrightarrow C_N$ where $C_A = \{h, v, p, r, c\}$ and $C_N = \{b, w\}$ are the color alphabets for arcs and nodes.*

$$C_A \xleftarrow{m_{GA}} G_A \underset{t_G}{\overset{s_G}{\rightrightarrows}} G_N \xrightarrow{m_{GN}} C_N$$

If we don't want to distinguish between arcs and nodes we use the notation $x \in G$ which means $x \in G_A$ or $x \in G_N$.

2.2 Definition (C–graph Morphism) *Given two C-graphs G and G' a C-graph morphism $f : G \longrightarrow G'$ is a pair of maps $(f_A : G_A \longrightarrow G'_A, f_N : G_N \longrightarrow G'_N)$ such that $f_N \circ s_G = s_{G'} \circ f_A$, $f_N \circ t_G = t_{G'} \circ f_A$, $m_{G'A} \circ f_A = m_{GA}$ and for all $x \in G_N$ $m_{G'N} \circ f_N(x) = m_{GN}(x)$ if $m_{GN}(x) \doteq b$.*

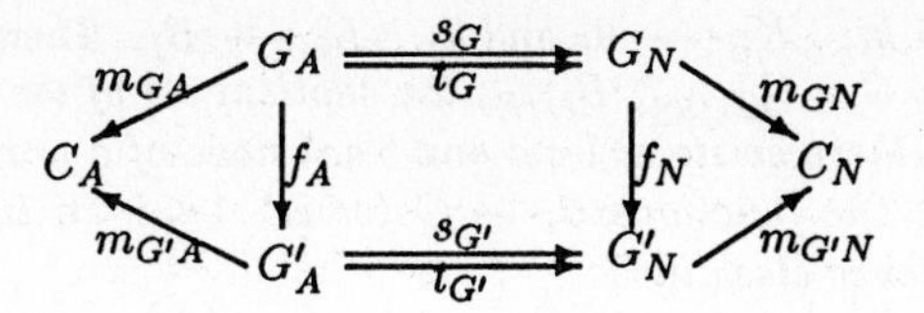

2.3 Definition (Graph Morphism) *A C–graph morphism is called graph morphism if* $m_{G'N} \circ f_N(x) = m_{GN}(x)$ *for all* $x \in G_N$.

A C–graph morphism f is said to be injective (surjective) if both f_A and f_N are injective (surjective) mappings.

In the graphical notation of C–graphs, nodes are presented as black or white circles and arcs as arrows. If the arcs are presented horizontally or vertically resp. then we omit the color "h" or "v" respectively. If an arc is represented as a part of a circular path then the color "p" is left out. Dashed arrows stand for arcs colored by "c" (connection arcs). A C–graph morphism between two C–graphs is represented by additional numbers that indicate the nodes which are mapped to each other. In the following examples it is not necessary to give the mappings for arcs explicitly because there is just one choice to define them such that they are C–graph morphisms.

2.4 Example

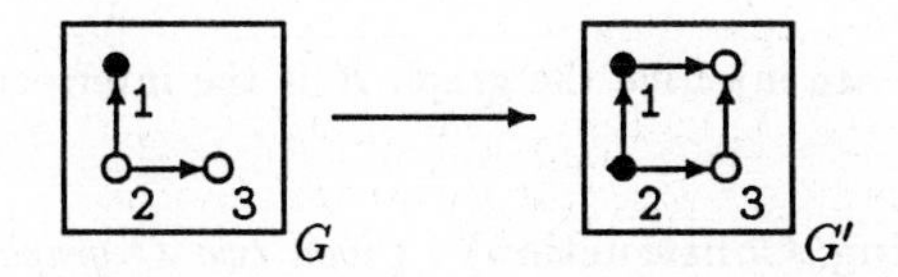

2.5 Definition (Pushout–star) *A C–graph star of degree* $n \geq 1$ *is a diagram* $S = (B_i \longleftarrow K_{ij} \longrightarrow B_j)_{1 \leq i < j \leq n}$ *consisting of C–graphs* B_i,B_j,K_{ij} *and C–graph morphisms* $K_{ij} \longrightarrow B_i$ *resp.* $K_{ij} \longrightarrow B_j$. *In the case* $n = 1$ *we only have the single C–graph* B_1. *A C–graph* G *together with the C–graph morphisms* $B_i \longrightarrow G$ *for* $1 \leq i \leq n$ *is called pushout–star of* S *(POS) if we have:*

1. *commutativity:* $K_{ij} \longrightarrow B_i \longrightarrow G = K_{ij} \longrightarrow B_j \longrightarrow G$ *for all* $1 \leq i < j \leq n$
2. *universal property: For all C–graphs* G' *and C–graph morphisms* $B_i \longrightarrow G'$ *satisfying* $K_{ij} \longrightarrow B_i \longrightarrow G' = K_{ij} \longrightarrow B_j \longrightarrow G'$ *for all* $i < j$ *there is a unique C–graph morphism* $\psi : G \longrightarrow G'$ *such that* $B_i \longrightarrow G \longrightarrow G' = B_i \longrightarrow G'$ *for all* $1 \leq i \leq n$.

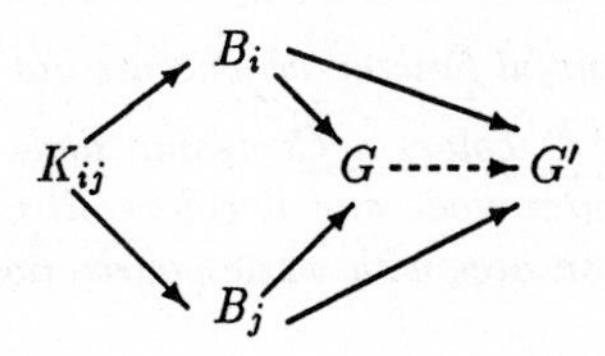

Remarks: In the definition above S can be denoted as the C–graph star of G: $S = \text{STAR}(G)$.

In the case $n = 2$ the pushout–star is called pushout. So all properties of pushout–stars are also valid for pushouts.

2.6 Proposition (Star Gluing Construction) *Given a C–graph star* $S = (B_i \longleftarrow K_{ij} \longrightarrow B_j)_{1 \leq i < j \leq n}$. *Let* $\bar{K}$ *be the equivalence relation generated by the relation* $K = \{(k_{ij}(x), k_{ji}(x)) \mid$

$x \in K_{ij},\ 1 \leq i < j \leq n\}$ *with* $k_{ij} : K_{ij} \longrightarrow B_i$ *and* $k_{ji} : K_{ij} \longrightarrow B_j$. *Then the pushout-star graph* $G = JOIN(S)$ *is defined by* $G = (\coprod_{1 \leq i \leq n} B_i)_{/K}$, *the quotient set of the disjoint union of all* B_i. *A node* x *in* G_N *is black if there exists at least one black node of a graph* B_i *in the equivalence class otherwise it is white. The functions* $B_i \longrightarrow G$ *for all* $1 \leq i \leq n$ *in the diagram send each element of* B_i *to its equivalence class in* G.

2.7 Definition (Pullback) *Given two C–graph morphisms* $B \longrightarrow G$ *and* $D \longrightarrow G$ *the pullback of these morphisms consists of a C–graph* K *together with two C–graph morphisms* $K \longrightarrow B$ *and* $K \longrightarrow D$ *such that:*

1. *commutativity:* $K \longrightarrow B \longrightarrow G = K \longrightarrow D \longrightarrow G$
2. *universal property: For all C–graphs* K' *and C–graph morphisms* $K' \longrightarrow B$ *und* $K' \longrightarrow D$ *satisfying* $K' \longrightarrow B \longrightarrow G = K' \longrightarrow D \longrightarrow G$ *there is a unique C–graph morphism* $K' \longrightarrow K$ *such that* $K' \longrightarrow K \longrightarrow D = K' \longrightarrow D$ *and* $K' \longrightarrow K \longrightarrow B = K' \longrightarrow B$.

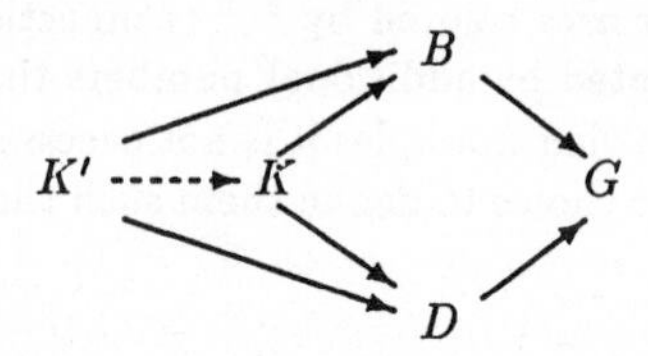

If $B \longrightarrow G$ and $D \longrightarrow G$ are injective the graph K is the intersection of B and D in G, in general we have:

2.8 Proposition (Splitting Construction) *Given two C–graph morphisms* $g : B \longrightarrow G$ *and* $c : D \longrightarrow G$ *the pullback graph* $K = SPLIT(B \longrightarrow G \longleftarrow D)$ *is defined by* $K = \{(x, y) \in B \times D | g(x) = c(y)\}$ *and the C–graph morphisms* $K \longrightarrow B$ *and* $K \longrightarrow D$ *are projections. A node* (x, y) *in* K *is black if both nodes* $x \in B$ *and* $y \in D$ *are black otherwise it is white.*

2.9 Definition (Functors in C–GRAPH)

1. *The remove functor* $\mathcal{R}$ *colors all nodes of a C–graph white.*
2. *The black search functor* $\mathcal{B}$ *is a forgetful functor which forgets that part of a C–graph that consists of white nodes and all arcs from and to these nodes.*
3. *The Cartesian functor* $\mathcal{C}$ *is a forgetful functor which cuts out the Cartesian part of a C–graph.*
4. *The polar functor* $\mathcal{P}$ *is a forgetful functor which cuts out the polar part of a C–graph.*
5. *The Cartesian object functor* $\mathcal{O}$ *colors a Cartesian node black if it is a target node of a connection arc with black source node and it colors all Cartesian nodes white if they are only target nodes of connection arcs with white source node.*

2.10 Definition (C–production) *A C–production* r *is a pair of C–graph morphisms* $r = (L \longleftarrow K \longrightarrow R)$ *such that* $\mathcal{R}(L) = \mathcal{R}(K) = K = \mathcal{R}(R)$ *(i.e. that the removed C–graphs* $\mathcal{R}(L)$, $\mathcal{R}(K)$ *and* $\mathcal{R}(R)$ *are isomorphic to the gluing graph* K*) and the C–graphs* L, R *and* K *are called left side, right side and gluing graph.*

2.11 Definition (Parallel Graph Transformation) *Given a C-graph G with the C-graph star* $STAR(G) = (L_i \longleftarrow K_{ij} \longrightarrow L_j)_{1 \le i < j \le n}$, *for each* L_i *with* $1 \le i \le n$ *a C-production* $r_i = (L_i \longleftarrow K_i \longrightarrow R_i)$ *and a unique factorization* $K_{ij} \longrightarrow L_p = K_{ij} \longrightarrow K_p \longrightarrow L_p$ *for* $1 \le i < j \le n$ *and* $p = i$ *or* $p = j$. *Then* $K_{ij} \longrightarrow R_p = K_{ij} \longrightarrow K_p \longrightarrow R_p$ *is the induced graph star defined by:*

$$STAR(G') = (R_i \longleftarrow K_{ij} \longrightarrow R_j)_{1 \le i < j \le n}$$

A parallel graph transformation $G \Longrightarrow G'$ *via C-productions* $r_1, \cdots, r_n$ *consists of the following two pushout-stars where G and G' are pushout star graphs of* $STAR(G)$ *and* $STAR(G')$:

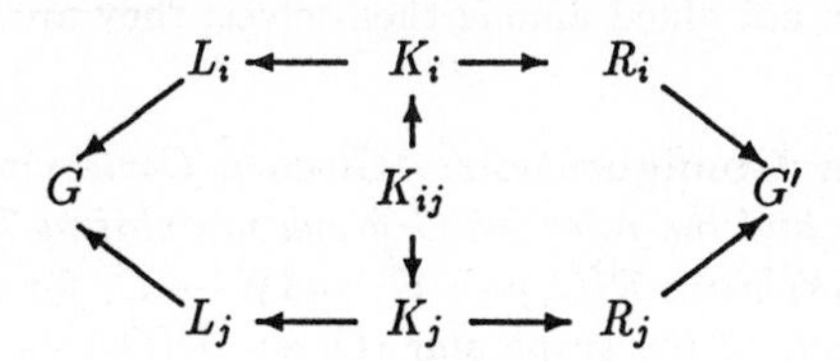

3 Configurations

Now we are able to model configuration spaces by C-graphs and C-graph stars. To do this a Cartesian grid structured graph is defined which represents the empty space and its Cartesian structure. If this grid structured graph is glued with some Cartesian object graphs which model arbitrary objects then the result of such a gluing construction is a Cartesian configuration graph that describes the position of these objects in the space and in relation to each other. If an object is able to rotate, the orientation of this object is modelled additionally to the position within a polar configuration. The Cartesian configuration is then extended by such polar configurations to a general configuration. A configuration is consistent if the objects don't overlap with each other.

3.1 Definition (Cartesian Grid Structured Graph) *A Cartesian grid structured graph S of the size* $n \times m$ *is a C-graph such that:*
$S_N = \{x_{ij} \mid 0 \le i \le n \wedge 0 \le j \le m\}$ $\quad S_A = \{(x_{ij}, x_{kl}) \mid \forall x_{ij}, x_{kl} \in S_N : (k = i + 1 \wedge j = l) \vee (i = k \wedge l = j + 1)\}$, $s_S((x,y)) = x$, $t_S((x,y)) = y$, $m_{SA}((x_{ij}, x_{kl})) = h$ *if* $k = i + 1 \wedge j = l$, $m_{SA}((x_{ij}, x_{kl})) = v$ *if* $i = k \wedge l = j + 1$ *for all* $(x_{ij}, x_{kl}) \in S_A$ *and* $m_{SN}(x) = w$ *for all* $x \in S_N$.

3.2 Definition (Cartesian Object Graph) *A Cartesian object graph O is a connected C-graph where all nodes are black and there exists an injective C-graph morphism* $\mathcal{R}(O) \longrightarrow S$ *if S is a Cartesian grid structured graph.*

3.3 Definition (Polar Graph) *A polar graph G of the size* $n \times m$ *is a C-graph such that:*
$G_N = \{x_{ij} \mid 0 \le i \le n \wedge 0 \le j \le m,\ x_{i0} = x_{im}\}$ $\quad G_A = \{(x_{ij}, x_{kl}) \mid \forall x_{ij}, x_{kl} \in G_N, 0 \le j < m : (j = l \wedge k = i + 1) \vee (i = k \wedge l = j + 1)\}$, $s_G((x,y)) = x$, $t_G((x,y)) = y$, $m_{GA}((x_{ij}, x_{kl})) = p$ *if* $i = k \wedge l = j + 1$ *and* $m_{GA}((x_{ij}, x_{kl})) = r$ *if* $j = l \wedge k = i + 1$ *for all* $(x_{ij}, x_{kl}) \in G_A$.

3.4 Definition (Cartesian Polar Grid Graph) *A Cartesian polar grid graph consists of a polar graph P of the size* $n \times m$, *a Cartesian graph which is a section of a Cartesian grid structured graph and at least* $n \times m$ *connection arcs such that each polar node is at least once source node and each Cartesian node is at least once target node. For each connected section* P_i *of P there is a connected Cartesian graph* C_i *which contains all target nodes of that connection arcs which have a source node in* P_i *and all arcs which are between these target nodes. All nodes have to be white. (See* PR_1 *in example 4.10.)*

3.5 Definition (Polar Object Graph) *A polar object graph PO is a connected black C–graph such that there exists an injective C–graph morphism $\mathcal{R}(PO) \longrightarrow \mathcal{P}(PR)$ if $\mathcal{P}(PR)$ is the polar graph of a Cartesian polar grid graph.*

A configuration is modelled by a C–graph star and its parallel gluing that yields a configuration graph. Such a graph star consists of a Cartesian grid structured graph S which represents the configuration space without any object and a set of Cartesian object graphs O_i. The position of an object is modelled by mapping a removed object graph $\mathcal{R}(O_i)$ into the grid structured graph $\mathcal{R}(O_i) \longrightarrow S$. If an object is able to rotate the orientation is modelled by a polar configuration. The object graphs are not glued among themselves; they are only related indirectly via the grid structured graph S.

3.6 Definition (Cartesian Configuration) *Given a Cartesian grid structured graph S, n Cartesian object graphs O_i and the injective C–graph morphisms $\mathcal{R}(O_i) \longrightarrow S$ for $1 \leq i \leq n$. Together with the C–graph morphisms $\mathcal{R}(O_i) \longrightarrow O_i$ and $\emptyset \longrightarrow O_i$ for each $1 \leq i \leq n$, a Cartesian configuration CCONF consists of the graph star $(O_i \longleftarrow \mathcal{R}(O_i) \longrightarrow S)_{1 \leq i \leq n}$ and its pushout-star graph C called Cartesian configuration graph. We use the denotation:*

$$CCONF : (O_i \longleftarrow \mathcal{R}(O_i) \longrightarrow S)_{1 \leq i \leq n}$$

Remark: The C–graph morphisms $\mathcal{R}(O_i) \longrightarrow O_i$ and $\emptyset \longrightarrow O_i$ for each $1 \leq i \leq n$ always exist.

3.7 Definition (Polar Configuration) *Given a Cartesian polar grid graph PR, a polar object graph PO and an injective C–graph morphism $\mathcal{R}(PO) \longrightarrow PR$. Together with the C–graph morphism $\mathcal{R}(PO) \longrightarrow PO$, a polar configuration PCONF consists of the gluing construction of these morphisms and the polar configuration graph $PK = \mathcal{O}(JOIN(PR \longleftarrow \mathcal{R}(PO) \longrightarrow PO))$.*

$$PCONF : (PR \longleftarrow \mathcal{R}(PO) \longrightarrow PO)$$

3.8 Definition (Configuration) *A configuration CONF consists of a Cartesian configuration CCONF, m polar configurations $PCONF_i$ if the objects i from 1 to m are able to rotate and the inclusions $O_i \longrightarrow PK_i$ $1 \leq i \leq m$ if O_i is a Cartesian object graph of CCONF and PK_i the polar configuration graph of $PCONF_i$ with $\mathcal{B}(\mathcal{C}(PK_i)) = O_i$. The configuration graph C is the configuration graph of the Cartesian configuration CCONF.*

$$CCONF : (CONF, \{PCONF_i | \ 1 \leq i \leq m\})$$

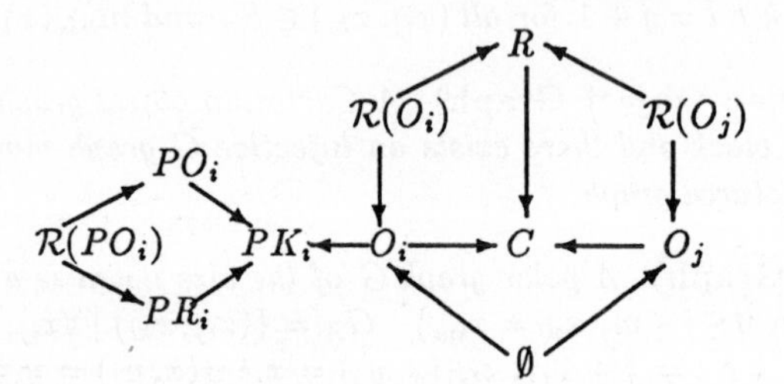

The configuration graph C has the same structure as the Cartesian grid structured graph S and all Cartesian object graphs O_i.

3.9 Definition (Consistency of Configurations) *A configuration with the configuration graph $C = JOIN(O_i \longleftarrow \mathcal{R}(O_i) \longrightarrow S)$ is called consistent if all pullbacks $SPLIT(O_i \longrightarrow C \longleftarrow O_j)$ are empty.*

(Examples of consistent configurations are shown in section 4 within the example of a configuration transformation.)

4 Basic motions

Basic motions are modelled by a two level configuration transformation consisting of the generation of an object production for arbitrary objects and the application of such an object production to a configuration.

4.1 Example of a Set of Elementary Productions for the Translation for one Step to the Right

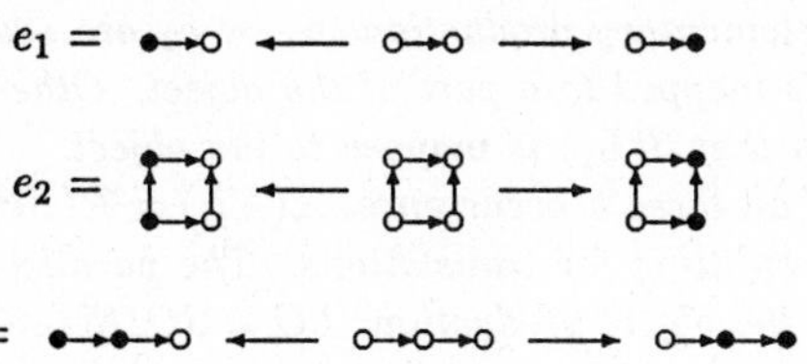

$e_1 =$

$e_2 =$

$e_3 =$

4.2 Example of a Set of Elementary Productions for the Positive Rotation about an Elementary Angle

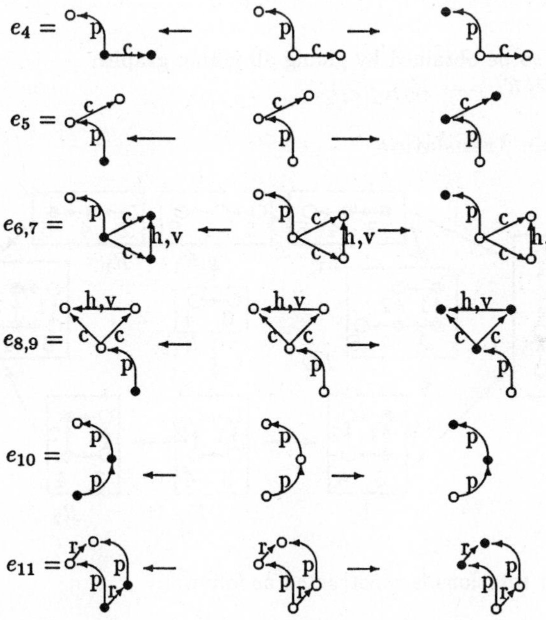

$e_4 =$

$e_5 =$

$e_{6,7} =$

$e_{8,9} =$

$e_{10} =$

$e_{11} =$

There are similar sets of elementary productions modelling other kinds of basic translations and rotations.

4.3 Definition (Object Production) *A C-production $o = (LO \longleftarrow KO \longrightarrow RO)$ is called an object production if we have:*

1. *The black Cartesian part of the right side $\mathcal{B}(\mathcal{C}(RO))$ is a Cartesian object graph.*
2. *The black polar part of the left side is equal to the black polar part of the right side: $\mathcal{B}(\mathcal{P}(LO)) = \mathcal{B}(\mathcal{P}(RO))$.*

An object production is called an object production for translations if all graphs LO, KO and RO are Cartesian and the black left side is equal to the black right side: $\mathcal{B}(LO) = \mathcal{B}(RO)$.

4.4 Proposition (Generation of Object Productions for Basic Translations) *Given a set of elementary productions* $\{e_i = (L_i \longleftarrow K_i \longrightarrow R_i) | \ 1 \leq i \leq 3\}$ *(see example 4.1) and a configuration graph C that models the configuration space with the object to be moved. An object production* $o = (LO \longleftarrow KO \longrightarrow RO)$ *is generated as follows:*

1. *If the object is modelled by a graph with more than just one node then all occurrences of the left sides of the elementary productions* e_2 *or* e_3 *are searched for such that the black part of the left side is mapped to a part of the object. Otherwise you have to look for an occurrence of* L_1 *such that* $\mathcal{B}(L_1)$ *is mapped to the object.*
   The white pullback of all these n occurrences $\mathcal{R}(K_{ij}) = \mathcal{R}(SPLIT(L_i \longrightarrow C \longleftarrow L_j)) \ \forall 1 \leq i \leq n$ *is called object splitting for translations. The parallel gluing of this object splitting yields the* <u>*left side*</u> *of the object production:* $LO = JOIN(L_i \longleftarrow \mathcal{R}(K_{ij}) \longrightarrow L_j)_{1 \leq i < j \leq n}$.
2. *The* <u>*right side*</u> *of the object production can be produced by a parallel graph transformation* $LO \Longrightarrow RO$: $RO = JOIN(R_i \longleftarrow \mathcal{R}(K_{ij}) \longrightarrow R_j)_{1 \leq i < j \leq n}$.
3. *The* <u>*gluing graph*</u> *of the object production KO is equal to the removed left side:* $KO = \mathcal{R}(LO)$.

Remark: KO could also be obtained by gluing all gluing graphs:
$KO = \text{JOIN}(K_i \longleftarrow \mathcal{R}(K_{ij}) \longrightarrow K_j)_{1 \leq i < j \leq n}$

4.5 Example of a Basic Translation

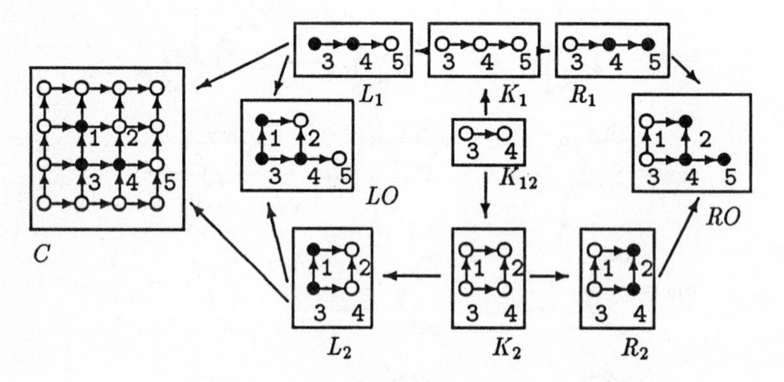

An object splitting for rotations is constructed as follows:

1. If the polar object graph PO for the moved object consists of more than one node then all occurrences of the left sides of the elementary productions e_{10} and e_{11} are searched for such that the black part of a left side is mapped to a part of the object.
2. For all black polar nodes or polar nodes of PK which are adjacent to the black ones in rotation direction we have the following:

   If this polar node is the source node of exactly one connection arc, all occurences of L_4 or L_5 are searched for and otherwise all occurrences of $L_6 - L_9$ are searched for such that one polar node of L_i is mapped to the polar node under consideration.

Now an object productions for basic rotations can be generated like those for basic translations.

4.6 Example of a Basic Rotation

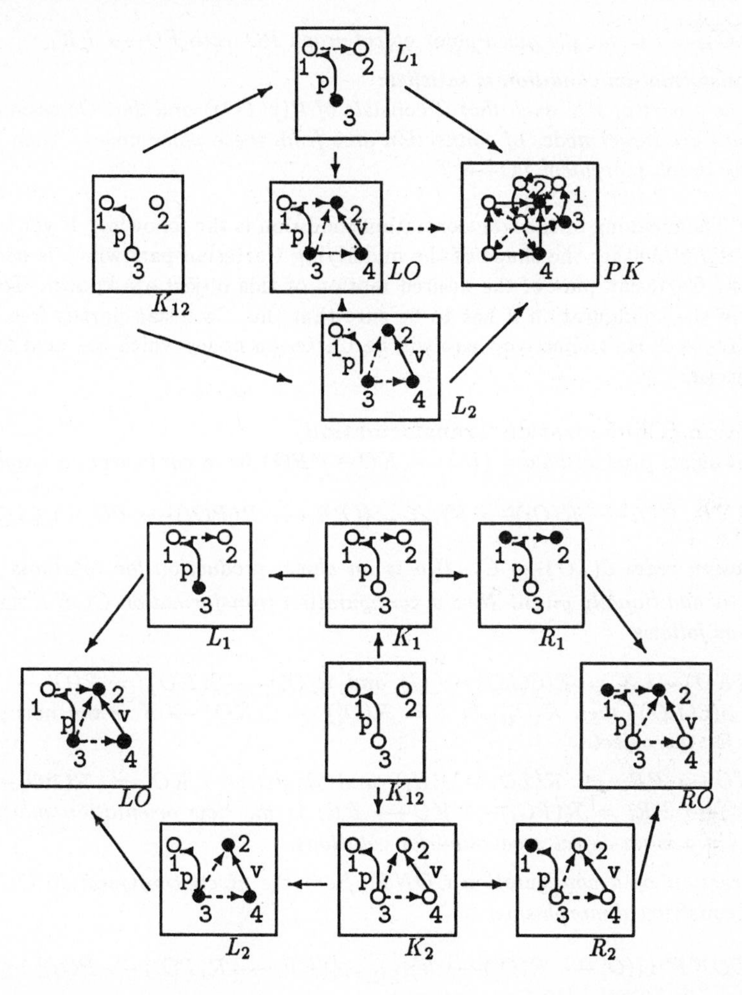

The object production obtained can now be applied to a configuration with an appropriate object. This is done by a configuration transformation changing the position morphism for the object and also transforming the polar configuration for this object if an object production for rotations is applied.

4.7 Definition (Cartesian Redex)

A Cartesian redex for an object production $o = (LO \longleftarrow KO \longrightarrow RO)$ in a configuration graph C is an injective graph morphism $\mathcal{C}(LO) \longrightarrow C$ such that there exists an object graph O with $\mathcal{B}(\mathcal{C}(LO)) = O$ and a C-graph morphism $O \longrightarrow C$ with $O \longrightarrow \mathcal{C}(LO) \longrightarrow C = O \longrightarrow C$.

Remark: The condition $O \longrightarrow \mathcal{C}(LO) \longrightarrow C = O \longrightarrow C$ has to be part of the definition because otherwise an object graph $O_1 = \mathcal{B}(\mathcal{C}(LO))$ could exist, but however there is another object graph $O_2 \neq \mathcal{B}(\mathcal{C}(LO))$ which should be moved without breaking into parts.

4.8 Definition (Polar Redex)

A polar redex for an object production for rotations $o = (LO \longleftarrow KO \longrightarrow RO)$ *in the polar configuration graph* PK *of a given polar configuration is an injective graph morphism* $p: LO \longrightarrow PK$ *such that*

1. $\mathcal{B}(\mathcal{P}(LO)) = PO$ *for the given polar object graph* PO *with* $PO \longrightarrow PK$
2. *the transformation condition is satisfied:*
   Let T *be a part of* PK *such that it consists of* $\mathcal{P}(p(LO))$ *and that Cartesian graph where all nodes are target nodes of connection arcs from these polar nodes. Then there exists a bijective graph morphism* $LO \longrightarrow T$.

Remark: The meaning of the transformation condition is the following: If you want to rotate a polar object PO neither the shape of the underlying Cartesian part which is occupied by the object nor the Cartesian part of the desired motion of this object are known. To preserve the consistency of the configuration it has to be sure that this Cartesian part is free. If the graph morphism $LO \longrightarrow T$ isn't bijective there will be Cartesian nodes which are used for the motion but not reserved.

4.9 Definition (Configuration Transformation)

Given an object production $o = (LO \longleftarrow KO \longrightarrow RO)$ *for a configuration graph* C *within*

$$CONF: ((O_i \longleftarrow \mathcal{R}(O_i) \longrightarrow S)_{1 \leq i \leq n}, \{(PR_i \longleftarrow \mathcal{R}(PO_i) \longrightarrow PO_i) \mid 1 \leq i \leq m\})$$

and a Cartesian redex $\mathcal{C}(LO) \longrightarrow C$. *If* o *is an object production for rotations a polar redex* $LO \longrightarrow PK$ *is additionally given. Now a configuration transformation* $CONF \overset{o}{\Longrightarrow} CONF'$ *is constructed as follows:*

1. *Let* $\mathcal{C}(KO) \longrightarrow S = \mathcal{R}(\mathcal{C}(LO) \longrightarrow C)$ *and* $\mathcal{R}(O_i') \longrightarrow \mathcal{C}(KO) = \mathcal{R}(O_i' \longrightarrow \mathcal{C}(RO))$ *with* $O_i' = \mathcal{B}(\mathcal{C}(RO))$ *then* $\mathcal{R}(O_i') \longrightarrow S = \mathcal{R}(O_i') \longrightarrow \mathcal{C}(KO) \longrightarrow S$ *is the new position morphism for the object* i.
2. *Let* $KO \longrightarrow PR_i = \mathcal{R}(LO \longrightarrow PK_i)$ *and* $\mathcal{R}(PO_i) \longrightarrow KO = \mathcal{R}(PO_i \longrightarrow RO)$ *then* $\mathcal{R}(PO_i) \longrightarrow PR_i = \mathcal{R}(PO_i) \longrightarrow KO \longrightarrow PR_i$ *is the new orientation morphism for the object* i *if* o *is an object production for rotations.*
3. *Construction of a configuration* $CONF'$ *from the given configuration* $CONF$ *and the newly constructed morphisms:*

$$CONF': ((O_i' \longleftarrow \mathcal{R}(O_i') \longrightarrow S)_{1 \leq i \leq n}, \{(PR_i \longleftarrow \mathcal{R}(PO_i) \longrightarrow PO_i) \mid 1 \leq i \leq m\})$$

with

$$O_i' = \begin{cases} O_i, & \text{if no Cartesian redex for } O_i \text{ exists.} \\ \mathcal{B}(\mathcal{C}(RO)), & \text{if a Cartesian redex for } O_i \text{ exists.} \end{cases}$$

$$\mathcal{R}(O_i') \longrightarrow S = \begin{cases} \mathcal{R}(O_i) \longrightarrow S, \\ \quad \text{if no Cartesian redex for } O_i \text{ exists.} \\ \mathcal{R}(\mathcal{B}(\mathcal{C}(RO))) \longrightarrow \mathcal{C}(KO) \longrightarrow S, \\ \quad \text{if a Cartesian redex for } O_i \text{ exits.} \end{cases}$$

$$\mathcal{R}(PO_i) \longrightarrow PR_i = \begin{cases} \mathcal{R}(PO_i) \longrightarrow PR_i, \\ \quad \text{if no polar redex for } O_i \text{ exists.} \\ \mathcal{R}(PO_i) \longrightarrow KO \longrightarrow PR_i, \\ \quad \text{if a polar redex for } O_i \text{ exists.} \end{cases}$$

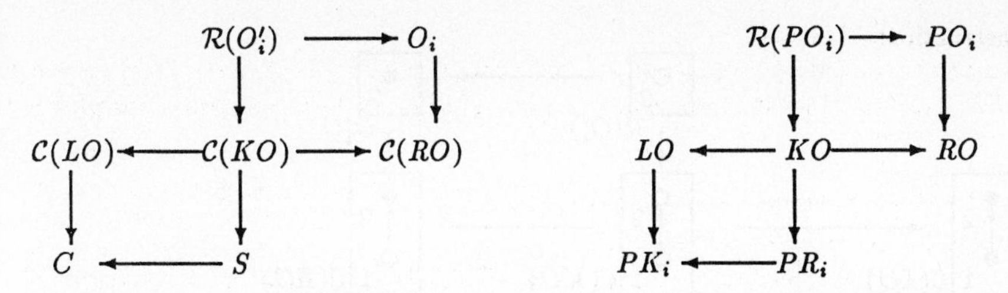

Remark: For each Cartesian object graph O_i' exists a removed Cartesian object graph $\mathcal{R}(O_i')$, a unique C–graph morphism $\mathcal{R}(O_i') \longrightarrow O_i'$ and a unique C–graph morphisms $\emptyset \longrightarrow O_i'$ $\forall\, 1 \leq i \leq n$. For each object that hasn't been moved we have $O_i' = O_i$ and $\mathcal{R}(O_i') \longrightarrow S = \mathcal{R}(O_i) \longrightarrow S$. The parallel gluing of all these morphisms together with the newly constructed ones yields the new configuration graph C'.

There exist always the C–graph morphisms $\mathcal{C}(KO) \longrightarrow S = \mathcal{R}(\mathcal{C}(LO) \longrightarrow C)$ because of the Cartesian redex and $\mathcal{R}(O_i') \longrightarrow \mathcal{C}(KO) = \mathcal{R}(O_i' \longrightarrow \mathcal{C}(RO))$ in consequence of $O_i' = \mathcal{B}(\mathcal{C}(RO))$. (The same can be stated for the orientation morphism.)

4.10 Example of a Configuration Transformation for a Rotation:

Object Production modelling a Basic Rotation:

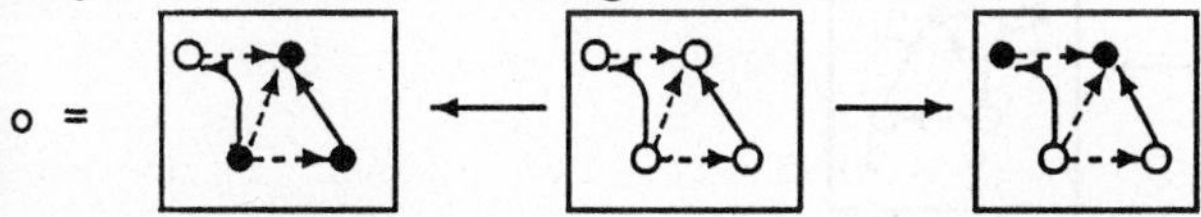

Configuration $CONF$:

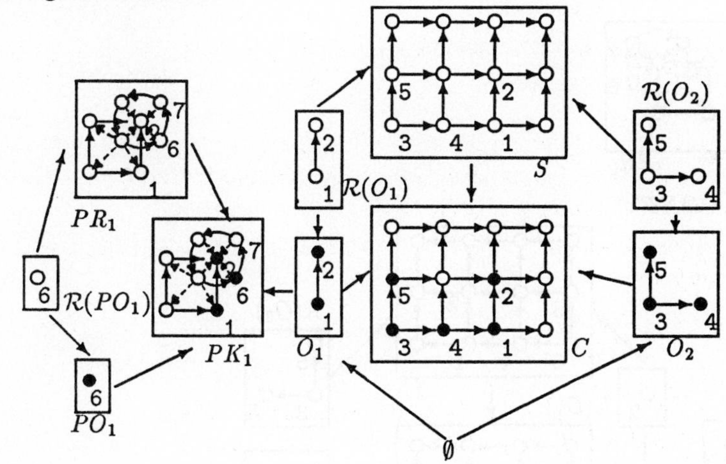

Position Construction:

Orientation Construction:

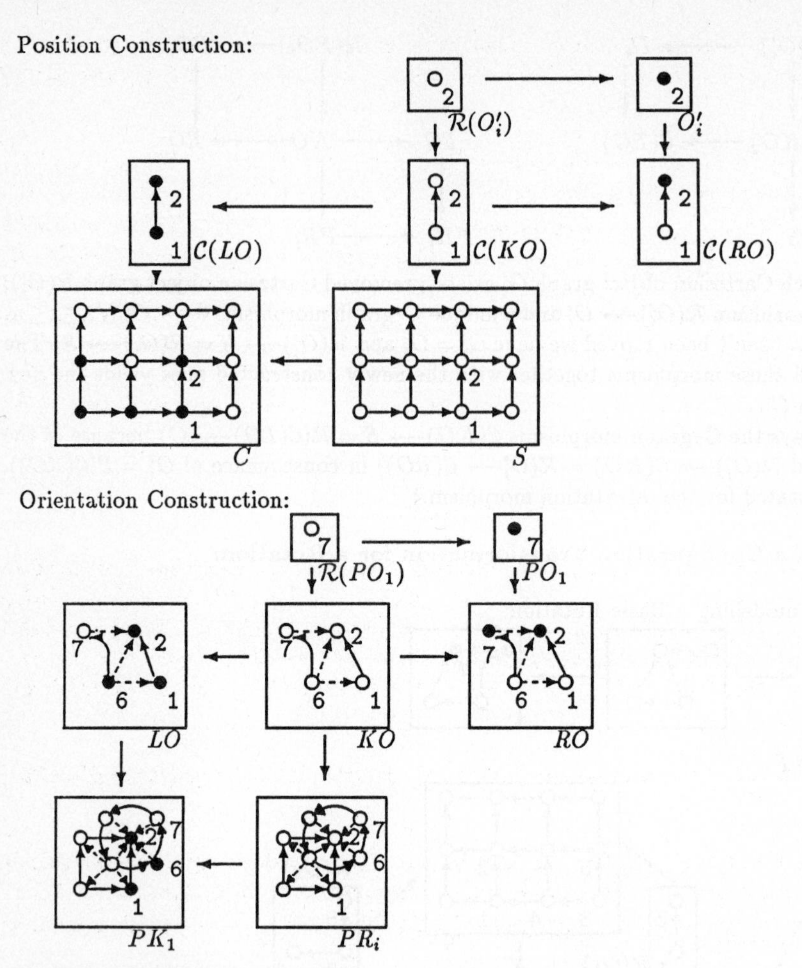

The New Configuration $CONF'$:

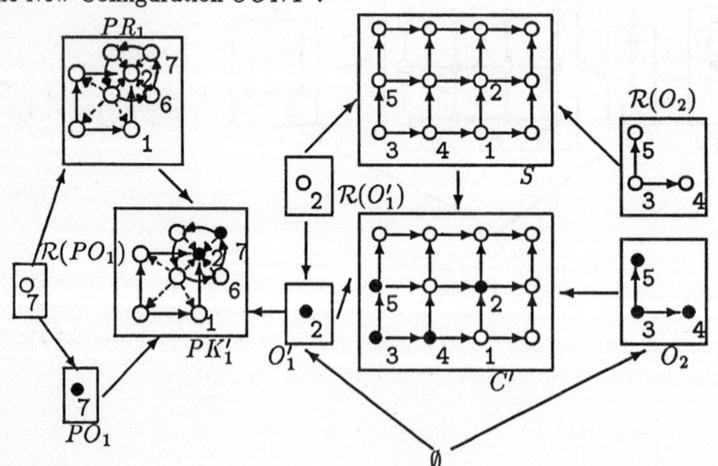

4.11 Consistency Theorem *Given a configuration transformation* $CONF \stackrel{o}{\Longrightarrow} CONF'$ *via the object production* $o = (LO \longleftarrow KO \longrightarrow RO)$ *the consistency of* $CONF$ *implies that of* $CONF'$.

Proof sketch: Let C or C' be the configuration graph of $CONF$ or $CONF'$ resp. and O_x the object graph of the moved object then we have $\text{SPLIT}(O_x \longrightarrow C \longleftarrow O_j) = \emptyset \quad \forall 1 \leq j \leq n$ because of the consistency of $CONF$. Since there exits a Cartesian redex $\mathcal{C}(LO) \longrightarrow C$ with $O_x \longrightarrow \mathcal{C}(LO) \longrightarrow C = O_x \longrightarrow C$ we have $\text{SPLIT}(O_x \longrightarrow \mathcal{C}(LO) \longrightarrow C \longleftarrow O_j) = \emptyset$. $\text{SPLIT}(\mathcal{C}(LO) \longrightarrow C \longleftarrow O_j) = \emptyset \quad \forall 1 \leq j \leq n$ can be obtained because $\mathcal{C}(LO) \longrightarrow C$ is a graph morphism and $\mathcal{B}(\mathcal{C}(LO)) = O_x$. So the area that is needed to execute the motion has been free and is now occupied which means that the object cannot collide with other during the motion. So we obtain $\text{SPLIT}(\mathcal{C}(RO) \longrightarrow C' \longleftarrow O_j) = \emptyset \quad \forall 1 \leq j \leq n$ and then $\text{SPLIT}(O'_x \longrightarrow C' \longleftarrow O_j) = \emptyset \quad \forall 1 \leq j \leq n$ because of $\mathcal{B}(\mathcal{C}(RO)) = O'_x$. None of the other objects are moved, so nothing changes concerning the consistency. □

5 Sequential composition of motions

5.1 Definition (R–Concurrent Object Production) *Given two object productions* $o_1 = (LO_1 \longleftarrow KO_1 \longrightarrow RO_1)$ *and* $o_2 = (LO_2 \longleftarrow KO_2 \longrightarrow RO_2)$ *with the object-relation* $R = SPLIT(RO_1 \longrightarrow C' \longleftarrow LO_2)$ *if* C' *is a configuration graph and* $\mathcal{B}(RO_1) = \mathcal{B}(R) = \mathcal{B}(LO_2)$ *a R-concurrent object production* $o = o_1 *_R o_2 = (LO \longleftarrow KO \longrightarrow RO)$ *is given by the construction of* $LO = JOIN(LO_1 \longleftarrow \mathcal{R}(R) \longrightarrow KO_2)$, $RO = JOIN(KO_1 \longleftarrow \mathcal{R}(R) \longrightarrow RO_2)$ *and* $KO = \mathcal{R}(LO)$.

A general construction of concurrent productions is given in [EhRo80] for graphs or [EHLP91] for PO-graphs.

5.2 Concurrency Theorem *Given an object-relation* R *for a pair of object productions* o_1 *and* o_2 *and the corresponding R-concurrent object production* $o = o_1 *_R o_2$ *there is a bijective correspondence between R-related transformations* $CONF \stackrel{o_1}{\Longrightarrow} CONF' \stackrel{o_2}{\Longrightarrow} CONF''$ *and direct transformations* $CONF \stackrel{o}{\Longrightarrow} CONF''$.

The proof of the concurrency theorem in [EhRo80] can be adapted to our case.

For the configuration transformation $CONF \stackrel{o}{\Longrightarrow} CONF''$ the consistency of $CONF$ is preserved because it can be shown that the R-concurrent object production $o = o_1 *_R o_2$ is an object production in the sense of definition 4.3. So the consistency theorem is also valid for this kind of configuration transformations.

6 Parallel motions

6.1 Definition (Parallel Independence) *Given two object productions* $o_1 = (LO_1 \longleftarrow KO_1 \longrightarrow RO_1)$ *and* $o_2 = (LO_2 \longleftarrow KO_2 \longrightarrow RO_2)$ *and a consistent configuration* $CONF$ *with the configuration graph* C. *The two configuration transformations* $CONF \stackrel{o_1}{\Longrightarrow} CONF_1$ *and* $CONF \stackrel{o_2}{\Longrightarrow} CONF_2$ *are parallel independent if*

$$SPLIT(\mathcal{C}(LO_1) \longrightarrow C \longleftarrow \mathcal{C}(LO_2)) = \emptyset.$$

Remark: Two motions are parallel independent if the areas where they take place are completely different. So the motions do not overlap each other during their whole execution.

The usual criterion of parallel independence for color preserving graph productions sharing only gluing items is not valid in our case because the image of $L \longrightarrow G$ consists of gluing items only and the productions are not color preserving.

Remark: Given a configuration $CONF$ and k object productions o_i with $1 \leq i \leq k$ such that each pair of productions is parallel independent a parallel configuration transformation $CONF \stackrel{\sum o_i}{\Longrightarrow} CONF'$ can be constructed similarily to the sequential configuration transformation in defintion 4.9.

6.2 Parallel Consistency Theorem *Given a parallel configuration transformation $CONF \stackrel{\sum o_i}{\Longrightarrow} CONF'$ via k object productions o_i $(1 \leq i \leq k)$ which are parallel independent for all pairs the consistency of $CONF$ implies that of $CONF'$.*

Proof sketch: Both objects modelled by the graphs O_x and O_y are moved by the productions o_i and o_j. SPLIT$(\mathcal{C}(LO_i) \longrightarrow C \longleftarrow \mathcal{C}(LO_j)) = \emptyset$ is satisfied because the productions o_i and o_j are parallel independent. So the areas that are needed are completely different and now occupied for each production which means that these objects can't collide during the motion. So we state SPLIT$(\mathcal{R}(\mathcal{C}(LO_i)) \longrightarrow S \longleftarrow \mathcal{R}(\mathcal{C}(LO_j))) = \emptyset =$ SPLIT$(\mathcal{R}(\mathcal{C}(RO_i)) \longrightarrow S \longleftarrow \mathcal{R}(\mathcal{C}(RO_j)))$ and then SPLIT$(\mathcal{C}(RO_i) \longrightarrow C' \longleftarrow \mathcal{C}(RO_j)) = \emptyset$. This implies SPLIT$(O'_x \longrightarrow C' \longleftarrow O'_y) = \emptyset$ because $\mathcal{B}(\mathcal{C}(RO_i)) = O'_x$ and $\mathcal{B}(\mathcal{C}(RO_j)) = O'_y$.

If O_x is the graph of an object which is moved by o_i and O_j the graph of an object which isn't moved we have the same situation as in theorem 4.11.

No other objects are moved so nothing changes concerning the consistency. □

Two parallel independent motions are not only parallel but also sequentially executable. (Church-Rosser-Property in e.g [EHLP91]) If the motions are not parallel independent they overlap in a certain part of that area where the motions take place. If their objects don't share this area part at the same time they can also be executed in parallel but possibly not sequentially. This can occur for example when one object moves to a part of that area which was occupied by another before. Some criteria to decide whether two motions can be executed in parallel without any collision are presented in [ST90].

7 Outlook

There are many open questions concerning an efficient modelling of collision-free motions in configuration spaces considered in this paper:

- In order to reduce the costs of modelling rather large configuration spaces a refinement concept for the grids used would be desirable. Then a rather coarse grid could be used to model the configuration space and might be refined at places where collisions could occur under certain circumstances. One approach to such a refinement concept is to have various grids with different degrees of fineness and the description of relations between these grids via graph morphisms. Then the coarsening and refining of configurations could be modelled by configuration transformations and might yield a configuration graph that models a space with different degrees of fineness.

- For modelling spatial configurations spaces the model has to be extendable to three dimensions. This seems to be straight forward as far as the Cartesian part of the modelling is concerned. The Cartesian grid would no longer consist of squares but rather cubes which are modelled by nodes. To model the vicinity of cubes an additional label for arcs is required. As far as the rotation is concerned the extension to the third dimension is dependent on the different possibilities of rotations. A rotation point for example could be modelled by a kind of globular grid, a rotation axis might be modelled by various polar grids which would be placed at a fixed distance to each other.

- Further investigations of application areas like object-orientated graphical user interfaces, work spaces of robots, railway systems, assembly lines, etc, are planned in future work.

(For more details to the introduced ideas and concepts as well as the proofs of the theorems and propositions presented see [ST90].)

References

[ArMa75] M.A. Arbib, E.G. Manes
Arrows, Structures and Functors
Academic Press, New York – San Francisco – London,1975

[BEHL87] P. Boehm, H. Ehrig, U. Hummert, M. Loewe
Towards distributed graph-grammars
Proc. 3rd Int. Workshop on Graph Grammars, Springer LNCS 291(1987),p.86–98

[Eh79] Hartmut Ehrig
Introduction to the Algebraic Theory of Graph Grammars
LNCS 73, Springer Verlag, Berlin, 1979

[EHLP91] H. Ehrig, A. Habel, M. Loewe, F. Parisi-Presicce
From Graph Grammars to High-Level Replacement Systems
This volume

[EhRo80] Hartmut Ehrig,Barry K. Rosen
Parallelism and Concurrency of Graph Manipulations
North-Holland Publishing Company, 1980

[PEM87] F. Parisi–Presicce, H. Ehrig, U. Montanari
Graph Rewriting with Unification and Composition
Proc. 3rd Int. Workshop on Graph Grammars, Springer LNCS 291(1987),p.496–511

[ST90] Holger Schween, Gabriele Taentzer
Anwendungen und Erweiterungen der Theorie von Graphgrammatiken auf Bewegungsprobleme in der Robotik
Diplomarbeit, TU Berlin, 1990

[Wi81] A. Wilharm
Anwendung der Theorie von Graphgrammatiken auf die Spezifikation der Prozesssteuerung von Eisenbahnsystemen
Comp. Sci. Report 81-15,TU Berlin,1981

Recognizing Edge Replacement Graph Languages in Cubic Time

Walter Vogler

Institut für Informatik
Technische Universität
Arcisstr. 21
D-8000 München 2

ABSTRACT: While in general the recognition problem for (hyper-)edge replacement grammars is *NP*-complete, there are polynomial algorithms for restricted graph classes. The degree of the corresponding polynomial depends on the size of the right hand sides of the grammar, and this size cannot be restricted without restricting the generative power.

In this paper we show that for the class of cyclically connected graphs the recognition problem for edge replacement grammars can be solved in cubic time, i.e. the degree of the polynomial does not depend on the size of the right hand sides of the grammar. For this result we give a suitable normal form for an edge replacement grammar. The algorithm uses the idea of the Cocke-Kasami-Younger algorithm and depends crucially on an algorithm of Hopcroft and Tarjan, which can be used to determine the form of a derivation tree for the given graph.

Keywords: Edge replacement, parsing, triconnected components

CONTENTS

0 Introduction

Graph grammars as generalizations of the usual string grammars have been defined in various ways, see [CER79,ENR83,ENRR87] for overviews. In order to have a class of graph grammars which is at the same time expressive and tractable one has developed graph grammars which one could call context-free. Context-free grammars where one edge or hyperedge is replaced in each step are defined in [BC87,Hab89,HK87a,HK87b], context-free grammars where one vertex is replaced in each step are defined in [RW86]. A comparison of these approaches can be found in [ER90] and [Vog90], and a specific discussion of context-freeness in [Cou87].

In this paper we will deal with the recognition problem of edge replacement grammars. In general this problem is *NP*-complete [LW87], thus we (probably) have to restrict the graphs that we can accept as inputs, if we want to give a polynomial recognition algorithm. This has been done in some papers, and the most general approach – it is dealing with hyperedge replacement – can be found in [Lau88]. The algorithm given there has the drawback that although it is polynomial, the degree of the polynomial depends on the size of the right hand sides of the grammar under consideration. Already in the case of edge replacement this size cannot be restricted without restricting the expressive power [HK87a]. In this paper we will give a polynomial recognition algorithm for edge replacement grammars, where the degree of the polynomial is fixed, namely 3 – or 2.376, since the worst case depends only on the worst case of context-free string recognition (see [CW87]).

A derivation step in an edge replacement grammar (ERG) consists in the removal of an edge, the addition of a graph with two distinguished vertices and the merging of these vertices with the source and the target of the deleted edge. We distinguish the two vertices implicitly: Each graph in this paper has a special edge, called the virtual edge, whose source and target are the two distinguished vertices. This allows to represent a rule of an ERG as one graph, where the (label of the) virtual edge is the left hand side, the rest of the graph the right hand side of the rule, compare [ER90].

As a consequence, a string $x_1 \dots x_n$ is not represented as a path whose edges are labelled $x_1, \dots, x_n$ as in [HK87a], but the representing graph consists of such a path plus a virtual edge from the first to the last vertex of the path; hence a string is represented by a cycle. Also, e.g. series-parallel graphs, which can be generated by an ERG, have an additional edge from their begin- to their end-vertex; hence they are cyclically connected. These remarks are important, since the class of graphs we can deal with is the class of cyclically connected graphs – thus we cover such prominent cases as strings and series-parallel graphs.

Let us compare this class of graphs with the class of graphs the algorithm of [Lau88] can deal with in the case of edge replacement: We require that the input graph itself as well as this graph after removing one arbitrary vertex have one component only – in [Lau88] a bounded number of components is allowed in both cases. On the other hand, removal of two vertices from a cyclically connected graph may result in any number of components, whereas in [Lau88] the number of components has to be bounded in this case, too.

Our recognition algorithm consists of a preprocessing phase where the grammar under consideration is brought into some normal form, which we call Chomsky Normal form: A rule is either 3-connected without multiple edges or its right hand side has two edges only.

Then the input graph is decomposed in linear time by the triconnected components algorithm of [HT73]. The result is more or less the only possible derivation tree of the input graph, except that the nonterminal labels are missing. Finally these labels are constructed in a bottom-up fashion, very similar to the Cocke-Kasami-Younger algorithm.

Section 1 contains the basic definitions and Section 2 introduces the triconnected components of graphs and edge replacement grammars. Section 3 exhibits how an ERG can be brought into Chomsky Normal form. Section 4 prepares the algorithm: It is shown how we will deal with edge directions; also the analysis of strings, i.e. elements of a free monoid, and of vectors, i.e. elements of a free commutative monoid, is prepared here. Finally, Section 5 gives the algorithm itself.

1 Preliminaries

Since we will replace directed edges by graphs via identifying one vertex of the graph with the source of the edge, another vertex with the target of the edge, our graphs have two distinguished vertices. We do not distinguish these vertices explicitly as in [HK87a], but implicitly by introducing a special edge between them, which we call the virtual edge. This representation is especially suitable for the use of results from [HT73] and can directly be used to represent productions of edge replacement systems – this representation of productions is used e.g. in [ER90].

Let Σ be a fixed countably infinite alphabet, let $\overline{\Sigma}$ be a disjoint copy of Σ and let $\overline{} : \Sigma \cup \overline{\Sigma} \to \Sigma \cup \overline{\Sigma}$ be the mapping that takes each element of one of the sets Σ and $\overline{\Sigma}$ to its copy in the other set. Let $\tau \notin \Sigma \cup \overline{\Sigma}$ and define $\overline{\tau} = \tau$.

A *graph* $G = (V, E, s, t, l, vir)$ consists of disjoint finite sets V and E, two functions $s, t : E \to V$ assigning to each edge e distinct vertices $s(e)$ and $t(e)$, called the *source* and the *target* of e, a *labelling* function $l : E \to \Sigma \cup \overline{\Sigma} \cup \{\tau\}$ and a distinguished edge $vir \in E$, called the *virtual edge* of G.

Note that a graph is loop-free, but may have multiple edges.

Two graphs G, G' are *weakly isomorphic* if there are bijections $\varphi : V \to V'$, $\eta : E \to E'$ such that $\eta(vir) = vir'$ and for all edges e of E we have

i) $\{\varphi(s(e)), \varphi(t(e))\} = \{s'(\eta(e)), t'(\eta(e))\}$

ii) $l(e) = l'(\eta(e))$ if $\varphi(s(e)) = s'(\eta(e))$

iii) $\overline{l(e)} = l'(\eta(e))$ if $\varphi(s(e)) = t'(\eta(e))$ (i.e. if (φ, η) *reverses* e)

Thus a weak isomorphism may reverse the direction of an edge if it changes the label according to $\overline{}$.

If additionally for all edges e $\varphi(s(e)) = s'(\eta(e))$, then G and G' are *isomorphic.*

If for a vertex v and an edge e of a graph we have $v \in \{s(e), t(e)\}$, then we call v *incident* with e. If for vertices v, w there is an edge e with $\{v, w\} = \{s(e), t(e)\}$, then v and w are called *adjacent.* A vertex is called *isolated* if it is not incident to any edge. A graph has *multiple edges* if there are distinct edges e, e' with $\{s(e), t(e)\} = \{s(e'), t(e')\}$.

When talking about connectivity or cycles we refer to the underlying undirected graph:

A *path* is an alternating sequence $v_0 e_1 v_1 \ldots e_n v_n$ of distinct vertices v_i and edges e_i such that $\{v_{i-1}, v_i\} = \{s(e_i), t(e_i)\}$ for $i = 1, \ldots, n$. Such a path is a v_0, v_n-path, the vertices v_i, $i = 1, \ldots, n-1$ are the *interior* vertices. Paths are *openly disjoint* if they have no interior vertices in common. A *cycle* is defined as a path except that we have $v_0 = v_n$ and $n \geq 3$ here. A path is *directed* if $(v_{i-1}, v_i) = (s(e_i), t(e_i))$, $i = 1, \ldots, n$. A graph G is n-*connected* if it has at least $n+1$ vertices and for all distinct vertices v, w there are at least n openly disjoint v, w-paths.

G is a *bond* if $|V| = 2$, thus necessarily for all $e \in E$ we have $\{s(e), t(e)\} = V$; it is a *triple bond* if additionally $|E| = 3$. A graph is *cyclically connected* if it is 2-connected or a bond with at least 3 edges.

2 Merge, split, and edge replacement grammars

Let G, G' be graphs that are disjoint except that G contains vir', i.e. $V \cap V' = \{s'(vir'), t'(vir')\}$, $E \cap E' = \{vir'\}$, $s(vir') = s'(vir')$, $t(vir') = t'(vir')$, $l(vir') = l'(vir')$, $vir \neq vir'$. Then

$$merge(G, G') = (V \cup V'\ ,\ E''\ ,\ (s \cup s')\mid_{E''}\ ,\ (t \cup t')\mid_{E''}\ ,\ (l \cup l')\mid_{E''}\ ,\ vir),$$

where $E'' = (E \cup E') - \{vir'\}$.

In words: We *merge* G and G' along vir' and delete this edge afterwards. Note that reversing vir' in both graphs and merging the modified graphs gives the same result.

Viewed the other way, we can describe how to separate a graph at two vertices: We say that $merge(G, G')$ is *split* at $s'(vir')$ and $t'(vir')$ into G and G', provided that $l'(vir') = \tau$ and $|E|, |E'| \geq 3$. This last condition is a non-triviality condition: It ensures that we cannot split a graph into a single edge and the remainder. (Remember that the virtual edge of G' is newly created and appears in G, too.) Thus G and G' have both strictly less edges than $merge(G, G')$.

If a graph is not 2-connected, one should decompose it into components and blocks before considering the separation at two vertices.

If we split a cyclically connected graph into G and G', split G and G' further etc. we obtain a *split decomposition* of the initial graph. The graphs obtained are called *split parts*. Formally, a split decomposition of G is a sequence of graphs defined inductively by:

- (G) is a split decomposition of G.
- Let $(G_1, \ldots, G_n)$ be a split decomposition of G and let, for some $i \in \{1, \ldots, n\}$, G_i be split into G' and G'' in such a way, that vir'' appears in G' but in no G_j, $j \in \{1, \ldots, n\} - \{i\}$. Then $(G_1, \ldots, G_{i-1}, G', G'', G_{i+1}, \ldots, G_n)$ is a split decomposition of G.

If a split part G_i contains an edge e that is the virtual edge of some other split part G_j we say that e is a *virtual edge somewhere*. If we view G_i as the father of G_j, then we have a tree structure on the split parts and could also call the split decomposition a decomposition tree.

By a result of [HT73] we have a linear dependence between the number of edges of a graph and the total number of edges in a split decomposition:

Lemma 2.1 ([HT73]) *Let G be a graph with at least 3 edges and let $(G_1, \ldots, G_n)$ be a split decomposition of G. Then the sum of the number of edges of $G_1, \ldots, G_n$ is at most $3 \cdot |E_G| - 6$.*

If $(H_1, \ldots, H_n)$ is a split decomposition of some graph G we define *result*(H_i) as the graph obtained from H_i as follows: If the graph constructed so far contains an edge that is a virtual edge somewhere, say in H_j, then merge the graph with H_j; repeat this as long as possible. In other words, we merge H_i with all its descendents in the decomposition tree. Obviously, the order of merging does not matter, hence *result*(H_i) is well-defined. Especially, if $vir_{H_i} = vir_G$, then *result*$(H_i) = G$.

If in a split decomposition no split part can be split further we call the split decomposition *total* and its parts *split components*. Due to the non-triviality condition any splitting of a graph will result in a total split decomposition after finitely many steps.

Proposition 2.2 ([HT73]) *Every split component of a cyclically connected graph is either a triple bond or a triangle or 3-connected without multiple edges.*

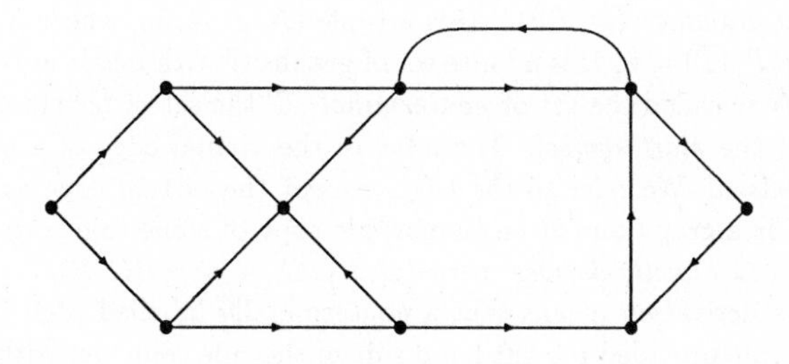

Figure 1

The split components of a graph are not uniquely determined. If we repeatedly merge each bond of a total split decomposition that contains the virtual edge of another bond with this bond until no bond contains the virtual edge of another bond, and in the same way repeatedly merge the (undirected) cycles starting with the triangles until no cycle contains the virtual edge of another

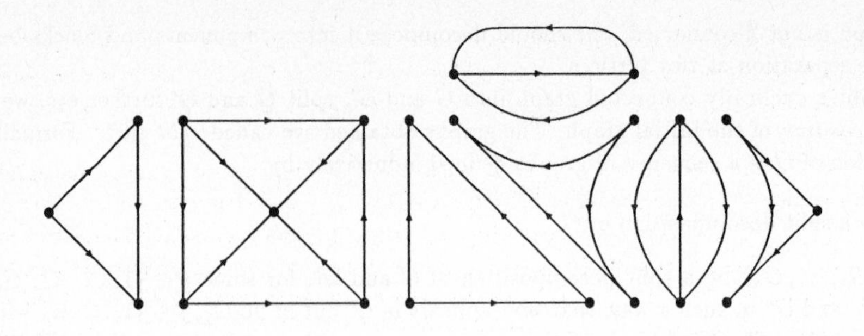

Figure 2

cycle, then we obtain a split decomposition which is called *collapsed*. Its split parts are called the *triconnected components* of the initial graph.

Figure 1 shows a graph without its edge labels and without distinguishing the virtual edge. Figure 2 shows a total split decomposition of this graph. To obtain a collapsed split decomposition one simply has to merge the two triangles in the middle.

The triconnected components of a graph are essentially unique by a result of [Mac37]; we rephrase this result in terms of directed graphs:

Theorem 2.3 *Let $(G_1, \ldots, G_n)$ and $(H_1, \ldots, H_m)$ be collapsed split decompositions of a graph G. Then $n = m$ and $(H_1, \ldots, H_n)$ can be reordered such that there exist weak isomorphisms (φ_i, η_i) : $G_i \to H_i$, $i = 1, \ldots, n$, with the properties:*

1) $\varphi_i = id$

2) $\eta_i(e) = e$ and $s_{G_i}(e) = s_{H_i}(e)$ for all $e \in E_G \cap E_{G_i}$

3) If for some $e \notin E_G$ we have $e \in E_{G_i} \cap E_{G_j}$ then: (φ_i, η_i) reverses e if and only if (φ_j, η_j) does.

Proof: The proof is by induction on the number of split components. One has to consider two ways of splitting G and has to show that they can result in the same collapsed split decomposition in the sense of the theorem. This requires a lengthy and tedious case analysis which we omit since the result is just the more detailled version for directed graphs of an old result. □

An *edge replacement grammar* (an ERG) $\mathcal{G}$ is a tuple $(N, T, \mathcal{R}, S)$, where N, T are finite subsets of $\Sigma \cup \overline{\Sigma}$ with $(N \cup \overline{N}) \cap (T \cup \overline{T}) = \emptyset$, $\mathcal{R}$ is a finite set of graphs G with labels in $N \cup T$ and $l(vir) \in N$, and $S \in N$. As usual, N is called the set of *nonterminals*, T the set of *terminals*, $\mathcal{R}$ the set of *rules* or *productions* and S is the *start symbol*. The label of the virtual edge of a production is the *left hand side* of the production. We refer to the edges except the virtual edge as the right hand side of the production. If G is a graph and R an isomorphic copy of some rule such that $merge(G, R)$ is defined then we say that G *directly derives* $merge(G, R)$, $G \to merge(G, R)$.

As in [HK87a] direct derivation means that a nonterminally labelled edge e of G is replaced by the right hand side of a rule provided the left hand side of the rule coincides with the label of e. Note that by the definition of *merge* we cannot replace the virtual edge of G (even if it is nonterminally labelled.)

The reflexive, transitive closure of $\to$ is denoted by $\xrightarrow{*}$, and G *derives* G' if $G \xrightarrow{*} G'$.

For $A \in N$ each graph G which can be derived from a rule with left hand side A is an *A-form*, an S-form is also called a *sentential form*, the set of S-forms is denoted by $\mathcal{S}(\mathcal{G})$. We say that G is *generated* from A by $\mathcal{G}$, and also that a sentential form is generated by $\mathcal{G}$.

The *language* $\mathcal{L}(\mathcal{G})$ of $\mathcal{G}$ consists of all sentential forms G with $l(e) \in T$ for all $e \in E - \{vir\}$.

More precisely, sentential forms are isomorphism classes of graphs; but since the merge-operation is defined on representatives, we will – as usual – not distinguish between an isomorphism class and a representative.

If for an A-form G we collect the rule we started with and the isomorphic copies of rules we used in the direct derivation steps we get the *derivation tree* of this derivation for G (compare [Kre86]) – the collection of graphs is the vertex set of the tree and a graph which contains the virtual edge of some other graph is the father of this graph. Edge replacement is considered context-free because, if we merge the graphs of our collection in any order, the result will always be G. Now for every graph of our collection we change the label of every edge that is not in G to τ. This way we obtain the *split decomposition induced* by this derivation of G. If we repeatedly merge each bond which contains the virtual edge of another bond with this other bond and each cycle which contains the virtual edge of another cycle with this other cycle, then we obtain the *collapsed split decomposition induced* by this derivation of G.

Lemma 2.4 *The split decomposition induced by a derivation of a graph G is a split decomposition of G, provided all its graphs have at least 3 edges.*

Note that the collapsed split decomposition induced by a derivation of G is not necessarily a collapsed split decomposition of G, since the split decomposition induced by a derivation does not have to be total.

We can consider a usual context-free string grammar without λ-rules as an ERG as follows: A nonempty *string* $x_1, \ldots, x_n$ is *represented* as a directed path with n edges which are labelled $x_1, \ldots, x_n$ in the order they appear, plus a virtual edge from the first to the last vertex of the path. Thus strings are cycles, and a rule $A \to x_1 \ldots x_n$ is a cycle representing $x_1 \ldots x_n$ where the virtual edge is labelled A.

A collapsed split decomposition induced by a derivation is in the string case simply a cycle representing the derived string.

If a string grammar is in Chomsky Normal Form, all rules are triangles or bonds with two edges. If we only want to derive strings of length at least 2, we can transform this easily into a normal form where all rules are triangles. This normal form we will generalize to ERG's and call it Chomsky Normal Form.

This generalization is influenced by Proposition 2.2, and it is chosen in order to use the results of [HT73]. Let us remark that we cannot expect a generalization where the size of the rules is bounded by some constant. In [HK87a] it is shown that for each k there exists an ERG such that no ERG can generate the same language if all its rules have at most k vertices.

The string case is the case of free monoids. We can also treat the case of free commutative monoids, i.e. the vector case. A graph G *represents* a *vector* over an alphabet A, i.e. a mapping $v : A \to \mathbb{N}_0$, if it is a bond, for all edges e we have $s_G(e) = s_G(vir_G)$, and for each $a \in A$ there are exactly $v(a)$ edges in $E_G - \{vir_G\}$ with label a.

3 Normal forms

In this section we will introduce two normal forms. First we will eliminate chain productions and erasing rules. A rule is a *chain production* if it is a bond with two edges. (I.e. we also consider the replacement of a nonterminal by a terminal edge as a chain production.) An *erasing* rule is a bond with one edge (the virtual edge). An ERG $\mathcal{G}$ is in *weak normal form* if

i) $\mathcal{G}$ has no chain productions, or every chain production has the start symbol as left hand side and the start symbol does not appear in any right hand side

ii) no rule is erasing, or the only erasing rule has the start symbol as left hand side and the start symbol does not appear in any right hand side.

Proposition 3.1 *For each ERG $\mathcal{G}$ there exists an ERG $\mathcal{G}'$ in weak normal form with $\mathcal{L}(\mathcal{G}) = \mathcal{L}(\mathcal{G}')$.*

Proof: Application of the usual techniques. □

Lemma 3.2 *Let $\mathcal{G}$ be an ERG in weak normal form, let G be a cyclically connected A-form of $\mathcal{G}$ and R a rule that is used in a derivation of G. Then R is cyclically connected.*

Proof: Assume R is not cyclically connected. Since G is cyclically connected and $\mathcal{G}$ is in weak normal form, R cannot be a bond with one or two edges, i.e. it has at least 3 vertices. Since R is not 2-connected, we conclude that in R there exists a vertex $v \notin \{s_R(vir_R), t_R(vir_R)\}$ such that there do not exist two openly disjoint paths from v to $s_R(vir_R)$ or $t_R(vir_R)$. Thus using R in a derivation results in a graph which is not cyclically connected, and further derivation steps do not change this, a contradiction. □

In this paper we are interested in recognizing cyclically connected graphs. In view of the above results we will ignore rules which are not cyclically connected when transforming an ERG into *Chomsky Normal Form* (CNF). An ERG is in CNF if all rules are 3-connected without multiple edges, or are triangles or triple bonds.

Theorem 3.3 *For each ERG $\mathcal{G}$ there exists an ERG $\mathcal{G}'$ in CNF such that $\mathcal{L}(\mathcal{G}') = \{G \in \mathcal{L}(\mathcal{G}) \mid G$ is cyclically connected $\}$.*

Proof: By 3.1 we can assume that $\mathcal{G}$ is in weak normal form, by 3.2 we can assume that all rules of $\mathcal{G}$ are cyclically connected. This implies that all $G \in \mathcal{L}(\mathcal{G})$ are cyclically connected. Iteratively, for each rule R of $\mathcal{G}$ we consider a total split decomposition of R; for each newly created virtual edge we introduce a new nonterminal and label both occurences of that edge with this nonterminal; then we replace R by all the graphs in this modified split decomposition. Since the rules of $\mathcal{G}$ are replaced one after the other, we will always introduce new nonterminals; in the resulting grammar $\mathcal{G}'$ each new nonterminal appears exactly twice, once in some right hand side and once as left hand side.

By 2.2 $\mathcal{G}'$ is in CNF.

By induction on the derivation length we see that $\mathcal{S}(\mathcal{G}) \subseteq \mathcal{S}(\mathcal{G}')$, since any application of a rule R in $\mathcal{G}$ can be simulated by applying all the rules of $\mathcal{G}'$ that replaced R. Vice versa, consider the derivation of some $G \in \mathcal{L}(\mathcal{G}')$: When applying the first of the rules that replaced some R (necessarily the only one whose virtual edge has an 'old' nonterminal label) the only way to get rid of the new nonterminals is to apply in every case the only rule that has such a nonterminal as left hand side. Thus we have to apply all the rules that replaced R. By context-freeness the result is the same as applying R. We conclude that $\mathcal{L}(\mathcal{G}) = \mathcal{L}(\mathcal{G}')$. □

The idea of our normal form is that either the right hand side of a rule has at most two edges or the rule is 3-connected without multiple edges. This idea can also be applied when we consider the generation of terminal graphs which are not cyclically connected, but such a generalization is not studied here.

4 Edge directions, bonds and cycles

Grammars in CNF are related to split decompositions as follows:

Lemma 4.1 *Let $\mathcal{G}$ be an ERG in CNF, and let for some nonterminal A the graph G be an A-form. Then the split decomposition induced by some derivation of G is a total split decomposition of G. The collapsed split decomposition induced by some derivation of G is a collapsed split decomposition of G.*

Proof: In view of 2.2 the first claim follows from 2.4. Now the second claim is immediate from the definitions. □

The collapsed split decomposition of a graph is unique in the sense of Theorem 2.3 and can be computed in linear time [HT73]. Thus for a graph generated by an ERG in CNF we can determine the form of the derivation tree with the following exceptions:

- directions of the virtual edges
- triangles
- bonds.

To deal with the first problem we have introduced the "reverse labels" $\overline{\Sigma}$, the second problem is just the recognition problem for context-free string grammars, and the third leads to the recognition problem of semilinear sets.

First we will deal with the edge directions. If for a graph grammar $\mathcal{G}$ we have $N \cup T \subseteq \Sigma$, then we call $\mathcal{G}' = (N \cup \overline{N}, T \cup \overline{T}, \mathcal{R}', S)$ the *completion* of $\mathcal{G}$, if $\mathcal{R}'$ consists of all graphs weakly isomorphic to some graph in $\mathcal{R}$. (More precisely, we take a representative of each isomorphism class of such graphs, hence $\mathcal{R}'$ is finite.)

We have the following relation between $\mathcal{G}$ and its completion:

Lemma 4.2 *Let $\mathcal{G}$ be an ERG with $N \cup T \subseteq \Sigma$, let $\mathcal{G}'$ be its completion and let G be a graph with labels in $N \cup T$. Then G is an A-form in $\mathcal{G}$ if and only if every (or some) graph that is weakly isomorphic to G with unreversed virtual edge is an A-form in $\mathcal{G}'$ if and only if every (or some) graph that is weakly isomorphic to G with reversed virtual edge is an $\overline{A}$-form in $\mathcal{G}'$.*

Proof: Let G' be a graph that is weakly isomorphic to an A-form G in $\mathcal{G}$ with unreversed virtual edge. Take a derivation tree of G, and whenever a graph R in this tree contains an edge of G that is reversed in G', then reverse it in R (and change the label to its $\overline{\ }$-image). This way we obtain a derivation tree of G' in $\mathcal{G}'$. Vice versa, we can change a derivation tree of G' in $\mathcal{G}'$ to a derivation tree of G in $\mathcal{G}'$. Whenever some edge is labelled $\overline{A} \in \overline{\Sigma}$ in this derivation tree, we reverse both occurrences of this edge and change the label to A. Thus we get a derivation tree of G in $\mathcal{G}$.

The other equivalence is similar. □

For the derivation of cycles we will need the following connection to string derivation:

Lemma 4.3 *Let $\mathcal{G}'$ be the completion of some CNF grammar $\mathcal{G}$, let G be a cycle and an A-form in $\mathcal{G}'$. Then:*

i) There is a unique graph G' which is weakly isomorphic to G, represents a string and has $l'(vir') = A$.

ii) G' can be derived in $\mathcal{G}'$ by using only rules that represent strings.

Proof:

i) Since $l'(vir') = A$ the weak isomorphism cannot reverse vir. Thus to get a graph G' that represents a string we have to direct the other edge e_1 incident to $s'(vir')$ away from this vertex. Now the edge $e_2 \neq e_1$ that is incident to $t'(e_1)$ has to be directed away from $t'(e_1)$ etc.

ii) G' consists of its virtual edge and a directed path; let $v_0, \ldots, v_n$ be the vertices of this path in this order.
By 4.2 G' can be derived in $\mathcal{G}'$. Consider a derivation tree for G'. Since $\mathcal{G}$, and thus also $\mathcal{G}'$ is in CNF all graphs in this tree are triangles. Now we direct all edges of graphs in this tree such that the source has a smaller index than the target. If we have to reverse an edge e we change its label B to $\overline{B}$. No edge of G' has to be reversed, and if some edge is reversed, then both occurrences of this edge are reversed. Hence we obtain a derivation tree for G' as desired. □

Analogously, one obtains a version of the last lemma for the case of vectors:

Lemma 4.4 *Let $\mathcal{G}'$ be the completion of some CNF grammar $\mathcal{G}$, let G be a bond and an A-form in $\mathcal{G}'$. Then*

i) There is a unique graph G' which is weakly isomorphic to G, represents a vector and has $l'(vir') = A$.

ii) G' can be derived in $\mathcal{G}'$ by using only rules that represent vectors.

5 The algorithm

Now we are ready to present the algorithm for the *recognition problem* of edge replacement graph languages:

We have a fixed ERG $\mathcal{G} = (N, T, \mathcal{R}, S)$ with $N \cup T \subseteq \Sigma$. Given a cyclically connected graph G with labels in $N \cup T$ we want to decide whether $G \in \mathcal{L}(\mathcal{G})$.

In a preprocessing phase we transform $\mathcal{G}$ into CNF and construct its completion – by 3.1, 3.3 and 4.2 we may call the result $\mathcal{G}$ again. (It should be remarked that constructing the completion is very time consuming, and in fact, it could be avoided. But using the completion the algorithm is easier to understand.)

Now we apply the algorithm of Hopcroft and Tarjan to determine a collapsed split decomposition of G in linear time [HT73]: This algorithm takes an undirected graph without edge labels represented as adjacency lists as input. Thus we have to modify the algorithm in order to keep track of edge labels. Furthermore we modify G as follows: For each edge e we have an entry $(l(e), t(e))$ in the list of $s(e)$, now we add an entry $(\overline{l(e)}, s(e))$ to the list of $t(e)$. This is more or less the usual way of representing undirected graphs as directed graphs. The algorithm of [HT73] performs a depth-first-search on the input graph, which turns it into a directed graph G', i.e. deletes for each edge e of G either the entry $(l(e), t(e))$ in the list of $s(e)$ or the entry $(\overline{l(e)}, s(e))$ in the list of $t(e)$. If the original virtual edge is deleted, G' has the corresponding reverse edge as virtual edge. G' is weakly isomorphic to G.

The algorithm of [HT73] produces a total split decomposition first. Whenever a split is performed, a new virtual edge is created; keeping track of these edges gives the tree of the decomposition and allows the algorithm of [HT73] to produce a collapsed split decomposition $(H_1, \ldots, H_n)$ of G'. If the virtual edge of G has been reversed in G' we can reverse it again in G' and in the corresponding triconnected component, say H_1. Hence we can assume that $l'(vir') = S$ (otherwise we have $G \notin \mathcal{L}(\mathcal{G})$ immediately). By 4.2 we now have to check whether $G' \in \mathcal{L}(\mathcal{G})$.

By 2.3 $(H_1, \ldots, H_n)$ is essentially the only collapsed split decomposition of G', and by 4.1 and 4.2 there exists a derivation for G' in $\mathcal{G}$ if and only if we can assign to each τ-labelled edge e

in $(H_1, \ldots, H_n)$ a label from N (the same label to both occurrences of e) such that each of the modified graphs $H'_1, \ldots, H'_n$ is an A-form of $\mathcal{G}$ for some $A \in N$. Hence we will process the tree of the decomposition bottom-up and assign to the virtual edge e_i of each H_i the set $poslab(e_i)$ of nonterminal labels A such that $result(H_i)$ is an A-form (if we change the label of vir_{H_i} to A).

To process a triconnected component $H = H_i$ we can assume that we already have determined $poslab(e)$ for all edges e of H that are a virtual edge somewhere. For each other edge e – except the virtual edge – put $poslab(e) = \{l_H(e)\}$, and put $poslab(vir_H) = \emptyset$.

Case a) H is 3-connected.

Consider every 3-connected $R \in \mathcal{R}$ and all bijections $\varphi : V_R \to V_H$, $\eta : E_R \to E_H$ such that

$$\forall e \in E_R : s_H(\eta(e)) = \varphi(s_R(e)) \wedge t_H(\eta(e)) = \varphi(t_R(e)).$$

If for every $e \in E_R - \{vir_R\}$ we have $l_R(e) \in poslab(\eta(e))$, then we add $l_R(vir_R)$ to $poslab(vir_H)$.

If in the end $poslab(vir_H) = \emptyset$, then $G \notin \mathcal{L}(\mathcal{G})$; otherwise we proceed with the next triconnected component.

Since R is a fixed finite set of graphs this step requires constant time; H has at least 4 edges in this case.

Case b) H is a cycle.

We determine H' according to 4.3i), i.e. H' is weakly isomorphic to H with unreversed virtual edge and represents a string. For each edge e that is reversed in H' we replace $poslab(e)$ by $\{\overline{A} | A \in poslab(e)\}$. By 4.3ii) we are now in the context-free string case and can determine $poslab(vir_H)$ by the Cocke-Kasami-Younger algorithm [HU79]. (Actually, this algorithm takes as input a string of terminals, and transforms this in a first step into a string of sets of nonterminals. This is where we start).

This step requires $O(|E_H|^3)$ time – or even only $O(|E_H|^{2.376})$ time (see [CW87]).

Case c) H is a bond.

We determine H' according to 4.4i), and modify each $poslab(e)$ as above, if necessary. By 4.4ii) and Parikh's Theorem [Gin66] it is not hard to see that we have to decide for each $A \in N$ whether H' represents a vector in some fixed semi-linear set, which corresponds to the A-forms of $\mathcal{G}$ that represent vectors. This can be done in time linear in $|E_H|$, see [FMR68].

Finally, we check whether $S \in poslab(vir_G)$. If so, $G \in \mathcal{L}(\mathcal{G})$ otherwise $G \notin \mathcal{L}(\mathcal{G})$.

Since by 4.4 the number of edges of the triconnected components is linear in $|E_G|$ we have the desired result. (Obviously, we can drop the assumption $N \cup T \subseteq \Sigma$, since $N \cup T$ is finite.)

Theorem 5.1 *Let $\mathcal{G}$ be a fixed ERG. Then we can decide whether a given cyclically connected graph G is in $\mathcal{L}(\mathcal{G})$ in time $O(|E|^3)$.*

Remark: Actually, the above proof shows that the algorithm can be implemented to run in time $O(|E|^{2.376})$. Furthermore, as pointed out by one of the referees, linear time suffices if the grammar in CNF has no triangles as rules. Note, that taking the ERG $\mathcal{G}$ as part of the input makes the graph isomorphism problem a subproblem (compare Case a) above).

6 Conclusion

We have presented a (sub-)cubic algorithm that decides whether a cyclically connected graph is generated by an edge replacement grammar.

This indicates that the recognition problem for hyperedge replacement grammars may be polynomial where the degree is not dependent on the size of the right hand sides of the grammar as in

[Lau88], but possibly on the maximal number of vertices incident to a nonterminal hyperedge. To confirm or disprove this is a challenging problem. Very recently a solution to this problem has been announced by Frank Drewes (Bremen).

In this paper not only rules, but also the graphs generated by an ERG contain a virtual edge. This edge gives some essential information for the recognition. If a cyclically connected graph is given without its virtual edge, we can of course solve the recognition problem in time $O(|E|^3 \cdot |V|^2)$, $O(|E|^{2.376} \cdot |V|^2)$ resp. by considering each pair of vertices as possible source and target of the virtual edge. An interesting question is whether there exists a more efficient solution.

Also it would be desirable to determine more precisely for which graph classes and ERG's the recognition problem can be solved in polynomial time where the polynomial is of fixed degree, and under which circumstances the problem is *NP*-complete. E.g. it is known that for ERG's that generate trees only (but may have other graphs as sentential forms) recognition is *NP*-complete. Some of these ERG's have rules with empty right hand side; if we additionally forbid this, the complexity is not known despite of some efforts. For a result that determines for the case of node rewriting one borderline between polynomial and *NP*-complete recognition see [Bra88].

Acknowledgement

I thank Hans-Jörg Kreowski for long discussions which prepared the ground for this paper, and my colleagues who helped me with the literature on semi-linear sets; I also thank the referees for their valuable comments.

References

[BC87] M. Bauderon and B. Courcelle. Graph expressions and graph rewritings. *Math. Systems Theory*, 20:83–127, 1987.

[Bra88] F.J. Brandenburg. On polynomial time graph grammars. In R. Cori et al., editors, *STACS '88, Proc. of the 5th Annual Symposium on Theor. Aspects of Comp. Sci., Bordeaux*, Lect. Notes Comp. Sci. 294, pages 227–236. Springer, 1988.

[CER79] V. Claus, H. Ehrig, and G. Rozenberg, editors. *Graph-Grammars and Their Application to Computer Science and Biology*, Lect. Notes Comp. Sci. 73. Springer, 1979.

[Cou87] B. Courcelle. An axiomatic definition of context-free rewriting and its application to NLC graph grammars. *Theor. Comp. Sci.*, 55:141–181, 1987.

[CW87] D. Coppersmith and S. Winograd. Matrix multiplication via arithmetic progressions. *Proc. 19th ACM Symp. on Theory of Computing,* New York City, pages 1–6, 1987.

[ENR83] H. Ehrig, M. Nagl, and G. Rozenberg, editors. *Graph Grammars and Their Application to Computer Science*, Lect. Notes Comp. Sci. 153. Springer, 1983.

[ENRR87] H. Ehrig, M. Nagl, G. Rozenberg, and A. Rosenfeld, editors. *Graph-Grammars and Their Application to Computer Science*, Lect. Notes Comp. Sci. 291. Springer, 1987.

[ER90] J. Engelfriet and G. Rozenberg. A comparison of boundary graph grammars and context-free hypergraph grammars. *Inf. and Computation*, 84:163–206, 1990.

[FMR68] P. C. Fischer, A. R. Meyer, and A. L. Rosenberg. Counter machines and counter languages. *Math. Systems Theory,* 2:265–283, 1968.

[Gin66] S. Ginsburg. *The Mathematical Theory of Context-Free Languages.* McGraw-Hill, New York, 1966.

[Hab89] A. Habel. Hyperedge replacement: Grammars and languages. Diss., FB Mathem. / Inform., Uni. Bremen, 1989.

[HK87a] A. Habel and H.-J. Kreowski. Characteristics of graph languages generated by edge replacement. *Theor. Comp. Sci.*, 51:81–115, 1987.

[HK87b] A. Habel and H.-J. Kreowski. Some structural aspects of hypergraph languages generated by hyperedge replacement. In F.J. Brandenburg et al., editors, *STACS '87, Proc. of the 4th Annual Symposium on Theor. Aspects of Comp. Sci.*, Passau, Lect. Notes Comp. Sci. 247, pages 207–219. Springer, 1987.

[HT73] J. E. Hopcroft and R. E. Tarjan. Dividing a graph into triconnected components. *SIAM J. Comp.*, 2:135–158, 1973.

[HU79] J. E. Hopcroft and J. D. Ullman. *Introduction to Automata Theory, Languages and Computation.* Addison-Wesley, Reading, Mass., 1979.

[Kre86] H.-J. Kreowski. Rule trees represent derivations in edge replacement systems. In G. Rozenberg and A. Salomaa, editors, *The Book of L*, pages 217–232. Springer, Berlin, 1986.

[Lau88] C. Lautemann. Efficient algorithms on context-free languages. In T. Lepistö et al., editors, *Automata, Languages and Programming* Tampere, Lect. Notes Comp. Sci. 317, pages 362–378. Springer, 1988.

[LW87] K.-J. Lange and E. Welzl. String grammars with disconnecting or a basic root of the difficulty in graph grammar parsing. *Discrete Appl. Math.*, 16:17–30, 1987.

[Mac37] S. MacLane. A structural characterization of planar combinatorial graphs. *Duke Math. J.*, 3:460–472, 1937.

[RW86] G. Rozenberg and E. Welzl. Boundary nlc graph grammars – basic definitions, normal forms and complexity. *Inf. and Control*, 69:136–167, 1986.

[Vog90] W. Vogler. On hyperedge replacement and bnlc graph grammars. In M. Nagl, editor, *Graph-Theoretic Concepts in Comp. Sci., Proc. 15th Int. Workshop WG '89,* Castle Roduc, Lect. Notes Comp. Sci. 411, pages 78–93. Springer, 1990.

Computing by Graph Transformation: Overall Aims and New Results

ESPRIT Basic Research Working Group No. 3299

Hartmut Ehrig, Michael Löwe, Editors
Technical University Berlin
Computer Science Department
Franklinstraße 28/29, D-1000 Berlin 10

0. Contents

1. Scope of this Document

This document gives an overview over the activities in the ESPRIT Basic Research Working Group No. 3299 "Computing by Graph Transformation (GraGra)". It provides the interested researcher with information about the overall aim (section 3), the partners (section 2), and the scientific background (section 6) of the project. Researchers who are interested in the technicallities of the project find a summary of the work which has been done in the project's first year in section 4. Section 5 contains a complete list of the publications of project members within this period and a list of papers which are currently in preparation and are going to be published soon (most of them in the proceedings of the 4th Int. Workshop on Graph Grammars, Bremen, March 5-9, 1990, appearing probably end of 1990).

2. The GraGra Working Group

Technische Universität Berlin
Hartmut Ehrig, P. Boehm, M. Korff, M. Löwe, H. Schween, and G. Taentzer.
European Cooperators: G. Engels, Braunschweig; F. Parisi-Presicce, L'Aquila;
H.-J. Schneider, Erlangen; U. Grude, Berlin.

Université de Bordeaux I
Bruno Courcelle, K. Barbar, M. Bauderon, M. Billaud, Y. Metivier, M. Mosbach, P. Lafou,
A. Raspaud, M. Sefiani, E. Sopena, R. Straudh, and S. Yoccoz (Talence).
European Cooperators: J.C. Raoult, Rennes; M. Dauchet, Lille.

Universität Bremen
Hans-Jörg Kreowski, A. Habel, and E. Jeltsch.
European Cooperators: C. Lautemann, Mainz; F.J. Brandenburg, Passau; M. Nagl, Aachen.

Freie Universität Berlin
Emo Welzl, and G. Lackner.
European Cooperators: T. Lengauer, Paderborn.

Rijksuniversiteit te Leiden
Grzegorz Rozenberg, J. Engelfriet, and L. Heyker.
European Cooperators: D. Janssens, Diepenbeck.

Università di Pisa
Ugo Montanari, A. Maggiolo-Schettini, P. Degano, A. Corradini, R. Gorrieri, G.M. Prima,
and F. Rossi.
European Cooperators: R. De Nicola, Pisa; A. D'Atri, Roma; F. Parisi-Presicce, L'Aquila.

3. Overall Aim

3.1 Relationship to the State of the Art

Graph transformations are very common in many areas of Computer Science and related fields, especially they are used in many different kinds of non-numeric computation. Moreover,since the early beginnings of Computer Science graphical representation have played a fundamental role to explain complex situations on an intuitive level. Today this fundamental role of graphs for visual understanding is one of the main reasons that graphical computer interfaces are those which are the most useful ones for human machine interaction.

On the other hand it is widely believed that graphs are only of limited use for internal machine representation because classical graph theory offers only little help and most graph algorithms are highly inefficient in general.

In contrast to most graph algorithms which are globally defined the basic idea of graph transformations is of local nature. This means that the transformation from one graph into another one is based on local changes where only certain subgraphs are transformed and the complement remains unchanged.

One of the systematic lines of studies towards the thorough understanding of the nature of such graph transformations was initiated about 20 years ago by Rosenfeld et. al. Their research was motivated by practical applications in different areas, in particular by applications to syntactic pattern recognition. Since the local transformations were given in a form similar to productions in Chomsky grammars which are used to specify natural and programming languages this new direction initiated by Rosenfeld et. al. was called "graph grammars".

In the early 70's graph grammars were independently introduced in Europe by Schneider (Erlangen/Berlin), Ehrig (Berlin), and Montanari (Pisa).

In Berlin (Technical University) the "algebraic approach" to graph grammars was developed by H. Ehrig and his group in cooperation with H. J. Schneider (who had initiated graph grammars in Erlangen) where the basic notions were formalized in the framework of algebra and category theory. The main line of research was concerned with the notions of parallelism and concurrency in the framework of graph grammars. When H.-J. Kreowski moved to Bremen, a lot of research on graph transformation mainly concentrated on hyperedge replacement and non-sequential graph grammar processes was pursuited by his group in Bremen.

In the early days of graph grammars, Erlangen has been an important center of research. Here H. J. Schneider and his group were following the algebraic approach and also developing a more traditional grammatical approach with applications directed towards software engineering and, in particular, compiler construction. This line of research was continued by M. Nagl. By today, he and his group applied graph grammars successfully to the whole range of problems in the area of software engineering.

In the late 70's, G. Rozenberg in the Netherlands initiated the "Node-Label controlled (NLC) approach" to graph grammars. Through the years, this approach was consolidated providing a solid mathematical framework for a considerable part of research in the theory of graph grammars. This research has been pursuited in Leiden (The Netherlands) and Boulder (Colorado, U.S.A.). This line of research has been taken up by E. Welzl in Graz. Today he is in Berlin (Free University) where his research is concentrated mostly on algorithmic and implementational aspects of graph grammars.

Since about three years a new group on graph grammars was built up in France by B. Courcelle in Bordeaux combining methods from logic and algebra. Main intentions are to improve efficiency of term rewriting by use of graphs and to build up a logical and combinatorial theory of graphs.

Research and development projects concerning distributed computer systems and computer networks have shown a significant lack of theoretical foundations. This problem has influenced U. Montanari in Pisa to define graph grammars for concurrent and distributed systems and to build up a corresponding group in graph grammars in Italy.

In addition to these groups mentioned above several other activities concerning graph grammars have been started during the last decade, especially in the U.S.A. and Australia. In the USA important contributions to theory and application of the algebraic approach were given by B.K. Rosen (IBM Yorktown Heights). Today the different European groups seem to have a leading role. Especially they have organized already three very successful International Workshops on Graph Grammars and their Applications to Computer Science, 1978 and 1982 in Germany, and 1986 - due to the growing international interest - together with Rosenfeld (Maryland) in the U.S.A.

By today, many of the approaches mentioned above are quite successful. However, even for experts in graph grammars it is difficult to relate the concepts and results of different approaches to each other in a clean formal way. The field is far away from a unified theory which can be represented in text books and applied to the wide range of Computer Science problems which are addressed by the different approaches up to now.

3.2 Novelty of the Approach

Computing originally was done on the level of the "von Neumann Machine" which is based on machine instructions and registers. This kind of low level computing was considerably improved by assembler and high level imperative languages. From the conceptual - but not yet from the efficiency point of view these languages were further improved by functional and logical programming languages. This newer kind of computing is mainly based on term rewriting, which - in the terminology of graphs and graph transformations - can be considered as a concept of tree transformations. Trees, however, don't allow sharing of common substructures, which is one of the main reasons for efficiency problems concerning functional and logical programs. This leads to consider graphs rather than trees as the fundamental structure of computing.

The novelty of our approach is to advocate graph transformations for the whole range of computing. Our concept of **Computing by Graph Transformations** is not limited to programming but includes also specification and implementation by graph transformations, as well as graph algorithms and computational models and computer architectures for graph transformations.

Thus it is supposed to extend current graph grammar techniques in view of specification, programming and implementation concepts.
One of the main objections against graphs as fundamental data structure is the high complexity of most graph algorithms (typically NP-complete). Most recently, however, it was shown how to use the theory of graph grammars to reduce the complexity of graph algorithms from intractable (NP-complete) to tractable (polynomial) and even efficient (linear) cases.

On the other hand the efficiency of imperative programming techniques in contrast to functional and logical ones is the reason - as pointed out above - that most of our computers are still based on the von Neumann architecture with its concept of references and pointers. Efficiency of many well-known algorithms is based on that capability. The new idea is to use the well-known concept of references and pointers to allow efficient representation and transformation of quite arbitrary graphs and to overcome the restrictions of tree transformations which are mainly used up to now.

Combining these aspects there is really a good chance for a breakthrough in several areas of computer sciences using graph transformations.

3.3 Research Goals

The main research goals corresponding to the four research areas of this action are the following:

FOUNDATIONS

The main research goal in this area is a comparative study, unification and the strengthening of the concepts developed in the different approaches which are mentioned in A.1.1. This should lead to a much better understanding of common goals and results of different approaches and, eventually, to a theory that could provide a uniform framework for a big part of research in this area. In particular, the research on foundations should provide a uniform and solid basis for the investigations pursuited in this project as discussed in II., III., and IV.

CONCURRENT COMPUTING

A high level concept for concurrent and distributed computation based on graph transformations should be developed which is suitable to express operational semantics of concurrent computations as well as specifications of distributed systems. Moreover modular graph grammars and modular specification techniques for distributed systems based on graph transformations should be developed.

EXECUTABILITY OF ALGEBRAIC SPECIFICATIONS AND
GRAPH TRANSFORMATIONS IN OTHER AREAS

Algebraic specification techniques for data types and software systems which are currently based on term rewriting should be extended to graph transformations. This extension should provide an adequate level for efficient implementation of algebraic specifications, and formal verification of such implementations against abstract algebraic specifications. It is desirable that these techniques are extendible from algebraic specifications to functional programs. It should be further investigated how advances in the theory of graph transformations can be applied to other areas like production systems, logical programming, analysis of programs, and syntactical pattern generation leading to significant improvements in these areas.

ALGORITHMIC AND IMPLEMENTATIONAL ASPECTS

The parsing problem for graph grammars should be solved in a tractable way for suitable classes extending those of precedence graph grammars. In a similar way further algorithms for searching in graphs and decidability of graph properties should be restricted from the general case - which is known to be intractable for most problems - to interesting classes of graphs generated by some grammar such that the restricted problem becomes tractable.
On the other hand graph transformation systems should be implemented in an efficient way in order to obtain useful tools to support computation by graph transformations for various applications.

3.4 Perspectives and Envisionaged Advances

Today functional and logical programming languages - although very important for various kinds of applications - are still significantly less efficient than most imperative programming languages. As pointed out in section A.1.2 one main reason for this problem is the fact that functional and logical programming is based on term rewriting and hence on tree transformations. A theory of graph transformations - as outlined in section A.1.2 - combined with a suitable hardware which directly supports graph transformations (graph reduction machine) may lead to a breakthrough concerning efficiency of functional and logical programming.
In a similar way we can expect a breakthrough concerning efficient implementation of formal specifications, like different kinds of algebraic specifications, once we have an effective procedure to transform specifications into graph transformations. This applies to software specification of centralized systems, but it may be even more important for concurrent and distributed systems concerning efficient reasoning about and execution of such specifications. Of course, we cannot

expect to reach these goals within this action, but we can extend the theory of graph transformations considerably in order to achieve such perspectives in subsequent R & D projects within ESPRIT III. As a basis for such projects we will organize workshops and publish state of the art reports concerning computing by graph transformations within this action.

3.5 Relationship to Other Basic Research Actions

It is one of the main aims of this action to cooperate not only with other experts in this research area (see cooperators of each partner) but also with other proposed ESPRIT Basic Research Actions, especially with the following ones:

- A Comprehensive Algebraic Approach to System Specification and Development; ESPRIT Basic Research Working Group No. 3264
- Models, Languages, and Logics for Concurrent Distributed Systems; ESPRIT Basic Research Action No. 3011
- Algebraic and Syntactic Methods in Computer Science; ESPRIT Basic Research Working Group No. 3166
- Semantics and Pragmatics of Generalized Graph Rewriting; ESPRIT Basic Research Action No. 3074.

4. New Results

This section provides a brief summary of the new results which have been achieved within the first year of the project. Some of the work has been published already (see 5.1). References to papers which are currently in preparation are marked with an asterisk and listed in 5.2. Most of this work has been reported on at the 4th International Workshop on Graph Grammars which took place in Bremen on March 5-9, 1990. The proceedings of the workshop is going to contain most of these papers (appears end of 1990).

4.1 Foundations

4.1.1 Grammatical Models

There are two types of graph grammars, the context-free and the non-context-free ones.
The Hyperedge Replacement Grammars (HR) are the best known among the context-free ones. A fundamental reference is Habel's dissertation [Hab 1] that covers most of the known results concering them.

The C-ed NCE Grammars, based on vertex replacement, are more difficult and less well-known than the previous ones, but they emerge as a robust context-free class for which one has several independet characterizations [Eng 2, Eng*].
Non-context-free graph rewritings can be defined by double pushouts (now axiomatized and generalized to other high level replacement systems by [EHKP*]) or by single pushouts. The single pushout approach is attractive because it is potentially simpler and more general than the previous one, and is investigated by [LE*], after Kennaway and Raoult. Other types of non-context-free rewriting systems are introduced in [MBLS 1], [MBLS 2], and [ML*] (motivated by distributed computations) and in [ACPS] (for the recognition of sets of graphs, see below).

4.1.2 Algebraic and Logical Techniques

Context-free graph grammars can be considered as systems of equations based on graph operations. This is very useful for manipulations of graph grammars, and allows to compare context-free sets of

graphs with recognizable ones (defined from finite congruences), and regular ones (defined by regular expressions).
Monadic second-order logic has also proved very useful to express graph properties and to define sets of graphs. Such sets are regular [Eng 1], close to be recognizable [Cou 3], and definable by graph reduction the inverse of grammatical generation [ACPS]. [Cou 4] gives conditions under which monadic second-order formulas with edge quantifications can be translated into similar formulas with vertex quantification only.
An important result by Engelfriet gives a logical characterization of C-ed NCE, independent of graph rewriting rules [Eng*]. This tends to prove that this class is definitely a good, robust one.

4.1.3 Comparative Study and Unification

There are many notions of graphs and graph grammars. [KR] gives a common framework for discussing all the existing graph rewriting approaches.
[CER*] unifies edge replacement and vertex replacement into a common class for which appropriate graph operations permit an algebraic treatment. [EH*] shows that the boundedness degree restriction reduces C-ed NCE to hyperedge replacement.
[HKL] compares various notions of compatibility between graph operations and graph properties, either combinatioral or algebraic and prove them all equivalent.
By considering systems of equations generating sets of finite relational structures, [Cou*] unifies several results on the logic of C-ed NCE and HR sets of graphs.
All these results tend to unify various notions into single mathematically sound ones.
The categorical approach to high level replacement systems in [EHKP*] unifies different graph grammar approaches based on double pushouts and includes also structure grammars and algebraic specification grammars [EP*] recently introduced to model rule based modular system design (see 4.3.1).
Another related field of research is to unify graph grammars with string and tree grammars, that is to consider the string- or tree-generating power of graph grammars. This is done in [EH 1], [EH 2], and [ML*].

4.1.4 Graph Theoretic Aspects

[ER 1] deals with 2-structures, a combinatorial notion extending that of graphs in a way close to graph grammars.
The effects of restrictions like bounded degree or bounded tree width on the grammatical structure of graph sets are considered in [EH*], [Cou 3], [Cou 4], and [ACPS].

4.1.5 New Notions

No good notion of a graph automaton is known up to now. [DBW*] introduces them in the special case of planar dags. [EV*] and [Rao *] consider tree transductions based on graph grammar notions.
Infinite graphs described by systems of equations and/or logical formulas are investigated in [Cou 1], [Cou 2], and [Bau].

4.? Concurrent Computing

4.2.1 Models of Concurrent System

The model of actor system based on graph grammar has been further studied and improved in [JR], [JLR].
In particular, (concurrent) computation processes are represented in great detail as directed acyclic graphs (computation graphs). A notion of applicability of computation graphs to graphs is introduced: a pair <g, C>m where g is a graph and C, a computation graph, is a description of a possible computation history of the system if and only if C is applicable to g. It is shown that the two ways of describing

derivation processes (i.e., by structured transformation and by computation graphs) are consistent with each other :each pair <g,C> such that C is applicable to g corresponds to a structured transformation that describes its external effect, and vice versa. The use of computation graphs is demonstrated by proving that actor grammars rewriting only destination-complete graphs can be simulated (step-by-step) by actor grammars without restriction on the rewritten graphs.

4.2.2 Models for Synchronized Graph Rewriting and Concurrent Logic Languages

As a first step towards modelling the behavior of concurrent logic grograms through graph grammars,[RCMEL*] proposes a categorical interpretation of logic programs as particular classes of graph grammars.
More in detail, a logic program is shown to be equivalent to a context-free graph grammar with an empty set of terminals. Each clause can be seen as a production and the application of a clause to a goal is naturally described via the "double push-out" construction used in graph grammar. Also, the constraints imposed by unification to the applicability of clauses are faithfully represented in this framework. Finally, a refutation is described as the derivation of a discrete graph, in the corresponding grammar. This mapping is developed in three levels that consider logic programs with variables only first, then with variables and constants, and finally with functions.
Further development of the above sketched results will consider aspects of logic programs related to concurrency, such as synchronization mechanisms, committed-choice and so on.

4.2.3 Structure Transformation and Pr/T nets

The purpose of the research was to develop tools for describing concurrent behaviors, based on structure rewriting systems. In particular, connections were to be established between structure derivation and processes in Predicate/Transition nets (Pr/T nets) in order to borrow concepts and proof methods from one theory to develop the other. The research has followed two lines.
The first provided to a translation of structure rewriting systems into Pr/T nets and vice versa, and to a method for studying composition and simplification of nets by reasoning on the correspondent rewriting systems [MPW], [PM*].
The second line has introduced a general concept of transformation system which includes Pr/T nets and grammars for transforming relational structures as special cases [MW 1*, MW 2*]. An operational and a denotational semantics for open systems have been given and compared. Both these semantics are compositional. A way of defining and proving behavioral equivalences of the considered systems has been also described. Derivations in grammars for transforming relational structures have a semantics which is equivalent to that given in A. Maggiolo Schettini, J. Winkowski, "Processes of Transforming Structures" Journal of Computer and System Sciences 24 (1982) pp. 245-282.

4.3 Executability of Algebraic Specifications and Graph Transformation in Other Areas

4.3.1 Execution of Algebraic Specifications

In this field, research has foccused on two aspects: Graph transformation as an implementational level for algebraic specifications offering some efficiency speed up and graph transformation as a rule based system which helps in the design of modular algebraic specifications.
With respect to the implementational level, jungle evaluation is investigated in [HP 1] and [HP 2]. Jungle evaluation provides an efficient graph grammatical way of the evaluation of functional expressions and, in particular, of terms w.r.t. many-sorted algebraic specifications. Jungle evaluation employs the Berlin approach to graph grammars so that the theory of derivations available in this approach, especially w.r.t. possible parallel evaluations, can be applied. New results in [Plu*] show that jungle evaluation is appropriate as an implementational device since termination is preserved if terminating systems of jungle evaluation rules are combined, even in those cases where term rewriting systems fail to have this property.

While jungle evaluation focusses on standard and automatic implementation on the graph level, [Löw] starts to introduce an implementation method of algebraic specifications by graph transformation systems. The algebraic semantics given to certain graph transformation systems (again the Berlin approach is used) allows to combine algebraic and graphical implementation steps within a common framework. A special problem of expression evaluation, namely garbage collection, can be advantagously tackled by applying single pushout derivations [LE*].

A survey about the techniques and methods used in graphic implementations of expression evaluation is in preparation [HKLP *].

[Par] and [EP*] apply graph transformation within the design of modular algebraic specification. The intend is to tackle the problem of designing modular systems which realizes a given goal specification. This is done by treating the interfaces of modular specifications as graph grammar productions and by converting direct derivations and operations on graph productions into system design. This connection between the theory of algebraic specifications and graph grammars gives new insight into both fields such that both theories can benefit from each other.

4.3.2 Syntactic Pattern Generation and Recognition

Hans-Jörg Kreowski and Eric Jeltsch [JK] have developed a grammatical inference algorithm which gets finite sets of sample graphs as inputs and constructs hyperedge replacement grammars as output. The inference mechanism is based on the decomposition of rules using edge-disjoint coverings of the right-hand sides of the rules.

4.3.3 Production Systems

[Kor*] applies graph transformation techniques to production systems. Elaborated syntactical independency criteria, based on well-known properties of the Berlin approach to graph transformations, like parallel and sequential independence, lead to a considerable optimization of standard search algorithms for production systems.

4.3.4 Application of Graph Grammar Models to Robotics

[ST*] model robot configurations, their movements (translations and rotations), and the analysis of dependencies of movements with a two level graph grammar model based on the Berlin approach. The first level describes elementary movements, which are composed on the second level to describe movement of complex and structured objects rsp. robots. It is the abstract categorical description of the Berlin approach to graph transformation which facilitates this two level method and leads to a very intuitive and well-structured discrete model for robotics.

4.4 Algorithmic and Implementational Aspects

The research in this area concentrates on hyperedge replacement grammars (and more general on the algebraic approach).

[ALR] investigates graph transformation datastructures and algorithms based on graph grammars. A new graph grammar model (related to the algebraic approach) is introduced, together with an implementation based on that model. The design of the model allows unified data structures in the implementation, and, moreover, new possibilities for structuring graphs are used (related to refinement techniques in programming languages).

The Bremen group emphasizes the investigation of compatible graph-theoretic properties; (compatible with the derivation process of graph grammars). In this way complexity and decidability results on a very general level can be obtained. In particular

1. properties of compatible properties [Hab 2]
2. decidability of the existence and the non-existence of members in generated languages with given compatible properties [HKV 1]

3. decidability of boundedness problems [HKV 2, HKV 3]
4. comparison with other properties and concepts [HKL, Lau 2], and
5. filtering members with given compatible properties from generated languages [HK]

are studied.

Moreover, [Lau 1] investigates special classes of hyperedge replacement grammars with efficient solution of the membership problem.

Similarly, [LWW*] tries to clarify the boundary between NP-completeness and polynomial time for the membership problem for context-free edge replacement grammars. It is shown that under certain assumptions, membership can be decided in close to cubic time (getting close to the bound for context-free string grammars).
Finally, in [ER 2], Ehrenfeucht and Rozenberg demonstrate how to use the notion of a clan to establish the complexity of T-labelings of graphs and hence the complexity of dependence graphs.

5. References

5.1 Finished and Published Papers (period March 1989 - February 1990)

[ACPS] S. Arnborg, B. Courcelle, A. Proskurowski, D. Seese: An Algebraic Theory of Graph Reduction, Techn. Report 90-02, University Bordeaux (1990).

[ALR] R. Arlt, M. Löwe, M. Röder: Basic Data Structures and Algorithms for Graph Grammar Implementations, Internal Technical Report, TU Berlin (1989).

[Bau] M. Bauderon: Infinite Hypergraphs, to appear in: Theoretical Computer Science.

[Cou 1] B. Courcelle: The Definability of Equational Graphs in Monadic 2nd Order Logic, in: ICALP'89, Springer, LNCS 372, 207-221 (1989).

[Cou 2] B. Courcelle: The Monadic 2nd Order Logic of Graphs IV: Definability Properties of Equational Graphs, to appear in: Annals of Pure and Applied Logic.

[Cou 3] B. Courcelle: The Monadic 2nd Order Logic of Graphs V: On Closing the Gap between Recognicability and Definability, Techn. Report 89-91, Univ. Bordeaux (1989). To appear in: Theoretical Computer Science.

[Cou 4] B. Courcelle: The Monadic 2nd Order Logic of Graphs VI: On Several Representations of Graphs by Relational Structures, Techn. Report 89-99, Univ. Bordeaux (1989). To appear in: Theoretical Computer Science.

[EH 1] J. Engelfriet, L.M. Heyker: The String Generating Power of Context-free Hypergraph Grammars, Techn. Report 89-05, Univ. Leiden (1989). To appear in JCSS.

[EH 2] J. Engelfriet, L.M. Heyker: The Term-generating Power of Context-free Hypergraph Grammars and Attribute Grammars, Techn. Report 89-17, Univ. Leiden (1989).

[EL] H. Ehrig, M. Löwe (ed.): Computing by Graph Transformation: Objectives, State-of-the-art, References, Techn. Report 89-14, Techn. Univ. Berlin (1989).

[Eng 1] J. Engelfriet: A Regular Characterization of Graph Languages Definable in Monadic Second-order Logic, Techn. Report 89-03, Univ. Leiden (1989).

[Eng 2] J. Engelfriet: Context-free NCE Graph Grammars, in: FCT'89, Springer, LNCS 380, 148-161 (1989).

[ER 1] A. Ehrenfeucht, G. Rozenberg: Angular 2-structures, Techn. Report 89-10, Univ. Leiden (1989).

[ER 2] A. Ehrenfeucht, G. Rozenberg: Clans and the Complexity of Dependence Graphs, Report 89-02, Leiden.

[Hab 1] A. Habel: Hyperedge Replacement: Grammars and Languages, Dissertation, University Bremen, FB Math. / Informatik (1989).

[Hab 2] A. Habel: Graph-Theoretic Properties Compatible with Graph Derivations, in WG'88, Springer, LNCS 344, 11-29 (1989).

[HK] A. Habel, H.-J. Kreowski: Filtering Hyperedge-Replacement Languages Trough Compatible Properties, Techn. Report, Univ. Bremen 1989, 12. pages

[HKL] A. Habel, H.-J. Kreowski, C. Lautemann: A Comparison of Compatible, Finite, and Inductive Graph Properties, Techn. Report 7/89, Univ. Bremen (1989).

[HKV 1] A. Habel, H.-J. Kreowski, W. Vogler: Metatheorems for Decision Problems on Hyperedge Replacement Languages, Acta Informatica 26, 657-677 (1989).

[HKV 2] A. Habel, H.-J. Kreowski, W. Vogler: Decidable Boundedness Problems for hyperedge-Replacement Graph Grammars, In J. Diaz, F. Orejas (Eds.): TAPSOFT '89, Proc. of the International Joint Conference on Theory and Practice of Software Development, Vol. 1, Springer-Verlag, Berlin, LNCS 351, 275-289 (1989).

[HKV 3] A. Habel, H.-J. Kreowski, W. Vogler: Decidable Boundedness Problems for Sets of Graphs Generated by Hyperedge-Replacement, to appear in: Theoretical Computer Science (1990)

[HP 1] B. Hoffmann, D. Plump: Jungle Evaluation for Efficient Term Rewriting, in Grabowski et al.: Algebraic and Logic Programming, Springer, Berlin, LNCS 351, 275-289 (1989).

[HP 2] B. Hoffmann, D. Plump: Jungle Evaluation for Efficient Term Rewriting, to appear in: RAIRO Theoretical Informatics and Applications (1990).

[JK] E. Jeltsch, H.-J. Kreowski: Grammatikalische Inferenz auf der Basis von Hyperkantenersetzung, Techn. Report 9/89, Univ. Bremen (1989).

[JLR] D. Janssens, M. Lens, G. Rozenberg: Computation Graphs for Actor Grammars, Techn. Report, Univ. Leiden (1989)

[JR] D. Janssens, G. Rozenberg: Actor Grammars, in Mathematical System Theory 22, 75-107 (1989).

[KR] H.-J. Kreowski, G. Rozenberg: On Structured Graph Grammars, Part I and II, to appear in: Information Sciences (1990).

[Lau 1] C. Lautemann: The Complexity of Graph Languages generated by Hyperedge Replacement, Techn. Report 4/89, Univ. Bremen 1989.

[Lau 2] C. Lautemann: Tree Decomposition of Graphs and Tree Automata, Techn. Report, Univ. Bremen 1989.

[Löw] M. Löwe: Implementing Algebraic Specifications by Graph Transformation Systems, Techn. Report 89-26, Techn. Univ. Berlin (1989).

[MBLS 1] Y. Metiviet, M. Billaud, P. Lafou, E. Sopena: Graph Rewriting Systems with Priorities, in: WG'89, Springer, LNCS 411 (1990).

[MBLS 2] Y. Metiviet, M. Billaud, P. Lafou, E. Sopena: Graph Rewriting Systems with Priorities: Definitions and Applications, Techn. Report 8908, Univ. Bordeaux (1989).

[MPW] A. Maggiolo Schettini, G.M. Pinna, J. Winkowski: A Compositional Semantics for Unmarked Predicate/Transition Nets, Fund. Infor. (1990)

[Par] F. Parisi-Presicce: Modular System Design Applying Graph Grammars Techniques, in: ICALP'89, Springer, LNCS 372 (1989). To appear in: Theoretical Computer Science.

5.2 Papers in Preparation

[CER*] B. Courcelle, J. Engelfriet, G. Rozenberg: Handle-rewriting Hypergraph Grammars.

[Cou*] B. Courcelle: Graphs as Relational Structures: An Algebraic and Logical Approach.

[DBW*] M. Dauchet, M. Bossu, B. Warin: Rational and Recognizable Sets of Planar Directed Acyclic Graphs.

[EH*] J. Engelfriet, L.M. Heyker: Context-free Hypergraph Languages of Bounded Degree.

[EHKP*] H. Ehrig, A. Habel, H.-J. Kreowski, F. Parisi-Presicce: High-level Replacement Systems.

[Eng*] J. Engelfriet: A Characterization of Context-free Graph Languages Using Monadic Second-order Logic.

[EP*] H. Ehrig, F. Parisi-Presicce: Algebraic Specification Grammars: A Junction between Module Specifications and Graph Grammars.

[EV*] J. Engelfriet, H. Vogler: Context-free Hypergraph Grammars and Macro Tree Transducers.

[HKLP*] B. Hoffmann, H.-J. Kreowski, M. Löwe, D. Plump: Evaluation of Expressions by Graph Transformation: A Survey.

[Kor*] M. Korff: Application of Graph Grammars to Rule-based Systems.

[LE*] M. Löwe, H. Ehrig: Algebraic Approach to Graph Transformation Based on Single Pushout Derivations.

[LWW*] G. Lackner, E. Wanke, E. Welzl: On a Class of Polynomial Passable Edge Replacement Grammars.

[ML*] Y. Metivier, I. Litovsky: Words and Trees Rewriting Systems with Priorities.

[MW1*] A. Maggiolo Schettini, J. Winkowski: A Generalization of Predicate / Transition Nets.

[MW2*] A. Maggiolo Schettini, J. Winkowski: Grammars for Transforming Relational Structures and Predicate /Transition Nets.

[Plu*] D. Plump: Implementing Term Rewriting by Graph Reduction: Termination of Combined Systems.

[PM*] G.M. Pinna, A. Maggiolo Schettini: Transformations of Pr/T Nets via Translation into Structure Grammars.

[Rao*] J.C. Raoult: Tree Transduction Based on Hyperedge Rewriting.

[RCMEL*] F. Rossi,A. Corradini, U. Montanari, H. Ehrig, M. Löwe: Graph Grammars and Logic Programming.

[ST*] H. Schween, G. Taentzer: Application of Graph Grammars to Robotics.

6. Bibliography of Selected Papers by Project Members and Cooperators

[Aal Roz 1] IJ. J. Aalbersberg, G. Rozenberg: Traces, dependency graphs and DNLC grammars; Discr. Appl. Math. 13 (1986), 79-85

[Aal Roz 2] IJ. J. Aalbersberg, G. Rozenberg: Theory of traces; Report 86-16, Leiden 1986, to appear in TCS

[Bau Cou] M. Bauderon, B. Courcelle: Graph expressions and graph rewritings; Math. Syst. Theory 20 (1987), 83-127

[Boe Ehr Hum Löw] P. Boehm, H. Ehrig, U. Hummert, M. Löwe: Towards distributed graph grammars; in [Ehr Nag Roz Ros], 86-98

[Boe Fon Hab 1] P. Boehm, H. Fonio, A. Habel: Amalgamation of graph transformations with applications to synchronization; LNCS 185, 267-285 (1985)

[Boe Fon Hab 2] P. Boehm, H. Fonio, A. Habel: Amalgamation of graph transformations: A synchronization mechanism, JCSS 34 (1987), 307-408

[Bra 1] F.J. Brandenburg: On partially ordered graph-grammars; in [Ehr Nag Roz Ros], 99-111

[Bra 2] F. J. Brandenburg: On polynomial time graph grammars; Proc. STACS 1988, LNCS 294, Springer-Verlag, 1988, 227-236

[Cas Mon] I. Castellani, U. Montanari: Graph grammars for distributed systems; in [Ehr Nag Roz], 20-38

[Cla Ehr Roz] V. Claus, H. Ehrig, G. Rozenberg (eds.): Graph grammars and their application to computer Science and biology, LNCS 73, Springer-Verlag, 1979

[Cou 1] B. Courcelle: Recognizability and second order definability for sets of finite graphs; Report I-8634, Bordeaux, 1986. See also [Ehr Nag Roz Ros], 112-146

[Cou 2] B. Courcelle: On using context-free graph grammars for analyzing recursive definitions; to appear

[Cou 3] B. Courcelle: Equivalences and transformations of regular systems. Applications to recursive program schemes and grammars; Theor. Comp. Sci. 42 (1988), 1-122

[Cou 4] B. Courcelle: an axiomatic definition of context-free rewriting and its application to NLC graph-grammars; Theor. Comp. Sci. 55 (1987), 141-181

[Deg Mon 1] P. Degano, U. Montanari: A model of distributed systems based on graph rewriting; Journal of the ACM Vol. 34, No 2, April 1987, 411-449

[Deg Mon 1] P. Degano, U. Montanari: Concurrent histories: A basis for observing distributed systems; Journal of Computer and System Sciences, 34, 442-461 (1987)

[Ehr Roz] A. Ehrenfeucht, G. Rozenberg: On the structure of dependence graphs; in "Concurrency and Nets" (K. Voss, H.J. Genrich, G. Rozenberg, eds.), Springer-Verlag, 1987, 141-170

[Ehr 1] H. Ehrig: Introduction to the algebraic theory of graph grammars (a survey); in [Cla Ehr Roz], 1-69

[Ehr 2] H. Ehrig: Aspects of concurrency in graph grammars; in [Ehr Nag Roz], 58-81

[Ehr 3] H. Ehrig: Tutorial introduction to the algebraic approach of graph grammars; in [Ehr Nag Roz Ros], 3-14

[Ehr Hab] H. Ehrig, A. Habel: Graph grammars with application conditions; in "The Book of L", G. Rozenberg, A. Salomaa (eds.), Springer-Verlag, 1985, 87-100

[Ehr Kre 1] H. Ehrig, H.-J. Kreowski: Categorical theory of graphical systems and graph grammars; Conf. Report Algebraic System Theory, Udine 1975, Springer-Verlag, Lecture Notes Econ. Math. Syst. 131 (1976), 323-351

[Ehr Kre 2] H. Ehrig, H.-J. Kreowski: Applications of graph grammar theory to consistency, synchronization and scheduling in data base systems; Inform. Syst. 5, 225-238 (1980)

[Ehr Kre Mag Ros Win] H. Ehrig, H.-J. Kreowski, A. Maggiolo-Schettini, B.K. Rosen, J. Winkowski: Transformation of structures: An algebraic approach; Math. Syst. Theory 14 (1981), 305-334

[Ehr Nag Roz] H. Ehrig, M. Nagl, G. Rozenberg (eds.): Graph-grammars and their application to computer science, LNCS 153, Springer-Verlag, 1983

[Ehr Nag Roz Ros] H. Ehrig, M. Nagl, G. Rozenberg, A. Rosenfeld (eds.): Graph-grammars and their application to computer science, LNCS 291, Springer-Verlag, 1987

[Ehr Ros 2] H. Ehrig, B.K. Rosen: Concurrency of manipulation in multidimensional information structures; LNCS 64, 165-176 (1978)

[Eng 2] J. Engelfriet: Generating strings with hypergraph grammars; in "Essays on concepts, formalisms, and tools" (eds. P.R.J. Asveld, A. Nijholt), CWI Tract 42, Amsterdam, 1987, 43-58

[Eng Lei Roz] J. Engelfriet, G. Leih, G. Rozenberg: Apex graph grammars and attribute grammars; Report 87-04, Leiden, 1987, to appear in Acta Informatica. See also [Ehr Nag Roz Ros], 167-185

[Fab Mag] W. Fabbri, A. Maggiolo-Schettini: A note on structure rewriting systems and Pr/T nets; Dipartimento di Informatica Pisa, Nota Scientifica S-35-84, November 1984

[Gen Jan Roz Thi1] H.J. Genrich, D. Janssens, G. Rozenberg, P.S. Thiagarajan: Petri nets and their relation to graph grammars; in [Ehr Nag Roz], 115-129

[Gen Jan Roz Thi2] H.J. Genrich, D. Janssens, G. Rozenberg, P.S. Thiagarajan: Generalized handle grammars and their relation to Petri nets; EIK 20 (1984), 179-206

[Hab Kre 1] A. Habel, H.-J. Kreowski: Characteristics of graph languages generated by edge replacement; TCS 51 (1987), 81-115

[Hab Kre 2] A. Habel, H.-J. Kreowski: Some structural aspects of hypergraph languates generated by hyperedge replacement; Proc. STACS 1987, LNCS 247, Springer-Verlag, 1987, 207-219

[Hab Kre 3] A. Habel, H.-J. Kreowski: Pretty pattern produced by hyperedge replacement, to appear in Proc. WG 1987

[Hab Kre Plu] A. Habel, H.-J. Kreowski, D. Plump: Jungle Evaluation, to appear in Proc. 5th Workshop on Specification of Abstract Data Types.

[Jan Roz 1] D. Janssens, G. Rozenberg: On the structure of node-label-controlled graph languages; Information Sciences 20 (1980), 191-216

[Jan Roz 2] D. Janssens, G. Rozenberg: Neighborhood-uniform NLC grammars; Comput. Vision, Graphics, Image Proc. 35 (1986), 131-151

[Jan Roz 3] D. Janssens, G. Rozenberg: Hypergraph systems and their extensions; RAIRO Informatique Theorique 17 (1983), 163-196

[Jan Roz 4] D. Janssens, G. Rozenberg: A characterization of context-free string languates by directed node-label controlled graph grammars; Acta Informatica 16 (1981), 63-85

[Jan Roz 5] D. Janssens, G. Rozenberg: Basic notions of actor grammars: A graph grammar model for actor computation; in [Ehr Nag Roz Ros], 280-298

[Kre 1] H.-J. Kreowski: A comparison between Petri-nets and graph grammars, Springer LNCS 100 (1981), 306-317

[Kre 2] H.-J. Kreowski: Graph grammar derivation processes, Proc. Graphtheoretic Concepts in Comp. Sci. 1983, Trauner Verlag, Linz 1983, 136-150

[Kre 3] H.-J. Kreowski: Is parallelism already concurrency? - Part I: Derivations in graph grammars; in: [Ehr Nag Roz], 343-360

[Kre Roz] H.-J. Kreowski, G. Rozenberg: On the constructive descriptiion of graph languages accepted by finite automata, Proc. MFCS 1981, LNCS 118, 1981, 398-409

[Kre Wil 1] H.-J. Kreowski, A. Wilharm: Solving conflicts in graph grammar derivation processes, Proc. Workshop, Graphtheoretic Concepts in Comp. Sci. 1985, Trauner Verlag, Linz 1985, 161-180

[Kre Wil 2] H.-J. Kreowski, A. Wilharm: Net processes correspond to derivation processes in graph grammars, Theoret. Comp. Sci. 44, 275-305 (1986)

[Kre Wil 3] H.-J. Kreowski, A. Wilharm: Is parallelism already concurrency? - Part II: Non-sequential processes in graph grammars; in: [Ehr Nag Roz], 361-377

[Lau] C. Lautemann: Efficient algorithms on context-free graph languages; to appear in Proc. ICALP 1988

[Len Wag] T. Lengauer, K.W. Wagner: The correlation between the complexities of the non-hierarchical and hierarchical version of graph problems; LNCS 247, (1987), 100-113

[Len Wan] T. Lengauer, E. Wanke: Efficient solution of connectivity problems on hierarchically defined graphs; to appear in SIAM J. Comput. (1988)

[Mag Nap Tor] A. Maggiolo-Schettini, M. Napoli, G. Tortora: Web structure: A tool for representing and manipulating programs; IEEE T. on Soft. Eng., 1988, to appear

[Mag Win 1] A. Maggiolo-Schettini, J. Winkowski: Processes of transforming structures; J. Comp. Syst. Sci. 24 (1982), 245-238

[Mag Win 2] A. Maggiolo-Schettini, J. Winkowski: Towards a programming language for manipulating relational data bases; in: D. Bjorner (ed.), Formal Description of Programming Concepts II, North-Holland, Amsterdam, 1983, 265-280

[Mar Mon] A. Martelli, U. Montanari: Nonserial Dynamic Programming: On the Optimal Strategy of Variable Elimination for the Rectangular Lattice, Journal of Mathematical Analysis and Application, 40, No 1, (Oct 1972), 226-242

[Mes Mon] J. Meseguer, U. Montanari: Petri nets are monoids; to appear in Proc. Logics in Comp. Sci. 1988

[Mol Par] D. Moldovan, F. Parisi-Presicce: Parallelism analysis in rule-based systems using graph grammars; in: [Ehr Nag Roz Ros], 427-439

[Mon Ros] U. Montanari, F. Rossi: An efficient algorithm for the solution of hierarchical networks of constraints; in [Ehr Nag Roz Ros], 440-457

[Roz Wel 1] G. Rozenberg, E. Welzl: Boundary NLC graph grammars - basic definitions, normal forms, and complexity; Inform. Contr. 69 (1986), 136-167

[Roz Wel 2] G. Rozenberg, E. Welzl: Graph theoretic closure properties of the family of boundary NLC graph languages; Acta Informatica 23 (1986), 289-309

[Win Mag] J. Winkowski, A. Maggiolo-Schettini: An algebra of processes; J. Comp. Syst. Sci. 35 (1987), 206-228

QMW LIBRARY
(MILE END)

Lecture Notes in Computer Science

For information about Vols. 1–454
please contact your bookseller or Springer-Verlag

Vol. 455: C.A. Floudas, P.M. Pardalos, A Collection of Test Problems for Constrained Global Optimization Algorithms. XIV, 180 pages. 1990.

Vol. 456: P. Deransart, J. Maluszyński (Eds.), Programming Language Implementation and Logic Programming. Proceedings, 1990. VIII, 401 pages. 1990.

Vol. 457: H. Burkhart (Ed.), CONPAR '90 – VAPP IV. Proceedings, 1990. XIV, 900 pages. 1990.

Vol. 458: J.C.M. Baeten, J.W. Klop (Eds.), CONCUR '90. Proceedings, 1990. VII, 537 pages. 1990.

Vol. 459: R. Studer (Ed.), Natural Language and Logic. Proceedings, 1989. VII, 252 pages. 1990. (Subseries LNAI).

Vol. 460: J. Uhl, H.A. Schmid, A Systematic Catalogue of Reusable Abstract Data Types. XII, 344 pages. 1990.

Vol. 461: P. Deransart, M. Jourdan (Eds.), Attribute Grammars and their Applications. Proceedings, 1990. VIII, 358 pages. 1990.

Vol. 462: G. Gottlob, W. Nejdl (Eds.), Expert Systems in Engineering. Proceedings, 1990. IX, 260 pages. 1990. (Subseries LNAI).

Vol. 463: H. Kirchner, W. Wechler (Eds.), Algebraic and Logic Programming. Proceedings, 1990. VII, 386 pages. 1990.

Vol. 464: J. Dassow, J. Kelemen (Eds.), Aspects and Prospects of Theoretical Computer Science. Proceedings, 1990. VI, 298 pages. 1990.

Vol. 465: A. Fuhrmann, M. Morreau (Eds.), The Logic of Theory Change. Proceedings, 1989. X, 334 pages. 1991. (Subseries LNAI).

Vol. 466: A. Blaser (Ed.), Database Systems of the 90s. Proceedings, 1990. VIII, 334 pages. 1990.

Vol. 467: F. Long (Ed.), Software Engineering Environments. Proceedings, 1969. VI, 313 pages. 1990.

Vol. 468: S.G. Akl, F. Fiala, W.W. Koczkodaj (Eds.), Advances in Computing and lnformation – lCCI ´90. Proceedings, 1990. VII, 529 pages. 1990.

Vol. 469: I. Guessarian (Ed.), Semantics of Systeme of Concurrent Processes. Proceedings, 1990. V, 456 pages. 1990.

Vol. 470: S. Abiteboul, P.C. Kanellakis (Eds.), ICDT ´90. Proceedings, 1990. VII, 528 pages. 1990.

Vol. 471: B.C. Ooi, Efficient Query Processing in Geographic Information Systems. VIII, 208 pages. 1990.

Vol. 472: K.V. Nori, C.E. Veni Madhavan (Eds.), Foundations of Software Technology and Theoretical Computer Science. Proceedings, 1990. X, 420 pages. 1990.

Vol. 473: I.B. Damgård (Ed.), Advances in Cryptology – EUROCRYPT ´90. Proceedings, 1990. VIII, 500 pages. 1991.

Vol. 474: D. Karagiannis (Ed.), Information Syetems and Artificial Intelligence: Integration Aspects. Proceedings, 1990. X, 293 pages. 1991. (Subseries LNAI).

Vol. 475: P. Schroeder-Heister (Ed.), Extensions of Logic Programming. Proceedings, 1989. VIII, 364 pages. 1991. (Subseries LNAI).

Vol. 476: M. Filgueiras, L. Damas, N. Moreira, A.P. Tomás (Eds.), Natural Language Processing. Proceedings, 1990. VII, 253 pages. 1991. (Subseries LNAI).

Vol. 477: D. Hammer (Ed.), Compiler Compilers. Proceedings, 1990. VI, 227 pages. 1991.

Vol. 478: J. van Eijck (Ed.), Logics in Al. Proceedings, 1990. IX, 562 pages. 1991. (Subseries in LNAI).

Vol. 480: C. Choffrut, M. Jantzen (Eds.), STACS 91. Proceedings, 1991. X, 549 pages. 1991.

Vol. 481: E. Lang, K.-U. Carstensen, G. Simmons, Modelling Spatial Knowledge on a Linguistic Basis. IX, 138 pages. 1991. (Subseries LNAI).

Vol. 482: Y. Kodratoff (Ed.), Machine Learning – EWSL-91. Proceedings, 1991. XI, 537 pages. 1991. (Subseries LNAI).

Vol. 483: G. Rozenberg (Ed.), Advances In Petri Nets 1990. VI, 515 pages. 1991.

Vol. 484: R. H. Möhring (Ed.), Graph-Theoretic Concepts In Computer Science. Proceedings, 1990. IX, 360 pages. 1991.

Vol. 485: K. Furukawa, H. Tanaka, T. Fullsaki (Eds.), Logic Programming '89. Proceedings, 1989. IX, 183 pages. 1991. (Subseries LNAI).

Vol. 486: J. van Leeuwen, N. Santoro (Eds.), Distributed Algorithms. Proceedings, 1990. VI, 433 pages. 1991.

Vol. 487: A. Bode (Ed.), Distributed Memory Computing. Proceedings, 1991. XI, 506 pages. 1991.

Vol. 488: R. V. Book (Ed.), Rewriting Techniques and Applications. Proceedings, 1991. VII, 458 pages. 1991.

Vol. 489: J. W. de Bakker, W. P. de Roever, G. Rozenberg (Eds.), Foundations of Object-Oriented Languages. Proceedings, 1990. VIII, 442 pages. 1991.

Vol. 490: J. A. Bergstra, L. M. G. Feljs (Eds.), Algebraic Methods 11: Theory, Tools and Applicatlons. VI, 434 pages. 1991.

Vol. 491: A. Yonezawa, T. Ito (Eds.), Concurrency: Theory, Language, and Architecture. Proceedings, l989. VIII, 339 pages. 1991.

Vol. 492: D. Sriram, R. Logcher, S. Fukuda (Eds.), Computer-Aided Cooperative Product Development. Proceedings, 1989 VII, 630 pages. 1991.

Vol. 493: S. Abramsky, T. S. E. Maibaum (Eds.), TAPSOFT '91. Volume 1. Proceedings, 1991. VIII, 455 pages. 1991.

Vol. 494: S. Abramsky, T. S. E. Maibaum (Eds.), TAPSOFT '91. Volume 2. Proceedings, 1991. VIII, 482 pages. 1991.

Vol. 495: 9. Thalheim, J. Demetrovics, H.-D. Gerhardt (Eds.), MFDBS '91. Proceedings, 1991. VI, 395 pages. 1991.

Vol. 496: H.-P. Schwefel, R. Männer (Eds.), Parallel Problem Solving from Nature. Proceedings, 1991. XI, 485 pages. 1991.

Vol. 497: F. Dehne, F. Fiala. W.W. Koczkodaj (Eds.), Advances in Computing and Intormation - ICCI '91 Proceedings, 1991. VIII, 745 pages. 1991.

Vol. 498: R. Andersen, J. A. Bubenko jr., A. Sølvberg (Eds.), Advanced Information Systems Engineering. Proceedings, 1991. VI, 579 pages. 1991.

Vol. 499: D. Christodoulakis (Ed.), Ada: The Choice for '92. Proceedings, 1991. VI, 411 pages. 1991.

Vol. 500: M. Held, On the Computational Geometry of Pocket Machining. XII, 179 pages. 1991.

Vol. 501: M. Bidoit, H.-J. Kreowski, P. Lescanne, F. Orejas, D. Sannella (Eds.), Algebraic System Specification and Development. VIII, 98 pages. 1991.

Vol. 502: J. Bārzdiņš, D. Bjørner (Eds.), Baltic Computer Science. X, 619 pages. 1991.

Vol. 503: P. America (Ed.), Parallel Database Systems. Proceedings, 1990. VIII, 433 pages. 1991.

Vol. 504: J. W. Schmidt, A. A. Stogny (Eds.), Next Generation Information System Technology. Proceedings, 1990. IX, 450 pages. 1991.

Vol. 505: E. H. L. Aarts, J. van Leeuwen, M. Rem (Eds.), PARLE '91. Parallel Architectures and Languages Europe, Volume I. Proceedings, 1991. XV, 423 pages. 1991.

Vol. 506: E. H. L. Aarts, J. van Leeuwen, M. Rem (Eds.), PARLE '91. Parallel Architectures and Languages Europe, Volume II. Proceedings, 1991. XV, 489 pages. 1991.

Vol. 507: N. A. Sherwani, E. de Doncker, J. A. Kapenga (Eds.), Computing in the 90's. Proceedings, 1989. XIII, 441 pages. 1991.

Vol. 508: S. Sakata (Ed.), Applied Algebra, Algebraic Algorithms and Error-Correcting Codes. Proceedings, 1990. IX, 390 pages. 1991.

Vol. 509: A. Endres, H. Weber (Eds.), Software Development Environments and CASE Technology. Proceedings, 1991. VIII, 286 pages. 1991.

Vol. 510: J. Leach Albert, B. Monien, M. Rodríguez (Eds.), Automata, Languages and Programming. Proceedings, 1991. XII, 763 pages. 1991.

Vol. 511: A. C. F. Colchester, D.J. Hawkes (Eds.), Information Processing in Medical Imaging. Proceedings, 1991. XI, 512 pages. 1991.

Vol. 512: P. America (Ed.), ECOOP '91. European Conference on Object-Oriented Programming. Proceedings, 1991. X, 396 pages. 1991.

Vol. 513: N. M. Mattos, An Approach to Knowledge Base Management. IX, 247 pages. 1991. (Subseries LNAI).

Vol. 514: G. Cohen, P. Charpin (Eds.), EUROCODE '90. Proceedings, 1990. XI, 392 pages. 1991.

Vol. 515: J. P. Martins, M. Reinfrank (Eds.), Truth Maintenance Systems. Proceedings, 1990. VII, 177 pages. 1991. (Subseries LNAI).

Vol. 516: S. Kaplan, M. Okada (Eds.), Conditional and Typed Rewriting Systems. Proceedings, 1990. IX, 461 pages. 1991.

Vol. 517: K. Nökel, Temporally Distributed Symptoms in Technical Diagnosis. IX, 164 pages. 1991. (Subseries LNAI).

Vol. 518: J. G. Williams, Instantiation Theory. VIII, 133 pages. 1991. (Subseries LNAI).

Vol. 519: F. Dehne, J.-R. Sack, N. Santoro (Eds.), Algorithms and Data Structures. Proceedings, 1991. X, 496 pages. 1991.

Vol. 520: A. Tarlecki (Ed.), Mathematical Foundations of Computer Science 1991. Proceedings, 1991. XI, 435 pages. 1991.

Vol. 521: B. Bouchon-Meunier, R. R. Yager, L. A. Zadek (Eds.), Uncertainty in Knowledge-Bases. Proceedings, 1990. X, 609 pages. 1991.

Vol. 522: J. Hertzberg (Ed.), European Workshop on Planning. Proceedings, 1991. VII, 121 pages. 1991. (Subseries LNAI).

Vol. 523: J. Hughes (Ed.), Functional Programming Languages and Computer Architecture. Proceedings, 1991. VIII, 666 pages. 1991.

Vol. 524: G. Rozenberg (Ed.), Advances in Petri Nets 1991. VIII, 572 pages. 1991.

Vol. 525: O. Günther, H.-J. Schek (Eds.), Large Spatial Databases. Proceedings, 1991. XI, 471 pages. 1991.

Vol. 526: T. Ito, A. R. Meyer (Eds.), Theoretical Aspects of Computer Software. Proceedings, 1991. X, 772 pages. 1991.

Vol. 527: J.C.M. Baeten, J. F. Groote (Eds.), CONCUR '91. Proceedings, 1991. VIII, 541 pages. 1991.

Vol. 528: J. Maluszynski, M. Wirsing (Eds.) Programming Language Implementation and Logic Programming. Proceedings, 1991. XI, 433 pages. 1991.

Vol. 529: L. Budach (Ed.) Fundamentals of Computation Theory. Proceedings, 1991. XII, 426 pages. 1991.

Vol. 530: D. H. Pitt, P.-L. Curien, S. Abramsky, A. M. Pitts, A. Poigné, D. E. Rydeheard (Eds.), Category Theory and Computer Science. Proceedings, 1991. VII, 301 pages. 1991.

Vol. 531: E. M. Clarke, R. P. Kurshan (Eds.), Computer-Aided Verification. Proceedings, 1990. XIII, 372 pages. 1991.

Vol. 532: H. Ehrig, H.-J. Kreowski, G. Rozenberg (Eds.), Graph Grammars and Their Application to Computer Science. Proceedings, 1990. X, 703 pages. 1991.

Lecture Notes in Computer Science 532

Edited by G. Goos and J. Hartmanis

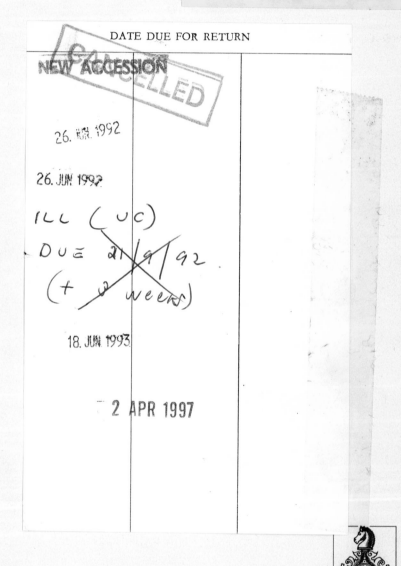

Advisory Board: W. Brauer D. Gries J. Stoer